GRAPHENE

GRAPHENE

Fundamentals, Devices, and Applications

Serhii Shafraniuk

Pan Stanford Publishing

Published by

Pan Stanford Publishing Pte. Ltd.
Penthouse Level, Suntec Tower 3
8 Temasek Boulevard
Singapore 038988

Email: editorial@panstanford.com
Web: www.panstanford.com

British Library Cataloguing-in-Publication Data
A catalogue record for this book is available from the British Library.

Graphene: Fundamentals, Devices, and Applications

ISBN 978-981-4613-47-7 (Hardcover)
ISBN 978-981-4613-48-4 (eBook)

Printed in the USA

Contents

Preface

The book was motivated by my work which I started back in 2005 studying Andreev reflection at the interface between a normal metal and a carbon nanotube. Experiments revealed interesting phenomena which required better understanding. Furthermore, questions which emerged from the experiments initialized the development of theory.

The book focuses mostly on electric charge and thermal transport properties of gated graphene. Graphene, which has a strong inherent connection with carbon nanotubes, started attracting strong attention in recent years and became the next logical step in that work.

Graphene is the first example of two-dimensional materials and is the most important areas of contemporary research. It forms the basis for new nanoelectronic applications. Graphene, which comprises field-effect structures, has remarkable physical properties. Unconventional electronic properties of graphene constitute the key factor forming a strong potential of the new material in a variety of applications.

Many ideas proposed in graphene research are frontier and futuristic, although some have immediate technological applications. The core scientific principles of all graphene applications, however, are grounded in physics and chemistry. There are currently numerous specialized graphene and graphene-related texts or monographs at the graduate and senior undergraduate levels. This book is targeted at a wide audience ranging between graduate students and postdoctoral fellows to mature researchers and industrial engineers. It has evolved from the author's own research experience and from interaction with other scientists at other institutions.

The book includes the following topics: an introduction and a historical perspective of graphene; basic physical principles and the simplest mathematical models; physics of Dirac fermions in graphene; Klein tunneling, anomalous Landau levels; transport in gated graphene; quantized states; phonon transport and the Raman effect; scattering on defects, ripples, and phonons; screening and plasma excitations; many-body effects; Andreev reflection; nonequilibrium effects on the nanoscale; the thermoelectric effect; sensing and emission of electromagnetic THz waves; and other atomic monolayers.

This book is suitable for graduate students, postdoctoral fellows, researchers, and engineers at academic and industrial institutions.

The book is practical and user friendly and can be adopted for teaching of graduate students in nanoscience and nanotechnology, materials science and engineering, and physics and chemistry.

The book can be an instructor-friendly textbook for research fellows and industrial engineers.

The book offers numerous interesting, well-illustrated, and recent examples from the author's research work.

The book also has a combined and unique multidisciplinary flavor needed when working in the field of nanoscale science.

I am grateful to Dr. H. Weinstock for valuable discussions over the course of eight years. I would like to thank Dr. R. Rosenstein for valuable discussions and comments during the preparation of this book. Professors P. Barbara, V. Chandrasekhar, I. Nevirkovets, and J. B. Ketterson have also contributed with significant discussions and valuable advice. Several chapters of this book (i.e., Chapters 2, 3, 5, 6, 8, 9, and 10) are based on my research work supported by AFOSR grants FA9550-06-1-0366, FA9550-09-1-0685, and FA9550-11-1-0311. The work discussed in Chapter 7 has been supported by the NSF grant DMR-0239721.

Serhii Shafraniuk

Introduction

The class of carbon-based crystals comprises a large variety of different systems with an equally large diversity of physical properties. Carbon is considered as a basic element of life because it forms a ground of all organic chemistry. The dimensionality of carbon-based systems and flexibility of the carbon atom bonding determine the variety of physical properties of these structures.

There are a few very important structures which are composed of only carbon atoms. A special case is graphene (**G**), depicted in Fig. 1, which is a 2D allotrope of carbon, forming an atomic monolayer sheet. A single-atomic graphene layer looks like a composition of benzene rings stripped of their hydrogen atoms (Pauling 1972). Its honeycomb structure is composed of hexagons. To explore the

Figure 1 A monoatomic sheet of graphene.

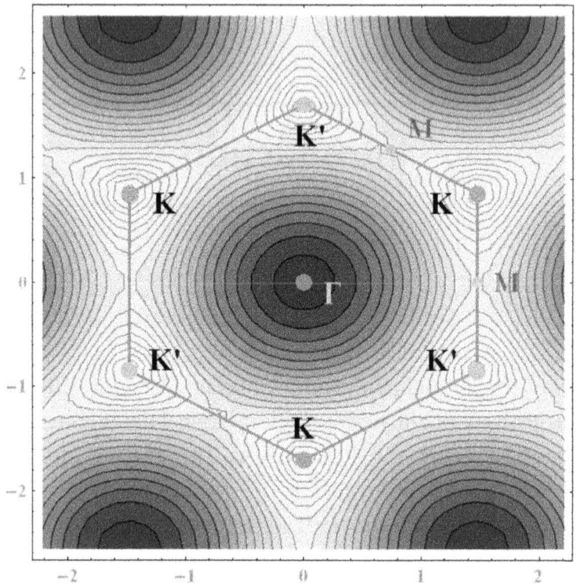

Figure 2 Brillouin zone of graphene.

properties of graphene is to help to understand the electronic properties of all the other allotropes shown in Fig. 1.6 (Geim 2009).

It is generally acknowledged that a stable monoatomic carbon sheet (graphene) might form a basis for promising new nanoelectronic applications (Novoselov et al. 2004). Graphene which comprises field-effect structures has remarkable physical properties. Unconventional electronic properties of graphene owing to its unusual Brillouin zone shown in Fig. 2 constitute a key factor forming the strong potential of the new material in a variety of applications. Major material parameters de facto are dependent from the chemical element number and from valence since they eventually determine periodicity of the lattice and the strength of Coulomb interaction in the sample. The electric transport and optical properties of the solid state crystal essentially depend on details of the electronic band structure (Kittel 1962, Abrikosov 1988), where the valence band is responsible for binding of the carbon atoms in a crystal lattice. The electrons can be either tightly bound to the ions or they move freely in the periodic lattice (Kittel

1962, Abrikosov 1988). Depending on this structure, the material is either insulating or conducting. In conventional semiconductors a phase transition between the insulating and conducting states occurs when doping them with specific impurities which act as donors or acceptors of electrons. Conventional semiconductors are extremely sensitive not only to imperfections and impurities but also to electron excitation by light. The knowledge of how to dope the semiconductors using appropriate impurities opened the path for developing of semiconductor electronics such as diodes and transistors. Also, in semiconductors there is a rich variety of optical properties based on the specific electron band structure of these materials. Graphene can also be doped. This is done in two different ways either by chemical doping or gate electrode doping. The latter approach benefits from the 2D structure of graphene which makes it possible to attach the back and top gate electrodes. The gate electrodes control the electric charge carrier concentration. Furthermore, the gate electrodes might have a variety of geometries which set a certain spatial profile of the electrochemical potential of the graphene sheet. The doping of the gate electrode allows us to significantly change the electron transport by mere shifting the Fermi level between the valence and conductance bands. In this case, the gated graphene sheet actually acts as a field-effect transistor (FET). The transport and low-energy spectral properties of such a FET are very sensitive to the values of gate voltages.

Another nontrivial feature of graphene is its large number of allotropes shown in Fig. 1.6 and it also indicates a remarkable diversity. There are several other materials which are closely related to graphene since they are composed of carbon atoms. However, they show striking differences of mechanical, electron transport and optical properties. Along with graphene, one might consider fullerenes, carbon nanotubes (CNTs), diamond, graphite, and coal. The differences between these carbon materials are not chemical but rather geometrical since the carbon atoms are arranged in variety of ways. If carbon atoms are arranged spherically, they form a molecule called fullerene (Andreoni and Curioni 1996). From the physical standpoint such molecules are 0D objects, their electron excitation spectrum consists of discrete levels. Fullerenes are derived from graphene with the introduction of pentagons.

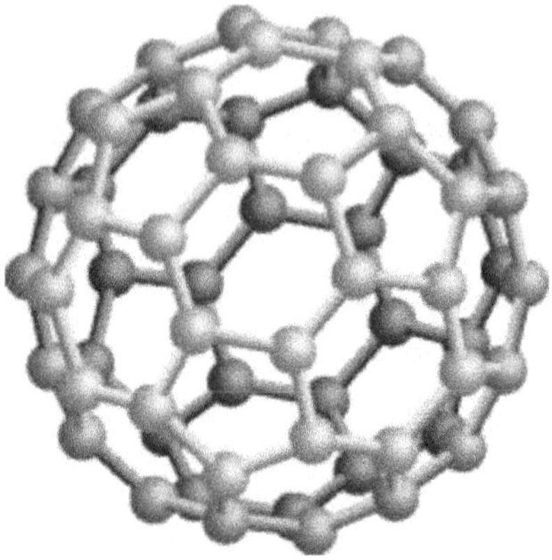

Figure 3 A fullerene buckyball.

The pentagons create defects which result in positive curvature; thus, the fullerenes can be represented as graphene wrapped in the shape of a ball (Fig. 3). One more relevant object, a carbon nanotube (CNT), is derived if one wraps a graphene sheet along certain directions, connecting the carbon bonds at opposite edges. CNTs consist of only hexagons and can be treated as 1D objects. The 3D allotrope of carbon, graphite, is widely known since 1564, when pencils were invented. Its capability to writing emerges from the fact that graphite has a layered intrinsic structure and represents stacks of graphene layers which are coupled by weak van der Waals forces. During the writing, a pencil is pressed against the paper, where the graphite stacks are actually unloaded from the pencil. There is a high probability that somewhere among those stacks there are monoatomic graphene layers. It is a remarkable fact that finally after 440 years of using pencils, the monoatomic graphene layers have been identified and isolated. Such a prolonged period of unawareness persisted because no one anticipated the fact that graphene might be in a free state. Another complication was the absence of experimental instruments and methods to find the single-atom-thick fragments among the pencil debris covering macroscopic areas (Geim and MacDonald 2007). Graphene is not very difficult to make but is sometimes hard to find. The circumstance which allowed the discovery was a subtle optical visibility of the graphene sheet

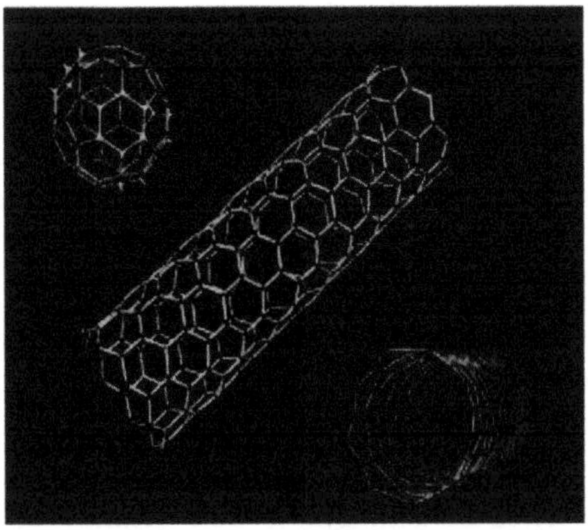

Figure 4 Carbon nanotubes.

resting on top of a chosen SiO_2 substrate. Graphene eventually had been spotted because it has a different color which was detectable by a conventional optical microscope (Novoselov et al. 2004, Abergel et al. 2007, Blake et al. 2007, Casiraghi et al. 2007).

The 2D graphene sheets can either be rolled creating CNTs or they can form carbon balls, the fullerene. Shin needles of graphite which are CNTs have been the subject of extensive research. CNT, in the form of concentric tubes (Figs. 4, 1.6) were discovered by Iijima (Iijima 1991) where configuration of the 2D graphite is arranged in a helical fashion representing multiwall nanotubes. Because the interwall spacing in the multiwall CNT is bigger than the spacing among the nearest neighbor atoms in a graphite sheet, the electronic properties of multiwall CNT therefore are dominated by those of a single carbon lattice layer. In the following years, single-wall nanotubes were also obtained (Kiang et al. 1995, Bethune 1996, Bethune 2002). The CNT can be either a metal or a semiconductor, depending on its diameter and helical arrangement. The electronic structure of a CNT and whether it is either metallic or semiconducting can be derived from the band structure of a graphene sheet when periodic boundary conditions are applied around the tube in the angular direction. This calculation had initially been done within a tight-binding model and neglecting the curvature of the nanotube (Sawada and Hamada

1992, Hamada 1993, Ando 2005). The low-energy electron spectral properties of graphene and CNTs were also obtained using the **k-p** method, and an effective-mass approximation (Ajiki and Ando 1993). The simplicity of the effective mass approximation made it very useful when exploring the rich variety of electronic properties of graphene and CNTs. That method had been successful in providing an understanding of magnetic properties (Ajiki and Ando 1993), influence of the Aharonov–Bohm effect on the band gap, the optical absorption spectra (Ajiki and Ando 1994), exciton effects (Ando 1997, Ando 2004, Ando 2004), interaction effects on band gaps (Sakai et al. 2004), lattice instabilities in the absence (Viet et al. 1994) and presence of a magnetic field (Ajiki and Ando 1995), magnetic properties of nanotube ensembles, effects of spin–orbit interaction (Ando 2000), and of electronic properties of nanotube caps (Yaguchi and Ando 2001, Yaguchi and Ando 2002, Yaguchi and Ando 2003). The effective mass model (**k-p** approximation) also helps to describe the electron transport properties which involve low-energy excitations. It successfully describes the suppression of backward scattering (Ando et al. 1998, Ando and Suzuura 2002), the origin of the ideally conducting channel, the influence of strong and short-range scattering centers like lattice defects (Ando et al. 1999), topological defects and junctions, (Matsumura and Ando 1998), and also the Stone–Wales defects (Matsumura and Ando 2001). The **k-p** approximation is also very helpful when considering the electron–phonon scattering involving long-wavelength phonons (Suzuura and Ando 2000). In this book we will consider mostly electronic states and transport properties of graphene described in the **k-p** approximation which allows a simple analytical derivation and also provides a good illustration.

Graphene shows remarkable structural diversity pronounced in its electronic properties. For instance, a trigonal planar structure with a bond between carbon atoms that are separated by 1.42 Å is the result of sp_2 hybridization of one s orbital and two p orbitals. We will discuss this effect in Chapter 1. The electronic band corresponding to this bond ensures the robustness of the lattice structure for all the allotropes (Fig. 5). This follows from the Pauli principle: these bands form a deep valence band since their shell is filled. There is a third p orbital which is directed

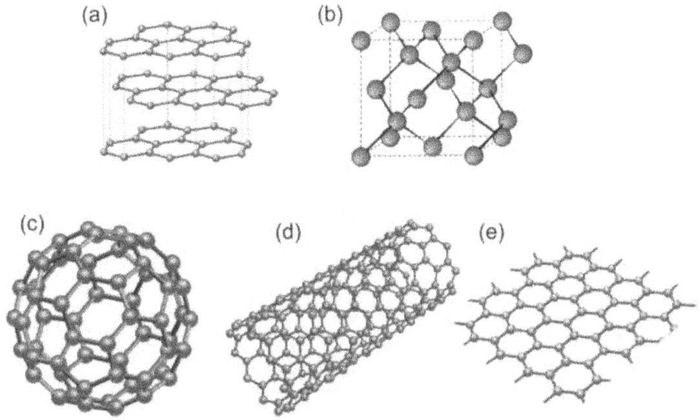

Figure 5 Various allotropes of graphene with geometrical transformations. Adapted from NPG.

perpendicular to the planar structure which is not affected by hybridization and its binding with neighboring carbon atoms has a covalent nature (see Fig. 1.3, Chapter 1). That coupling leads to formation of a σ-band which is half-filled because each p orbital has one extra electron. If electrons bands are half-filled, they have a well-articulated tight-binding character. Therefore one might expect that strong correlation effects can play an important role in graphene due to the presence of half-filled bands. The role of electron–electron interactions in graphene currently is the subject of intense study. Another remarkable feature has been emphasized by P. R. Wallace in 1946 who noticed an unusual semimetallic behavior in graphene. During those early years, the study of graphene was regarded as a purely theoretical and abstract exercise, since no one expected the 2D systems to exist in reality. Nevertheless, this work served as an important building block in deriving the properties of graphene allotropes. Another important theoretical task was to establish the nature of coupling between the stacked graphene layers. The interlayer coupling can change their properties considerably. The use of a simple band structure model failed to properly describe the interaction between the graphene planes which is due to van der Waals forces. The problem had been resolved by utilizing many-body methods. This approach has provided a consistent theory of interacting graphene layers.

However, the band model works just fine for the monoatomic graphene. Existing experimental data suggest that the low-energy excitations in graphene are essentially the massless, chiral, Dirac fermions. If the electron and hole concentrations in graphene are equal to each other, that is, graphene is neutral, the chemical potential crosses exactly the Dirac point. The low-energy region is located near so-called K-points in the energy space and is described by the Dirac wave equation for the four-component wavefunction. The excitations are propagated with a constant speed $v_F = c/300$, 300 times slower than light speed c. The relativistic properties of the Dirac excitations in graphene had been confirmed in many experiments. They include, for example, the anomalous integer quantum Hall effect which persists even at room temperatures owing a large value the cyclotron energies for "relativistic" electrons. A distinguished feature of the Weyl equation is reflected by presence of Landau level at $\varepsilon = 0$. The zero-energy level is independent versus the magnetic field and is observed by measuring of divergences in magnetic susceptibility (Geim and Novoselov 2007). The unconventional quantum Hall effect is quantitatively different from the regular quantum Hall effect taking place in Si and Ga-Al-As heterostructures (Stone et al. 1992) which poses itself as a distinguished feature of chiral fermions. Another remarkable feature of the Dirac fermions shows up in the so-called Klein paradox see Fig. 1.12 which permits Dirac fermions to be transmitted with probability equal to one through a classically forbidden region. Originally, the Klein paradox (McKellar and Stephenson 1987) was predicted only for relativistic particles. For the practical observation of the Klein paradox very high energies which had not been experimentally achievable are necessary. Surprisingly, graphene provided the opportunity for the Klein paradox to be observed in achievable experimental conditions (Geim and Novoselov 2007).

Many of the remarkable electron transport properties of graphene are due to an "intrinsic purity" of the material which ensures a ballistic propagation of the charge carriers. The physical basis of this "intrinsic purity" is related to the coherence of the electrons. In graphene this is in turn due to the chirality of elementary low-energy excitations which behave as Dirac fermions. The chirality of Dirac fermions is caused by a specific symmetry of the electron wavefunctions which is determined by graphene's honeycomb crystal lattice. One outstanding consequence of the Klein paradox is the intrinsic "purines" of graphene (Geim and Novoselov 2007). More specifically, atomic impurities in graphene are serving as a source of random potentials as in regular materials. In conventional metals and semiconductors, the random potential has a large effect on the motion of charge carriers. When the impurity concentration is altered from zero to large, the electron transport changes from ballistic to diffusive. In contrast, in graphene, the electron transport remains ballistic up to large impurity concentrations since the impurities are "invisible" to chiral electrons and holes (Geim 2012).

The fabrication of graphene devices having nanoscopic dimensions is a challenging technological problem. In particular, it involves the sample's purity, quality of interfaces, presence of atomic and structural defects, etc. Currently, fabrication technology is still in early stage and many contemporary graphene devices are "dirty" rather than "clean." Numeric simulations of such "dirty" graphene devices exploits the diffusion–drift approximation. This approximation is valid when the concentration of lattice defects and impurity is sufficiently high. An intensive scattering of charge carriers inside a channel or from the substrate insulator and intrinsic scattering on phonons in graphene field-effect transistors (GFETs) influences the whole scenario of electric charge transport. Many approaches to modeling of field-effect transistors tend to disregard an important element of the transport description, that is, satisfying the continuity condition for diffusion–drift current in channels. An ignoring of the continuity condition complicates the definition of component of the diffusion current and also introduces an undesired ambiguity computing of the current–voltage characteristics at the edges of the operation modes (linear and saturation, subthreshold and above threshold). A variety of important details (e.g., spatial variations of the electrostatic and chemical potentials, compressibility of 2D electron system, and the form of the Einstein relation in charge-confined channels, etc.) are also often disregarded in the simulations of devices. All this encourages us to elaborate an analytical self-consistent treatment of the transport properties of graphene devices. These devices are examples of new and interesting physical systems, the electrostatics of which has a contribution of quantum capacitance, and this requires new insights for correct modeling and simulation. This task is in large degree devoted to development of such an analysis of the carrier transport in the emerging graphene devices. The novel approach is based on the explicit solution of the microscopic equations which contain specific and new aspects of the charge and heat transport problem. The role of charged defects near or at the interfaces between graphene and insulated layers will be also discussed in this book.

The electron transport properties of graphene nanoribbons depend critically on their edges. As has been noticed by Neto et al. (2006), Peres (2009), and Peres (2010), the basic low-energy electronic properties depend drastically on the boundary conditions imposed on the wavefunctions in mesoscopic samples with various types of edges. There are two distinct basic types of edges in graphene stripes which preserve the electric neutrality. They are referred to as the zigzag and armchair edge shape. As we will see in Chapter 6, due to the chiral nature of the excitations in graphene, the boundary conditions at the ribbon edge determine the electron excitation spectrum (Nakada et al. 1996, Wakabayashi et al. 1999, Brey and Fertig 2006, Brey and Fertig 2006). We will see that graphene ribbons with two different types of edges, zigzag and armchair, show remarkably different electron transport properties. For instance, the graphene ribbons with zigzag edges contain surface states which are absent in the ribbons with the armchair edges.

Another unconventional property of graphene which follows from its 2D nature, is its similarity to a membrane which can bend in the out-of-plane direction (Kosmrlj and Nelson 2013). Graphene simultaneously has also metallic properties, since the chiral fermions propagate over the locally curved surface. This represents a deep analogy with problems of quantum gravity. Because the graphene sheet bends in the out-of-plane direction, there are corresponding graphene lattice oscillations which are regarded as phonons. The out of plane flexural modes have no analogues in regular 3D solids. They are responsible for absence of

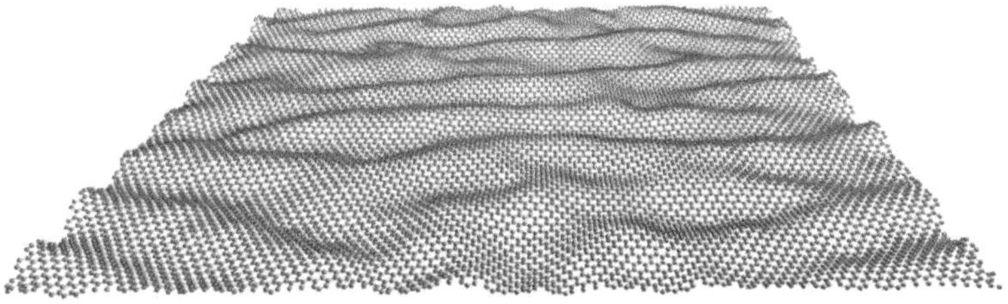

Figure 6 Out-of-plane bending of the graphene sheet.

the long range structural order in soft membranes leading to the phenomenon of crumpling as has been noticed (Kosmrlj and Nelson 2013). However, if a graphene sheet rests on a substrate or on a special rack which holds the graphene sheet in place, it partially stabilizes. But even in the last case there is a residual ripple which resembles the frozen flexural mode (see Fig. 6).

A lot of interesting phenomena take place when one attaches superconducting electrodes to graphene. In that case, besides the conventional transmission and reflection, the electron transport also involves the so-called Andreev reflection. In addition to changing the electron transport, the electrodes serve as probes which provide very useful information about the interfaces between graphene and the electrodes attached to it (Heersche et al. 2007). Graphene has a robust intrinsic coherence which is greatly conducive to propagation of Cooper pairs.

There are also many other quantum interference effects which have been measured experimentally. Among confirmed experimental observations were weak localization, universal conductance fluctuations (Morozov, Novoselov et al. 2006), the Aharonov–Bohm effect in graphene rings (Recher, Trauzettel et al. 2007), the anomalous Hall effect, and many others. Graphene can be readily utilized in a variety of applications. Examples are single-molecule detection (Schedin et al. 2007, Wehling et al. 2008), and spin injection (Blake et al. 2007, Cho et al. 2007). Because of its unusual structural and electronic flexibility, graphene can be tailored chemically and/or structurally in many different ways: deposition of metal atoms (Calandra and Mauri 2007, Uchoa et al. 2008) or molecules (Schedin et al. 2007, Leenaerts et al. 2008a,

2008b, 2008c) on top, intercalation as done in graphite intercalated compounds (Tanuma 1985, Yosida and Sato et al. 1985, Yosida and Tanuma et al. 1985, Dresselhaus and Dresselhaus 1994), incorporation of nitrogen and/or boron in its structure (Martins et al. 2007, Peres et al. 2007) in analogy with what has been done in nanotubes (Stephan et al. 1994), and use of different substrates that modify the electronic structure (Calizo et al. 2007, Giovannetti et al. 2007, Varchon et al. 2007, Zhou et al. 2007, Das et al. 2008, Faugeras et al. 2008). A new way of controlling the basic parameters of graphene materials and devices inspire creation of the graphene-based systems with magnetic and superconducting features (Uchoa and Neto 2007).

In recent years a lot of attention has been paid to the electromagnetic properties of graphene (Mele et al. 2000, Syzranov et al. 2008) and CNTs (Maksimenko et al. 2006, Nemilentsau et al. 2006, Slepyan et al. 2006, Fuse et al. 2007, Fuse, Kawano, and Yamaguchi et al. 2007a, Fuse, Kawano, and Yamaguchi et al. 2007b, Kawano et al. 2009, Kawano and Ishibashi 2010, Kawano 2011, Kawano 2012, Rinzan et al. 2012, Kawano 2013a, Kawano 2013b). Relativistic properties and chirality of low-energy excitations in graphene represent new fundamental features (Ando 2005, Katsnelson and Novoselov 2007). They present many new exciting opportunities beyond ordinary semiconductors where the electrons behave as spinless excitations. In contrast, electrons in the new carbon materials are characterized by chirality and by two one-half pseudospins. This changes the whole scenario of electron transport in graphene. Electron chirality and the pseudospin conservation have a large impact on the transport properties. The scattering time of chiral electrons on impurities, phonons, and other electrons becomes much longer because the pseudospin conservation imposes additional selection rules on the elementary scattering processes. This makes the electron propagation in pristine graphene more ballistic than diffusive. Thus, the electron mean free path l_e ($l_e = \min\{l_{ee}, l_{ep}, l_{ei}\}$, where $l_{ee,ep,ei} = v\tau_{ee,ep,ei}$, and $\tau_{ee,ep,ei}$ are scattering times of an electron on the other electrons, phonons, and impurities) might exceed the sample length L_T ($l_e > L_T$) (Ando 2005, Shafranjuk 2007, Shafranjuk 2009). Such a long scattering length $l_e > L_T$ in graphene actually means that the 2D material is intrinsically "purified." Intrinsic "pureness" of graphene emerges from a relativistic-like nature of the electrons (Ando 2005, Katsnelson and Novoselov 2007). Ballistic transport of the quasiparticle excitations in graphene is in contrast to the electron transport in conventional semiconductors. In ordinary 2D semiconductors, where the electron–electron collisions are frequent, the electron system is transformed into a fluid which is characterized by a finite viscosity where the ballistic transport is then impossible. The observable consequence of the electron fluid behavior is the excitation of "shallow water" waves. Such plasma excitations with an acoustic dispersion law has been examined for regular semiconductors (Cheremisin et al. 1998). The story is greatly changed for the "intrinsically pure" graphene where the low-energy electron excitations have much longer scattering times. Because the electron–electron scattering length $l_{ee} = v\tau_{ee}$ in "clean" suspended graphene can be remarkably long (Ando 2005, Katsnelson and Novoselov 2007), the chiral electrons behave rather as a diluted gas and not as a fluid. It is difficult for the "shallow water" plasma waves to emerge in graphene, because the nature of low-energy electron excitations in the two distinct systems is quite different.

The mechanical properties of graphene and CNTs are also unique owing to their huge Young's modulus. CNTs are capable of sustaining high strains without fracture since they have the biggest tensile strength of any material known. Graphene and CNTs are under consideration as functional units for the constructing of future molecular-scale machines. Applications of graphene and CNTs enable robust mechanical structures to be used on an atomic scale as probes in scanning tunneling microscopy and atomic force microscopy and tweezers used for atomic-scale manipulation, etc.

This book consists of 11 chapters. Chapter 1 is devoted to the concept of chiral fermions which in graphene serve as a basis for understanding of the unique properties. We begin by discussing the low-energy electron excitations, derive the Dirac equation for chiral fermions, formulate the tight-binding scheme, and compute the density of electron states in pristine graphene. The electronic states of 2D graphite or a graphene sheet are discussed first in the nearest-neighbor tight-binding model, then an effective-mass equation equivalent to that of a 2D neutrino model is introduced, and a characteristic topological anomaly is discussed. The formation of Landau levels in a magnetic field is discussed together with effects of trigonal warping, finite curvature, and strain. The band structure of nanotubes is discussed first from a general point of view and then on the basis of the effective-mass scheme. We also discuss the consequences of the chiral properties of the elementary excitations. This involves the Berry phase and topological singularity, the Klein paradox and chiral tunneling, and the Landau levels. We will make explicit the basic assumptions that are admitted when modeling graphene devices.

In Chapter 2 we consider phenomena which are related to the intrinsic coherence in graphene and their impact on the properties of field-biased graphene junctions. We distinguish two limiting cases of electron transport. One limit corresponds to a ballistic propagation of chiral particles over long distances which occurs in "clean" samples at relatively low temperatures when the electron–phonon collisions are weak. The other limit takes place in "dirty" monolayers resting on dielectric substrates. Other contributions into the electron scattering emerge from the presence of ripples (shown in Fig. 6), interface roughness, and changed impurity atoms. We shall discuss the propagation of electron and hole excitations in graphene, the quantum capacitance, the Einstein relation, the electrostatics of gated graphene devices, the charge traps near the graphene/oxide interface, and the steady-state electrostatics of the GFETs. Besides we estimate the characteristic scales of gated graphene, solve the electrostatic equation, compute the capacitance of the channel and of the gate, and find the diffusion–drift current, the ratio of the diffusion and drift currents, the electric current continuity, and the inhomogeneous behavior of the chemical and electrostatic potential along the channel. In the same Chapter 2 we formulate the microscopic model of electron transport through FET. Other subjects are the chiral tunneling through a rectangular barrier, the role of the armchair and zigzag edges, deviation of an electron inside a wide chiral barrier, the electric current density across the chiral barrier, gate voltage-controlled quantization, and a hybrid graphene/CNT junction. The modeling of current channel, computing of the electric current characteristics, the saturation regime (pinch-off), the linear behavior in low fields, the transit time through the channel, the diffusion–drift approximation, effects in the high field, the generalized boundary conditions, the pseudodiffusive dynamics, the role of confinement, and *Zitterbewegung* will be discussed as well.

In Chapter 3 we study the quantized states in graphene ribbons. At the beginning of the chapter we formulate the tight-binding model for bilayer graphene and then consider the bilayer graphene junction. Using the symmetry properties of the chiral excitations and, in particular, the PT-invariance of the Dirac Hamiltonian, we obtain heavy chiral fermion (HCF) states in a graphene stripe with zigzag edges.

Chapter 4 is dedicated to phonons and Raman scattering, phonon spectra in the monoatomic sheets and nanoribbons, and the phonon transport in 2D graphene crystals. We consider the momentum diagram of phonon transport and discuss the thermal conductivity due to phonons in nanoribbons. The role of the degrees of freedom and the connection between the molecular vibrations and infrared radiation will also be discussed. Other subjects of interest in this chapter are Raman scattering, Stokes and anti-Stokes scattering, Raman scattering versus fluorescence, selection rules for Raman scattering, Raman amplification and stimulated Raman scattering, and the requirement of coherence. We also describe practical applications of the Raman spectroscopy, higher-order Raman spectra, Raman spectroscopy of graphene, Kohn anomalies, double resonance, origin of the D and G peaks, how to deduce the strength of electron–phonon coupling from the Raman line width, the Raman spectroscopy of graphene and graphene layers, failure of the adiabatic Born–Oppenheimer approximation and the Raman spectrum of doped graphene, the effect of disorder on the Raman peaks, and the role of edges in graphene ribbons.

In Chapter 5 we focus our attention on the following issues. We discuss the electron scattering on atomic defects and phonons in graphene, scattering of electrons on charged impurities in graphene and CNTs, scattering of electrons on phonons and phonon drag, random phase approximation (RPA), and derivation of Lindhard formula. Effects of electron–impurity scattering are discussed and an absence of backward scattering is predicted. The latter occurs when the scatterer size is larger than the lattice constant. Further, the presence of a perfectly conducting channel and its sensitivity to the presence of various symmetry breaking effects such as inelastic scattering, magnetic field and flux, short-range scatterers, and trigonal warping are discussed. We also compute the electron–phonon scattering time, polarization operator, dielectric function of the valence electrons, electron–phonon coupling strength, susceptibility of pristine graphene, and electron–impurity scattering time. It allows us to analyze the screening by interacting electrons, plasma oscillations, scattering of phonons in a few-layer graphene, phonon drag, and electron–phonon coupling. Furthermore, as we will discuss bellow in Chapter 5, the chiral particles are also immune to some degree against scattering on the lattice oscillations (phonons). Therefore, chiral fermions in graphene can propagate without scattering over large distances exceeding a few micrometers (Novoselov et al. 2005). Another remarkable phenomenon is that the chiral fermions are experiencing a jittery behavior of their wavefunction which had been called *Zitterbewegung* (Katsnelson 2007, Katsnelson and Novoselov 2007, Rusin and Zawadzki 2007). It appears that the chiral electrons in graphene can propagate as far as a few micrometers without scattering which has been established experimentally by Geim and Novoselov (2007), Tan et al. (2007), Bolotin, Sikes, Hone et al. (2008), and Bolotin, Sikes, Jiang et al. (2008).

In Chapter 6 we examine many body effects and excitations in graphene, excitons, excitonic states, observation of excitons in graphene, and electron scattering on indirect excitons. The importance of exciton effects and electron–electron interaction on band gaps is emphasized and some related experiments are discussed.

Chapter 7 is devoted to Andreev reflection at the graphene/metal interface, conversion between electrons and holes at the N/S interface, the BTK model of Andreev reflection, experimental study of the Andreev reflection in graphene, interpretation of Andreev reflection in graphene-based junctions, composite Andreev reflection amplitude, and Andreev retroreflection versus the specular reflection. We also suggest a simple theoretical model to describe the electronic transport at Van Hove singularities in graphene stripes and CNTs.

Chapter 8 is dedicated to nonequilibrium effects in graphene devices. In this chapter we discuss relevance of nonequilibrium effects in graphene junction and validity of the "homogeneous" approximation inside the chiral barrier. It allows us to compute the tunneling rates for a gated graphene field effect transistor (GFET). We also compute the Keldysh and advanced Green functions for chiral fermions, the AC field power which is absorbed by tunneling electrons, and also the nonequilibrium electric current. We derive the quantum kinetic equation for distribution function of chiral fermions to examine the effect of energy pumping into the quantum well by using the δ-function approximation. In a simple limiting case of symmetric junction we compute the electric current and also determine the nonequilibrium contribution. Besides we implement the quantum Jarzynski equality to examine dissipation dynamics of the one-half pseudospin which is taking place in graphene and CNT systems.

In Chapter 9 we consider thermoelectric effect in graphene and CNT quantum dots (QDs). Departing from the quantum theory of electronic thermal transport, we discuss the graphene thermoelectric nanocoolers and electricity cogenerators, performance of the thermoelectric setup, the role of electron transport and elastic collisions, the reversible Peltier effect in carbon nanojunctions, Fourier law, the potential cooling power, electron thermal conductivity, and the thermoelectric figure of merit. By comparing the recent experiments for measuring the thermal conductivity of graphene with predictions of microscopic model of the thermoelectric effect, we review the phonon transport and thermal conductivity, converting of heat to electricity by a graphene stripe with heavy chiral fermions, blocking the phonon flow by multilayered electrodes, molecular dynamics simulations, nonequilibrium thermal injection, and the role of interfaces in the thermoelectricity. We complete the chapter with summarizing the perspectives of the thermoelectric research in graphene.

In Chapter 10 we describe sensing of external electromagnetic field by graphene and CNT QDs. We formulate key assumptions of the microscopic model of the CNT sensor and then compute basic characteristics of the THz sensor based on a CNT QD. We analyze the role of chiral transport, the effect of the electromagnetic field and its influence on responsivity, and quantum efficiency of the THz detector, noise-equivalent power, the intrinsic noise, frequency range, and operation temperature. We address existing experiments and technological approaches (Mele et al. 2000, Fuse et al. 2007, Fuse, Kawano, and Yamaguchi et al. 2007a, Fuse, Kawano, and Yamaguchi et al. 2007b, Rinzan et al. 2012) and, the critical theoretical concepts (Akturk et al. 2007, Shafranjuk 2007, Shafranjuk 2008, Syzranov et al. 2008, Kienle and Leonard 2009, Shafranjuk 2009, Shafranjuk 2011, Shafranjuk 2011), and we also present results of the T-ray sensing by the graphene and/or CNT devices (Rinzan et al. 2012). We will also describe physical mechanisms which determine how an AC field changes the electron transport through the QD as compared to the steady-state case. Major theoretical results will be scrutinized within contents of their relevance in respect to existing experimental data. We also illustrate how the theory (Akturk et al. 2007, Shafranjuk 2007, Shafranjuk 2008, Syzranov et al. 2008, Kienle and Leonard 2009, Shafranjuk 2009, Shafranjuk 2011b, Shafranjuk 2011b) helps to extract essential nanodevice parameters from the experiments. They include the level spacing and width, the level population, the effective electron temperature, etc. Other useful characteristics are noise, level broadening due to the electron–phonon interaction, and the spectrum of thermal phonons generated due to electron–phonon collisions. The last microscopic process serves as a major source of the intrinsic thermal noise (Shafranjuk 2009, Shafranjuk 2011a, 2000b). This allows us to establish necessary requirements for the QD to keep it "quiet." Besides we also discuss the Johnson–Nyquist noise and the shot noise.

A lot of attention nowadays is being devoted to QD detectors where an external AC field induces the interlevel transitions (Fuse et al. 2007, Fuse, Kawano, and Yamaguchi et al. 2007a, Fuse, Kawano, and Yamaguchi et al. 2007b, Rinzan et al. 2012). The QD detectors utilize a change in the electron transport properties under influence of AC external fields. If the electron energy in the source and drain electrodes ε matches the quantized level position E_{n}, provided $\varepsilon \to E_{\mathrm{n}}$, the tunneling probability through the QD is high. Relevant resonant peaks in the DC conductance G across QD are measured experimentally. An external AC field alters electron energy ε during the tunneling by the photon energies $\pm m\hbar\omega$ (m is the number of absorbed/emitted photons; $\hbar\omega$ is the photon energy), so the energy matching condition changes as $\varepsilon \pm m\hbar\omega \rightarrow E_{\mathrm{n}}$. The AC field modified resonance condition ensures that the side peaks in the DC conductance arise at new values $V_{\mathrm{SD}} \rightarrow V_{\mathrm{SD}} \pm m\hbar\omega/e$ of the source–drain voltage V_{SD}. At low temperatures, the side peaks may be a few orders of magnitude taller as compared to its steady-state value. The sharp changes in the QD conductance constitute an idea of the AC field sensing. The sharp and crisp resonances ensure the high accuracy of the QD THz detectors. By measuring the resonance width, spacing, and magnitude, one immediately determines the frequency, intensity, and even polarization of the external THz field. Another important issue is reducing of the intrinsic and parasitic noises which are limiting the sensitivity of the device (Shafranjuk 2009, Shafranjuk 2011a, 2011b). In this respect, the GFET and/or CNT FET (Tselev et al. 2009, Yang et al. 2012, Yang et al. 2013) as well as the THz detectors (Mele et al. 2000, Fuse et al. 2007, Fuse, Kawano, and Yamaguchi et al. 2007a, Fuse, Kawano, and Yamaguchi et al. 2007b, Rinzan et al. 2012) might have many basic advantages over regular semiconducting devices (Shafranjuk 2008, Shafranjuk 2009, Sizov and Rogalski 2010, Shafranjuk 2011a, Shafranjuk 2011b).

In the same Chapter 10 we will consider a THz sensor based on graphene and a CNT FET QD. The sensor is composed of graphene and/or CNT where quantized electron states are formed. The CNT FET QD is exposed to the THz field which induces transitions between the quantized states. We will see that the interlevel transitions are affecting the DC differential conductance which becomes sturdily dependent on the THz field parameters. The huge change in the DC conductance indicates a high sensitivity of the CNT/graphene FET transport versus the external THz field. Functioning of the CNT/graphene QD utilizes a variety of processes. This includes transmissions of the particles from attached electrodes into and out the CNT/graphene QD, interelectrode transitions of the particles through the QD via the discrete energy levels, and the field-induced interlevel transitions. We describe the experimental CNT/graphene QD setup and the measurement in conditions of nonequilibrium cooling caused by the photon-assisted single electron tunneling (PASET) through the CNT QD. The theoretical model of that phenomenon is also presented.

Chapter 11 is devoted to description of other atomic monolayers, monolayer, and a few-layered materials. The renewed interest in inorganic 2D materials with unique electronic and optical attributes has been motivated by the remarkable properties of graphene. Most of attention in this respect is paid to transition metal dichalcogenides (TMDCs) which are layered materials with strong in-plane bonding and weak out-of-plane interactions. That structural property enables exfoliation into 2D layers of single unit cell thickness. Despite the fact that TMDC materials have been studied for decades, only recently the advances in nanoscale materials' characterization and device fabrication have opened up new opportunities. The motivation emerges from great perspective for the 2D layers of thin TMDCs in nanoelectronics and opto-electronics. Many TMDCs such as MoS_2, $MoSe_2$, WS_2, and WSe_2 demonstrate considerable band gaps that change from indirect to direct in single layers, allowing applications such as transistors, photodetectors, and electroluminescent devices. In Chapter 11 we review the origin of TMDCs, methods for preparing of atomically thin layers, and their electronic and optical properties. The fabrication methods, bottom-up fabrication, electronic bandstructure, electric transport in nanodevices, and electronic transport versus scattering mechanisms also represent the subject of this chapter. Besides we discuss the TMDC transistors, perspectives of the TMDC electronics, optoelectronics, vibrational and optical properties of TMDCs, transparent and flexible optoelectronics, photodetection and photovoltaics, emission of light, orbit, spin, and valley interactions, molecular sensing, and the prospects for future advances in electronics and optoelectronics.

References

Abergel, D. S. L., A. Russell, and V. I. Fal'ko (2007). Visibility of graphene flakes on a dielectric substrate. *Appl Phys Lett*, **91**(6): 063125.

Abrikosov, A. A. (1988). *Fundamentals of the Theory of Metals*. Amsterdam, New York, North Holland, sole distributors for USA and Canada, Elsevier Science.

Ajiki, H., and T. Ando (1993). Magnetic-properties of carbon nanotubes. *J Phys Soc Jpn*, **62**(7): 2470–2480.

Ajiki, H., and T. Ando (1994). Magnetic-properties of carbon nanotubes [**62**, 2470, (1993)]. *J Phys Soc Jpn*, **63**(11): 4267–4267.

Ajiki, H., and T. Ando (1995). Magnetic-properties of ensembles of carbon nanotubes. *J Phys Soc Jpn*, **64**(11): 4382–4391.

Akturk, A., N. Goldsman, G. Pennington, and A. Wickenden (2007). Terahertz current oscillations in single-walled zigzag carbon nanotubes. *Phys Rev Lett*, **98**(16): 166803.

Ando, T. (1997). Excitons in carbon nanotubes. *J Phys Soc Jpn*, **66**(4): 1066–1073.

Ando, T. (2000). Spin-orbit interaction in carbon nanotubes. *J Phys Soc Jpn*, **69**(6): 1757–1763.

Ando, T. (2004). Excitons in carbon nanotubes revisited: dependence on diameter, Aharonov-Bohm flux, and strain. *J Phys Soc Jpn*, **73**(12): 3351–3363.

Ando, T. (2004). Theory of electronic states and optical absorption in carbon nanotubes. *P Soc Photo-Opt Ins*, **5349**: 1–10.

Ando, T. (2005). Theory of electronic states and transport in carbon nanotubes. *J Phys Soc Jpn*, **74**(3): 777–817.

Ando, T., T. Nakanishi, and M. Igami (1999). Effective-mass theory of carbon nanotubes with vacancy. *J Phys Soc Jpn*, **68**(12): 3994–4008.

Ando, T., T. Nakanishi, and R. Saito (1998). Berry's phase and absence of back scattering in carbon nanotubes. *J Phys Soc Jpn*, **67**(8): 2857–2862.

Ando, T., and H. Suzuura (2002). Presence of perfectly conducting channel in metallic carbon nanotubes. *J Phys Soc Jpn*, **71**(11): 2753–2760.

Andreoni, W., and A. Curioni (1996). Freedom and constraints of a metal atom encapsulated in fullerene cages. *Phys Rev Lett*, **77**(5): 834–837.

Bethune, D. S. (1996). Adding metal to carbon: production and properties of metallofullerenes and single-layer nanotubes. *NATO Adv Sci I E-App*, **316**: 165–181.

Bethune, D. S. (2002). Carbon and metals: a path to single-wall carbon nanotubes. *Phys B*, **323**(1–4): 90–96.

Blake, P., E. W. Hill, A. H. C. Neto, K. S. Novoselov, D. Jiang, R. Yang, T. J. Booth, and A. K. Geim (2007). Making graphene visible. *Appl Phys Lett*, **91**(6): 063124.

Bolotin, K. I., K. J. Sikes, J. Hone, H. L. Stormer, and P. Kim (2008). Temperature-dependent transport in suspended graphene. *Phys Rev Lett*, **101**(9): 096802.

Bolotin, K. I., K. J. Sikes, Z. Jiang, M. Klima, G. Fudenberg, J. Hone, P. Kim, and H. L. Stormer (2008). Ultrahigh electron mobility in suspended graphene. *Solid State Commun*, **146**(9–10): 351–355.

Brey, L., and H. A. Fertig (2006). Edge states and the quantized Hall effect in graphene. *Phys Rev B*, **73**(19): 195408.

Brey, L., and H. A. Fertig (2006). Electronic states of graphene nanoribbons studied with the Dirac equation. *Phys Rev B*, **73**(23): 235411.

Calandra, M., and F. Mauri (2007). Electronic structure of heavily doped graphene: the role of foreign atom states [**76**, art no 161406, (2007)]. *Phys Rev B*, **76**(19).

Calizo, I., W. Z. Bao, F. Miao, C. N. Lau, and A. A. Balandin (2007). The effect of substrates on the Raman spectrum of graphene: graphene-on-sapphire and graphene-on-glass. *Appl Phys Lett*, **91**(20).

Casiraghi, C., A. Hartschuh, E. Lidorikis, H. Qian, H. Harutyunyan, T. Gokus, K. S. Novoselov, and A. C. Ferrari (2007). Rayleigh imaging of graphene and graphene layers. *Nano Lett*, **7**(9): 2711–2717.

Cheremisin, M. V., M. I. Dyakonov, M. S. Shur, and G. Samsonidze (1998). Influence of electron scattering on current instability in field effect transistors. *Solid State Electron*, **42**(9): 1737–1742.

Cho, S. J., Y. F. Chen, and M. S. Fuhrer (2007). Gate-tunable graphene spin valve. *Appl Phys Lett*, **91**(12).

Das, A., B. Chakraborty, and A. K. Sood (2008). Raman spectroscopy of graphene on different substrates and influence of defects. *B Mater Sci*, **31**(3): 579–584.

Dresselhaus, M. S., and G. Dresselhaus (1994). New directions in intercalation research. *Mol Cryst Liq Cryst Sci Technol Sect a: Mol Cryst Liq Cryst*, **244**: 1–12.

Faugeras, C., A. Nerriere, M. Potemski, A. Mahmood, E. Dujardin, C. Berger, and W. A. de Heer (2008). Few-layer graphene on SiC, pyrolitic graphite, and graphene: a Raman scattering study. *Appl Phys Lett*, **92**(1).

Fuse, T., Y. Kawano, M. Suzuki, Y. Aoyagi, and K. Ishibashi (2007). Coulomb peak shifts under terahertz-wave irradiation in carbon nanotube single-electron transistors. *Appl Phys Lett*, **90**(1).

Fuse, T., Y. Kawano, T. Yamaguchi, Y. Aoyagi, and K. Ishibashi (2007a). Quantum response of carbon nanotube quantum dots to terahertz wave irradiation. *Nanotechnology*, **18**(4).

Fuse, T., Y. Kawano, T. Yamaguchi, Y. Aoyagi, and K. Ishibashi (2007b). Single electron transport of carbon nanotube quantum dots under THz laser irradiation. *AIP Conf Proc*, **893**: 1013–1014.

Geim, A. K. (2009). Graphene: status and prospects. *Science*, **324**(5934): 1530–1534.

Geim, A. K. (2012). Graphene prehistory. *Phys Scripta*, **T146**.

Geim, A. K., and A. H. MacDonald (2007). Graphene: exploring carbon flatland. *Phys Today*, **60**(8): 35–41.

Geim, A. K., and K. S. Novoselov (2007). The rise of graphene. *Nat Mater*, **6**(3): 183–191.

Giovannetti, G., P. A. Khomyakov, G. Brocks, P. J. Kelly, and J. van den Brink (2007). Substrate-induced band gap in graphene on hexagonal boron nitride: ab initio density functional calculations. *Phys Rev B*, **76**(7).

Hamada, N. (1993). Electronic Band-structure of carbon nanotubes: toward the 3-dimensional system. *Mater Sci Eng, B*, **19**(1–2): 181–184.

Heersche, H. B., P. Jarillo-Herrero, J. B. Oostinga, L. M. K. Vandersypen, and A. F. Morpurgo (2007). Bipolar supercurrent in graphene. *Nature*, **446**(7131): 56–59.

Iijima, S. (1991). Helical microtubules of graphitic carbon. *Nature*, **354**(6348): 56–58.

Katsnelson, M. I. (2007). Conductance quantization in graphene nanoribbons: adiabatic approximation. *Eur Phys J B*, **57**(3): 225–228.

Katsnelson, M. I., and K. S. Novoselov (2007). Graphene: new bridge between condensed matter physics and quantum electrodynamics. *Solid State Commun*, **143**(1–2): 3–13.

Kawano, Y. (2011). Highly sensitive detector for on-chip near-field THz imaging. *IEEE J Sel Top Quant*, **17**(1): 67–78.

Kawano, Y. (2012). Terahertz sensing and imaging based on nanostructured semiconductors and carbon materials. *Laser Photonics Rev*, **6**(2): 246–257.

Kawano, Y. (2013a). Terahertz nano-devices and nano-systems. *Handbook of Terahertz Technology for Imaging, Sensing and Communications*, Vol. 34: 403–422.

Kawano, Y. (2013b). Terahertz waves: a tool for condensed matter, the life sciences and astronomy. *Contemp Phys*, **54**(3): 143–165.

Kawano, Y., and K. Ishibashi (2010). On-chip near-field terahertz detection based on a two-dimensional electron gas. *Phys E*, **42**(4): 1188–1191.

Kawano, Y., T. Uchida, and K. Ishibashi (2009). Terahertz sensing with a carbon nanotube/two-dimensional electron gas hybrid transistor. *Appl Phys Lett*, **95**(8).

Kiang, C. H., W. A. Goddard, R. Beyers, and D. S. Bethune (1995). Carbon nanotubes with single-layer walls. *Carbon*, **33**(7): 903–914.

Kienle, D., and F. Leonard (2009). Terahertz response of carbon nanotube transistors. *Phys Rev Lett*, **103**(2).

Kittel, C. (1962). *Elementary Solid State Physics: A Short Course*. New York, Wiley.

Kosmrlj, A., and D. R. Nelson (2013). Mechanical properties of warped membranes. *Phys Rev E*, **88**(1).

Leenaerts, O., B. Partoens, and F. M. Peeters (2008a). Adsorption of H(2)O, NH(3), CO, NO(2), and NO on graphene: a first-principles study. *Phys Rev B*, **77**(12).

Leenaerts, O., B. Partoens, and F. M. Peeters (2008b). Graphene: a perfect nanoballoon. *Appl Phys Lett*, **93**(19).

Leenaerts, O., B. Partoens, and F. M. Peeters (2008c). Paramagnetic adsorbates on graphene: a charge transfer analysis. *Appl Phys Lett*, **92**(24).

Maksimenko, S. A., A. A. Khrushchinsky, G. Y. Slepyan, and O. V. Kibis (2006). Nonlinear interaction of electromagnetic waves with chiral carbon nanotubes. Helical parametrization. *Nanomodeling II*, **6328**.

Martins, T. B., R. H. Miwa, A. J. R. da Silva, and A. Fazzio (2007). Electronic and transport properties of boron-doped graphene nanoribbons. *Phys Rev Lett*, **98**(19).

Matsumura, H., and T. Ando (1998). Effective-mass theory of carbon nanotube junctions. *J Phys Soc Jpn*, **67**(10): 3542–3551.

Matsumura, H., and T. Ando (2001). Conductance of carbon nanotubes with a Stone-Wales defect. *J Phys Soc Jpn*, **70**(9): 2657–2665.

McKellar, B. H., and G. J. Stephenson, Jr. (1987). Klein paradox and the Dirac-Kronig-Penney model. *Phys Rev A*, **36**(6): 2566–2569.

Mele, E. J., P. Kral, and D. Tomanek (2000). Coherent control of photocurrents in graphene and carbon nanotubes. *Phys Rev B*, **61**(11): 7669–7677.

Morozov, S. V., K. S. Novoselov, M. I. Katsnelson, F. Schedin, L. A. Ponomarenko, D. Jiang, and A. K. Geim (2006). Strong suppression of weak localization in graphene. *Phys Rev Lett*, **97**(1).

Nakada, K., M. Fujita, G. Dresselhaus, and M. S. Dresselhaus (1996). Edge state in graphene ribbons: Nanometer size effect and edge shape dependence. *Phys Rev B*, **54**(24): 17954–17961.

Nemilentsau, A. M., A. A. Khrutchinskii, G. Y. Slepyan, and S. A. Maksimenko (2006). Third-order nonlinearity and plasmon properties in carbon nanotubes. *Carbon Nanotubes*, **222**: 175–176.

Neto, A. C., F. Guinea, and N. M. R. Peres (2006). Drawing conclusions from graphene. *Phys World*, **19**(11): 33–37.

Novoselov, K. S., A. K. Geim, S. V. Morozov, D. Jiang, M. I. Katsnelson, I. V. Grigorieva, S. V. Dubonos, and A. A. Firsov (2005). Two-dimensional gas of massless Dirac fermions in graphene. *Nature*, **438**(7065): 197–200.

Novoselov, K. S., A. K. Geim, S. V. Morozov, D. Jiang, Y. Zhang, S. V. Dubonos, I. V. Grigorieva, and A. A. Firsov (2004). Electric field effect in atomically thin carbon films. *Science*, **306**(5696): 666–669.

Pauling, L. (1972). Stirling approximation. *Chem Brit*, **8**(10): 447.

Peres, N. M. R. (2009). The transport properties of graphene. *J Phys: Condens Mater*, **21**(32).

Peres, N. M. R. (2010). Colloquium: the transport properties of graphene: an introduction. *Rev Mod Phys*, **82**(3): 2673–2700.

Peres, N. M. R., F. D. Klironomos, S. W. Tsai, J. R. Santos, J. M. B. L. dos Santos, and A. H. C. Neto (2007). Electron waves in chemically substituted graphene. *EPL*, **80**(6).

Recher, P., B. Trauzettel, A. Rycerz, Y. M. Blanter, C. W. J. Beenakker, and A. F. Morpurgo (2007). Aharonov-Bohm effect and broken valley degeneracy in graphene rings. *Phys Rev B*, **76**(23).

Rinzan, M., G. Jenkins, H. D. Drew, S. Shafranjuk, and P. Barbara (2012). Carbon nanotube quantum dots as highly sensitive terahertz-cooled spectrometers. *Nano Lett*, **12**(6): 3097–3100.

Rusin, T. M., and W. Zawadzki (2007). Transient Zitterbewegung of charge carriers in mono- and bilayer graphene, and carbon nanotubes. *Phys Rev B*, **76**(19).

Sakai, H., H. Suzuura, and T. Ando (2004). Effective-mass approach to interaction effects on electronic structure in carbon nanotubes. *Phys E*, **22**(1–3): 704–707.

Sawada, S., and N. Hamada (1992). Energetics of carbon nanotubes. *Solid State Commun*, **83**(11): 917–919.

Schedin, F., A. K. Geim, S. V. Morozov, E. W. Hill, P. Blake, M. I. Katsnelson, and K. S. Novoselov (2007). Detection of individual gas molecules adsorbed on graphene. *Nat Mater*, **6**(9): 652–655.

Shafranjuk, S. E. (2007). Sensing an electromagnetic field with photon-assisted Fano resonance in a two-branch carbon nanotube junction. *Phys Rev B*, **76**(8).

Shafranjuk, S. E. (2008). Probing the intrinsic state of a one-dimensional quantum well with photon-assisted tunneling. *Phys Rev B*, **78**(23).

Shafranjuk, S. E. (2009). Directional photoelectric current across the bilayer graphene junction. *J Phys: Condens Mater*, **21**(1).

Shafranjuk, S. E. (2011). Electromagnetic properties of the graphene junctions. *Eur Phys J B*, **80**(3): 379–393.

Shafranjuk, S. E. (2011). Resonant transport through a carbon nanotube junction exposed to an ac field. *J Phys: Condens Mater*, **23**(49).

Sizov, F., and A. Rogalski (2010). THz detectors. *Prog Quant Electron*, **34**(5): 278–347.

Slepyan, G. Y., M. V. Shuba, S. A. Maksimenko, and A. Lakhtakia (2006). Theory of optical scattering by achiral carbon nanotubes and their potential as optical nanoantennas. *Phys Rev B*, **73**(19).

Stephan, O., P. M. Ajayan, C. Colliex, P. Redlich, J. M. Lambert, P. Bernier, and P. Lefin (1994). Doping graphitic and carbon nanotube structures with boron and nitrogen. *Science*, **266**(5191): 1683–1685.

Stone, M., H. W. Wyld, and R. L. Schult (1992). Edge waves in the quantum Hall effect and quantum dots. *Phys Rev B: Condens Matter*, **45**(24): 14156–14161.

Suzuura, H., and T. Ando (2000). Huge magnetoresistance by phonon scattering in carbon nanotubes. *Phys E*, **6**(1–4): 864–867.

Syzranov, S. V., M. V. Fistul, and K. B. Efetov (2008). Effect of radiation on transport in graphene. *Phys Rev B*, **78**(4).

Tan, Y. W., Y. Zhang, K. Bolotin, Y. Zhao, S. Adam, E. H. Hwang, S. Das Sarma, H. L. Stormer, and P. Kim (2007). Measurement of scattering rate and minimum conductivity in graphene. *Phys Rev Lett*, **99**(24).

Tanuma, S. (1985). Proceedings of the international-symposium on graphite-intercalation compounds, Tsukuba, Japan, 27–30 May 1985. Opening address. *Synth Met*, **12**(1–2): 1–3.

Tselev, A., Y. F. Yang, J. Zhang, P. Barbara, and S. E. Shafranjuk (2009). Carbon nanotubes as nanoscale probes of the superconducting proximity effect in Pd-Nb junctions. *Phys Rev B*, **80**(5).

Uchoa, B., C. Y. Lin, and A. H. C. Neto (2008). Tailoring graphene with metals on top. *Phys Rev B*, **77**(3).

Uchoa, B., and A. H. C. Neto (2007). Superconducting states of pure and doped graphene. *Phys Rev Lett*, **98**(14).

Varchon, F., R. Feng, J. Hass, X. Li, B. N. Nguyen, C. Naud, P. Mallet, J. Y. Veuillen, C. Berger, E. H. Conrad, and L. Magaud (2007). Electronic structure of epitaxial graphene layers on SiC: effect of the substrate. *Phys Rev Lett*, **99**(12).

Viet, N. A., H. Ajiki, and T. Ando (1994). Lattice instability in metallic carbon nanotubes. *J Phys Soc Jpn*, **63**(8): 3036–3047.

Wakabayashi, K., M. Fujita, H. Ajiki, and M. Sigrist (1999). Electronic and magnetic properties of nanographite ribbons. *Phys Rev B*, **59**(12): 8271–8282.

Wehling, T. O., K. S. Novoselov, S. V. Morozov, E. E. Vdovin, M. I. Katsnelson, A. K. Geim, and A. I. Lichtenstein (2008). Molecular doping of graphene. *Nano Lett*, **8**(1): 173–177.

Yaguchi, T., and T. Ando (2001). Topological effects in capped carbon nanotubes. *J Phys Soc Jpn*, **70**(12): 3641–3649.

Yaguchi, T., and T. Ando (2002). Electronic states in capped carbon nanotubes [**70**, 1327, (2001)]. *J Phys Soc Jpn*, **71**(11): 2824–2824.

Yaguchi, T., and T. Ando (2003). Cap states in capped carbon nanotubes by effective-mass theory. *Phys E*, **18**(1–3): 220–222.

Yang, Y., G. Fedorov, P. Barbara, S. E. Shafranjuk, B. K. Cooper, R. M. Lewis, and C. J. Lobb (2013). Coherent nonlocal transport in quantum wires with strongly coupled electrodes. *Phys Rev B*, **87**(4).

Yang, Y. F., G. Fedorov, J. Zhang, A. Tselev, S. Shafranjuk, and P. Barbara (2012). The search for superconductivity at van Hove singularities in carbon nanotubes. *Supercond Sci Tech*, **25**(12).

Yosida, Y., K. Sato, S. Tanuma, and Y. Iye (1985). Magnetic-susceptibility of Sbcl5-graphite intercalation compounds. 2. *Synth Met*, **12**(1–2): 319–324.

Yosida, Y., S. Tanuma, and Y. Iye (1985). Magnetic-susceptibility of Sbcl5-graphite intercalation compounds. *J Phys Soc Jpn*, **54**(7): 2635–2640.

Zhou, S. Y., G. H. Gweon, A. V. Fedorov, P. N. First, W. A. De Heer, D. H. Lee, F. Guinea, A. H. C. Neto, and A. Lanzara (2007). Substrate-induced bandgap opening in epitaxial graphene. *Nat Mater*, **6**(11): 916–916.

Chapter 1

Chiral Fermions in Graphene

Electronic transport properties of conventional metals and semiconductors are described in terms of quasiparticle excitations, electrons, and holes. Simple analytical models of Fermi gas and Fermi liquid provide an illustrative depiction of various electronic phenomena, and they also allow a deeper understanding the properties of composite systems in terms of elementary excitations (Abrikosov 1988). They include transport in the inhomogeneous structures, like tunneling junctions and constrictions (Wolf 2012), and also account for influence the external DC magnetic (Ando and Seri 1997, Arimura and Ando 2012) and AC electromagnetic fields (Shafranjuk 2009, Shafranjuk 2011) on the electron transport of charged particles in the conducting materials.

1.1 Low-Energy Electron Excitations in Graphene

The discovery of graphene (Geim 2011, Geim 2012) invoked the question whether a similar handy model can be formulated for the new material. In response, it motivated creating of a simple and illustrative mathematical model (Ando 2005, Katsnelson et al. 2006, Katsnelson 2012) capable of systematically covering the electronic

Graphene: Fundamentals, Devices, and Applications
Serhii Shafraniuk

ISBN 978-981-4613-47-7 (Hardcover), 978-981-4613-48-4 (eBook)
www.panstanford.com

Figure 1.1 A sheet of monoatomic graphene with the honeycomb crystal lattice.

low-energy properties of graphene and adequately understanding the obtained experimental data. The honeycomb lattice symmetry (we sketch it in Fig. 1.1) and appropriate value of the intersite transfer integral make graphene to be unique and different from the conventional conductors (Abrikosov 1988).

Electrons and holes in regular semiconductors and metals are described by a Schrödinger equation for spinless quasiparticles with a finite effective mass (Abrikosov 1988). This approach appears to be incomplete for the atomically monolayer graphene; its low-energy elementary excitations are characterized by additional degrees of freedom (Ando 2005, Katsnelson et al. 2006, Katsnelson 2012). Furthermore, the elementary excitations in graphene act like particle with two one-half pseudospins. Thus, an adequate and handy description of the low-energy electronic properties in graphene is based on the relativistic Dirac equation for charged particles with two one-half pseudospins. Such relativistic model captures basic features of the low-energy electron spectrum and describes the transport properties of graphene. In this chapter we consider a model of electron system in terms of chiral noninteracting relativistic particles. Many-body interaction effects will be addressed in the consecutive chapters. Below we derive the wave equation which is based on the simplest symmetry assumptions and contains minimum parameters. Nevertheless, it nicely and adequately captures the most important features of the electron system in graphene. The honeycomb graphene atomic lattice is depicted in Figs. 1.1 and 1.2. The lattice is characterized by primitive translation vectors $\mathbf{a} = a(1,\ 0)$ and $\mathbf{b} = a(-1/2,\ \sqrt{3}/2)$,

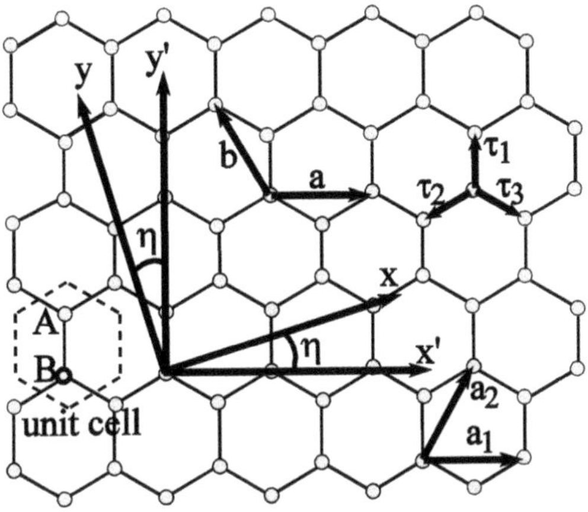

Figure 1.2 The coordinate system and the lattice structure of graphene. Here **a** and **b** are two primitive translation vectors. The hexagonal unit cell represented by a dashed line contains two carbon atoms denoted by A and B. Three vectors directed from a B site to the nearest-neighbor A sites are given by $\boldsymbol{\tau}_l$ ($l = 1, 2, 3$). The coordinates x' and y' are fixed onto the graphene, and x and y are along the circumference and axis, respectively, and denote a chiral angle. Another choice of primitive translation vectors is a couple of $\mathbf{a}_1$ and $\mathbf{a}_2$.

where the lattice constant is $a = 0.246$ nm. The elementary lattice cell contains two carbon atoms, which are denoted as A and B. The nearest-neighbor carbon atoms are connected by the vectors $\boldsymbol{\tau}_1 = a(0,\ 1/\sqrt{3})$, $\boldsymbol{\tau}_2 = a(-1/2, -1/2\sqrt{3})$, and $\boldsymbol{\tau}_3 = a(1/2, -1/2\sqrt{3})$. One easily calculates the area of a hexagonal unit cell as $\Omega_0 = a^2\sqrt{3}/2$ (Kittel 2005).

The primitive reciprocal lattice vectors $\mathbf{a}^*$ and $\mathbf{b}^*$ characterizing the first Brillouin zone (Kittel 2005) are $\mathbf{a}^* = (2\pi/a)(1,\ 1/\sqrt{3})$ and $\mathbf{b}^* = (2\pi/a)(0,\ 2/\sqrt{3})$, respectively (see Figs. 2, 1.3). Then one immediately computes the area of the hexagonal first Brillouin zone as

$$\Omega_0^* = \left(2/\sqrt{3}\right)(2\pi/a)^2 \tag{1.1}$$

The corners K and K′ of the Brillouin zone (see Figs. 2, 1.3) are located in the reciprocal momentum space at the points with coordinates $K = (2\pi/a)(1/3,\ 1/\sqrt{3})$ and $K' = (2\pi/a)(1/3,\ 1/\sqrt{3})$. We will see that K and K′ points play a special role in the transport

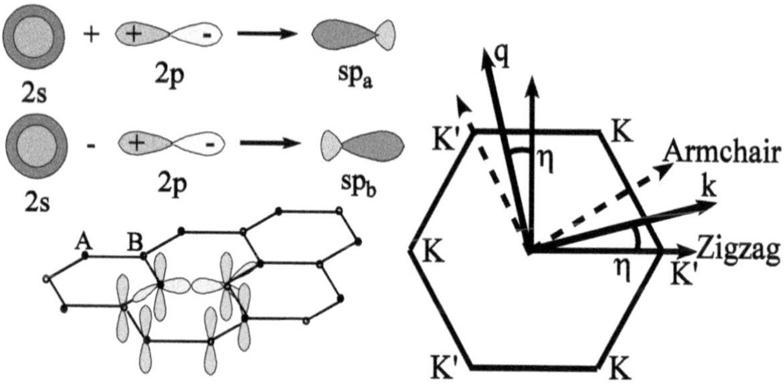

Figure 1.3 Left: Bonding in the x–y atomic plane due to the sp^2-hybridized orbitals in graphene. The remaining $2p$ orbitals (cyan) are directed perpendicular to the plane. They provide one conducting electron per C atom. Right: The first Brillouin zone of graphene has the honeycomb shape. The neutrality points K and K′, which are positioned at the hexagon's vertices, are connecting via the cone-shaped valence and conductance bands.

electronic properties of graphene. One also utilizes the useful relationships $\exp(i\mathbf{K}\cdot\boldsymbol{\tau}_1) = \omega$, $\exp(i\mathbf{K}\cdot\boldsymbol{\tau}_2) = 1/\omega$, $\exp(i\mathbf{K}\cdot\boldsymbol{\tau}_3) = 1$, $\exp(i\mathbf{K}'\cdot\boldsymbol{\tau}_1) = 1$, $\exp(i\mathbf{K}'\cdot\boldsymbol{\tau}_2) = 1/\omega$, and $\exp(i\mathbf{K}'\cdot\boldsymbol{\tau}_3) = \omega$, where $\omega = \exp(2\pi i/3)$. The above simple relationships can be used to obtain a low-energy spectrum of elementary excitations starting from a tight-binding model. The composite ground-state electronic wavefunction, for sites A and B, is decomposed into a sum of product terms expressing the donation of conduction electrons from p_z localized orbitals, $\varphi(\mathbf{r} - \mathbf{R}_{A(B)})$ and translational plane-wave-like functions $\psi_{A(B)}$. Then one obtains

$$\psi(\mathbf{r}) = \sum_{R_A} \psi_A(\mathbf{R}_A)\,\varphi(\mathbf{r} - \mathbf{R}_A) + \sum_{R_B} \psi_B(\mathbf{R}_B)\,\varphi(\mathbf{r} - \mathbf{R}_B) \qquad (1.2)$$

where the A and B atomic site coordinates are $\mathbf{R}_A = n_a\mathbf{a} + n_b\mathbf{b} + \boldsymbol{\tau}_1$ and $\mathbf{R}_B = n_a\mathbf{a} + n_b\mathbf{b}$. We set the zero energy at the carbon p_z level and denote the transfer integral between nearest-neighbor carbon atoms as $-\gamma_0$. From the Schrödinger equation

$$\mathbf{H}_{ij}\psi_j = \varepsilon\psi_i \qquad (1.3)$$

where the indices $i,\ j = A,\ B$. In the matrix form, the Hamiltonian is written as

$$\hat{\mathbf{H}} = E_0 \begin{pmatrix} 1 & 0 \\ 0 & 1 \end{pmatrix} + \begin{pmatrix} 0 & -\gamma_0 \sum_{l=1}^{3} \exp\left(-i\mathbf{k}\cdot\boldsymbol{\tau}_l\right) \\ -\gamma_0 \sum_{l=1}^{3} \exp\left(i\mathbf{k}\cdot\boldsymbol{\tau}_l\right) & 0 \end{pmatrix} \tag{1.4}$$

which gives

$$\begin{aligned} E_0\psi_A - \gamma_0 \textstyle\sum_{l=1}^{3} \exp\left(-i\mathbf{k}\cdot\boldsymbol{\tau}_l\right)\, \psi_B &= \varepsilon\psi_A \\ -\gamma_0 \textstyle\sum_{l=1}^{3} \exp\left(i\mathbf{k}\cdot\boldsymbol{\tau}_l\right)\, \psi_A + E_0\psi_B &= \varepsilon\psi_B \end{aligned} \tag{1.5}$$

or is rewritten in the matrix form as

$$\begin{pmatrix} E_0 & -\gamma_0 \sum_{l=1}^{3} \exp\left(-i\mathbf{k}\cdot\boldsymbol{\tau}_l\right) \\ -\gamma_0 \sum_{l=1}^{3} \exp\left(i\mathbf{k}\cdot\boldsymbol{\tau}_l\right) & E_0 \end{pmatrix} \begin{pmatrix} \psi_A \\ \psi_B \end{pmatrix} = \varepsilon \begin{pmatrix} \psi_A \\ \psi_B \end{pmatrix} \tag{1.6}$$

Henceforth, we set $E_0 = 0$. Then we merely obtain

$$\begin{aligned} \varepsilon\psi_A\left(\mathbf{R}_A\right) &= -\gamma_0 \textstyle\sum_{l=1}^{3} \Psi_B\left(\mathbf{R}_A - \boldsymbol{\tau}_l\right) \\ \varepsilon\psi_B\left(\mathbf{R}_B\right) &= -\gamma_0 \textstyle\sum_{l=1}^{3} \Psi_A\left(\mathbf{R}_B + \boldsymbol{\tau}_l\right) \end{aligned} \tag{1.7}$$

We assume that the electron wavefunction is modulated by an envelope functions $f_{A,B}$ and thereby takes the form $\psi_A\left(\mathbf{R}_A\right) \propto f_A\left(\mathbf{k}\right) \exp\left(i\mathbf{k}\cdot\mathbf{R}_A\right)$ and $\psi_B\left(\mathbf{R}_B\right) \propto f_B\left(\mathbf{k}\right) \exp\left(i\mathbf{k}\cdot\mathbf{R}_B\right)$. In this way we separate "slower" effect of the "honeycomb" crystal symmetry from the faster plane wave factors $\exp\left(i\mathbf{k}\cdot\mathbf{R}_{A(B)}\right)$. Then the above Eq. 1.7 is rewritten as

$$\begin{pmatrix} 0 & h_{AB}\left(\mathbf{k}\right) \\ h_{AB}^{*}\left(\mathbf{k}\right) & 0 \end{pmatrix} \begin{pmatrix} f_A\left(\mathbf{k}\right) \\ f_B\left(\mathbf{k}\right) \end{pmatrix} = \varepsilon \begin{pmatrix} f_A\left(\mathbf{k}\right) \\ f_B\left(\mathbf{k}\right) \end{pmatrix} \tag{1.8}$$

In Eq. 1.8 we have introduced the notation

$$\begin{aligned} h_{AB}\left(\mathbf{k}\right) = -\gamma_0 \sum_{l=1}^{3} \exp\left(-i\mathbf{k}\cdot\boldsymbol{\tau}_l\right) &= -\gamma_0 (\exp\left(-i\mathbf{k}\cdot\boldsymbol{\tau}_1\right) \\ &+ \exp\left(-i\mathbf{k}\cdot\boldsymbol{\tau}_2\right) + \exp\left(-i\mathbf{k}\cdot\boldsymbol{\tau}_3\right)) \end{aligned} \tag{1.9}$$

with the newly introduced vectors

$$\begin{aligned} &\boldsymbol{\tau}_1 = a\left(0,\ 1/\sqrt{3}\right),\ \boldsymbol{\tau}_2 = a\left(-1/2,\ -1/2\sqrt{3}\right), \\ &\boldsymbol{\tau}_3 = a\left(1/2,\ -1/2\sqrt{3}\right) \end{aligned} \tag{1.10}$$

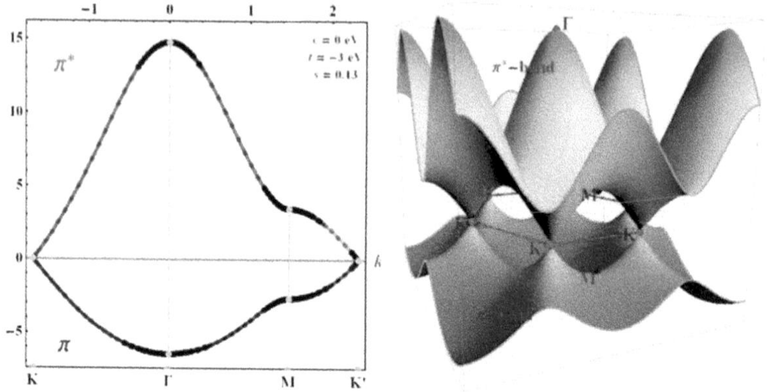

Figure 1.4 The energy dispersion law in pristine graphene according to the Dirac equation model.

The matrix equation Eq. 1.8 has a nontrivial solution if the determinant vanishes:

$$\left\| \begin{matrix} \varepsilon & h_{\mathrm{AB}}(\mathbf{k}) \\ h_{\mathrm{AB}}(\mathbf{k})^* & \varepsilon \end{matrix} \right\| = \varepsilon^2 - \gamma_0^2 e^{-ik\tau_1} e^{ik\tau_1} - \gamma_0^2 e^{-ik\tau_1} e^{ik\tau_2}$$
$$- \gamma_0^2 e^{ik\tau_1} e^{-ik\tau_2} - \gamma_0^2 e^{-ik\tau_1} e^{ik\tau_3} - \gamma_0^2 e^{ik\tau_1} e^{-ik\tau_3}$$
$$- \gamma_0^2 e^{-ik\tau_2} e^{ik\tau_2} - \gamma_0^2 e^{-ik\tau_2} e^{ik\tau_3} - \gamma_0^2 e^{ik\tau_2} e^{-ik\tau_3}$$
$$- \gamma_0^2 e^{-ik\tau_3} e^{ik\tau_3} = 0 \tag{1.11}$$

The last algebraic equation Eq. 1.11 has the following roots:

$$\varepsilon_{1,2} = \pm\gamma_0 \sqrt{e^{-ik\tau_1} + e^{-ik\tau_2} + e^{-ik\tau_3}} \sqrt{e^{ik\tau_1} + e^{ik\tau_2} + e^{ik\tau_3}} \tag{1.12}$$

The values of $\varepsilon_{1,2}$ in Eq. 1.12 actually represent the energy branches of excitations in the electron system shown in Fig. 1.4.

1.2 Dirac Equation for Chiral Fermions

1.2.1 *Dirac Equation for Monoatomic Graphene*

Electronic transport in conducting materials is determined by electron states near the Fermi energy (Abrikosov 1988, Kittel 2005). In an undoped pristine graphene (see Fig. 1.5), such states are localized near K and K′ points in the momentum space (Geim 2011). An awkward nearest-neighbor tight-binding model can be

Figure 1.5 A monoatomic sheet of graphene.

reformatted in a simpler and more illustrative fashion to describe electron states in vicinity of K and K′ points (Ando 2005). Below we derive an equation for an envelope electron wavefunction in the effective mass approximation which describes the states in vicinity of K and K′ points. Such an equation explicitly accounts for the symmetry of the system and provides an accurate description of the electron transport properties of graphene if its parameters are defined correctly. The honeycomb lattice of graphene is sketched in Figs. 1.1 and 1.2. Near the K and K′ points seating on the Fermi level $\varepsilon = 0$ we compose the electron wavefunctions of the plane waves modulated by envelope functions as

$$\begin{aligned}\psi_{\mathrm{A}}(\mathbf{R}_{\mathrm{A}}) &= e^{i\mathbf{K}\cdot\mathbf{R}_{\mathrm{A}}} F_{\mathrm{A}}^{K}(\mathbf{R}_{\mathrm{A}}) + e^{i\eta} e^{i\mathbf{K}'\cdot\mathbf{R}_{\mathrm{A}}} F_{\mathrm{A}}^{K'}(\mathbf{R}_{\mathrm{A}}) \\ \psi_{\mathrm{B}}(\mathbf{R}_{\mathrm{B}}) &= -\omega e^{i\eta} e^{i\mathbf{K}\cdot\mathbf{R}_{\mathrm{B}}} F_{\mathrm{B}}^{K}(\mathbf{R}_{\mathrm{B}}) + e^{i\mathbf{K}'\cdot\mathbf{R}_{\mathrm{B}}} F_{\mathrm{B}}^{K'}(\mathbf{R}_{\mathrm{B}}) \end{aligned} \qquad (1.13)$$

where $F_{\mathrm{A(B)}}^{K(K')}$ denote the envelope functions. By introducing auxiliary spinor matrices

$$\begin{aligned} a(\mathbf{R}_{\mathrm{A}}) &= \begin{pmatrix} e^{-i\mathbf{K}\cdot\mathbf{R}_{\mathrm{A}}} \\ e^{-i\eta} e^{-i\mathbf{K}'\cdot\mathbf{R}_{\mathrm{A}}} \end{pmatrix} \\ b(\mathbf{R}_{\mathrm{B}}) &= \begin{pmatrix} -\omega e^{-i\eta} e^{-i\mathbf{K}\cdot\mathbf{R}_{\mathrm{B}}} \\ e^{-i\eta} e^{-i\mathbf{K}'\cdot\mathbf{R}_{\mathrm{B}}} \end{pmatrix} \end{aligned} \qquad (1.14)$$

the above equations in Eq. 1.13 are rewritten as

$$\begin{aligned} \psi_{\mathrm{A}}(\mathbf{R}_{\mathrm{A}}) &= a(\mathbf{R}_{\mathrm{A}})^{\dagger} F_{\mathrm{A}}(\mathbf{R}_{\mathrm{A}}) \\ \psi_{\mathrm{B}}(\mathbf{R}_{\mathrm{B}}) &= b(\mathbf{R}_{\mathrm{B}})^{\dagger} F_{\mathrm{B}}(\mathbf{R}_{\mathrm{B}}) \end{aligned} \qquad (1.15)$$

where we also use

$$F_{\mathrm{A}} = \begin{pmatrix} F_{\mathrm{A}}^{K} \\ F_{\mathrm{A}}^{K'} \end{pmatrix}, F_{\mathrm{B}} = \begin{pmatrix} F_{\mathrm{B}}^{K} \\ F_{\mathrm{B}}^{K'} \end{pmatrix} \qquad (1.16)$$

In our new notations, equations for the envelope functions $F_{\mathrm{A(B)}}^{K(K')}$ are written as

$$\begin{aligned} \varepsilon a(\mathbf{R}_\mathrm{A})^\dagger F_\mathrm{A}(\mathbf{R}_\mathrm{A}) &= -\gamma_0 \textstyle\sum_{l=1}^{3} \left[b(\mathbf{R}_\mathrm{A} - \boldsymbol{\tau}_l)^\dagger F_\mathrm{B}(\mathbf{R}_\mathrm{A} - \boldsymbol{\tau}_l) \right] \\ \varepsilon b(\mathbf{R}_\mathrm{B})^\dagger F_\mathrm{B}(\mathbf{R}_\mathrm{B}) &= -\gamma_0 \textstyle\sum_{l=1}^{3} \left[a(\mathbf{R}_\mathrm{B} + \boldsymbol{\tau}_l)^\dagger F_\mathrm{A}(\mathbf{R}_\mathrm{B} + \boldsymbol{\tau}_l) \right] \end{aligned} \tag{1.17}$$

It is technically instructive to introduce the smoothing function $g(\mathbf{r}' - \mathbf{R}_\mathrm{A})$ which satisfies the following relationships:

$$\sum_{R_\mathrm{A}} g(\mathbf{r}' - \mathbf{R}_\mathrm{A}) = \sum_{R_\mathrm{B}} g(\mathbf{r}' - \mathbf{R}_\mathrm{B}) = 1 \tag{1.18}$$

and

$$\int dr' g(r' - \mathbf{R}_\mathrm{A}) = \int d\mathbf{r}' g(\mathbf{r}' - \mathbf{R}_\mathrm{B}) = \Omega_0 \tag{1.19}$$

Then

$$\begin{aligned} &\sum_{\mathbf{R}_\mathrm{A}} g(r' - \mathbf{R}_\mathrm{A})\; a(\mathbf{R}_\mathrm{A}) \times \varepsilon a(\mathbf{R}_\mathrm{A})^* F_\mathrm{A}(\mathbf{R}_\mathrm{A}) \\ &= -\gamma_0 \sum_{\mathbf{R}_\mathrm{A}} g(r' - \mathbf{R}_\mathrm{A})\; a(\mathbf{R}_\mathrm{A}) \\ &\quad \times \sum_{l=1}^{3} \left[b(\mathbf{R}_\mathrm{A} - \boldsymbol{\tau}_l)^* F_\mathrm{B}(\mathbf{R}_\mathrm{A} - \boldsymbol{\tau}_l) \right] \varepsilon b(\mathbf{R}_\mathrm{B})^* F_\mathrm{B}(\mathbf{R}_\mathrm{B}) \\ &= -\gamma_0 \sum_{l=1}^{3} \left[a(\mathbf{R}_\mathrm{B} + \boldsymbol{\tau}_l)^* F_\mathrm{A}(\mathbf{R}_\mathrm{B} + \boldsymbol{\tau}_l) \right] \end{aligned}$$

Here we approximate the smoothing function by

$$g(\mathbf{r}' - \mathbf{R}) \approx \Omega_0 \delta(\mathbf{r}' - \mathbf{R}) \tag{1.20}$$

which gives

$$\begin{aligned} &\varepsilon \sum_{R_\mathrm{A}} g(\mathbf{r}' - \mathbf{R}_\mathrm{A})\; a(\mathbf{R}_\mathrm{A})\; a(\mathbf{R}_\mathrm{A})^\dagger\; F_\mathrm{A}(\mathbf{R}_\mathrm{A}) \\ &= -\gamma_0 \sum_{R_\mathrm{A}} \sum_{l=1}^{3} g(\mathbf{r}' - \mathbf{R}_\mathrm{A})\; a(\mathbf{R}_\mathrm{A})\; b(\mathbf{R}_\mathrm{A} - \boldsymbol{\tau}_l)^\dagger\; F_\mathrm{B}(\mathbf{R}_\mathrm{A} - \boldsymbol{\tau}_l). \end{aligned} \tag{1.21}$$

Besides, we use the expansion

$$\begin{aligned} &\varepsilon \sum_{R_\mathrm{A}} g(\mathbf{r}' - \mathbf{R}_\mathrm{A})\; a(\mathbf{R}_\mathrm{A})\; a(\mathbf{R}_\mathrm{A})^\dagger F_\mathrm{A}(\mathbf{r}') \\ &= -\gamma_0 \sum_{R_\mathrm{A}} \sum_{l=1}^{3} g(\mathbf{r}' - \mathbf{R}_\mathrm{A})\; a(\mathbf{R}_\mathrm{A})\; b(\mathbf{R}_\mathrm{A} - \boldsymbol{\tau}_l)^\dagger F_\mathrm{B}(\mathbf{r}' - \boldsymbol{\tau}_l) \\ &= -\gamma_0 \sum_{R_\mathrm{A}} \sum_{l=1}^{3} g(\mathbf{r}' - \mathbf{R}_\mathrm{A})\; a(\mathbf{R}_\mathrm{A})\; b(\mathbf{R}_\mathrm{A} - \boldsymbol{\tau}_l)^\dagger \\ &\quad \left[F_\mathrm{B}(\mathbf{r}') - \boldsymbol{\tau}_l \cdot (i\nabla')\; F_\mathrm{B}(\mathbf{r}') + \ldots \right] \end{aligned} \tag{1.22}$$

Also we take into account that

$$\sum_{R_A} g\left(\mathbf{r}' - \mathbf{R}_A\right) a\left(\mathbf{R}_A\right) a\left(\mathbf{R}_A\right)^{\dagger} \approx \begin{pmatrix} 1 & 0 \\ 0 & 1 \end{pmatrix} \tag{1.23}$$

along with

$$\sum_{R_A} g\left(\mathbf{r}' - \mathbf{R}_A\right) a\left(\mathbf{R}_A\right) b\left(\mathbf{R}_A - \boldsymbol{\tau}_l\right)^{\dagger} \approx \begin{pmatrix} -\omega e^{i\eta} e^{-i\mathbf{K}\cdot\boldsymbol{\tau}_l} & 0 \\ 0 & e^{-i\eta} e^{-i\mathbf{K}'\cdot\boldsymbol{\tau}_l} \end{pmatrix} \tag{1.24}$$

Furthermore we are using

$$\sum_{l} e^{-i\mathbf{K}\cdot\boldsymbol{\tau}_l} \left(\tau_l^x \ \tau_l^y\right) = \frac{\sqrt{3}}{2}\omega^{-1} a\left(+i \ +1\right) \tag{1.25}$$

$$\sum_{l} e^{-i\mathbf{K}'\cdot\boldsymbol{\tau}_l} \left(\tau_l^x \ \tau_l^y\right) = \frac{\sqrt{3}}{2} a\left(-i \ +1\right) \tag{1.26}$$

which gives a concise equation connecting A and B atomic sites

$$\varepsilon F_A(\mathbf{r}) = \gamma \begin{pmatrix} \hat{k}_x - i\hat{k}_y & 0 \\ 0 & \hat{k}_x + i\hat{k}_y \end{pmatrix} F_B(\mathbf{r}) \tag{1.27}$$

In the above Eq. 1.27 we have replaced $\mathbf{r}' \rightarrow \mathbf{r}$, $\hat{\mathbf{k}}' \rightarrow -i\nabla'$ and $\hat{\mathbf{k}} \rightarrow -i\nabla$. An analogous equation which connects the B and A atomic sites is derived in the form

$$\varepsilon F_B(\mathbf{r}) = \gamma \begin{pmatrix} \hat{k}_x + i\hat{k}_y & 0 \\ 0 & \hat{k}_x - i\hat{k}_y \end{pmatrix} F_A(\mathbf{r}) \tag{1.28}$$

and can be simply written as

$$\mathbf{H}_0 \mathbf{F}(\mathbf{r}) = \varepsilon \mathbf{F}(\mathbf{r}) \tag{1.29}$$

The above two Eqs. 1.27 and 1.28 correspond to Hamiltonian, which can be rewritten in the compact matrix form:

$$\mathbf{H}_0 = \begin{array}{c} \begin{matrix} KA & KB & K'A & K'B \end{matrix} \\ \begin{pmatrix} 0 & \gamma\left(\hat{k}_x - i\hat{k}_y\right) & 0 & 0 \\ \gamma\left(\hat{k}_x + i\hat{k}_y\right) & 0 & 0 & 0 \\ 0 & 0 & 0 & \gamma\left(\hat{k}_x + i\hat{k}_y\right) \\ 0 & 0 & \gamma\left(\hat{k}_x - i\hat{k}_y\right) & 0 \end{pmatrix} \end{array}, \tag{1.30}$$

whereas the envelope function matrices are

$$\mathbf{F}(\mathbf{r}) = \begin{pmatrix} \mathbf{F}^K(\mathbf{r}) \\ \mathbf{F}^{K'}(\mathbf{r}) \end{pmatrix}, \quad \mathbf{F}^{K(K')}(\mathbf{r}) = \begin{pmatrix} \mathbf{F}_A^{K(K')}(\mathbf{r}) \\ \mathbf{F}_B^{K(K')}(\mathbf{r}) \end{pmatrix} \tag{1.31}$$

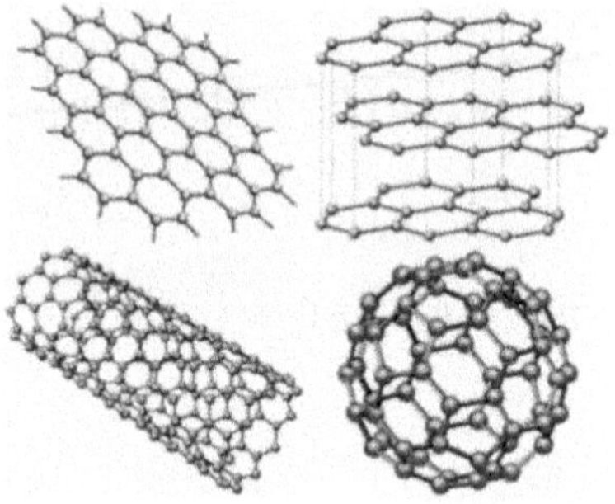

Figure 1.6 Graphene, graphite, nanotube, and buckyball. Adapted from Novoselov et al. (2005) © NPG.

Envelope functions now satisfy the following equations:

$$\begin{aligned}\gamma\left(\hat{\mathbf{k}}\cdot\sigma\right)F^{K}\left(\mathbf{r}\right)&=\varepsilon F^{K}\left(\mathbf{r}\right)\\ \gamma\left(\hat{\mathbf{k}}'\cdot\sigma\right)F^{K'}\left(\mathbf{r}\right)&=\varepsilon F^{K'}\left(\mathbf{r}\right)\end{aligned}\qquad(1.32)$$

where $\sigma = \left(\sigma_x, \sigma_y\right)$ are the Pauli matrices, $\hat{k}'_x = \hat{k}_x$, and $\hat{k}'_y = -\hat{k}_y$. Formulas in Eq. 1.32 represent the relativistic Dirac equation with the vanishing electron mass, which is also known as Weyl's equation for neutrino. We emphasize that Eq. 1.32 for the matrix envelope functions in Eq. 1.31 $\mathbf{F}\left(\mathbf{r}\right)$ follows directly from the nearest-neighbor tight-binding approximation. Nonetheless, equations in Eq. 1.32 are pretty general and remain valid for more general band structure of graphene.

Using the above-derived equations in Eq. 1.32, which ideally describe a suspended pristine graphene sheet, one finds the energy dispersion law in the form

$$\varepsilon_s\left(k\right) = s\gamma\left|k\right| \qquad (1.33)$$

where $s = \pm 1$ correspond to conduction and valence bands, respectively. From Eq. 1.33 one can see that parameter γ is actually the velocity v of chiral quasiparticle excitations in the system. The density of electron states in pristine graphene then is obtained as

$$D\left(\varepsilon\right) = \frac{1}{S}\sum_{s,k}\delta\left(\varepsilon - \varepsilon_k\right) = \frac{\left|\varepsilon\right|}{2\pi\gamma^2} \qquad (1.34)$$

where S is the sample area. In Fig. 1.7 (right) we have sketched conic shaped dispersion law and the electron density of states $D\left(\varepsilon\right)$. The

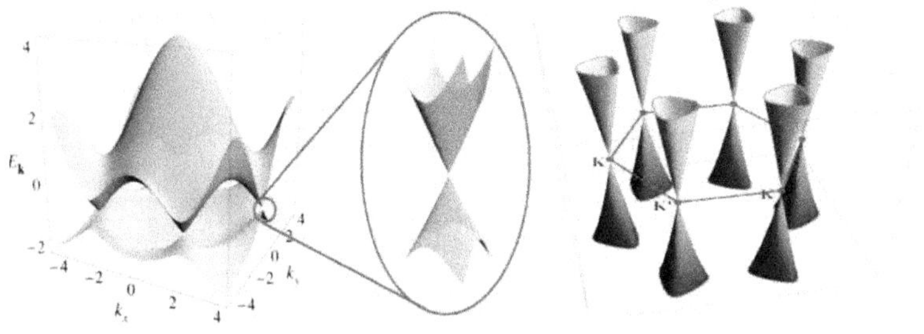

Figure 1.7 Left: Electronic dispersion in the honeycomb lattice: (left): Energy spectrum (in units of t) for finite values of t and t', with $t = 2.7$ eV and $t' = -0.2t$; (right): zoom-in of the energy bands close to one of the Dirac points. Adapted from Novoselov et al. (2005) © NPG. Right: The cones at the corners of the Brillouin zone of graphene.

density of states $D(\varepsilon)$ depends on excitation energy linearly and vanishes at zeroth energy $\varepsilon = 0$.

1.2.2 *Tight-Binding Scheme*

We assume that electrons can hop to both nearest- and next-nearest-neighbor atoms on graphene lattice. Then the tight-binding Hamiltonian acquires the form

$$H = -t \sum_{i,\,j,\,\sigma} \left(a^{\dagger}_{\sigma,\,i} b_{\sigma,\,j} + h.c. \right) - t' \sum_{i,\,j,\,\sigma} \left(a^{\dagger}_{\sigma,\,i} a_{\sigma,\,j} + b^{\dagger}_{\sigma,\,i} b_{\sigma,\,j} + h.c. \right) \tag{1.35}$$

where we use units such that $\hbar = 1$, and a_i, $a_i^{\dagger}$ are the annihilation and creation operators acting on an electron with spin $\sigma = \uparrow, \downarrow$ localized on an atomic site $\mathbf{R}_i$ of sublattice A (see Fig. 1.2). In the same way we define also operators b_i, $b_i^{\dagger}$ for sublattice B, t is the nearest-neighbor hopping energy between different sublattices, and t' is the intersite next-nearest-neighbor hopping energy. The energy parameter t actually coincides with the parameter $-\gamma_0$ introduced earlier in Section 1.1. The new notation $-\gamma_0 \to t$ for the electron transfer is introduced to keep a correspondence with traditional notations when describing graphite. Numeric estimations give $t = 2.8$ eV, whereas for hopping within the same sublattice one respectively obtains $0.02t < t' < 0.2t$. Tight-binding fit to cyclotron resonance experiments (Deacon et al. 2007) gives $t' \simeq 0.1$ eV.

Energy bands obtained from the above Hamiltonian are (Wallace 1947)

$$E_{\pm}(\mathbf{k}) = \pm t\sqrt{3 + f(\mathbf{k})} - t' f(\mathbf{k}) \qquad (1.36)$$

where we have introduced an auxiliary function

$$f(\mathbf{k}) = 2\cos\left(\sqrt{3}k_y a\right) + 4\cos\left(\frac{\sqrt{3}}{2}k_y a\right)\cos\left(\frac{3}{2}k_x a\right) \qquad (1.37)$$

The plus sign in Eq. 1.36 corresponds to the upper π^* band, whereas the minus sign to the lower π band. As it follows from Eq. 1.36, electron spectrum is symmetric in respect to zero energy in the absence of the electron hopping between different cells, that is, if $t' = 0$. A finite intercell hopping between two carbon atoms belonging to the same sublattice, which corresponds to $t' \neq 0$, brakes the electron–hole symmetry since it causes π and π^* bands to become asymmetric. The full band structure of graphene is shown in Fig. 1.4 where both t and t' are finite. The 3D electron excitation spectrum in the vicinity of the K point is represented in Fig. 1.7. We also show details of the band structure in vicinity of Dirac points K (or K′) in the Brillouin zone. The dispersion law has been obtained by Wallace in 1947, who expanded $E_{\pm}(\mathbf{k})$ near the vector $\mathbf{k} = \mathbf{K} + \mathbf{q}$ by assuming that $|q| << |K|$. Thus, close to the K and K′ points one gets (Wallace 1947)

$$E_{\pm}(\mathbf{q}) \approx \pm v|\mathbf{q}| + O\left[(q/K)^2\right] \qquad (1.38)$$

In the last formula, Eq. 1.38, $\mathbf{q}$ is the electron momentum measured relatively to the Dirac points and $v = 3ta/2$ is the Fermi velocity; its value is estimated as $v \simeq 1 \times 10^6$ m/s. The Fermi velocity v coincides with the parameter γ introduced in the former subsection (see Eq. 1.33). One can immediately notice a remarkable difference between the above formula and the case of a conventional conductor where $\varepsilon_q = q^2/2m$. The electron mass m, entering the last formula, does not appear in Eq. 1.38, whereas the Fermi velocity is independent of energy or momentum. In conventional conductors one normally gets $v = k/m = \sqrt{2E/m}$ which indicates a substantial dependence of velocity versus energy. Including the intercell hopping ($t' \neq 0$) along with the second order over q/K for the spectrum near the Dirac point one arrives at

$$E_{\pm}(\mathbf{q}) \approx 3t' \pm v|\mathbf{q}| - \left(\frac{9t'a^2}{4} \pm \frac{3ta^2}{8}\sin(3\theta_q)\right)|\mathbf{q}|^2 \qquad (1.39)$$

where the angle θ_q in the momentum space is

$$\theta_q = \arctan\left(\frac{q_x}{q_y}\right) \quad (1.40)$$

From Eq. 1.39 one can notice that presence of t' inflicts the energy shift of the Dirac point, and it also breaks the electron–hole symmetry. Besides, the electron spectrum appears to be anisotropic in the momentum space, and it has threefold symmetry. The last fact is quoted as a *trigonal warping* of the electronic spectrum (Ando et al. 1998, Dresselhaus and Dresselhaus 2002).

According to Eq. 1.39, the elementary excitations in pristine graphene behave as massless fermions which are described by the relativistic Dirac equation, Eq. 1.32 (Geim 2011). Therefore one might introduce a cyclotron mass which depends on the electronic density (Novoselov et al. 2005, Zhang et al. 2005). Following Ashcroft and Mermin (1976) we define the cyclotron mass in the semiclassical approximation as

$$m^* = \frac{1}{2\pi}\left[\frac{\partial A\,(E)}{\partial E}\right]_{E=E_{\mathrm{F}}} \quad (1.41)$$

where $A\,(E)$ is area in the momentum space delineated by the electron orbit. One obtains

$$A\,(E) = \pi q\,(E)^2 = \pi\frac{E^2}{v^2} \quad (1.42)$$

From the above Eq. 1.41, with aid of Eq. 1.42, one obtains

$$m^* = \frac{E_{\mathrm{F}}}{v^2} = \frac{k_{\mathrm{F}}}{v} \quad (1.43)$$

The last quantity is expressed via the electronic density n taking into account contributions from two Dirac points K and K′ and the spin. Then, when $n = k_{\mathrm{F}}^2/\pi$ where k_{F} is the Fermi momentum one also finds

$$m^* = \frac{\sqrt{\pi}}{v}\sqrt{n} \quad (1.44)$$

The last Eq. 1.44 can be used for fitting to the experimental data as shown in Fig. 1.8. The Fermi velocity and hopping parameter extracted in such a way are $v_{\mathrm{F}} \approx 10^6$ m/s and $t \approx 3$ eV, respectively. The square root dependence $\sqrt{n}$ of the cyclotron mass versus electron density serves as an experimental evidence of the massless Dirac fermion excitations in graphene (Novoselov et al. 2005, Zhang et al. 2005, Jiang et al. 2007). We emphasize that cyclotron mass in conventional conductors which corresponds to the ordinary parabolic Schrödinger dispersion is constant.

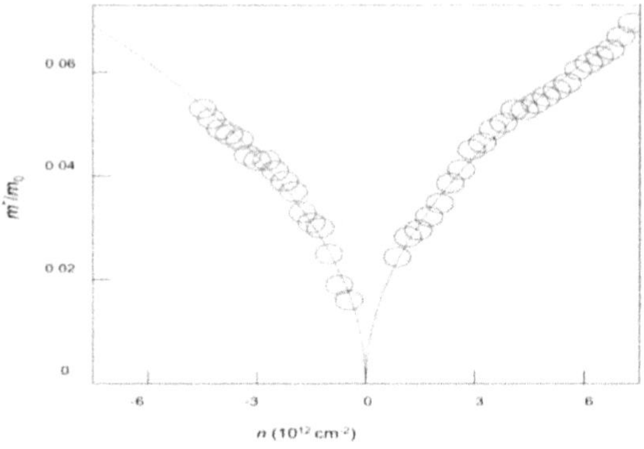

Figure 1.8 The ratio of cyclotron mass m^* of charge carriers in graphene to the free-electron mass m_0 as a function of concentration n. For electrons n is positive, whereas for holes it is negative. The experimental data are shown as symbols and are extracted from the temperature dependence of Shubnikov–de Haas (SdH) oscillations. Solid curves correspond to the fit by Eq. 1.41. Adapted from Novoselov et al. (2005) © NPG.

1.2.3 *Density of Electron States in Graphene*

More precise expression for the density of states per unit cell is derived from Eq. 1.39. We show the corresponding functions in Fig. 1.9 for both $t' = 0$ and $t' \neq 0$ where one might notice a semimetallic behavior in both cases (Wallace 1947, Bena and Kivelson 2005). A simple analytical expression for the density of states per unit cell is derived in the limit $t' = 0$ (Hobson and Nierenberg 1953):

$$\rho(E) = \frac{4}{\pi^2}\frac{|E|}{t^2}\frac{1}{\sqrt{Z_0}}\mathbf{F}\left(\frac{\pi}{2}, \sqrt{\frac{Z_1}{Z_0}}\right) \tag{1.45}$$

where $\mathbf{F}(\pi/2, x)$ is the complete elliptic integral of the first kind,

$$Z_0 = \begin{pmatrix} \left(1+\left|\frac{E}{t}\right|\right)^2 - \frac{1}{4}\left(1-\left|\frac{E}{t}\right|\right)^2 \text{ if } -t \leq E \leq t \\ 4\left|\frac{E}{t}\right| \quad \text{if} \quad -3t \leq E \leq -t \quad \text{or} \quad t \leq E \leq 3t \end{pmatrix} \tag{1.46}$$

and

$$Z_1 = \begin{pmatrix} 4\left|\frac{E}{t}\right| \text{ if } -t \leq E \leq t \\ \left(1+\left|\frac{E}{t}\right|\right)^2 - \frac{1}{4}\left(1-\left|\frac{E}{t}\right|\right)^2 \text{ if } -3t \leq E \leq -t \,\text{or}\, t \leq E \leq 3t \end{pmatrix} \tag{1.47}$$

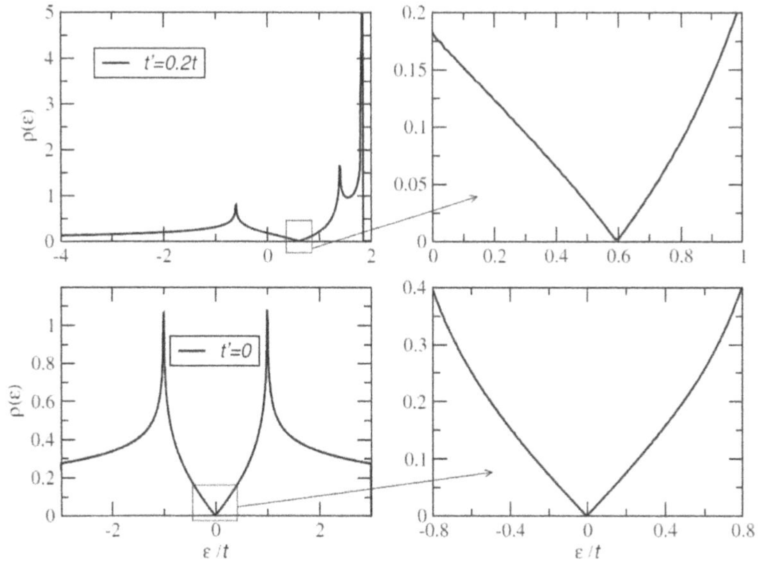

Figure 1.9 Density of electron states per unit cell as a function of energy (in units of intersite nearest-neighbor hopping energy (t) computed from the energy dispersion for two values of the intersite next-nearest-neighbor hopping energy $t' = 0.2t$ (top) and $t' = 0$ (bottom). On the right side we also show zoom-ins of the vicinity of the neutrality point of one electron per site. Adapted from Hobson and Nierenberg (1953).

In vicinity of the Dirac point, the dispersion is approximated by Eq. 1.39 and the electron density of states per unit cell Eq. 1.45 is simplified as

$$\rho(E) = \frac{2A_c}{\pi}\frac{|E|}{v^2} \tag{1.48}$$

where we have included the degeneracy of 4, A_c is the unit cell area given by $A_c = 3\sqrt{3}a^2/2$. We emphasize that the density of states in graphene is different from the density of states in carbon nanotubes (Saito et al. 1992a, 1992b, 1992c). In the latter case (see Fig. 1.10), the electron motion is quantized in the lateral direction perpendicular to the nanotube axis. It causes the so-called Van Hove singularities $1/\sqrt{E}$ in the electron density of states emerging due to the 1D nature of electronic spectrum. That type of quantization does not occur in pristine graphene which is considered to have infinite dimensions. A similar quantization in the lateral direction

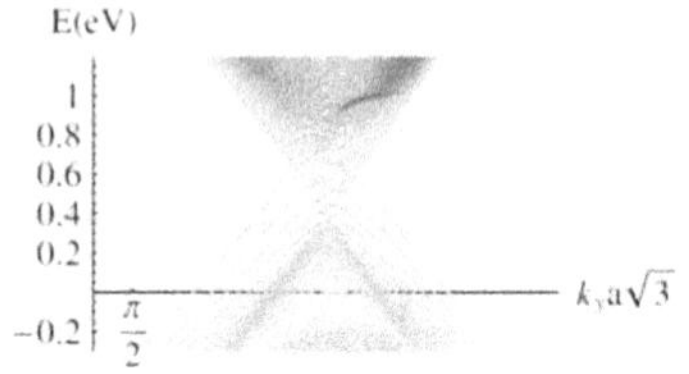

Figure 1.10 Energy spectrum in units of t' for a graphene ribbon 600 α wide, as a function of the momentum **k** along the ribbon (in units of $1/\sqrt{3}a$), in the presence of confining potential with the height of $V_0 = 1$ eV and width $180a$. Adapted from Chen et al. (2013) and Ziatdinov et al. (2013).

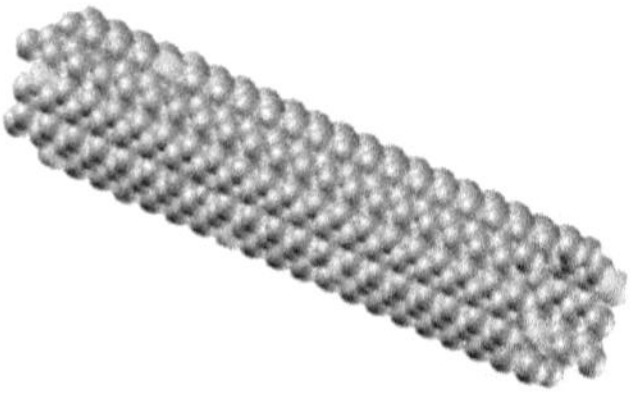

Figure 1.11 Carbon nanotube represents a rollover of a graphene sheet. Atomic impurities are indicated by different colors.

to the ribbon length takes place in graphene ribbons: quantization of the electron motion in the lateral direction thus is responsible for the appearance of subband structure and an energy gap at the Fermi energy as shown in Fig. 1.11 (Chen et al. 2013, Ziatdinov et al. 2013). The density of states can be expressed as

$$\rho(E) = \frac{\sqrt{3}a^2}{2\pi R} \sum_i dk \int dk \delta(k - k_i) \left| \frac{\partial \varepsilon(k_i)}{\partial k} \right|^{-1} \tag{1.49}$$

where R is the radius of the nanotube, $\varepsilon(k_i) = E$. The unique feature here is the presence of Van Hove singularities at the band edges. Basic features of the DOS are understood with using of the dispersion relation around the Fermi point. This gives

$$\rho(E) = \frac{\sqrt{3}a}{\pi^2 R\gamma} \sum_{m=1}^{N} \frac{|E|}{\sqrt{E^2 - \varepsilon_m^2}} \tag{1.50}$$

where $\varepsilon_m = |3m + 1| (a\gamma/2R)$ for semiconducting nanotubes and $\varepsilon_m = |3m| (a\gamma/2R)$ for metallic nanotubes. The semiconducting type of the energy spectrum is very important for a variety of applications including creating the graphene field-effect transistors which will be discussed in the following chapters.

1.3 Berry Phase and Topological Singularity in Graphene

The steady-state zero-field Dirac Hamiltonian Eq. 1.30, which describes electronic states in the absence of the magnetic field characterized by the eigenfunctions

$$\mathbf{F}_{sk}(\mathbf{r}) = \frac{1}{\sqrt{LA}} e^{i\mathbf{k}\cdot\mathbf{r}} \mathbf{F}_{sk} \tag{1.51}$$

is valid in vicinity of K(K′) points. The eigenvector $\mathbf{F}_{sk}$ is generally presented as

$$\mathbf{F}_{s\mathbf{k}} = e^{i\phi_s(\mathbf{k})} R^{-1}[\theta(\mathbf{k})][s] \tag{1.52}$$

where $\theta(\mathbf{k})$ is the angle between electron momentum $\mathbf{k}$ and the k_y axis, that is, $k_x + ik_y = +|\mathbf{k}| e^{i\theta(\mathbf{k})}$, $k_x - ik_y = -|\mathbf{k}| e^{-i\theta(\mathbf{k})}$ and we have introduced an arbitrary phase $\phi_s(\mathbf{k})$. Furthermore, we have introduced the spin-rotation operator $R(\theta)$ as

$$R(\theta) = e^{i\frac{\theta}{2}\sigma_z} = \begin{pmatrix} e^{i\frac{\theta}{2}} & 0 \\ 0 & e^{-i\frac{\theta}{2}} \end{pmatrix} \tag{1.53}$$

In Eq. 1.53, σ_z is the Pauli matrix. The eigenvector $[s]$ for the electron state $\mathbf{k}$ and $k_y > 0$ is

$$[s] = \frac{1}{\sqrt{2}} \begin{pmatrix} -is \\ 1 \end{pmatrix} \tag{1.54}$$

Direct calculation using Eqs. 1.52–1.54 gives

$$\begin{aligned} R^{-1}(\theta) &= R(-\theta) \\ R(\theta_1 + \theta_2) &= R(\theta_1) R(\theta_2) \end{aligned} \tag{1.55}$$

Furthermore, one obtains

$$R(\theta \pm 2\pi) = -R(\theta) \tag{1.56}$$

which in particular gives

$$R(-\pi) = -R(+\pi) \tag{1.57}$$

Function $R(\theta)$ in fact describes rotations of the one-half spin. The flip of sign in Eqs. 1.56 and 1.57 occurring after the spin rotation by

2π around $\mathbf{k} = 0$ is related to a topological singularity at $\mathbf{k} = 0$. The mentioned topological singularity at $\mathbf{k} = 0$ is understood in terms of Berry's phase (Simon 1983, Berry 1984).

Let us assume that the Hamiltonian depends on certain time-dependent parameter $s(t)$. If time changes from $t = 0$ to $t = T$ then $s(t)$ changes from $s(0)$ to $s(T)$. Besides, we request that Hamiltonian should become the same at $t = T$ again as it was at $t = 0$, that is, $\mathcal{H}[s(t)] = \mathcal{H}[s(0)]$ though $s(0) \neq s(T)$ not necessarily. If the process is sufficiently slow and adiabatic, and there is no degeneracy, the new state at $t = T$ is essentially the same as it was at $t = 0$. This means that apart from the phase multiplier $e^{-i\varphi}$, the electron wavefunction $\psi_s(\mathbf{k})$ is also the same at the two different time moments $t = 0$ and $t = T$. The extra phase φ is called Berry's phase (Simon 1983, Berry 1984) and is computed as

$$\varphi = -i \int_0^T dt \left\langle \psi[s(t)] \right| \left. \frac{\partial \psi[s(t)]}{\partial t} \right\rangle \tag{1.58}$$

Let us see what happens when we rotate momentum $\mathbf{k}$ adiabatically in the anticlockwise direction as a function of time from $t = 0$ to $t = T$. We start at $t = 0$ from a trial wavefunction where we select the "spin" part of eigenfunction in the particular form

$$\psi_s(\mathbf{k}) = \frac{1}{\sqrt{2}} \begin{pmatrix} -is \cdot e^{-i\theta(\mathbf{k})} \\ 1 \end{pmatrix} \tag{1.59}$$

The wavefunction Eq. 1.59 corresponds to setting the phase $\phi_s(\mathbf{k}) = -\theta(\mathbf{k})/2$ to maintain the wavefunction to be continuous versus $\theta(\mathbf{k})$. After rotating the momentum $\mathbf{k}$ in the anticlockwise direction adiabatically as a function of time t around the origin during the time interval $0 < t < T$, the wavefunction Eq. 1.59 is transformed into $\psi_s(\mathbf{k}) \cdot e^{-i\varphi}$. In the last expression φ is Berry's phase given by

$$\varphi = -i \int_0^T dt \left\langle \psi_s[\mathbf{k}(t)] \right| \left. \frac{\partial \psi_s[\mathbf{k}(t)]}{\partial t} \right\rangle = -\pi \tag{1.60}$$

which means the change of sign when rotating $\mathbf{k}$ by 2π around the origin $\mathbf{k} = 0$. The spin-rotation operation $R^{-1}[\theta(\mathbf{k})][s]$ is performed in Eq. 1.59 when continuously changing the direction of $\mathbf{k}$ including Berry's phase. The change of the sign happens only if the origin $\mathbf{k} = 0$ is enclosed inside the closed contour. If the closed contour does not enclose the origin $\mathbf{k} = 0$, the wavefunction sign remains the same. The topological singularity at $\mathbf{k} = 0$ causes various zero mode anomalies in observable characteristics (Shon and Ando 1998, Ando et al. 2002a, 2002b). A remarkable example is a finite value of the electric conductivity at $\varepsilon = 0$ which would not be the case for a zero gap semiconductor. Calculations performed using the kinetic Boltzmann equation indicate a metallic behavior when extrapolated to $\varepsilon = 0$. A more strict approach based on self-consistent Born approximation for the conductivity yields the universal value

$$G = \frac{e^2}{h} \tag{1.61}$$

when the electron energy vanishes, $\varepsilon \to 0$, whereas G coincides with the Boltzmann equation result at finite ε (Shon and Ando 1998). Similar anomalies take place also in the magnetoconductivity and in the dynamic conductivity, which serve as a clear indication of special properties of graphene (Ando et al. 2002a, 2002b, Zheng and Ando 2002).

1.4 Klein Paradox and Chiral Tunneling

The most remarkable difference between ordinary 2D conductors and graphene comes up when scattering of the chiral fermions on a potential barrier (Ando and Nakanishi 1998, McEuen et al. 1999, Katsnelson et al. 2006, Katsnelson and Novoselov 2007) in two dimensions see Fig. 1.12. The unconventional character

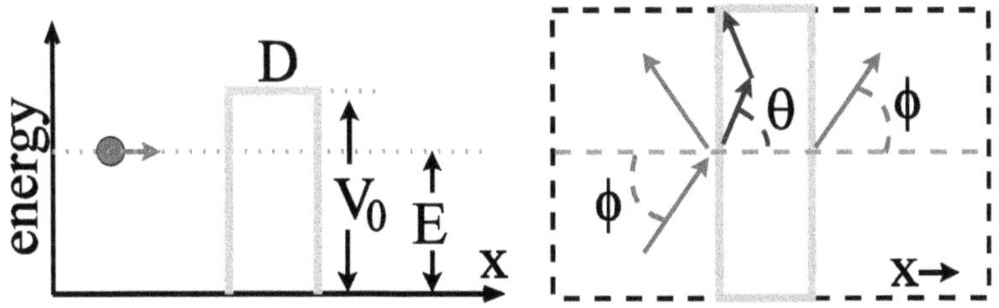

Figure 1.12 Klein tunneling in graphene. Top: Scattering the Dirac electrons by a square potential. Bottom: Definition of the angles ϕ and θ used in the scattering formalism in regions I, II, and III.

of transmission of the chiral fermion is well illustrated when considering a piece-wise geometry with a rectangular barrier with thickness D. The propagation of the particles is described by the Dirac equation, Eq. 1.32. Unlikely to the case of conventional tunneling, the transmission of chiral particles through the potential barrier must take into account their two one-half pseudospins. It is accomplished by using matrix representation of the chiral fermion's wavefunctions having the structure, as given by Eq. 1.31. Spinor representation of the chiral fermion wavefunction by a gauge transformation takes the form

$$\psi_K(k) = \frac{1}{\sqrt{2}} \begin{pmatrix} 1 \\ \pm e^{i\theta_K} \end{pmatrix} \tag{1.62}$$

The particle's momentum in our geometry is not mixed up during scattering in the vicinity of K and K' points. The simplest trial wavefunction is composed of incident and reflected plane waves which propagate in three different regions. On the left side from the barrier ($x < 0$) one writes

$$\psi_L(k) = \frac{1}{\sqrt{2}} \begin{pmatrix} 1 \\ se^{i\phi} \end{pmatrix} e^{i(k_x x + k_y y)} + \frac{r}{\sqrt{2}} \begin{pmatrix} 1 \\ se^{i(\pi-\phi)} \end{pmatrix} e^{i(-k_x x + k_y y)} \tag{1.63}$$

where $\phi = \arctan(k_y/k_x)$, $k_x = k_F \cos(\phi)$, $k_y = k_F \sin(\phi)$, and $\mathbf{k}_\text{F}$ is the Fermi momentum. Inside the barrier region, one uses

$$\psi_\text{B}(k) = \frac{a}{\sqrt{2}} \begin{pmatrix} 1 \\ s'e^{i\theta} \end{pmatrix} e^{i(q_x x + k_y y)} + \frac{b}{\sqrt{2}} \begin{pmatrix} 1 \\ s'e^{i(\pi-\theta)} \end{pmatrix} e^{i(-q_x x + k_y y)} \tag{1.64}$$

where $\theta = \arctan(k_y/q_x)$ and

$$q_x = \sqrt{(V_0 - E)^2/v^2 - k_y^2} \tag{1.65}$$

On the right hand side of the barrier there is a transmitted wave only,

$$\psi_R(k) = \frac{t}{\sqrt{2}} \begin{pmatrix} 1 \\ se^{i\phi} \end{pmatrix} e^{i(k_x x + k_y y)} \tag{1.66}$$

where $s = \text{sign}(E)$ and $s' = \text{sign}(E - V_0)$. The requirement of continuity of the wavefunction

$$\psi_L(x = 0, y) = \psi_\text{B}(x = 0, y) \tag{1.67}$$

and

$$\psi_\text{B}(x = D, y) = \psi_R(x = D, y) \tag{1.68}$$

provides us with linear equations to compute the coefficients r, a, b, and t. The solution gives us the transmission through the barrier as $T(\phi) = tt^*$ which is

$$T(\phi) = \frac{\cos^2\theta\cos^2\phi}{[\cos(Dq_x)\cos\phi\cos\theta]^2 + \sin^2(Dq_x)(1 - ss'\sin\phi\sin\theta)^2} \quad (1.69)$$

Here we neglect the contribution from evanescent waves in the barrier region. Such approximation is justified at finite $V_0 \neq 0$, that is, when chemical potential in the barrier region coincides with the Dirac energy. As follows from the obtained expression for $T(\phi)$, there is an angular symmetry $T(\phi) = T(-\phi)$. Furthermore, the barrier is ideally transparent ($T(\phi) = 1$) as soon as a standing wave is formed inside it at $q_x = n\pi/L$, with n being integer. In the latter case, the barrier transparency $T(\phi)$ turns to be independent versus the incidence angle ϕ. Besides, one gets the ideal barrier transparency $T(0) = 1$ also for the normal incidence (i.e., when $\phi \to 0$ and $\theta \to 0$) irrespectively of Dq_x.

The above unconventional tunneling characterized by ideal transparency for chiral relativistic fermions constitutes the Klein paradox sketched in Fig. 1.12 (Calogeracos and Dombey 1999). It does not take place for conventional nonrelativistic electrons which are described by the Schrödinger equation, which always gives $T(\phi) < 1$. One can further simplify Eq. 1.69 considering the limit $|V_0| >> |E|$, which gives

$$T(\phi) \simeq \frac{\cos^2\phi}{1 - \cos^2(Dq_x)\sin^2\phi} \quad (1.70)$$

Angular dependence of $T(\phi)$ for two distinct barrier heights V_0 is shown in Fig. 1.13 and the 3D plot is shown in Fig. 1.14. As it follows from the plot, the ideal barrier transparency $T(\phi) = 1$ takes place for a few values of the incidence angle. In graphene composed of two atomic layers (Katsnelson et al. 2006, Shafranjuk 2009) the straightforward (i.e., when $\phi \to 0$ and $\theta \to 0$) tunneling probability vanishes. However, the angular dependence is pronounced at finite incidence angles.

Klein tunneling through the simplest rectangular barrier as shown in Fig. 1.12 serves as a good illustration for understanding electron transport in inhomogeneous systems composed of

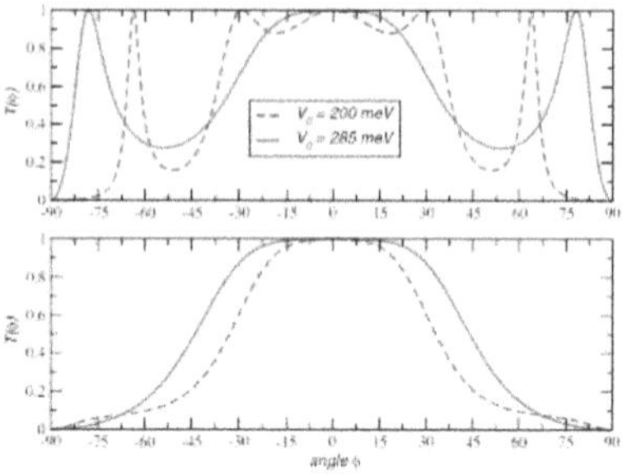

Figure 1.13 Angular behavior of $T\ (\phi)$ for two different values of V_0: $V_0 =$ 200 meV, dashed line; $V_0 = 285$ meV, solid line. The remaining parameters are $D = 110$ nm (top), $D = 50$ nm (bottom), $E = 80$ meV, $k_F = 2\pi/\delta\lambda$, and $\lambda = 50$ nm. Adapted from Castro Neto et al. (2009).

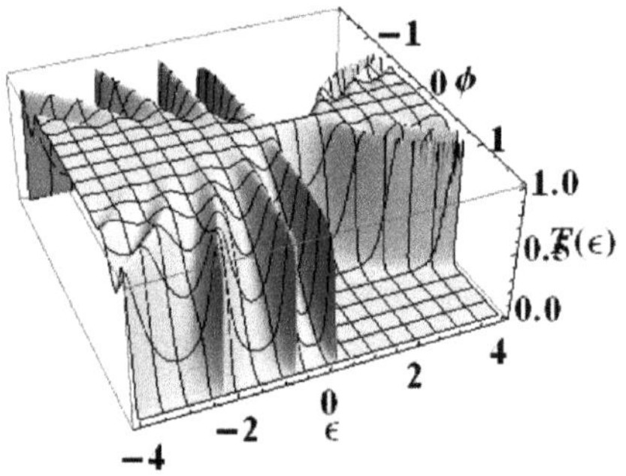

Figure 1.14 Chiral tunneling probability T (E, ϕ) versus electron energy E and incidence angle ϕ.

graphene. An ideal tunneling probability and the absence of reflection at normal incidence occur when one neglects the inter-valley scattering and the symmetry change between two different sublattices (Katsnelson et al. 2006). An interesting consequence of the Klein paradox takes place when tunneling occurs between two graphene regions with different type of charge carriers (e.g., one region is n-doped, while the other is p-doped). In particular, if we have electron conductivity on one side of the barrier, whereas the hole conductivity is on the other side, the normally incident chiral electrons are fully converted into the chiral holes. Here there is a similarity of chiral tunneling between the p-

and *n*-graphene regions and photons propagating in a medium with negative reflection index (Cheianov et al. 2007). Because of opposite sign the effective mass (or relationship between the momentum and velocity) for electrons and holes, the momentum for a hole created in the *p*-region is the inverse of that for an electron incoming from the *n*-region. If, additionally, we request the conservation of momentum parallel to the $n-p$ interface, the velocity of the tunneling particle is inverted. Consequently, a bunch of transmitted holes in the *p*-region is actually focused on the source of incident electrons located in the *n*-region. The angular dependence of the chiral tunneling probability $T(\phi)$ depends on the shape of the barrier. According to Cheianov et al. (2007), the transmission at finite angles becomes better for the barriers which vary slowly versus x. It was suggested that such an effect can take place in graphene quantum dots wired, for example, with the *n*-type source and *p*-type drain electrodes (Cserti et al. 2007). Another interesting issue is about whether it is possible to form the graphene quantum dots or not: The normally incident electrons experience no reflection from the dot's walls; thereby they escape from the dot immediately. The answer is affirmative since one can exploit electrons and holes with a finite parallel momentum which indeed are reflected from the quantum dot walls and are prone to creating sharp quantized levels in the lateral direction. Hence one might engineer graphene quantum dots where there are potential barriers in the perpendicular direction (Silvestrov and Efetov 2007). Another interesting phenomenon occurs in a disordered undoped graphene where the $p-n$ puddles arise (Katsnelson et al. 2006, Martin et al. 2008). The electron transport in such an inhomogeneous system proceeds via transmissions through the $p-n$ junctions between the puddles (Cheianov et al. 2007, Fogler et al. 2007, Shklovskii 2007).

Experimental evidence was obtained (Stander et al. 2009, Young and Kim 2009) when transport through potential barriers in graphene was investigated using a set of metallic gates capacitively coupled to graphene to modulate the potential landscape. When a gate-induced potential step is steep enough, disorder becomes less important and the resistance across the step is in quantitative agreement with predictions of Klein tunneling of Dirac fermions up to a small correction. The authors Stander et al. (2009) and Young and Kim (2009) have performed magnetoresistance measurements at low magnetic fields and compared them to recent predictions.

The internal symmetry in graphene can be broken by applying the external field or due to fluctuations. In particular, a DC magnetic field breaks the time reversal symmetry, while the atomic defects and impurities break the inversion symmetry of the crystal lattice. Combining influence of the disorder and magnetic field causes the violation of symmetry between the valleys. It causes dependence of the transmission coefficients versus the valley index which results in different paths for electrons propagating along different valleys. That effect is similar to controlling the electron spin in mesoscopic devices (Tworzydlo et al. 2007). If a large DC magnetic field is applied to a *p-n* junction which separates regions with distinct quantized Hall conductivities, the chiral currents can flow at both edges (Abanin and Levitov 2007). In turn it induces back scattering between the Hall currents at the edges of the graphene stripe. The aforementioned phenomena are observed experimentally by measuring the transport properties of the graphene ribbons with additional top gates that play the role of tunable potential barriers (Han et al. 2007, Huard et al. 2007, Lemme et al. 2007, Ozyilmaz et al. 2007, Williams et al. 2007).

1.5 Landau Levels in Graphene

When a finite DC magnetic field B is applied perpendicular to a graphene sheet, it causes quantization of electron motion and appearance of Landau levels. The electron properties are described by the same equation as Eq. 1.30, where the electron momentum $\hat{\mathbf{k}} = -i\nabla$ is replaced with $\hat{\mathbf{k}} = -i\nabla + (e/c\hbar)\,\mathbf{A}$. In the last formula, $\mathbf{A}$ is the vector potential of the magnetic field

$$B_z = \frac{\partial A_y}{\partial x} - \frac{\partial A_x}{\partial y} \tag{1.71}$$

We first consider the Landau quantization for electron states near the K point. Commutator of the x and y components of electron momentum $\hat{\mathbf{k}}$ is

$$\left[\hat{k}_x, \hat{k}_y\right] = -i\frac{eB}{c\hbar} = -il^{-2} \tag{1.72}$$

where l is the magnetic length

$$l = \sqrt{\frac{c\hbar}{eB}}. \tag{1.73}$$

By introducing new operators

$$\begin{aligned} a &= \left(l/\sqrt{2}\right)\left(\hat{k}_x - i\hat{k}_y\right) \\ a^{\dagger} &= \left(l/\sqrt{2}\right)\left(\hat{k}_x + i\hat{k}_y\right) \end{aligned} \tag{1.74}$$

and using the anticommutator $\left[a, a^{\dagger}\right] = 1$ one can rewrite Hamiltonian in vicinity of K(K′) points as

$$\mathcal{H}_0 = \frac{\sqrt{2}\gamma}{l}\begin{pmatrix} 0 & a \\ a^{\dagger} & 0 \end{pmatrix} \tag{1.75}$$

Landau quantization of the electron motion in graphene is conveniently described in terms of Hermite polinomial functions

$$h_n(x, y) = \frac{\left(a^{\dagger}\right)^n}{\sqrt{n!}} h_0(x, y) \tag{1.76}$$

The functions $h_n(x, y)$ entering Eq. 1.76 have the following properties

$$a h_0(x, y) = 0 \tag{1.77}$$

and

$$\begin{aligned} a^{\dagger} h_n &= \sqrt{n+1} h_{n+1} \\ a h_{n+1} &= \sqrt{n+1} h_{n+1} \\ a^{\dagger} a h_n &= n h_n \end{aligned} \tag{1.78}$$

One consequence for graphene is that there exists a Landau level with zero energy $\varepsilon_0 = 0$ described by the wavefunction

$$F_0^K = \begin{pmatrix} 0 \\ h_0 \end{pmatrix} \tag{1.79}$$

The excited states correspond to Landau levels at finite energies

$$\varepsilon_n = \text{sign}(n)\frac{\sqrt{2}\gamma}{l}\sqrt{|n|} \tag{1.80}$$

where the corresponding wavefunction is given by

$$F_n^K = \frac{1}{\sqrt{2}}\begin{pmatrix} \text{sign}(n)\, h_{|n|-1} \\ h_{|n|} \end{pmatrix} \tag{1.81}$$

In the above formula $n = \pm 1, \pm 2 \ldots,$ is an integer and sign(n) denotes the sign of n. Similar expressions are also derived for the electron states in vicinity of K′ point. The remarkable feature of the Weyl equation is reflected by presence of the Landau level at $\varepsilon = 0$. This zero-energy level is independent versus the magnetic field and is observed by measuring divergences in magnetic susceptibility (Geim and Novoselov 2007).

1.6 Modeling of Graphene Devices

Basic transport properties of an ideal, pristine single-atomic-layer graphene indicate numerous anomalies which are not present in regular conducting materials (Katsnelson et al. 2006, Tworzydlo et al. 2006, Castro Neto et al. 2009, Shafranjuk 2009). As it follows from the aforesaid, basic unconventional features of graphene emerge from its honeycomb crystal symmetry and appropriate coupling between the atomic sites. More specifically, peculiarities of the low-energy chiral transport in graphene originate from the electron states $\psi_{\mathbf{A(B)}}^{K(K')}$, which belong to the A and B sublattices in the vicinity of K and K′ points. At those points, the conduction and valence bands touch each other. Because the four states are intrinsically connected (they couple to each other via transfer integrals between nearest-neighbor carbon atoms) they are combined into a four-component wavefunction $\hat{\psi}$. The four-component state $\hat{\psi}$ corresponds to a single chiral particle characterized by two one-half pseudospins (Ando 2005). Under certain conditions the particle may represent a single electron or a hole. For instance, a particle described by the spinor wavefunction $\hat{\psi}$ with only one "↑" nonvanishing component is related to a single electron propagating along an atomic chain on the A sublattice. In this case the propagation angle $\chi = 0$ where

$$\chi = \arctan \frac{q}{k} \tag{1.82}$$

is the azimuthal angle between the 2D electron momentum $\mathbf{p} = (\hbar k, \hbar q)$ and the $\hat{x}$ axis (see Fig. 1.15). A single hole moving in the opposite direction along the atomic chain on the B sublattice has only the spin "↓" component (in this case $\chi = 0$ also). However, if four components of $\hat{\Psi}$ are finite they pose as electrons and

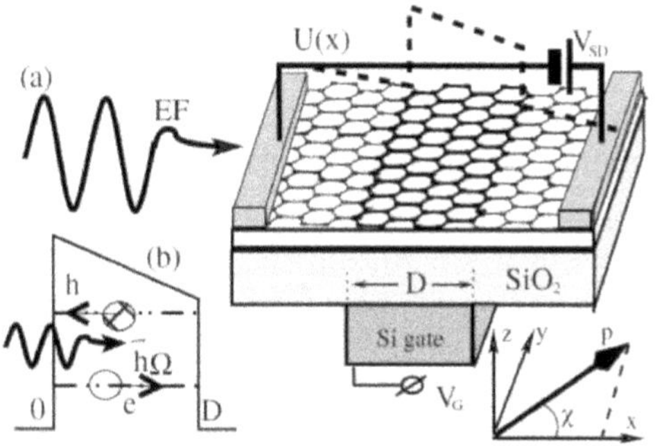

Figure 1.15 (a) The monolayer graphene junction (GJ) exposed to the external electromagnetic field (EF) and controlled by the gate voltage V_G. The chiral barrier region is denoted by darker hexagons, while the barrier profile $U(x)$ is shown by the dashed line. (b) Quantum *e–h* interference inside the chiral barrier.

holes which are hopping between the A and B sublattices and are converted into each other. This intermediate case with electron and hole hopping between the A and B sublattices is interpreted as tilting the one-half pseudospin. The pseudospin tilting degree depends on the value χ. In graphene samples of finite size the *quantized states* emerge not only due to interference between incident and reflected pure electrons or pure holes, but for finite propagation angles $\chi \neq 0$, the eigenstate is rather a hybrid of the electron and hole states.

This yields a complex interference pattern formed by the chiral particles with tilted pseudospins. One interesting aspect of the external field influence on the chiral transport through a graphene junction had been addressed by Syzranov et al. (2008). Using the rotating wave approximation (RWA), the authors Syzranov et al. (2008) considered the effect of an external AC field. They have found that the AC field induces a finite energy gap in the excitation spectrum of graphene which henceforth is posing as a two band conductor. Noteworthy, that depending on the AC field amplitude and frequency, the AC field may either suppress or stimulate the DC current throughout the graphene junction. In Chapter 10 we will consider a different aspect of the field influence on the chiral transport through the graphene junction. In particular, we address a consequence of the direct coupling between the AC field and the pseudospin. Since the propagation direction of a chiral fermion depends on the pseudospin tilting, the AC field which tilts the pseudospin also deflects the chiral particles from their steady state trajectories. Another consequence of these phenomena is that we go far beyond the rotating wave approximation formerly used by Syzranov et al. (2008): We will not introduce restrictions on the AC field amplitude nor on the field frequency. Our approach allows comparing of the weak and strong AC field limits and of different frequency regions. The simplest GJ setup is a graphene ribbon with two metallic source (S) and drain (D) electrodes attached to its ends (see Fig. 1.15). The electrodes are distanced by $\sim (L - D)/2 \geq l_{\mathrm{i}}$ (L being the length of the whole ribbon, D is the barrier length, and l_{i} is the electron elastic mean free path in graphene) from the potential barrier induced by a finite gate voltage $V_{\mathbf{G}} \neq 0$ on the same ribbon as shown in Fig. 1.15. In the setup there is no material separation between the barrier and the adjacent electrodes: The chiral barrier is induced in the graphene locally by the back gate voltage $V_{\mathbf{G}}$ as shown in Fig. 1.15. Besides, the gate voltage $V_{\mathbf{G}}$ controls the ratio between the electron and hole concentrations within the barrier region.

Relative crystallographic misorientation of two adjacent graphene regions is characterized by a finite angle $\kappa \neq 0$ between chains of carbon atoms belonging to either A or B sublattices. If the crystal lattice in the adjacent regions of electrodes and the barrier match each other ($\kappa = 0$), and there is no scattering on phonons and impurities, the pseudospins conserve. We call the transmission through the gate voltage-induced barrier the chiral tunneling when the pseudospins are conserved. If there are crystallographic misorientations ($\kappa \neq 0$), the pseudospins may flip and the tunneling becomes nonchiral. The nonchiral contribution emerges also from scattering on the lattice imperfections (atomic impurities and defects) and from the electron–phonon scattering. In our model setup shown in Fig. 1.15 we neglect the nonchiral transport and the nonchiral tunneling. It means that we assume the pseudospins are always conserved. Then the transport of chiral particles is supposed to be ballistic inside the electrode regions and also inside the barrier. The barrier length D is determined by extent of the Si gate region as sketched in Fig. 1.15a. So, it arises on the same monoatomic graphene ribbon (marked as darker hexagons) where the electrodes reside. The gate voltage V_G is applied via the n^{++} Si gate placed beneath the dielectric SiO_2 substrate.

Maximum size of the graphene sample where the electron propagation is ballistic is restricted by several factors. One restriction is due to electron scattering on corrugations of the graphene sheet (Katsnelson and Geim 2008, Kim and Neto 2008), disorder (Chen, Jang, and Adam et al. 2008, Chen, Jang, and Xiao et al. 2008, Jang et al. 2008), and charged impurities (Hwang et al. 2007a, 2007b, Tan et al. 2007). All the mentioned scattering processes are prone to limit the electron elastic mean free path l_i. It motivates obtaining of graphene samples where the scattering centers are minimized. "Clean" graphene sheets with $l_i \geq L$ were recently reported in a series of publications (Tan et al. 2007, Bolotin, Sikes, and Hone et al. 2008, Bolotin, Sikes, and Jiang et al. 2008). In the annealed graphene ribbons one achieves $l_i \geq 1 - 5$ μm, which means that the condition

$$l_i > \max\{D,\ W\} \tag{1.83}$$

is fulfilled in those samples (Tan et al. 2007, Bolotin, Sikes, and Hone et al. 2008, Bolotin, Sikes, and Jiang et al. 2008). In Eq. 1.81 W is the width of the middle electrode of ballistic propagation. The annealing makes the electron mean free path l_i to be considerably longer, so l_i exceeds the micron scale. Nevertheless, most of available samples have a submicron size. Another restriction comes from the electron–phonon collisions which at the temperature $T = 10$ K provides the inelastic electron mean free path $l_{ep} \simeq 3$ μm. Besides there is a restriction on D due to a requirement that the AC field must be fairly homogeneous within the barrier. This gives $D \leq 2\pi c/\Omega$ (Ω is the field frequency) which for $D \leq 5$ μm is equivalent to $\Omega \geq 0.5$ THz.

The ballistic chiral transport addressed in this chapter is pronounced in various unconventional DC electronic and magnetic properties of the monoatomic layer graphene (Novoselov et al. 2004). We will see that angular dependence of the chiral tunneling determines anisotropy of the electric current density across the junction. The resonant tunneling originates from a constructive quantum interference taking place between an electron (e) and a hole (h) as indicated in Fig. 1.15b. Inside the graphene barrier, the e and h chiral particles are characterized by the same pseudospin. They also propagate in opposite directions, as sketched in Fig. 1.14, which is a consequence of Klein paradox (Krekora et al. 2004a, 2004b, 2004c). The pseudospin conservation enforces special quantum mechanical selection rules for tunneling through the chiral barrier. When a finite gate voltage $V_G \neq 0$ induces a chiral barrier in a "clean" ribbon, the tunneling becomes dependent on the angle χ. Probability of the chiral tunneling is ideal ($T = 1$) in the straightforward direction $\chi = 0$, while it experiences finite maximums at some selected angles $\chi = \chi_n$ (Katsnelson et al. 2006).

For the sake of simplicity we also assume that the energy which is pumped into the system is too low to cause the electron distribution be deviating from equilibrium. The energy supplied by the external field into the barrier region is taken away due to quasiparticle diffusion into the adjacent electrodes and due to escape of phonons into the electrodes and substrate. Namely, we assume that diffusion of the quasiparticle excitations from the barrier region into electrodes and diffusion of phonons created in the electron–phonon collisions and then escaping into the electrodes and substrate drive the energy from the system away faster than an external field supplies it. Under the above assumption, the electron distribution function remains nearly equilibrium in the barrier region and is firmly equilibrium in the electrodes.

The field-induced directional resonant chiral tunneling across the graphene junction which is directly related to the Klein paradox will be considered in Chapter 10. We will see that an external field modifies the angular dependence of the electric current density through the graphene ribbon. Physically, the field causes an additional wavefunction e–h phase shift ϕ inside the graphene junction. In this way, it impacts the quantization of the electron motion in finite size samples. The quantized states are formed in both the lateral $\hat{x}$ and longitudinal $\hat{y}$ directions. The quantization in the longitudinal $\hat{x}$ direction is physically distinct from the quantization in the lateral $\hat{y}$ direction: The lateral $\hat{y}$ quantization comes from regular reflections at the ribbon edges. The longitudinal $\hat{x}$ quantization originates from electron–hole conversions (Ando 2005, Katsnelson and Geim 2008) at the chiral barrier edges. Details of the lateral quantization are sensitive to shape of the ribbon edges. The edge reflections can mix electron states from different A(B) sublattices and K(K′) valleys which corresponds to flipping of the two corresponding one-half pseudospins.

Quantized energy levels are pronounced in the electron transport across the junction. Energy $E_n(\phi)$ of the quantized state $|n\rangle$ depends on the wavefunction phase ϕ explicitly. Therefore the external field shifts the energy levels

$$E_n \to E_n^{(0)} + \delta_n \tag{1.84}$$

by δ_n relatively to their steady state position $E_n^{(0)}(\phi)$. For such reasons, the averaged transmission probability $\overline{T(\chi)}$ (where χ is the propagation angle) across the chiral barrier is also changed. The corresponding altering of $\overline{T(\chi)}$ is quoted as a directional shift of the transmission resonances (deflection). The deflection degree depends on the frequency and amplitude of the external AC field. This constitutes a new feature of the photon-assisted chiral tunneling.

Later in Chapter 10 we will compute the angular dependence of steady state differential conductivity σ (φ, $V_{\mathbf{SD}}$) (φ is the azimuthal angle between the $\hat{x}$ axis and the line connecting two tiny S and D electrodes) for a monolayer graphene junction biased by the source–drain voltage $V_{\mathbf{SD}}$. We will see that it causes the AC field–induced directional shift of the transmission resonances. The dependence σ versus φ can be measured experimentally in two different setups which we will discuss further in the following Sections. We will also consider two different shapes of the ribbon edges, that is, zigzag and armchair. The stationary solution obtained in this chapter is then utilized for studying the AC properties of graphene junction in the following chapters. We will also discuss the experimental setup where the directional photoelectric effect can be observed. In Chapter 10 we examine the AC field–induced intrinsic noise in the graphene junction and consider the deflection phenomena when the intensity of the AC field is arbitrary.

Problems

1-1. Calculate the area of a hexagonal unit cell and the area of the hexagonal first Brillouin zone of the honeycomb graphene atomic lattice.

1-2. Consider how the electron dispersion law (Eq. 1.12) is changed if the intersite next-nearest-neighbor hopping is not neglected, that is, if one retains terms with $t' \neq 0$ in Eqs. 1.8 and 1.9, where t' is the intersite next-nearest-neighbor hopping energy.

1-3. Compute the density of electron states in pristine graphene without and with *trigonal warping* of the electronic spectrum. Analyze how trigonal warping affects the electron density of states.

1-4. What is the difference between the mass of elementary excitation in graphene and the cyclotron mass?

1-5. Explain why the dispersion laws of electron excitations in pristine graphene and carbon nanotubes are different from each other, although the materials have the same crystal lattice symmetry, spacing, and intersite coupling. What is the origin of Van Hove singularities in carbon nanotubes?

1-6. Explain why the electric conductivity of pristine graphene has a finite value $G = e^2/h$ at $\varepsilon = 0$, which should not be in a zero-gap semiconductor where it is supposed to vanish.

1-7. Why does the tunneling of chiral relativistic particles lead to the Klein paradox with an ideal transmission probability and angular dependence on the angle of incidence? What is the influence of the intervalley scattering and symmetry change between the two different sublattices?

References

Abanin, D. A., and L. S. Levitov (2007). Quantized transport in graphene p-n junctions in a magnetic field. *Science*, **317**(5838): 641–643.

Abrikosov, A. A. (1988). *Fundamentals of the Theory of Metals*. Amsterdam, New York, North Holland, sole distributors for USA and Canada, Elsevier Science.

Ando, T. (2005). Theory of electronic states and transport in carbon nanotubes. *J Phys Soc Jpn*, **74**(3): 777–817.

Ando, T., and T. Nakanishi (1998). Impurity scattering in carbon nanotubes: absence of back scattering. *J Phys Soc Jpn*, **67**(5): 1704–1713.

Ando, T., T. Nakanishi, and R. Saito (1998). Berry's phase and absence of back scattering in carbon nanotubes. *J Phys Soc Jpn*, **67**(8): 2857–2862.

Ando, T., and T. Seri (1997). Quantum transport in a carbon nanotube in magnetic fields. *J Phys Soc Jpn*, **66**(11): 3558–3565.

Ando, T., Y. S. Zheng, and H. Suzuura (2002a). Dynamical conductivity and zero-mode anomaly in honeycomb lattices. *J Phys Soc Jpn*, **71**(5): 1318–1324.

Ando, T., Y. S. Zheng, and H. Suzuura (2002b). Exotic transport properties of two-dimensional graphite. *Microelectron Eng*, **63**(1–3): 167–172.

Arimura, Y., and T. Ando (2012). Diamagnetism of graphene with gap in nonuniform magnetic field. *J Phys Soc Jpn*, **81**(2).

Ashcroft, N. W., and N. D. Mermin (1976). *Solid State Physics*. New York, Holt.

Bena, C., and S. A. Kivelson (2005). Quasiparticle scattering and local density of states in graphite. *Phys Rev B*, **72**(12).

Berry, M. V. (1984). Quantal phase-factors accompanying adiabatic changes. *Proc R Soc London, A*, **392**(1802): 45–57.

Bolotin, K. I., K. J. Sikes, J. Hone, H. L. Stormer, and P. Kim (2008). Temperature-dependent transport in suspended graphene. *Phys Rev Lett*, **101**(9).

Bolotin, K. I., K. J. Sikes, Z. Jiang, M. Klima, G. Fudenberg, J. Hone, P. Kim, and H. L. Stormer (2008). Ultrahigh electron mobility in suspended graphene. *Solid State Commun*, **146**(9–10): 351–355.

Calogeracos, A., and N. Dombey (1999). Klein tunnelling and the Klein paradox. *Int J Mod Phys A*, **14**(4): 631–643.

Castro Neto, A. H., F. Guinea, N. M. R. Peres, K. S. Novoselov, and A. K. Geim (2009). The electronic properties of graphene. *Rev Mod Phys*, **81**(1): 109–162.

Cheianov, V. V., V. Fal'ko, and B. L. Altshuler (2007). The focusing of electron flow and a Veselago lens in graphene p-n junctions. *Science*, **315**(5816): 1252–1255.

Chen, J. H., C. Jang, S. Adam, M. S. Fuhrer, E. D. Williams, and M. Ishigami (2008). Charged-impurity scattering in graphene. *Nat Phys*, **4**(5): 377–381.

Chen, J. H., C. Jang, S. D. Xiao, M. Ishigami, and M. S. Fuhrer (2008). Intrinsic and extrinsic performance limits of graphene devices on SiO2. *Nat Nanotechnol*, **3**(4): 206–209.

Chen, X. B., Y. Xu, X. L. Zou, B. L. Gu, and W. H. Duan (2013). Interfacial thermal conductance of partially unzipped carbon nanotubes: linear scaling and exponential decay. *Phys Rev B*, **87**(15).

Cserti, J., A. Palyi, and C. Peterfalvi (2007). Caustics due to a negative refractive index in circular graphene p-n junctions. *Phys Rev Lett*, **99**(24).

Deacon, R. S., K. C. Chuang, R. J. Nicholas, K. S. Novoselov, and A. K. Geim (2007). Cyclotron resonance study of the electron and hole velocity in graphene monolayers. *Phys Rev B*, **76**(8).

Dresselhaus, M. S., and G. Dresselhaus (2002). Intercalation compounds of graphite. *Adv Phys*, **51**(1): 1–186.

Fogler, M. M., D. S. Novikov, and B. I. Shklovskii (2007). Screening of a hypercritical charge in graphene. *Phys Rev B*, **76**(23).

Geim, A. K. (2011). Nobel Lecture: Random walk to graphene. *Rev Mod Phys*, **83**(3): 851–862.

Geim, A. K. (2012). Graphene prehistory. *Phys Scripta*, **T146**.

Geim, A. K., and K. S. Novoselov (2007). The rise of graphene. *Nat Mater*, **6**(3): 183–191.

Han, M. Y., B. Ozyilmaz, Y. B. Zhang, and P. Kim (2007). Energy band-gap engineering of graphene nanoribbons. *Phys Rev Lett*, **98**(20).

Hobson, J. P., and W. A. Nierenberg (1953). The statistics of a 2-dimensional, hexagonal net. *Phys Rev*, **89**(3): 662–662.

Huard, B., J. A. Sulpizio, N. Stander, K. Todd, B. Yang, and D. Goldhaber-Gordon (2007). Transport measurements across a tunable potential barrier in graphene. *Phys Rev Lett*, **98**(23).

Hwang, E. H., S. Adam, and S. Das Sarma (2007a). Carrier transport in two-dimensional graphene layers. *Phys Rev Lett*, **98**(18).

Hwang, E. H., S. Adam, and S. Das Sarma (2007b). Transport in chemically doped graphene in the presence of adsorbed molecules. *Phys Rev B*, **76**(19).

Jang, C., S. Adam, J. H. Chen, D. Williams, S. Das Sarma, and M. S. Fuhrer (2008). Tuning the effective fine structure constant in graphene: opposing effects of dielectric screening on short- and long-range potential scattering. *Phys Rev Lett*, **101**(14).

Jiang, Z., E. A. Henriksen, L. C. Tung, Y. J. Wang, M. E. Schwartz, M. Y. Han, P. Kim, and H. L. Stormer (2007). Infrared spectroscopy of Landau levels of graphene. *Phys Rev Lett*, **98**(19).

Katsnelson, M. I. (2012). *Graphene: Carbon in Two Dimensions*. Cambridge, Cambridge University Press.

Katsnelson, M. I., and A. K. Geim (2008). Electron scattering on microscopic corrugations in graphene. *Philos Trans R Soc A*, **366**(1863): 195–204.

Katsnelson, M. I., and K. S. Novoselov (2007). Graphene: New bridge between condensed matter physics and quantum electrodynamics. *Solid State Commun*, **143**(1–2): 3–13.

Katsnelson, M. I., K. S. Novoselov, and A. K. Geim (2006). Chiral tunnelling and the Klein paradox in graphene. *Nat Phys*, **2**(9): 620–625.

Kim, E. A., and A. H. C. Neto (2008). Graphene as an electronic membrane. *EPL*, **84**(5).

Kittel, C. (2005). *Introduction to Solid State Physics*. Hoboken, NJ, Wiley.

Krekora, P., Q. Su, and R. Grobe (2004a). Klein paradox in spatial and temporal resolution. *Phys Rev Lett*, **92**(4).

Krekora, P., Q. Su, and R. Grobe (2004b). Relativistic electron localization and the lack of Zitterbewegung. *Phys Rev Lett*, **93**(4).

Krekora, P., Q. Su, and R. Grobe (2004c). Transitions into the negative-energy Dirac continuum. *Phys Rev A*, **70**(5).

Lemme, M. C., T. J. Echtermeyer, M. Baus, and H. Kurz (2007). A graphene field-effect device. *IEEE Electron Device Lett*, **28**(4): 282–284.

Martin, J., N. Akerman, G. Ulbricht, T. Lohmann, J. H. Smet, K. Von Klitzing, and A. Yacoby (2008). Observation of electron-hole puddles in graphene using a scanning single-electron transistor. *Nat Phys*, **4**(2): 144–148.

McEuen, P. L., M. Bockrath, D. H. Cobden, and J. G. Lu (1999). One dimensional transport in carbon nanotubes. *Microelectron Eng*, **47**(1–4): 417–420.

Novoselov, K. S., A. K. Geim, S. V. Morozov, D. Jiang, M. I. Katsnelson, I. V. Grigorieva, S. V. Dubonos, and A. A. Firsov (2005). Two-dimensional gas of massless Dirac fermions in graphene. *Nature*, **438**(7065): 197–200.

Novoselov, K. S., A. K. Geim, S. V. Morozov, D. Jiang, Y. Zhang, S. V. Dubonos, I. V. Grigorieva, and A. A. Firsov (2004). Electric field effect in atomically thin carbon films. *Science*, **306**(5696): 666–669.

Ozyilmaz, B., P. Jarillo-Herrero, D. Efetov, and P. Kim (2007). Electronic transport in locally gated graphene nanoconstrictions. *Appl Phys Lett*, **91**(19).

Saito, R., M. Fujita, G. Dresselhaus, and M. S. Dresselhaus (1992a). Electronic-structure of carbon-fibers based on C-60. *Electrical, Optical, and Magnetic Properties of Organic Solid State Materials*, Vol. 247: 333–338.

Saito, R., M. Fujita, G. Dresselhaus, and M. S. Dresselhaus (1992b). Electronic-structure of chiral graphene tubules. *Appl Phys Lett*, **60**(18): 2204–2206.

Saito, R., M. Fujita, G. Dresselhaus, and M. S. Dresselhaus (1992c). Electronic-structure of graphene tubules based on C-60. *Phys Rev B*, **46**(3): 1804–1811.

Shafranjuk, S. E. (2009). Directional photoelectric current across the bilayer graphene junction. *J Phys: Condens Mater*, **21**(1).

Shafranjuk, S. E. (2011). Electromagnetic properties of the graphene junctions. *Eur Phys J B*, **80**(3): 379–393.

Shklovskii, B. I. (2007). Simple model of Coulomb disorder and screening in graphene. *Phys Rev B*, **76**(23).

Shon, N. H., and T. Ando (1998). Quantum transport in two-dimensional graphite system. *J Phys Soc Jpn*, **67**(7): 2421–2429.

Silvestrov, P. G., and K. B. Efetov (2007). Quantum dots in graphene. *Phys Rev Lett*, **98**(1).

Simon, B. (1983). Holonomy, the quantum adiabatic theorem, and Berry phase. *Phys Rev Lett*, **51**(24): 2167–2170.

Stander, N., B. Huard, and D. Goldhaber-Gordon (2009). Evidence for Klein tunneling in graphene p-n junctions. *Phys Rev Lett*, **102**(2).

Syzranov, S. V., M. V. Fistul, and K. B. Efetov (2008). Effect of radiation on transport in graphene. *Phys Rev B*, **78**(4).

Tan, Y. W., Y. Zhang, K. Bolotin, Y. Zhao, S. Adam, E. H. Hwang, S. Das Sarma, H. L. Stormer, and P. Kim (2007). Measurement of scattering rate and minimum conductivity in graphene. *Phys Rev Lett*, **99**(24).

Tworzydlo, J., I. Snyman, A. R. Akhmerov, and C. W. J. Beenakker (2007). Valley-isospin dependence of the quantum Hall effect in a graphene p-n junction. *Phys Rev B*, **76**(3).

Tworzydlo, J., B. Trauzettel, M. Titov, A. Rycerz, and C. W. J. Beenakker (2006). Sub-Poissonian shot noise in graphene. *Phys Rev Lett*, **96**(24).

Wallace, P. R. (1947). The band theory of graphite. *Phys Rev*, **71**(7): 476–476.

Williams, J. R., L. DiCarlo, and C. M. Marcus (2007). Quantum hall effect in a gate-controlled p-n junction of graphene. *Science*, **317**(5838): 638–641.

Wolf, E. L. (2012). *Principles of Electron Tunneling Spectroscopy*. Oxford, New York, Oxford University Press.

Young, A. F., and P. Kim (2009). Quantum interference and Klein tunnelling in graphene heterojunctions. *Nat Phys*, **5**(3): 222–226.

Zhang, Y. B., Y. W. Tan, H. L. Stormer, and P. Kim (2005). Experimental observation of the quantum Hall effect and Berry's phase in graphene. *Nature*, **438**(7065): 201–204.

Zheng, Y. S., and T. Ando (2002). Hall conductivity of a two-dimensional graphite system. *Phys Rev B*, **65**(24).

Ziatdinov, M., S. Fujii, K. Kusakabe, M. Kiguchi, T. Mori, and T. Enoki (2013). Visualization of electronic states on atomically smooth graphitic edges with different types of hydrogen termination. *Phys Rev B*, **87**(11).

Chapter 2

Intrinsic Coherence of Graphene

Chapter 1 deals with steady-state chiral tunneling. We have concluded that the transmission probability is ideal ($T = 1$) if the momentum of an incident electron is directed perpendicular to the chiral barrier, which constitutes the Klein paradox. Another interesting moment is that an incident electron is converted into a hole inside the chiral barrier. However, if the angle of incidence is finite, the elementary excitation represents a combination of the electron and the hole which propagate over two different A and B sublattices.

2.1 Field-Biased Graphene Junctions

In this section we consider the effect of external field which takes place in the field-effect transistor (FET). The FET setup involves a voltage biased graphene ribbon which is enclosed between metallic source (S) and drain (D) electrodes as shown in Fig. 2.1. In this configuration, the bias voltage V_{SD} is applied between the S and D electrodes and it drops entirely on the ribbon (see Fig. 2.1). The local gate electrode can be placed beneath the dielectric substrate which is made, for example, from dielectric SiO_2. The gate electrode

Graphene: Fundamentals, Devices, and Applications
Serhii Shafraniuk

ISBN 978-981-4613-47-7 (Hardcover), 978-981-4613-48-4 (eBook)
www.panstanford.com

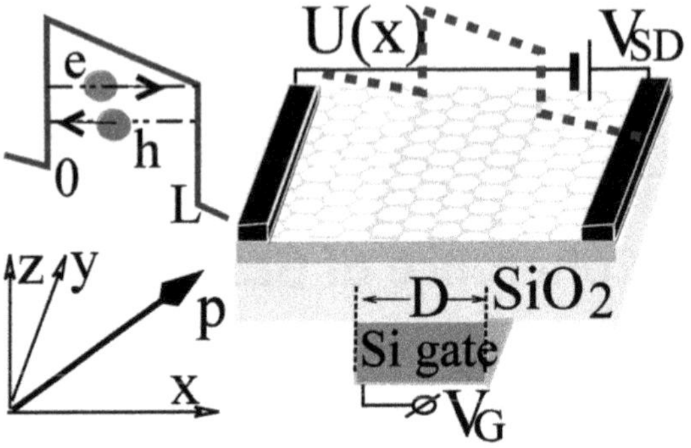

Figure 2.1 Chiral tunneling through the graphene junction with metal electrodes. Left upper panel: Tunneling through the chiral barrier tilted by the source–drain bias voltage. When the incidence angle φ is finite, the chiral particle represents a combination of an electron and a hole. Main panel: Graphene junction with metal source and drain electrodes (black) and the doped Si gate located underneath the dielectric SiO_2 substrate. Left lower panel shows the coordinate system.

itself is fabricated from, for example, p-doped Si, as shown in the right panel of Fig. 2.1. An alternative FET geometry might involve local gate electrodes placed on the top of graphene sheet. In either case, when a local gate voltage V_G is applied to the local gate electrode it alters the spatial profile of the electrochemical potential $\mu(x)$ by inducing a potential barrier in a limited region $0 < x < L$, as shown in the upper-left panel of Fig. 2.1. Formally, the profile U (x) of potential barrier is computed from electroneutrality condition which describes the coordinate-dependent altering $\mu(x)$ in the graphene ribbon for $0 < x < D$ (see Fig. 2.1). However, for the sake of simplicity, we model the barrier as having a mere rectangular shape. When a finite-bias source–drain voltage $V_{SD} \neq 0$ is applied, the U (x) barrier is tilted, as sketched in Fig. 2.1. For a "clean" graphene junction with finite dimensions, the ballistic motion of electrons and holes is quantized. We will see that the quantization imposes additional constrains on the directional tunneling diagram (Katsnelson et al. 2006, Katsnelson and Novoselov 2007, Shafranjuk 2009, Shafranjuk 2011). There are two carbon atoms per unit cell in graphene, forming two different triangular sublattices, A and B. This allows to represent a charge carrier eigenstate in graphene as a two-component vector (Ando 2005, Katsnelson et al. 2006, Katsnelson and Novoselov 2007). Because the fundamental electron transport properties are determined by low-energy excitations, here we focus our attention mostly on eigenstates in vicinity of the K point (quoted as the Dirac point where the excitation energy vanishes) in the Brillouin zone of graphene. In practice, the K point itself produces practically no contribution into the electron transport. It happens due to various reasons. For instance, the K point can be hardly achievable, because the electron states in graphene ribbon interact with impurities and phonons, as well as with evanescent states in the substrate (Cheianov et al. 2007, Giovannetti et al. 2007, Hwang et al. 2007a, 2007b, Nomura and MacDonald 2007, Tan et al. 2007, Ando 2008, Bolotin, Sikes, and Hone et al. 2008, Bolotin, Sikes, and Jiang et al. 2008, Chen et al. 2008, Katsnelson and Geim 2008, Kim and Neto 2008). Although the point does not immediately affect the transport properties, its presence is very important for the whole scenario of the electron transport in graphene ribbons and in carbon nanotubes (CNTs). The K point determines not only the electron excitation spectrum but it also contributes to quantum mechanical probabilities of electron scatterings and chiral tunneling (Ando 2005, Katsnelson et al. 2006, Katsnelson and Novoselov 2007) (see Fig. 2.2).

According to Cheianov et al. (2007), Giovannetti et al. (2007), Hwang et al. (2007a, 2007b), Nomura and MacDonald (2007), Tan et al. (2007), Ando (2008), Bolotin, Sikes, and Hone et al. (2008), Bolotin, Sikes, and Jiang et al. (2008), Chen et al. (2008), Katsnelson and Geim (2008), and Kim and Neto (2008), the elastic electron-impurity and inelastic electron–phonon scatterings both depend on the Coulomb screening. According to Cheianov et al. (2007), Giovannetti et al. (2007), Hwang et al. (2007a, 2007b), Nomura and MacDonald (2007), Tan et al. (2007), Ando (2008), Bolotin, Sikes, and Hone et al. (2008), Bolotin, Sikes, and Jiang et al. (2008), Chen et al. (2008), Katsnelson and Geim (2008), and Kim and Neto (2008), electron scattering on charged impurities may essentially impact the whole scenario of charge carrier's transport in graphene ribbons. When the electron–impurity interaction is strong, the electron mean free path in graphene may become as short as a few nanometers. The strong scattering actually happens in samples with charged

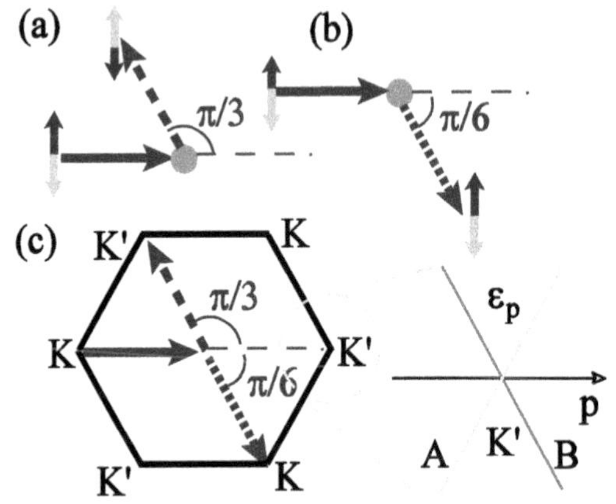

Figure 2.2 An elastic scattering of a chiral electron shown by a brown arrow with two smaller one-half pseudospin arrows on atomic impurities. (a) The electron scattering on a charged impurity (red circle) is strong. Let us assume that the electron momentum deviates by angle $\varphi = \pi/3$ emerging from the K-point and getting into the K′ point. Then the intervalley one-half pseudospin must flip; thus such a process is prohibited. (b) A weaker electron–impurity scattering on a neutral impurity (smaller magenta circle). This process is permitted since the scattering between the two different K points conserves the pseudospins. (c) The (a) and (b) cases in terms of directions of the electron momentum in the Brillouin zone during the scattering processes.

impurities. The local electric charge accumulated on foreign atomic impurities (see processes scetched in Fig. 2.2) and lattice defects is not well screened in graphene as it happens, for example, in metals where the electron concentration is considerably higher (in 3D metal samples typically $n = 10^{21}$–10^{23} cm^3).

Another electron scattering mechanism involves ripples of graphene sheets (Fig. 2.3). Despite the presence of scattering (Cheianov et al. 2007, Giovannetti et al. 2007, Hwang et al. 2007a, 2007b, Nomura and MacDonald 2007, Tan et al. 2007, Ando 2008, Bolotin, Sikes, and Hone et al. 2008, Bolotin, Sikes, and Jiang et al. 2008, Chen et al. 2008, Katsnelson and Geim 2008, Kim and Neto 2008), the electron mean free path can be long enough to exceed the sample size. Recent technological and experimental advances in fabricating and transport measurements of “clean” graphene sheets suggest that the mean free path of the charge carriers might exceed

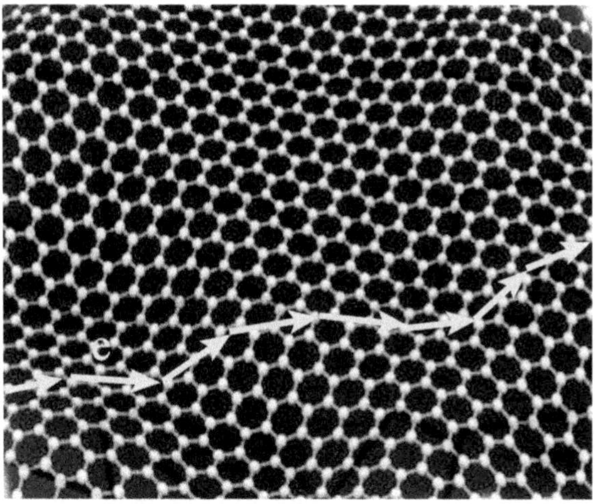

Figure 2.3 The electron scattering on out-of-plane ripples in a graphene sheet. Graphene ripples originate from the out-of-plane bending.

as much as 5 μm (Tan et al. 2007, Bolotin, Sikes, and Hone et al. 2008, Bolotin, Sikes, and Jiang et al. 2008. Therefore the electric transport of charge carrier in small graphene flakes by size ≤ 5 μm can be regarded as ballistic rather than as diffusive. When considering the spatial scale of interest $\leq l_{\mathrm{i,ep}}$, one might ignore the electron scattering on defects and impurities. This corresponds to the ballistic limit which is easily handled in terms of the Dirac equation (Ando 2005, Katsnelson et al. 2006, Tworzydlo et al. 2006, Katsnelson and Novoselov 2007, Tworzydlo et al. 2007, Shafranjuk 2009).

Here emerges a key difference between graphene on the one hand and conventional conductors (i.e., regular semiconductors and normal metals) on the other hand: Owing to unique symmetry of the crystal lattice, the envelope wavefunctions in graphene satisfy rather the Dirac equation than the Schrödinger equation. The matrix components of the 4×4 Hamiltonian respectively are ascribed to the A and B sublattices and also to K and K′ points in the momentum space. Coupling between electron states which belong to the A and B sublattices causes the electron and holes to be intrinsically connected with each other. Thus, the electron and hole waves serve as components of more general four-component wavefunction. The

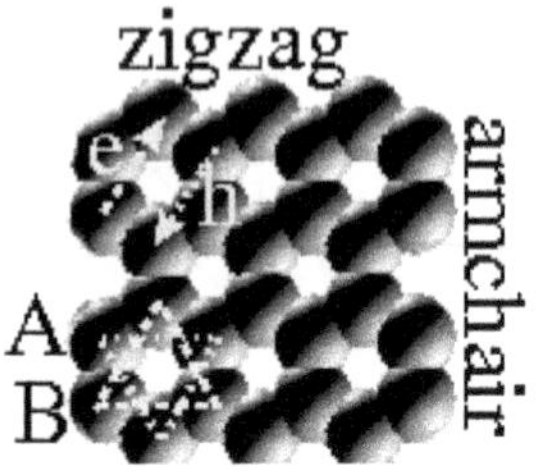

Figure 2.4 A graphene fragment with zigzag and armchair edges. Electrons with pseudospin "↑" propagate over the A sublattice, while holes with pseudospin "↓" move over the B sublattice.

envelope wavefunction in graphene is composed of electron and hole states ascribed to A and B sublattices and also to the K and K′ points. In other words, an elementary excitation in our model is characterized by two additional quantum numbers which are one-half pseudospins "↑" and "↓." One pseudospin indicates that the particle belongs to a certain sublattice, either A or B, whereas the other one-half pseudospin serves to ascribe the particle either to K or K′ point. For instance, one quotes an electron with momentum **p** propagating on the A sublattice near the K point and having a positive energy as an excitation with two one-half pseudospins "↑↑." Consequently, a hole with the same momentum **p** having a negative energy and propagating on the B sublattice in the opposite direction is characterized by two one-half pseudospins "↓↓," as illustrated in Fig. 2.4. In the more general case, when the propagation direction is not parallel to chains of the carbon atoms belonging to the A and B sublattices, an elementary excitation is neither electron nor hole, but instead a hybrid composition of them. The one-half pseudospin of a chiral particle in graphene may be directed in four possible ways: one pseudospin $s_{A(B)} = \uparrow$ $(\downarrow)$ ascribed to the A and B sublattices, and another pseudospin $s_{K(K')} = \uparrow$ $(\downarrow)$ for the same particle ascribed to K and K′ points. However, in either case, the direction of the one-half pseudospin coincides with the direction of the propagation which is opposite for the A and B sublattices. One remarkable feature of the chiral transport is that the particle's pseudospins are conserved during their propagation. In an ideal case, a chiral particle with the one-half pseudospin propagates in a sheet of the

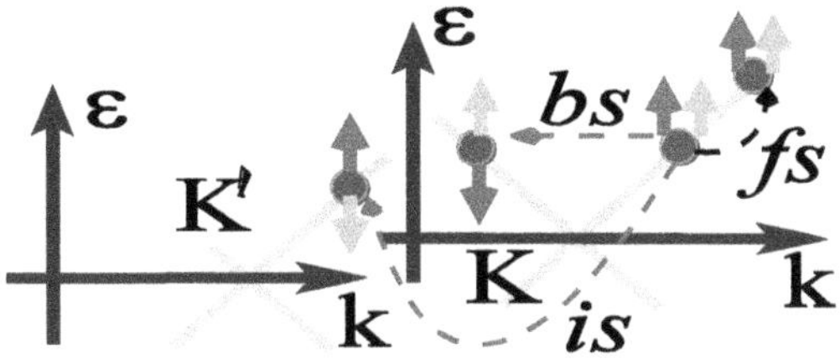

Figure 2.5 A flip of the one-half pseudospins during electron scatterings. The back scattering presumes the flip of the *intravalley* one-half pseudospin when the electron (hole) velocity is reversed. During the *intervalley* scattering process the electron momentum **p** must be reversed, which is accompanied by a flip of the *intervalley* one-half pseudospin.

homogeneous, pristine, suspended graphene without any collisions and back scattering. However, if a chiral fermion crosses a border between two adjacent graphene regions characterized by different crystallographic orientation, the pseudospin (one or both of them) can flip, since the orientation of it may change from one region to another (Fig. 2.5). An external AC field is coupled to the pseudospins directly, so it may tilt or flip the pseudospins as well.

In this chapter we employ quasiclassical approximation to describe transport of the charge carriers in graphene. It allows us to make a simple analytical treatment. Such an approach is well tractable and captures basic electron properties determined by the wavefunction symmetry and by spectrum of the low-energy excitations. The quasiclassical approach (Ando 2005, Katsnelson et al. 2006, Tworzydlo et al. 2006, Katsnelson and Novoselov 2007, Tworzydlo et al. 2007, Shafranjuk 2009) is valid on the spatial scale comparable to or larger than the wavefunction envelope size $\delta x \approx \hbar v/\bar{\varepsilon}$ ($\bar{\varepsilon}$ is the average electron energy). For instance, if one takes the graphene ribbon width $\sim$10 nm, the uncertainty of the electron energy achieves $\sim$7 meV (Brey and Fertig 2006a, 2006b). This means that the larger is the graphene sample the better precision is achieved when using the quasiclassical Dirac equation approximation. The quasiclassical method is accurate when describing the transport properties of graphene and CNTs. Similar ideas have been widely implemented for describing the transport properties of superconductors.

2.2 Electron and Hole Excitations in Graphene

One computes the 2D electron density of states $g_{2D}(\varepsilon)$ in graphene by taking into account the degeneracy of valley, pseudospin, and over the angle as

$$g_{2D}(\varepsilon)\, d\varepsilon = 4\frac{dp_x dp_y}{(2\pi\hbar)^2} = \frac{4}{(2\pi\hbar)^2}\frac{\varepsilon}{v_F^2}2\pi d\varepsilon \tag{2.1}$$

One can see that $g_{2D}(\varepsilon)$ corresponds to the number of discrete eigenenergy levels per unit energy within the unit area (states/eV cm^2). By substituting the dispersion law of gapless graphene $\varepsilon = v_F\sqrt{p_x^2 + p_y^2}$ into Eq. 2.1 one obtains

$$g_{2D}(\varepsilon) = \frac{2\varepsilon}{\pi\hbar^2 v_F^2}\text{sign}\varepsilon = \frac{2|\varepsilon|}{\pi\hbar^2 v_F^2} \tag{2.2}$$

In the last formula $\hbar$ is Plank's constant and $v_F = 10^6$ m/s is the Fermi velocity in graphene. The density of electrons per unit area is obtained as

$$\begin{aligned} n_e(\mu) &= \int_0^\infty d\varepsilon g_{2D}(\varepsilon) f_{FD}(\varepsilon - \mu) \\ &= \frac{2}{\pi}\left(\frac{k_B T}{\hbar v_F}\right)^2 \int_0^\infty du \frac{u}{1+\exp\left(u - \frac{\mu}{k_B T}\right)} \\ &= -\frac{2}{\pi}\left(\frac{k_B T}{\hbar v_F}\right)^2 Li_2\left(-e^{\frac{\mu}{k_B T}}\right) \end{aligned} \tag{2.3}$$

where we have used the equilibrium Fermi–Dirac function $f_{FD}(\varepsilon - \mu)$, μ is the chemical potential, T is the temperature, k_B is the Boltzmann constant, $Li_n(x)$ is the polylogarithm function of the order of n defined as

$$Li_n(z) = \sum_{k=1}^{\infty}\frac{z^k}{k^n} \tag{2.4}$$

The density of holes n_h is obtained by using the electron-hole symmetry $\gamma(\varepsilon) = \gamma(-\varepsilon)$, which gives

$$\begin{aligned} n_h(\mu) &= \int_{-\infty}^0 d\varepsilon g_{2D}(\varepsilon)(1 - f_{FD}(\varepsilon - \mu)) \\ &= -\frac{2}{\pi}\left(\frac{k_B T}{\hbar v_F}\right)^2 Li_2\left(-e^{-\frac{\mu}{k_B T}}\right) = n_e(-\mu) \end{aligned} \tag{2.5}$$

The above Eqs. 2.3 and 2.5 describe the full charge density per unit area or the charge imbalance in the form

$$n_S = n_e - n_h = \int_0^\infty d\varepsilon g_{2D}(\varepsilon)\left(f(\varepsilon-\mu) - f(\varepsilon+\mu)\right)$$
$$= \frac{2}{\pi}\left(\frac{k_B T}{\hbar v_F}\right)^2 \left[Li_2\left(-e^{-\frac{\mu}{k_B T}}\right) - Li_2\left(-e^{\frac{\mu}{k_B T}}\right)\right] \tag{2.6}$$

The total carrier density of charged sheet of the pristine graphene determines the bipolar conductivity is obtained as

$$N_S = n_e + n_h = -\frac{2}{\pi}\left(\frac{k_B T}{\hbar v_F}\right)^2 \left[Li_2\left(-e^{-\frac{\mu}{k_B T}}\right) + Li_2\left(-e^{\frac{\mu}{k_B T}}\right)\right] \tag{2.7}$$

In the vicinity of the charge neutrality point (NP) one obtains the intrinsic density with equal densities of electrons and holes. That NP is characterized by the vanishing chemical potential $\mu = 0$ which gives

$$N_S(\mu = 0) = n_e + n_h = -\frac{2}{\pi}\left(\frac{k_B T}{\hbar v_F}\right)^2 2Li_2(-1)$$
$$= -\frac{2}{\pi}\left(\frac{k_B T}{\hbar v_F}\right)^2 2\sum_{k=1}^{\infty}\frac{(-1)^k}{2^k} = \frac{\pi}{3}\left(\frac{k_B T}{\hbar v_F}\right)^2 \tag{2.8}$$

At room temperature $T = 300$ K, the intrinsic carrier density roughly corresponds to $n_i \cong 8 \times 10^{10}$ cm^{-2}, which exceeds the corresponding value of n_i for silicon. Expanding the polylogarithm function in the vicinity of the $\mu = 0$ one obtains

$$-Li_2\left(-e^{\frac{\mu}{k_B T}}\right) \cong \frac{\pi^2}{12} + \ln 2\frac{\mu}{k_B T} + \frac{1}{4}\left(\frac{\mu}{k_B T}\right)^2 \tag{2.9}$$

which serves as a good approximation as soon as $\mu < 5k_B T$. It gives

$$n_e(\mu) \cong \frac{2}{\pi}\left(\frac{k_B T}{\hbar v_F}\right)^2\left[\frac{\pi^2}{12} + \frac{1}{2}\left(\frac{\mu}{k_B T}\right)^2\right]$$
$$= \frac{\pi}{6}\left(\frac{k_B T}{\hbar v_F}\right)^2 + \frac{\mu^2}{\pi(\hbar v_F)^2} = n_i + \frac{\mu^2}{\pi(\hbar v_F)^2} \tag{2.10}$$

where we have used the following asymptotic

$$-Li_2(-e^z) \cong \frac{\pi^2}{12} + \frac{z^2}{2}, \quad z >> 1 \tag{2.11}$$

$$-Li_2\left(-e^{\frac{\mu}{k_BT}}\right) \cong \frac{\pi^2}{12} + \frac{1}{2}\left(\frac{\mu}{k_BT}\right)^2 \tag{2.12}$$

The above approximation, Eqs. 2.11 and 2.12, is valid at the charge NP and allows a simple expression for electron charge concentration, Eq. 2.10, since it gives correct asymptotic at $\mu >> k_BT$ and in the intermediate region $\mu \approx k_BT$. However, one should be careful when using that approximation for computing the capacitance at $\mu \approx 0$ because the number of linear terms in μ is insufficient in the latter case. In the vicinity of point with zero chemical potential ($\mu \approx 0$), the mentioned approximation fails because of the disorder. One also obtains the formula for the channel electron density per unit area for degenerate system at $\mu >> k_BT$ as

$$n_S = n_e - n_h = \int_0^{\mu} d\varepsilon g_{2D}(\varepsilon) \simeq \frac{\mu^2}{\pi (\hbar v_F)^2} \tag{2.13}$$

2.3 Quantum Capacitance of Graphene

An important characteristic of the intrinsic properties of graphene is the quantum capacitance which is computed as follows. It is determined by derivatives of Eqs. 2.3 and 2.5, which read

$$\frac{dn_e}{d\mu} = \frac{2}{\pi}\left(\frac{k_BT}{\hbar v_F}\right)^2 \ln\left(1 + e^{\frac{\mu}{k_BT}}\right) \tag{2.14}$$

$$\frac{dn_h}{d\mu} = -\frac{2}{\pi}\left(\frac{k_BT}{\hbar v_F}\right)^2 \ln\left(1 + e^{-\frac{\mu}{k_BT}}\right) \tag{2.15}$$

Then one obtains an expression for the quantum capacitance of the graphene charge in the form

$$C_Q = e\frac{d(n_e - n_h)}{d\mu} = e^2 \int_{-\infty}^{\infty} d\varepsilon g_{2D}(\varepsilon)\left(-\frac{\partial f_{FD}}{\partial \varepsilon}\right)$$
$$= \frac{2k_BT}{\pi}\left(\frac{e}{\hbar v_F}\right)^2 \ln\left(2 + 2\cos h\frac{\mu}{k_BT}\right) \tag{2.16}$$

In particular, for an unbiased case ($\mu = 0$), the quantum capacitance is

$$C_Q^{\min} = 2\ln 2\frac{k_BT}{\pi}\left(\frac{e}{\hbar v_F}\right)^2 \tag{2.17}$$

When the doping level is high (i.e., assuming $\mu >> k_B T$) one obtains an *approximate* formula for the quantum capacitance

$$C_Q \cong e^2 \frac{dn_S}{d\mu} = \frac{2|\mu|}{\pi} \left(\frac{e}{\hbar v_F} \right)^2 \tag{2.18}$$

In expression for the total density of free carriers one obtains an expression, which is valid for arbitrary μ

$$\frac{d(n_e + n_h)}{d\mu} = \frac{2|\mu|}{\pi} \left(\frac{1}{\hbar v_F} \right)^2 \tag{2.19}$$

The last Eq. 2.19 represents an *exact* for ideal graphene for any chemical potential which was not the case for formerly introduced Eqs. 2.17 and 2.18. That is a consequence of their relationship with the exact form of the Einstein relation.

2.4 Einstein Relation in Graphene

Recent advances in fabricating graphene allow us to obtain fairly "clean" samples where the concentration of charged impurities is low. The charged impurities are removed, for example, by a proper annealing. Other scattering mechanisms involve riffles of graphene sheets, corrugations, dislocations, and other lattice defects. Then, the electron mean free path l_e is determined by scattering on the lattice oscillations, phonons. The electron scattering on phonons is reduced due to chiral properties of electron excitations in graphene which are characterized by one-half pseudospin. Since the electron–phonon scattering presumes flip-flops of the pseudospin they are inhibited by the pseudospin conservation law. The reduced scattering of chiral electrons results in a prolonged scattering length which exceeds several micrometers. The intrinsic "cleanness" of pristine graphene can be exported when fabricating the FETs (Fig. 2.6).

In this subsection we consider another limit of "dirty" graphene which takes place in many practical graphene devices since their fabrication technology is still in a premature stage. Electron scattering on the charged impurities in the gate insulating oxide along with the elastic electron–phonon scattering are major factors which determine the transport properties of "dirty" graphene. Then,

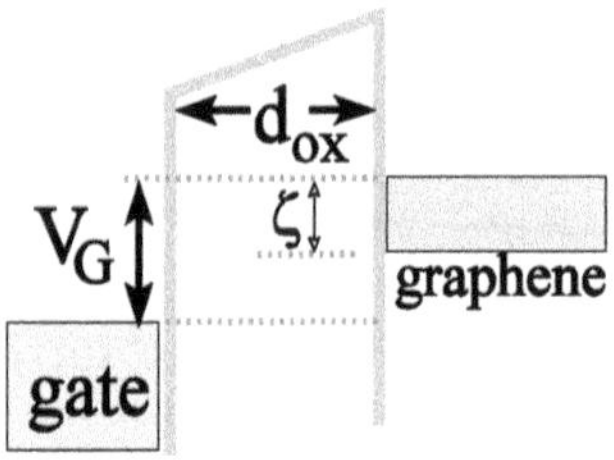

Figure 2.6 Energy diagram of a graphene FET.

properties of the graphene FET resemble those of the silicon metal–oxide–semiconductor field-effect transistors (MOSFETs). One can compute the diffusion constant of the 2D graphene sheet by using the Fermi velocity v_{F} and transport relaxation time τ_{tr} or mean free path $l = \varpi_{\Phi}\tau_{\mathrm{tr}}$

$$D = \frac{v_{\mathrm{F}}^2\tau_{\mathrm{tr}}}{2} = \frac{v_{\mathrm{F}} l}{2} \tag{2.20}$$

The Einstein relation yields the electron and hole mobility $\mu_{\mathrm{e/h}}$ in the form

$$\mu_{\mathrm{e/h}} = \frac{eD_{\mathrm{e/h}}}{n_{\mathrm{e/h}}}\frac{dn_{\mathrm{e/h}}}{d\mu} \equiv \frac{eD_{\mathrm{e/h}}}{\varepsilon_{\mathrm{D}}} \tag{2.21}$$

where $e = |e|$ and we also have introduced the diffusion energy (Ando et al. 1982)

$$\varepsilon_{\mathrm{D}} \equiv \frac{n_{\mathrm{e/h}}}{dn_{\mathrm{e/h}}/d\mu} \tag{2.22}$$

When the device regime is far from the graphene charge NP E_0 and we find $\varepsilon_{\Delta} = \varepsilon_{\mathrm{F}}/2$. One also obtains the bipolar conductivity as a the sum of electron and hole components

$$\sigma_0 = e^2\left(D_{\mathrm{e}}\frac{dn_{\mathrm{e}}}{d\mu} + D_{\mathrm{h}}\frac{dn_{\mathrm{h}}}{d\mu}\right) \tag{2.23}$$

Under an assumption of the electron–hole symmetry, when the electron and hole diffusion coefficients become equal to each other $D_{\mathrm{e}} = D_{\mathrm{h}} = D_0$, from Eq. 2.19 one finds the total bipolar conductivity as

$$\sigma_0 = e^2 D_0\frac{d\left(n_{\mathrm{e}} + n_{\mathrm{h}}\right)}{d\mu} = \frac{2e^2}{h}\frac{\varepsilon_{\mathrm{F}}\tau_{\mathrm{tr}}}{\hbar} = \frac{2e^2}{h}k_{\mathrm{F}}l \tag{2.24}$$

where we have used the dispersion law of the gapless pristine graphene $\hbar v_F k_F = \mu \cong \varepsilon_F$ to define the Fermi wavevector k_F. The above formula in Eq. 2.24 also allows us to obtain the carrier mobility in highly doped ($\mu >> k_B T$) graphene

$$\mu_0 = \frac{e v_F \tau_{tr}}{p_F} = \frac{el}{p_F} \tag{2.25}$$

where we have implemented the Einstein relation in a different form as being expressed via conductivity and quantum capacitance

$$D_0 C_Q = e\mu_0 N_S = \sigma_0 \tag{2.26}$$

Notice that in fact $l \propto p_F$, and therefore μ_0 weakly depends on the Fermi energy of graphene.

2.5 Electrostatics of Gated Graphene Devices: Charge Traps Near the Graphene/Oxide Interface

The presence of charged oxide defects at the interface separating graphene and an insulated oxide layer (G/O interface) strongly affects the general transport properties of a graphene FET. Geometrically, the defects are situated either exactly at the G/O interface or in an immediate vicinity of it inside the oxide on a short distance 1–3 nm from the interface. The defects serve as traps of electric charge whose mechanism is as follows. The defects generally can have different charge states, which opens the possibility for the charge carriers (electrons and holes) to recharge the traps within the FET channel. The charge traps are coupled to the conducting charge carriers by means of the tunneling exchange interaction. Therefore, the electric charge on a certain trap depends on the Fermi level position: If the Fermi energy ε_F exceeds the defect energy E_{ct}, then trap levels are filled; when $E_{ct} > \varepsilon_F$ they are empty. One can distinguish between two kinds of traps which are either acceptors or donors. The former acceptor traps are charged negatively when their energy level is filled, whereas they are neutral if the level is empty ($-/0$). The latter donor traps are charged positively when their energy level is empty but become negatively charged if the level is filled ($0/+$). In either case, the Fermi energy diminishes when the gate voltage V_G increases, while the traps are getting

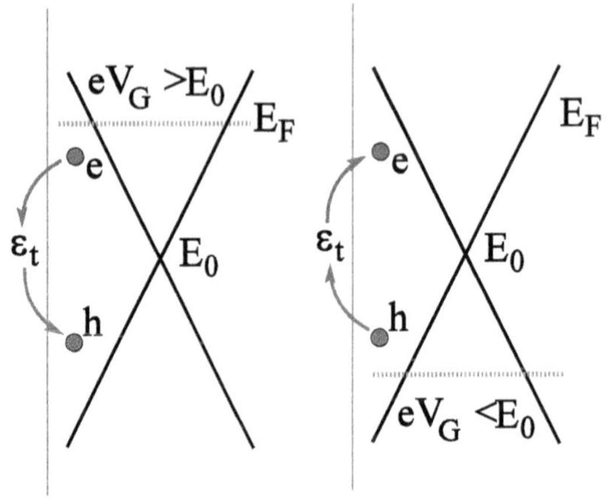

Figure 2.7 Coupling of charge traps to the conducting electrons due to the tunneling exchange at the G/O interface, which corresponds either to (a) filling or to (b) emptying.

filled up and the negative charging prevails (see Fig. 2.7). The Fermi energy ε_F is adjusted by applying a finite electric voltage V_G to the gate electrode. Each ε_F value at the G/O interface corresponds to certain "equilibrium" filling which also causes a finite density of equilibrium-trapped charge $Q_t(\mu) = eN_t(\mu)$ which is considered as positive. If the recharging time of a trap is short, they quickly come into the equilibrium with the substrate. Such traps with the short recharging time are regarded as interface traps (N_{it}). Otherwise, if the coupling time between a trap and the substrate is longer than the measurement time they are quoted as oxide trapped traps (N_{ot}). There is no clear distinction between the two kinds of the charge traps. It depends on the temperature of measurement and on how fast the gate voltage sweeps. One introduces the interface trap capacitance per unit area C_{it} as

$$C_{it} \equiv \frac{d}{d\mu}\left(-eN_t\left(\mu\right)\right) > 0 \tag{2.27}$$

In the last Eq. 2.27, the trap charge $Q_t(\mu) = eN_t(\mu)$ which depends on the Fermi-level position is determined by the electric charge of all the traps. However, because of a finite value of the voltage sweep time t_s, only the "interface traps" with low recharging time constants $\tau_r < t_s$ are being recharged. The capacitance of

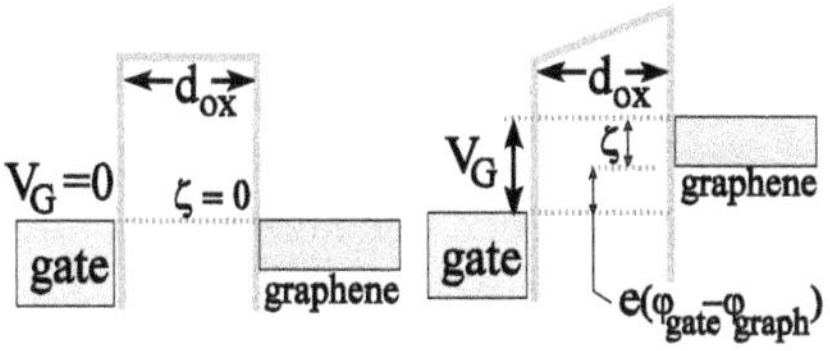

Figure 2.8 Schematics of the energy diagram of gate-oxide-graphene structure at $V_G = 0$ (on the left) and $V_G > 0$ (on the right). For simplicity we set $\varphi_{gg} = 0$.

interface trap corresponds to the energy density of the defect levels $D_{it}(\mathrm{cm}^{-2}\mathrm{eV}^{-1})$. This is illustrated by the formula

$$C_{it} = e^2 D_{it}(\mu) \tag{2.28}$$

In modern silicon MOSFETs, typical capacitance of interface trap is of the order of $D_{it} \approx 10^{11}$–10^{12} cm^{-2}eV^{-1}. That parameter is changed under the ionizing radiation impact.

2.6 Steady-State Electrostatics of Graphene Field-Effect Transistors

The simplest design of the graphene FET has the geometry of a two-plate capacitor. Electric charges of the opposite signs are accumulated in such a device on the gate electrode and graphene sheet, respectively. The level of charge neutrality E_{NP} is typically selected as the chemical potential in graphene. We also define the work function for Dirac point of graphene (electron affinity) as measured in respect the vacuum energy level E_{vac} as

$$\chi_g = E_{vac} - E_{NP} \tag{2.29}$$

The graphene work function is evaluated as $\chi_g \approx 4.5$ eV. The source–drain bias voltage across our FET causes the difference of electrochemical potentials. For the sake of simplicity we account here only for two contributions to the electrochemical potential

$$\mu = \zeta + U = \zeta - e\varphi \tag{2.30}$$

$$\mu_{graphene} = -\chi_g + \zeta - e\varphi_{graphene} = E_{NP} + \zeta \tag{2.31}$$

In the above Eqs. 2.21 and 2.31, $\varphi_{\text{graphene}}$ is electrostatic potential of graphene sheet, ζ is the chemical potential which is independent on the electric charge, whereas U and φ are the electrostatic energy and potential $\Upsilon = -\varepsilon\varphi$, respectively. Because the quantum capacitance per unit area is extremely large in such device, we can neglect by the voltage drop in the gate which typically is fabricated good 3D conductors

$$\mu_{\text{gate}} = -e\varphi - W_{\text{gate}} \tag{2.32}$$

Here $E_{\text{NP}} = -\chi_{\text{g}} - \varepsilon\phi_{\text{graphene}}$ is the Dirac (or charge neutrality) point which determines zero of energy and W_{gate} is the work function of the gate material. If the gate voltage is, for example, positive, it increases both the chemical potential and electrostatic potential of the graphene sheet. The two potentials compensate each other exactly, while keeping the electrochemical potential of the graphene sample unchanged (see Fig. 2.6).

The difference between the electrochemical potential of graphene μ_{graphene} and of the gate μ_{gate} determines the bias between the graphene sheet and the conducting gate

$$eV_{\text{G}} = \mu_{\text{graphene}} - \mu_{\text{gate}} = e\varphi_{\text{gg}} + \zeta + e\left(\varphi_{\text{gate}} - \varphi_{\text{graphene}}\right) \tag{2.33}$$

where the work function difference between the gate and graphene is

$$e\varphi_{\text{gg}} = W_{\text{gate}} - \chi_{\text{g}} \tag{2.34}$$

If the charged oxide defects are located straight on the insulator–graphene interface, whereas there are no electric charges inside the oxide layer, the electric field E_{ox} is fairly uniform across the insulating layer:

$$\varphi_{\text{gate}} - \varphi_{\text{graphene}} = E_{\text{ox}} d_{\text{ox}} = \frac{eN_{\text{gate}}}{\varepsilon_{\text{ox}}\varepsilon_0} d_{\text{ox}} \equiv \frac{eN_{\text{gate}}}{C_{\text{ox}}} \tag{2.35}$$

In the above formulas $N_{\text{gate}}\,(V_{\text{G}})$ is the charge carrier density (per unit area) inside the gate/metallic gate and C_{ox} is the oxide (insulator) capacitance per unit area:

$$C_{\text{ox}} = \frac{\varepsilon_{\text{ox}}\varepsilon_0}{d_{\text{ox}}} \tag{2.36}$$

where ε_{ox} is the dielectric constants of the insulator.

2.7 Characteristic Scales of Gated Graphene

The gated structure is characterized by the electric neutrality condition

$$N_{\mathrm{G}} + N_{\mathrm{t}} = n_{\mathrm{S}} \tag{2.37}$$

where N_{G} is the gate charge density, n_{s} is the charge imbalance density, and N_{t} is the defect density. In Eq. 2.37, $N_{\mathrm{t}} > 0$ and $N_{\mathrm{G}} > 0$, whereas n_{s} can be either positive or negative and all the densities are per unit area. Using the above charge neutrality condition (2.37) one overwrites Eq. 2.33 as

$$eV_{\mathrm{G}} = e\varphi_{\mathrm{gg}} + e\varphi + \frac{e^2}{C_{\mathrm{ox}}}\left(n_{\mathrm{S}}\left(\zeta\right) - N_{\mathrm{t}}\left(\zeta\right)\right) \tag{2.38}$$

The above value of the gate voltage corresponding to the charge NP V_{NP} now can be defined as

$$V_{\mathrm{NP}} \equiv V_{\mathrm{G}}\left(\zeta = 0\right) = \varphi_{\mathrm{gg}} - \frac{eN_{\mathrm{t}}\left(\zeta = 0\right)}{C_{\mathrm{ox}}} \tag{2.39}$$

Taking into account that the sign of the chemical potential is determined by sign(V_{G}- V_{NP}) one also obtains

$$e\left(V_{\mathrm{G}} - V_{\mathrm{NP}}\right) = \zeta + \frac{e^2 n_S}{C_{\mathrm{ox}}} + \frac{e^2\left(N_t\left(\zeta = 0\right) - N_{\mathrm{t}}\left(\zeta\right)\right)}{C_{\mathrm{ox}}} \tag{2.40}$$

Equation 2.40 is simplified in the limit $V_{\mathrm{NP}} = 0$ and when the interface trap charges at the NP are absent. Then one gets

$$e^2\left(N_{\mathrm{t}}\left(\zeta = 0\right) - N_{\mathrm{t}}\left(\zeta\right)\right) \cong C_{\mathrm{it}}\zeta \tag{2.41}$$

Taking into account Eq. 2.13 one obtains the basic equation of graphene planar electrostatics as

$$eV_{\mathrm{G}} = \varepsilon_{\mathrm{F}} + \frac{e^2 n_S}{C_{\mathrm{ox}}} + \frac{C_{\mathrm{it}}}{C_{\mathrm{ox}}}\varepsilon_{\mathrm{F}} \equiv m\varepsilon_{\mathrm{F}} + \frac{\varepsilon_{\mathrm{F}}^2}{2\varepsilon_{\mathrm{a}}} \tag{2.42}$$

where we have replaced ζ by ε_{F} and

$$m \equiv 1 + \frac{C_{\mathrm{it}}}{C_{\mathrm{ox}}} \tag{2.43}$$

is the "ideality factor." The characteristic energy scale

$$\varepsilon_{\mathrm{a}} = \frac{\pi\left(\hbar v_{\mathrm{F}}\right)^2 C_{\mathrm{ox}}}{2e^2} = \frac{\varepsilon_{\mathrm{ox}}}{8\alpha_{\mathrm{G}}}\frac{\hbar v_{\mathrm{F}}}{d_{\mathrm{ox}}} \tag{2.44}$$

which appears in Eq. 2.42 reflects specifics of the graphene–insulator–gate FET electrostatics and therefore one can also introduce the graphene "fine structure constant" as (in SI units):

$$\alpha_{\mathrm{G}} = \frac{e^2}{4\pi\varepsilon_0 \hbar v_{\mathrm{F}}} \tag{2.45}$$

The dielectric constants of SiO_2 $\varepsilon_{\mathrm{ox}} = 4$ and $\varepsilon_{\mathrm{ox}} = 16$ for HfO_2. The value of ε_{a} actually determines a natural spatial scale

$$a_{\mathrm{Q}} \equiv \frac{\hbar v_{\mathrm{F}}}{\varepsilon_{\mathrm{a}}} = \frac{2e^2}{\pi \hbar v_{\mathrm{F}} C_{\mathrm{ox}}} = \frac{8\alpha_G}{\varepsilon_{\mathrm{ox}}} d_{\mathrm{ox}} \tag{2.46}$$

which is specific to the graphene gated structures and corresponding characteristic density. The characteristic length a_{Q} appears to be of the order of the oxide thickness for the insulator HfO_2 with $\varepsilon_{\mathrm{ox}} \approx 16$ only because the graphene "fine structure constant" has the value $\alpha_{\mathrm{G}} \approx 2.0$–$2.2$. We note that the energy scale ε_{a} is also represented as the functions of Fermi energy and wavevector k_{F}, quantum capacitance and charge density

$$\frac{\varepsilon_{\mathrm{a}}}{\varepsilon_{\mathrm{F}}} = \frac{C_{\mathrm{ox}}}{C_{\mathrm{Q}}} = \frac{1}{k_{\mathrm{F}} a_{\mathrm{Q}}} = \frac{1}{\sqrt{\pi n_{\mathrm{S}} a_{\mathrm{Q}}^2}} \tag{2.47}$$

2.8 Solving the Electrostatic Equation

An explicit dependence of the electron Fermi energy versus the gate voltage at $V_{\mathrm{G}}>0$ is obtained as a solution of the algebraic Eq. 2.42:

$$\varepsilon_{\mathrm{F}} = \sqrt{m^2\varepsilon_{\mathrm{a}}^2 + 2\varepsilon_{\mathrm{a}} e V_{\mathrm{G}}} - m\varepsilon_{\mathrm{a}} \tag{2.48}$$

It provides the following relation for the graphene charge density versus the gate voltage

$$\frac{e^2 n_{\mathrm{S}}}{C_{\mathrm{ox}}} = eV_{\mathrm{G}} - m\varepsilon_{\mathrm{F}} = eV_{\mathrm{G}} + m^2\varepsilon_{\mathrm{a}} - m\sqrt{m^2\varepsilon_{\mathrm{a}}^2 + 2\varepsilon_{\mathrm{a}} e V_{\mathrm{G}}} \tag{2.49}$$

The above formula can be rewritten in a different form (Fang et al. 2007, Zebrev 2008)

$$en_{\mathrm{S}}(V_{\mathrm{G}}) = C_{\mathrm{ox}} \left(|V_{\mathrm{G}} - V_{\mathrm{NP}}| + V_0 \left(1 - \sqrt{1 + 2\frac{|V_{\mathrm{G}} - V_{\mathrm{NP}}|}{V_0}} \right) \right) \tag{2.50}$$

where we have introduced the characteristic voltage $V_0 \equiv m^2\varepsilon_\alpha/e$, which is defined when the interface trap capacitance is accounted for.

General tendency in behavior of the charge density versus the gate voltage depends on the characteristic values. If the gate voltage is high, $|V_G - V_{NP}| >> V_0$, which is also the case when the oxide layer is "thick," one arrives to the linear behavior

$$en_S \cong C_{ox}\left(|V_G - V_{NP}| - \sqrt{2V_0\,|V_G - V_{NP}|}\right) \tag{2.51}$$

In the last case, the gate voltage drops mostly on the oxide layer which in particular happens for the typical oxide thickness $d_{ox} = 300$ nm. If one reduces the thickness of the gate oxide for the graphene FET down to a few or tens of nanometers, then one arrives to a quadratic law

$$en_S \cong C_{ox}\left((V_G - V_{NP})\left(\frac{V_G - V_{NP}}{V_0}\right)\right), \quad V_G - V_{NP} < V_0 \tag{2.52}$$

Similar tendency also takes place if the gate bias $C_{ox}\,|V_G - V_{NP}| < en_Q$ is relatively small.

2.9 Capacitance of the Channel and of the Gate

Many valuable parameters of the FETs can be obtained from direct measurements of the device's capacitance. Using Eq. 2.38 one obtains

$$\frac{dV_G}{d\mu} = 1 + \frac{C_Q + C_{it}}{C_{ox}} \tag{2.53}$$

We can define the low-frequency gate capacitance as

$$C_G = e\left(\frac{\partial N_G}{\partial V_G}\right) = e\frac{dN_G/d\mu}{dV_G/d\mu} = \frac{C_Q + C_{it}}{1 + \frac{C_Q + C_{it}}{C_{ox}}} = \left(\frac{1}{C_{ox}} + \frac{1}{C_Q + C_{it}}\right)^{-1} \tag{2.54}$$

The relevant equivalent electric circuit which corresponds to Eq. 2.54 is sketched in Fig. 2.9. Another useful characteristic is the intrinsic channel capacitance defined as

$$C_{CH} = e\left(\frac{\partial N_S}{\partial V_G}\right) = e\frac{dN_S/d\mu}{dV_G/d\mu} = \frac{C_Q}{1 + \frac{C_Q + C_{it}}{C_{ox}}} = \frac{C_{ox}}{1 + \frac{C_Q + C_{it}}{C_Q}} \tag{2.55}$$

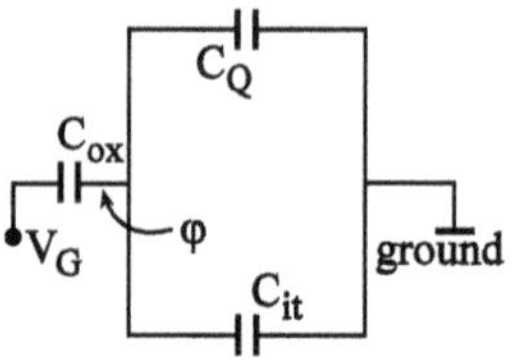

Figure 2.9 The equivalent circuit for computation of gate capacitance $C_G(V_G)$.

In the above Eq. 2.55, all the participating capacitances are nonvanishing and positive for any arbitrary V_G. Frequently the channel capacitance C_{CH} is regarded as the total gate capacitance C_{tot}, while the interface trap capacitance C_{it} is sometimes ignored. In the latter case, when $C_{it} = 0$, the gate and the channel capacitances coincide, that is, $C_{CH} = C_G$. In a more general case, when $C_{it} \neq 0$, C_G and C_{CH} are related to each other as

$$\frac{C_G}{C_{CH}} = 1 + \frac{C_{it}}{C_Q} \tag{2.56}$$

The above formulas for differential capacitances are preserved for any form of interface trap energy spectrum. In a particular oversimplified situation, when the interface traps are characterized by a constant energy density spectrum, the capacity–voltage characteristics $C_{CH}(V_G)$ are symmetric in respect to the NP. Then one derives the channel capacity by a mere direct differentiation of Eq. 2.50 which gives

$$C_{CH} = e\frac{dn_S}{dV_G} = C_{ox}\left[1 - \frac{1}{\sqrt{1 + 2\frac{|V_G - V_{NP}|}{V_0}}}\right] \tag{2.57}$$

The capacitance–voltage characteristics $C_G(V_G)$ is influenced by the interface trap capacitance. An equivalent circuit for computing of $C_G(V_G)$ is shown in Fig. 2.9.

If the interface trap capacitance vanishes, that is, $C_{it} = 0$ when $m = 1$, the capacitance–voltage dependencies are regarded as universal curves. They only depend on thickness and permittivity of the gate oxide through the parameter ε_a. A problematic point is that in practice one should distinguish between the quantum and the interface trap capacitances which in an equivalent circuit are connected in parallel. One can extract parameters of the interface trap spectra by comparing the "ideal" capacitance–voltage characteristics with real measurement data (Nicollian and Brews 1984).

2.10 Ratio of the Diffusion and Drift Currents

The electron current density J_S in the channel is represented by the sum of drift and diffusion parts

$$J_S = J_{dr} + J_{diff} = e\mu_0 n_S \frac{d\varphi}{dy} + eD_0 \frac{dn_S}{dy} \tag{2.58}$$

where μ_0 and D_0 are the electron mobility and diffusivity, respectively and y is a coordinate along the channel. An equivalent form of the above expression reads

$$J_D = \sigma_0 E \left[1 - \frac{D_0}{\mu_0} \frac{dn_S}{n_S d\zeta} \frac{d\zeta}{d\varphi}\right] \tag{2.59}$$

where we have introduced the electric field along the channel $E = -d\varphi/dy$ and the graphene sheet conductivity $\sigma_0 = e\mu_0 n_s$. Besides, in Eq. 2.59, $\zeta(y)$ and $\varphi(y)$ are the local chemical and electrostatic potentials in the graphene channel, respectively. Furthermore, using the Einstein relation for 2D system of noninteracting carrier, one can rewrite the above Eq. 2.58 for the diffusion–drift current in a different form as (Zebrev and Useinov 1990)

$$J_S \equiv e\mu_0 n_S E \left(1 - \frac{1}{e}\frac{d\zeta}{d\varphi}\right) = e\mu_0 n_S E\,(1+\kappa) \tag{2.60}$$

As it follows from the above formula, the ratio of gradients of chemical (ζ) and electrostatic (φ) potentials along the channel actually acts as a ratio of diffusion to the drift current:

$$\kappa = -\frac{1}{e}\frac{d\zeta}{d\varphi} = \frac{J_{dif}}{J_{dr}} \tag{2.61}$$

where ζ and φ are the components of electrochemical potential (or local Fermi energy for a high-doping case). The steady-state electrochemical potential does not depend on the coordinate ($\mu = \zeta - \varepsilon\varphi = \text{const.}$), whereas the derivative $d\zeta/d\varphi$ is identically equal to unity. It corresponds to the fact that the diffusion–drift current components exactly compensate each other:

$$\frac{1}{e}\frac{d\zeta}{d\varphi}\bigg|_{\mu} = -\frac{1}{e}\frac{\partial\mu/\partial\varphi|_{\zeta}}{\partial\mu/\partial\zeta|_{\varphi}} = 1 \tag{2.62}$$

A different scenario takes place in the nonequilibrium case. Both, the diffusion and drift currents flow in the same direction ($d\zeta/d\phi < 0$), and therefore one obtains $\kappa > 0$. Contrary to the equilibrium limit, nonequilibrium electrostatic and chemical potentials now are independent on each other because the chemical potential which determines the density of electrons is irrelevant to the electric charge density and the electrostatic potential. The electron density $n_s(\zeta)$ inside the 2D channel depends mostly on the local value of chemical potential ζ, whereas the dependences of it on the electrostatic φ and total electrochemical potentials μ are negligibly weak. A key point is that the distribution of electrochemical potential along the channel differs from the distribution of the electrostatic potential. Therefore the ratio κ must be computed under conforming to the electric neutrality condition. Here one assumes that the gradual channel approximation along the channel length is justified under the nonequilibrium condition $V_{DS}>0$. Considering $\varphi(y)$ and $\zeta(y)$ in Eq. 2.38 as independent variables and differentiating Eq. 2.38 with respect to chemical potential ζ, one obtains

$$\kappa = -\frac{1}{e}\frac{\partial\zeta}{\partial\varphi}\bigg|_{V_G} = -\frac{1}{e}\frac{\partial V_G/\partial\varphi|_{\zeta}}{\partial V_G/\partial\zeta|_{\varphi}} = \frac{C_{ox}}{C_Q + C_{it}} \tag{2.63}$$

where we have also assumed that V_G does not depend on y, whereas n_S depends only on chemical potential ζ. Another assumption is that the dimensionless parameter κ remains constant along the channel for a given electric bias and is expressed via the ratio of characteristic capacitances. If one neglects the interface trap density ($C_{it} = 0$), which corresponds to an idealized graphene channel, then the κ ratio depends solely on ε_α and the Fermi energy

$$\kappa\,(C_{it} = 0) = \frac{C_{ox}}{C_Q} = \frac{\varepsilon_a}{\varepsilon_F} = \frac{1}{k_F a_Q} = \frac{1}{\sqrt{\pi n_S a_Q^2}} \tag{2.64}$$

Where again a_Q is the natural spatial scale defined with Eq. 2.46. If capacitance C_Q is large, which corresponds to a high-doped regime, whereas simultaneously C_{ox} is low, that is, when the gate oxide layer is thick, one gets $C_Q >> C_{ox}$ giving $\kappa << 1$. It means that the drift component of the current dominates over the diffusion component or vice versa in the opposite limit.

2.11 Continuity of the Electric Current

When considering the electric transport in conducting materials, the basic starting point is the condition of continuity of the electric current. Using the simplified assumptions, below we obtain explicit analytical solution of continuity equation for the current density of the channel. It can be written in the form

$$\frac{dJ_S}{dy} = 0 \Leftrightarrow \frac{d\,(n_S E)}{dy} = 0 \tag{2.65}$$

where $J_s = J_{dr} + J_{diff}$ is the total drain current which is conserved along the channel. It allows us to write an equation which describes the electric field distribution along the channel (Zebrev and Useinov 1990) in the form

$$\frac{dE}{dy} = \frac{e}{n_S}\frac{dn_S}{d\zeta}\frac{-1}{e}\frac{d\zeta}{d\varphi}\frac{d\varphi}{dy} = \frac{e\kappa}{\varepsilon_D}E^2 \tag{2.66}$$

where κ and ε_D are considered to be independent on the drain-source bias, whereas their dependence on the position along the channel and the gate voltage is retained. The above Eq. 2.66 is solved immediately as

$$E\,(y) = \frac{E\,(0)}{1 - \frac{\kappa e E(0)}{\varepsilon_D}y} \tag{2.67}$$

where one should implement appropriate boundary conditions (BCs) to define the electric field near the source $E(0)$. It is accomplished with imposing of finite source-drain voltage bias V_{SD} which, for example, gives

$$V_{SD} = (1+\kappa)\int_0^L E\,(y)dy \tag{2.68}$$

where L is the length of the channel. Explicit expressions for $E(0)$ and $E(y)$ are obtained from Eqs. 2.67 and 2.68 as

$$E\,(0) = \frac{\Upsilon\,(\kappa)}{e}\frac{\varepsilon_D}{\kappa L} \tag{2.69}$$

$$E(y) = \frac{\frac{\Upsilon(\kappa)}{e}\frac{\varepsilon_D}{\kappa L}}{1 - \frac{y}{L}\Upsilon(\kappa)} \tag{2.70}$$

where the last formula. Eq. 2.70, describes the electric field distribution along the channel. We have introduced the auxiliary function

$$\Upsilon(\kappa) = 1 - \exp\left\{-\frac{\kappa}{1+\kappa}\frac{eV_D}{\varepsilon_D}\right\} \tag{2.71}$$

2.12 Inhomogeneous Behavior of Chemical and Electrostatic Potentials along the Channel

Coordinate dependence of the chemical and electrostatic potentials along the channel is obtained by integrating Eq. 2.70. It gives

$$\begin{aligned} \varphi(y) - \varphi(0) &= -\frac{\varepsilon_D}{\kappa e}\Xi(\kappa) \\ \zeta(y) - \zeta(0) &= \varepsilon_D \Xi(\kappa) \\ \mu(y) - \mu(0) &= \varepsilon_D \frac{1+\kappa}{\kappa}\Xi(\kappa) \end{aligned} \tag{2.72}$$

where we have introduced an auxiliary function

$$\Xi(\kappa) = \ln\left\{1 - \frac{y}{L}\Upsilon(\kappa)\right\} \tag{2.73}$$

In the above formulas, Eqs. 2.72 and 2.73, the quantities $\zeta(0)$, $\mu(0)$, and $\varphi(0)$ are the potentials nearby the source which depend on the gate source bias V_{GS}. For any value of the gate voltage V_{GS}(and respectively $\kappa(V_G)$), there is a certain full drop of electrochemical potential μ on the channel length which in turn is given by the source–drain bias V_D:

$$e(\varphi(L) - \varphi(0)) + \zeta(0) - \zeta(L) = \frac{eV_{DS}}{1+\kappa} + \frac{e\kappa V_{DS}}{1+\kappa} = eV_{DS} \tag{2.74}$$

Expanding the last expression at low drain bias and for high carrier density case ($\kappa < 1$) we arrive at the known dependence of electrostatic potential on the coordinate along the channel:

$$\varphi(y) - \varphi(0) \cong \frac{y}{L}V_D \tag{2.75}$$

One can see that the above dependence has a trivial linear form as in any good conductor, whereas the corresponding spatial change of the chemical potential along the channel length is negligible $\Delta\zeta = \kappa\Delta\varphi << \phi$. Therefore, as compared to electrostatic potential, the full drop of chemical potential in the high-doped channel is small. However its importance pops up in the regime of saturation.

2.13 Microscopic Model of Electron Transport through a Field-Effect Transistor

Applicability of the relativistic Dirac model to graphene depends on the temperature, purity (characterized by the electron–impurity scattering time τ_i), and on the geometrical dimensions (i.e., the chiral barrier length D and width W). The wavefunction phase is broken when electrons inside the barrier collide with atomic defects, impurities, and phonons. During the transmission of a particle through the chiral barrier the shape of which is modulated by the external field, an electron absorbs the field energy. The absorbed energy is ultimately dissipated due to emission of phonons during the electron–phonon collisions. Absorption of the field energy with electrons and the energy dissipation due to the phonon emission cause different impacts on the electron wavefunction phase inside the barrier. Typically, the field strength is quite homogeneous over the sample and also the field acts on all the electrons simultaneously. On the contrary, the electron–phonon absorption and emission processes occur randomly and they only involve an electron and a phonon. Thus, the field acts on all the electrons coherently, whereas the phonon emission is rather a random process. Therefore the electron wavefunction phase becomes randomized as soon as electrons and holes emit the thermal phonons. The randomness limits the applicability of mere Dirac equation and it also requires of more general method to study the electron transport. The electron phases are getting randomized either inside or outside the chiral barrier: If the ratio $\tau_{ep}/\tau_{dw} < 1$, the phonons are emitted inside the barrier and the scattering approach does not work. However, if the electron–phonon collision time τ_{ep} is long enough assuring that $\tau_{ep}/\tau_{dw} > 1$, the electron wavefunction phase is preserved inside the barrier region, whereas it is randomized outside the barrier. In the latter case, the electron motion retains the coherence and is ballistic inside the barrier since the time dependence of the electron wavefunction is the same for all the electrons. Outside the barrier,

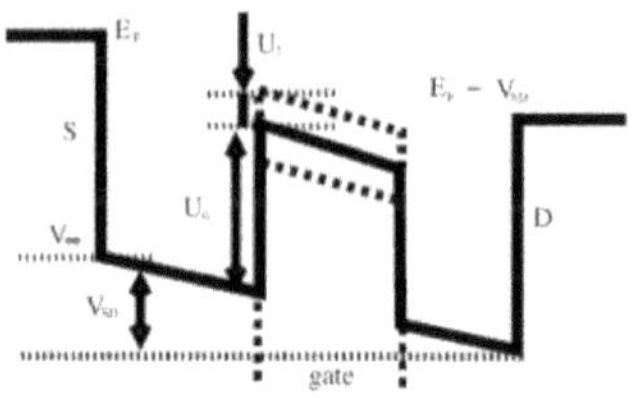

Figure 2.10 Coordinate dependence of the electric potential across the voltage-biased double-barrier graphene junction.

where the electrons lose their energy by emitting the phonons, the phase is eventually randomized on the distance $v\tau_{\rm ep} \geq D$. Then, the Dirac equation can be utilized on the spatial scale $\sim v\tau_{\rm ep}$ only. On the longer scale ($\geq v\tau_{\rm ep}$), the electron transport is better described with method of kinetic equation. The ratio $\tau_{\rm ep}/\tau_{\rm dw}$ depends on the electron energy, temperature, and the barrier length. For instance, for the electron energies $\overline{\varepsilon} \simeq 1 - 10$ meV at $T = 10$ K one gets $\tau_{\rm ep} \simeq 3 \times 10^{-12}$s and $v\tau_{\rm ep} \simeq 3$ μm, while the dwell time for $D = 1$ μm is estimated as $\tau_{\rm dw} \simeq 10^{-12}$s, which gives $\tau_{\rm ep}/\tau_{\rm dw} \simeq 5$. It changes at higher temperatures. For instance, at $T = 100$ K, when $\tau_{\rm ep} \simeq 3 \times 10^{-13}$s and $\overline{\varepsilon} \simeq 1-10$ meV, one obtains $v\tau_{\rm ep} \simeq 0.3$ μm and $\tau_{\rm ep}/\tau_{\rm dw} \simeq 0.5$. In the latter case, the barrier must be shorter than $v\tau_{\rm ep} \leq 0.3$ μm; otherwise, the phonon emission randomizes the electron phase inside the chiral barrier. The mentioned randomization also destroys quantized states inside the well. Therefore in this chapter we pay our attention to chiral barriers with length $D \leq v\tau_{\rm ep}$ where $\tau_{\rm ep}/\tau_{\rm dw} \geq 1$ and the quantized states exist.

Following Schuessler et al. (2009a, 2009b), we model the source (S) and drain (D) metallic contacts as highly doped graphene regions. The dependence of the electric potential on the x coordinate across the junction is sketched in Fig. 2.10. The electrochemical potential $E_{\rm F}$ in the metallic S and D leads is shifted far from the steady-state electrochemical potential V_∞ counted from the Dirac point in graphene (see Fig. 2.10). Below, for the sake of simplicity we set $V_\infty = 0$ in graphene for $x < 0$, while $V_\infty = V_{\rm SD}$ for $x > D$. The electron transverse momentum q_n is preserved in the "clean" system and its value is obtained from the quantization condition in the y direction which are considered in the following subsections.

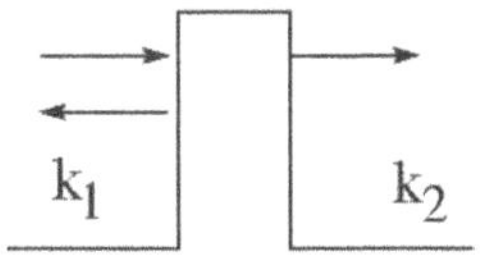

Figure 2.11 Simplest model of the rectangular barrier, where the blue arrows show the incident, transmitted, and reflected electrons, respectively.

2.14 Conventional Tunneling through a Rectangular Barrier

The electron wavefunction satisfies the Schrödinger wave equation

$$\left(-\frac{\hbar^2\nabla^2}{2m}+V(x)\right)\psi=E\psi \tag{2.76}$$

where the electron wavevector is

$$k=\sqrt{\frac{2m}{\hbar^2}(E-V_0)} \tag{2.77}$$

When the barrier is high $V_0 >> E$, the wavevector inside the barrier is

$$k\approx\frac{i}{\hbar}\sqrt{2mV_0} \tag{2.78}$$

To compute the tunneling probability through a simple rectangular barrier (Fig. 2.11) we use the following trial electron wavefunction. For the electron wave emerging from the left ($x < 0$)

$$\Psi_{\mathrm{L}}=e^{iqy}\left(e^{ikx}+re^{-ikx}\right)e^{i\varphi_1} \tag{2.79}$$

On the right hand side ($x > D$)

$$\Psi_{\mathrm{R}}=te^{iqy}e^{ikx}e^{i\varphi_3} \tag{2.80}$$

Inside the barrier we use

$$\Psi_{\mathrm{B}}=e^{iqy}\left(ae^{ik_2x}+be^{-ik_2x}\right)e^{i\varphi_2} \tag{2.81}$$

The BCs for a double-barrier junction (Fig. 2.10) assume that wavefunctions are continuous

$$\begin{aligned}\Psi_{\mathrm{L}}(0,\ 0)&=\Psi_{\mathrm{B}}(0,\ 0)\\ \Psi_{\mathrm{B}}(D,\ 0)&=\Psi_{\mathrm{R}}(D,\ 0),\end{aligned} \tag{2.82}$$

whereas the derivatives are not continuous. Then at $x = 0$ we write

$$\frac{\partial \Psi_{\mathrm{L}}(0,\ 0)}{\partial x} - \frac{\partial \Psi_{\mathrm{B}}(0,\ 0)}{\partial x} = Z_1 \Psi_{\mathrm{B}}(0,\ 0) \tag{2.83}$$

where for technical reasons we've introduced a wavefunction jump $\propto Z_1$. At the other interface $x = D$ we have

$$\frac{\partial \Psi_{\mathrm{B}}(D,\ 0)}{\partial x} - \frac{\partial \Psi_{\mathrm{R}}(D,\ 0)}{\partial x} = Z_2 \Psi_{\mathrm{R}}(D,\ 0)$$

which gives transmission amplitude in the form

$$t = \frac{2ie^{-i(Dk+\phi_1-\phi_3)}kk_2}{k_2(2ik + Z_1 + Z_2)\cos(Dk_2) + (k^2 - i(Z_1+Z_2)k + k_2^2 - Z_1 Z_2)\sin(Dk_2)} \tag{2.84}$$

We emphasize that the transmission coefficient t depends on the phase difference $\phi_1 - \phi_3$, while the reflection coefficient has no such a multiplier

$$r = \frac{e^{2iDk_2}(k + k_2 + iZ_1)(k - k_2 - iZ_2) - (k - k_2 + iZ_1)(k + k_2 - iZ_2)}{e^{2iDk_2}(-k + k_2 + iZ_1)(-k + k_2 + iZ_2) - (k + k_2 - iZ_1)(k + k_2 - iZ_2)} \tag{2.85}$$

which at $Z_1 = 0$ and $Z_2 = 0$ becomes

$$t_0 = \frac{2ikk_2 e^{-iDk} e^{i(\varphi_1 - \varphi_3)}}{2ikk_2\cos(Dk_2) + \left(k^2 + k_2{}^2\right)\sin(Dk_2)} \tag{2.86}$$

where we introduced the *phase difference* φ across the whole junction.

$$r_0 = \frac{(k^2 - k_2^2)\sin(Dk_2)}{2ikk_2\cos(Dk_2) + \left(k^2 + k_2^2\right)\sin(Dk_2)} \tag{2.87}$$

When a DC voltage is applied, then one sets $\varphi = \varphi_3 - \varphi_1$ and $\varphi - \varphi' = eV(t - t')$:

$$\begin{aligned}
\varphi_{\mathrm{R}}(x)\,\varphi_{\mathrm{R}}^{*}\left(x'\right)\,e^{i\varepsilon_k(t-t')} &= t_{kk'} \cdot t'_{kk'} \cdot e^{iqy} e^{ik(x-x')} e^{i\varepsilon_k(t-t')} \\
&= |t_{kk'}|^2\, e^{iqy} e^{ik(x-x')} e^{i(\varphi-\varphi')} e^{i\varepsilon_k(t-t')} \\
&= |t_{kk'}|^2\, e^{iqy} e^{ik(x-x')} e^{i(\varepsilon_k - eV)(t-t')}
\end{aligned} \tag{2.88}$$

where we used the time-dependent interface transparency $t_{kk'}(t) = t_{kk'}e^{i\varphi(t)}$, $t'_{kk'} = t_{kk'}(t') = t_{kk'}e^{i\varphi(t')}$, since the phase difference $\varphi(t)$ across the whole junction is time dependent.

2.15 Chiral Tunneling through a Rectangular Barrier

We depart from the Dirac equation completed by appropriate BCs and constrains. The steady-state Dirac equation for low-energy quasiparticles in graphene (Ando 2005, Katsnelson et al. 2006, Tworzydlo et al. 2006, Katsnelson and Novoselov 2007, Tworzydlo et al. 2007, Shafranjuk 2009) is

$$-i\hbar v\left[\left(\hat{\sigma}_x \otimes \hat{1}\right)\partial_x + \left(\hat{\sigma}_y \otimes \hat{\tau}_z\right)\partial_y\right]\Psi + eU(x)\Psi = \varepsilon_{\mathrm{p}}\Psi \quad (2.89)$$

where $\otimes$ is the Kronecker product, $v \simeq c/300$ is the massless fermion speed, $\hat{\sigma}_i$ and $\hat{\tau}_k$ are the Pauli matrices, $\{i, k\} = x, y, z$, and the barrier potential $U(x)$ is induced by the gate voltage $V_{\mathbf{G}}$ (for typical gate voltage $V_{\mathbf{G}} = 1\mathrm{V}$ and for the $\mathrm{SiO_2}$ thickness $d = 300$ nm, one finds (Novoselov et al. 2005) $U_0 = 2$ meV, where U_0 is the height of chiral barrier at $V_{\mathbf{SD}} = 0$). If the source–drain bias voltage $V_{\mathbf{SD}} = 0$, then U_0 corresponds to a local stationary shift of the electrochemical potential V_∞ by a finite gate voltage $V_{\mathbf{G}}$ applied to the section of the graphene ribbon marked in Fig. 2.1a by darker hexagons (see also Fig. 2.10). Thus, U_0 is proportional to $V_{\mathbf{G}}$ with a proportionality constant α, which is called the gate efficiency, that is, $U_0 = e\alpha V_{\mathbf{G}}$. The simplest steady-state model implements a piecewise geometry of the graphene junction when the gate voltage-induced barrier is approximated by a rectangular shape

$$U(x) = U_0\theta(x)\,\theta(D - x) \quad (2.90)$$

and $V_{\mathbf{SD}} = 0$. In this case, the chiral fermion wavefunctions $\hat{\Psi}$ are expressed as combinations of plane waves. This allows the chiral fermion excitation energy ε_p to be written in terms of the steady-state energy eigenvalues ε_p for the infinite space. In the graphene electrode regions just outside the chiral barrier (i.e., $x < 0$ and $x > D$) one gets

$$\varepsilon_p = eV_\infty \pm \hbar v\sqrt{k^2 + q_n^2} = \hbar v\left(k_{\mathrm{F}m} \pm \sqrt{k^2 + q_n^2}\right) \quad (2.91)$$

where zero energy at $\varepsilon_{\mathrm{p}} = 0$ is associated with the energy of the Dirac point

$$V_\infty = \frac{\hbar v k_{\mathrm{Fm}}}{e} \quad (2.92)$$

which in our setup serves as an equilibrium background electrochemical potential in the bulk of graphene ribbon outside the barrier, k_{Fm} is the Fermi momentum in the leads, q_n is the transverse momentum, $\mathbf{p} = (\hbar k, \hbar q_n)$ is the 2D electron momentum. In most practical cases the leads are well coupled to the external environment. Therefore the electron and hole states in the leads can be regarded as equilibrium. In the piece-wise approximation, one finds the electron energy ε_p as an eigenvalue of the Hamiltonian in Eq. 2.89. The electron excitation energy inside the barrier is given by

$$\varepsilon_p = eU_0 \pm \hbar v \tilde{p} \tag{2.93}$$

where the absolute values of electron momentum in the chiral electrodes is

$$\tilde{p} = \sqrt{\tilde{k}^2 + q_n^2} \tag{2.94}$$

We use the fact that $V_\infty \to U_0$ inside the gated region, as shown in Fig. 2.10. The last formula can also be rewritten as

$$\tilde{k} = \sqrt{\tilde{p}^2 - q_n^2} \tag{2.95}$$

The honeycomb symmetry of the graphene lattice determines structure of the electron and hole wavefunctions. In vicinity of the K (or K′) point an elementary excitation with a certain finite momentum $\mathbf{p}$ may acquire either positive (attributed to an electron) or a negative (ascribed to a hole) energy $\pm v|\mathbf{p}|$.

The steady-state tunneling amplitude t_n across the chiral barrier is obtained by matching of the BCs at $x = 0$ and $x = D$ for the trial chiral fermion wavefunctions

$$\begin{aligned} \Psi_{\mathbf{L}}(x=0) &= \Psi_{\mathbf{B}}^{(0)}(x=0) \\ \Psi_{\mathbf{B}}^{(0)}(x=D) &= \Psi_{\mathbf{R}}(x=D) \end{aligned} \tag{2.96}$$

where $\Psi_{\mathbf{L},\mathbf{B},\mathbf{R}}$ are the electron wavefunctions in the left, chiral barrier, and the right graphene regions, respectively. In Eq. 2.96 index "0" stands for the stationary steady-state solution in absence of the AC field. We note that both $\Psi_{\mathbf{L}}$ and $\Psi_{\mathbf{R}}$ coincide with the steady-state functions, even if the external field is applied. This is due to strong coupling of the electrodes to the external environment. The transverse modes are determined by the BCs at $y = 0$ and

$y = W$. The rectangular barrier configuration, Eq. 2.90, is obtained by setting $V_{SD} \to 0$. The scattering state is constructed as

$$\Psi = \begin{cases} \Phi_{\mathbf{L}} \text{ if } x < 0 \\ \Phi_{\mathbf{B}} \text{ if } 0 < x < D \\ \Phi_{\mathbf{R}} \text{ if } x > D \end{cases} \tag{2.97}$$

where

$$\begin{aligned} \Phi_{\mathbf{L}} &= \chi_{n,\,k} e^{ikx} + r_n \chi_{n,\,-k} e^{-ikx} \\ \Phi_{\mathbf{B}} &= \alpha_n \chi_{n,\,\tilde{k}} e^{i\tilde{k}x + i\phi} + \beta_n \chi_{n,\,-\tilde{k}} e^{-i\tilde{k}x + i\phi} \\ \Phi_{\mathbf{R}} &= t_n \chi_{n,\,k} e^{ik(x-D)} \end{aligned} \tag{2.98}$$

In the above formulas in Eqs. 2.97 and 2.98 k is positive for the conductance band, while it is negative for the valence band. In Eqs. 2.97 and 2.98 we have introduced the auxiliary functions

$$\begin{aligned} \chi_{n,\,k}(y) = & \, a_n \left|\uparrow\right\rangle \otimes \left(\left|\uparrow\right\rangle + z_{n,\,k} \left|\downarrow\right\rangle\right) e^{iq_n y} \\ & + a_n' \left|\downarrow\right\rangle \otimes \left(z_{n,\,k} \left|\uparrow\right\rangle + \left|\downarrow\right\rangle\right) e^{iq_n y} \\ & + b_n \left|\uparrow\right\rangle \otimes \left(z_{n,\,k} \left|\uparrow\right\rangle + \left|\downarrow\right\rangle\right) e^{-iq_n y} \\ & + b_n' \left|\downarrow\right\rangle \otimes \left(\left|\uparrow\right\rangle + z_{n,\,k} \left|\downarrow\right\rangle\right) e^{-iq_n y} \end{aligned} \tag{2.99}$$

where

$$\left|\uparrow\right\rangle = \begin{pmatrix} 1 \\ 0 \end{pmatrix} \text{ and } \left|\downarrow\right\rangle = \begin{pmatrix} 0 \\ 1 \end{pmatrix} \tag{2.100}$$

are 1×2 matrices,

$$z_{n,\,k} = s_{n,\,k}(k + iq_n)/p \tag{2.101}$$

where $s_{n,\,k} = \pm 1$ is applied to conductive (valence) bands

$$p = \frac{\varepsilon_p}{\hbar v} = \sqrt{k^2 + q_n^2} \tag{2.102}$$

Here k is positive for the conductance band and negative for the valence band. The factor $z_{n,\,k}$ conforms the identity

$$z_{n,\,k} z_{n,\,-k} \equiv -1 \tag{2.103}$$

Permitted values of the angle $\tilde{\chi}$ inside the graphene ribbon of a finite width W are obtained from BCs along the y direction (see the next subsections).

A simple analytical expression for the tunneling amplitude is available for a rectangular barrier configuration, Eq. 2.90, and a wide ribbon ($W > l_{\mathrm{i}}$) in the steady state as

$$t_{p\tilde{p}}^{(0)} = \frac{e^{-i\phi}k\tilde{k}s\tilde{s}}{k\tilde{k}s\tilde{s}\cos(\tilde{k}D) - i\left(p\tilde{p} - s\tilde{s}q_{\mathrm{n}}^{2}\right)\sin(\tilde{k}D)} \tag{2.104}$$

where

$$\tilde{p} = \mp\frac{\varepsilon_p - eU_0}{\hbar v} \tag{2.105}$$

$s = \mathbf{sign}(\varepsilon_p)$, $\tilde{s} = \mathbf{sign}(\varepsilon_p - eU_0)$. Equation 2.104 involves the transversal momentum q_n which indicates the angular dependence of the chiral tunneling amplitude. The dependence on lateral momentum component is a remarkable property of chiral tunneling. This makes a difference from the nonchiral tunneling case where

$$t_{p\tilde{p}} = \frac{4k\tilde{k}}{e^{-i\tilde{k}D}\left(k+\tilde{k}\right)^2 - e^{i\tilde{k}D}\left(k-\tilde{k}\right)^2} \tag{2.106}$$

The above Eq. 2.106 for the conventional tunneling amplitude contains no explicit dependence on the lateral electron momentum q_n. For narrow ribbons with $W < l_{\mathrm{i}}$, the lateral quantization becomes important. The quantized states strongly depend on the BCs at the edges which are determined by their shape (Fujita et al. 1996, Nakada et al. 1996a, 1996b, Brey and Fertig 2006a, 2006b).

2.15.1 *Tilted Chiral Barrier*

If the source–drain voltage is finite, $V_{\mathbf{SD}} \neq 0$, the simple analytical expressions for $t_{p\tilde{p}}$ like Eqs. 2.104 and 2.106 are available no more. For $V_{\mathbf{SD}} \neq 0$ one sets

$$U_x = U_0 + \mathcal{E}_1 x \tag{2.107}$$

where $\mathbf{E}_1 = V_{\mathbf{SD}}/L$ is the electric field and x is the coordinate along the ribbon (see Fig. 2.1). When computing the tunneling amplitude for the tilted barrier, for the sake of simplicity, we neglect the terms $\sim x^2$, while keeping the linear $\sim x$ terms only. Then the scattering state for the trapezoidal barrier configuration is expressed via linear combinations of Airy functions $\mathbf{Ai}(\varsigma_x)$ and $\mathbf{Bi}(\varsigma_x)$

$$\begin{aligned}\Psi = {} & \theta(-x)[\chi_{n,k}\mathbf{Ai}(\varsigma_x) + r_n\chi_{n,-k}\mathbf{Bi}(\varsigma_x)] \\ & + \theta(x-D)t_n\chi_{n,k}\mathbf{Ai}(\varsigma_{x-D}) + \theta(x)\theta(D-x) \\ & \times[\alpha_n\chi_{n,\tilde{k}}\mathbf{Ai}(\tilde{\varsigma}_x)e^{i\phi} + \beta_n\chi_{n,-\tilde{k}}\mathbf{Bi}(\tilde{\varsigma}_x)e^{i\phi}]\end{aligned} \tag{2.108}$$

where we have introduced the auxiliary functions

$$\begin{aligned}\varsigma_x &= -(k/\upsilon)^2 - \upsilon x \\ \tilde{\varsigma}_x &= -(k/\tilde{\upsilon})^2 - \tilde{\upsilon} x\end{aligned} \tag{2.109}$$

where $\upsilon = \sqrt[3]{2\mathcal{E}_1 p}$, $\tilde{\upsilon} = \sqrt[3]{2\mathcal{E}_1 \tilde{p}}$. We emphasize that in a stronger field one should keep the terms $\sim x^2$ in Eq. 2.89. Then the trial wavefunction is a linear combination of the hypergeometric functions rather than of the Airy functions taking place in Eq. 2.108. Thus, for a trapezoidal barrier one uses Airy functions at the place of the plane waves

$$\begin{aligned}e^{ikx} &\to \mathbf{Ai}(k, x) \\ e^{-ikx} &\to \mathbf{Bi}(k, x)\end{aligned} \tag{2.110}$$

$$\Psi_{n,k} = \chi_{n,k}(y)\,\mathbf{Ai}(k, x) \tag{2.111}$$

where again

$$\begin{aligned}\chi_{n,k}(y) = a_n &\begin{pmatrix} 1 \\ z_{n,k} \\ 0 \\ 0 \end{pmatrix} e^{iq_n y} + a'_n \begin{pmatrix} 0 \\ 0 \\ z_{n,k} \\ 1 \end{pmatrix} e^{iq_n y} \\ +b_n &\begin{pmatrix} z_{n,k} \\ 1 \\ 0 \\ 0 \end{pmatrix} e^{-iq_n y} + b'_n \begin{pmatrix} 0 \\ 0 \\ 1 \\ z_{n,k} \end{pmatrix} e^{-iq_n y}\end{aligned} \tag{2.112}$$

$$z_{n,k} = \pm \frac{k + iq_n}{\sqrt{k^2 + q_n^2}} \tag{2.113}$$

In the above formulas, the sign "+" corresponds to the conductance band, while "−" corresponds to the valence band. The above Eq. 2.112 is rewritten in another form

$$\begin{aligned}\chi_{n,k}(y) = a_n &|\uparrow\rangle \otimes (|\uparrow\rangle + z_{n,k}|\downarrow\rangle)\, e^{iq_n y} \\ &+ a'_n |\downarrow\rangle\, (\otimes (z_{n,k}|\uparrow\rangle + |\downarrow\rangle))\, e^{iq_n y} \\ &+ b_n |\uparrow\rangle\, (\otimes (z_{n,k}|\uparrow\rangle + |\downarrow\rangle))\, e^{-iq_n y} \\ &+ b'_n |\downarrow\rangle \otimes (|\uparrow\rangle + z_{n,k}|\downarrow\rangle)\, e^{-iq_n y}\end{aligned} \tag{2.114}$$

2.15.2 *Tunneling through a Graphene Quantum Well*

First we consider the tunneling through a graphene quantum well (GQW) where two metallic leads are attached to a graphene stripe. Energy of the fermion state in the leads ($x < 0$ and $x > L$) reads

$$\varepsilon = eV_\infty \pm \hbar v_{\mathrm{F}m}\sqrt{k^2 + q_n^2} = \hbar v_{\mathrm{F}m}\left(k_{\mathrm{F}m} \pm \sqrt{k^2 + q_n^2}\right) \quad (2.115)$$

By setting $\varepsilon = 0$ for a normal incidence ($q_n = 0$) one gets $eV_\infty \pm \hbar v_{\mathrm{F}m}k_{\mathrm{F}m} = 0$ or $eV_\infty = \hbar v_{\mathrm{F}m}k_{\mathrm{F}m}$. In the metallic lead one also finds

$$k = \sqrt{\left(\frac{\varepsilon - \mu}{\hbar v_{\mathrm{F}m}}\right)^2 - q_n^2} \quad (2.116)$$

Inside the graphene stripe the energy is

$$\varepsilon = eV_{\mathbf{G}} \pm \hbar v\sqrt{\tilde{k}^2 + q_n^2} \quad (2.117)$$

assuming that the modes inside the strip may be either propagating or evanescent. In the above formulas $v = c/300$ is the massless fermion speed. Eq. 2.117 also gives

$$\tilde{k} = \sqrt{\left(\frac{\varepsilon - eV_{\mathbf{G}}}{\hbar v}\right)^2 - q_n^2} \quad (2.118)$$

The transverse modes are determined by the BCs at $y = 0$ and $y = W$. Permitted values of the angle χ inside the graphene stripe are obtained from the BCs along the y direction, $p_y = q_n = n\pi/W$. In the triple graphene–graphene–graphene G/G/G junction, where the electrochemical potential of the middle G section is adjusted, for example, by applying a local gate voltage, one may use the angular representation

$$\begin{aligned} k &= \frac{\varepsilon}{v\hbar}\cos\varphi \\ q &= \frac{\varepsilon}{v\hbar}\sin\varphi \\ k_1 &= \frac{1}{v\hbar}\sqrt{(\varepsilon - eV_0)^2 - \varepsilon^2\sin^2\varphi} \end{aligned} \quad (2.119)$$

$$\begin{aligned} \varphi &= \arctan\frac{q}{k} \\ \varphi_1 &= \arctan\frac{q}{k_1} = \arctan\frac{\sin\varphi}{\sqrt{((\varepsilon - eV_0)/\varepsilon)^2 - \sin^2\varphi}} \end{aligned} \quad (2.120)$$

This gives the quantized angle χ_n inside the graphene strip as

$$\varphi_n = \arcsin\frac{n\pi}{W}\frac{v\hbar}{\varepsilon}, \quad (2.121)$$

provided that

$$k_n = \frac{\varepsilon}{v\hbar}\left(1 - \left(\frac{n\pi}{W}\cdot\frac{v\hbar}{\varepsilon}\right)^2\right)$$
$$q_n = \frac{n\pi}{W}$$
$$k_1 = \frac{1}{v\hbar}\sqrt{(\varepsilon - eV_0)^2 - \left(\frac{n\pi}{W}\right)^2} \qquad (2.122)$$

The quantized angle is introduced as follows

$$\varphi_n \to \frac{1}{\pi}\int d\varphi\delta\left(\varphi - \varphi_n\right) \to \frac{1}{\pi}\int d\varphi\frac{\gamma}{\left(\varphi - \varphi_n\right)^2 + \gamma^2} \qquad (2.123)$$

since the y component of the fermion momentum q does not change during the tunneling. The scattering state is constructed using the piece-wise trial wavefunction

$$\Psi = \begin{cases} \Phi_{\mathbf{L}} & \text{if} \quad x < 0 \\ \tilde{\Phi} & \text{if} \quad 0 < x < L \\ \Phi_{\mathbf{R}} & \text{if} \quad x > L \end{cases} \qquad (2.124)$$

where

$$\Phi_{\mathbf{L}} = \chi_{n,\,k}e^{ikx} + r_n\chi_{n,-k}e^{-ikx} \qquad (2.125)$$

In the right hand side electrode

$$\Phi_{\mathbf{R}} = t_n\chi_{n,\,k}e^{ik(x-L)} \qquad (2.126)$$

and inside the middle graphene section which serves as a chiral barrier

$$\tilde{\Phi} = \alpha_n\chi_{n,\,\tilde{k}}e^{i\tilde{k}x+i\phi} + \beta_n\chi_{n,-\tilde{k}}e^{-i\tilde{k}x+i\phi} \qquad (2.127)$$

where k is positive for the conductance band, while being negative for the valence band. Direct calculation of the transmission coefficient for the chiral tunneling gives

$$t_n = \frac{e^{-i\phi}kk_1ss_1}{kk_1ss_1\cos(k_1L) - i\left(\sqrt{k_1^2 - q^2}\sqrt{k^2 - q^2} - ss_1q^2\right)\sin(k_1L)}, \qquad (2.128)$$

while for a normal incidence ($q = 0$) it becomes

$$t_n = \frac{2e^{-i\phi}kk_1ss_1}{2kk_1ss_1\cos(k_1L) - 2i\,|k|\cdot|k_1|\,\sin(k_1L)} \qquad (2.129)$$

From the last Eq. 2.129 one can see that the tunneling amplitude t_n reverses the sign due to flipping the sign of s_1. For the semiconducting nanotubes we use

$$\varepsilon = \hbar v_F \sqrt{q_\nu (n)^2 + k^2} \text{ or } k = \sqrt{\frac{\varepsilon^2}{\hbar^2 v_F^2} - q_\nu (n)^2} \quad (2.130)$$

and $\gamma = \hbar v$, where the transverse electron momentum is quantized due to a finite width of the graphene stripe $W \neq 0$. As we will see in the next subsection, the lateral quantization of the electron motion which occurs in graphene stripes of finite width also depends on the shape of the edges.

2.16 Role of Edges: Armchair Edges

Termination of the electron wavefunction at the armchair edges consists of a line of dimers. Therefore, the wavefunction amplitude vanishes on the two edges at the K and K′ points in both sublattices A and B (Fujita et al. 1996, Nakada et al. 1996a, 1996b, Brey and Fertig 2006a, 2006b). This leads to the following BCs along the $\hat{y}$ axis

$$\begin{aligned} &\Psi_A(0) = \Psi_{A'}(0), \Psi_B(0) = -\Psi_{B'}(0) \text{ at } y = 0 \\ &\Psi_A(W) = \Psi_{A'}(W), \Psi_B(W) = -\Psi_{B'}(W) \text{ at } y = W \end{aligned} \quad (2.131)$$

The armchair edges mix the two valleys and cause scattering of a wave propagating on a certain valley to the opposite valley. The BCs define the quantized transverse momentum

$$q_n = \frac{2\pi n}{2W + a_0} + \frac{2\pi}{3a_0} \quad (2.132)$$

where a_0 is the lattice constant. One may see that the electron confinement is similar in both the armchair and smooth edge configurations. Due to intervalley mixing while reflecting from the edges, the electron wavefunction of each mode inside the ribbon is a mixture the two valleys:

$$\begin{aligned} \chi^{ac}_{q,k}(y) = &[a \cdot |\uparrow\rangle \otimes (|\uparrow\rangle + z_{q,k} |\downarrow\rangle) \\ &+ a' \cdot |\downarrow\rangle \otimes (z_{q,k} |\uparrow\rangle + |\downarrow\rangle)]e^{iqy} \\ &+ [b \cdot |\uparrow\rangle \otimes (z_{q,-k} |\uparrow\rangle + |\downarrow\rangle) \\ &+ b' \cdot |\downarrow\rangle \otimes (|\uparrow\rangle + z_{q,-k} |\downarrow\rangle)]e^{-iqy} \end{aligned} \quad (2.133)$$

One also introduces an auxiliary function

$$\breve{\rho}_{q,k} = |\uparrow\rangle \otimes \left(|\uparrow\rangle + z_{q,k}\,|\downarrow\rangle\right) e^{iqy} + |\downarrow\rangle \otimes \left(|\uparrow\rangle + z_{q,-k}\,|\downarrow\rangle\right) e^{-iqy} \tag{2.134}$$

which allows for Eq. 2.133 to have a compact form

$$\chi^{ac}_{q,k}(y) = a \cdot \breve{\rho}_{q,k} + a' \cdot \breve{\rho}_{q,-k} + b \cdot \breve{\rho}_{q,-k} + b' \cdot \breve{\rho}_{-q,-k} \tag{2.135}$$

The net envelope wavefunction has the same chirality for both valleys contributions, as a result of the symmetric BCs. Also one should note that the wavefunctions $\Psi(q)$ and $\Psi(-q)$ represent distinct solutions in which the components of the two valleys are interchanged. This gives the lowest zero mode with zero transverse momentum which exists inside the channel, even if the width of it becomes smaller than the Fermi wavelength. The above formulas allow to compute the tunneling amplitude which has a conventional form given by Eq. 2.106.

2.17 Role of Edges: Zigzag Edges

The electron envelope wavefunction for a narrow graphene ribbon with zigzag edges (Fig. 2.12) terminates at one edge for the A sublattice and at the opposite edge for the B sublattice. The BCs along the $\hat{y}$ direction read

$$\Psi_{\mathrm{A}}(0) = \Psi_{\mathrm{A'}}(0) = 0 \text{ at } y = 0$$
$$\text{and } \Psi_{\mathrm{B}}(W) = \Psi_{\mathrm{B'}}(W) = 0 \text{ at } y = W \tag{2.136}$$

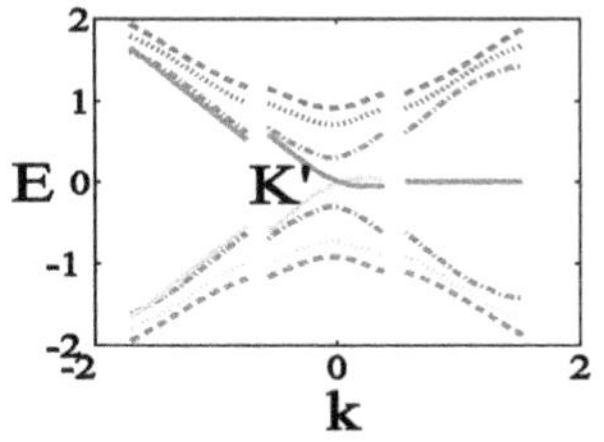

Figure 2.12 Electron energy dispersion law for the graphene stripe with zigzag edges. The flat zero-energy band corresponds to edge states of the stripe.

The above BCs, Eq. 2.136, lead to the dispersion relationship in the form of transcendental algebraic equation

$$k = \frac{q_n}{\tan(q_n W)} \tag{2.137}$$

which constitutes the quantization condition for the transversal electron momentum q_n. Instead of Eqs. 2.99, 2.133, and 2.135 one now implements

$$\begin{aligned} \chi^{zz}_{n,k}(x) = a_n \cdot |\uparrow\rangle \otimes \left(\zeta^{A}_{q}(x) \cdot |\uparrow\rangle + \zeta^{B}_{q,k,s}(x) \cdot |\downarrow\rangle\right) \\ + a'_n \cdot |\downarrow\rangle \otimes \left(\zeta^{B}_{q,k,s}(x) \cdot |\uparrow\rangle + \zeta^{A}_{q}(x) \cdot |\downarrow\rangle\right) \\ + b_n \cdot |\uparrow\rangle \otimes \left(\zeta^{B}_{q,k,s}(x) \cdot |\uparrow\rangle + \zeta^{A}_{q}(x) \cdot |\downarrow\rangle\right) \\ + b'_n \cdot |\downarrow\rangle \otimes \left(\zeta^{A}_{q}(x) \cdot |\uparrow\rangle + \zeta^{B}_{q,k,s}(x) \cdot |\downarrow\rangle\right) \end{aligned} \tag{2.138}$$

where we have introduced auxiliary functions related to the A and B sublattices

$$\begin{aligned} \zeta^{A}_{q}(x) &= \sin(q \cdot x) \\ \zeta^{B}_{q,k,s}(x) &= \frac{s \cdot i\,[-q \cdot \cos(q \cdot x) + k \cdot \sin(q \cdot x)]}{p} \end{aligned} \tag{2.139}$$

where p is the absolute value of the electron momentum, $s = \pm 1$ for conductance (valence) bands. In the same assumption as when deriving Eqs. 2.104 and 2.106, the transmission coefficient for a chiral barrier in the graphene ribbon with zigzag edges is obtained as

$$t^{zz} = 2e^{-i\phi} \frac{\beta_{k,\tilde{k}}}{\beta_{\tilde{k},k} + \beta_{k,\tilde{k}}} \tag{2.140}$$

where

$$\beta_{k,\tilde{k}} = k(\tilde{k}\cos(\tilde{k}D) - q\sin(\tilde{k}D)) \tag{2.141}$$

$$\beta_{\tilde{k},k} = \tilde{k}\,(k\cos(kD) - q\sin(kD)) \tag{2.142}$$

where $q = \sqrt{p^2 - k^2}$, and $p = \varepsilon/\hbar v$. One can see that tunneling amplitude for a narrow ribbon with zigzag edges has quite a different form than the former Eq. 2.104. That is because the electron wavefunction for a ribbon with zigzag edges has a distinct intrinsic structure as compared to the armchair edge case.

2.18 Deviation of an Electron inside a Wide Chiral Barrier

The lateral component of the electron momentum is

$$q = k\tan\chi = p\sin\chi \tag{2.143}$$

where the full electron momentum

$$p = \sqrt{k^2+q^2} = k\sqrt{1+\tan^2\chi} = \frac{\varepsilon}{\hbar v} \tag{2.144}$$

is conserved. The electron propagation angle inside the barrier becomes

$$\tilde{\chi} = \arctan\frac{q_n}{\tilde{k}} = \arctan\left(\sin\frac{\chi}{\sqrt{(1-\tilde{\upsilon})^2-\sin^2\chi}}\right) \tag{2.145}$$

where

$$\tilde{p} = \frac{\varepsilon - eU}{\hbar v} \text{ and } \tilde{\upsilon} = \frac{eU}{\varepsilon} \tag{2.146}$$

One can see that a finite DC gate voltage $V_G \neq 0$ causes a deviation of an electron from its steady-state trajectory. In particular, for a small barrier the deviation is

$$\delta\chi \simeq \tilde{\upsilon}\frac{\tan\chi}{\cos 2\chi} \tag{2.147}$$

2.19 Electric Current Density across the Chiral Barrier

In this subsection we compute the electric current density through the chiral barrier B between the source (S) and drain (D) electrodes. The S and D electrodes belong to the same graphene ribbon as B does. The difference between the S, D, and B is that in the S and D

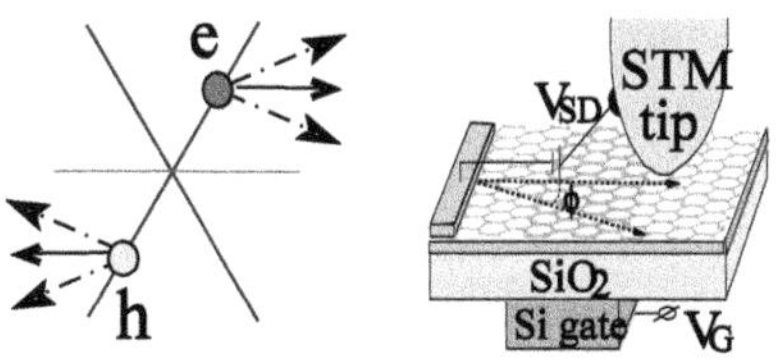

Figure 2.13 Detecting angular distribution of the tunneling probability through the chiral barrier with an STM tip.

electrodes the gate voltage vanishes ($V_{\mathbf{G}} \equiv 0$), while inside B the finite gate voltage $V_{\mathbf{G}} \neq 0$ induces a finite shift of electron energy $\varepsilon \rightarrow \varepsilon + eU_0$. Then, the electric current density is expressed by the Landauer–Büttiker formula

$$j(\varphi,\ V_{\mathbf{SD}}) = \frac{4ev}{W} \int \left| t_{U_0,\,\varepsilon}(\varphi) \right|^2 \cdot [n(\varepsilon) - n(\varepsilon - eV_{\mathbf{SD}})] \qquad (2.148)$$

The density of electric current depends on the azimuthal angle φ between $\hat{x}$ axis and the line connecting two tiny S and D probe electrodes as we discuss below. Anisotropy of the electric current density (current) is pronounced when the size $d_{\mathbf{el}}$ of the S and D probe electrodes is sufficiently small $d_{\mathbf{el}} << D_{\mathbf{el}} \sin \varphi_{\mathbf{ir}}$ where $D_{\mathbf{el}}$ is the distance between the S and D electrodes and $\varphi_{\mathbf{ir}}$ is the angle between two closest maximums. In Eq. 2.148 we set

$$n(\varepsilon) = \frac{1}{\exp(\varepsilon/T) + 1} \qquad (2.149)$$

and for a long monoatomic graphene ribbon

$$N(\varepsilon) = \frac{WD|\varepsilon|}{2\pi\hbar^2 v^2} \qquad (2.150)$$

where W is the ribbon width and D is the chiral barrier length. Because of the electron density of states $N(\varepsilon) \propto W$, the density of electric current j in Eq. 2.148 is practically independent versus the graphene stripe width W. When the sample dimensions are finite, the electron density of states $N(\varepsilon)$ acquires a more complex form: It accounts also for singularities originating from geometrical quantization and interaction between the graphene and substrate. We emphasize that the chiral barrier height U_0 and the source–drain bias voltage $V_{\mathbf{SD}}$ enter Eq. 2.148 in an essentially different way, which also depends on a particular setup. The form of $t_{U_0,\,\varepsilon}(\varphi)$ entering Eq. 2.148 differs for the wide ribbon, and for ribbons with zigzag and armchair edges—compare, that is, Eqs. 2.104, 2.106, and 2.140. From Eq. 2.148 one infers that the angle-dependent transmission coefficient $t_\varepsilon(\varphi)$ essentially determines anisotropy of the electric current density. The resonances of $t_\varepsilon(\varphi)$ indicate preferable directions of chiral tunneling of the electrons and holes. We will see that the strong angular dependence is proclaimed in "clean" graphene ribbons where the electron propagation is ballistic. Initially we computed the steady-state differential conductivity

$$\sigma(\varphi,\ V_{\mathbf{SD}}) = \frac{\partial I}{\partial V_{\mathbf{SD}}} \qquad (2.151)$$

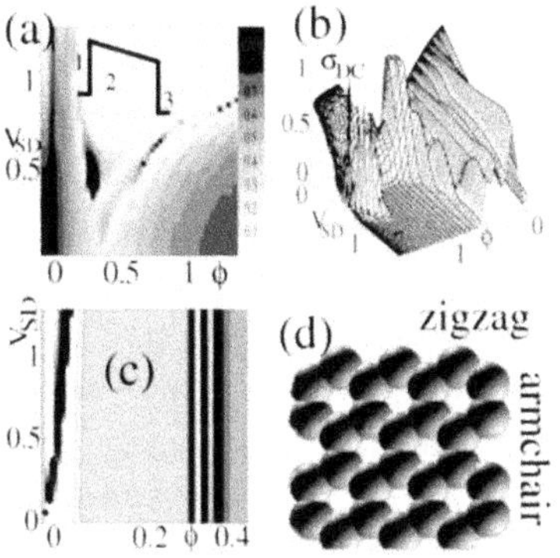

Figure 2.14 (a) Contour plot of the steady-state differential conductivity σ_{SD} versus V_{SD} and φ (in radians) for a trapezoidal barrier formed on a wide ($W >> l_i$) ribbon. (b) The 3D plot of σ_{SD} (in units of $8e^2\ |V_{SD}|/(vh^2)$) versus V_{SD} (in units of the barrier height U_0) and the azimuthal angle φ (in radians) for $D = 15$ and $U_0 = 1$. (c) Contour plot $\sigma_{SD}(V_{SD}, \varphi)$ for a ribbon with zigzag edges for $W = 15$, $D = 20$, and $U_0 = 1$. (d) The corner of a graphene ribbon where the upper edge has a zigzag shape, while the right-side edge has the armchair shape.

versus the azimuthal angle φ and the source–drain voltage $V_{\mathbf{SD}}$. The contour plot in Fig. 2.14a and the 3D plot in Fig. 2.14b both suggest that in the range $0 < V_{\mathbf{SD}} < 1.5$ (in units of U_0) and $0 < \varphi < \pi/2$, the conductivity $\sigma(\varphi, V_{\mathbf{SD}})$ experiences pronounced singularities and resonances. In particular, $\sigma(\varphi, V_{\mathbf{SD}})$ becomes almost ideal at $\varphi \leq 0.3$ and at arbitrary $V_{\mathbf{SD}}$. As the deviation angle φ exceeds $\varphi_{\max} \approx 0.3$, the barrier transparency sharply decreases and then vanishes above certain threshold value $V_{\mathbf{SD}} > V_{\mathbf{SD}}^{\mathbf{th}}(\varphi)$. Besides, in Fig. 2.14a, one can notice dark dots and spots which correspond to an ideal transparency of the chiral barrier. The ideal transparency occurs when the chiral particle tunnels via the resonant scattering energy levels. This is the case if one of the levels matches the electrochemical potential of the attached electrodes. The numeric data presented in Fig. 2.14a,b correspond to the chiral barrier height $U_0 = 1$ and length $D = 15$. The quantization depending on directions of the electric charge propagation between S and D electrodes which abelong to the same graphene ribbon is a remarkable feature of the contour plot (a) and of 3D plot shown in Fig. 2.14b. The dependence versus the azimuthal angle φ follows from a fundamental property of the Dirac equation, Eq. 2.89, which describes the chiral fermions with an angle-dependent pseudospin. The 3D plot in Fig. 2.14b shows numerous resonances of ideal transparency [related to the dark areas and spots visible in Fig. 2.14a where $T = |t_n|^2 \equiv 1$]. They occur below the barrier as soon as $V_{\mathbf{SD}} > V_{\mathbf{SD}}^{\mathbf{th}}(\varphi)$ at finite azimuthal angles $\varphi < \pi/2$. In Fig. 2.14c we also show contour plot of the differential conductivity $\sigma_{\mathbf{SD}}(V_{\mathbf{SD}}, \varphi)$ for a narrow ribbon (by width $W = 15$ and length $D = 20$) with zigzag edges computed for $U_0 = 1$. One can notice sharp features which emerge from electron scattering on the zigzag edges. Fig. 2.14d shows a fragment of a graphene ribbon where the upper edge has a zigzag shape, while the right-side edge has the armchair shape. Corresponding probability for electron transmissions through the chiral barrier formed by a graphene ribbon with zigzag edges is shown in Fig. 2.14 for the ribbon width $W = 15$, barrier length $D = 20$, and the barrier height $U_0 = 1$. Curve 1 in Fig. 2.14 shows $\mathcal{T}_{00}^{\mathbf{zz}}(E)$ which corresponds to a transition between the lowest bands. In the same figure curve 2 corresponds to $\mathcal{T}_{01}^{\mathbf{zz}}(E)$ for a transition between the lowest band in the electrode and first band inside the barrier, and curve 3 is the dependence $\mathcal{T}_{02}^{\mathbf{zz}}(E)$ for a transition between the lowest band in the electrode and second band inside the barrier. The electron energy bands $\varepsilon_k^{\mathbf{zz}}$ for a narrow graphene ribbon with zigzag edges near the K point are shown in the right inset.

2.20 Gate Voltage–Controlled Quantization

In graphene stripes with certain geometry one can control the angle-dependent quantized energy level spacing inside the barrier by applying the gate voltage $V_{\mathbf{G}}$. When the chiral barrier is induced applying a gate voltage, the dependence of electron wavevector on the barrier height U is determined by

$$\tilde{k}^2 = \frac{q_n^2}{\tan^2 \chi} = \left(\frac{\varepsilon - eU}{\hbar v}\right)^2 - q_n^2 \qquad (2.152)$$

Then, for a ribbon with armchair edges one writes the energy-level positions as

$$\varepsilon_n = \hbar v \left(\frac{2\pi n}{2W + a_0} + \frac{2\pi}{3a_0} \right) \sqrt{\frac{1}{\tan^2 \chi} + 1} + eU \qquad (2.153)$$

The above formulas show that the gate voltage V_G shifts the energy level positions in the graphene ribbons with armchair edges. However V_G does not affect the level spacing itself. A different situation occurs inside the ribbon with zigzag edges. According to Eq. 2.153 the longitudinal and the transversal electron momentum components are coupled to each other. Therefore any modification of electron motion in the $\hat{x}$ direction also affects the electron properties in the $\hat{y}$ direction. The deviation angle of an electron populating the quantized level n inside the chiral barrier is found to be dependent on the gate voltage as

$$\tilde{\chi}_n = \arctan \frac{q_n}{\tilde{k}} = \arctan \frac{q_n}{\sqrt{\left(\left(\varepsilon - eU \right) / \hbar v \right)^2 - q_n^2}} \qquad (2.154)$$

2.21 A Hybrid Graphene/CNT Junction

When placing a CNT over the graphene sheet one can form a hybrid graphene/CNT junction (Fig. 2.15). The interface transparency between the CNT and graphene is less than ideal due to the following reasons. When the graphene sheet contacts a semiconducting nanotube, one gets a Schottky barrier which emerges at the interface due to distinct charge carrier concentrations in the two materials. An extra electric charge accumulated near the interface may produce Schottky barriers. The interface transparency is also reduced because of distinction of the electron energy dispersion laws $E_{\mathrm{p}}^{\mathrm{G,\,T}}$

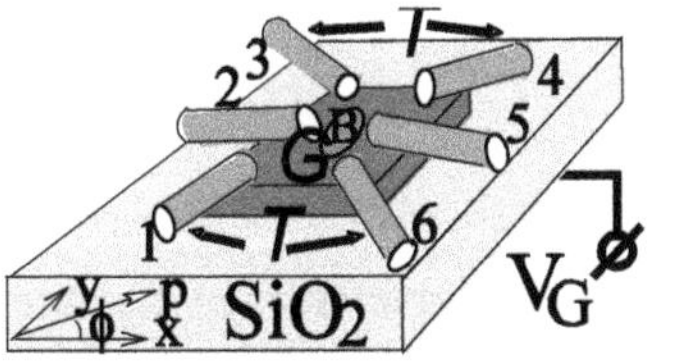

Figure 2.15 A hybrid graphene/CNT junction where the directional diagram of chiral tunneling is detected.

in the two materials. When an electron is injected, for example, from the CNT into the graphene sheet, its energy must be preserved. However, due to different dispersion laws, the momentum conservation requirement cannot be fulfilled. This decreases the transmission probability through the graphene/CNT interface. The mathematical expression for the transmission coefficient through the hybrid GT junction is obtained from the BC as

$$t_{\mathrm{GT}} = \frac{2k_{\mathrm{i}}s_{\mathrm{i}}e^{-i\phi}(k_{\mathrm{i}} - iq_{\mathrm{i}})\kappa_{\mathrm{t}}}{\kappa_{\mathrm{i}}\left(s_{\mathrm{i}}\kappa_{\mathrm{i}}\kappa_{\mathrm{t}} + s_{\mathrm{t}}(k_{\mathrm{i}} - iq_{\mathrm{i}})(k_{\mathrm{t}} - iq_{\mathrm{t}})\right)} \tag{2.155}$$

where $\mathbf{p}_{\mathrm{i}} = \{k_{\mathrm{i}}, q_{\mathrm{i}}\}$ and $\mathbf{p}_{\mathrm{t}} = \{k_{\mathrm{t}}, q_{\mathrm{t}}\}$ are incoming and transmitted electron moments correspondingly.

$$\kappa_{\mathrm{i,\,t}} = \sqrt{k_{\mathrm{i,\,t}}^2 + q_{\mathrm{i,\,t}}^2} \tag{2.156}$$

The quantization of the transversal momentum $q_{\mathrm{i,\,t}}$ proceeds differently for the wide graphene ribbon, narrow graphene ribbon, and in the CNT, where

$$q = \frac{2}{d_{\mathrm{T}}}\left(n - \frac{\nu}{3}\right) \tag{2.157}$$

d_{T} is the CNT diameter, n is the subband index, and ν is the chirality index (see, for example, Ref. [Ando 2005]). For this reason, the transmission coefficient, Eq. 2.155, differs for various pairs of the electrodes. The effect of transmission of a chiral particle from a wide graphene ribbon into the adjacent CNT is illustrated supposing that in graphene $q_{\mathrm{i}} = 0$ (i.e., the incident chiral particle in graphene propagates along the AB chains). After penetrating into the CNT from graphene, the particle preserves its energy, while its transversal momentum in the CNT with, for example, $n = 0$, $\nu = 1$ must be $q_{\mathrm{t}} = -(2/d_{\mathrm{T}})\nu/3$. Then, from Eq. 2.155 one gets

$$t_{\mathrm{GT}} = \frac{\sqrt{\varepsilon^2 + \Delta^2} + \varepsilon - i\Delta}{2\sqrt{\varepsilon^2 + \Delta^2}} \tag{2.158}$$

where $\Delta = 2v\nu/(3d_{\mathrm{T}})$. Substituting Eq. 2.155 or Eq. 2.158 in Eq. 2.148 one computes the electric current density across the hybrid GT interface without the Schottky barrier. One may notice that the graphene/CNT interface transparency explicitly depends on the graphene ribbon width and edges as well as on the CNT diameter and chirality which determine the values of $q_{i,t}$ in Eq. 2.155 or Δ in Eq. 2.158.

2.22 Electric Current Characteristics

We write the expression for the total drain current by taking the gradient of the electrochemical potential in the source vicinity

$$I_D = -W\mu_0 n_S(0)\left.\frac{d\mu}{dy}\right|_{y=0} = eW\mu_0 n_S(0)(1+\kappa)E(0)$$
$$= e\frac{W}{L}D_0 n_S(0)\frac{1+\kappa}{\kappa}\Upsilon(\kappa) \qquad (2.159)$$

where W is the channel width, and we have used the Einstein formula $D_0 = \mu_0\varepsilon_D/e$ under assumption that the temperature is constant. One can see outside the charge NP the total 2D charge density $eN_s \approx en_s$ is almost equals to charge imbalance density. The saturated source–drain voltage V_{DSAT} is defined as

$$V_{DSAT} = 2\frac{1+\kappa}{\kappa}\frac{\varepsilon_D}{e} = \frac{1+\kappa}{\kappa}\frac{\varepsilon_F}{e} \qquad (2.160)$$

where ε_F is the Fermi energy which coincides with the chemical potential in vicinity of the source since $\varepsilon_D \approx \varepsilon_F/2$ for $\zeta = \varepsilon_F >> k_B T$. The chemical potential in the drain vicinity can also be written as

$$\zeta(L) = \left(1 - \frac{V_D}{V_{DSAT}}\right) \qquad (2.161)$$

Because of the electrostatic blocking, the condition $V_D = V_{DSAT}$ reflects the fact that the chemical potential and current both vanish which is regarded as a pinch-off for silicon MOSFETs. The channel current thus can be written as

$$I_D = \frac{W}{L}\sigma_0(0)\frac{V_{DSAT}}{2}\left(1 - \exp\left\{-2\frac{V_D}{V_{DSAT}}\right\}\right) \qquad (2.162)$$

where σ_0 is the low-field conductivity nearby the source. As it follows from Eq. 2.162, V_{DSAT} actually determines the saturation onset of the drain current. Besides, Eq. 2.162 in fact represents I–V characteristics of the electric current along the channel in the gated graphene device.

2.23 The Saturation Regime (Pinch-Off)

As it follows from above Eqs. 2.106, 2.63, and 2.64

$$eV_{\text{DSAT}} = \varepsilon_{\text{F}} \frac{1+\kappa}{\kappa} = m\varepsilon_{\text{F}} + \frac{\varepsilon_{\text{F}}^2}{\varepsilon_{\text{a}}} \tag{2.163}$$

Besides, we take into account that $|V_{\text{G}} - V_{\text{NP}}| = m\varepsilon_{\text{F}} + \varepsilon_{\text{F}}^2/2\varepsilon_{\text{a}}$, which gives

$$V_{\text{DSAT}} = |V_{\text{G}} - V_{\text{NP}}| + \frac{\varepsilon_{\text{F}}^2}{2\varepsilon_{\text{a}}} = |V_{\text{G}} - V_{\text{NP}}| + \frac{en_{\text{S}}}{C_{\text{ox}}} \tag{2.164}$$

The last formula reflects a specific character of the graphene FET. The pinch-off saturation of the graphene FET is not observed if the dielectric layer is thick, and thus one gets very large $V_{\text{DSAT}} \cong 2\,|V_{\text{G}} - V_{\text{NP}}| >> 1$. The saturation voltage V_{DSAT} depends parametrically versus ε_{a} and the interface trap capacitance C_{it}. The regime of the electric current saturation caused by electrostatic pinch-off is achieved when the source–drain bias is high, that is, $V_{\text{D}} > V_{\text{DSAT}}$. It is described by Eq. 2.162 rewritten as

$$I_{\text{DSAT}} \cong \frac{W}{L} D_0 n_{\text{S}}(0) \frac{1+\kappa}{\kappa} = \frac{W}{L} \sigma_0(0) \frac{V_{\text{DSAT}}}{2} \tag{2.165}$$

2.24 Linear Behavior in Low Fields

The triode exhibits a linear behavior when

$$V_{\text{D}} << V_{\text{DSAT}} = \varepsilon_{\text{F}} \frac{1+\kappa}{\kappa} \tag{2.166}$$

The electric current through the channel is determined by the drift component which dominates in conditions of high doping $\kappa <<1$. Such behavior is typical for metals. On the other side, the diffusion component of the current emerges in the other limit when $\kappa >>1$. The crossover, when the two electric current components become equal to each other, takes place if $C_{\text{it}} = 0$ and at $\varepsilon_{\text{F}} = \varepsilon_{\text{a}}$ which is rather an idealized situation. Similar crossover might also occur when the characteristic channel density is $n_{\text{s}} = n_{\text{Q}}$.

The diffusion current with significant current saturation predominates for fractions of the curves below the dashed lines, whereas the drift current which linearly depends on the drain bias occurs in the region above the dashed lines. The region of diffusion current is small for "dirty" samples with thick oxide layers. When the drain bias is small one obtains a regular linear expression by expanding Eq. 2.162 in series over V_D

$$I_D \cong e\frac{W}{L}\mu_0 n_S V_D \tag{2.167}$$

If one assumes that the mobility μ_0 does not depend on the gate voltage, then the small-signal transconductance in the linear mode takes the form

$$g_m \equiv \left.\frac{\partial I_D}{\partial V_G}\right|_{V_D} = \frac{W}{L}\mu_0 C_{CH} V_D \tag{2.168}$$

where C_{CH} is the channel capacitance; see also Eq. 2.55. One can also rewrite Eq. 2.168 to introduce the field-effect mobility μ_{FE} as

$$g_m = \frac{W}{L}\mu_0 C_{CH} V_D \equiv \frac{W}{L}\mu_{FE} C_{ox} V_D \tag{2.169}$$

The field-effect mobility μ_{FE} depends on the partial mobility μ_0 determined by purely "microscopic" scattering mechanisms and also on the exchange by electric charges with extrinsic traps (defects in the gate oxides, chemical dopants, etc.)

$$\mu_{FE}\left(C_{it} + \Delta C_{it}\right) = \frac{\mu_{FE}\left(C_{it}\right)}{1 + \frac{\Delta C_{it}}{C_{ox}+C_Q+C_{it}}} \tag{2.170}$$

The field-effect mobility extracted from the slope of experimental conductivity curves versus the gate voltage is smaller than the actual microscopic mobility. It is also considerably reduced in the vicinity of the charge NP. The agreement between μ_{FE} and μ_0 occurs only if $C_Q >> mC_{ox}$, which means that $\varepsilon_F >> m\varepsilon_a$ taking place in conditions of high doping. An electric stress, a wear-out, or ionizing radiation all cause degradation of the transconductance due to interface trap buildup in FETs. When changing the capacitance $C_{it} \rightarrow C_{it} + \Delta C_{it}$ due to the externally induced interface trap one also affects the field-effect mobility. The process is described by formula which involves the initial value $\mu_{FE}(C_{it})$:

$$\mu_{FE}\left(C_{it} + \Delta C_{it}\right) = \frac{\mu_{FE}\left(C_{it}\right)}{1 + \frac{\Delta C_{it}}{C_{ox}+C_Q+C_{it}}} \tag{2.171}$$

Equations (2.55) and (2.167) allow computing of the logarithmic swing which characterizes the $I_{\mathrm{ON}}/I_{\mathrm{OFF}}$ ratio and equals numerically to the gate voltage alteration needed for current change by an order as

$$S \equiv \frac{1}{\frac{d\left(\log_{10} I_{\mathrm{D}}\right)}{dV_{\mathrm{G}}}} = n_{\mathrm{S}} \ln 10 / \left(\frac{dn_{\mathrm{S}}}{dV_{\mathrm{G}}}\right) = en_{\mathrm{S}} \ln 10 / C_{\mathrm{CH}}\left(V_{\mathrm{G}}\right) \quad (2.172)$$

The last equation can also be rewritten to resemble the formula which had been formerly derived for the silicon MOSFET:

$$\begin{aligned} S &= \ln 10 \frac{en_{\mathrm{S}}}{C_{\mathrm{Q}}}\left(1 + \frac{C_{\mathrm{it}} + C_{\mathrm{Q}}}{C_{\mathrm{ox}}}\right) = \ln 10 \frac{en_{\mathrm{S}}}{C_{\mathrm{ox}}}\left(1 + \frac{C_{\mathrm{it}} + C_{\mathrm{ox}}}{C_{\mathrm{Q}}}\right) \\ &= \ln 10 en_{\mathrm{S}}\left(\frac{1}{C_{\mathrm{ox}}} + \frac{m}{C_{\mathrm{Q}}}\right) \end{aligned} \quad (2.173)$$

Further we take into account that for the nondegenerate subthreshold mode of the silicon FETs operation the diffusion energy $\varepsilon_{\mathrm{D}} = e^2 n_{\mathrm{s}}/C_{\mathrm{Q}} \approx \varepsilon_{\mathrm{F}}/2$ acts here as the thermal potential $e\varphi_D = k_{\mathrm{B}} T$. The subthreshold swing depends versus the gate voltage which was not the case for the silicon FET. Outside the small vicinity of the charge NP and for the "clean" interface $C_{\mathrm{Q}} >> mC_{\mathrm{ox}}$ one gets an assessment of the logarithmic swing for a "thick" oxide layer as $S \geq en_{\mathrm{S}} \ln 10/C_{\mathrm{ox}} >> 1\mathrm{V/decade}$. Otherwise, for "thin" oxide layer with $C_{\mathrm{Q}} << mC_{\mathrm{ox}}$ the corresponding assessment is $S \cong m\varepsilon_{\mathrm{F}} \ln 10/2e$.

2.25 Transit Time through the Channel

Equation 2.70, which describes the distribution of electric field versus coordinate y along the channel, allows us to compute the transit time as

$$\tau_{\mathrm{TT}} = \int_0^L \frac{dy}{\mu_0 (1+\kappa) E(y)} \quad (2.174)$$

An immediate integration gives

$$\tau_{\mathrm{TT}} = \frac{L^2}{2D_0} \frac{\kappa}{1+\kappa} \cos h\left(\frac{\kappa}{1+\kappa} \frac{eV_{\mathrm{G}}}{2\varepsilon_{\mathrm{D}}}\right) = \frac{L^2}{\mu_0 V_{\mathrm{DSAT}}} \cos h\left(\frac{V_{\mathrm{D}}}{V_{\mathrm{DSAT}}}\right) \quad (2.175)$$

This expression for the drift flight time is valid in the linear regime when $\varsigma_{\mathrm{D}} << \varsigma_{\mathrm{DSAT}}$

$$\tau_{\mathrm{TT}} = \frac{L^2}{\mu_0 V_{\mathrm{D}}} \tag{2.176}$$

The diffusion time in the opposite limit $\varsigma_{\mathrm{D}} > \varsigma_{\mathrm{DSAT}}$ and for low carrier density $\kappa >> 1$ becomes

$$\tau_{\mathrm{TT}} = \frac{L^2}{2D_0}\frac{\kappa}{1+\kappa} \cong \frac{L^2}{2D_0} \tag{2.177}$$

2.26 The Diffusion–Drift Approximation

The basic assumption above was about the validity of macroscopic diffusion–drift approximation which serves as a departure point for practical simulation of gated devices. Because of the semiclassical origin of the diffusion–drift approximation its validity is limited only to high carrier density and small wave lengths. Validity of the Boltzmann equation and of the diffusion–drift approach depends on the correspondence between basic spatial scales which involve mean free path l, the channel length Λ, the carrier's De Broglie wavelength $\lambda_{\mathrm{D}} = \hbar v_{\mathrm{F}}/\varepsilon_{\mathrm{D}} \lambda_{\mathrm{D}} = v_{\mathrm{F}}/\varepsilon_{\mathrm{D}}$, and the Fermi energy ε_{F} in graphene. The ballistic transport takes flare for $\Lambda < l$. The semiclassical case with weak scattering and well-defined dispersion law conditions takes place at $\lambda_{\Phi} < \lambda < \Lambda$. Because the electron mobility in graphene is independent of the charge carrier density, the electron wavelength must be small. It can be reformulated as

$$\lambda_{\mathrm{F}} < l \longleftrightarrow n_S > \frac{2e}{\hbar\mu_0} \cong 3\times 10^{15}\frac{1}{\mu_0\cdot \mathrm{cm}^2} \tag{2.178}$$

where we have used Eq. 2.25. It means that the diffusion–drift approximation is valid for not too small carrier densities and in low electric fields. Due to presence of unavoidable disorder at the Dirac point with smooth potential relief the semiclassical approach remains valid even in vicinity of the NP. A strong electric field in transverse direction close to the drain electrode causes the semiclassical approximation to break because the local charge density is diminished. The electric field which acts near the drain electrode might split the electron–hole pairs and violate the equilibrium balance between generation and recombination. It results in an additional nonequilibrium generation of the drain current which is induced by the electric field. When the carrier density is low there emerge significant quantum effects of interband interaction which are regarded as "trembling" or *Zitterbewegung* (Katsnelson 2006). Such a phenomenon resembles generation and recombination of virtual electron–hole pairs.

2.27 Effects in the High Field

A strong electric field when being applied to the field-effect gated device accelerates the electric charge carriers until their drift velocity tends to become saturated when the electric field is high enough. Saturation of the electric current resulting from the velocity saturation has been suggested for explaining the recent electronic transport experiments in graphene transistors (Meric et al. 2008). The relevance of diffusion–drift equations to existing experimental data can be improved introducing a field-dependent mobility. They account for the velocity saturation in high electric fields:

$$\mu_0(E) = \frac{\mu_0}{1 + E/E_C} \quad (2.179)$$

Here μ_0 is the mobility in a low field, v_{SAT} is the saturation velocity, the value of which is determined by the emission of optical phonons, $E_{SAT} \approx v_{SAT}/\mu_0 = (1\text{–}5) \times 10^4$ V/cm. The formerly mentioned correspondence between the electrostatic pinch-off and the saturation of electron velocity is described by the dimensionless ratio (Zebrev 1992)

$$a = \frac{V_{DSAT}}{2E_C L} = \frac{|V_G - V_{NP}| + en_S/C_{ox}}{2E_C L} \quad (2.180)$$

One can distinguish different mechanisms of saturation of the electric current. If the channel is long, while C_{ox} is large ($\alpha << 1$), then the electrostatically induced current pinch-off dominates. On the other hand, in the gated devices with a short channel and thick gate oxides ($\alpha >> 1$), the channel current saturation $I_\Delta = \Omega \cdot \varepsilon \cdot n_s \cdot v_{SAT}$ occurs due to limitations of the drift velocity. If the diffusion–drift approximation is applicable, the basic difference between the electric charge transport in graphene and in conventional silicon MOSFET emerges from the specific dispersion law in graphene. It causes an unconventional behavior of graphene FET the statistics and electrostatics of which appear to be different from conventional semiconductors.

2.28 Generalized Boundary Conditions

BCs for the electron wavefunction in graphene samples of finite size can be reformulated in a concise form (McCann and Fal'ko 2004, Akhmerov and Beenakker 2007a, 2007b). At the edge of graphene sheet one should request that the solution of Dirac equation satisfies (McCann and Fal'ko 2004):

$$\Psi = \mathcal{M}\Psi \tag{2.181}$$

In graphene ribbons or flakes the edges are typically abrupt on the atomic scale. Electron scattering on and reflection from the edges therefore couples the valleys. The intervalley coupling is accounted for by implementing the appropriate BC. At the moment we neglect a possible local magnetization, and we also assume that the BC itself does not break time-reversal symmetry which formally means that M commutes with T. Those requirements lead to the following BCs (Akhmerov and Beenakker 2007a, 2007b):

$$\Psi = \mathcal{M}\Psi, \quad \mathcal{M} = (\boldsymbol{\nu} \cdot \boldsymbol{\tau}) \otimes (\mathbf{n} \cdot \boldsymbol{\sigma}) \tag{2.182}$$

where $\boldsymbol{\nu}$ and ν are the 3D unit vectors which parametrize BC. The requirement that no current flows throughout the boundary (with normal $\mathbf{n}_{\mathrm{B}}$, pointing outward) constrains the vector $\mathbf{n}$ as

$$\mathbf{n} \cdot \mathbf{n}_{\mathrm{B}} = 0 \tag{2.183}$$

There are three most instructive boundary geometries (Berry and Mondragon 1987, Brey and Fertig 2006a, 2006b):

- For a zigzag edge the row of missing atoms at the edge might be either on the A or B sublattice (see Fig. 2.16. It accounts for

  $$\Psi_{\mathrm{A}} = \Psi'_{\mathrm{A}} = 0 \tag{2.184}$$

  or

  $$\Psi_{\mathrm{B}} = \Psi'_{\mathrm{B}} = 0 \tag{2.185}$$

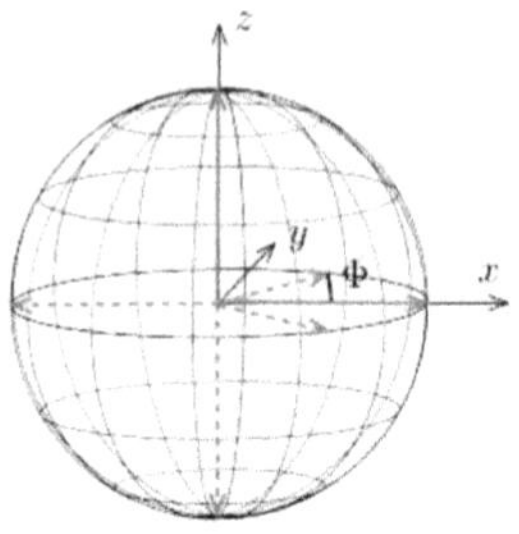

Figure 2.16 Schematics of the valley isospin located on the Bloch sphere for a zigzag edge. The arrows along the *z* axis and for an armchair edge (arrows in the *x*–*y* plane). The solid and dashed arrows correspond to opposite edges. Adapted from Akhmerov and Beenakker (2007a, 2007b).

In the relevant BC matrix *M* one should use

$$\boldsymbol{\nu} = \pm\hat{\mathbf{z}},\ \mathbf{n} = \hat{\mathbf{z}} \tag{2.186}$$

For opposite zigzag edges which involve atoms belonging to distinct A and B sublattices, the angle Φ between the vectors $\boldsymbol{\nu}$ are equal and they are independent of the separation W between the edges.

- An armchair shape of the edge of a graphene stripe or flake requires that

$$\Psi_X e^{i\mathbf{K}\cdot\mathbf{r}} + \Psi_{X'} e^{-i\mathbf{K}\cdot\mathbf{r}} = 0 \tag{2.187}$$

where $\Xi = A,\ B$. The wavefunction thus vanishes on the both A and B sublattices. This gives a prerequisite that

$$\boldsymbol{\nu}\cdot\hat{\mathbf{z}} = 0,\ \mathbf{n} = \hat{\mathbf{z}}\times\mathbf{n}_{\mathrm{B}} \tag{2.188}$$

Then the angle $\Phi = |\mathbf{K}|\,W + \pi$ depends on the edge separation W: if $W = 3an/2$ (n being an integer), then $\Phi = \pi$, otherwise $\Phi = \pm\pi/3$.

- If the quantum well is formed due to confinement by an infinite mass then the corresponding BC is

$$\boldsymbol{\nu} = \hat{\mathbf{z}},\ \mathbf{n} = \hat{\mathbf{z}}\times\mathbf{n}_{\mathrm{B}} \tag{2.189}$$

One obtains two eigenstates $|+\boldsymbol{\nu}\rangle$ and $|-\boldsymbol{\nu}\rangle$ of $\boldsymbol{\nu}\cdot\boldsymbol{\tau}$ (defined by $\boldsymbol{\nu}\cdot\boldsymbol{\tau}\,|\pm\boldsymbol{\nu}\rangle = \pm\,|\pm\boldsymbol{\nu}\rangle$) corresponding to the valley polarization which could be either parallel or antiparallel to the unit vector $\boldsymbol{\nu}$. Because vector $\boldsymbol{\nu}$ is transformed under

Figure 2.17 Graphene flake with armchair and zigzag edges.

rotations in the same way as the electron spin it is called the valley *isospin*. On the Bloch sphere, vector $\boldsymbol{\nu}$ can be represented by a certain point, as shown in Fig. 2.17.

For *zigzag edges* or for the *infinite mass confinement* vector $\boldsymbol{\nu}$ is directed along $\hat{\mathbf{z}}$, whereas the corresponding eigenstate belongs to a certain individual valley. For armchair edges vector $\boldsymbol{\nu}$ is located within the x-y plane, whereas the eigenstate is a coherent equal weight superposition of the two distinct valleys. There is a deep analogy between vector $\boldsymbol{\nu}$ and a strong magnetic field: the valley polarization of edge states is actually determined by direction of $\boldsymbol{\nu}$ in the boundary matrix M. In the lowest Landau level the edge states are valley-polarized although the Hall conductance

$$G_{\mathrm{H}} = g\frac{e^2}{h} \tag{2.190}$$

where g is the degeneracy factor of edge states, which is independent of the direction ν of the valley isospin (Abanin et al. 2006, Brey and Fertig 2006a, 2006b). The "half integer" Hall conductance $G_{\mathrm{H}} = (n+1/2)4e^2/h$ measured by Novoselov et al. (2005) and Zhang et al. (2005) suggests that the lowest ($n = 0$) Landau level has no valley degeneracy, while the spin degeneracy is preserved ($g = 2$ rather than $g = 4$). Thus, G_{H} does not depend on the directions of valley polarization. In the following chapters we will see that Andreev reflection and Klein tunneling allow measuring of the valley isospin in the quantum Hall effect.

The BCs for edges of graphene stripes or flakes (Fig. 2.18) impose a constraint on the electron–hole symmetry. In an unbounded system, when the electrostatic potential is zero ($U = 0$), the Dirac Hamiltonian anticommutes with $\tau_z \otimes \sigma_z$, that is,

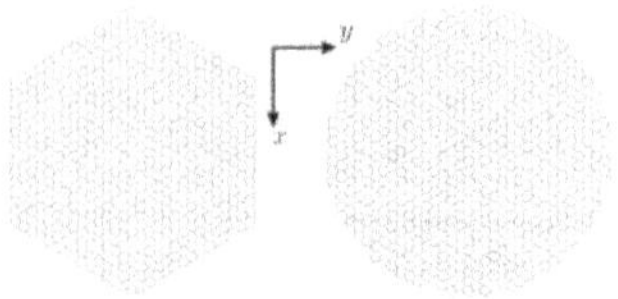

Figure 2.18 Two graphene flakes with the same zigzag boundary condition: $\Psi = \pm\tau_z \otimes \sigma_z \Psi$. The sign switches between + and − at the armchair orientation (when the tangent to the boundary has an angle with the y axis that is a multiple of 60°). From Akhmerov and Beenakker (2008) and Beenakker et al. (2008).

$$\begin{aligned} H_D \Psi &= E\Psi \\ H_D (\tau_z \otimes \sigma_z) \Psi &= -E (\tau_z \otimes \sigma_z) \Psi \end{aligned} \tag{2.191}$$

which implies the electron–hole symmetry of the spectrum. The requirement to preserve the electron–hole symmetry is fulfilled only for two types of edges, either zigzag or armchair. In terms of the boundary $\mathcal{M}$ matrix it assumes that $\mathcal{M}$ must commute with $\tau_z \otimes \sigma_z$. The electron–hole symmetry is preserved by any boundary that is simply a termination of the lattice where the potential of the edge is set to zero.

2.29 Pseudodiffusive Dynamics

If the Fermi level in a graphene sheet is at the NP K(K′), the electric conduction exhibits unconventional features. Electron density of states vanishes in the K(K′) point vicinity and therefore the transmission probability changes respectively. Microscopically, the transport occurs entirely via evanescent electron states in the direction of electric current. If the graphene strip is short and wide, the number of evanescent modes with transmission probability of order unity is large $W/L{>>}1$. Technically, the transmission probabilities appear to be the same for both, the evanescent modes in the graphene stripe and for diffusive modes in a disordered piece of metal with the same conductance (Tworzydlo et al. 2006).

Let us briefly consider the transmission of evanescent modes through the undoped graphene (Katsnelson et al. 2006, Tworzydlo et al. 2006). Because of an infinitely long wavelength in the K(K′)

point vicinity, we do not need to know the detailed shape of the electrostatic potential profile at the interface between the metal contacts and the graphene sheet. Therefore, one can model it simply as having a rectangular shape, as shown in Fig. 2.11. Here we assume that the Fermi energy coincides with energy of the K(K′) points. Then, the interband tunneling goes over into intraband tunneling. If $W/L >> 1$, we can disregard individual transmission probabilities of the evanescent modes. It allows us to ignore the dependence on the BC at $y = 0, W$ It is important to know how many modes $\rho(T)\,dT$ (counting all degeneracy factors) there are with transmission probabilities in the interval $(T, T+dT)$. The result is

$$\rho(T) = \frac{g}{2T\sqrt{1-T}} \tag{2.192}$$

where $g = 4W/(\pi L)$ is the conductance in units of e^2/h. The term "pseudodiffusive dynamics" is introduced because the above expression looks as the distribution (Dorokhov 1984) for diffusion modes in a disordered metal with the same conductance g_1. Experimental evidence of the validity of the above formula for g has been obtained by (Miao et al. 2007). Another relevant experiment which checks the bimodal shape of the distribution is the measurement of the shot noise at the K(K′) points. One also computes the Fano factor (ratio of shot noise power and mean current) as (Tworzydlo et al. 2006)

$$F = 1 - \frac{\int_0^1 T^2\rho(T)\,dT}{\int_0^1 T\rho(T)\,dT} = \frac{1}{3} \tag{2.193}$$

which also coincides with the corresponding expression for a disordered metal (Beenakker and Buttiker 1992). The experimental evidence that Fano factor is equal to 1/3 had been found in work (Danneau et al. 2008).

2.30 Confinement and *Zitterbewegung*

An attempt to confine the Dirac electrons causes *Zitterbewegung*, or jittery motion of the wavefunction of the Dirac problem. The value of electron momentum becomes uncertain due to the Heisenberg principle as a consequence of wave packet localization. In contrast to nonrelativistic limit, the momentum uncertainty of a Dirac particle with a zero rest mass leads to uncertainty in the particle's energy. That happens because the position–momentum uncertainty relation is independent from the energy–time uncertainty relation. One consequence is that for an ultrarelativistic particle, a particle-like state can have hole-like states in its time evolution. If one starts at some time $t = 0$, for example, from a Gaussian shape of the wave packet with width w and momentum close to $\mathbf{K}$

$$\psi_0(\mathbf{r}) = \frac{e^{-r^2/2w^2}}{\sqrt{\pi}w} e^{i\mathbf{K}\cdot\mathbf{r}}\phi \tag{2.194}$$

where ϕ is a spinor composed of positive energy states being associated with $\psi_{+,\mathrm{K}}$. The Dirac equation has an eigenfunction which is written as a Fourier series

$$\psi(\mathbf{r}, t) = \int \frac{d^2k}{(2\pi)^2} \sum_{a=\pm1} \alpha_{a,\mathbf{k}} \psi_{a,K}(\mathbf{k}) e^{-ia(\mathbf{k}\cdot\mathbf{r}+vpkt)} \tag{2.195}$$

where $\alpha_{a,\mathbf{k}}$ are coefficients. Then previous formula can be rewritten using the inverse Fourier transform, which gives

$$\alpha_{\pm,\mathbf{k}} = \sqrt{\pi} w e^{-k^2w^2/2} \psi_{\pm,K}(\mathbf{k})\phi \tag{2.196}$$

One can see that the ratio $\alpha_+/\alpha_- = 1$ which means the same relative weight for positive energy and negative energy states. Thus, in a wave packet, there are as many positive energy states as the negative energy states. The wavefunction therefore becomes quickly delocalized when $t_0 \neq 0$. Besides one can see that a wave packet of electron-like states has hole-like components which is a remarkable feature of massless fermions in relativistic quantum mechanics. To see how *Zitterbewegung* manifests itself in magnetotransport one can reformulate the problem in terms of the tight-binding model (Peres et al. 2006, Chen et al. 2007). In the simplest geometry the confining potential is 1D. We implement a potential which decays exponentially away from the edges into the bulk with a penetration depth λ. We add a potential V_i on site $\mathbf{R}_i$ as

$$H_\mathrm{e} = \sum_i V_i n_i \tag{2.197}$$

where n_i is the on-site electronic density. Here V_i vanishes in the bulk but becomes large at the edge of the sample. Then the confining potential is modeled by

$$V(x) = V_0 \left[e^{-(x-L/2)\lambda} + e^{(x-L/2)\lambda}\right] \tag{2.198}$$

where x is the direction of confinement and V_0 is the strength of the potential. The above simple model allows computing the electronic spectrum for a graphene ribbon. Figure 1.11 shows the electronic spectrum for a graphene ribbon of width $L = 600\alpha$. As follows from the plot, the confining potential breaks the electron–hole symmetry, whereas for $V_0 > 0$, the hole part of the spectrum is deformed. For electron states with k near the Dirac point, the hole branch dispersion becomes

$$E_{n,\sigma=-1}(k) \approx -\gamma_n k^2 - \zeta_n k^4 \tag{2.199}$$

where $n>0$ and is an integer and $\gamma_n < 0$ ($\gamma_n > 0$) for $n < N^*$ ($n > N^*$). Further, one can shift the chemical potential μ using a gate electrode to a negative value $\mu < 0$ which is the hole region of the spectrum. Then, as soon as $n = N^*$, one gets a divergence of the hole effective mass ($\gamma_{N^*} = 0$), which is a manifestation of *Zitterbewegung*. That anomaly can be observed in the Shubnikov–de Haas (SdH) oscillations.

Problems

2-1. Why are some processes of electron scattering in pristine graphene permitted, whereas others are prohibited?
2-2. Determine how the full charge density per unit area (or the charge imbalance) changes when a finite gate voltage is applied to a graphene sample.
2-3. Compute the bipolar conductivity $\sigma_0(\mu)$ using the same dependence for the total carrier density $N_s(\mu)$ of a charged sheet of pristine graphene which represents a graphene field-effect transistor (FET), that is, in conditions when a finite gate voltage is applied ($\mu \neq 0$).
2-4. Compute the quantum capacitance of a single-wall carbon nanotube (CNT) FET when a finite gate voltage $V_G \neq 0$ is

applied underneath the substrate. What is the effect of Van Hove singularities on the quantum capacitance?

2-5. Compute the bipolar conductivity of a CNT FET when a finite gate voltage $V_G \neq 0$ is applied. What is the contribution of Van Hove singularities?

2-6. Explain how the basic equation of graphene planar electrostatics (Eq. 2.42) changes if the value of the gate voltage corresponding to the charge neutrality point (NP) is finite, $V_{NP} \neq 0$.

2-7. Determine in which way the interface trap charges affect the dependence of the electron Fermi energy on the gate voltage.

2-8. What is the condition for the drift current component to dominate over the diffusion part or vice versa?

2-9. Explain why the Dirac equation fails to describe the electron transport in conditions when the electron scattering on the phonons, atomic impurities, and lattice imperfections in the graphene and CNT samples is strong.

References

Abanin, D. A., P. A. Lee, and L. S. Levitov (2006). Spin-filtered edge states and quantum hall effect in graphene. *Phys Rev Lett*, **96**(17).

Akhmerov, A. R., and C. W. J. Beenakker (2007a). Detection of valley polarization in graphene by a superconducting contact. *Phys Rev Lett*, **98**(15).

Akhmerov, A. R., and C. W. J. Beenakker (2007b). Pseudodiffusive conduction at the Dirac point of a normal-superconductor junction in graphene. *Phys Rev B*, **75**(4).

Akhmerov, A. R., and C. W. J. Beenakker (2008). Boundary conditions for Dirac fermions on a terminated honeycomb lattice. *Phys Rev B*, **77**(8).

Ando, T. (2005). Theory of electronic states and transport in carbon nanotubes. *J Phys Soc Jpn*, **74**(3): 777–817.

Ando, T. (2008). Carrier-density dependence of optical phonons in carbon nanotubes: art. no. 012006. *Int Symp Adv Nanodevices Nanotechnol*, **109**: 12006–12006.

Beenakker, C. W. J., A. R. Akhmerov, P. Recher, and J. Tworzydlo (2008). Correspondence between Andreev reflection and Klein tunneling in bipolar graphene. *Phys Rev B*, **77**(7).

Beenakker, C. W. J., and M. Buttiker (1992). Suppression of shot noise in metallic diffusive conductors. *Phys Rev B*, **46**(3): 1889–1892.

Berry, M. V., and R. J. Mondragon (1987). Neutrino billiards: time-reversal symmetry-breaking without magnetic-fields. *Proc R Soc London, A*, **412**(1842): 53–74.

Bolotin, K. I., K. J. Sikes, J. Hone, H. L. Stormer, and P. Kim (2008). Temperature-dependent transport in suspended graphene. *Phys Rev Lett*, **101**(9).

Bolotin, K. I., K. J. Sikes, Z. Jiang, M. Klima, G. Fudenberg, J. Hone, P. Kim, and H. L. Stormer (2008). Ultrahigh electron mobility in suspended graphene. *Solid State Commun*, **146**(9–10): 351–355.

Brey, L., and H. A. Fertig (2006a). Edge states and the quantized Hall effect in graphene. *Phys Rev B*, **73**(19).

Brey, L., and H. A. Fertig (2006b). Electronic states of graphene nanoribbons studied with the Dirac equation. *Phys Rev B*, **73**(23).

Cheianov, V. V., V. Fal'ko, and B. L. Altshuler (2007). The focusing of electron flow and a Veselago lens in graphene p-n junctions. *Science*, **315**(5816): 1252–1255.

Chen, H. Y., V. Apalkov, and T. Chakraborty (2007). Fock-Darwin states of dirac electrons in graphene-based artificial atoms. *Phys Rev Lett*, **98**(18).

Chen, J. H., C. Jang, S. Adam, M. S. Fuhrer, E. D. Williams, and M. Ishigami (2008). Charged-impurity scattering in graphene. *Nat Phys*, **4**(5): 377–381.

Danneau, R., F. Wu, M. F. Craciun, S. Russo, M. Y. Tomi, J. Salmilehto, A. F. Morpurgo, and P. J. Hakonen (2008). Evanescent wave transport and shot noise in graphene: ballistic regime and effect of disorder. *J Low Temp Phys*, **153**(5–6): 374–392.

Dorokhov, O. N. (1984). On the coexistence of localized and extended electronic states in the metallic phase. *Solid State Commun*, **51**(6): 381–384.

Fang, T., A. Konar, H. L. Xing, and D. Jena (2007). Carrier statistics and quantum capacitance of graphene sheets and ribbons. *Appl Phys Lett*, **91**(9).

Fujita, M., K. Wakabayashi, K. Nakada, and K. Kusakabe (1996). Peculiar localized state at zigzag graphite edge. *J Phys Soc Jpn*, **65**(7): 1920–1923.

Giovannetti, G., P. A. Khomyakov, G. Brocks, P. J. Kelly, and J. van den Brink (2007). Publisher's note: substrate-induced band gap in graphene on hexagonal boron nitride: ab initio density functional calculations [**76**, art. no. 073103, (2007)]. *Phys Rev B*, **76**(7).

Hwang, E. H., S. Adam, and S. Das Sarma (2007a). Carrier transport in two-dimensional graphene layers. *Phys Rev Lett*, **98**(18).

Hwang, E. H., S. Adam, and S. Das Sarma (2007b). Transport in chemically doped graphene in the presence of adsorbed molecules. *Phys Rev B*, **76**(19).

Katsnelson, M. I., and A. K. Geim (2008). Electron scattering on microscopic corrugations in graphene. *Philos Trans R Soc A*, **366**(1863): 195–204.

Katsnelson, M. I., and K. S. Novoselov (2007). Graphene: New bridge between condensed matter physics and quantum electrodynamics. *Solid State Commun*, **143**(1–2): 3–13.

Katsnelson, M. I., K. S. Novoselov, and A. K. Geim (2006). Chiral tunnelling and the Klein paradox in graphene. *Nat Phys*, **2**(9): 620–625.

Kim, E. A., and A. H. C. Neto (2008). Graphene as an electronic membrane. *Epl*, **84**(5).

McCann, E., and V. I. Fal'ko (2004). Symmetry of boundary conditions of the Dirac equation for electrons in carbon nanotubes. *J Phys: Condens Mater*, **16**(13): 2371–2379.

Miao, F., S. Wijeratne, Y. Zhang, U. C. Coskun, W. Bao, and C. N. Lau (2007). Phase-coherent transport in graphene quantum billiards. *Science*, **317**(5844): 1530–1533.

Nakada, K., M. Fujita, G. Dresselhaus, and M. S. Dresselhaus (1996a). Edge state in graphene ribbons: nanometer size effect and edge shape dependence. *Phys Rev B*, **54**(24): 17954–17961.

Nakada, K., M. Fujita, K. Wakabayashi, and K. Kusakabe (1996b). Localized electronic states on graphite edge. *Czech J Phys*, **46**: 2429–2430.

Nicollian, E. H., and J. R. Brews (1984). Instrumentation and analog implementation of the Q-C method for MOS measurements. *Solid State Electron*, **27**(11): 953–962.

Nomura, K., and A. H. MacDonald (2007). Quantum transport of massless dirac fermions. *Phys Rev Lett*, **98**(7).

Novoselov, K. S., A. K. Geim, S. V. Morozov, D. Jiang, M. I. Katsnelson, I. V. Grigorieva, S. V. Dubonos, and A. A. Firsov (2005). Two-dimensional gas of massless Dirac fermions in graphene. *Nature*, **438**(7065): 197–200.

Peres, N. M. R., A. H. C. Neto, and F. Guinea (2006). Dirac fermion confinement in graphene. *Phys Rev B*, **73**(24).

Schuessler, A., P. M. Ostrovsky, I. V. Gornyi, and A. D. Mirlin (2009a). Analytic theory of ballistic transport in disordered graphene. *Phys Rev B*, **79**(7).

Schuessler, A., P. M. Ostrovsky, I. V. Gornyi, and A. D. Mirlin (2009b). Ballistic transport in disordered graphene. *Adv Theor Phys*, **1134**: 178–186.

Shafranjuk, S. E. (2009). Directional photoelectric current across the bilayer graphene junction. *J Phys: Condens Mater*, **21**(1).

Shafranjuk, S. E. (2009). Reversible heat flow through the carbon tube junction. *EPL*, **87**(5).

Shafranjuk, S. E. (2011). Electromagnetic properties of the graphene junctions. *Eur Phys J B*, **80**(3): 379–393.

Tan, Y. W., Y. Zhang, K. Bolotin, Y. Zhao, S. Adam, E. H. Hwang, S. Das Sarma, H. L. Stormer, and P. Kim (2007). Measurement of scattering rate and minimum conductivity in graphene. *Phys Rev Lett*, **99**(24).

Tworzydlo, J., I. Snyman, A. R. Akhmerov, and C. W. J. Beenakker (2007). Valley-isospin dependence of the quantum Hall effect in a graphene p-n junction. *Phys Rev B*, **76**(3).

Tworzydlo, J., B. Trauzettel, M. Titov, A. Rycerz, and C. W. J. Beenakker (2006). Sub-Poissonian shot noise in graphene. *Phys Rev Lett*, **96**(24).

Zebrev, G. I. (2008). Graphene nanoelectronics: electrostatics and kinetics. *Proc. SPIE* **7025**, micro- and nanoelectronics 2007, 70250M.

Zebrev, G. I., and R. G. Useinov (1990). Simple-model of current-voltage characteristics of a metal-insulator semiconductor transistor. *Sov Phys Semicond: USSR*, **24**(5): 491–493.

Zhang, Y. B., Y. W. Tan, H. L. Stormer, and P. Kim (2005). Experimental observation of the quantum Hall effect and Berry's phase in graphene. *Nature*, **438**(7065): 201–204.

Chapter 3

Quantized States in Graphene Ribbons

In this chapter we consider a tight-binding model which formerly was suggested for graphite. We implement this model to describe the stack with a finite number of graphene layers and also to explore the quantized states in the ribbons of bilayer and single-layer graphene.

3.1 Tight-Binding Model of Bilayer Graphene

Electron properties of the simplest bilayer consisting of two monoatomic graphene sheets which are stacked one on top of another (see Fig. 3.1a) indicate new anomalies not visible in monolayers (McCann et al. 2006). Thus, the graphene bilayer is modeled

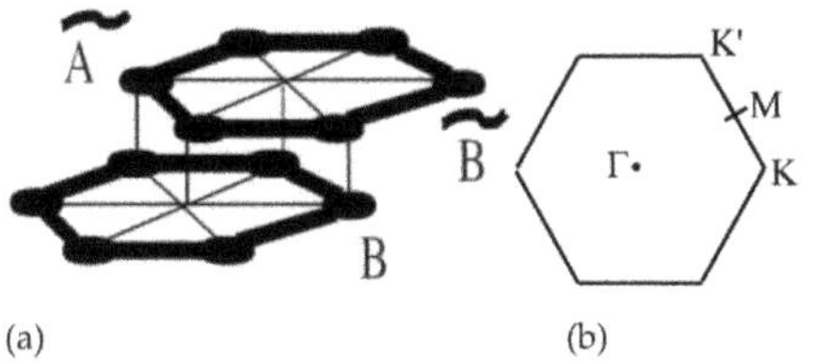

Figure 3.1 Lattice structure of (a) bilayer graphene and (b) the Brillouin zone.

Graphene: Fundamentals, Devices, and Applications
Serhii Shafraniuk

ISBN 978-981-4613-47-7 (Hardcover), 978-981-4613-48-4 (eBook)
www.panstanford.com

as two coupled hexagonal lattices consisting of four nonequivalent sites A, B and Ã, B̃ in the bottom and top layers, respectively (see Fig. 3.1b). One writes the tight-binding Hamiltonian for the bilayer graphene as

$$
\begin{aligned}
H_{t,b} = &-\gamma_0 \sum_{i,j,m,\sigma} \left(a^{\dagger}_{m,i,\sigma} b_{m,j,\sigma} + h.c.\right) - \gamma_1 \sum_{j,\sigma} \left(a^{\dagger}_{1,j,\sigma} a_{2,j,\sigma} + h.c.\right) \\
&-\gamma_4 \sum_{i,j} \left(a^{\dagger}_{1,j,\sigma} b_{2,j,\sigma} + a^{\dagger}_{2,j,\sigma} b_{1,j,\sigma} + h.c.\right) \\
&-\gamma_3 \sum_{j,\sigma} \left(b^{\dagger}_{1,j,\sigma} b_{2,j,\sigma} + h.c.\right)
\end{aligned}
\tag{3.1}
$$

where we have introduced annihilation operators $a_{m,i,\sigma}$ and $b_{m,i,\sigma}$ acting on an electron with spin σ on sublattices A and B, in plane $m = 1{,}2$, at site $\mathbf{R}_i$. Here for hopping parameters we use the graphite notations $\gamma_0 = t$ is the in-plane hopping energy. Other parameters are as follows: γ_1 is the hopping energy between atom $\tilde{A}_1$ and atom $\tilde{A}_2$ (see Fig. 3.1), $\gamma_1 = t_{\perp} \approx 0.4$ eV in graphite (Brandt et al. 1988, Brandt et al. 1988, Dresselhaus and Dresselhaus 2002), γ_4 is the hopping energy between atom $\tilde{A}_1$ ($\tilde{A}_2$) and atom $\tilde{B}_2$ ($\tilde{B}_1$), $\gamma_4 \approx 0.04$ eV in graphite (Brandt et al. 1988, Brandt et al. 1988, Dresselhaus and Dresselhaus 2002) and γ_3 connects $\tilde{B}_1$ and $\tilde{B}_2$, $\gamma_3 \approx 0.3$ eV in graphite (Brandt et al. 1988, Brandt et al. 1988, Dresselhaus and Dresselhaus 2002). In the continuum limit, by expanding the electron momentum in the vicinity of K and K′ points in the Brillouin zone, one obtains the model Hamiltonian

$$
H = \sum_k \Psi_k^{\dagger} H_k \Psi_k \tag{3.2}
$$

where the chiral fermion Hamiltonian H operates in the space of the four-component wavefunctions $\hat{\Psi}$

$$
\Psi_k^{\dagger} = \left(b_1^{\dagger}(\mathbf{k})\; a_1^{\dagger}(\mathbf{k})\; a_2^{\dagger}(\mathbf{k})\; b_2^{\dagger}(\mathbf{k}) \right) \tag{3.3}
$$

is a four-component spinor

$$
H_K \equiv \begin{pmatrix} -V & vk & 0 & 3\gamma_3 a k^* \\ vk^* & -V & \gamma_1 & 0 \\ 0 & \gamma_1 & V & vk \\ 3\gamma_3 a k & 0 & vk^* & V \end{pmatrix} \tag{3.4}
$$

and $k = k_x + ik_y$ is a complex number. For the sake of simplicity, in Eq. 3.4 we set $\gamma_4 = 0$. Furthermore, we introduce a bias potential V which is applied between the two graphene sheets. The bias here is a half of the shift of electrochemical potential between the two layers. In the limiting case of zero bias $V = 0$ and small γ_3, $v_F k << 1$, an effective Hamiltonian is written as

$$H_K \equiv \begin{pmatrix} 0 & \frac{v^2 k^2}{\gamma_1} + 3\gamma_3 a k^* \\ \frac{v^2 (k^*)^2}{\gamma_1} + 3\gamma_3 a k & 0 \end{pmatrix} \quad (3.5)$$

where we have eliminated the high-energy states perturbatively. The momentum-dependent coupling between the sublattices of different layers originates from the electron hopping $\sim\gamma_4$. Another part of coupling comes from a small renormalization of γ_1 and also from a nonequivalence between sublattices within the layer. If in Eq. 3.5 one sets $\gamma_3 = 0$, the electron–hole spectrum is symmetric and it consists of two parabolic bands

$$\varepsilon_{k,\pm} \approx \pm\frac{v^2 k^2}{t_\perp} \quad (3.6)$$

Merging with each other at $\varepsilon = 0$ (as shown in Fig. 3.2 on the left). Furthermore, the additional two bands springing at finite $\varepsilon = \pm t_\perp$. The approximation yields a metallic bilayer with a constant density of states (DOS). The term γ_3 represents a source of *trigonal distortion*, or *warping*. Thus, the qualitative change of the energy bands represents a rebuilding of the electron excitation spectrum. The trigonal distortion, unlike the one introduced by a large momentum in Eq. 3.4, occurs at low energies. Although the electron–hole symmetry is preserved, instead of two bands which touch each

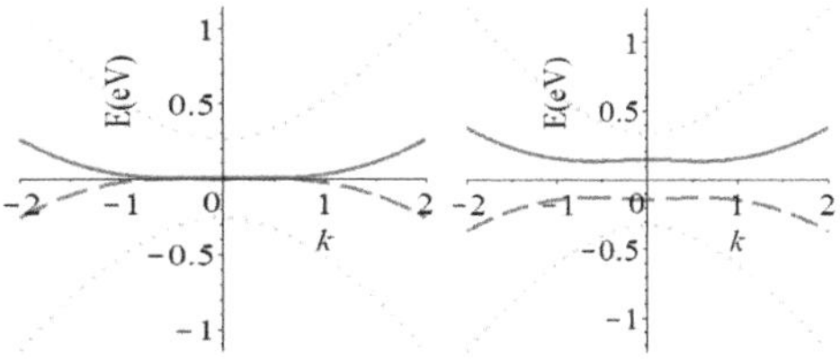

Figure 3.2 Left: Band structure of bilayer graphene for $V = 0$ and $\gamma_3 = 0$. The electron wavevector is in units of nanometers $^{-1}$. Right: Band structure of bilayer graphene for $V \neq 0$ and $\gamma_3 = 0$.

other at $k = 0$, we obtain three sets of Dirac-like linear bands. One Dirac point is positioned at $\varepsilon = 0$ and $k = 0$, whereas three other Dirac points lie at the three equivalent points also at $\varepsilon = 0$ but with a finite momentum. Stability of the Dirac points, where the electron bands touch each other, is studied using topological arguments (Manes et al. 2007). The winding number of the point, where the two parabolic bands merge with each other for $\gamma_3 = 0$, is equal $+2$. The winding number here illustrates topological properties of the electron spectral branches. Thus, the winding number is introduced for a closed curve in the plane which encircles a given point. Such number represents the total number of times when the curve encircles the point counterclockwise with no change of the wavefunction. The trigonal warping term γ_3 causes splitting of the Dirac point at $k = 0$ which is characterized by the winding number -1. Furthermore, it also splits the three Dirac points at $k \neq 0$ which are described with the winding numbers $+1$. One can also split the $\gamma_3 = 0$ degeneracy into two Dirac points with winding number $+1$ by applying an in-plane magnetic field or slightly rotating one graphene layer in respect to another. The inversion symmetry and equivalence of the two layers is broken with the term V in Eq. 3.4. Then the dispersion relation becomes

$$\varepsilon_{\pm,k}^2 = V^2 - v^2k^2 + \frac{t_\perp^2}{2} \pm \sqrt{4V^2v^2k^2 + t^2v^2k^2 + \frac{t_\perp^2}{4}} \tag{3.7}$$

From the last formula one can see that V actually induces an energy gap in the vicinity of the K point (but not directly at it). If $V << t$, and the electron moments k are small, one expands the above formula for the energy of the conduction band as

$$\varepsilon_k \approx V - \frac{2Vv^2k^2}{t_\perp} + \frac{v^4k^4}{2t_\perp^2 V} \tag{3.8}$$

With replacing ε_k by $-\varepsilon_k$ one obtains the dispersion law for the valence band. The energy gap in the bilayer's spectrum thus emerges at

$$k^2 \approx \frac{2V^2}{v^2} \tag{3.9}$$

We emphasize that the gap in the bilayer graphene system under bias depends on the applied voltage and therefore it can be measured experimentally (McCann et al. 2006, Novoselov et al. 2006). The appearance of an energy gap in the bilayer graphene spectrum makes it very interesting for technological applications.

3.2 A Bilayer Graphene Junction

When neglecting many-body effects one might describe the electron properties of bilayer graphene in terms of envelope wavefunctions (Shafranjuk 2009). The main part of the effective Hamiltonian in the single-electron approximation is

$$\hat{H} = -\hbar^2 \left(\pi_-^2 \hat{\sigma}_+ + \pi_+^2 \hat{\sigma}_-\right)/2m + U(x) \tag{3.10}$$

where

$$\pi_\pm = (k - e\mathrm{A}_x(t)/\hbar) \pm i\left(q - e\mathrm{A}_y(t)/\hbar\right) \tag{3.11}$$

$\hat{\sigma}_\pm = \hat{\sigma}_x \pm \hat{\sigma}_y$, $\hat{\sigma}_i$ are the Pauli matrices, $i = \{x, y, z\}$, the effective mass m is expressed via coupling strength $\gamma_{\tilde{\mathrm{A}}\mathrm{B}}$ between Ã and B as

$$m = \frac{\gamma_{\tilde{\mathrm{A}}\mathrm{B}}}{2v^2} = 0.054 \tag{3.12}$$

where

$$v = (\sqrt{3}/2)a\gamma_{\mathbf{AB}} \tag{3.13}$$

$a = 0.246$ nm is the lattice constant, $\gamma_{\mathbf{AB}} \approx 0.4$ eV, $A_{x,y}(t)$ are the corresponding x-y components of the vector potential $\mathbf{A}(t)$, which generally might be time dependent, and $U(x)$ is the electric potential in graphene, which might be controlled by the gate voltage $V_{\mathbf{G}}$. In the latter case, the graphene structure represents a field-effect transistor (FET). Equation 3.10 describes interlayer coupling via a dimer state formed by pairs of carbon A atoms located in the bottom and top layers, respectively, as shown in Fig. 3.1b. A weak direct A coupling and a small interaction due to the bottom and top layer asymmetry (which opens a minigap in the electron spectrum (McCann et al. 2006, Novoselov et al. 2006)) are both hereafter neglected.

For graphene devices of finite dimensions, the motion of chiral fermions is quantized. The quantization imposes additional constrains on the directional tunneling diagram. Permitted values of the angle $\tilde{\varphi}_n$ inside the graphene barrier are obtained from boundary conditions (BCs) along the y direction, so the y component of the electron momentum $\mathbf{p} = (\hbar k, \hbar q)$ is quantized as

$$\tilde{q}_n = \frac{n\pi}{W} \tag{3.14}$$

where W is the barrier width, which gives

$$\tilde{\varphi}_n = \arctan\frac{n\pi}{k'_\varepsilon W} \tag{3.15}$$

where

$$k'_\varepsilon = \sqrt{\frac{2m}{\hbar^2}}\sqrt{|\varepsilon - U_0| - |\varepsilon|\left(\frac{1-\cos 2\varphi}{2}\right)} \tag{3.16}$$

The last formula also means that $\tilde{q}_n$ depends on the electron energy variable ε.

Probability of the electron tunneling through a chiral barrier in bilayer graphene is computed as follows. The chirality has no significance for $E > V$ (i.e., outside the barrier). On the left side from the barrier ($x < 0$), one uses the following trial function:

$$\begin{aligned}\Psi_{1L} &= e^{iqy}\left(e^{ikx} + b_1 e^{-ikx} + c_1 e^{\chi\kappa}\right)\\ \Psi_{2L} &= e^{iqy}\left(e^{ikx+2i\varphi} + b_1 e^{-ikx-2i\varphi} - c_1 h_1 e^{\chi\kappa}\right) s_1\end{aligned} \tag{3.17}$$

At positive ($x > 0$) one implements

$$\begin{aligned}\Psi_{1R} &= e^{iqy}\left(e^{ikx}a_3 + d_3 e^{-\chi\kappa}\right)\\ \Psi_{2R} &= e^{iqy}\left(e^{ikx+2i\varphi}a_3 - \frac{d_3 e^{-\chi\kappa}}{h_3}\right) s_3\end{aligned} \tag{3.18}$$

Chirality is pronounced inside the barrier of height V, when the electron excitation energy is $E < V$ (which corresponds to a finite $\varphi \neq 0$).

$$\begin{aligned}\Psi_{1B} &= e^{iq_2y}\left(e^{ik_2x}a_2 + b_2 e^{-ik_2x} + d_2 e^{-\chi\kappa_2} + c_2 e^{\chi\kappa_2}\right)\\ \Psi_{2B} &= e^{iq_2y}\left(e^{ik_2x+2i\varphi}a_2 + b_2 e^{-ik_2x-2i\varphi} - c_2 e^{\chi\kappa_2}h - \frac{d_2 e^{-\chi\kappa_2}}{h_2}\right) s_2\end{aligned} \tag{3.19}$$

where the incidence angle φ is given at the beginning. Then, after the electron enters the graphene barrier, we set the constrain that the y component q is conserved

$$\varphi_{\mathrm{B}} = \arcsin\left(\sqrt{\frac{|E|}{|E - V_{\mathrm{B}}|}}\sin\varphi\right) \tag{3.20}$$

The electron wavevector inside the barrier is

$$\begin{aligned}\hbar k_{\mathrm{B}} &= \sqrt{2m\,|E-V_{\mathrm{B}}|}\cos\varphi_{\mathrm{B}} \\ &= \sqrt{m\,|E-V_{\mathrm{B}}|}\sqrt{\frac{|E|}{|E-V_{\mathrm{B}}|}(\cos 2\varphi - 1) + 2} \\ &= \sqrt{2m}\sqrt{|E-V_{\mathrm{B}}| - |E|\,(1-\cos 2\varphi)/2} \end{aligned} \quad (3.21)$$

Thus, the electron wavevectors in the sections of our piece-wise model are expressed as

$$\begin{aligned}\hbar k_i &= \sqrt{2m\,|E-V_i|}\cos\varphi_i \\ \hbar q_i &= \sqrt{2m\,|E-V_i|}\sin\varphi_i \\ \kappa_i &= \sqrt{k_i^2 + 2q_i^2} \\ h_i &= \left(\sqrt{1+\sin^2\varphi_i} - \sin\varphi_i\right) \\ s_i &= \mathbf{sign}\,(V_i - E) \end{aligned} \quad (3.22)$$

Using the fact that the charge carrier velocity in the bilayer graphene

$$v_E = \frac{\hbar k_i}{m} = \sqrt{\frac{2\,|E-V_i|}{m}}\cos\varphi_i \quad (3.23)$$

We set

$$\frac{\hbar^2 k_i^2}{2m} \to \frac{m v_E^2}{2} \quad (3.24)$$

The barrier thickness D in units of $2\pi/\kappa_0$ where

$$\kappa_0 = \frac{\sqrt{2mU_0}}{\hbar} \quad (3.25)$$

Then D is in units $h/\sqrt{2mU_0}$. The simplified steady-state solution for a rectangular barrier gives

$$t_{\mathbf{2GW}} = -\frac{2k(k_2-k)s e^{2i(Dk_2+2\varphi)}}{e^{2iDk_2}(k-k_2)^2 s_2 - (k+k_2)^2 s_2}, \quad (3.26)$$

whereas the conventional expression is

$$t_{\mathrm{cl}} = \frac{k k_2 e^{-iDk} e^{i\varphi}}{k k_2 \cos(Dk_2) - i\left(k^2 + {k_2}^2\right)\sin(Dk_2)/2} \quad (3.27)$$

The electric current density $j = I(V_{\mathbf{sd}})/W$ (I is the electric current, $V_{\mathbf{sd}}$ is the bias voltage, and W is the graphene stripe width) between the electrodes 1 and 3 is computed as

$$j = 2\pi e \int d\varepsilon\, \chi_\varepsilon \left[G_3^K(\varepsilon) - G_1^K(\varepsilon)\right] \quad (3.28)$$

where we introduced the factor χ_ε. If 1 and 3 electrodes are made of a monolayer graphene or are metallic, then

$$\chi_\varepsilon = v_F N(0) \tag{3.29}$$

where v_F and $N(0)$ are the relevant Fermi velocity and the electron DOS at the Fermi level. However, if the 1 and 3 electrodes are made of the bilayer graphene itself, which case we consider in details below, then $\chi_\varepsilon = v_\varepsilon N(\varepsilon)$ where

$$v_\varepsilon = \frac{\hbar |k|}{m} = \sqrt{\frac{2|\varepsilon|}{m}} \tag{3.30}$$

and

$$N(\varepsilon) = \frac{m}{\pi \hbar^2} \sum_k \theta(\varepsilon - E_k) \tag{3.31}$$

are the energy-dependent velocity and the two-dimensional electron DOS in bilayer graphene, E_k is the kth electron energy level in the graphene barrier stripe,

$$G_r^K(\varepsilon) = -i \sum_{\mathbf{p}} \left|t_{\mathbf{p}}\right|^2 e^{iqy} e^{ikD} (2n_{\mathbf{p}} - 1)\, \delta\left(\varepsilon - \varepsilon_{\mathbf{p}} + \delta_{r,3} e V_{\mathbf{sd}}\right) \tag{3.32}$$

is the Keldysh–Green function (Keldysh 1965), r is the electrode index, $\delta_{r,3}$ is the Kronecker symbol, and $n_{\mathbf{p}}$ is the distribution function of electrons with momentum $\mathbf{p}$. A straightforward calculation using the methods of Keldysh (1965) and Datta (1992) gives

$$\begin{aligned} j &= (\pi/2)e \int d\varepsilon \chi_\varepsilon \sum_{\mathbf{p}} \left|t_{\mathbf{p}}\right|^2 [(2n_{\mathbf{p}} - 1)\cdot\delta(\varepsilon - \varepsilon_{\mathbf{p}} + eV) \\ &\quad -(2n_{\mathbf{p}} - 1)\cdot\delta(\varepsilon - \varepsilon_{\mathbf{p}})] \\ &= \pi e \int d\varepsilon \chi_\varepsilon |t_\varepsilon|^2 (n_{\varepsilon - eV} - n_\varepsilon) \end{aligned} \tag{3.33}$$

Taking for simplicity $N(\varepsilon) = m/\left(\pi\hbar^2\right)$, from Eq. 3.33 one finds the zero-temperature steady-state conductance in the form

$$G_0 = \frac{e^2}{\hbar^2} \overline{T} W \sqrt{2meV_{\mathbf{sd}}} = \frac{2e^2}{h} \overline{T} N_{\mathbf{ch}}(V_{\mathbf{sd}}) \tag{3.34}$$

where $\overline{T} = \left|t_{eV_{\mathbf{sd}}}\right|^2$ is the transparency of the chiral graphene barrier. In Eq. 3.34 we introduced the voltage-dependent dimensionless number of conducting channels

$$N_{\mathbf{ch}}(V_{\mathbf{sd}}) = \pi W \sqrt{2meV_{\mathbf{sd}}} \tag{3.35}$$

The dependence $N_{\mathbf{ch}}$ versus $V_{\mathbf{sd}}$ stems from the energy dependence of the electron velocity in bilayer graphene v_ε. Equation 3.34 coincides with the well-known Landauer formula with a number of conducting channels $N_{\mathbf{ch}}$. The calculation results will be convenient to normalize to an auxiliary conductivity defined as

$$\tilde{\sigma}_0 = \frac{G_0}{W}\left(V_{\mathbf{sd}} = \frac{U_0}{e}\right) = \frac{2e^2}{h}\pi\sqrt{2mU_0} \qquad (3.36)$$

(where we use $\overline{T} \simeq 1$ at $V_{\mathbf{sd}} = U_0/e$, U_0 is the graphene barrier height). The transmission amplitude t_ε across the voltage biased junction is obtained within a simple model which represents the chiral fermion wavefunctions via Airy functions. The Hamiltonian Eq. 3.10 yields a gapless semiconductor with massive chiral electrons and holes having a finite mass m. Let us consider tunneling of those fermions with the energy E incident on the barrier under the angle φ. Since the potential barrier is formed in the longitudinal direction, the y component $\hbar q$ of the momentum $\mathbf{p}$ is conserved, while the x component $\hbar k$ is not. The trial chiral fermion wavefunction takes a piece-wise form (Katsnelson and Novoselov 2007). An analytical steady-state solution (Katsnelson et al. 2006) is obtained at $V_{\mathbf{sd}} = 0$ for a rectangular barrier expressing the electron and hole wavefunctions via combinations of plane waves. Matching the continuous BCs one finds (Novoselov et al. 2005, Geim and Novoselov 2007, McCann et al. 2007a, 2007b) the tunneling amplitude $t_{\mathbf{2GW}}$ for a normal electron incidence ($\varphi = 0$) as

$$t_{\mathbf{2GW}} = -\frac{2k(k' - k)se^{2i(Dk'+2\phi)}}{e^{2iDk'}(k - k')^2 s' - (k + k')^2 s'} \qquad (3.37)$$

where the electron wavevector in the electrode is

$$k = \frac{\sqrt{2m|E|}}{\hbar} \qquad (3.38)$$

and inside the barrier is

$$k' = \frac{\sqrt{2m(E - U_0)}}{\hbar} \qquad (3.39)$$

ϕ is the phase drop across the graphene barrier, $s' = \mathbf{sign}(U_0 - E)$. For a classic rectangular barrier one obtains

$$t_{\mathrm{cl}} = \frac{kk' e^{-iDk} e^{i\phi}}{kk' \cos(Dk') - i\left(k^2 + k'^2\right)\sin(Dk')/2} \qquad (3.40)$$

Although Eqs. 3.37 and 3.40 are instructive, the experimentally measured characteristics are relevant rather to a finite bias voltage ($V_{\mathbf{sd}} \neq 0$) across the graphene barrier and finite incidence angles $\varphi \neq 0$. The electric field $\mathbf{E}$ in the latter case penetrates inside the bilayer graphene barrier and electrodes, forcing the charge carriers to accelerate. The simplest electron and hole wavefunctions in that case are represented via the Airy functions rather than via plane waves. The CF wavefunction $\hat{\Psi}(x)$ is obtained from the Dirac equation $\hat{H}\hat{\Psi} = E\hat{\Psi}$, where E is the electron energy. For calculations one uses the tilted barrier potential

$$U(x) = -\mathcal{E} \cdot x\,[\theta(-x) + \theta(x - D)] + [U_0 - \mathbf{E} \cdot x]\,\theta(x)\,\theta(D - x) \tag{3.41}$$

where

$$\mathcal{E} = \frac{V_{\mathbf{sd}}}{D} \tag{3.42}$$

is the electric field, which penetrates into the graphene barrier. Then, components of the fermion momentum $\mathbf{p} = (\hbar k, \hbar q)$ are written as

$$\hbar q = \sqrt{2m\,|E|}\sin\varphi \tag{3.43}$$

and

$$\hbar k(x) = i\sqrt{2m(U(x) - E)}\cos\varphi(x) \tag{3.44}$$

$$\varphi(x) = \arcsin\left(\frac{q\sin\varphi}{k(x)}\right) \tag{3.45}$$

where D is the barrier thickness and φ is the electron incidence angle in the electrode 1. The corresponding trial wavefunction is

$$\begin{aligned}
\hat{\Psi} &= \hat{\Psi}_1\theta(-x) + \hat{\Psi}_2\theta(D - x) + \hat{\Psi}_3\theta(x - D)\\
\hat{\Psi}_1 &= e^{iqy}[\lambda\mathbf{Bi}(\zeta_{k,\,x}) + b_1\tilde{\lambda}\mathbf{Bi}(\zeta_{k,\,x}) + c_1\lambda^{\dagger}\mathbf{Ai}(\zeta_{ik,\,x})]\\
\hat{\Psi}_2 &= e^{iqy}[a_2\mathbf{Ai}(\zeta_{k',\,x})\mu + b_2\mathbf{Bi}(\zeta_{k',\,x})\tilde{\mu}\\
&\quad + d_2\mathbf{Bi}(\zeta_{ik',\,x})\mu^{\dagger} + c_2\mathbf{Ai}(\zeta_{ik',\,x})\mu^{\ddagger}]\\
\hat{\Psi}_3 &= e^{iqy}[a_3\mathbf{Ai}(\zeta_{k,\,x})\,\nu + d_3\mathbf{Bi}(\zeta_{ik,\,x})\,\tilde{\nu}]
\end{aligned} \tag{3.46}$$

where

$$\zeta_{k,\,x} = -\frac{k^2 + \mathcal{E}x}{(-\mathcal{E})^{2/3}} \tag{3.47}$$

$$k = \frac{\sqrt{2m|E|}\cos\varphi}{\hbar} \tag{3.48}$$

is the electron wavevector in the electrode

$$k' = \frac{\sqrt{2m(E - U_0)}\cos\varphi'}{\hbar} \tag{3.49}$$

is the electron wavevector inside the graphene barrier. In the above formulas we have used the following notations:

$$\varphi' = \arcsin\left(\frac{q}{k'}\sin\varphi\right) \tag{3.50}$$

$$s_1 = -1,\ s_2 = \text{sign}\,(U_0 - E) \tag{3.51}$$

$$s_3 = \text{sign}\,(-V_{\mathbf{sd}} - E) \tag{3.52}$$

Furthermore, we use

$$h' = \sqrt{1 + \sin^2\varphi'} - \sin\varphi' \tag{3.53}$$

$$\lambda = (|\uparrow\rangle + s_1 e^{2i\varphi}\,|\downarrow\rangle) \tag{3.54}$$

$$\tilde{\lambda} = (|\uparrow\rangle + s_1 e^{-2i\varphi}\,|\downarrow\rangle) \tag{3.55}$$

$$\lambda^{\dagger} = (|\uparrow\rangle + s_1 h_1\,|\downarrow\rangle) \tag{3.56}$$

$$\begin{aligned} \nu &= (|\uparrow\rangle + s_3 e^{2i\varphi}\,|\downarrow\rangle) \\ \tilde{\nu} &= (|\uparrow\rangle - s_3/h_3\,|\downarrow\rangle) \end{aligned} \tag{3.57}$$

and

$$\begin{aligned} \mu &= (|\uparrow\rangle + s_2 e^{2i\varphi'}\,|\downarrow\rangle) \\ \tilde{\mu} &= (|\uparrow\rangle + s_2 e^{-2i\varphi'}\,|\downarrow\rangle) \\ \mu^{\dagger} &= (|\uparrow\rangle - s_2/h_2\,|\downarrow\rangle) \\ \mu^{\ddagger} &= (|\uparrow\rangle - s_2 h_2\,|\downarrow\rangle) \end{aligned} \tag{3.58}$$

In the above equations we have introduced auxiliary "spin up" and "spin down" matrices as

$$|\uparrow\rangle = \begin{pmatrix} 1 \\ 0 \end{pmatrix} \tag{3.59}$$

and

$$|\downarrow\rangle = \begin{pmatrix} 0 \\ 1 \end{pmatrix} \tag{3.60}$$

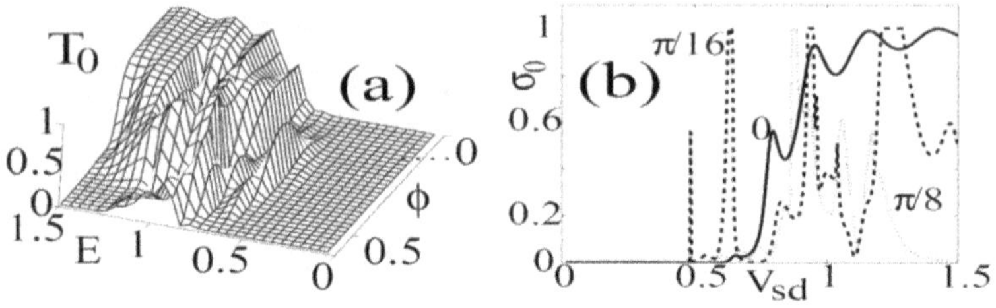

Figure 3.3 (a) The steady-state tunneling transparency T_0 versus the electron energy E (in units of the graphene barrier height U_0) and the azimuthal angle ϕ (in radians). (b) The corresponding steady-state differential conductance σ_0 (in units of $\tilde{\sigma}_0 = \left(2e^2/h\right)\pi\sqrt{2mU_0}$) versus the source–drain bias voltage V_{sd} (in units of U_0/e) for three angles of incidence ϕ. The sharp peaks at $V_{\mathrm{sd}} < U_0/e$ when $\varphi \neq 0$ originate from the electron–hole interference inside the barrier. Adapted from Shafranjuk (2009).

As we have learned in the previous chapter, the chiral tunneling through the barrier in the monolayer graphene has a strong angular dependence, whereas the tunneling probability T for the normal incidence is ideal ($T = 1$). The tendency is different for the tunneling in the bilayer graphene. In latter case, the chirality of the particles is pronounced only at finite incidence angles $\varphi \neq 0$.

Probability of the steady-state tunneling T_0 for a normally incident chiral particle vanishes below the barrier ($E < U_0$), while it is finite above the barrier (when $E \geq U_0$). In Fig. 3.3a we plot T_0 versus the energy E of an electron incident to the barrier under the angle ϕ. In Fig. 3.3b we show the steady-state tunneling differential conductance $\sigma_0(V_{\mathbf{sd}})$ for different incidence angles ϕ. Both the plots in Fig. 3.3 are related to $U_0 = 2$ meV, which corresponds to the surface charge density $n = 10^{11}$ cm^{-2} induced by the gate voltage $V_{\mathbf{G}} = 1$ V across the SiO_2 substrate with thickness $d = 300$ nm.

3.3 Heavy Chiral Fermion State in a Graphene Stripe

Let us consider a nonequilibrium thermal injection of "hot" electrons and holes from a source of heat into the heavy chiral fermion (HCF) energy levels which are formed inside a narrow (10–100 nm) graphene stripe. The steady state of the graphene stripe

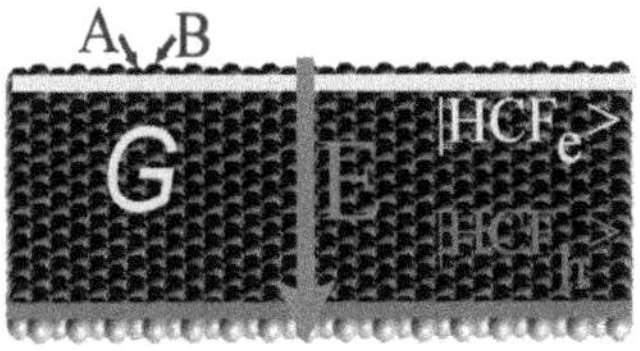

Figure 3.4 Graphene stripe G with heavy fermion states $|HCF_e >$ and $|HCF_h >$ polarized by a transversal electric field $\mathbf{E} = \{0, E_y, 0\}$. Zigzag edges of G are terminated by carbon atoms belonging to two different sublattices A and B.

had been studied by Brey and Fertig (2006a, 2006b) for the case of equal population of the electron and hole states. The electrochemical potential μ is then positioned at zero energy $\mu = 0$. In a graphene stripe G with zigzag edges a sharp energy level emerges at the energy $\varepsilon = 0$ (Brey and Fertig 2006a, 2006b). Here we examine transport properties of a graphene stripe polarized in the lateral y direction with an electric field $\mathbf{E} = \{0, E_y, 0\}$. We will see that the zero-energy steady state is modified as soon as $\mathbf{E} \neq 0$. The modified state is characterized by two opposite one-half pseudospins localized on the opposite edges of G as shown in Fig. 3.4. Pseudospin "↑" corresponds to electrons (e) localized on one edge, whereas holes (h) with pseudospin "↓" are accumulated on the opposite edge. Sharp distinctive $\mathrm{HCF}_{e(h)}$ levels at energies $\varepsilon = \pm\Delta$ result from electron reflections at the atomic zigzag edges (Brey and Fertig 2006a, 2006b) of the graphene stripe G as schematically shown in Fig. 3.4. The $\mathrm{HCF}_{e(h)}$ level spacing

$$2\Delta = 2e|\mathbf{E}|W \tag{3.61}$$

is controlled with the electric field $\mathbf{E} \neq 0$. The origin of the HCF resonances is illustrated in terms of a mere mean field approach (Ando 2005, Shafranjuk 2009). The method (Ando 2005, Shafranjuk 2009) describes the electronic excitation spectrum E_p of the graphene stripe in terms of the Dirac equation $\mathcal{H}\Psi = E\Psi$ for bipartite sublattices. Below we compute the DOS and evaluate the effective electron mass m^* at the HCF levels.

3.4 Quantum-Confined Stark Effect

A graphene stripe with zigzag edges actually represents an electric dipole with momentum $\mathbf{D} = e\mathbf{W}$ and energy $\Delta = \mathbf{D}E_y$ when an electric field $\mathbf{E} = \{0,\ E_y,\ 0\}$ is applied. One important property of a dipole is that its energy Δ changes sign under a parity transformation **P**. Since **D** is a vector, its expectation value in a state $|\psi>$ must be proportional to $<\psi|\mathbf{D}|\psi>$. Thus, under time reversal operation, **T**, an invariant state must have a vanishing dipole momentum $\mathbf{D} = 0$. In other words, a nonvanishing $\mathbf{D} \neq 0$ indicates both $\mathcal{P}$ and $\mathcal{T}$ symmetry breaking, whereas the $\mathcal{PT}$ symmetry keeps maintained. We emphasize that the graphene stripe represents a one-dimensional quantum well in the lateral y direction with quantized energy subbands. In absence of an external electric field, electrons and holes within the quantum well may only occupy the Van Hove singularity (VHS) energy subbands. A finite electric field $\mathbf{E} = \{0,\ E_y,\ 0\}$ causes the *quantum-confined Stark effect* (QCSE), which is expressed in the VHS splitting. Therefore, electron states are moved to higher energies, while the hole states are shifted to lower energies. Besides, the external electric field pulls electrons and holes to opposite sides of the G stripe in the perpendicular direction, thus, separating them spatially. The spatial separation between the electrons and holes is limited with presence of the confining potential barriers fringing the graphene stripe. It means that the two $\mathrm{HCF}_{\mathrm{e(h)}}$ edge states characterized with the opposite $\pm$ one-half pseudospins and opposite electric charges $\pm e$ exist in the system subjected to the transverse electric field. The QCSE is used in QCSE optical modulators, which allow optical communication signals to be switched on and off rapidly.

Polarization of the G stripe by the electric field $\mathbf{E} \neq 0$ is accounted for by the term $\propto \Delta$ in the Hamiltonian

$$\mathcal{H} = -i\hbar v \left(\left(\hat{\sigma}_x \otimes \hat{1} \right) \partial_x + \left(\hat{\sigma}_y \otimes \hat{\tau}_z \right) \partial_y \right) + U(x) \left(\hat{1} \otimes \hat{1} \right) + \Delta \left(\hat{\sigma}_z \otimes \hat{\tau}_z \right) \quad (3.62)$$

where $v = 8.1 \times 10^5$ m/s $\simeq c/300$ is the massless fermion speed, $\hat{\sigma}_i$ and $\hat{\tau}_k$ are the Pauli matrices, and $\otimes$ is the Kronecker product, $\{i,\ k\} = 1 \ldots 3$. Given an $m \times n$ matrix A and a $p \times q$ matrix B, their Kronecker product $C = A \otimes B$, also called their matrix direct product, is an $(mp) \times (nq)$ matrix with elements defined by $c_{\alpha\beta} = a_{ij} b_{kl}$ where $\alpha = p\,(i-1) + k$ and $\beta = q\,(j-1) + l$. Besides, in Eq. 3.62

$$E_y = \frac{V_{\mathbf{SG}}}{W} = \frac{V_{\mathbf{GA}} - V_{\mathbf{GB}}}{W} \tag{3.63}$$

is the magnitude of transverse electric field, V_{SG} is the split gate voltage. The dipole energy term $\propto \Delta = DE_y$ in Eq. 3.62 violates either the $\mathcal{T}$ or $\mathcal{P}$ symmetry when taken separately, whereas simultaneously it maintains the combined $\mathcal{PT}$ symmetry (see the next section of this chapter). We assume that the barrier potential $U\,(x)$ depends on the coordinate x in a piece-wise form

$$U\,(x) = U_0 \left[\theta\left(x - \frac{L_0}{2}\right) + \theta\left(-x - \frac{L_0}{2}\right)\right] \tag{3.64}$$

which is defined by the back gates $G_{\mathbf{L(R)}}$. The Pauli matrices $\hat{\sigma}_z \otimes \hat{\tau}_z$ in the Δ term of Eq. 3.62 account for the opposite electric charges $\pm e$ of electrons and holes located at the opposite stripe's edges when the electrochemical potential $\mu = 0$. However, if both edge state levels $\mathrm{HCF}_{\mathrm{e(h)}}$ are occupied by quasiparticles with the same sign of electric charge, either by electrons or holes, a conventional form of the Δ term with the unity matrices $\hat{\sigma}_0 \otimes \hat{\tau}_0$ must be used instead. The replacement

$$\hat{\sigma}_3 \otimes \hat{\tau}_3 \rightarrow \hat{\sigma}_0 \otimes \hat{\tau}_0 \tag{3.65}$$

in Eq. 3.62 should be done, for example, when the back gates $G_{\mathbf{L(R)}}$ shift the electrochemical potential μ above ($\mu > \Delta$ for HCF_e or below ($\mu < -\Delta$ for $\mathrm{HCF}_{\mathrm{h}}$) the HCF levels.

3.5 PT Invariance of the Dirac Hamiltonian

The two-dimensional Dirac equation (Ando 2005) describes states with wavevector $\mathbf{k}$ in the valley centered at the corner of the Brillouin zone with wavevector

$$\mathbf{K} = \frac{4\pi}{3a}\hat{\mathbf{x}} \tag{3.66}$$

The valley at the opposite corner at $-\mathbf{K}$ produces an independent set of states with amplitudes $\Psi'_{\mathrm{A}}(r)e^{-i\mathbf{K}\cdot\mathbf{r}}$ and $\Psi'_{\mathrm{B}}(r)e^{-i\mathbf{K}\cdot\mathbf{r}}$ on the A and B sublattices. The two components $\Psi'_{\rm A}$ and $\Psi'_{\rm B}$ satisfy the same Dirac equation Eq. 3.4 with $p_x \to -p_x$. The spinor

$$\Psi = \begin{pmatrix} \Psi_{\rm A} \\ \Psi_{\rm B} \\ -\Psi'_{\rm B} \\ \Psi'_{\rm A} \end{pmatrix} \tag{3.67}$$

containing both valleys therefore satisfies the four-dimensional Dirac equation (Ando 2005)

$$\begin{pmatrix} v\mathbf{p}\cdot\sigma & 0 \\ 0 & v\mathbf{p}\cdot\sigma \end{pmatrix} \Psi = E\Psi \tag{3.68}$$

where

$$\mathbf{p} = -i\hbar\left(\frac{\partial}{\partial x}, \frac{\partial}{\partial y}\right) \tag{3.69}$$

The above Dirac equation Eq. 3.68 can be rewritten introducing the second set of Pauli matrices $\tau = (\tau_x, \tau_y, \tau_z)$ which act on the valley degree of freedom as

$$\hat{H}\hat{\Psi} = E\hat{\Psi} \tag{3.70}$$

where

$$\hat{H} = v\left[(\mathbf{p} + e\mathbf{A})\cdot\sigma\right]\otimes\tau_0 + U\sigma_0\otimes\tau_0 \tag{3.71}$$

The intervalley time reversal operator (which also flips the one-half pseudospin) is introduced as

$$\mathcal{T}_K = \begin{pmatrix} 0 & 0 & 0 & 1 \\ 0 & 0 & -1 & 0 \\ 0 & -1 & 0 & 0 \\ 1 & 0 & 0 & 0 \end{pmatrix} \mathcal{C} \tag{3.72}$$

where **C** is the operator of complex conjugation. This gives

$$\mathcal{T}_K\hat{\Psi} = \mathcal{C}\begin{pmatrix} 0 & 0 & 0 & 1 \\ 0 & 0 & -1 & 0 \\ 0 & -1 & 0 & 0 \\ 1 & 0 & 0 & 0 \end{pmatrix}\begin{pmatrix} \Psi_{\rm A} \\ \Psi_{\rm B} \\ -\Psi'_{\rm B} \\ \Psi'_{\rm A} \end{pmatrix} = \begin{pmatrix} \Psi'_{\rm A} \\ \Psi'_{\rm B} \\ -\Psi_{\rm B} \\ \Psi_{\rm A} \end{pmatrix} \tag{3.73}$$

that is, it swaps $K \leftrightarrow K'$. The intravalley time reversal operator

$$\mathcal{T}_{\rm AB} = \begin{pmatrix} 0 & 1 & 0 & 0 \\ 1 & 0 & 0 & 0 \\ 0 & 0 & 0 & -1 \\ 0 & 0 & -1 & 0 \end{pmatrix} \mathcal{C} \tag{3.74}$$

gives

$$\mathcal{T}_{\rm AB}\hat{\Psi} = \mathbf{C}\begin{pmatrix} 0 & 1 & 0 & 0 \\ 1 & 0 & 0 & 0 \\ 0 & 0 & 0 & -1 \\ 0 & 0 & -1 & 0 \end{pmatrix}\begin{pmatrix} \Psi_{\rm A} \\ \Psi_{\rm B} \\ -\Psi'_{\rm B} \\ \Psi'_{\rm A} \end{pmatrix} = \mathcal{C}\begin{pmatrix} \Psi_{\rm B} \\ \Psi_{\rm A} \\ -\Psi'_{\rm A} \\ \Psi'_{\rm B} \end{pmatrix} \tag{3.75}$$

that is, it swaps A $\leftrightarrow$ B. One can immediately note that the above Dirac Hamiltonian Eq. 3.62 is not invariant in respect to the time reversals either $\mathcal{T}_K$ or $\mathcal{T}_{\rm AB}$ because neither

$$\mathcal{T}_K \hat{H} \mathcal{T}_K^{-1} \neq \hat{H} \tag{3.76}$$

nor $\mathcal{T}_{\rm AB}\hat{H}\mathcal{T}_{\rm AB}^{-1} \neq \hat{H}$. However, after flipping both the pseudospins one gets

$$\mathcal{T}_K\mathcal{T}_{\rm AB}\hat{H}\mathcal{T}_{\rm AB}^{-1}\mathcal{T}_K^{-1} = \hat{H}\,(-\mathbf{p}, -\mathbf{A}) = -\hat{H} \tag{3.77}$$

which means that mere time invariance is broken for the $\mathcal{T}_K\mathcal{T}_{\rm AB}\hat{H}\mathcal{T}_{\rm AB}^{-1}\mathcal{T}_K^{-1}$ antiunitarian transformation. Nevertheless, the parity–time (**PT**) invariance is indeed maintained, which becomes evident after an additional use of the parity transformation as

$$CP\hat{H}\,(-\mathbf{p}, -e\mathbf{A})\;C^{-1}P^{-1} = \hat{H}\,(\mathbf{p}, -\mathbf{A}) \tag{3.78}$$

If in Eq. 3.78 one sets the magnetic field to zero (i.e., $\mathbf{A} = 0$) Hamiltonian (H-Dirac) appears actually to be the $\mathcal{PT}$ invariant. Below we follow Brey and Fertig (2006a, 2006b) and consider the edge states of a graphene stripe with zigzag edges subjected to a transverse electric field $\mathbf{E} = \{0,\ E_y,\ 0\}$. The edge states are formed due to interference of evanescent waves for which the electron momentum is represented as $\mathbf{p} = \{k,\ iq\}$. Besides, we add into the Dirac Hamiltonian Eq. 3.62 the symmetry-breaking term $\Delta\,(\hat{\sigma}_z \otimes \hat{\tau}_z)$

$$H_{\rm d} = H_0 + H_\Delta = v\begin{pmatrix} 0 & -q+ik & 0 & 0 \\ -q-ik & 0 & 0 & 0 \\ 0 & 0 & 0 & q+ik \\ 0 & 0 & q-ik & 0 \end{pmatrix} + \Delta\begin{pmatrix} 1 & 0 & 0 & 0 \\ 0 & -1 & 0 & 0 \\ 0 & 0 & -1 & 0 \\ 0 & 0 & 0 & 1 \end{pmatrix} \tag{3.79}$$

The above additional term causes opening an *energy gap* 2Δ in the excitation spectrum

$$\varepsilon_{1,2} = v\mathbf{p}_{1,2} = \pm v\sqrt{k^2 + q^2 + \left(\frac{\Delta}{v}\right)^2} \tag{3.80}$$

The $\mathcal{PT}$ inversion acts on H_d as follows. First we apply the $\mathcal{T}$ inversion only which flips both the one-half pseudospins

$$\mathbf{T}_K\mathbf{T}_\text{AB}\hat{H}_\text{d}\mathbf{T}_\text{AB}^{-1}\mathbf{T}_K^{-1} = \mathbf{T}_K\mathbf{T}_\text{AB}\begin{pmatrix} \Delta & -v(q-ik) & 0 & 0 \\ -v(q+ik) & \Delta & 0 & 0 \\ 0 & 0 & \Delta & v(q+ik) \\ 0 & 0 & v(q-ik) & \Delta \end{pmatrix}\mathbf{T}_\text{AB}^{-1}\mathbf{T}_K^{-1}$$

$$= \begin{pmatrix} -\Delta & -v(ik+q) & 0 & 0 \\ -v(q-ik) & \Delta & 0 & 0 \\ 0 & 0 & \Delta & v(q-ik) \\ 0 & 0 & v(q+ik) & -\Delta \end{pmatrix}$$

At the next step, we perform the parity inversion which gives $k \to -k$, and $W \to -W$ (whereas the sign of the y component of the electron momentum $\mathbf{q}$ corresponding to evanescent waves remains unchanged). Then the combined $\mathcal{PT}$ symmetry is restored as

$$\mathbf{PT}_K\mathbf{T}_\text{AB}\hat{H}_\text{d}\mathbf{T}_\text{AB}^{-1}\mathbf{T}_K^{-1}\mathbf{P}^{-1}$$

$$= \begin{pmatrix} \Delta & -v(-ik+\mathbf{q}) & 0 & 0 \\ -v(\mathbf{q}+ik) & -\Delta & 0 & 0 \\ 0 & 0 & -\Delta & v(\mathbf{q}+ik) \\ 0 & 0 & v(\mathbf{q}-ik) & \Delta \end{pmatrix} = H_\text{d} \tag{3.81}$$

In the above Eq. 3.82 we have taken into account that momentum $\mathbf{D} = e\mathbf{W}$ of a dipole exposed to an electric field $\mathbf{E} = \{0,\ E_y,\ 0\}$ is accompanied by an energy increase $\Delta = eWE_y$ for electrons and the energy decrease $-\Delta = -eWE_y$ for holes. The presence of the term $\propto \Delta$ breaks the $\mathbf{T}$ symmetry of the Dirac Hamiltonian, which is then restored after the parity inversion. The combined $\mathbf{PT}$ symmetry is indeed maintained, though the time reversal and parity symmetry are broken when $\mathbf{P}$ or $\mathbf{T}$ acts separately from the other.

3.6 Heavy Chiral Fermions at Zigzag Edges of a Graphene Stripe

Electron excitation spectrum is computed as follows. The opposite zigzag edges of the G stripe are terminated by carbon atoms belonging to two distinct sublattices A and B (Gosalbez-Martinez et al. 2012, Soriano and Fernandez-Rossier 2012). In this geometry the electron wavefunction Ψ satisfies the BCs

$$\Psi_{\mathbf{A}}(0) = \Psi_{\mathbf{A}'}(0) = 0 \tag{3.82}$$

at $y = 0$ and

$$\Psi_{\mathbf{B}}(W) = \Psi_{\mathbf{B}'}(0) = 0 \tag{3.83}$$

at $y = W$. Here A and B refer to two different A and B sublattices, while the prime denotes the K′ point in the momentum space. At $V_{\mathbf{SG}} = 0$, there are two flat bands giving rise to a large value of the DOS directly at the Fermi energy and associated with the zigzag edge states (Brey and Fertig 2006a, 2006b). When $V_{\mathbf{SG}} \neq 0$, the resulting pseudospin polarization driven by the $\propto \Delta$ term in Eq. 3.82 yields an excitation spectrum

$$\varepsilon_{\pm} = \pm v\sqrt{k^2 + q^2 + \left(\frac{\Delta}{v}\right)^2} \pm U \tag{3.84}$$

where $\mathbf{k}$ and $\mathbf{q}$ are the longitudinal and transverse electron momentums. The sign of U in the last formula is defined by the $G_{\mathbf{R(L)}}$ back gates, whereas $U \equiv 0$ for $-L_0/2 < x < L_0/2$. Thus, excitation spectrum of the electrons propagating along the G stripe ($q = 0$) acquires the energy gap 2Δ. The gap separates excitations with opposite $\pm$ one-half pseudospin polarization (Gosalbez-Martinez et al. 2012, Soriano and Fernandez-Rossier 2012). Physically it means that the electric charge is depleted on one zigzag edge, while it is simultaneously accumulated on the other. The field increases the electron energy by Δ, while the energy of the hole is decreased by the same amount. As a result, an electric dipole is formed as soon as $V_{\mathbf{SG}} = E_y W \neq 0$. Then, solving Eq. 3.68 in conjunction with the BC yields the transversal quantization condition in the form

$$k = \frac{q_\varepsilon}{\tan q_\varepsilon W} \tag{3.85}$$

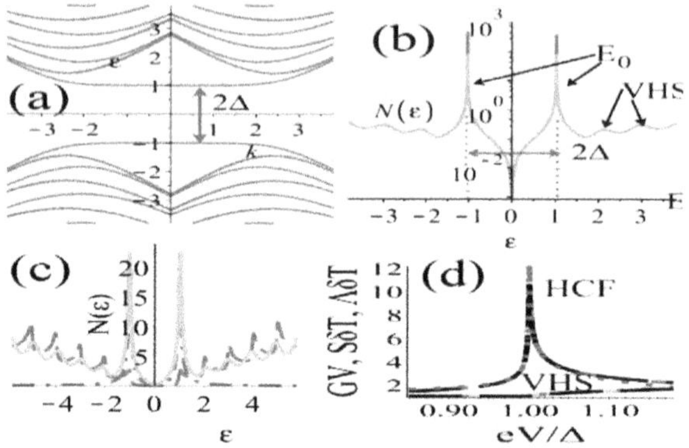

Figure 3.5 (a) The energy subbands $E(k)$ in the electron spectrum of a graphene stripe G with zigzag edges, as computed from Eqs. 3.2 and 3.3. (b) Density of electron states N_ε for three inelastic scattering rates $\gamma_{1,2,3} = 0.01$, 0.002, and 0.0004 in units of Δ. The two sharp peaks constitute the heavy chiral fermion (HCF) excitations arising at energies $\pm\Delta$. The corresponding peak height $N(\varepsilon = \pm\Delta)$ depends on γ. It exceeds the Van Hove singularity (VHS) peaks N_{VHS} by the big factors $N(\varepsilon = \pm\Delta)/N_{\text{VHS}} = \sqrt{m^*/m_e} = 10$ and 100. (c) The HCF singularity in the electron DOS $N(\varepsilon)$ at energy $\varepsilon = \Delta$ (green solid curve) as compared to the VHS singularities at $\varepsilon = \Delta_{\text{VHS}}^{(n)}$ (red dash curve) for $\gamma = 0.05\Delta$. Blue dot-dash curve shows the fitting function $1/(\varepsilon - \Delta)^{1.3}$. (d) A comparison plot showing an increase of thermoelectric current $G(V)$, the Seebeck voltage $S\delta T$, and the electron heat current $\Lambda_e\delta T$ when the source–drain voltages V_{SD} match either HCF (correspondingly shown by cyan, red, and black curves) or VHS (shown by magenta, green, and blue curves) singularities.

where

$$q_\varepsilon = \pm\sqrt{\left(\frac{\varepsilon \mp U}{v}\right)^2 - k^2 - \left(\frac{\Delta}{v}\right)^2} \tag{3.86}$$

Equations 3.86 and 3.87 actually determine the energy subbands when they are solved with respect to the excitation energy ε. Thus, for $V_{\text{SG}} \neq 0$, the $\text{HCF}_{\text{e(h)}}$ level splitting 2Δ is finite and the graphene zigzag stripe G is a band insulator with the pseudospin polarization. The steady-state energy spectrum of the G stripe numerically computed from Eqs. 3.86 and 3.87 is shown in Fig. 3.5a. One can see that $E(k)$ curves at $k = 0$ feature two flat bands $\varepsilon = \pm\Delta$, which represent the highest occupied and lowest unoccupied bands. The two flat bands cause very sharp HCF singularities in the DOS, which emerge at energies $\varepsilon = \pm\Delta$ (see Fig. 3.5a–c). An approximate analytical expression for the DOS is obtained from Eqs. 3.86 and 3.87 in the vicinity of HCF singularities $\varepsilon = \pm\Delta$ in the form $\propto 1/(\varepsilon - \Delta)^{\alpha}$. Here $\alpha \simeq 1.3$ is obtained by using the last formula to fit the numeric solution of Eqs. 3.86 and 3.87. Series of weaker periodic peaks in Fig. 3.5b taking place at $|\varepsilon| > \Delta$ represent VHSs. The DOS divergences are smoothed by setting $\varepsilon \to \varepsilon + i\gamma$. The effective HCF bandwidth γ at the HCF singularity is actually the rate of inelastic electron–phonon collisions, which is typically low inside the graphene stripe. The flat bands correspond to a very high effective electron mass

$$m^* = (10^2 - 10^5) \times m_{\mathrm{e}} >> m_{\mathrm{e}} \tag{3.87}$$

which we evaluate as

$$m^* \simeq \frac{\hbar^2}{\gamma a^2} \tag{3.88}$$

where a is the graphene lattice constant (Kittel 2005). The transport experiments (Tan et al. 2007, Bolotin et al. 2008a, 2008b) performed on annealed graphene samples give the electron mean free path $l_{\varepsilon} = \hbar v_{\mathrm{F}}/\gamma = 3.3$ μm, which corresponds to $\gamma = 0.2$ meV or $\gamma = (0.01 - 0.001) \times \Delta$. In principle, in "clean" graphene samples one can get the electron mean free path $l_{\varepsilon} \geq 15$ μm, which corresponds to $\gamma \leq 0.04$ meV. Then, applying the split gate voltage $V_{\mathbf{SG}} \neq 0$, one might achieve, for example, $\Delta = 100$ meV. This gives $\gamma/\Delta = 0.0004$ which in turn yields as much as

$$\frac{m^*}{m} \simeq 3 \times 10^4 \tag{3.89}$$

The VHS spacing

$$\Delta_{\mathbf{VHS}} \simeq \frac{h v_{\mathbf{F}}}{\pi W} \tag{3.90}$$

in graphene stripes by width $W \simeq 10$ nm is ~0.4 eV. This corresponds to temperatures ~4600 K. Such magnitude of spacing much exceeds the temperature broadening, which for temperatures of interest $T \approx 300 - 600$K is always much narrower than 4600 K. Therefore, the main contribution at temperatures $T < 600$ K comes from the lowest subband. It might be different for wider graphene stripes with width $W \geq 100$ nm where the VHS spacing is below ~300 K. However, the contribution of HCF state into ZT is always several orders of magnitude stronger than the impact of the VHSs of higher energy bands.

The aforesaid suggests that a G stripe with zigzag edges produces much stronger HCF singularities in the DOS than conventional VHS peaks can do, as illustrated in Fig. 3.5c. On the one hand, the VHS peaks in the DOS have the form

$$\rho(\varepsilon) \sim \frac{1}{\sqrt{\varepsilon - n\Delta_{\mathbf{VHS}}}} \tag{3.91}$$

where n is an integer. On the other hand, the HCF peaks are considerably sharper

$$\rho(\varepsilon) \sim \frac{1}{(\varepsilon - \Delta)^{1.3}} \tag{3.92}$$

Thus, a conventional VHS in the DOS causes only minor anomalies in the electric conductance and Seebeck coefficient, as illustrated by Fig. 3.5d. However, when the source–drain bias voltage $V_{\mathbf{SD}}$ matches Δ/e, a much stronger HCF singularity gives a sizable increase of the electric transport coefficients G, $\mathbf{S}$ and Λ_e along the G stripe, as is illustrated in Fig. 3.5d. Thus, the condition $V_{\mathbf{SD}} = \Delta/e$ ensures a strong thermoelectric effect.

Problems

3-1. Explain why trigonal warping in bilayer graphene occurs at low energies.

3-2. Explain which properties of bilayer graphene are characterized by winding numbers.

3-3. Determine what is the effect of rotation of one graphene layer on the other in bilayer graphene? How is the mutual rotating of the two atomic planes in bilayer graphene expressed in the electron excitation spectrum?

3-4. Discuss the basic difference between two cases of chiral tunneling in monolayer graphene on the one hand and bilayer graphene on the other hand.

3-5. How should we measure the energy gap in the electron excitation spectrum of bilayer graphene when a finite bias voltage is applied between the atomic layers?

3-6. Explain why for chiral fermions propagating in a stripe of monolayer graphene the time reversal symmetry is violated when applying a lateral electric field. What type of symmetry is preserved instead?
3-7. What makes heavy chiral fermion (HCF) singularities shown in Fig. 3.5b,c sharper than the Van Hove singularities (VHSs) in the electron density of states (DOS)?

References

Ando, T. (2005). Theory of electronic states and transport in carbon nanotubes. *J Phys Soc Jpn*, **74**(3): 777–817.

Bolotin, K. I., K. J. Sikes, J. Hone, H. L. Stormer, and P. Kim (2008). Temperature-dependent transport in suspended graphene. *Phys Rev Lett*, **101**(9).

Bolotin, K. I., K. J. Sikes, Z. Jiang, M. Klima, G. Fudenberg, J. Hone, P. Kim, and H. L. Stormer (2008). Ultrahigh electron mobility in suspended graphene. *Solid State Commun*, **146**(9–10): 351–355.

Brandt, N. B., V. N. Davydov, V. A. Koulbachinskii, and O. M. Nikitina (1988). Galvanomagnetic properties of graphite-intercalation compounds of acceptor type. *Fiz Nizk Temp*, **14**(4): 348–356.

Brandt, N. B., V. A. Kulbachinskii, O. M. Nikitina, V. V. Avdeev, V. Y. Akim, S. G. Ionov, and K. N. Semenenko (1988). Electron properties of heterointercalated graphite compounds of the 1st stage C10cucl2.3,6 Ici. *Vestn Mosk Univ, Ser 3: Fiz, Astron*, **29**(3): 64–68.

Brey, L., and H. A. Fertig (2006a). Edge states and the quantized Hall effect in graphene. *Phys Rev B*, **73**(19).

Brey, L., and H. A. Fertig (2006b). Electronic states of graphene nanoribbons studied with the Dirac equation. *Phys Rev B*, **73**(23).

Datta, S. (1992). Exclusion-principle and the Landauer–Büttiker formalism. *Phys Rev B*, **45**(3): 1347–1362.

Dresselhaus, M. S., and G. Dresselhaus (2002). Intercalation compounds of graphite. *Adv Phys*, **51**(1): 1–186.

Geim, A. K., and K. S. Novoselov (2007). The rise of graphene. *Nat Mater*, **6**(3): 183–191.

Gosalbez-Martinez, D., D. Soriano, J. J. Palacios, and J. Fernandez-Rossier (2012). Spin-filtered edge states in graphene. *Solid State Commun*, **152**(15): 1469–1476.

Katsnelson, M. I., and K. S. Novoselov (2007). Graphene: New bridge between condensed matter physics and quantum electrodynamics. *Solid State Commun*, **143**(1–2): 3–13.

Katsnelson, M. I., K. S. Novoselov, and A. K. Geim (2006). Chiral tunnelling and the Klein paradox in graphene. *Nat Phys*, **2**(9): 620–625.

Keldysh, L. V. (1965). Diagram technique for nonequilibrium processes. *Sov Phys JETP: USSR*, **20**(4): 1018.

Kittel, C. (2005). *Introduction to Solid State Physics*. Hoboken, NJ, Wiley.

Manes, J. L., F. Guinea, and M. A. H. Vozmediano (2007). Existence and topological stability of Fermi points in multilayered graphene. *Phys Rev B*, **75**(15).

McCann, E., D. S. L. Abergel, and V. I. Fal'ko (2007a). Electrons in bilayer graphene. *Solid State Commun*, **143**(1–2): 110–115.

McCann, E., D. S. L. Abergel, and V. I. Fal'ko (2007b). The low energy electronic band structure of bilayer graphene. *Eur Phys J*, **148**: 91–103.

McCann, E., K. Kechedzhi, V. I. Fal'ko, H. Suzuura, T. Ando, and B. L. Altshuler (2006). Weak-localization magnetoresistance and valley symmetry in graphene. *Phys Rev Lett*, **97**(14).

Novoselov, K. S., A. K. Geim, S. V. Morozov, D. Jiang, M. I. Katsnelson, I. V. Grigorieva, S. V. Dubonos, and A. A. Firsov (2005). Two-dimensional gas of massless Dirac fermions in graphene. *Nature*, **438**(7065): 197–200.

Novoselov, K. S., E. McCann, S. V. Morozov, V. I. Fal'ko, M. I. Katsnelson, U. Zeitler, D. Jiang, F. Schedin, and A. K. Geim (2006). Unconventional quantum Hall effect and Berry's phase of 2 pi in bilayer graphene. *Nat Phys*, **2**(3): 177–180.

Shafranjuk, S. E. (2009a). Directional photoelectric current across the bilayer graphene junction. *J Phys: Condens Mater*, **21**(1).

Shafranjuk, S. E. (2009b). Reversible heat flow through the carbon tube junction. *EPL*, **87**(5).

Soriano, D., and J. Fernandez-Rossier (2012). Interplay between sublattice and spin symmetry breaking in graphene. *Phys Rev B*, **85**(19).

Tan, Y. W., Y. Zhang, K. Bolotin, Y. Zhao, S. Adam, E. H. Hwang, S. Das Sarma, H. L. Stormer, and P. Kim (2007). Measurement of scattering rate and minimum conductivity in graphene. *Phys Rev Lett*, **99**(24).

Chapter 4

Phonons and Raman Scattering in Graphene

Phonons, the quantized oscillations of a crystal lattice, have attracted strong interest among physics and engineering communities who study graphene. On the one hand, attention is paid to acoustic phonons which are responsible for the heat transport in graphene at room temperatures. On the other hand, optical phonons are utilized in Raman experiments with few-layer graphene when counting the number of atomic planes. Elucidation of the fundamental difference between the phonon spectra, energy dispersion, and scattering rates in 2D and 3D systems represents another subject of an intensive research work. There is a remarkable distinction between the phonons which exist in the 2D crystals on the one side and the phonons existing in the 3D bulk crystals on the other side. The former system is represented with graphene or with the basal planes of graphite, while the latter case is related to the most of the 3D solid-state crystals. In this chapter we consider phonon transport in graphene which is subjected to unconventional boundary conditions. We also address the roles of in-plane and cross-plane phonon oscillations and discuss experiments which measure the thermal conductivity. We will focus our attention on defects introduced by various lattice imperfections like strains,

Graphene: Fundamentals, Devices, and Applications
Serhii Shafraniuk

ISBN 978-981-4613-47-7 (Hardcover), 978-981-4613-48-4 (eBook)
www.panstanford.com

defects, and impurity atoms which influence the phonon transport in graphene and in few-layer graphene by changing the thermal conductivity of those systems.

4.1 Phonon Modes in 2D Graphene

There are three acoustic phonon modes in the 2D graphene. Two of them are in-plane modes, longitudinal acoustic (LA) and transverse acoustic (TA) with a linear dispersion relation, whereas the third out-of-plane acoustic (ZA) mode has a quadratic dispersion relation. One consequence is that at low temperatures the out of plane mode which causes the temperature-dependent contribution ($\sim T^{1.5}$) to dominate over the T^2-dependent thermal conductivity contribution of the linear modes (Mingo and Broido 2005, Graf et al. 2007). Another unconventional feature of graphene is the negative value of Grüneisen parameters for the lowest TA–ZA modes (Mounet and Marzari 2005). This phenomenon was predicted by Lifshitz in 1952 and is quoted as a "membrane effect." In simply speaking it can be illustrated as behavior of a string which, when stretched, will vibrate with a smaller amplitude and a higher frequency. According to this analogy, the phonon frequencies for lowest TA–ZA modes increase when the in-plane lattice parameter become larger because atoms in the layer upon stretching will be less free to move in the out of plane direction. Contribution into the phonon modes characterized by a negative Grüneisen parameter emerges at low temperatures $T \leq T^*$ when the optical modes with positive Grüneisen parameters remain not excited. Then, as soon as $T \leq T^*$, the contribution from the phonon modes with negative Grüneisen parameters will dominate. Since the thermal expansion coefficient of a crystal is directly proportional to the Grüneisen parameter, it allows observing of the contribution of anomalous modes experimentally.

4.2 Phonon Spectra in Graphene and Graphene Nanoribbons

Many critically important parameters of solid-state crystals like the sound velocity, phonon density of states, phonon–phonon or electron–phonon scattering rates, and lattice heat capacity, as well as the phonon thermal conductivity, are determined by the phonon energy spectrum. Therefore, an adequate understanding of thermal and electrical properties of graphene, few-layer graphene (FLG), and graphene nanoribbons (GNRs) (Evans et al. 2010, Munoz et al. 2010, Savin and Kivshar 2010, Savin et al. 2010) motivate investigations of phonon energy spectra in these materials and structures (Maultzsch et al. 2004). Raman measurements serve as a powerful tool for studying the optical phonon properties. Raman spectroscopy not only provides values of parameters of the optical phonon spectra but it also allows us to determine the number of graphene layers, their quality, and their stacking order (Ghosh et al. 2010, Lui et al. 2011, Cong, Yu, Saito et al. 2011, Cong, Yu, Sato et al. 2011). Crystal lattice oscillations distort the graphene Brillouin zone which is sketched in Fig. 4.1a. The carbon atoms deviate from their equilibrium positions, whereas the interatomic interaction acts to restore the steady-state shape of the lattice cell. It causes crystal lattice oscillations elementary quanta of which are quoted as phonons. Energy dispersion of phonon oscillations in graphene and its allotropes had been explored by numerous research groups. The curves of the phonon energy dispersion along the Γ–M–K–Γ directions in graphite had been measured by X-ray inelastic scattering (Maultzsch et al. 2004). The phonon spectra in graphite were determined using Raman spectroscopy (Maultzsch et al. 2004, Mounet and Marzari 2005, Mohr et al. 2007). The same method was adopted for graphene (Lindsay and Broido 2011, Yan et al. 2008, Wang, Cao et al. 2009, Wang, Wang et al. 2009, Mazzamuto et al. 2011) and GNRs. (Qian et al. 2009, Aksamija and Knezevic 2011, Droth and Burkard 2011, Mazzamuto et al. 2011). The Raman spectroscopy approach is well suited for revealing specific features of particular phonon modes in graphene and its allotropes. Various theoretical approaches had been implemented by many research groups to compute the phonon energy dispersion in graphite, graphene, and GNRs. The methods included the continuum model (Qian et al. 2009, Droth and Burkard 2011), the Perdew–Burke–Ernzerhof model, generalized gradient approximation (GGA) (Maultzsch et al. 2004, Mounet and Marzari 2005, Mohr et al. 2007), first-order local density function approximation (LDA) (Dubay and

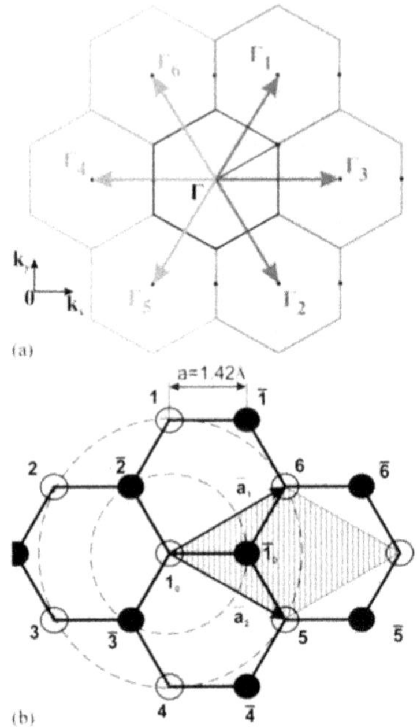

Figure 4.1 (a) Reciprocal graphene lattice. (b) Graphene crystal lattice. The unit cell has a rhombic shape shown as a shaded region. Adapted from Nika, Pokatilov et al. (2009) © APS.

Kresse 2003, Yan et al. 2008), fourth- and fifth-nearest-neighbor force constant (4NNFC and 5NNFC) approaches (Wirtz and Rubio 2004, Mohr et al. 2007), Born-von Karman or valence force field (VFF) model of the lattice dynamics (Falkovsky 2008, Nika, Pokatilov et al. 2009, Perebeinos and Tersoff 2009), and the Tersoff and Brenner potentials (Lindsay and Broido 2010) or the Tersoff and Lennard–Jones potentials (Lindsay et al. 2011). The aforementioned models have utilized different sets of the fitting parameter. The values of parameters were obtained by comparing the numeric results with the phonon dispersion derived in the independent experiments (Maultzsch et al. 2004, Mounet and Marzari 2005, Mohr et al. 2007).

A number of the parameters in theoretical models varies between 5 (Wirtz and Rubio 2004) and 23 (Wang, Wang et al. 2009). It depends on specifics of the model and on the number

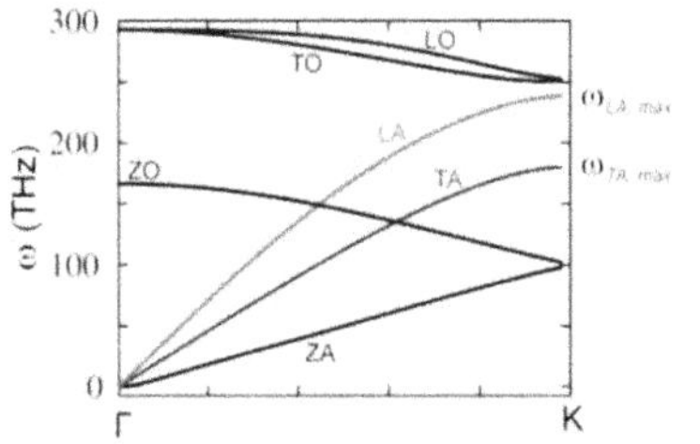

Figure 4.2 Phonon dispersion law $\omega(q)$ in graphene according to the valence force field model. Adapted from Ghosh et al. (2009) © DPG.

of the considered atomic neighbors. For example, VFF model of graphene suggested by Nika, Pokatilov et al. (2009) implements only six parameters. In this model, all interatomic forces are interpreted either as bond-stretching or bond-bending forces (Martin 1970). The model accounts for stretching and bending interactions with two out-of-plane and two in-plane atomic neighbors along with the doubled stretching–stretching interactions with the nearest in-plane neighbors. The honeycomb crystal lattice of graphene utilized in the model (Martin 1970) is presented in Figs. 4.2 and 4.3. The rhombic unit cell of graphene, which is shown as a dashed region in Fig. 4.1b, contains two atoms and is defined by two basis vectors $a_1 = a(3, \sqrt{3})\ /2$ and $a_2 = a\ (3,-\sqrt{3})\ /2$. Here $a = 0.142$ nm is the distance between two nearest carbon atoms. Six phonon polarization branches $s = 1...6$ in SLG are shown in Fig. 4.1b. Currently, there is only a qualitative agreement between different theoretical models. Furthermore, distinct models predict different phonon dispersion laws in the vicinity of Γ, M, or K points in the Brillouin zone. However, there is a certain general consensus concerning the phonon spectral branches as follows: (*i*) ZA and out-of-plane optical (ZO) phonons with the displacement vector along the Z axis; (*ii*) TA and transverse optical (TO) phonons, which correspond to the transverse vibrations within the graphene plane; and (*iii*) LA and longitudinal optical (LO) phonons, which correspond to the longitudinal vibrations within the graphene plane. Nonetheless, some of the models give the same results for of the LO–LA phonon frequencies (Wirtz and Rubio 2004, Mohr et al. 2007) and the ZO–TA phonon frequencies (Maultzsch et al. 2004, Mounet

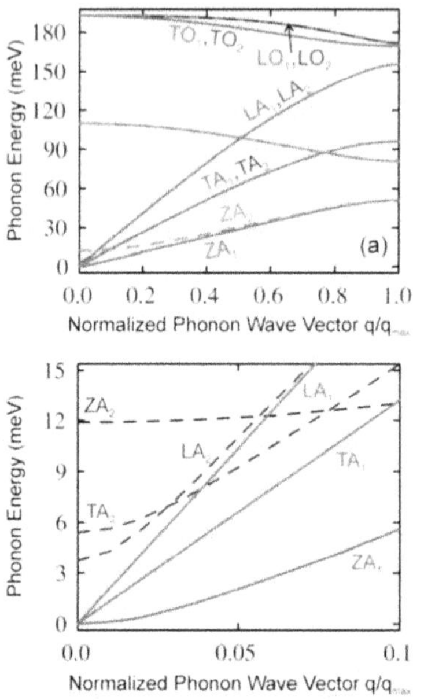

Figure 4.3 Bilayer graphene phonons computed within the valence force field model for the Γ–*M* direction (upper panel) and near the Brillouin zone center (lower panel). Adapted from Ghosh et al. (2010) © NPG.

and Marzari 2005, Mohr et al. 2007) at M point. Nevertheless, other models, predict nonequal frequencies for the mentioned phonon branches at M point (Falkovsky 2008, Nika, Pokatilov et al. 2009, Perebeinos and Tersoff 2009). The difference between computed phonon frequencies is pronounced in the thermal conductivity characteristics which vary in a wide range versus the phonon frequencies obtained numerically.

There are 6·*n* quantized phonon branches in *n*-layer graphene, because the unit cell of it contains 2·*n* atoms. Figure 4.3 shows the phonon dispersions in bilayer graphene. Coupling the long-wavelength phonons and quantization the low-energy part of spectrum with $q < 0.1q_{max}$ for LA, TA, LO, TO, and ZO phonons (see the lower panel of Fig. 4.3) and with $q < 0.4\ q_{max}$ for ZA phonons is caused by a weak van der Waals interaction between carbon lattice monolayers. *n*-layer graphene has much lower intrinsic thermal conductivity as compared with the atomic monolayer. It happens due to the change of the phonon spectrum as compared to single-layer graphene.

4.3 Phonon Transport in 2D Crystals

Phonon transport in the 2D materials has certain essentials. According to theoretical models (Saito and Dhar 2010), the intrinsic thermal conductivity exhibits a logarithmic divergence in the 2D systems, $K \approx \ln(N)$, and the power law divergence in 1D systems, $K \approx N\alpha$, where N is the number of atoms and $0 < \alpha < 1$ (Saito and Dhar 2010). Technically, the logarithmic divergence is avoided by introduction of the *extrinsic* scattering mechanisms such as scattering on defects or coupling to the substrate (Saito and Dhar 2010). Presence of a formal logarithmic divergence in the intrinsic thermal conductivity of the 3D material makes a difference from the 3D case. In the latter 3D case thermal conductivity is limited by the crystal anharmonicity and therefore has a finite value (Saito and Dhar 2010). The ambiguity in definition of the intrinsic thermal conductivity for graphene (Ghosh et al. 2008, Ghosh et al. 2009) and CNTs (Kim et al. 2001) is resolved if one considers a crystal of a given size. Because graphene atoms are oscillating in all three directions, it makes graphene not an ideal 2D crystal. However, in single-atomic-layer graphene there is a weak scattering of low-energy phonons by other phonons within the sheet. The thermal conductivity, therefore, depends on the graphene sheet size.

In this way, scattering of phonons on the sheet boundaries contributes to their relaxation in graphene. Another mechanism, which limits the mean free path of low-energy phonons, is related to the higher-order anharmonic processes. They involve Umklapp scattering and inclusion of the normal phonon processes with participation of more than three phonons. The normal phonon processes do not contribute to thermal resistance directly. However, they affect distribution of the phonon modes (Lindsay et al. 2010, Singh et al. 2011). In either case, even simplified models suggest that the graphene sample should be quite large (>10 μm) to make the thermal conductivity size independent. The thermal conductivity due to phonons is

$$K_{\mathrm{p}} = \sum_{j} C_J(\omega) v_j^2(\omega) \tau_j(\omega)\, d\omega \tag{4.1}$$

where j is the index of the phonon polarization branches, $\tau_j(\omega)$ is the phonon relaxation time, $v_j(\omega)$ is the phonon group velocity of the jth branch, $C_J(\omega)$ is the contribution to heat capacity from the jth branch, and $N_0(\omega_j) = 1/\left(\exp\left[\hbar\omega_j/k_{\mathrm{B}}T\right] - 1\right)$ is the Bose–Einstein phonon equilibrium distribution function. In Eq. 4.1 one should take into account the relevant phonon branches. They include two TA and one LA branches. If the thermal conductivity is affected by extrinsic processes like scattering of phonons on a rough boundary or by the phonon–defect scattering, it is regarded as extrinsic. Because the acoustic phonons are spatially confined in the nanostructures, this causes the phonon energy spectra to be quantized. Such discreteness of the phonon energy spectra generally causes a decreasing of the phonon group velocity. Altering the phonon energies, group velocities, and the density of states, which are complemented with the phonon scattering from boundaries, altogether modifies the thermal conductivity of nanostructures.

Specific details of phonon transport in graphene are well illustrated by the Klemens equation, which had been derived by Klemens (2001). According to Klemens (2001), the thermal conductivity of graphene limited by Umklapp processes is written as

$$K = \frac{\rho_{\mathrm{m}}}{2\pi\gamma^2} \frac{\bar{\upsilon}^4}{f_{\mathrm{m}} T} \ln\left(\frac{f_{\mathrm{m}}}{f_{\mathrm{B}}}\right) \tag{4.2}$$

Here f_{m} is the upper limit of the phonon frequencies defined by the phonon dispersion, $\bar{\upsilon}$ is an average value of the phonon group velocity, and

$$f_{\mathrm{B}} = \sqrt{\frac{M\bar{\upsilon}^3 f_{\mathrm{m}}}{4\pi\gamma^2 k_{\mathrm{B}} T L}} \tag{4.3}$$

is the low-bound cutoff frequency of acoustic phonons, which depends on the sample size. The sample's size L actually limits the phonon mean free path inside the graphene layer.

Equation 4.2 can be refined further by taking into account the actual maximum phonon frequencies and actual Grüneisen parameters γ_s (where index s = TA, LA marks the TA and the LA branches of phonons, respectively), that is, when they are determined separately for the TA and LA phonon branches, respectively. The Umklapp scattering rates depend on the Grüneisen parameter which determines degree of the lattice anharmonicity. When the restoring force is nonlinear in the displacement, the phonon frequencies ω_i change with the volume V. The Grüneisen parameter γ_s of an individual vibrational mode s then can be defined as (the negative of) the logarithmic derivative of the corresponding frequency ω_s

$$\gamma_s = -\frac{V}{\omega_s}\frac{\partial \omega_s}{\partial V} \tag{4.4}$$

The Grüneisen parameters of the sample are obtained by averaging the phonon mode-dependent $\gamma_s(\mathbf{q})$ (see Eq. 4.4) for all the relevant phonons (here $\mathbf{q}$ is the wavevector). The thermal conductivity of graphene is

$$K = \frac{1}{4\pi T \hbar} \sum_{s=\text{TA,LA}} \int_{q_{\min}}^{q_{\max}} \left[E_q^s v_s(q)\right]^2 \tau_{\text{U},s}^K(q) \left(-\frac{\partial N_0}{\partial E_q^s}\right) q dq \tag{4.5}$$

In the last Eq. 4.5 $E_q^s = \hbar\omega_s(q)$ is the phonon energy, $\hbar = 0.335$ nm is the thickness of graphene layer, and $v_s(q) = d\omega_s(q)/dq$ is the phonon group velocity. The steady-state Bose–Einstein function entering Eq. 4.5 is

$$N_0\left(E_q\right) = \frac{1}{e^{\frac{E_q^s}{k_B T}} - 1}, \tag{4.6}$$

whereas the Umklapp relaxation time $\tau_{\text{U},s}^K(q)$ depends on the three phonon modes. The time $\tau_{\text{U},s}^K(q)$ is determined with taking into account the different lifetimes for the LA and TA phonon branches as

$$\tau_{\text{U},s}^K(q) = \frac{1}{\gamma_s^2}\frac{M\bar{\upsilon}_s^2}{k_B T}\frac{\omega_{s,\max}}{\omega^2} \tag{4.7}$$

where s = TA, LA, $\bar{\upsilon}_s$ is the average phonon velocity for the branch s, $\omega_{s,\max} = \omega(q_{\max})$ is the maximum cutoff frequency for a given branch, M is the corresponding atomic mass. The description is simplified further considering that ZA phonons have low group velocity and a large (negative) Grüneisen parameter γ_{ZA} (Klemens 2001, Nika, Ghosh et al. 2009). Therefore contribution of the ZA phonons to thermal transport might be neglected (Nika, Ghosh et al. 2009). Equation 4.5 can be utilized to compute the thermal conductivity (see Figs. 4.4, 4.5) by incorporating an actual dependence of the phonon frequency $\omega_s(q)$ and the phonon velocity $v_s(q) = d\omega_s(q)/dq$ on the phonon wave number $\mathbf{q}$. The next simplification of the model implies the linear dispersion $\omega_s(q) = \bar{\upsilon}_s q$ which gives

$$K_{\mathrm{U}} = \frac{1}{4\pi T\hbar} \sum_{s=\mathrm{TA,LA}} \int_{\omega_{\min}}^{\omega_{\max}} \omega^3 \tau^K_{\mathrm{U},s}(\omega) \left(-\frac{\partial N_0(\omega)}{\partial \omega}\right) d\omega \tag{4.8}$$

The integration over ω is performed by substituting the model expression Eq. 4.7 into Eq. 4.8. Then one arrives at

$$K_{\mathrm{U}} = \frac{M}{4\pi T\hbar} \sum_{s=\mathrm{TA,LA}} \frac{\omega_{s,\max}\bar{\upsilon}_s^2}{\gamma_s^2} F(\omega_{s,\min}, \omega_{s,\max}) \tag{4.9}$$

where we have introduced the auxiliary function

$$\begin{aligned} F(\omega_{s,\min}, \omega_{s,\max}) &= \int_{\hbar\omega_{s,\min}}^{\hbar\omega_{s,\max}} \xi \frac{\exp(\xi)}{[\exp(\xi)-1]^2} d\xi \\ &= \left[\ln\{\exp(\xi)-1\} + \frac{\xi}{1-\exp(\xi)} - \xi\right]\Bigg|_{\hbar\omega_{s,\min}/k_{\mathrm{B}}T}^{\hbar\omega_{s,\max}/k_{\mathrm{B}}T} \end{aligned} \tag{4.10}$$

We also utilize the actual phonon dispersion in graphene (see Fig. 4.3) to define the dimensionless variable $\xi = \hbar\omega/k_{\mathrm{B}}T$, and the upper cutoff frequencies $\omega_{s,\max}$ in Eq. 4.10. We also use that

$$\omega_{\mathrm{LA,max}} = 2\pi f_{\mathrm{LA,max}}(\Gamma M) = 241\mathrm{THz} \tag{4.11}$$

and

$$\omega_{\mathrm{TA,max}} = 2\pi f_{\mathrm{TA,max}}(\Gamma M) = 180\mathrm{THz} \tag{4.12}$$

Equation 4.9 is even more simplified when assuming that $\hbar\omega_{s,\max} > k_{\mathrm{B}}T$ which is the case near room temperature. It gives

$$F(\omega_{s,\min}) \approx -\ln\left|e^{\Omega}-1\right| + \Omega\frac{e^{\Omega}}{e^{\Omega}-1} \tag{4.13}$$

where $\Omega = \hbar\omega_{\text{s,min}}/(k_{\text{B}}T)$. The heat transport in single-layer graphene differs from the heat transport in basal planes of the bulk graphite. In the latter case, there is some low-bound cutoff frequency ω_{min} above which the heat transport is approximately 2D. On the contrary, below ω_{min} there is a strong coupling with the cross-plane phonon modes. It leads to propagation of heat in all directions. Besides, it also reduces the contributions of these low-energy modes to heat transport along basal planes to negligible values. Onset of the cross-plane coupling, which is the ZO′ phonon branch near ~4 THz visible in the spectrum of bulk graphite serves as a reference point. During calculations, one takes into account that $\omega_{\text{min}} = \omega_{\text{ZO}'}\ (q = 0)$ for the ZO′ branch which helps to avoid the logarithmic divergence in the Umklapp-limited thermal conductivity.

The 2D nature of the phonon transport in atomically monolayer graphene causes the heat conduction here to have a different scenario compared with the bulk graphite. In particular, the onset of the cross-plane heat transport in the long-wavelength limit does not occur in graphene, thus, there is no ZO′ branch in the phonon dispersion of graphene (see Fig. 4.2). For such reasons, the low-bound cutoff frequencies $\omega_{s,\text{min}}$ for each s are determined from the condition that the mean free path of phonons is shorter than the flake size L (see Fig. 4.4)

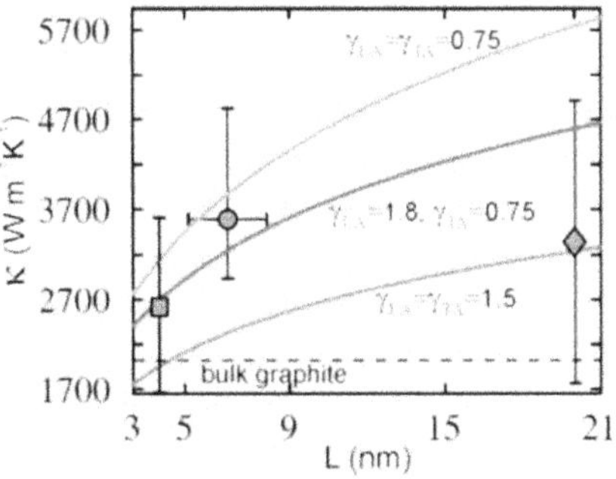

Figure 4.4 Numeric data for the room-temperature thermal conductivity of graphene versus the lateral size for different magnitudes of the Grüneisen parameter. The points are the experimental data from Balandin et al. (2008) and Ghosh et al. (2008) (circle) and Jauregui et al. 2010) (rhombus).

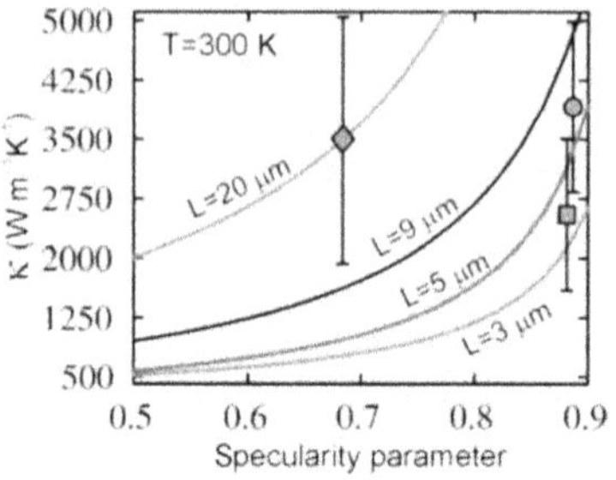

Figure 4.5 Numeric results for the room-temperature thermal conductivity of suspended graphene versus the specularity parameter p for phonon scattering from the flake edges. One can see that K_U is sharply dependent on the size of the graphene flakes. The points are experimental data from Balandin et al. (2008) and Ghosh et al. (2008) (circle) and Jauregui et al. (2010) (rhombus).

$$\omega_{s,\min} = \frac{\bar{\upsilon}_s}{\gamma_s}\sqrt{\frac{M\bar{\upsilon}_s}{k_B T}\frac{\omega_{s,\max}}{L}} \tag{4.14}$$

Another important factor is the dependence of the thermal conductivity K versus the graphene sample size like it takes place in Eq. 4.14 for $\omega_{s,\min}$. The sample size dependence shown in Fig. 4.4 is useful because it helps to avoid numeric errors when computing the thermal conductivity. The last equation Eq. 4.14 represents an example of a simple analytical approximation which allows computing of the thermal conductivity of graphene layer. Despite a simplicity of the approximation, the numeric results retain most significant features of the graphene phonon spectra including LA and TA branches characterized by distinct $\bar{\upsilon}_s$ and γ_s. The model also mimics well the 2D nature of heat transport in graphene in the whole frequency interval including zero phonon frequency $\omega_{\min} = \omega_{ZO'}(q = 0)$.

In Fig. 4.4 we show dependence of the graphene thermal conductivity versus the flake size L. To illustrate sensitivity of the result to Grüneisen parameters we plot the dependence for the averaged values of Grüneisen parameter $\gamma_{LA} = 1.8$ and $\gamma_{TA} = 0.75$ which are obtained from ab initio calculations along with several other close sets of $\gamma_{LA,TA}$. If the graphene flake size is small the dependence of the thermal conductivity K versus L is strong. It becomes less significant for flakes as soon as $L \geq 10$ μm. Computed characteristics agree well with known experimental data for the suspended, exfoliated, and CVD graphene. The horizontal line in Fig. 4.4 corresponds to thermal conductivity of the bulk graphite measured experimentally. One can see that in smaller L graphene flakes K becomes bigger than in the bulk graphite. The thermal conductivity at room temperature shown in Fig. 4.5 represents an *intrinsic* value which is restricted only by the three-phonon Umklapp scattering. Nevertheless it also accounts for the graphene flake size L which implicitly depends on Umklapp scattering via the long-wavelength (low-bound) cutoff frequency according to Eq. 4.14. In reality there is also phonon scattering on various lattice defects, as well as the scattering on the polycrystalline grain boundaries. Those additional processes contribute to the phonon part of thermal conductivity which can be merely included into the model using appropriate effective flake size $L^* \neq L$. An uncontrolled increase of a graphene flake with size L is prevented by scattering on lattice defects.

4.4 Momentum Diagram of Phonon Transport in Graphene

A simple model which was sketched in the former Subsection is not sufficiently accurate to describe the thermal transport in graphene. The Klemens approximation of the relaxation time defined by Eq. 4.7 ignores the 2D three-phonon Umklapp processes. The Umklapp processes in graphene can be of two types. One type represents an absorption of a phonon with the wavevector $\mathbf{q}'(\omega')$ belonging to heat flux by another phonon with a different wavevector $\mathbf{q}(\omega)$. The result of such an absorption process which involves two phonons is that the latter phonon leaves the state $\mathbf{q}$. The momentum and energy conservation laws in this case are written as follows:

$$\left.\begin{aligned} \mathbf{q} + \mathbf{q}' = \mathbf{b}_i + \mathbf{q}'' \\ \omega + \omega' = \omega'' \end{aligned}\right\}, i = 1 \ldots 3 \tag{4.15}$$

where $\mathbf{b}_i = \Gamma\Gamma_i, i = 1 \ldots 6$ is the vector of the reciprocal lattice (see Fig. 4.1a). There is also a different type of the phonon-scattering processes which corresponds to splitting of a single phonon with the wavevector $\mathbf{q}$ into two separate phonons with wavevectors $\mathbf{q}'$ and $\mathbf{q}''$. Another possibility is merging of two phonons having the wavevectors $\mathbf{q}'$ and $\mathbf{q}''$ with a consequent creation of a single phonon characterized by the wavevector $\mathbf{q}$. Then the momentum and energy conservation laws for the latter type of phonon scattering are written as

$$\left.\begin{array}{l} \mathbf{q}+\mathbf{b}_i=\mathbf{q}'+\mathbf{q}'' \\ \omega=\omega'+\omega'' \end{array}\right\}, i=4\ldots6 \qquad (4.16)$$

Thermal conductivity of graphene is computed considering all the possible three-phonon Umklapp processes which obey the Eqs. 4.15 and 4.16, whereas the actual phonon dispersions are observed (Nika, Pokatilov et al. 2009). The next step is to use Eqs. 4.15 and 4.16 for elucidating the pairs of phonon modes with wavevectors and frequencies ($\mathbf{q}'$, s') and ($\mathbf{q}''$, s'') for each phonon mode ($\mathbf{q}_i$, s) for which the conditions Eqs. 4.15 and 4.16 are observed.

In this way one obtains a *phase diagram* for all the three-phonon transitions which are permitted in the $\mathbf{q}'$ space (Nika, Pokatilov et al. 2009). Another approximation is to use the matrix element of the three-phonon interaction for the Umklapp scattering rates in a long-wave limit (Nika, Pokatilov et al. 2009)

$$\frac{1}{\tau_{ij}^{I,II}(s,q)}=\frac{\hbar\gamma_s^2(\mathbf{q})}{3\pi\rho\upsilon_s^2(\mathbf{q})}\sum_{s',s''}\iint\omega_s(\mathbf{q})\,\omega'_{s'}(\mathbf{q}')\,\omega''_{s''}(\mathbf{q}'')$$
$$\left\{N_0\left[\omega'_{s'}(\mathbf{q}')\right]\mp N_0\left[\omega''_{s''}(\mathbf{q}'')\right]+\frac{1}{2}\mp\frac{1}{2}\right\}$$
$$\times\delta\left[\omega_s(\mathbf{q})\pm\omega'_{s'}(\mathbf{q}')-\omega''_{s''}(\mathbf{q}'')\right]dq'_l dq'_\perp \qquad (4.17)$$

In Eq. 4.17, $\gamma_s(\mathbf{q})$ is the mode-dependent Grüneisen parameter, which is determined for each phonon wavevector and polarization branch, ρ is the surface mass density, $\mathbf{q}'_l$ and $\mathbf{q}'_\perp$ are the components of the vector $\mathbf{q}'$ parallel or perpendicular to the lines defined by Eqs. 4.15 and 4.16. The upper signs in Eq. 4.17 correspond to the first type of the phonon–phonon scattering processes, whereas the lower sign is related to the second type. Integration over the phonon momentum $\mathbf{q}_l$ is performed along the segment of curves defined by the conditions Eqs. 4.15 and 4.16, whereas $q_\perp$ is taken perpendicular to those segments. The straightforward integration along $q_{\perp}$ in Eq. 4.17 gives the 1D quadrature

$$\frac{1}{\tau_U^{I,II}(s,\mathbf{q})} = \frac{\hbar\gamma_s^2(\mathbf{q})\,\omega_s(\mathbf{q})}{3\pi\rho\upsilon_s^2(\mathbf{q})} \sum_{s',s''} \int_l \frac{\pm\left(\omega''_{s''} - \omega_s\right)\omega''_{s''}}{\upsilon_{\perp,s'}\left(\omega'_{s'}\right)} \times \left(N'_0 \mp N''_0 + \frac{1}{2} \mp \frac{1}{2}\right) dq'_l \tag{4.18}$$

Besides, one can also take into account the phonon scattering on the rough edges of graphene. It is accomplished with introducing an effective scattering time

$$\frac{1}{\tau_{B,j}} = \frac{\upsilon_j}{L}\frac{1-p}{1+p} \tag{4.19}$$

where p is the *specularity parameter* defined as probability of the phonon specular scattering at the boundary. Then, according to (Ziman 1965), the total phonon relaxation rate is given by

$$\frac{1}{\tau_{\text{tot}}(s,q)} = \frac{1}{\tau_{\text{U}}(s,q)} + \frac{1}{\tau_{\text{B}}(s,q)} \tag{4.20}$$

One can see how the phonon boundary scattering affects the room temperature thermal conductivity in Fig. 4.5. In Fig. 4.5 we show κ versus the specularity parameter of phonon boundary scattering for different sizes of graphene flakes as computed by using Eqs. 4.17 and 4.18. For comparison we also show the experimental data points for suspended, exfoliated, and CVD graphene.

4.5 Thermal Conductivity Due to Phonons in Graphene Nanoribbons

The nature of thermal transport in GNRs has attracted a strong attention of theoreticians and experimentalists. The role of lateral dimensions of the GNRs in the phonon transport properties is important both for fundamental knowledge and practical applications. Recent theoretical research had focused on phonon transport and heat conduction in GNRs with various lengths, widths, edge roughness, and defect concentrations. Along with the nonequilibrium Green's function method (Zhai and Jin 2011) and molecular dynamics (MD) simulations (Evans et al. 2010, Munoz et al. 2010), the BTE approach (Aksamija and Knezevic 2011) had been used as well. The obtained theoretical results suggest that the thermal conductivity is increased from ~1000 W/(mK) to 7000 W/(mK) for graphene ribbons with constant $L = 10$ nm and width W changing between 1 and 10 nm. The effect of roughness of GNR edges causes a sharp decrease of the thermal conductivity by a few orders of magnitude down from the value of K for graphene ribbons with perfect edges (Savin et al. 2010). The phonon part of thermal conductivity is also substantially decreased due to an isotopic superlattice modulation of a GNR or due to defects of the crystal lattices (Ouyang et al. 2009). However, uniaxial stretching of a GNR applied in the longitudinal direction causes an opposite effect. It might improve K as compared to its steady-state value. In particular, for the 5 nm armchair or zigzag GNR one might get an increase of the low-temperature thermal conductance up to 36% caused by stretching-induced convergence of phonon spectra to the low-frequency region (Zhai and Jin 2011).

One example of calculation results for the thermal conductivity of GNR with rough edges versus temperature is shown in Fig. 4.6 (Aksamija and Knezevic 2011). The roughness of the ribbon's edges is characterized by parameter Δ. The data in Fig. 4.6 correspond to the ribbon width $W = 5$ nm and the edge roughness parameter

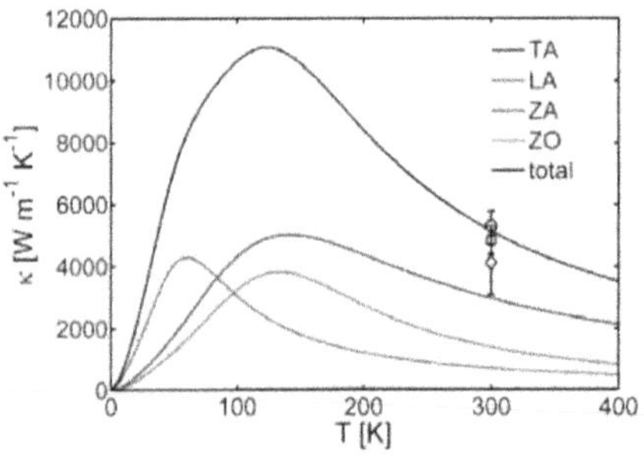

Figure 4.6 The contributions of TA, LA, ZA, and ZO phonon branches into the total thermal conductivity of a graphene nanoribbon. The ribbon width is $W = 5$ nm and the edge roughness parameter $\Delta = 1$ nm. The points are data of experiments by Balandin et al. (2008) and Ghosh et al. (2008). Adapted from Vidano et al. (1981) and Aksamija and Knezevic (2011) © APS.

$\Delta = 1$ nm. The model (Aksamija and Knezevic 2011) has taken into account the three-phonon Umklapp processes, mass defect, and rough edge scatterings. The model predicts the room temperature thermal conductivity $K \approx 5500$ W/(mK) for the GNR. Nonlinear thermal transport effects emerging from large temperature biases of GNRs which have rectangular and triangular shapes had been studied by Hu et al. (2011). They found that the short ($\sim$6 nm) GNRs with a rectangular shape exhibit a negative differential thermal conductance in a certain interval of the applied temperature difference. The nonlinear effect diminishes as soon as the length of the rectangular GNRs increases. A numeric computing (Xie et al. 2011) suggests that the thermal transport in zigzag GNRs is determined mostly by the cumulative influence of the edge roughness and by local defects. Unfortunately, there are not many experiments concerning the thermal transport in GNRs yet. As reported by Murali, Yang et al. (2009) where an electrical self-heating method had been used, the thermal conductivity at 700–800 K of sub-20-nm GNRs might exceed 1000 W/(mK).

4.6 Raman Scattering

Raman scattering is related to inelastic scattering of a photon. The effect had been predicted theoretically by A. Smekal in 1923 (Smekal 1923) and had been discovered experimentally by C. V. Raman and K. S. Krishnan in liquids (Raman and Krishnan 1928a, 1928b) and by G. Landsberg and L. I. Mandelstam in crystals (Landsberg and Mandelstam 1928a, 1928b).

Scattering of a photon on atoms or molecules typically is an elastic process which is quoted as Rayleigh scattering. It means that the photon has the same energy before and after the scattering. However, some photons (approximately 1 in 10 million) experience an inelastic scattering since the photon frequency after the scattering becomes different from the frequency of the incident photon. Raman scattering of a photon on a gas molecule is accompanied by the change of molecule energy. During the inelastic scattering process, the photon induces a transition between two quantized energy levels of that particular molecule. This kind of

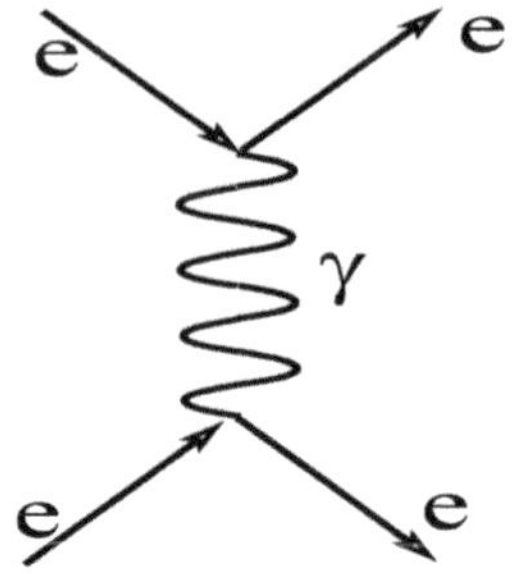

Figure 4.7 Feynman diagram which describes emitting a virtual photon due to scattering of the two electrons.

Raman scattering attracts attention of chemists who implement it in various spectroscopy applications.

Historically, the inelastic scattering of photons had been theoretically suggested by A. Smekal in 1923 (Smekal 1923). Experimental observation of the "molecular diffraction of light" in 1922 by Indian physicist C. V. Raman eventually led to discovery on 28 February 1928 of the radiation effect which bears his name. Initially the Raman effect had been reported by Raman and Krishnan (1928a, 1928b). The same phenomena also had independently been discovered on February 21, 1928, by Landsberg and Mandelstam. In 1930 Raman was awarded the Nobel Prize for the work on the scattering of light. In recognition of significance as a tool for analyzing the composition of liquids, gases, and solids, the Raman effect in 1998 was designated as the National Historic Chemical Landmark.

4.7 Role of the Degrees of Freedom

A chemical compound which consists of N atoms has $3N$ degrees of freedom in total. Number $3N$ corresponds to ability of each atom to be displaced in three different directions (x, y, and z). When considering molecules, the movement of the molecule is generally regarded as a whole. Thus, it is convenient to distinguish different portions of the $3N$ degrees of freedom. A common convention is to separate the molecular degrees of freedom into translational, rotational, and vibrational motions. Translational motion of the molecule as a whole along each of the three spatial dimensions also corresponds to the three degrees of freedom. Consequently, other three degrees of freedom are related to molecular rotations around the x, y, and z axes. However, there are certain exceptions from the above scheme. For instance, linear molecules can only have two rotations since the rotations along the bond axis do not change the positions of atoms in the molecule. Other remaining molecular degrees of freedom are related with vibrational modes. The latter modes correspond to molecular stretching and bending motions of chemical bonds. One evaluates the number of vibrational modes of a linear molecule as

$$3N - 3 - 2 = 3N - 5 \tag{4.21}$$

In the case of a nonlinear molecule the number of vibrational modes is calculated as

$$3N - 6 \tag{4.22}$$

4.8 Molecular Vibrations and Infrared Radiation

Molecular vibrations occur with frequencies which range between 10^{12} and 10^{14} Hz. These frequencies correspond to the THz domains (or far-infrared region) of the electromagnetic field spectrum. Every molecule of the matter has a finite mass and thus is characterized by a certain amount of vibrational energy. During collisions and interactions with other molecules belonging to the same sample, the vibrational energy of the molecule changes versus time. At the room temperature or below, the majority of molecules occupy the lowest energy state, which is also quoted as the ground state. Quantum states having the higher energies are regarded as excited states and they are populated by the minority of molecules. The occupation probability of a certain state by a molecule with a given energy is determined by the Boltzmann distribution. Simple evaluation using the Boltzmann distribution function confirms that at temperatures of typical spectroscopic experiments most of molecules belong to the ground vibrational state. One might excite the molecule from the ground state to an excited state when the molecule absorbs a photon having the corresponding energy. In the relevant experiment of infrared (IR) spectroscopy one illuminates a sample using the IR beam. Some of the IR photons are absorbed in the sample, whereas the others are transmitted through it. The ratio of the transmitted beam intensity to the incident beam intensity at a given wavelength indicates that some photons of the incident beam were absorbed by molecules of the sample during the vibrational transition. The vibrational energy of those molecules was increased to excite them into the higher energy state at the expense of the absorbed photon energy

$$E = h\nu \tag{4.23}$$

In the above Eq. 4.23, h is Planck's constant and ν is the frequency of radiation. Thus, if the frequency ω of the incident beam is known, one determines the vibrational transition energy E.

4.9 Various Processes of Light Scattering

The inelastic scattering of light can serve as an efficient probe of molecular vibrations. During an inelastic scattering process, the absorbed photon is then re-emitted as another photon with lower energy. In Raman spectroscopy, the high-intensity laser radiation with wavelengths either in visible or near-IR regions of the spectrum shines on a sample. Photons from the laser beam produce an oscillating polarization in the molecules, exciting them to a virtual energy state. Thus, Raman scattering corresponds to a molecule excitation by an incident photon to a higher vibrational mode. In Raman process, the difference in energies of the incident and scattered phonons is equal to the excitation energy of the new vibration mode.

The light-induced oscillation of molecular polarization couples with other polarizations of the same molecule, for example, vibrational and electronic excitations. The coupling between the molecular polarizations actually represents a mechanism which initializes the Raman scattering. Furthermore, the Raman scarring does not occur if the molecular polarization does not couple to the other possible polarization. In the last case, the vibrational state will not change since the scattered photon will have the same energy as the incident photon. The elastic scattering of that type is quoted as Rayleigh scattering.

When the molecular polarization couples to a vibrational state with a higher energy than the initial state, the energy of the incident photon differs from the energy of the scattered photon. The difference of the energies causes the molecular vibrations to be excited. In terms of the perturbation theory, the Raman effect is regarded as consequent absorption and subsequent emission of a photon via an intermediate quantum state of a material. The intermediate state can be either a "real," that is, stationary state or a virtual state.

4.10 Stokes and Anti-Stokes Scattering

There are different scenarios of the light scattering on gas, liquid, or solid-state molecules shown in Fig. 4.8. When energies of incident and scattered photons are the same, the scattering process is regarded as the Rayleigh scattering which happens with no energy exchange between the light and the molecules. When the scattered photon has lower energy than the incident photon, that is, the atom or molecule absorbs energy the process corresponds to the Stokes scattering. However, if the scattered photon has larger energy than the incident photon, then the process is quoted as the anti-Stokes scattering in which the atom or molecule loses energy (see energy diagram in Fig. 4.8).

Raman interaction yields two possible outcomes: (i) the material absorbs energy so that the emitted photon has a lower energy than the absorbed photon; this event is quoted as the Stokes Raman scattering; and (ii) the material loses energy during scattering the photon, which means that the emitted photon has a bigger energy than the absorbed photon. The latter scenario corresponds to the anti-Stokes Raman scattering. In either case the difference between energy of the absorbed and emitted photons corresponds to the energy difference between two resonant states of the material and is independent of the absolute energy of the photon.

Spectrum of scattered photons is called Raman spectrum. The optical spectrum shows intensity of the scattered light versus the

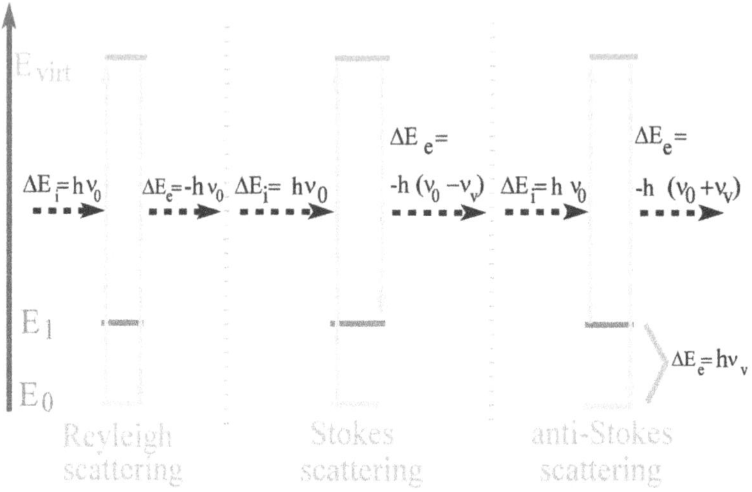

Figure 4.8 Various processes of light scattering. The first diagram on the left depicts Rayleigh scattering, when the photon energy is conserved because the incident and scattered photons have the same energy. Another process is Stokes scattering, when an atom or molecule absorbs energy. In the latter case the scattered photon has lower energy than the incident photon. The last process corresponds to the anti-Stokes scattering, during which an atom or a molecule loses its energy since the scattered photon has larger energy than the incident photon.

frequency difference $\Delta\nu$ between the transmitted and incident photons. Locations of the corresponding Stokes and anti-Stokes peaks are creating a symmetric pattern around $\Delta\nu = 0$. The frequency shifts are symmetric around zero frequency $\nu = 0$ because they are related to energy difference between the corresponding upper and lower resonant states. The difference in intensities is observed only for the pairs of features which are also temperature dependent. That happens because the corresponding intensities depend on populations of initial states of the material, which are altered versus the temperature. In the steady state, the upper state is less populated than the lower state. For these reasons, the Stokes transitions have the transition rate from the lower to the upper state higher than the anti-Stokes transitions which occur in the opposite direction. It causes the Stokes scattering peaks to be stronger than the anti-Stokes scattering peaks. Because the ratio of the two opposite processes is temperature dependent, it can be practically exploited for temperature measurement.

4.11 Raman Scattering vs. Fluorescence

There is a difference between the Raman effect on the one hand and fluorescence on the other hand. In the latter case the light of incidence is completely absorbed, therefore the system initially goes to an excited state, and then arrives at various lower states only after a certain resonance lifetime. Actually, both processes yield the same result. In either case, a photon with a frequency different from that of the incident photon is created, provided that a molecule either is excited to a higher- or is recombined to a lower-energy level. A fundamental feature of the Raman effect is that it occurs at any frequency of the incident light. Therefore, the Raman effect is not a resonant phenomenon as is the case of fluorescence. During experimental observations, the Raman peaks are separated from the excitation frequency by certain fixed intervals of frequency, whereas the fluorescence peak occurs at the specific frequency.

4.12 Selection Rules for Raman Scattering

Raman transitions between two quantum states are initialized only if a nonzero derivative of polarizability α with respect to the normal coordinate Q (i.e., the vibration or rotation) remains finite. Mathematically, it is formulated as

$$\frac{\partial \alpha}{\partial Q} \neq 0 \tag{4.24}$$

The above Eq. 4.24 means that Raman shift is optically activated only if the molecule is polarized. The distortion α of a molecule in an electric field, and therefore the vibrational Raman cross section are determined with the molecular polarizability. Molecular vibrations and rotations, which participate in Raman scattering, occur for molecules or crystals. The respective symmetries are determined using appropriate character table for the symmetry group. They are described in a variety of textbooks on quantum mechanics or group theory for chemistry.

4.13 Raman Amplification and Stimulated Raman Scattering

The spontaneous Raman scattering which has been described above occurs within random time intervals. It happens when a small part of incoming photons is scattered by the material. Another Raman scattering scenario takes place when a fraction of Stokes photons have previously been generated by spontaneous Raman scattering. If those Stokes photons remain inside the sample then the Raman scattering rate is increased above the spontaneous Raman scattering. That happens because the pump photons are converted more rapidly into additional Stokes photons. The same outcome happens if one deliberately illuminates the sample with a "signal light" consisting of the Stokes photons along with the original "pump light." It stimulates the Raman scattering as well. In either case the total Raman-scattering rate is increased. Thus, the more Stokes photons are present in the sample, the faster they multiply. The aforementioned processes tend to amplify the Stokes light in the presence of the pump light, which is utilized in Raman amplifiers and Raman lasers. The nonlinear-optical effect of the stimulated Raman scattering is described, for example, in terms of the third-order nonlinear susceptibility $\chi^{(3)}$.

4.14 A Requirement of Coherence

Let us consider two points A and B inside the light beam which are spaced by d. The external beam with wavelength λ excites Raman photons with a wavelength λ', which in general is different from λ. The corresponding phase difference between the two points is $\theta = 2\pi d(1/\lambda - 1/\lambda')$. If $\theta = \pi$, the amplitude of scattered light, becomes opposite to the incoming light amplitude. In this case the scattered beam of Raman photons is weak. The Raman scattering amplitude might be increased, for example, using appropriately oriented optically anisotropic crystals to ensure that $\lambda' = \lambda$ since the crossing of two beams limits the spacing d. Another method is regarded as impulsive stimulated Raman scattering (ISRS). In the latter approach, the pulse duration is shorter than any of the relevant time constants in the system. Although the spatial interference between the incident and Raman beams becomes impossible on such a short timescale, it nevertheless yields a finite frequency shift which appears in the form of the Hubble's redshift. In practice, because the ISRS is very weak, it requires to use femtosecond laser pulses. If the pulses are very short, the ISRS becomes observable, which cannot be accomplished using ordinary incoherent light.

4.15 Practical Applications

A very useful implementation of Raman spectroscopy is the material's analysis. One can identify the materials by measuring spectrum of the Raman-scattered beams. Thus, one determines, for example, the presence of molecular constituents and their state. The method of Raman spectroscopy is suitable for a precise analysis of a variety of materials, including gases, liquids, and solids. Furthermore, even such complex materials as biological organisms and human tissues can also be studied by the Raman spectroscopy approach. Besides, Raman spectroscopy allows us to measure, for example, the high-frequency phonon and magnon excitations in solid materials. Stimulated Raman transitions are also widely used for manipulating by the trapped ion's energy levels, and thus constitute the basic qubit states, whereas Raman amplification is used in optical amplifiers. Other applications of Raman spectroscopy include measuring the force constant and bond length for molecules which do not have an IR absorption spectrum.

Raman light detection and ranging (lidar) is used in atmospheric physics to measure the atmospheric extinction coefficient and vertical distribution of water vapor. The Raman lidar is an active, ground-based laser instrument for remote sensing, which measures vertical profiles of water vapor–mixing ratio and of several cloud- and aerosol-related quantities. Lidar is an optical analog of radar, which uses the pulses of laser radiation to probe the atmosphere.

The lidar system is fully computer automated, and can run unattended many days following a brief (~5 min) startup period.

4.16 Higher-Order Raman Spectra

Implementing the high-intensity continuous wave lasers allows us to obtain the broad bandwidth spectra by means of stimulated Raman spectroscopy. That approach exploits the process of the four-wave mixing, when wavelengths of the two incident photons are equal to each other, whereas the emission spectra emerge in two different bands which are separated from the incident beams with spacing of the photon energies. Initially, the Raman spectrum is composed of a spontaneous emission and then is amplified thereafter. Raman spectra of higher order are then generated by using Raman spectrum which serves as an initial input, thereby building a chain of new spectra with a decrease of their amplitude. The four-wave mixing is accomplished using an intensive pumping in long fibers. One complication is the intrinsic noise which emerges during an initial spontaneous process. Intrinsic noise is reduced, for example, either by using a feedback loop as in a resonator to stabilize the process, or by simply seeding a spectrum at the beginning. Raman amplification and spectrum generation are benefiting from a rapid evolution of the fiber laser field, and also from the progress in the development of transversal coherent high-intensity light sources which are used in the broadband telecommunication and imaging applications.

4.17 Raman Spectroscopy of Graphene

An ability to identify and characterize carbon materials serves as a major requirement during carbon research. That condition is critically important also for a variety of applications as well as for the lab and at mass production. Additionally, the characterization must be precise, instant, and nondestructive, as well as it should provide the maximum of structural and electronic information. All the listed requirements to the analyzing tool are satisfied in Raman spectroscopy. Therefore, Raman spectroscopy actually represents a critical research tool in diverse fields including physics, engineering, chemistry, and biology. Certainly, a vast majority of nowadays publications on carbon matter contain Raman analysis (Ferrari and Robertson 2004). Thus, Raman spectroscopy tends to become a mandatory element of the research activity within the field of graphene. In particular, the Raman approach helps to identify and determine the quality of graphene samples already at earliest stages of fabrication. When producing graphene, either with mechanical cleavage or with the epitaxy, chemical exfoliation, there appear all sorts of carbon species. The same problem emerges during the growth of carbon nanotubes. Unwanted side products and structural damages also emerge during structuring graphene into devices. To avoid the aforementioned complications it is required to monitor structural details. The latter task is well accomplished by Raman spectroscopy. The Raman spectroscopy method therefore serves as the standard tool when comparing the materials produced by different research groups. This is a commonly accepted practice in the field of nanotubes and amorphous and diamond-like carbons (Ferrari and Robertson 2004). For other carbon-based materials, the researchers implement various methods of optical characterization. In particular, carbon nanotubes are studied by the method of photoluminescence excitation spectroscopy, whereas amorphous and diamond-like carbons are examined using ellipsometry or X-ray photoelectron spectroscopy (XPS) (Patsalas et al. 2001). One can expect that in the growing field of graphene research, there will be a variety of alternative optical approaches including those which are more advanced than Raman spectroscopy.

Strong motivation to favoring the Raman method when studying carbon materials is a simplicity and handiness of the spectral interpretation. A few of very remarkable features of the Raman spectra emerge for all the carbon systems, regardless of whether it is a conjugated polymer or fullerene (Ferrari and Robertson 2004). One could mention several very intense bands in the 1000–2000 cm^{-1} region along with a few modulations of the second order. It is easy to distinguish hard amorphous carbon and metallic nanotubes just by comparing the relative positions, intensity, and shapes of spectral features (Ferrari and Robertson 2004). Definitely, a fundamental reason why Raman spectroscopy is always resonant for carbon materials, including graphene, is related to the peculiar dispersion of electrons. It makes the Raman method an excellent tool to accomplish the probing of electronic properties along with the lattice vibrations (Ferrari and Robertson 2004). For the aforementioned reasons, the carbon materials (Ferrari and Robertson 2004) and FLG (Gogotsi et al. 2000, Cancado, Pimenta, Neves, Dantas et al. 2004, Cancado, Pimenta, Neves, Medeiros-Ribeiro et al. 2004) have been studied by the method of Raman spectroscopy for about 40 years.

4.18 Kohn Anomalies, Double Resonance, and D and G Peaks

A similarity between Raman spectra of different carbon materials emerges as the so-called G and D peaks which for visible excitation are positioned near 1560 and 1360 cm^{-1}, respectively. One can notice how the similarity emerges in Fig. 4.9, where we show Raman spectra of graphite, metallic and semiconducting nanotubes, and high and low sp^3 amorphous carbons, all measured for visible excitation. The D peak position versus the excitation energy for graphite with defects is represented in Fig. 4.10 (Pocsik et al. 1998). A Raman spectrum peak close to 1060 cm^{-1} (T peak) had been

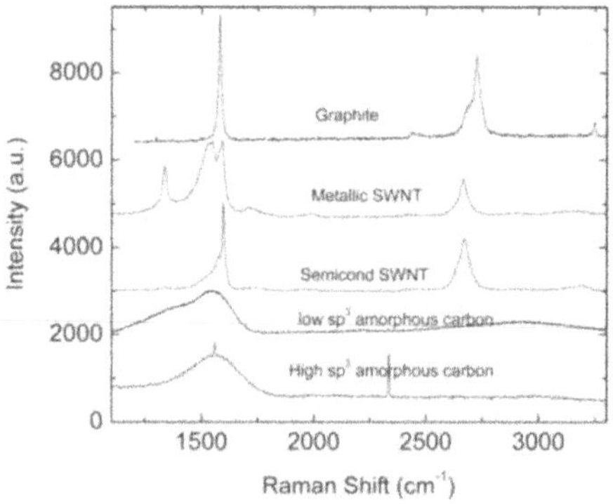

Figure 4.9 Raman spectra of graphite, metallic and semiconducting carbon nanotubes, and low and high sp^3 amorphous carbons. Adapted from Pocsik et al. (1998).

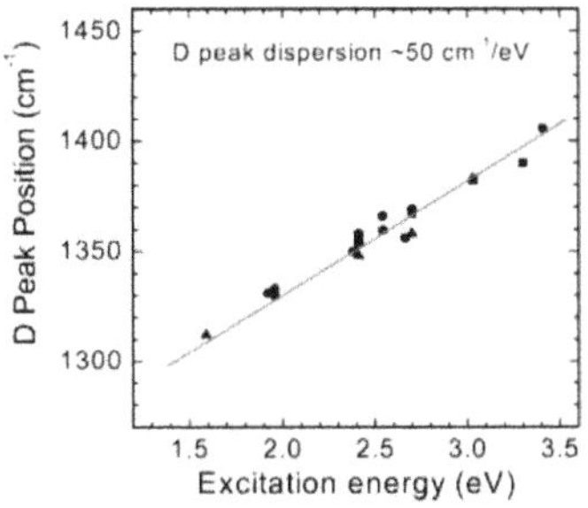

Figure 4.10 D peak dispersion versus excitation energy. Adopted from Pocsik et al. (1998).

reported by Ferrari and Robertson (2001a, 2001b) in amorphous carbons in ultraviolet (UV) excitation. In carbon films, due to a resonance taking place between the visible excitation and π states, the Raman spectra are also dominated by the sp^2 sites. Therefore, sp^2 vibrations strongly contribute into the visible Raman spectra even in highly sp^3 amorphous carbon samples. The diamond sp^3 peak at 1332 cm^{-1} is observed only for diamond or samples containing a sizable diamond phase (Ferrari and Robertson 2001a, 2001b). The Raman signature of the C–C vibrations for the amorphous sp^3 can only be observable for UV excitation because its cross section is negligible for visible excitations. The cross section for graphite at 514 nm is about 55 times larger as compared to the analogous cross section for diamond, whereas for hydrogenated amorphous carbon the cross section is about 230 times bigger than for the diamond.

The correspondence between the D and G peaks and the electronic bands which are present in all polyaromatic hydrocarbons is understood in terms of the so-called molecular picture of carbon material (Castiglioni et al. 2001, Castiglioni et al. 2004). According to the mentioned model, the G peak originates from the bond stretching of all pairs of sp^2 atoms in both rings and chains. The other D peak represents a signature of the breathing modes of sp^2 atoms in rings (Ferrari and Robertson 2000, Castiglioni et al. 2001). Nevertheless, there are still doubts concerning applicability of the "solid state"-like handling of the relevant bands. Initially the D peak was related to an A_{1g} breathing mode at K. It was assumed that the mode is activated because violation of the fundamental Raman selection rule $\mathbf{q} = 0$. The next step was to relate it with maxima in the vibrational density of states of graphite at M and K points (Nemanich and Solin 1979). However, the last assumption does not incorporate the dispersion of the D position consistently with photon energy (see Fig. 4.10; Pocsik et al. 1998). It fails to explain the presence of dispersive D peak overtone at about 2710 cm^{-1} taking place even when the D peak is absent. Neither it explains why the I_D/I_G ratio is dispersive (Pocsik et al. 1998). Besides, it is not clear why the D mode is more intense than other modes with smaller Δq despite the fact that they are closer to the Γ points and why the D mode is also observed in disordered graphite where the plane correlation length $L_a \approx 30$ nm. That is inconsistent with the Heisenberg indetermination principle, which limits participation of phonons by squeezing them to a much narrower Δq range around Γ (see, for example, Ferrari and Robertson 2000). In Fig. 4.11 we've sketched the double resonance for D_0 peak (close to Γ) and D peak (close to K) (Reich and Thomsen 2000a, 2000b, Thomsen et al. 2000, Thomsen and Reich 2000). Another hypothesis is that the D peak originates from a resonant Raman coupling (see, that is, Pocsik et al. 1998), which ensures a strong increase of Raman cross section for the phonon wavevector $\mathbf{q}$ immediately when $\mathbf{k} = \mathbf{q}$ (regarded as the "quasiselection rule") where $\mathbf{k}$ is the wavevector of electronic transition which is excited by the incident photon (Ferrari and Robertson 2000). Nonetheless, the "quasiselection rule" cannot explain why it works only for phonons which belong to a particular optical branch, whereas the other phonons are not involved in the process. The double resonance as an activation mechanism was suggested by Baranov et al. (1987), Reich and Thomsen (2000a, 2000b), and Thomsen and Reich (2000). In terms of the double resonance, the Raman scattering is represented as a fourth-order process, as sketched in Fig. 4.12 (Baranov et al. 1987, Reich and Thomsen 2000a, 2000b, Thomsen and Reich 2000). In Fig. 4.12 we show an activation process for the D peak: (i) an electron/hole pair is excited under the laser radiation; (ii) exchange of momentum $\mathbf{q} \approx \mathbf{K}$ comes from the electron–phonon scattering; (iii) alternatively it originates from scattering on defects; and (iv) the recombination of electrons and holes occurs. The condition of double resonance is fulfilled if the energy is conserved during all the transitions (Baranov

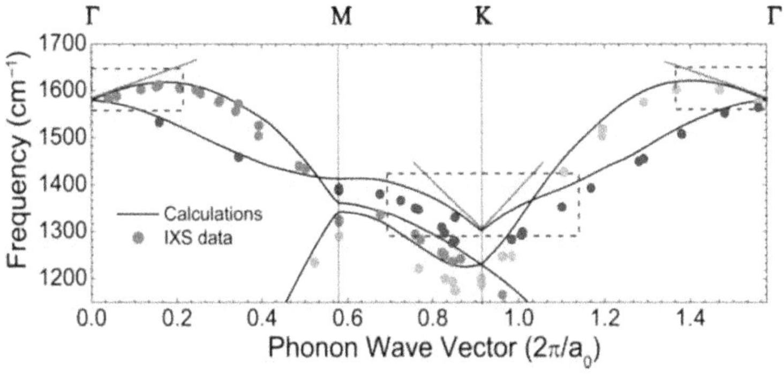

Figure 4.11 Calculated phonon dispersion of graphene (Piscanec, Lazzeri et al. 2004) compared to the experimental data on graphite from Maultzsch et al. (2004).

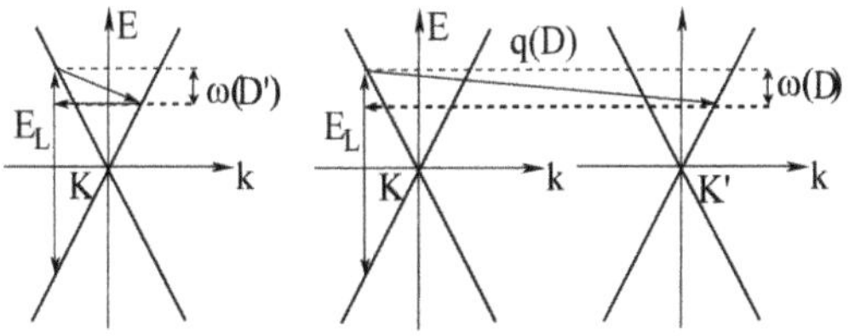

Figure 4.12 The scheme of double resonance for the D_0 peak (which is close to Γ) and for the D peak (which is close to *K*). The incident laser energy is equal to E_L.

et al. 1987, Reich and Thomsen 2000a, 2000b, Thomsen and Reich 2000). Another mechanism describes formation of the so-called D_0 peak due to activation of phonons with small **q**. It becomes possible due to intravalley scattering which is shown in Fig. 4.12. The D_0 peak is observed at $\sim$1620 cm^{-1} in graphite. An important circumstance for an adequate understanding of the Raman D peak is also a precise knowledge of the phonon dispersion near the K point. The exact shape of phonon branches determines their energy dispersion around the K point. In principle, if the electron–phonon coupling (EPC) is neglected, there should be four Raman active phonon branches around K point in graphene. In addition to the three branches shown in Fig. 4.11 there is a lower lying optical branch at $\sim$800 cm^{-1} which intersects the K point at lower lying **q**.

However, contrary to general expectations, only the D peak actually has a visible intensity (Tuinstra and Koenig 1970). According to the results obtained within the molecular approach (Mapelli et al. 1999, Ferrari and Robertson 2000, Ferrari and Robertson 2001a, 2001b), the D peak is related to the phonon branch starting from the **K**–A'_1 mode. The last assumption is based on the symmetry considerations and also on the fact that Raman cross section in aromatic molecules with enhancing size is large. Another possible origin of the D peak might emerge from the doubly degenerate linearly dispersive 1200 cm^{-1} E' mode at the **K** point as had been suggested by Piscanec et al. (2004). Thus, the D peak is actually related to the highest optical branch starting from the **K**–A'_1 mode (Tuinstra and Koenig 1970, Mapelli et al. 1999, Ferrari and Robertson 2000, Ferrari and Robertson 2001a, 2001b). Definitely, amongst the K phonons, the A'_1 branch is characterized by the largest EPC (Lazzeri et al. 2006). An additional argument is that according to Fig. 4.13 the A'_1 branch has a linear dependence versus the phonon wavevector **q**. The linear dispersion physically is related to the Kohn anomaly at **K**, which is consistent with the experimentally obtained dispersion of D peak in Fig. 4.10 (Pocsik et al. 1998). Generally speaking, the lattice vibrations are screened by the electronic states. Such

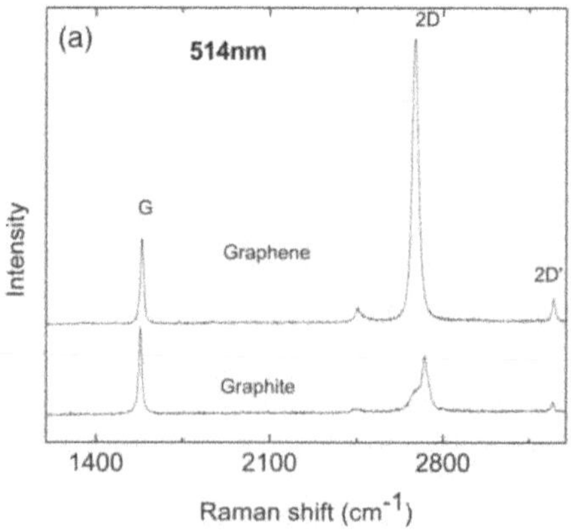

Figure 4.13 Raman spectra of graphene versus similar data for graphite measured at 514.5 nm. Adapted from Nemanich and Solin (1979) and Vidano et al. 1981).

screening of lattice oscillations by the conducting electrons can be dramatically changed close to certain points of the Brillouin zone in metals, which is determined by shape of the Fermi surface. The relevant modification of the phonon dispersion is regarded as the Kohn anomaly (Kohn 1959). The Kohn anomaly is determined by the condition $\mathbf{k}_2 = \mathbf{k}_1 + \mathbf{q}$ where the wavevector $\mathbf{q}$ links two electronic states $\mathbf{k}_1$ and $\mathbf{k}_2$, both positioned on the Fermi surface (Kohn 1959). The gap separating the occupied and empty electronic states in graphene vanishes at the two Brillouin zone points $\mathbf{K}$ and $\mathbf{K}'$. Taking into account the condition $\mathbf{K}' = 2\mathbf{K}$, the two mentioned points are connected by vector $\mathbf{K}$. This means that the Kohn anomalies take place either for $\mathbf{q} = \Gamma$ or when $\mathbf{q} = \mathbf{K}$. According to Piscanec et al. (2004), there are two essential Kohn anomalies in graphene at Γ–E_{2g} and K–A'_1, and both are visible in Fig. 4.11. The presence of two Kohn anomalies introduces an ambiguity when deriving the phonon branches at Γ and K points with the force constant approaches. Therefore, deduction of phonon branches at Γ and K points using their Raman spectra which makes a difference from semiconducting tubes (Piscanec et al. 2004). Besides, the Kohn anomalies in metallic nanotubes are much sharper than in graphite. In addition, the mentioned anomalies also cause a phonon softening, which is not the case for the phonon dispersions of folded graphite on the one hand and the metallic tubes on the other hand (Dubay and Kresse 2003, Piscanec et al. 2004, Lazzeri et al. 2006). One can conclude that the D peak originates from the LO phonons near the K point (Tuinstra and Koenig 1970, Tuinstra and Koenig 1970, Baranov et al. 1987, Thomsen and Reich 2000), it is active by double resonance, whereas a strong dispersion of the D peak versus the excitation energy is caused by the Kohn anomaly near the K point (Piscanec et al. 2004, Piscanec et al. 2007).

4.19 Derivation of Electron–Phonon Coupling from the Raman Line Width

Basic properties of graphene and carbon nanotubes like superconductivity, excited state dynamics, Raman spectra, phonon dispersions, and transport essentially depend on the EPC. In particular, the interaction between electrons and optical phonons in carbon nanotubes actually restricts the high-field ballistic transport (Yao et al. 2000, Park et al. 2004, Lazzeri et al. 2005, Perebeinos et al. 2005a, 2005b). Unfortunately, the known numeric results concerning the interaction of electrons with optical phonons in carbon nanotubes contradict each other (see, for example, Table II of Piscanec et al. 2007). Here we

The electron interaction with the optical phonons in graphene along with the contribution of Kohn anomalies can be described using a simple model to analyzing of known experimental data (Piscanec et al. 2004). Such a model includes the EPC for optical phonons of graphene, and, most importantly, it present a strategy for the experimental determination of EPCs. On the basis of density functional theory (DFT) calculations, Piscanec et al. (2004) suggested that the anomalies are pronounced as two sharp kinks in the phonon dispersion shown in Fig. 4.13. According to the model (Piscanec et al. 2004) the slope of the kinks $\eta^{\mathrm{LO}}(\Gamma, K)$ is proportional to the ratio of the square of the EPC matrix element and slope of the π bands at the K point (Piscanec et al. 2004)

$$\eta^{LO}(\Gamma) = \frac{\sqrt{3}a_0^2}{8M\omega_\Gamma}\frac{\alpha_{ep}^2(\Gamma)}{v_F} \tag{4.25}$$

$$\eta^{\mathrm{LO}}(K) = \frac{\sqrt{3}a_0^2}{8M\omega_K}\frac{\alpha_{\mathrm{ep}}^2(K)}{v_{\mathrm{F}}} \tag{4.26}$$

where $\alpha_{\mathrm{ep}}(\Gamma, K)$ is the EPC constant near the Γ or K points, respectively, M is the mass of carbon atom, $a_0 = 2.46$ Å is the graphite lattice spacing, $v_{\mathrm{F}} \simeq 10^6$ m/s is the graphite Fermi velocity, and $\hbar\omega_\Gamma = 196$ meV (Piscanec et al. 2004). Besides, the ratio of two slopes is

$$\frac{\eta^{\mathrm{LO}}(\Gamma)\,\omega_\Gamma}{\eta^K(K)\,\omega_K} = 2 \tag{4.27}$$

The experimentally measured phonon dispersion is compared with the numeric results (Piscanec et al. 2004) in Fig. 4.13. From Fig. 4.13 one can conclude that the inelastic X-ray scattering data (Maultzsch et al. 2004, Telg et al. 2004) for phonons around Γ agree well with the calculated phonon dispersion by Piscanec et al. (2004). One can fit the X-ray scattering data (Maultzsch et al. 2004, Telg et al. 2004) using Eq. 4.25. Then, on the one hand, the corresponding quadratic fit gives $\eta^{\mathrm{LO}}(\Gamma) \approx 133\ \mathrm{cm}^{-1}$ Å, which yields the experimental value $\alpha_{\mathrm{ep}}(\Gamma) \approx 39\ (\mathrm{eV/Å})^2$. On the other hand, the numeric DFT gives a close value $\sim 46\ (\mathrm{eV/Å})^2$; thus, the agreement between the experiment (Maultzsch et al. 2004, Telg et al. 2004) and theory (Piscanec et al. 2004) appears to be very good. Unlike to the Γ point vicinity, the experimental phonon dispersions near the K point are less definite. Then, Eq. 4.27 is utilized to evaluate $\alpha_{\mathrm{ep}}(K)$ from $\alpha_{\mathrm{ep}}(\Gamma)$. Another approach for obtaining α_{ep} consists in considering the G peak line width. The electron coupling with the optical phonons tends to broaden the Raman G band in graphite, graphene, and the G peak in metallic nanotubes (Piscanec et al. 2004, Piscanec et al. 2007). In an ideal crystal the broadening γ of the phonon energy $\hbar\omega$ depends on interaction between phonons and other elementary excitations. One can write $\gamma = \gamma^{\mathrm{an}} + \gamma_0^{\mathrm{ep}}$, where γ^{an} is the phonon energy broadening which is due to interaction with other phonons and γ_0^{ep} with electron–hole pairs. The first term γ^{an} corresponds to the anharmonic terms in the interatomic potential. The additional term γ_0^{ep} comes from the electron–phonon interaction and appears only in systems where the electron gap is zero. If for some reasons $\gamma^{\mathrm{an}} \approx 0$, or its value is known, then one determines the electron–phonon interaction by measuring the line width. Implementing the Fermi golden rule, one finds contribution of the electron–phonon interaction to the G peak as

$$\gamma_0^{\mathrm{ep}} = \frac{\sqrt{3}a_0^2}{4M}\frac{\alpha_{\mathrm{ep}}^2(\Gamma)}{v_{\mathrm{F}}^2}, \tag{4.28}$$

which is valid when the phonon energy and momentum are conserved (i.e., $q \leq \omega_\Gamma / v_{\mathrm{F}}$). If not, then one gets $\gamma_0^{\mathrm{ep}} = 0$. The energy and momentum conservation laws are observed close to the G peak of graphene and of undoped graphite. However, the conservation laws are violated for the double resonant D′ mode close to Γ. For these reasons, the D′ peak is considerably sharper than the G peak (Tan et al. 2000).

The broadening of G peak was experimentally studied using a single crystal graphite around $\sim 13\ \mathrm{cm}^{-1}$ (Piscanec et al. 2007). The experimental data indicate no increase of broadening the G peak in the 2–900 K temperature range. Considering that the resolution of the Raman spectrometer is about 1.5 cm^{-1}, one can state that the anharmonic contribution γ^{an} is below the spectral resolution. It means that $\gamma^{\text{ep}}(G) = 11.5$ cm^{-1}. On the other hand, implementing Eq. 4.27, one obtains $\alpha^2_{\text{ep}}(\Gamma) \approx 47$ (eV/Å)^2, which is well consistent with the results of the numeric DFT and with the fact that $\gamma^{\text{an}}(G)$ is low. After combining Eq. 4.28 with Eq. 4.25, one obtains the expression for the Fermi velocity in the form

$$v_{\text{F}} = \frac{2\eta^{\text{LO}}(\Gamma)\,\omega_{\Gamma}}{\gamma^{\text{ep}}} \tag{4.29}$$

Using the last Eq. 4.29, one can actually measure the Fermi velocity directly, since it is expressed in terms of experimental quantities related to the phonon spectrum. The experimental data shown in Fig. 4.13 correspond to $v_{\text{F}} \approx 7 \times 10^5$ m/s, which agrees well with other experiments including ARPES or magnetotransport (Novoselov et al. 2005, Zhou et al. 2006).

One can generalize Eq. 4.28 for describing a graphite (or graphene) sample which is doped to get the Fermi level shifted ($\varepsilon_{\text{F}} \neq 0$, ε_{F} is the Fermi level)

$$\gamma^{\text{EPC}}(\varepsilon_{\text{F}}) = \gamma_0^{\text{EPC}}\{f(-\hbar\omega_{\Gamma}/2 - \varepsilon_{\text{F}}) - f(\hbar\omega_{\Gamma}/2 - \varepsilon_{\text{F}})\} \tag{4.30}$$

where $f(x)$ is the Fermi–Dirac distribution of argument x. As it follows from Eq. 4.30, even without doping, one might considerably decrease the electron–phonon interaction constant versus temperature. Because the anharmonic fraction γ^{an} of broadening is low compared to γ^{ep}, the G peak broadening tends to decrease as the temperature grows. It makes a remarkable difference from other materials, where the broadening always increases with temperature.

4.20 Raman Spectroscopy of Graphene and Graphene Layers

Raman spectra of graphene and of bulk graphite measured with excitation at 514.5 nm are compared in Figs. 4.14 and 4.15 (Ferrari et al. 2006). One can notice two most intense features represented by the G peak at 1580 cm^{-1} and G′ band positioned at ~2700 cm^{-1} which should be quoted as 2G peak because it corresponds to the

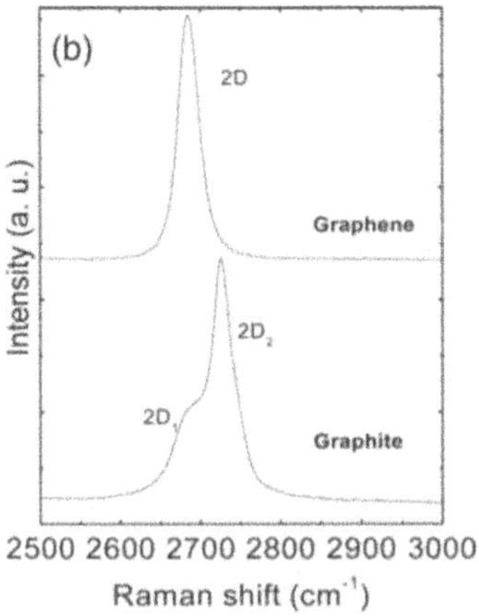

Figure 4.14 Comparison of the 2D peaks in graphene and graphite. Adapted from Nemanich and Solin (1979) and Vidano et al. (1981).

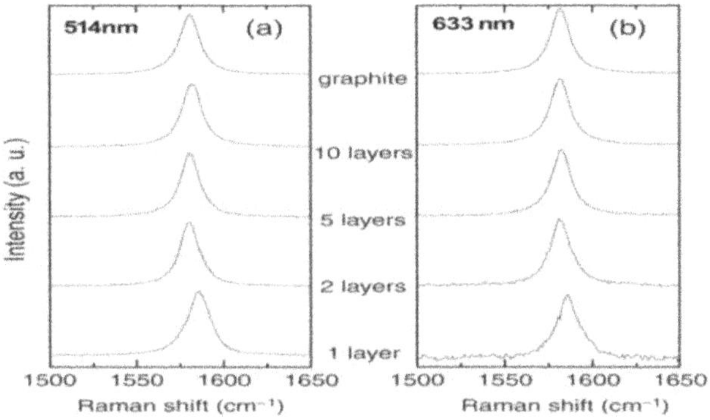

Figure 4.15 Evolution of the G peak as a function of the number of layers for (a) 514 nm and (b) 633 nm excitations. Adapted from Gupta et al. (2006).

second order of the D peak. There is also another peak in Figs. 4.14 and 4.15 which lies at $\sim$3250 cm^{-1}, and the frequency of which is higher than the position of the 2G peak. The latter peak is thus not 2G but rather it originates from the second order of the intravalley D′ peak of Fig. 4.12 (see the first sketch). Therefore one might quote the second peak as 2D′ peak. A remarkable change of the shape and intensity of the 2D peak in graphene compared to bulk graphite is evident from Fig. 4.15.

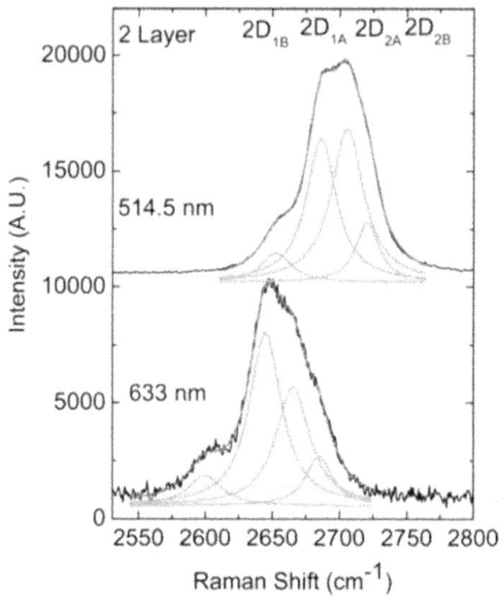

Figure 4.16 The four components of the 2D peak in bilayer graphene. Adopted from Gupta et al. (2006). Adapted from Nemanich and Solin (1979) and Vidano et al. (1981).

Let us consider the structure of the 2D peak in bulk graphite in more details. It consists of two components $2D_1$ and $2D_2$ (Nemanich and Solin 1979, Vidano et al. 1981) the relevant heights of which are measured as roughly 1/4 and 1/2 of the G peak height, respectively. On the contrary, graphene has a single, sharp 2D peak, the intensity of which is about four times stronger than of the G peak (Ferrari et al. 2006). It is interesting also to learn the evolution of Raman spectra versus the number of atomic layers in graphene which are represented in Figs. 4.15 and 4.16. In Figs. 4.15 and 4.16 one can observe the development of the 2D band versus the number of layers for 514.5 nm and 633 nm excitations. It reflects a prominent fact that the 2D band in bilayer graphene is much broader and is shifted up in energies as compared to monolayer graphene. The 2D band itself differs from bulk graphite and consists of four components: $2D_{1B}$, $2D_{1A}$, $2D_{2A}$, and $2D_{2B}$. The relative intensities of the two compounds, $2D_{1A}$ and $2D_{2A}$, are higher than those of the remaining two ones, which is evident from Fig. 4.16. Further increase of the number of layers causes considerable decrease of relative intensity of the lower frequency $2D_1$ peaks, as seen from Fig. 4.16. If the number of layers in graphene exceeds five, the Raman spectrum becomes similar to the spectrum of bulk graphite. Thus, Raman spectroscopy is able to distinguish between single-layer graphene, bilayer graphene, and FLG (less than five layers). The few-layer features of the Raman spectra are not pronounced in nanographite, which agrees with experiments. Similar 2D peak features are also observed in the Raman spectrum of single-wall carbon nanotubes (Jorio et al. 2002). However, there are also significant differences between the Raman spectra of the single-wall carbon nanotubes and graphene. Those differences allow us to clearly distinguish between the two materials. The basic difference between a monatomic sheet and a nanotube originates from confinement and curvature of the latter material. They result in splitting of the two degenerate modes of a single G peak into two distinct G+ and G− peaks (Jorio et al. 2002, Piscanec et al. 2007). Furthermore, the single 2D peak splits into the four components in bilayer graphene, whereas it is decomposed in only two components in bulk graphite. The evolution of the electronic bands with the number of layers serves as a key circumstance when trying to interpret the double structure of the 2D peak in graphite (Nemanich and Solin 1979, Vidano et al. 1981).

The scattering process which is responsible for the formation of the 2D peak in graphene involves two different phonons with opposite momentum in the highest optical branch near the K point (see Fig. 4.16). The double resonance relates wavevectors of the phonon to the electronic band structure (Thomsen and Reich 2000). Therefore, the 2D peak position depends on the excitation energy as much as twice faster than the first-order D peak does. It causes the 2D Raman frequency to become twice larger than frequency of the scattering phonon. The frequency value itself is obtained from the resonance condition. The energy diagram shown in Fig. 4.11b illustrates the origin of the 2D peak. In that diagram, one should assume that the electron–phonon scattering occurs with an exchange of momentum $-\mathbf{q}$. That exchange of momentum allows us to satisfy the Raman fundamental selection rule for the second-order scattering, since $\mathbf{q}+(-\mathbf{q}) = 0$. The last condition is applicable only to phonons whose momentum satisfies $\mathbf{q} > \mathbf{K}$, along the

Γ–K–M direction (**K** $<$ **q** $<$**M**), which conforms to conditions of the double resonance (Ferrari et al. 2006).

Another possible process, which might contribute to the double resonance, involves two phonons with wavevectors **q** $<$ **K** and **q** $\approx$ **K**. However, its impact is almost negligible in terms of the Raman intensity. That happens due to a trigonal warping of the band structure which yields no electron–phonon interaction for the **q-K** phonon during this particular transition. Phonons with **q** $<$ **K**, which participate in the process, occupy just a tiny region of the phase–space. One can consider various mechanisms which are responsible for the four components of the 2D peak in bilayer graphene. They could be either the splitting of phonon branches (Ferrari and Robertson 2000, Castiglioni et al. 2001) or the splitting of electronic bands (Ferrari and Robertson 2000, Castiglioni et al. 2001). The former mechanism had been numerically studied by the DFT by Piscanec et al. (2004) and Ferrari et al. (2006). The obtained numeric data suggest that the effect of electronic bands causes just a small splitting of the phonon branches, the frequency of which remains below ~1.5 cm^{-1}. This is much smaller than the 2D splitting observed experimentally. Interaction between the adjacent graphene planes in bilayer graphene splits the π and π^* bands into four bands, whereas splitting for electrons and holes is different. On the one hand, an external optical beam couples only two pairs of the four bands (Ferrari et al. 2006). On the other hand, in the highest optical branch, there are two almost degenerate phonons which couple the four electron bands with each other. The latter four processes occur with participating phonons with momentums $\mathbf{q}_{1B}$, $\mathbf{q}_{1A}$, $\mathbf{q}_{2A}$, and $\mathbf{q}_{2B}$. Because the phonon bands around **K** have a strong dispersion, the corresponding momentums $\mathbf{q}_{1B}$, $\mathbf{q}_{1A}$, $\mathbf{q}_{2A}$, and $\mathbf{q}_{2B}$ characterize phonons with different frequencies. Thus, different phonon frequencies cause the four peaks in the Raman spectrum of bilayer graphene. The phonon scattering with momentums $\mathbf{q}_{1A}$ and $\mathbf{q}_{2A}$ between the bands of the same type yields a larger Raman peak intensity than the phonon scattering with momentums $\mathbf{q}_{1B}$ and $\mathbf{q}_{2B}$. It happens because the former process occurs within a larger-phase space region where the double-resonance condition is satisfied as compared to the latter process, which happens in a smaller-phase space region (Ferrari et al. 2006).

4.21 Failure of Adiabatic Born–Oppenheimer Approximation and Raman Spectrum of Doped Graphene

One can move the Kohn anomaly aside when displacing it from $q = 0$ using the doping of graphene either chemically or electrically. Because the doping alters the electrochemical potential of the charge carriers in graphene, it also causes the Kohn anomaly to shift away from its steady-state position. The corresponding shift is detected by measuring the first order of nondouble resonant Raman scattering which involves phonons with zero wavevector $\mathbf{q} = 0$. Raman spectroscopy experiment was performed on graphene where the electrochemical potential was altered by applying a gate voltage (electric doping). The obtained Raman data show that the G peak position changes as a function of the electron and hole doping. Intuitively one should expect a stiffening of the G peak (Pisana et al. 2007).

Experimental results for the G peak position and full width at half maximum (FWHM) at temperature $T = 200$ K are represented in Fig. 4.17a,b. The properties of G peak are qualitatively described by Eq. 4.30. However, a more detailed understanding of the G peak behavior versus doping can be achieved on the basis of DFT calculations (Piscanec et al. 2004, Ferrari et al. 2006) which include beyond Born–Oppenheimer corrections to the dynamic matrix (Pisana et al. 2007). The adiabatic Born–Oppenheimer (ABO) approximation describes interaction between the electrons and nuclei under the assumption that the lighter electrons stay in their own instantaneous ground state at any time moment while following the motion of the heavier nuclei adiabatically (Born and Oppenheimer 1927, Ziman 1960, Motakabbir and Grimvall 1981). The key condition that such an approximation is justified corresponds to the case when the energy gap between ground and excited electronic states is larger than the energy scale of the nuclear motion. However, the aforementioned condition is not always fulfilled. In particular, the ABO approximation appears to be pretty efficient for many metals, since it provides an accurate prediction for chemical reactions, MD, and phonon frequencies.

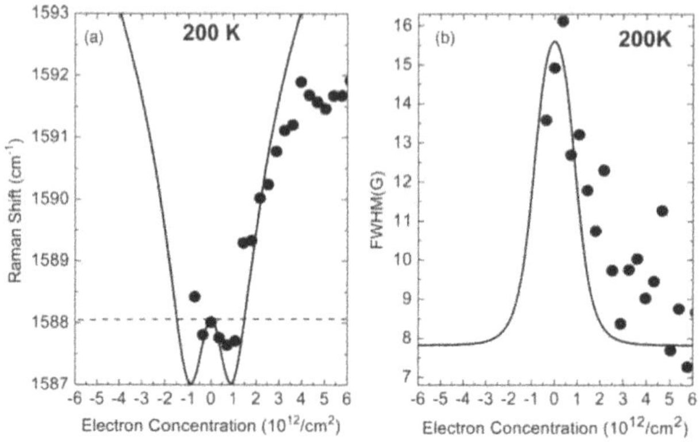

Figure 4.17 (a) The G peak position versus the concentration of electrons at 200 K. (Dots) measurements, (horizontal-dashed line) adiabatic Born–Oppenheimer, and (line) finite-temperature non-adiabatic calculation. When the Fermi energy equals half of the phonon energy there is a minimum observed in the calculations at $\sim 10^{12}$ cm^{-2}. (b) FWHM(G) at 200 K as a function of electron concentration. (dots) measured, (line) theoretical FWHM of a Voigt profile obtained from a Lorentzian component given by Eq. 4.31 and a constant Gaussian component of $\sim$8 cm^{-1}. Adapted from Pisana et al. (2007).

The ABO approximation remarkably fails in graphene (Pisana et al. 2007). The reason is that frequency of electron–impurity, electron–electron, and electron–phonon scattering with nonzero momentum phonons is comparable or higher than the frequency of nuclei oscillations. More specifically, typical frequency of the electron momentum relaxation in graphene is $\sim 10^{13}$ s^{-1}, whereas the width of G peak is much broader, $\sim 3.3 \times 10^{14}$ s^{-1}(Moos et al. 2001, Kampfrath et al. 2005, Zhang et al. 2006). Because of frequent collisions, the electrons do not have enough time to adjust their momentums to reach the instantaneous adiabatic ground state. It causes the violation of the ABO condition. At zero temperature, one can utilize an analytical formula for the G peak shifts (Lazzeri and Mauri 2006):

$$\hbar D\omega\,(G) = \frac{A \cdot \alpha_{\mathrm{ep}}^{2}\,(\Gamma)}{\pi\, M\omega_0\,(G)\,\hbar v_{\mathrm{F}}^{2}} \left[|\varepsilon_{\mathrm{F}}| + \frac{\hbar\omega_0\,(G)}{4} \ln \left(\left| \frac{|\varepsilon_{\mathrm{F}}| - \frac{\hbar\omega_0(G)}{2}}{|\varepsilon_{\mathrm{F}}| + \frac{\hbar\omega_0(G)}{2}} \right| \right) \right] \tag{4.31}$$

where $A = 5.24$ Å^2 is the graphene unit cell area, whereas $\omega_0(G)$ is the G peak frequency for an undoped graphene. According to Lazzeri and Mauri (2006) and Pisana et al. (2007), Eq. 4.31 can be generalized to a finite temperature and there is a very good agreement between the experiments and the nonadiabatic finite T calculations. Such a case corresponds to a special situation when the D peak phonons are away from K and the D peak is still described by the ABO approximation. The result, Eq. 4.31, remains valid even in the presence of moderate doping which does not happen for the G peak phonon since the latter is positioned always at $q = 0$. The situation changes for the levels of higher doping when there is a contribution from the charge transfer effects. Significant p-doping causes the phonon stiffening, whereas the high n-doping leads to the phonon softening (Lazzeri and Mauri 2006, Pisana et al. 2007). The result of doping can be represented as a sum of nonadiabatic effects and a charge transfer. If the doping level is below 0.6 eV, it causes just a little asymmetry of the G peak stiffening (Lazzeri and Mauri 2006). In principle, one can estimate the level of doping by considering the fact that the 2D peak responds differently to hole and electron doping, whereas the G peak always stiffens. At the same time the ratio of intensities I_{2D}/I_G is reduced versus the doping level. The above tendencies are illustrated by Fig. 4.18, which suggests that the G peak positions, FWHM, and ratio I_{2D}/I_G all are changing versus the doping level. It can be understood under an assumption that an inhomogeneous self-doping takes place. In

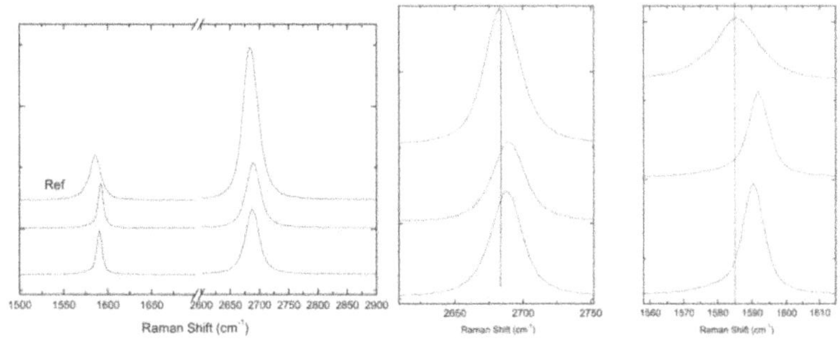

Figure 4.18 Raman spectra measured at different points of a graphene sample. Adapted from Ferrari, A.C. SSC (2007) © Elsevier.

that case, the inhomogeneity scale is smaller than the Raman spot dimensions; thus, an asymmetrical G peak can be observed. A similar tendency emerges when one compares the data obtained for suspended graphene with the spectra of graphene sample on a substrate (Ferrari et al. 2006). Although Raman spectra for both cases are similar, the G peak is broader and shifted down in the former case, whereas the intensity of the 2D peak is much higher. The mentioned results well agree with the decrease of self-doping when the substrate is removed. Another subtle feature is a tiny increase of disorder in the suspended graphene which becomes visible in the D peak.

4.22 Influence of Atomic and Structural Disorders

Degree of disorder represents an important parameter when measuring Raman scattering on different locations of the same sample. The disorder along with the turbostraticity must also be taken into account when comparing different multilayered samples or devices. It is convenient to implement a common strategy for classification of disorder. For the sake of simplicity, the classification of all the Raman spectra can be performed in three stages, which allow us to distinguish between, for example, graphite and amorphous carbons (Ferrari and Robertson 2004). It assumes that the Raman spectrum depends on the following

- ratio of the sp^2/sp^3 phases;
- presence of the rings or chains of the sp^2 phase;
- the sp^2 phase clustering; and
- disorder of bonds.

A useful illustrative characteristic is the amorphization trajectory (Ferrari and Robertson 2004) which links graphite and highly sp^3 amorphous carbon. It might be divided contingently in the following parts

Graphite → nanocrystalline graphite → low sp^3 amorphous carbon→ high sp^3 amorphous carbon.

The authors Tuinstra and Koenig (1970) have emphasized that the ratio of the D peak intensity to the G peak intensity can approximately be written as

$$\frac{I_D}{I_G} = \frac{C\,(\lambda)}{L_a} \tag{4.32}$$

where L_a is the in-plane correlation length and C (488 nm) is $\sim$4.4 nm (see Tuinstra and Koenig 1970, Knight and White 1989, Sinha and Menendez 1990, Matthews et al. 1999). One estimates an effective value of L_a for a system with mixed grain sizes as

$$\frac{1}{L_a^{\mathrm{Eff}}} = \sum_i^N X_i \frac{1}{L_{ai}} \tag{4.33}$$

where X_i are the fractions of volume and L_{ai} are the grain's dimensions. Furthermore, by assuming that I_D/I_G coincides with the integrated area ratios rather than ratio of the peak heights one can use an approximate expression (Tuinstra and Koenig 1970, Knight and White 1989, Sinha and Menendez 1990, Matthews et al. 1999):

$$C\,(\lambda) \approx 2.4 \times 10^{-10}\lambda^4 \tag{4.34}$$

The recent study (Cancado et al. 2006) has used the fitting $C(\lambda) \approx 2.4 \times 10^{-10}\lambda^4$, using I_D/I_G as an integrated area ratio, rather than the peak height ratio. Using the area ratio gives different results since FWHM(D) increases much more than FWHM(G) for decreasing L_a. Cuesta et al. 1998, Ferrari et al. 2003). We emphasize that the above Eqs. 4.32–4.34 must be used with caution since a complete theory for the Raman intensity of the G and D peaks is still in progress.

The origin of disorders and defects of the graphene lattice attracts a strong interest since they can alter the electron transport considerably. Evolution of the Raman spectrum is summarized as follows: (*i*) there is a distinguishable D peak, whereas an increase of the ratio I_D/I_G conforms to Eq. 4.32; (*ii*) another D′ peak emerges at $\sim$1620 cm^{-1}; (*iii*) the effect of disorder is pronounced in broadening FWHM of all the peaks; and (*iv*) there is no doublet structure of the D and 2D peaks. When the disorder is high, the G and D′ peaks merge into a single G line, whereas the average G peak position is gradually relocated from $\sim$1580 cm^{-1} to $\sim$1600 cm^{-1}. The doublet shape of the 2D peak indicates the turbostraticity due to the *c* axis ordering, whereas the disappearance of the doublet in the D peak and of its second-order signals the loss of 3D ordering (Lespade et al. 1984). It is well known that the turbostratic graphite shows just a single 2D peak without the planar AB staking (Lespade et al. 1984), whereas its FWHM is about twice bigger ($\sim$50 cm^{-1}) than for the 2D peak of graphene which is about 20 cm^{-1}. Besides, often there is a first-order D peak in the turbostratic graphite (Lespade et al. 1984). The next step represents an evolution from nanocrystalline graphite to mainly sp^2 amorphous carbon. When density of defects increases, they introduce a disorder of bond angles and lengths on the atomic scale. It causes softening of the phonon modes and thus in the broadening of the G peak. That process results in a fully disordered, sp^2-bonded a-C consisting of distorted sixfold rings or rings of other orders (e.g., with a 20% fraction of sp^3 at most). That happens, for example, when the amorphous carbon is sputtered (Li and Lannin 1990). Raman spectrum evolution consists of the following: (*i*) G peak decreases from 1600 to $\sim$1510 cm^{-1}; (*ii*) the intensity ratio $I\,(\mathrm{D})\,/I\,(\mathrm{G}) \propto M \propto L_a^2$; (*iii*) $I\,(\mathrm{D})\,/I\,(\mathrm{G}) \to 0$; and (*iv*) the G peak dispersion increases. Besides, as it follows from Fig. 4.19, there is a small modulated bump from $\sim$2400 to $\sim$3100 cm^{-1}, whereas the second-order Raman peaks are missing. The aforementioned tendency is noticeable from Fig. 4.19. One can see that the D peak intensity of amorphous carbons is much lower than it follows from Eq. 4.32. The reason for that discrepancy is understood when using a "molecular" model. When disorder increases, the ordered clusters diminish in size while the rings are getting distorted and starting to break up. During that process, the intensity I_G remains almost constant and is not affected by the disorder because the G peak is related to the relative motion of C sp^2 atoms. On the contrary, the sp^2 rings density decreases, which causes the intensity I_D to diminish. Therefore, Eq. 4.32 is not valid any more. In the limit of small in-plane correlation length L_a, the probability of finding a sixfold ring in the cluster is proportional to the cluster area which in turn is related to the D mode strength. It means that development of the D peak in amorphous carbons is caused by ordering but is not the case for graphite (Ferrari and Robertson 2000, Castiglioni et al. 2001). Then, one gets $I\,(\mathrm{D})\,/I\,(\mathrm{G}) \propto M \propto L_\mathrm{a}^2$ where M is the number of ordered rings. On this step of amorphization trajectory (see Fig. 4.20) one

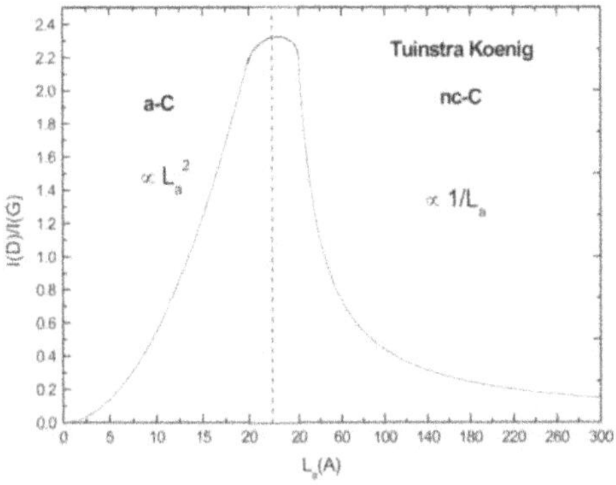

Figure 4.19 Schematic evolution of I (D)/I (G) as a function of disorder for visible excitation. The maximum of this curve is taken as the boundary between nanocrystalline graphite and amorphous carbons (Ferrari and Robertson 2004).

gets (Ferrari and Robertson 2000, Castiglioni et al. 2001)

$$\frac{I_{\mathrm{D}}}{I_{\mathrm{G}}} = C'(\lambda)\, L_{\mathrm{a}}^2 \tag{4.35}$$

The requirement of continuity between Eqs. 4.32 and 4.35 gives the value $C'(514\ \mathrm{nm}) \approx 0.0055$.

When the excitation wavelength diminishes, the G peak position in disordered carbons shifts to higher frequencies, from the IR to the UV part of optical spectrum (Ferrari and Robertson 2000, Castiglioni et al. 2001). In general, the dispersion rate of the G peak with disorder increases. However, there is no dispersion of the G peak in graphite, nanocrystalline graphite or glassy carbon (Ferrari and Robertson 2000, Castiglioni et al. 2001). The dispersion takes place only in the highly disordered carbons, where it is proportional to the degree of disorder. Roughly speaking, there are two types of materials. If only the sp^2 rings are present in the material, the G peak dispersion becomes saturated at the G position in nanocrystalline graphite the maximum of which is located at $\sim$1600 cm^{-1}. Otherwise, if the material also contains sp^2 chains in addition to the sp^2 rings, the G peak position might be well above 1600 cm^{-1} up to 1690 cm^{-1} at 229 nm excitation (Ferrari and Robertson 2000, Castiglioni et al. 2001). In all the carbons, in contrast to the conditional dispersion of the G peak, the other D peak

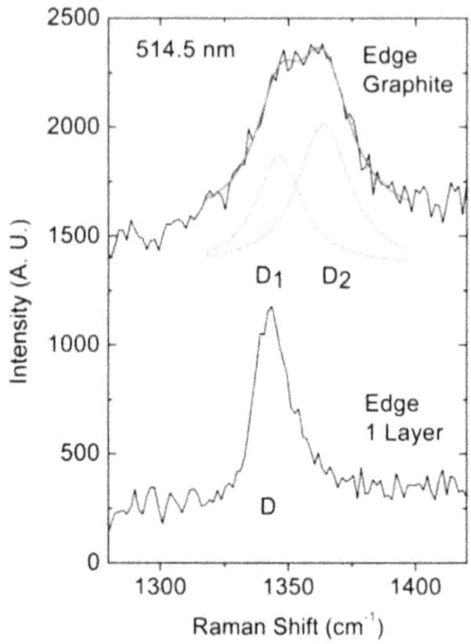

Figure 4.20 Raman spectra of graphite and graphene edges. Adapted from Nemanich and Solin (1979) and Vidano et al. (1981).

is always subject to dispersion versus excitation energy (Ferrari and Robertson 2000, Castiglioni et al. 2001). Nevertheless, in contrast to the G peak behavior, the dependence of dispersion versus disorder for the D peak is rather reciprocal: the stronger disorder causes the lower dispersion (Ferrari and Robertson 2000, Castiglioni et al. 2001).

4.23 Graphene Ribbons and Edges

If a laser spot includes an edge of the sample, the edge looks like a defect. Thus, despite the fact that the bulk of the sample is free of defects, a D peak in Raman spectrum will appear. In the Raman spectrum shown in Fig. 4.17 which was measured at the center of graphene layer, the D peak is absent. It is consistent with absence of a significant number of defects in the sample. However, if the laser spot involves the edge of graphene sample, one might notice the presence of a single D peak, as is evident from the Raman spectrum shown in Fig. 4.17. A different situation takes place for the D peak emerging at the edge of graphite which consists of two peaks D_1 and D_2 (Nemanich and Solin 1979, Vidano et al. 1981). The contribution of ribbon edges becomes more significant when the ribbon width diminishes. Another factor which is responsible for the D peak intensity is the level of disorder which also contributes to the magnitude of the ratio I_D/I_G. Besides, a ribbon of a finite width will cause the effect of confinement for both electrons and phonons. Raman spectra are also expected to depend on the polarization parallel to the edges, and could also be sensitive to chirality of the edges Cancado, Pimenta, Neves, Dantas et al. 2004, Cancado, Pimenta, Neves, Medeiros-Ribeiro et al. 2004).

One can conclude that the shapes, positions, and relative intensity of the G and 2D Raman peaks depend on the number of graphene layers which is caused by changes in the electronic structure and electron–phonon interactions. Both types of n- and p-doping result in a crisper G peak and in its moving to higher frequencies. Raman spectroscopy is a powerful tool for monitoring of the number of layers, layer quality, disorder, level of doping, and confinement in graphene samples.

Problems

4-1. Explain why the Grüneisen parameter for the lowest transverse acoustic (TA) out-of-plane acoustic (ZA) modes is negative.

4-2. Discuss why the contribution of ZA phonons to thermal transport in graphene might be neglected.

4-3. Explain why the acoustic phonon energy spectra in nanostructures are spatially quantized, and what impact this has on the phonon group velocity.

4-4. What are the basic assumptions for deriving the Klements equation, Eq. 4.2?

4-5. Explain why the heat transport in single-layer graphene differs from the heat transport in basal planes of bulk graphite.

4-6. Explain the role of 2D three-phonon Umklapp processes.

4-7. Explain why the number of vibrational modes $3N$-6 of nonlinear molecules is smaller than the number of vibrational modes $3N$-5 of linear molecules.

4-8. Explain why the locations of Stokes and anti-Stokes peaks create a symmetric pattern around the frequency difference $\Delta\nu = 0$ between transmitted and incident photons.

4-9. What is the physical difference between the Raman effect and fluorescence?

4-10. What is the necessary condition for a Raman shift to appear? How should molecular vibrations and rotations be classified?

4-11. How should the Fermi velocity and the electron–phonon coupling constant be measured directly from the Raman effect?

4-12. What is the difference between the structure of the 2D peak in bulk graphite and graphene (Kittel 2005, Klemens 2000, Bergman 2000)?

References

Aksamija, Z., and I. Knezevic (2011). Lattice thermal conductivity of graphene nanoribbons: anisotropy and edge roughness scattering. *Appl Phys Lett*, **98**(14).

Balandin, A. A., S. Ghosh, W. Z. Bao, I. Calizo, D. Teweldebrhan, F. Miao, and C. N. Lau (2008). Superior thermal conductivity of single-layer graphene. *Nano Lett*, **8**(3): 902–907.

Baranov, A. V., A. N. Bekhterev, Y. S. Bobovich, and V. I. Petrov (1987). Interpretation of some singularities in Raman-spectra of graphite and glass carbon. *Opt I Spektrosk*, **62**(5): 1036–1042.

Bergman, L., et al. (2000). Phonons, electron-phonon interactions, and phonon-phonon interactions in III-V nitrides. In *Ultrafast Phenomena in Semiconductors IV*, 100–111.

Born, M., and R. Oppenheimer (1927). Quantum theory of molecules. *Ann Phys*, **84**(20): 0457–0484.

Cancado, L. G., M. A. Pimenta, B. R. A. Neves, M. S. S. Dantas, and A. Jorio (2004). Influence of the atomic structure on the Raman spectra of graphite edges. *Phys Rev Lett*, **93**(24).

Cancado, L. G., M. A. Pimenta, B. R. A. Neves, G. Medeiros-Ribeiro, T. Enoki, Y. Kobayashi, K. Takai, K. Fukui, M. S. Dresselhaus, R. Saito, and A. Jorio (2004). Anisotropy of the Raman spectra of nanographite ribbons. *Phys Rev Lett*, **93**(4).

Cancado, L. G., K. Takai, T. Enoki, M. Endo, Y. A. Kim, H. Mizusaki, A. Jorio, L. N. Coelho, R. Magalhaes-Paniago, and M. A. Pimenta (2006). General equation for the determination of the crystallite size La of nanographite by Raman spectroscopy. *Appl Phys Lett*, **88**(16).

Castiglioni, C., F. Negri, M. Rigolio, and G. Zerbi (2001). Raman activation in disordered graphites of the A(1)(') symmetry forbidden k not equal 0 phonon: the origin of the D line. *J Chem Phys*, **115**(8): 3769–3778.

Castiglioni, C., M. Tommasini, and G. Zerbi (2004). Raman spectroscopy of polyconjugated molecules and materials: confinement effect in one and two dimensions. *Philos Trans R Soc A*, **362**(1824): 2425–2459.

Cong, C. X., T. Yu, R. Saito, G. F. Dresselhaus, and M. S. Dresselhaus (2011). Second-order overtone and combination Raman modes of graphene layers in the range of 1690–2150 cm(-1). *ACS Nano*, **5**(3): 1600–1605.

Cong, C. X., T. Yu, K. Sato, J. Z. Shang, R. Saito, G. F. Dresselhaus, and M. S. Dresselhaus (2011). Raman characterization of ABA- and ABC-stacked trilayer graphene. *ACS Nano*, **5**(11): 8760–8768.

Cuesta, A., P. Dhamelincourt, J. Laureyns, A. Martinez-Alonso, and J. M. D. Tascon (1998). Comparative performance of X-ray diffraction and Raman microprobe techniques for the study of carbon materials. *J Mater Chem*, **8**(12): 2875–2879.

Droth, M., and G. Burkard (2011). Acoustic phonons and spin relaxation in graphene nanoribbons. *Phys Rev B*, **84**(15).

Dubay, O., and G. Kresse (2003). Accurate density functional calculations for the phonon dispersion relations of graphite layer and carbon nanotubes. *Phys Rev B*, **67**(3).

Evans, W. J., L. Hu, and P. Keblinski (2010). Thermal conductivity of graphene ribbons from equilibrium molecular dynamics: effect of ribbon width, edge roughness, and hydrogen termination. *Appl Phys Lett*, **96**(20).

Falkovsky, L. A. (2008). Symmetry constraints on phonon dispersion in graphene. *Phys Lett A*, **372**(31): 5189–5192.

Ferrari, A. C., J. C. Meyer, V. Scardaci, C. Casiraghi, M. Lazzeri, F. Mauri, S. Piscanec, D. Jiang, K. S. Novoselov, S. Roth, and A. K. Geim (2006). Raman spectrum of graphene and graphene layers. *Phys Rev Lett*, **97**(18).

Ferrari, A. C., and J. Robertson (2000). Interpretation of Raman spectra of disordered and amorphous carbon. *Phys Rev B*, **61**(20): 14095–14107.

Ferrari, A. C., and J. Robertson (2001a). Origin of the 1150-cm(-1) Raman mode in nanocrystalline diamond. *Phys Rev B*, **63**(12).

Ferrari, A. C., and J. Robertson (2001b). Resonant Raman spectroscopy of disordered, amorphous, and diamondlike carbon. *Phys Rev B*, **64**(7).

Ferrari, A. C., and J. Robertson (2004). Raman spectroscopy in carbons: from nanotubes to diamond: preface. *Philos Trans R Soc A*, **362**(1824): 2269–2270.

Ferrari, A. C., S. E. Rodil, and J. Robertson (2003). Interpretation of infrared and Raman spectra of amorphous carbon nitrides. *Phys Rev B*, **67**(15).

Ghosh, S., W. Z. Bao, D. L. Nika, S. Subrina, E. P. Pokatilov, C. N. Lau, and A. A. Balandin (2010). Dimensional crossover of thermal transport in few-layer graphene. *Nat Mater*, **9**(7): 555–558.

Ghosh, S., I. Calizo, D. Teweldebrhan, E. P. Pokatilov, D. L. Nika, A. A. Balandin, W. Bao, F. Miao, and C. N. Lau (2008). Extremely high thermal conductivity of graphene: prospects for thermal management applications in nanoelectronic circuits. *Appl Phys Lett*, **92**(15).

Ghosh, S., D. L. Nika, E. P. Pokatilov, and A. A. Balandin (2009). Heat conduction in graphene: experimental study and theoretical interpretation. *New J Phys*, **11**.

Gogotsi, Y., J. A. Libera, N. Kalashnikov, and M. Yoshimura (2000). Graphite polyhedral crystals. *Science*, **290**(5490): 317–320.

Graf, D., F. Molitor, K. Ensslin, C. Stampfer, A. Jungen, C. Hierold, and L. Wirtz (2007). Spatially resolved raman spectroscopy of single- and few-layer graphene. *Nano Lett*, **7**(2): 238–242.

Gupta, A., G. Chen, P. Joshi, S. Tadigadapa, and P. C. Eklund (2006). Raman scattering from high-frequency phonons in supported n-graphene layer films. *Nano Lett*, **6**(12): 2667–2673.

Hu, J. N., Y. Wang, A. Vallabhaneni, X. L. Ruan, and Y. P. Chen (2011). Nonlinear thermal transport and negative differential thermal conductance in graphene nanoribbons. *Appl Phys Lett*, **99**(11).

Jauregui, L. A., Y. N. Yue, A. N. Sidorov, J. N. Hu, Q. K. Yu, G. Lopez, R. Jalilian, D. K. Benjamin, D. A. Delk, W. Wu, Z. H. Liu, X. W. Wang, Z. G. Jiang, X. L. Ruan, J. M. Bao, S. S. Pei, and Y. P. Chen (2010). Thermal transport in graphene nanostructures: experiments and simulations. *Graphene, Ge/III-V, Emerging Mater Post-CMOS Appl 2*, **28**(5): 73–83.

Jorio, A., C. Fantini, M. S. S. Dantas, M. A. Pimenta, A. G. Souza, G. G. Samsonidze, V. W. Brar, G. Dresselhaus, M. S. Dresselhaus, A. K. Swan, M. S. Unlu, B. B. Goldberg, and R. Saito (2002). Linewidth of the Raman features of individual single-wall carbon nanotubes. *Phys Rev B*, **66**(11).

Kampfrath, T., L. Perfetti, F. Schapper, C. Frischkorn, and M. Wolf (2005). Strongly coupled optical phonons in the ultrafast dynamics of the electronic energy and current relaxation in graphite. *Phys Rev Lett*, **95**(18).

Kim, P., L. Shi, A. Majumdar, and P. L. McEuen (2001). Thermal transport measurements of individual multiwalled nanotubes. *Phys Rev Lett*, **87**(21).

Kittel, C. (2005). *Introduction to Solid State Physics*, 8th ed., Wiley, Hoboken, NJ: xix, 680.

Klemens, P. G. (2001). Role of optical modes in thermal conduction. In *Thermal Conductivity 25: Thermal Expansion 13*, 291–299.

Klemens, P. G. (2001). Theory of thermal conduction in thin ceramic films. *Int J Thermophys*, **22**(1): 265–275.

Knight, D. S., and W. B. White (1989). Characterization of diamond films by raman-spectroscopy. *J Mater Res*, **4**(2): 385–393.

Kohn, W. (1959). Image of the Fermi surface in the vibration spectrum of a metal. *Phys Rev Lett*, **2**(9): 393–394.

Landsberg, G., and L. Mandelstam (1928a). A new appearance in the light diffusion in crystals. *Naturwissenschaften*, **16**: 772–772.

Landsberg, G., and L. Mandelstam (1928b). A new occurrence in the light diffusion of crystals. *Naturwissenschaften*, **16**: 557–558.

Lazzeri, M., and F. Mauri (2006). Nonadiabatic Kohn anomaly in a doped graphene monolayer. *Phys Rev Lett*, **97**(26).

Lazzeri, M., S. Piscanec, F. Mauri, A. C. Ferrari, and J. Robertson (2005). Electron transport and hot phonons in carbon nanotubes. *Phys Rev Lett*, **95**(23).

Lazzeri, M., S. Piscanec, F. Mauri, A. C. Ferrari, and J. Robertson (2006). Phonon linewidths and electron-phonon coupling in graphite and nanotubes. *Phys Rev B*, **73**(15).

Lespade, P., A. Marchand, M. Couzi, and F. Cruege (1984). Characterization of carbon materials with Raman microspectrometry. *Carbon*, **22**(4–5): 375–385.

Li, F., and J. S. Lannin (1990). Radial-distribution function of amorphous-carbon. *Phys Rev Lett*, **65**(15): 1905–1908.

Lindsay, L., and D. A. Broido (2010). Optimized Tersoff and Brenner empirical potential parameters for lattice dynamics and phonon thermal transport in carbon nanotubes and graphene [**81**, 205441, (2010)]. *Phys Rev B*, **82**(20).

Lindsay, L., and D. A. Broido (2011). Enhanced thermal conductivity and isotope effect in single-layer hexagonal boron nitride. *Phys Rev B*, **84**(15).

Lindsay, L., D. A. Broido, and N. Mingo (2010). Diameter dependence of carbon nanotube thermal conductivity and extension to the graphene limit. *Phys Rev B*, **82**(16).

Lindsay, L., D. A. Broido, and N. Mingo (2011). Flexural phonons and thermal transport in multilayer graphene and graphite. *Phys Rev B*, **83**(23).

Lui, C. H., Z. Q. Li, Z. Y. Chen, P. V. Klimov, L. E. Brus, and T. F. Heinz (2011). Imaging stacking order in few-layer graphene. *Nano Lett*, **11**(1): 164–169.

Mapelli, C., C. Castiglioni, G. Zerbi, and K. Mullen (1999). Common force field for graphite and polycyclic aromatic hydrocarbons. *Phys Rev B*, **60**(18): 12710–12725.

Martin, R. M. (1970). Eleastic properties of Zns structure semiconductors. *Phys Rev B*, **1**(10): 4005.

Matthews, M. J., M. A. Pimenta, G. Dresselhaus, M. S. Dresselhaus, and M. Endo (1999). Origin of dispersive effects of the Raman D band in carbon materials. *Phys Rev B*, **59**(10): R6585–R6588.

Maultzsch, J., S. Reich, C. Thomsen, H. Requardt, and P. Ordejon (2004). Phonon dispersion in graphite. *Phys Rev Lett*, **92**(7).

Mazzamuto, F., J. Saint-Martin, A. Valentin, C. Chassat, and P. Dollfus (2011). Edge shape effect on vibrational modes in graphene nanoribbons: a numerical study. *J Appl Phys*, **109**(6).

Mingo, N., and D. A. Broido (2005). Carbon nanotube ballistic thermal conductance and its limits. *Phys Rev Lett*, **95**(9).

Mohr, M., J. Maultzsch, E. Dobardzic, S. Reich, I. Milosevic, M. Damnjanovic, A. Bosak, M. Krisch, and C. Thomsen (2007). Phonon dispersion of graphite by inelastic x-ray scattering. *Phys Rev B*, **76**(3).

Moos, G., C. Gahl, R. Fasel, M. Wolf, and T. Hertel (2001). Anisotropy of quasiparticle lifetimes and the role of disorder in graphite from ultrafast time-resolved photoemission spectroscopy. *Phys Rev Lett*, **87**(26).

Motakabbir, K. A., and G. Grimvall (1981). Thermal-conductivity minimum of metals. *Phys Rev B*, **23**(2): 523–526.

Mounet, N., and N. Marzari (2005). First-principles determination of the structural, vibrational and thermodynamic properties of diamond, graphite, and derivatives. *Phys Rev B*, **71**(20).

Munoz, E., J. X. Lu, and B. I. Yakobson (2010). Ballistic thermal conductance of graphene ribbons. *Nano Lett*, **10**(5): 1652–1656.

Murali, R., Y. X. Yang, K. Brenner, T. Beck, and J. D. Meindl (2009). Breakdown current density of graphene nanoribbons. *Appl Phys Lett*, **94**(24).

Nemanich, R. J., and S. A. Solin (1979). 1st-order and 2nd-order Raman-scattering from finite-size crystals of graphite. *Phys Rev B*, **20**(2): 392–401.

Nika, D. L., S. Ghosh, E. P. Pokatilov, and A. A. Balandin (2009). Lattice thermal conductivity of graphene flakes: comparison with bulk graphite. *Appl Phys Lett*, **94**(20).

Nika, D. L., E. P. Pokatilov, A. S. Askerov, and A. A. Balandin (2009). Phonon thermal conduction in graphene: Role of Umklapp and edge roughness scattering. *Phys Rev B*, **79**(15).

Novoselov, K. S., A. K. Geim, S. V. Morozov, D. Jiang, M. I. Katsnelson, I. V. Grigorieva, S. V. Dubonos, and A. A. Firsov (2005). Two-dimensional gas of massless Dirac fermions in graphene. *Nature*, **438**(7065): 197–200.

Ouyang, T., Y. P. Chen, K. K. Yang, and J. X. Zhong (2009). Thermal transport of isotopic-superlattice graphene nanoribbons with zigzag edge. *EPL*, **88**(2).

Park, J. Y., S. Rosenblatt, Y. Yaish, V. Sazonova, H. Ustunel, S. Braig, T. A. Arias, P. W. Brouwer, and P. L. McEuen (2004). Electron-phonon scattering in metallic single-walled carbon nanotubes. *Nano Lett*, **4**(3): 517–520.

Patsalas, P., M. Handrea, S. Logothetidis, M. Gioti, S. Kennou, and W. Kautek (2001). A complementary study of bonding and electronic structure of amorphous carbon films by electron spectroscopy and optical techniques. *Diam Relat Mater*, **10**(3–7): 960–964.

Perebeinos, V., and J. Tersoff (2009). Valence force model for phonons in graphene and carbon nanotubes. *Phys Rev B*, **79**(24).

Perebeinos, V., J. Tersoff, and P. Avouris (2005a). Effect of exciton-phonon coupling in the calculated optical absorption of carbon nanotubes. *Phys Rev Lett*, **94**(2).

Perebeinos, V., J. Tersoff, and P. Avouris (2005b). Electron-phonon interaction and transport in semiconducting carbon nanotubes. *Phys Rev Lett*, **94**(8).

Pisana, S., M. Lazzeri, C. Casiraghi, K. S. Novoselov, A. K. Geim, A. C. Ferrari, and F. Mauri (2007). Breakdown of the adiabatic Born-Oppenheimer approximation in graphene. *Nat Mater*, **6**(3): 198–201.

Piscanec, S., M. Lazzeri, F. Mauri, A. C. Ferrari, and J. Robertson (2004). Kohn anomalies and electron-phonon interactions in graphite. *Phys Rev Lett*, **93**(18).

Piscanec, S., M. Lazzeri, J. Robertson, A. C. Ferrari, and F. Mauri (2007). Optical phonons in carbon nanotubes: Kohn anomalies, Peierls distortions, and dynamic effects. *Phys Rev B*, **75**(3).

Pocsik, I., M. Hundhausen, M. Koos, and L. Ley (1998). Origin of the D peak in the Raman spectrum of microcrystalline graphite. *J Non-Cryst Solids*, **227**: 1083–1086.

Qian, J., M. J. Allen, Y. Yang, M. Dutta, and M. A. Stroscio (2009). Quantized long-wavelength optical phonon modes in graphene nanoribbon in the elastic continuum model. *Superlattices Microstruct*, **46**(6): 881–888.

Raman, C. V., and K. S. Krishnan (1928a). A new type of secondary radiation. *Nature*, **121**: 501–502.

Raman, C. V., and K. S. Krishnan (1928b). The production of new radiations by light scattering: part I. *Proc R Soc London, A*, **122**(789): 23–35.

Reich, S., and C. Thomsen (2000a). Chirality dependence of the density-of-states singularities in carbon nanotubes. *Phys Rev B*, **62**(7): 4273–4276.

Reich, S., and C. Thomsen (2000b). Comment on "polarized Raman study of aligned multiwalled carbon nanotubes." *Phys Rev Lett*, **85**(16): 3544–3544.

Saito, K., and A. Dhar (2010). Heat conduction in a three dimensional anharmonic crystal. *Phys Rev Lett*, **104**(4).

Savin, A. V., and Y. S. Kivshar (2010). Surface solitons at the edges of graphene nanoribbons. *EPL*, **89**(4).

Savin, A. V., Y. S. Kivshar, and B. Hu (2010). Suppression of thermal conductivity in graphene nanoribbons with rough edges. *Phys Rev B*, **82**(19).

Singh, D., J. Y. Murthy, and T. S. Fisher (2011). On the accuracy of classical and long wavelength approximations for phonon transport in graphene. *J Appl Phys*, **110**(11).

Sinha, K., and J. Menendez (1990). 1st-order and 2nd-order resonant Raman-scattering in graphite. *Phys Rev B*, **41**(15): 10845–10847.

Smekal, A. (1923). Contribution to my work "Remarks on the quantisation of non determined periodic system." *Z Phys*, **15**: 58–60.

Tan, P. H., Y. Tang, C. Y. Hu, F. Li, Y. L. Wei, and H. M. Cheng (2000). Identification of the conducting category of individual carbon nanotubes from Stokes and anti-Stokes Raman scattering. *Phys Rev B*, **62**(8): 5186-5190.

Telg, H., J. Maultzsch, S. Reich, F. Hennrich, and C. Thomsen (2004). Chirality distribution and transition energies of carbon nanotubes [**93**, art. no. 177401, (2004)]. *Phys Rev Lett*, **93**(18).

Thomsen, C., P. M. Rafailov, H. Jantoljak, and S. Reich (2000). Resonant Raman scattering in carbon nanotubes. *Phys Status Solidi B*, **220**(1): 561–568.

Thomsen, C., and S. Reich (2000). Doable resonant Raman scattering in graphite. *Phys Rev Lett*, **85**(24): 5214–5217.

Tuinstra, F., and J. L. Koenig (1970). Raman spectrum of graphite. *J Chem Phys*, **53**(3): 1126.

Vidano, R. P., D. B. Fischbach, L. J. Willis, and T. M. Loehr (1981). Observation of Raman band shifting with excitation wavelength for carbons and graphites. *Solid State Commun*, **39**(2): 341–344.

Wang, H., B. W. Cao, M. Feng, Y. F. Wang, Q. H. Jin, D. T. Ding, and G. X. Lan (2009). Family behaviour of Raman-active phonon frequencies of single-wall nanotubes of C, BN and BC3. *Spectrochim Acta A*, **71**(5): 1932–1937.

Wang, H., Y. F. Wang, X. W. Cao, M. Feng, and G. X. Lan (2009). Vibrational properties of graphene and graphene layers. *J Raman Spectrosc*, **40**(12): 1791–1796.

Wirtz, L., and A. Rubio (2004). The phonon dispersion of graphite revisited. *Solid State Commun*, **131**(3–4): 141–152.

Xie, Z. X., K. Q. Chen, and W. H. Duan (2011). Thermal transport by phonons in zigzag graphene nanoribbons with structural defects. *J Phys: Condens Mater*, **23**(31).

Yan, J. A., W. Y. Ruan, and M. Y. Chou (2008). Phonon dispersions and vibrational properties of monolayer, bilayer, and trilayer graphene: density-functional perturbation theory. *Phys Rev B*, **77**(12).

Yao, Z., C. L. Kane, and C. Dekker (2000). High-field electrical transport in single-wall carbon nanotubes. *Phys Rev Lett*, **84**(13): 2941–2944.

Zhai, X. C., and G. J. Jin (2011). Stretching-enhanced ballistic thermal conductance in graphene nanoribbons. *EPL*, **96**(1).

Zhang, Y., Z. Jiang, J. P. Small, M. S. Purewal, Y. W. Tan, M. Fazlollahi, J. D. Chudow, J. A. Jaszczak, H. L. Stormer, and P. Kim (2006). Landau-level splitting in graphene in high magnetic fields. *Phys Rev Lett*, **96**(13).

Zhou, S. Y., G. H. Gweon, J. Graf, A. V. Fedorov, C. D. Spataru, R. D. Diehl, Y. Kopelevich, D. H. Lee, S. G. Louie, and A. Lanzara (2006). First direct observation of Dirac fermions in graphite. *Nat Phys*, **2**(9): 595–599.

Ziman, J. M. (1960). A note on the selection rules for optical transitions in alloys. *Philos Mag*, **5**(55): 757–758.

Ziman, J. M. (1965). Phonons and phonon interactions. *Philos Mag*, **11**(110): 438.

Chapter 5

Electron Scattering on Atomic Defects and Phonons in Graphene

In this chapter we consider the effect of chirality on conservation of electron one-half pseudospins in the metallic carbon nanotubes. It restricts probabilities of certain scattering processes when electron momentum is located in the vicinity of Dirac points K and K′. In experimental situations it corresponds to low values of the bias voltage V_{SD} and of temperature T, that is, when $\max\{V_{SD}, T\} \leq \Delta$. In particular, conservation of the pseudospin leads to suppression of electron–impurity and of electron–phonon back scattering and thus impacts the whole transport of electrons.

5.1 Pseudospin Conservation during the Scattering of Chiral Fermions

During the electron–impurity scattering process, the spatial part of the wavefunction remains unchanged. In pristine graphene and metallic nanotubes, summation over all probabilities of the possible electron-scattering processes causes the full cancellation of matrix elements, which results in an absence of the backward scattering.

Graphene: Fundamentals, Devices, and Applications
Serhii Shafraniuk

ISBN 978-981-4613-47-7 (Hardcover), 978-981-4613-48-4 (eBook)
www.panstanford.com

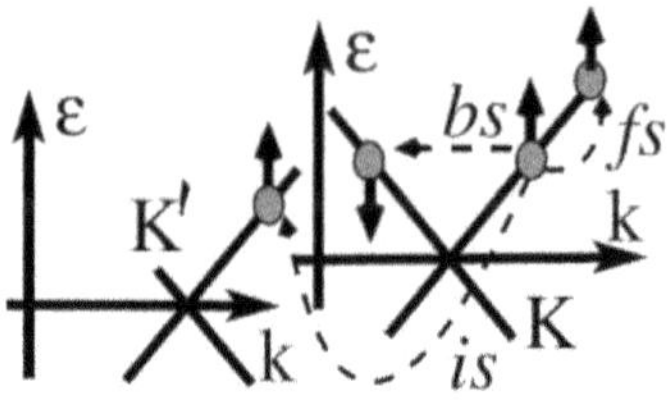

Figure 5.1 The forward scattering (fs), backward scattering (bs), and intervalley scattering (is) processes in the vicinity of K and K′ points. The pseudospins are indicated by the arrows.

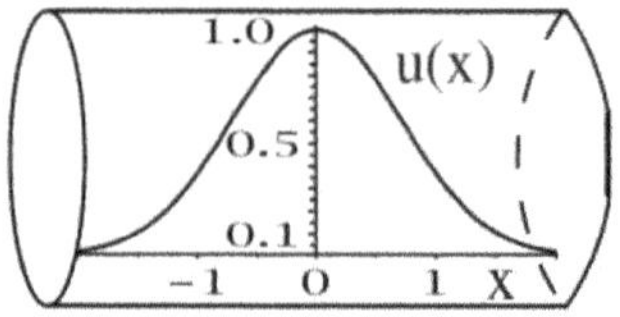

Figure 5.2 Spatial profile of the charged impurity potential inside a narrow graphene stripe.

The backward scattering of chiral electrons corresponds to a spin rotation by $+\pi\ (2m+1)$, where m is an integer, while the time reversal process corresponds to a spin rotation by $-\pi\ (2m+1)$(see sketch in Fig. 5.1). The spatial dependence of the charged impurity potential is shown schematically in Fig. 5.2. Quite a different scenario takes place in the semiconducting nanotubes. There, the wavefunction symmetry at the K and K′ points is destroyed because the states with the pseudospins "↑" and "↓" are hybridized. The wavefunction hybridization is conveniently described in terms of the fictitious flux $\varphi = -(\nu/3)\ \varphi_0$ (Ando 2005, Shafranjuk 2009). It allows us to compute the scattering probability which remains essentially finite for the semiconducting nanotubes. In the latter case, which differs from that of metallic nanotubes, the electron back scattering produces a finite contribution. It indicates an absence of perfectly conducting channel in semiconducting nanotubes. In the vicinity of K point for $n = 0$ and $\nu = \pm 1$, the component along the tube axis direction of electron wavevector is $\mathbf{k}_{\pm} = (-2\pi\nu/3L, \pm|k|)$ where $\pm$ correspond to the right (left) moving excitations. Since the net sign of the electron momentum now is not reversed to the opposite, there is no direct correspondence between the back scattering $\mathbf{k}_+ \rightarrow \mathbf{k}_-$ and the spin rotation either by $+\pi\ (2m+1)$ or by $-\pi\ (2m+1)$. Physically, it means that the one-half pseudospin is not a good quantum number anymore. Because of the contribution of finite back scattering, the semiconducting nanotubes have acquired the nonzero resistance.

Let us discuss the impact of electron–impurity and electron–phonon collisions on the electron transport. The electron transport properties of a material are characterized by the transport coefficients. For instance, Seebeck coefficient $\alpha_\pm$ describes an ability of a material or a device to generate a finite electric voltage when a temperature gradient is applied across the sample. The Seebeck coefficient $\alpha_\pm$ where "$\pm$" indicates the sign of charge carriers, is obtained as

$$\alpha_\pm = -\frac{1}{q_\pm T}\frac{L_\pm^{(1)}}{L_\pm^{(0)}} \tag{5.1}$$

where $q_\pm = \mp e$ is the electric charge of the electron (hole). Besides, using the same formulation, the conductivity of a nanotube is $\sigma_\pm = L_\pm^{(0)}$, whereas the corresponding electron (hole) thermal conductivity is

$$\kappa_\pm = \frac{1}{e^2 T}\left(L_\pm^{(2)} - \frac{\left[L_\pm^{(1)}\right]^2}{L_\pm^{(0)}}\right) \tag{5.2}$$

The functions $L_\pm^{(\alpha)}$ entering the above formulas are defined as

$$L_\pm^{(\alpha)} = \frac{L}{A}\frac{2e^2}{h}N_{\text{ch}}\int (\varepsilon - V_{\text{SD}})^\alpha D_\varepsilon \left|t_\varepsilon^\pm\right|^2 \left(-n_\varepsilon'\right) d\varepsilon \tag{5.3}$$

where

$$n_\varepsilon = \frac{1}{\exp\left(\frac{\varepsilon}{T}\right)+1} \tag{5.4}$$

is the Fermi–Dirac function, L is the nanotube length, $A = \pi d_{\text{T}}^2/4$ is the cross-section area, d_{T} is the nanotube diameter, and $N_{\mathbf{ch}}$ is the number of conducting channels. The functions $L_\pm^{(\alpha)}$ are computed from the microscopic model.

The impurity atoms create additional potential barriers randomly positioned along the tube. The electron–impurity scattering is described by the Boltzmann transport equation. Corresponding integrals $L_{\pm}^{(\alpha)}$ in Eq. 5.3 are modified as

$$L_{\pm}^{(\alpha)} \rightarrow L_{\pm}^{(\alpha)} + L_{\alpha}^{(\mathrm{i})} + L_{\alpha}^{(\mathbf{ep})} \tag{5.5}$$

where the additional terms $L_{\alpha}^{(\mathrm{i})}$ and $L_{\alpha}^{(\mathbf{ep})}$ are determined by the electron–impurity and electron–phonon scattering times $\tau_{\mathrm{i}}(\varepsilon)$ and $\tau_{\mathbf{ep}}(\varepsilon)$ respectively. Thus, one gets

$$L_{\alpha}^{(\mathrm{i,\,ep})} = \frac{e^2 v_{\mathrm{F}}^2}{2} \int d\varepsilon\, D_{\varepsilon} \tau_{\mathrm{i,\,ep}}(\varepsilon) \cdot \left(-n'_{\varepsilon}\right) \left(\varepsilon - V_{\mathbf{SD}}\right)^{\alpha} \tag{5.6}$$

The impurity scattering time of an electron from state k to another state k' is evaluated as

$$\tau_i^{-1}\left(k,\, k'\right) = \frac{2\pi}{\hbar} n_i \Sigma_{k'} |V_{kk'}|^2 \delta\left(\varepsilon_{k'} - \varepsilon_k\right) \tag{5.7}$$

where n_{i} is the number of impurity atoms per unit length and $V_{kk'}$ is the matrix element of the screened electron–impurity interaction. For metallic nanotubes, in the immediate vicinity of the K and K′ points, $k \simeq$ K and $k \simeq$ K′, and therefore one uses (Ando 2005) the following form of matrix elements:

$$V\left(K_{\pm},\, K_{+}\right) = V\left(K'_{\pm},\, K'_{+}\right) = \frac{\pm u_{\mathrm{A}} + u_{\mathrm{B}}}{2} \tag{5.8}$$

and

$$V\left(K_{\pm},\, K'_{+}\right) = V^*\left(K'_{\pm},\, K_{+}\right) = \frac{\mp e^{\mathrm{i}\eta} u'_{\mathrm{A}} - \omega^{-1} e^{-\mathrm{i}\eta} u'_{\mathrm{B}}}{2} \tag{5.9}$$

where the index $+(-)$ denotes forth (back) direction of the electron momentum, u_{A} and u_{B} are the effective potential strengths at the A and B atomic sublattice sites, $\omega = \exp(i\mathbf{K} \cdot \boldsymbol{\tau}_1)$, $\boldsymbol{\tau}_1 = a(0,\ 1/\sqrt{3})$, a is the carbon lattice spacing, and η is the chiral angle. The short-range electron–impurity interaction can be approximated by a Gaussian potential

$$u(x) = \frac{u}{\pi L_{\mathrm{i}}^2} \exp\left(-\frac{(x - x_{\mathrm{B}})^2}{L_{\mathrm{i}}^2}\right) \tag{5.10}$$

located at the B site x_{B}, spread over an effective range $\sim L_{\mathrm{i}}$, $L_{\mathrm{i}} = 1/n_{\mathrm{i}}$, and characterized by an integral intensity u (see sketch in Fig. 5.2). One finds that $u_{\mathrm{A}} \simeq 0$ and $u_{\mathrm{B}} \simeq 2$ at $L_{\mathrm{i}}/a < 1/2$, while $u_{\mathrm{A}} = u_{\mathrm{B}} = 1$ at $L_{\mathrm{i}}/a \geq 1/2$. The magnitude of the effective range L_{i} is actually determined by screening effects. Then the averaged matrix element amplitudes near K

$$\pi d_{\mathrm{T}} L_{\mathrm{i}} \sqrt{\langle |V\,(K_{\mp},\ K_{\pm})|^2 \rangle} \tag{5.11}$$

and near K′

$$\pi d_{\mathrm{T}} L_{\mathrm{i}} \sqrt{\left\langle \left|V\,\left(K'_{\mp},\ K'_{\pm}\right)\right|^2 \right\rangle} \tag{5.12}$$

characterize the intensity of back scattering. The probability of the latter process is sharply decreasing when the potential range L_{i}/a increases and is vanishing at $L_{\mathrm{i}}/a >> 1$. On the contrary, the forward scattering amplitudes

$$\pi d_{\mathrm{T}} L_{\mathrm{i}} \sqrt{\langle |V\,(K_{\pm},\ K_{\pm})|^2 \rangle} \tag{5.13}$$

are constant and at $L_{\mathrm{i}}/a > 1$ become ~ 1. It leads to a perfect electric conductivity of metallic nanotubes. The hybridization of the electron states which are characterized by the pseudospins "↑" and "↓" in the semiconducting nanotubes gives (Ando 2005)

$$V\,(K_{\pm},\ K_{+}) = \frac{\pm u_{\mathrm{A}} F_{-}^2\,(x_{\mathrm{i}}) + u_{\mathrm{B}} F_{+}^2\,(x_{\mathrm{i}})}{2L_{\mathrm{i}}} \tag{5.14}$$

$$V\,\left(K'_{\pm},\ K_{+}\right) = F_{+}\,(x_{\mathrm{i}})\ F_{-}\,(x_{\mathrm{i}})\ \frac{\mp e^{-i\eta} u_{\mathrm{A}}'^{*} - \omega e^{i\eta} u_{\mathrm{B}}'^{*}}{2L_{\mathrm{i}}} \tag{5.15}$$

$$V\,\left(K'_{\pm},\ K_{+}\right) = F_{+}\,(x_{\mathrm{i}})\ F_{-}\,(x_{\mathrm{i}})\ \frac{\mp e^{-i\eta} u_{\mathrm{A}}'^{*} - \omega e^{i\eta} u_{\mathrm{B}}'^{*}}{2L_{\mathrm{i}}} \tag{5.16}$$

$$V\,\left(K'_{\pm},\ K'_{+}\right) = \frac{\pm u_{\mathrm{A}} F_{+}^2\,(x_{\mathrm{i}}) + u_{\mathrm{B}} F_{-}^2\,(x_{\mathrm{i}})}{2L_{\mathrm{i}}} \tag{5.17}$$

and

$$V\,\left(K_{\pm},\ K'_{+}\right) = F_{+}\,(x_{\mathrm{i}})\ F_{-}\,(x_{\mathrm{i}})\ \frac{\mp e^{i\eta} u'_{\mathrm{A}} - \omega^{-1} e^{-i\eta} u'_{\mathrm{B}}}{2L_{\mathrm{i}}} \tag{5.18}$$

where x_{i} is the coordinate of the impurity atom position

$$F_{\pm}\,(x_{\mathrm{i}}) = \frac{1}{\sqrt{\pi d_{\mathrm{T}} I_0\,(2\zeta)}} \exp\left[\pm\zeta \cos\left(2\frac{x - x_{\mathrm{i}}}{d_{\mathrm{T}}}\right)\right] \tag{5.19}$$

and

$$\zeta = \frac{d_{\mathrm{T}}^2}{4L_{\mathrm{i}}^2} = \frac{\nu d_{\mathrm{T}}}{12L_{\mathrm{i}}} \tag{5.20}$$

where $I_0(2\zeta)$ is the modified Bessel function of the first kind. The above formulas illustrate the influence of the electron–impurity scattering on thermoelectric transport in semiconducting nanotubes. However we emphasize that the presence of impurities also causes a reduction of the phonon heat conductivity $\kappa_{\mathbf{ph}}$. It happens because the phonon mean free path becomes shorter, which eventually improves the figure of merit of the thermoelectric devices based on graphene and carbon nanotubes (CNTs), as we will see in Chapter 9.

5.2 The Phonon Drag Effect

Another intrinsic process which impairs the electron transport properties and, specifically, the thermoelectric effects is the phonon drag (Shafranjuk 2009). The phenomenon can have two possible scenarios (Bailyn 1967, Tsaousidou and Papagelis 2007): (*i*) Phonons flowing between the hot and cold ends of the nanotube generate electrons (or holes) moving in the same direction during phonon–electron collisions, and (*ii*) electrons and holes moving from the hot to the cold end of the nanotube generate the phonons. Both mechanisms eventually reduce the net Seebeck coefficient $\alpha_{\pm}$. The phonon drag contribution (Bailyn 1967) into the thermopower is

$$\alpha_{\mathrm{drag}} = \frac{2\,|e|\,k_{\mathrm{B}}}{\sigma}\,\Sigma_{kq}\frac{\partial N_q^0}{\partial T}\cdot\beta_{kq}\left(L_k - L_{k+q}\right)\cdot s_{\mathrm{g}} \tag{5.21}$$

where N_q^0 is the Bose–Einstein function, $L_k = v_k\tau_{\mathrm{ep}}(k)$ is the so-called coherence length (Perebeinos et al. 2005a, 2005b, 2005c), v_k is the electron band velocity, s_{g} is the group velocity of acoustic phonons with momentum $\mathbf{q}$ and dispersion $\omega_q = s_{\mathrm{g}}q$. Besides, for the sake of simplicity we have set the matrix elements β_{kq} of phonon–electron collisions $\beta_{kq} \simeq 1$. One can notice that electron back scattering contributes into the phonon drag as soon as $v_{k+q} = -v_k$. Nevertheless, since the back scattering processes flip the pseudospin, they are prohibited in metallic nanotubes. The back scattering is however partially permitted in semiconducting nanotubes where the states with the pseudospin "↑" and "↓" are hybridized as was noticed above. Long-wavelength acoustic phonons interact with electrons via the effective deformation potential

$$V_1 = g_1 \left(u_{xx} + u_{yy}\right) \tag{5.22}$$

where we have used the following notations

$$u_{xx} = \frac{\partial u_x}{\partial x} + \frac{2u_z}{d_\mathrm{T}} \tag{5.23}$$

$$u_{yy} = \frac{\partial u_y}{\partial y} \tag{5.24}$$

$$2u_{xy} = \frac{\partial u_x}{\partial y} + \frac{\partial u_y}{\partial x} \tag{5.25}$$

For graphene, one can use the numerically estimated value $g_1 = 30$ eV. The deformation potential constitutes a diagonal term V_1 of the matrix Hamiltonian. The nondiagonal part of the electron–phonon interaction is

$$V_2 = g_2 e^{3i\eta} \left(u_{xx} - u_{yy} + 2iu_{xy}\right) \tag{5.26}$$

where the term

$$g_2 = \frac{3}{4}\lambda_1\lambda_2\gamma_0 \tag{5.27}$$

originates from variations of the interatomic distance, which hereby change the magnitude of the transfer integral. For the s and p electrons one uses $\lambda_2 \simeq 2$ and $\lambda_1 \simeq 1/3$, which gives

$$g_2 \simeq \frac{\gamma_0}{2} \simeq 1.5 \text{ eV} \tag{5.28}$$

The relevant part of the electron–phonon interaction matrix element is

$$M_{ep} = \kappa_{\nu\phi}(n)\, V_1 \frac{\sigma^n_{\nu\phi}}{\left|\sigma^n_{\nu\phi}\right|^2} + s\mathrm{Re}\, V_2 \frac{\sigma^n_{\nu\phi}}{\left|\sigma^n_{\nu\phi}\right|} \tag{5.29}$$

where $s = \pm 1$ for conduction (valence) band, $\sigma^n_{\nu\phi} = \kappa_{\nu\phi}(n) - ik$ and

$$\kappa_{\nu\phi}(n) = 2\frac{n + \phi - \nu/3}{d_\mathrm{T}} \tag{5.30}$$

In metallic nanotubes, for the lower band with linear electron dispersion, $\kappa_{\nu\phi}(n) = 0$. Then, the first term in Eq. 5.29 for M_ep vanishes, while a much smaller second term still remains finite.

The electron–phonon scattering time is computed (Perebeinos et al. 2005a, 2005b, 2005c) as

$$\tau_{ep}^{-1}(i \to f) = \frac{2\pi}{\hbar} \cdot \Sigma_{f,q}\left|M_{ep}(i,\ f)\right|^2 \delta\left(E_f - E_i \pm \omega_q\right) \cdot \left[N_q + \delta_{\pm}\right] \tag{5.31}$$

where $\delta_+ = 1, \delta_- = 0$. According to Perebeinos et al. (2005a, 2005b, 2005c) the electron (hole) coupling to acoustic modes (which are relevant to the thermoelectric transport) is 2 orders of magnitude weaker than to the longitudinal optical (LO) phonons. The coherence length

$$L_k = v_k \tau_{\mathbf{ep}}(k) \tag{5.32}$$

is strongly energy dependent because

$$k = \sqrt{\left(\frac{\varepsilon - eU_0}{\hbar v}\right)^2 - q_\nu\,(n)^2} \tag{5.33}$$

The function Eq. 5.32 shows a broad maximum located between the energy of breathing mode and the LO optical phonon energy. The dependence (Eq. 5.32) shows a sharp downturn at energies $\varepsilon \le \varepsilon_c + \Omega_{LO}$ (Perebeinos et al. 2005a, 2005b, 2005c) where ε_c is the conduction band bottom and Ω_{LO} is the LO phonon frequency. Practically, it means that contribution of the acoustic phonons into the phonon drag involves a small fraction of electrons with energies within a very narrow range in the vicinity of $\varepsilon \approx \varepsilon_c + \Omega_{LO}$. For the aforementioned reasons, the phonon drag is diminished when the bias voltage V_{SD} is set outside the mentioned energy range. In the semiconducting nanotubes, the phonon drag is pronounced at arbitrary electron energies. It corresponds to arbitrary values of V_{SD} in transport experiments. For instance, as a result of the phonon drag effect, a relevant decrease of the reduced Seebeck coefficient $\alpha = (\alpha_{\pm} - \alpha_{\mathbf{drag}})/\alpha_{\pm}$ might achieve as much as $\sim 30\%$.

5.3 Screening by Interacting Electrons

Straightforward calculation of the electronic self-energy in the normal metals by using of perturbation series encounters a formal problem. The problem consists of a divergence of series

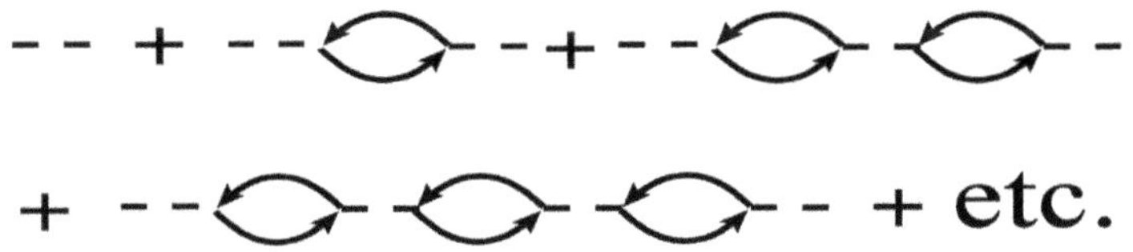

Figure 5.3 The series of bubbles which represents the effective electron interaction (dash lines) in random phase approximation (RPA). The electrons are represented by solid arrows.

at small momentum transfer **q**, which happens due to a singular nature of the Coulomb matrix elements. A fundamental reason of the formal divergence is the long-range nature of the Coulomb potential. The mentioned complications are solved using a technical trick which helps to avoid the singularity. It consists of the sum of a set of graphs which physically represents the screening of the Coulomb potential by the valence electrons. For the sake of simplicity we have disregarded the phonons at this point. According to Bohm and Pines (1953) and also to Gellmann and Brueckner (1957), the most significant screening effects in the long-wave limit $q \to 0$ is taken into account by summing the so-called bubble graphs. Following this path, one replaces the actual bare Coulomb interaction V with some effective potential V_c, which is presented by the series shown in Fig. 5.3. The corresponding approximation for computing the screening is referred to as the random phase approximation (RPA). The calculations are simplified significantly as soon as we approximate the Bloch states with the plane wave states and also neglect the processes in which the momentum is not conserved. Then, the matrix elements of the Coulomb interaction are represented as $V(q) = 4\pi e^2/q^2$. The diagrams of the Feynman series are evaluated as

$$V_c^{\mathrm{RPA}}(q) = V(q) - V(q)\, P^{\mathrm{RPA}}(q)\, V_c^{\mathrm{RPA}}(q) \qquad (5.34)$$

where we have introduced the RPA "irreducible polarizabi1ity" $P(q)$, which is evaluated as

$$P^{\mathrm{RPA}}(q) = 2i \int G_0(p+q) G_0(p) \frac{d^4 p}{(2\pi)^4} \qquad (5.35)$$

In the last Eq. 5.35, factor 2 comes from the sum over electron spins. Equation 5.34 can also be rewritten in the following form:

$$V_c^{\mathrm{RPA}}(q) = \frac{V(q)}{1 + V(q)\,P^{\mathrm{RPA}}(q)} = \frac{V(q)}{\kappa_0(q)} \tag{5.36}$$

where we have used the 4D vector $\mathbf{q} = (\mathbf{q}, q_0)$ and $V(q)$ depends only on $\mathbf{q}^2$ alone. As is evident from Eq. 5.36, the function $\kappa_0(\mathbf{q}, q_0)$, which represents the denominator, is understood as the dielectric function of the valence electrons. The dielectric function $\kappa_0(\mathbf{q}, q_0)$ is evaluated within the RPA and depends on the wavevector and energy (Hubbard 1957, Pines 1957). Using the standard diagram rules, one obtains the explicit form of the RPA polarization bubble as

$$\begin{aligned} \mathrm{Re}P^{\mathrm{RPA}}(\mathbf{q}, q_0) &= P\int \frac{d^3p}{(2\pi)^3} f_p\left(1 - f_{p+q}\right) \frac{2\left(\varepsilon_{p+q} - \varepsilon_p\right)}{q_0^2 - \left(\varepsilon_{p+q} - \varepsilon_p\right)^2} \\ \mathrm{Im}P^{\mathrm{RPA}}(\mathbf{q}, q_0) &= 2\pi \int \frac{d^3p}{(2\pi)^3} f_p\left(1 - f_{p+q}\right) \\ &\quad \times \left[\delta\left(q_0 - \varepsilon_{p+q} + \varepsilon_p\right) + \delta\left(q_0 + \varepsilon_{p+q} - \varepsilon_p\right)\right] \end{aligned} \tag{5.37}$$

where P indicates that the principal part of the integral is to be taken. It gives

$$\begin{aligned} \mathrm{Re}\,\kappa_0(\mathbf{q}, q_0) &= 1 + V(q)\,\mathrm{Re}P^{\mathrm{RPA}}(\mathbf{q}, q_0) \\ &= 1 - 2V(q)\,P\int \frac{d^3p}{(2\pi)^3} f_p\left(1 - f_{p+q}\right) \\ &\quad \times \frac{2\left(\varepsilon_{p+q} - \varepsilon_p\right)}{q_0^2 - \left(\varepsilon_{p+q} - \varepsilon_p\right)^2} \end{aligned} \tag{5.38}$$

and

$$\begin{aligned} \mathrm{Im}\,\kappa_0(\mathbf{q}, q_0) &= 2\pi V(q) \int \frac{d^3p}{(2\pi)^3} f_p\left(1 - f_{p+q}\right) \\ &\quad \times \left[\delta\left(q_0 - \varepsilon_{p+q} + \varepsilon_p\right) + \delta\left(q_0 + \varepsilon_{p+q} - \varepsilon_p\right)\right] \end{aligned} \tag{5.39}$$

The integrals entering the above Eqs. 5.38 and 5.39 were computed for the 3D case in an explicit form by Lindhard (1954):

$$\mathrm{Re}\kappa_0(\mathbf{q}, q_0) = 1 + \frac{2e^2 m k_F}{\pi q^2}\left\{1 + \frac{k_F}{2q}\left[1 - \left(\frac{mq_0}{qk_F} + \frac{q}{2k_F}\right)^2\right]\right.$$
$$\times \ln\left|\frac{1 + \left(\frac{mq_0}{qk_F} + \frac{q}{2k_F}\right)}{1 - \left(\frac{mq_0}{qk_F} + \frac{q}{2k_F}\right)}\right|$$
$$\left. - \frac{k_F}{2q}\left[1 - \left(\frac{mq_0}{qk_F} - \frac{q}{2k_F}\right)^2\right] \ln\left|\frac{1 + \left(\frac{mq_0}{qk_F} - \frac{q}{2k_F}\right)}{1 - \left(\frac{mq_0}{qk_F} - \frac{q}{2k_F}\right)}\right|\right\} \tag{5.40}$$

whereas the imaginary part is

$$Im\kappa_0(\mathbf{q}, q_0) = \begin{cases} 0 \quad \text{for} \quad 2m|q_0| > q^2 + 2qk_F \\ 0 \quad \text{for} \quad q > 2k_F \quad \text{and} \quad 2m|q_0| < q^2 + 2qk_F \\ 2\mathrm{e}^2\mathrm{m}^2\frac{q_0}{q^3} \text{ for } q < 2k_F \text{ and } 2m|q_0| < \left|q^2 - 2qk_F\right| \\ \frac{e^2 m k_F^2}{q^3}\left\{1 - \left(\frac{mq_0}{qk_F} - \frac{q}{2k_F}\right)^2\right\} \\ \text{for } \left|q^2 - 2qk_F\right| < 2m|q_0| < \left|q^2 + 2qk_F\right| \end{cases} \tag{5.41}$$

From the last Eq. 5.41, one obtains the static dielectric constant at zero energy by setting $q_0 = 0$, which gives $Im\kappa_0(\mathbf{q}, q_0) = 0$ and also

$$\kappa_0(\mathbf{q}, 0) = 1 + 0.66 r_s \left(\frac{k_F}{q}\right)^2 u\left(\frac{q}{2k_F}\right) \tag{5.42}$$

where

$$u(x) = \frac{1}{2}\left[1 + \frac{1 - x^2}{2x}\ln\left|\frac{1 + x}{1 - x}\right|\right] \tag{5.43}$$

and

$$\frac{4\pi r_s^3 a_0^3}{3} = \frac{1}{n} \tag{5.44}$$

In the last formula, Eq. 5.44, we have introduced the Bohr radius a_0, whereas r_s is a dimensionless measure of strength the Coulomb interaction. A comparison between $1/\kappa_0(\mathbf{q}, 0)$ given by Eqs. 5.42–5.44 and the Thomas–Fermi result is provided in Fig. 5.4 (left). From the plot in Fig. 5.4 (left) one can see the following features

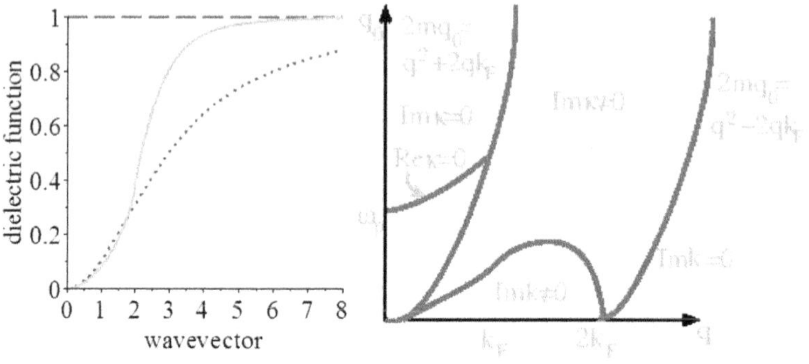

Figure 5.4 Left: The zero-frequency dielectric function $1/\mathrm{Re}\kappa\,(q, 0)$ versus the reduced wave number $q/2k_\mathrm{F}$ in RPA (solid line) and $1/\kappa_0\,(\mathbf{q}, q_0)$ in the Thomas–Fermi approximation (dotted line). Right: A diagram showing the behavior of the imaginary part of the RPA dielectric function for the general wave number q and frequency q_0.

1. In the limit $q \to 0$ one obtains the Thomas–Fermi result for κ_0

$$\kappa_0\,(\mathbf{q}, 0) = 1 + \frac{k_\mathrm{s}^2}{q^2} \tag{5.45}$$

   where the wave number of screening is defined as

$$k_\mathrm{s}^2 = \frac{6\pi n e^2}{E_\mathrm{F}} \tag{5.46}$$

2. In the other limit $q \to 2k_\mathrm{F}$ one arrives at the result, obtained by Langer and Vosko (1959) which corresponds to $d\kappa_0/dq \to \infty$. It corresponds to an asymptotic form of the screened Coulomb potential for large distance in the form of the oscillatory function

$$V\,(r) \propto \frac{\cos\,(2k_\mathrm{F}r + \phi)}{r^3} \tag{5.47}$$

   The above Eq. 5.47 differs from the Yukawa potential. The latter asymptotic Eq. 5.47 is confirmed in the experiments (Rowland 1962).
3. When the transfer of momentum is large, that is, $\mathbf{q} \to \infty$ one obtains $\kappa_0 \to 1$ which means that the screening becomes ineffective.

The imaginary part of the dielectric function κ becomes nonzero as soon as q_0 and q are interrelated, while the energy transfer is finite, $q_0 \neq 0$. In particular, if $|\mathbf{p}| < p_{\mathrm{F}}$ and $|\mathbf{p}+\mathbf{q}| > p_{\mathrm{F}}$, the delta function argument in Eq. 5.39 can vanish provided that

$$q^2 - 2qk_{\mathrm{F}} \leq 2m\,|q_0| \leq q^2 + 2qk_{\mathrm{F}} \tag{5.48}$$

That condition is illustrated by the diagram in Fig. 5.4 (right). If the transfer energy is large, that is, $|q_0| >> q^2 + (qk_{\mathrm{F}}/m)$, we arrive at the familiar limiting form

$$\mathrm{Re}\kappa_0\,(\mathbf{q}, q_0) = 1 - \frac{\omega_p^2}{q_0^2} \tag{5.49}$$

where we have defined the so-called *plasma frequency*

$$\omega_p^2 = \frac{4\pi n}{m} e^2 \tag{5.50}$$

The above two formulas, Eqs. 5.49 and 5.50, help to understand the effects of screening better. To include all the vacuum polarization processes, it is instructive to introduce an effective potential $V_c\,(q)$ (Hubbard 1957). Thus, we expand our model beyond the RPA limit, which is restricted by small transfers of electron momentum. It is achieved by introducing of the irreducible polarizability $P(q)$, for which a partial summation of the series is performed. The auxiliary function $P(q)$ is then defined as the sum of all graphs shown in Fig. 5.5, where a single interaction line enters and leaves the graph and which cannot be divided into any of two disconnected graphs by mere cutting of a single Coulomb line. Then we write

$$V_c\,(q) = \frac{V\,(q)}{1 + V\,(q)\,P\,(q)} \equiv \frac{V\,(q)}{\kappa\,(q)} \tag{5.51}$$

The RPA result corresponds to the first term in the series for $P(q)$. A compact expression for $\kappa\,(q)$ is obtained by considering the average

$$\langle 0|\, T\left\{\rho_{-q}\,(-\tau)\,\rho_q\,(0)\right\} |\rangle \tag{5.52}$$

The above average Eq. 5.52 is evaluated in the Heisenberg representation. In Eq. 5.52, ρ_q is the qth Fourier component of the electron density operator, which is defined as

$$\rho_q\,(t) = \sum_s \int e^{i\mathbf{q}\cdot\mathbf{r}} \psi_s^\dagger\,(\mathbf{r}, t)\,\psi_s\,(\mathbf{r}, t)\,d^3r = \sum_{p,s} c_{p,s}^\dagger\,(t)\,c_{p+q,s}\,(t) \tag{5.53}$$

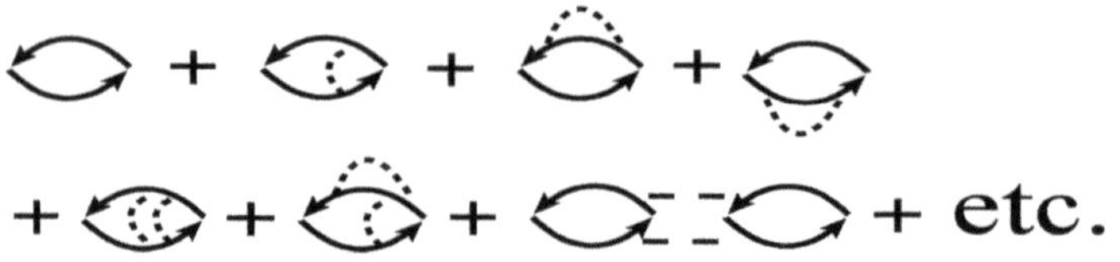

Figure 5.5 The perturbation series for the irreducible polarization propagator $P(q)$. RPA corresponds to the first term in the series.

At this stage the contribution of phonons is negligible. The Fourier transform of Eq. 5.52 over the time variable τ is related to the series for $V(q)$ when it is expanded as a perturbation series over the strength of Coulomb interaction

$$\frac{V_c(q) - V(q)}{V(q)} \equiv \frac{1}{\kappa(q)} - 1 = -iV(q)\int_{-\infty}^{\infty} e^{iq_0\tau}\langle 0|T\left\{\rho_{-q}(-\tau)\rho_q(0)\right\}|\rangle d\tau \tag{5.54}$$

The last Eq. 5.54 is instructive and helps to understand the analytical structure of $\kappa_0(\mathbf{q}, q_0)$ as a function of q_0 for fixed q. The Green function $G(\mathbf{p}, p_0)$ can be computed by inserting a complete set of eigenstates of the Hamiltonian between $\rho_{-q}(-\tau)$ and $\rho_q(0)$ in Eq. 5.52 which yields the spectral representation

$$\frac{1}{\kappa(q)} - 1 = \int_{-\infty}^{\infty} \frac{F(\mathbf{q}, \omega)}{q_0 - \omega + iq_0\delta} d\omega \tag{5.55}$$

where

$$F(\mathbf{q}, \omega) = V(q)\sum_n \left|\langle n|\rho_{-q}|\rangle\right|^2 \delta(\omega - \omega_{n0}) \quad \text{for} \quad \omega > 0 \tag{5.56}$$

and

$$F(\mathbf{q}, \omega) = -F(-\mathbf{q}, |\omega|) \text{ for } \omega < 0 \tag{5.57}$$

If the electron system is symmetric in respect to the inversion, Eq. 5.57 is equivalent to $F(\mathbf{q}, \omega) = -F(\mathbf{q}, -\omega)$ which for Eq. 5.55 gives

$$\frac{1}{\kappa(q)} - 1 = \int_{-\infty}^{\infty} F(\mathbf{q}, \omega)\left(\frac{1}{q_0 - \omega + iq_0\delta} - \frac{1}{q_0 + \omega + iq_0\delta}\right) d\omega \tag{5.58}$$

There is a similarity between the right-hand side of Eq. 5.58 and the spectral representation of the phonon (boson) Green's function. An essential difference between the two objects is that the weight function $F(\mathbf{q}, \omega)$ involves matrix elements of the electronic density fluctuation operator, whereas the longitudinal phonon weight function $\Phi(\mathbf{q}, \omega)$ involves matrix elements of the ionic density fluctuation operator. However, here we use an analogy between the two mentioned functions, because the poles of $1/\kappa(\mathbf{q}, q_0)$ correspond to the boson-type excitations of the electron gas which are excited by a longitudinal field.

5.4 Plasma Oscillations

The RPA model helps to understand the nature of the Bose excitations, introduced above. The imaginary part of Eq. 5.55 is

$$F(\mathbf{q}, q_0) = \begin{cases} -\frac{1}{\pi}\mathrm{Im}\frac{1}{\kappa(\mathbf{q}, q_0)} & \text{for } q_0 > 0 \\ \frac{1}{\pi}\mathrm{Im}\frac{1}{\kappa(\mathbf{q}, q_0)} & \text{for } q_0 < 0 \end{cases} \tag{5.59}$$

or in another form

$$F(q) = -\frac{1}{\pi}\frac{\kappa_2(q)\,\mathbf{sign}q_0}{\kappa_1^2(q) + \kappa_2^2(q)} \tag{5.60}$$

In Eqs. 5.59 and 5.60 the functions $\kappa_1(q)$ and $\kappa_2(q)$ are the real and imaginary parts of $\kappa(q)$, respectively. As it follows from the RPA model, for a given momentum $\mathbf{q}$, the interval

$$q^2 - 2qk_F \leq q_0 \leq q^2 + 2qk_F \tag{5.61}$$

corresponds to the continuum of boson excitations because $\kappa_2(q)$ is finite on this interval but vanishes elsewhere. The mentioned branch of boson-like excitations is identical to excitations of the noninteracting system. During the corresponding elementary process, an electron initially occupies state $\mathbf{p}$ within the Fermi sea, whereas the excited electron state is raised above the Fermi surface to a state $\mathbf{p} + \mathbf{q}$. Finite strength of correlations taking place

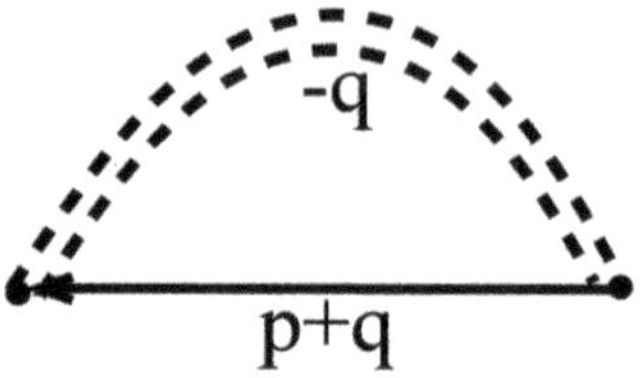

Figure 5.6 The screened exchange approximation for computing the electron self-energy. The electron momentum is $\mathbf{p}+\mathbf{q}$, whereas the Coulomb line transfers the momentum $-\mathbf{q}$. Thus, one obtains the most important contribution to the electron excitation energy $E\,(p)$, which comes from small momentum transfer processes ($|\mathbf{q}| << k_F$).

between the excited electron (or hole) and the background electrons introduces a difference between the "wavefunctions" of these excitations on the one hand and the single particle-like excitations as in the noninteracting case, on the other hand. Nevertheless, according to RPA, the corresponding energies are identical in the two cases. One can imagine it as the excited electron pushes the other adjacent electrons away, thus producing a "correlation hole," which accompanies the excited particle during its propagation through the system. Both the electron and the hole are combined into a Bose excitation. The local depletion of the background electrons creates the electric field. The field's lines are terminated at the fixed positive ions and are vanishing over long distances. According to the Gauss law, the absence of long-range electric field associated with the excitation, means that the net effective charge of the excitation is zero. In this scenario, when an excited electron moves through the system, a backflow of background electrons must accompany the excitation in such a way that the net transported charge remains zero. The whole picture is similar to the backflow of excitations in superfluid He^4, as was suggested by Feynman and Cohen. Technically, the backflow helps to obtain consistency when describing the response of the system to external fields. Without introducing the backflow, the continuity condition for electronic charge becomes violated which also causes violation of the gauge invariance. In the high-frequency and long-wavelength limit one obtains that $\kappa_1\,(q)$ vanishes as soon as

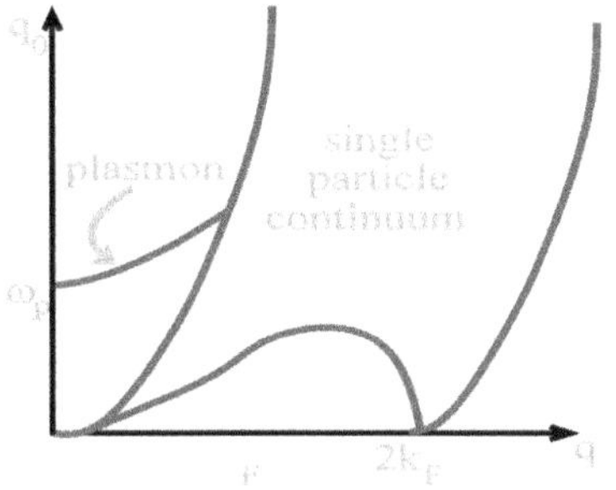

Figure 5.7 The RPA model of the excitation spectrum of the electron gas treated.

$$q_0^2 = \omega_{\mathrm{p}}^2 = \frac{4\pi ne^2}{m} \tag{5.62}$$

The above parameter ω_{p} is regarded as the electronic plasma frequency, Eq. 5.50, which is equal to the excitation energy and characterizes an excited state of the system. Such elementary Bose-type excitations had been named *plasmons* by Bohm and Pines. The plasmons correspond to the density–fluctuation waves of the electron system. If the transferred momentum **q** is sufficiently large, the plasmons are heavily damped and are transformed into the single particle-like continuum as sketched in Fig. 5.7. The plasmons actually determine the longitudinal response of the system.

Properties of electron excitations in the system with a finite interaction can be described introducing one-electron Green's function. In the electrically neutral and translation invariant system, the Hartree term in the self-energy Σ vanishes. The interaction effects are included using the Dyson equation. Thus, one takes into account the screening effect where the exchange self-energy shown in Fig. 5.6 is incorporated into the calculations within the RPA model with the screened potential $V_{\mathrm{c}}(q)$ instead of the bare Coulomb matrix element $V(q)$

$$\sum(p) = i\int G_0(p+q)\, V_{\mathrm{c}}(-q)\,\frac{d^4q}{(2\pi)^4} \tag{5.63}$$

Although the electron Green function G_0 and the screened potential V_{c} are available in explicit form, the computing of Eq. 5.63 for general p is not straightforward. Nevertheless, the mentioned calculation had been performed by Quinn and Ferrell. They have obtained the quasiparticle energy near the Fermi surface as

$$E_p = E_F \left\{ \frac{p^2}{k_F^2} - 0.166 \cdot r_s \left[\frac{p}{k_F} (\ln r_s + 0.203) + \ln r_s - 1.9 \right] \right\}, \quad (5.64)$$

whereas creating of the electron–hole pairs causes the damping rate

$$\left| 2\Gamma^{\text{pair}}(p) \right| = 2E_F \left(0.252 r_s^{1/2} \right) \left| \left(\frac{p}{k_F} \right)^2 - 1 \right| \quad (5.65)$$

One also obtains the effective mass of the quasiparticle at the Fermi surface by a mere differentiating of Eq. 5.64

$$\frac{1}{m^*} = \frac{1}{k_F} \frac{\partial E_p}{\partial p} \bigg|_{p=k_F} = \frac{1}{m} [1 - 0.083 \cdot r_s \cdot (\ln r_s + 0.203)] \quad (5.66)$$

Because the effective mass m^* actually determines the electronic specific heat C of a metal at low temperatures, one also obtains that

$$\frac{C}{C_{\text{free}}} = 1 + 0.083 \cdot r_s \cdot (\ln r_s + 0.203) \quad (5.67)$$

From the above formulas it is clear that because fulfillment of the condition $|E(p) - E_F| >> |\Gamma(p)|$, the quasiparticles are well defined, which holds valid at $|\varepsilon_p - E_F| \le E_F/5$. Strictly speaking, the above formulas in Eqs. 5.64 and 5.65 are correct in the limit of high electron density ($r_s < 1$). As is evident, such an assumption fails if the phonons are included. In real metals, r typically is $2 < r < 5$; thus one should be cautious when using Eqs. 5.64 and 5.65. One can successfully account for the Coulomb potential in the long wavelength limit by implementing the exchange graph screened within the RPA (see Fig. 5.8). The second-order diagram shown

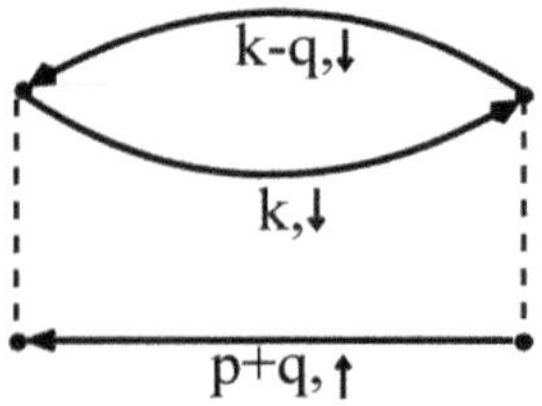

Figure 5.8 Leading contribution to the electron self-energy when the momentum transfers are large ($|\mathbf{q}| >> k_F$).

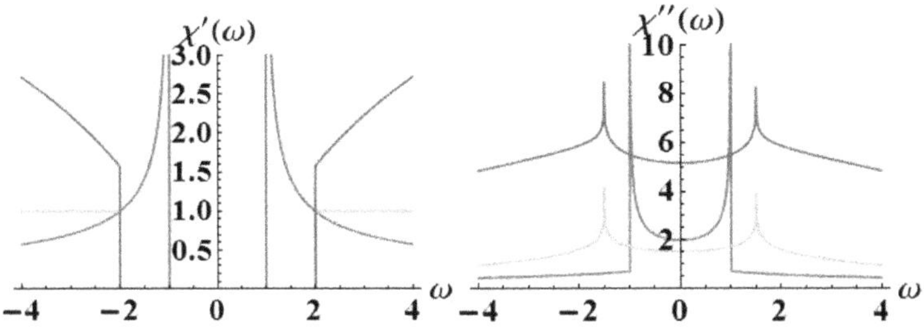

Figure 5.9 Frequency dependence of electron susceptibility. Left: Re$\chi(\omega)$ for $D = 1, 2,$ and 3, $\delta = 1$, $\mu = 0.5$, and $\Lambda = 16$. Right: Im$\chi(\omega)$ for $D = 1, 2,$ and 3, $\delta = 1$, $\mu = 0.5$, and $\Lambda = 16$.

in Fig. 5.9 describes the opposite limit of very short wavelengths, that is, the short-range interaction. In that case, the electrons with parallel spins do not interact with each other because they are spatially separated due to the Pauli principle.

5.5 Plasma Excitations in Graphene

5.5.1 *Electron Susceptibility for 3D and 2D Cases*

The optical susceptibility shown in Figs. 5.9–5.11 is expressed by

$$\chi(\omega) = -\sum_k \frac{|d_{cv}|^2}{L^3}(f_{v,k} - f_{c,k}) \times \left[\frac{1}{\varepsilon_{v,k} - \varepsilon_{c,k} + \omega + i\gamma} - \frac{1}{\varepsilon_{c,k} - \varepsilon_{v,k} + \omega + i\gamma}\right] \quad (5.68)$$

where indices v and c are attributed to the valence and conducting bands, respectively. The imaginary part of Eq. 5.68 is written in the form

$$\chi''(\omega) = -\pi \sum_k \frac{|d_{cv}|^2}{L^3}(f_{v,k} - f_{c,k})\,\delta(\varepsilon_{v,k} - \varepsilon_{c,k} + \omega) \quad (5.69)$$

The sum over the electron momentum **k** entering the above Eqs. 5.68 and 5.69 can be converted to integrals. The sum over the *D*-dimensional **k** vector is converted as

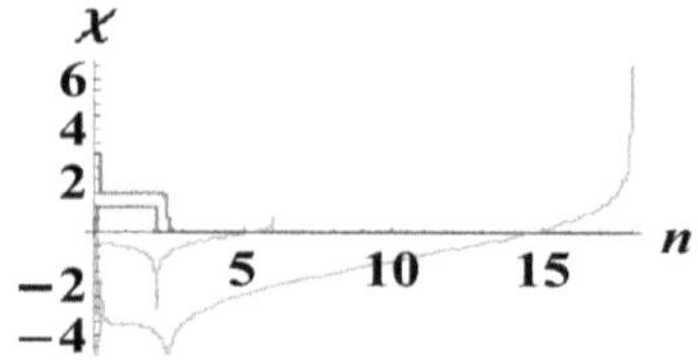

Figure 5.10 Electron susceptibility $\chi\,(n)$ versus electron concentration n for 2D graphene (red) and a 2D semiconductor (green).

$$\sum_k = \sum_k \frac{(\Delta k)^D}{(\Delta k)^D} \rightarrow \left(\frac{L}{2\pi}\right)^D \int dk^D = \left(\frac{L}{2\pi}\right)^D \int d\Omega_D \int_0^\infty dk{\cdot}k^{D-1} \tag{5.70}$$

where Ω_D is the space angle in D dimensions. If the integrand is isotropic, the space angle integral yields Ω_D which equals 4π in 3D, 2π in 2D, and 2 in 1D, respectively. The k integral is transformed to an integral over the energy variable E with $\hbar^2k^2/2m = E$ and $(\hbar^2/2m)\; 2kdk = dE$, which gives

$$\sum_k = \left(\frac{L}{2\pi}\right)^D \frac{1}{2} \int d\Omega_D \left(\frac{2m}{\hbar^2}\right)^{D/2} \int_0^\infty dE \cdot E^{D/2-1} \tag{5.71}$$

where the integrand $\rho_D\,(E) = E^{D/2-1}$ is the *density of states.*

5.5.2 *Optical Susceptibility*

The conversion from sums to integrals in Eqs. 5.68 and 5.69 gives

$$\chi\,(\omega) = \left(\frac{L}{2\pi}\right)^D \int d\Omega_D \int_0^\infty dk \cdot k^{D-1} \frac{|d_{cv}|^2}{L^3} \left(f_{\mathrm{v},\,k} - f_{\mathrm{c},\,k}\right)$$
$$\times \left[\frac{1}{\hbar\,(\varepsilon_{\mathrm{c},\,k} - \varepsilon_{\mathrm{v},\,k} + \omega + i\gamma)} - \frac{1}{\hbar\,(\varepsilon_{\mathrm{v},\,k} - \varepsilon_{\mathrm{c},\,k} + \omega + i\gamma)}\right] \tag{5.72}$$

The absorption is defined as

$$\alpha\,(\omega) = \frac{4\pi\omega}{n_{\mathrm{B}}c}\chi''\,(\omega) = \frac{8\pi^2\omega}{L^3 n_{\mathrm{B}}c} \sum_k |d_{\mathrm{cv}}|^2 \left(f_{\mathrm{v},\,k} - f_{\mathrm{c},\,k}\right) \delta\left(\varepsilon_{\mathrm{v},\,k} - \varepsilon_{\mathrm{c},\,k} + \omega\right), \tag{5.73}$$

while the refractive part is

$$\rho(\omega) = \frac{4\pi\omega}{n_{\rm B}c}\chi'(\omega) = \frac{8\pi^2\omega}{L^3 n_{\rm B}c}\sum_k |d_{cv}|^2 (f_{v,k} - f_{c,k}) \times \frac{E_k + E_{\rm g} + E_0^{(D)}}{\left(E_k + E_{\rm g} + E_0^{(D)}\right)^2 - \hbar^2\omega^2} \tag{5.74}$$

The negative electron mass $m_{\rm v} < 0$ in the valence band is avoided by introducing the holes with a negative effective mass $m_{\rm h} = -m_{\rm v} > 0$. Then

$$f_{{\rm v},k} \to 1 - f_{{\rm h},k} \tag{5.75}$$

and one uses

$$\frac{1}{\varepsilon_{{\rm c},k} - \varepsilon_{{\rm v},k} + \omega} - \frac{1}{\varepsilon_{{\rm v},k} - \varepsilon_{{\rm c},k} + \omega} = 2\frac{\varepsilon_{{\rm v},k} - \varepsilon_{{\rm c},k}}{\omega^2 - \left(\varepsilon_{{\rm v},k} - \varepsilon_{{\rm c},k}\right)^2} = 2\frac{E_k + E_{\rm g} + E_0^{(D)}}{\left(E_k + E_{\rm g} + E_0^{(D)}\right)^2 - \hbar^2\omega^2} \tag{5.76}$$

$$\hbar\left(\varepsilon_{{\rm c},k} - \varepsilon_{{\rm v},k}\right) = \frac{\hbar^2k^2}{2m_c} - \frac{\hbar^2k^2}{2m_v} + E_{\rm g} = \frac{\hbar^2k^2}{2m_{\rm r}} + E_{\rm g} \text{ where } \frac{1}{m_{\rm r}} = \frac{1}{m_{\rm e}} + \frac{1}{m_{\rm h}} \tag{5.77}$$

then one arrives at

$$\alpha(\omega) = \frac{8\pi^2\omega}{n_{\rm B}cL^3}\left(\frac{L}{2\pi}\right)^D \int d\Omega_D \int_0^\infty dk \cdot k^{D-1}|d_{\rm cv}|^2 \times\delta\left(\frac{\hbar^2k^2}{2m_{\rm r}} + E_{\rm g} + E_0^{(D)} - \hbar\omega\right)\left(1 - f_{{\rm e},k} - f_{{\rm h},k}\right) = \frac{8\pi^2\omega}{n_{\rm B}cL^{3-D}}\frac{|d_{\rm cv}|^2\,\Omega_D}{(2\pi)^D}S_D \tag{5.78}$$

and

$$\rho(\omega) = \frac{8\pi^2\omega}{n_{\rm B}cL^3}\left(\frac{L}{2\pi}\right)^D \int d\Omega_D \int_0^\infty dk \cdot k^{D-1}|d_{\rm cv}|^2 \times\frac{E_k + E_{\rm g} + E_0^{(D)}}{\left(E_k + E_{\rm g} + E_0^{(D)}\right)^2 - \hbar^2\omega^2}\left(1 - f_{{\rm e},k} - f_{{\rm h},k}\right) = \frac{8\pi^2\omega}{n_{\rm B}cL^{3-D}}\frac{|d_{\rm cv}|^2\,\Omega_D}{(2\pi)^D}\mathbf{L}_D \tag{5.79}$$

where

$$\begin{aligned}
S_D &= \int_0^\infty dk \cdot k^{D-1}\delta\left(\frac{\hbar^2 k^2}{2m_\mathrm{r}} + E_\mathrm{g} + E_0^{(D)} - \hbar\omega\right)(1 - f_{\mathrm{e},\,k} - f_{\mathrm{h},\,k}) \\
&= \int_0^\infty \sqrt{m_\mathrm{r}/\left(2E_k\hbar^2\right)}dE_k \cdot \left(\sqrt{2m_\mathrm{r}E_k/\hbar^2}\right)^{D-1} \\
&\quad \times\delta\left(E_k + E_\mathrm{g} + E_0^{(D)} - \hbar\omega\right)(1 - f_{\mathrm{e},\,k} - f_{\mathrm{h},\,k}) \\
&= \sqrt{\frac{m_\mathrm{r}}{2\hbar^2}}\int_0^\infty \frac{dE_k}{\sqrt{E_k}} \cdot \left(\frac{2m_\mathrm{r}E_k}{\hbar^2}\right)^{(D-1)/2} \\
&\quad \times\delta\left(E_k + E_\mathrm{g} + E_0^{(D)} - \hbar\omega\right)(1 - f_{\mathrm{e},\,k} - f_{\mathrm{h},\,k}) \\
&= \frac{1}{2}\left(\frac{2m_\mathrm{r}}{\hbar^2}\right)^{\frac{D}{2}}\int_0^\infty dE_k \cdot E_k^{\frac{1}{2}D-1} \\
&\quad \times\delta\left(E_k + E_\mathrm{g} + E_0^{(D)} - \hbar\omega\right)(1 - f_{\mathrm{e},k} - f_{\mathrm{h},\,k}) \qquad (5.80)
\end{aligned}$$

We also use the identity

$$\begin{aligned}
\int_0^\infty dE_k \cdot E_k^{\frac{1}{2}D-1}\delta\left(E_k + E_\mathrm{g} + E_0^{(D)} - \hbar\omega\right) &= \left(\hbar\omega - E_\mathrm{g} - E_0^{(D)}\right)^{\frac{1}{2}D-1} \\
&\quad \times\theta\left(\hbar\omega - E_\mathrm{g} - E_0^{(D)}\right) \qquad (5.81)
\end{aligned}$$

which gives

$$\begin{aligned}
L_D = &\sqrt{\frac{m_r}{2\hbar^2}}\int_0^\infty \frac{dE_k}{\sqrt{E_k}} \cdot \left(\frac{2m_r E_k}{\hbar^2}\right)^{(D-1)/2} \\
&\times\frac{E_k + E_g + E_0^{(D)}}{\left(E_k + E_g + E_0^{(D)}\right)^2 - \hbar^2\omega^2}(1 - f_{e,\,k} - f_{h,\,k}) \\
&\times\frac{1}{2}\left(\frac{2m_r}{\hbar^2}\right)^{\frac{D}{2}}\int_0^\infty dE_k E_k^{\frac{1}{2}D-1} \\
&\times\frac{E_k + E_g + E_0^{(D)}}{\left(E_k + E_g + E_0^{(D)}\right)^2 - \hbar^2\omega^2}(1 - f_{e,\,k} - f_{h,\,k}) \qquad (5.82)
\end{aligned}$$

where $E_k = \hbar^2 k^2/2m_r$, $\sqrt{2m_r E_k/\hbar^2} = k$, $dE_k = (\hbar^2/m_r)\ kdk$, and $\sqrt{m_r/\left(2E_k\hbar^2\right)}dE_k = dk$. Then, one gets

$$\begin{aligned} S_D &= \frac{1}{2}\left(\frac{2m_r}{\hbar^2}\right)^{\frac{D}{2}} \int_0^\infty dE_k E_k^{\frac{1}{2}D-1} \delta\left(E_k + \hbar\delta - \hbar\omega\right)\left(1 - f_{e,k} - f_{h,k}\right) \\ &= \frac{1}{2}\left(\frac{2m_r}{\hbar^2}\right)^{\frac{D}{2}} \left(\hbar\left(\omega - \delta\right)\right)^{\frac{1}{2}D-1} \theta\left(\hbar\left(\omega-\delta\right)\right) A\left(\omega\right) \end{aligned} \tag{5.83}$$

where $\delta = \left(E_g + E_0^{(D)}\right)/\hbar$ and the distribution factor

In the limit $T \to 0$ one also gets

$$S_D = \frac{1}{2}\left(\frac{2m_r}{\hbar^2}\right)^{\frac{D}{2}} \left(\hbar\omega - \hbar\delta\right)^{\frac{D}{2}-1} \cdot \theta\left(\hbar\omega - \hbar\delta\right) A\left(\omega\right) \tag{5.84}$$

5.5.3 *The 1D Case*

In the 1D case one simply gets

$$\begin{aligned} \mathbf{L}_D^{(0)} &= \frac{1}{2}\left(\frac{2m_r}{\hbar^2}\right)^{\frac{D}{2}} \int_0^\infty x^{\frac{1}{2}D-1} \frac{x + \hbar\delta}{\left(x+\hbar\delta\right)^2 - \left(\hbar\omega\right)^2} dx \\ &= \frac{\pi}{4}\sqrt{\frac{2m_r}{\hbar^3}}\left(\frac{1}{\sqrt{\delta-\omega}} + \frac{1}{\sqrt{\delta+\omega}}\right) \end{aligned} \tag{5.85}$$

However, for $D = 2, 3$, the corresponding integral diverges, and one needs a cutoff

$$L_D^{(0)} = \frac{1}{2}\left(\frac{2m_r}{\hbar^2}\right)^{\frac{D}{2}} \int_\mu^{v_F\Lambda} x^{\frac{1}{2}D-1} \frac{x + \hbar\delta}{\left(x+\hbar\delta\right)^2 - \left(\hbar\omega\right)^2} dx \tag{5.86}$$

where the lower limit mimics the distribution function. For the spatial dimension $D = 2$ one gets

$$\frac{2i}{\pi}\int_\mu^{v_F\Lambda} x^{\frac{1}{2}D-1} \frac{x + \hbar\delta}{\left(x+\hbar\delta\right)^2 - \left(\hbar\omega\right)^2} dx = -i\pi \log\left(\frac{\left(\Lambda+\delta\right)^2 - \omega^2}{\left(\mu+\delta\right)^2 - \omega^2}\right), \tag{5.87}$$

while for $D = 3$ it becomes

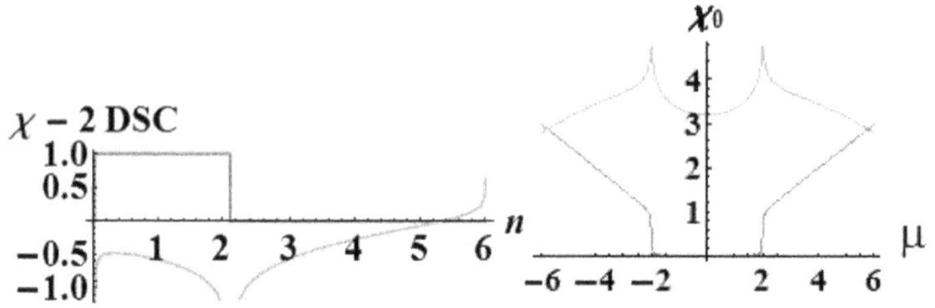

Figure 5.11 Left: The electron susceptibility $\chi\,(n)$ versus concentration for graphene (red) and a 2D semiconductor (green). Right: Similar plot $\chi\,(\mu)$ but versus the chemical potential μ.

$$
\begin{aligned}
&\frac{2i}{\pi}\int_{\mu}^{v_F\Lambda} x^{\frac{1}{2}D-1}\frac{x+\hbar\delta}{(x+\hbar\delta)^2-(\hbar\omega)^2}dx \\
&= i\sqrt{\delta}+\omega\left(\tan^{-1}\left(\frac{\sqrt{\mu}}{\sqrt{\delta+\omega}}\right)-\tan^{-1}\left(\frac{\sqrt{\Lambda}}{\sqrt{\delta+\omega}}\right)\right) \\
&\quad + 2i\left(\sqrt{\Lambda}-\sqrt{\mu}\right)+\sqrt{\omega}-\delta\tan^{-1}\left(\frac{\sqrt{\Lambda}}{\sqrt{\delta-\omega}}\right)-\sqrt{\delta} \\
&\quad -\omega\tan h^{-1}\left(\frac{\sqrt{\mu}}{\sqrt{\omega-\delta}}\right)
\end{aligned}
\tag{5.88}
$$

Above we've used that

$$
\tan^{-1}(a) = i\tanh^{-1}(-ia) \tag{5.89}
$$

and

$$
\sqrt{-(\delta-\omega)^2} = i\,|\delta-\omega| \tag{5.90}
$$

5.5.4 *The Relationship between n and μ*

In many practical applications it is very useful to know the relationship between the charge carrier density n and the chemical potential μ. The total number of particles is obtained from the sum rule

$$
N = \sum_{\mathbf{k},\,s} f_k = 2\sum_{\mathbf{k}} f_k \tag{5.91}
$$

which determines the chemical potential $\mu = \mu\,(n,\ T)$ as a function of the charged particle density n and temperature T. The sum is converted as

$$\sum_k \to \int_0^\infty d\varepsilon \rho^{(D)}(\varepsilon) \tag{5.92}$$

where the D-dimensional density of states is (see, for example, Eq. 4.7, Chapter 4)

$$\rho^{(D)}(\varepsilon) = \Omega_D \left(\frac{L}{2\pi}\right)^D \frac{1}{2}\left(\frac{2m}{\hbar^2}\right)^{D/2} \varepsilon^{(D-2)/2} \tag{5.93}$$

where Ω_D is the space angle in D dimensions.

5.5.5 *The 3D Case*

In the 3D case one obtains

$$\begin{aligned} N = 2\sum_k f_k &\to 4\pi\left(\frac{L}{2\pi}\right)^3\left(\frac{2m}{\hbar^2}\right)^{3/2}\int_0^\infty f_\varepsilon\sqrt{\varepsilon}d\varepsilon \\ &= 4\pi\left(\frac{L}{2\pi}\right)^3\left(\frac{2m}{\hbar^2}\right)^{3/2}\int_0^\infty \frac{1}{e^{\beta(\varepsilon-\mu)}+1}\sqrt{\varepsilon}d\varepsilon \end{aligned} \tag{5.94}$$

In the limit $T \to 0$, the Fermi gas becomes degenerate

$$f_k = \begin{pmatrix} 1 \\ 0 \end{pmatrix} \text{ for } \begin{pmatrix} \varepsilon < \mu \\ \varepsilon > \mu \end{pmatrix} \text{ or } f_k = \theta(\mu - \varepsilon) \tag{5.95}$$

where the *Fermi energy* is defined as

$$\mu(n, T = 0) = E_{\mathrm{F}} = \frac{\hbar^2 k_{\mathrm{F}}^2}{2m} \tag{5.96}$$

where k_{F} is the Fermi vector. In the *degenerate* limit (3D) one obtains

$$n = \frac{N}{L^3} = \frac{1}{2\pi^2}\left(\frac{2m}{\hbar^2}\right)^{3/2}\frac{2}{3}E_{\mathrm{F}}^{3/2} = \frac{1}{3\pi^2}k_{\mathrm{F}}^3 \tag{5.97}$$

and thus

$$k_{\mathrm{F}} = \left(3\pi^2 n\right)^{1/3} \tag{5.98}$$

which gives

$$\mu(n, T = 0) = E_{\mathrm{F}} = \frac{\hbar^2}{2m}\left(3\pi^2 n\right)^{2/3} \tag{5.99}$$

Under the L/M Padé approximation, one obtains an approximate expression which relates the chemical potential μ to the normalized charge carrier density $\nu = n/n_0$ for 3D (see Fig. 5.12)

$$\beta\mu \simeq \ln\nu + K_1\ln(K_2\nu + 1) + K_3\nu \tag{5.100}$$

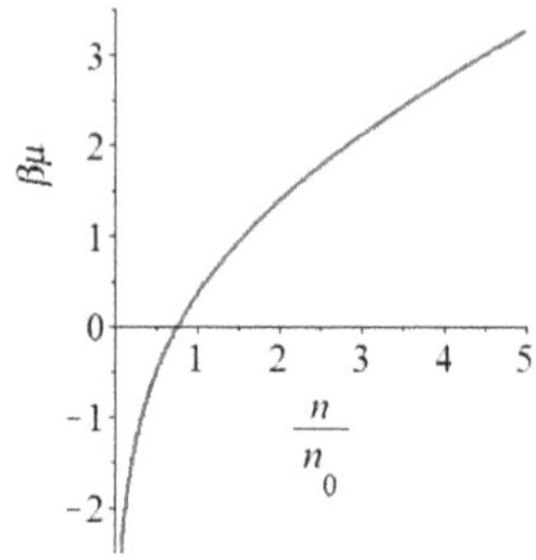

Figure 5.12 Chemical potential versus electron concentration for the 3D case.

where $K_1 = 4.897$, $K_2 = 0.045$, and $K_3 = 0.133$. The charge carrier concentration is

$$n_0 = \frac{1}{4}\left(\frac{2m}{\hbar^2 \pi \beta}\right)^{3/2} \tag{5.101}$$

For the 2D case one finds instead

$$n = \frac{N}{L^2} = \frac{1}{2\pi}\left(\frac{2m}{\hbar^2 \beta}\right)\int_0^\infty dx \frac{1}{e^x e^{-\beta\mu} + 1} \tag{5.102}$$

From the above formula (6.103) one obtains

$$\begin{aligned} n &= \frac{1}{2\pi}\left(\frac{2m}{\hbar^2 \beta}\right)\left(\frac{\Lambda}{T} + \log\left(\frac{e^{-\frac{\mu}{T}} + 1}{e^{\frac{\Lambda-\mu}{T}} + 1}\right)\right) \\ &\simeq \frac{mT}{\pi \hbar^2}\log\left(e^{-\frac{\mu}{T}} + 1\right) \end{aligned} \tag{5.103}$$

or, in another way

$$\beta\mu\,(n,\ T) = \ln\left(\exp\left(\frac{\hbar^2 \pi \beta n}{m} - 1\right)\right) \tag{5.104}$$

which plot is shown in Fig. 5.13. For a 1D case one gets

$$N = 2\sum_k f_k \to 2\int_0^\infty d\varepsilon \rho^{(1)}(\varepsilon)\ f_\varepsilon = \frac{L}{2\pi}\sqrt{\frac{2m}{\hbar^2}}\int_0^\infty d\varepsilon \frac{1}{\sqrt{\varepsilon}} f_\varepsilon \tag{5.105}$$

which gives the electron concentration in the form

$$\begin{aligned} n(\mu) &= \frac{N}{L} = \frac{1}{2\pi}\sqrt{\frac{2m}{\hbar^2}}\int_0^\infty dx \frac{1}{\sqrt{x}} \frac{1}{e^x e^{-\beta\mu} + 1} \\ &= -\sqrt{\pi}\frac{1}{2\pi}\sqrt{\frac{2m}{\hbar^2}}\mathrm{Li}_{\frac{1}{2}}\left(-e^b\right) \end{aligned} \tag{5.106}$$

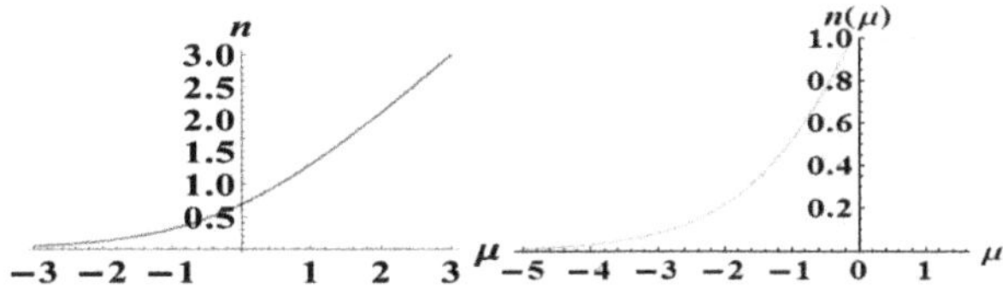

Figure 5.13 Left: Reciprocal dependence $\mu(n)$ for the 2D case. Right: Similar dependence of electron concentration versus electrochemical potential for the 1D case.

and

$$n(\mu) = -\frac{1}{2\sqrt{\pi}}\sqrt{\frac{2m}{\hbar^2}}\mathrm{Li}_{\frac{1}{2}}\left(-e^{\beta\mu}\right) \tag{5.107}$$

The absorption/generation instability is determined by the sign of the distribution factor

$$\tan h\left[\frac{\beta}{2}\left(\hbar\omega - E_{\mathrm{g}} - \mu\right)\right] \tag{5.108}$$

that is, by sign of

$$x = \hbar\omega - E_{\mathrm{g}} - \mu \tag{5.109}$$

where the total chemical potential is

$$\mu = \mu_{\mathrm{e}} + \mu_{\mathrm{h}} \tag{5.110}$$

5.6 Coupling between Electrons and Phonons

The long-range Coulomb forces between ions on the one hand and between ions and electrons on the other hand are screened out by the conduction electrons. Therefore, the dressed electron–phonon interaction largely depends on screening. The leading corrections in the long wavelength limit are described by the RPA. However, the situation becomes different for shorter wavelengths when the processes formerly neglected in the RPA become important. Nevertheless, the RPA provides roughly correct answers for any wavelength, because in the region where the RPA is least accurate it still gives small corrections to the bare quantities. Thus, the RPA provides an illustrative and satisfactory description of the entire electron–phonon system. For simplicity, we also neglect

Figure 5.14 The polarization operator.

the Umklapp processes in both the Coulomb interaction and in the electron–phonon interaction. The series of bubbles shown in Fig. 5.14 corresponds to the irreducible longitudinal phonon self-energy $\Pi_l(q)$. That string of bubbles represents the screening of the ion–ion potential in the RPA model. One might notice two remarkable differences between $V_c(q)$ and $\Pi_l(q)$: (a) In $\Pi_l(q)$ we have replaced the incoming and outgoing bare Coulomb lines which were in $V_c(q)$ by electron–phonon matrix elements, and (b) there is no leading term $V(q)$ in $\Pi_l(q)$, which was formerly present in the series for $V_c(q)$. Therefore, now we arrive at

$$\frac{V_c(q) - V(q)}{V^2(q)} = \frac{1}{V(q)}\left[\frac{1}{\kappa(q)} - 1\right] = \frac{1}{|g_{el}|^2}\Pi_l(q) \tag{5.111}$$

where we have used Eq. 5.36 in the left equality and have assumed that $|g_{el}|^2$ is a function only of the momentum transfer for longitudinal phonons. Equation 5.112 can be rewritten as follows

$$\Pi_l(q) = \frac{|g_{el}|^2}{V(q)}\left[\frac{1}{\kappa(q)} - 1\right] \tag{5.112}$$

The full Green function of phonons is obtained from the Dyson's equation

$$D_l^{-1}(q) = D_{0l}^{-1}(q) - \Pi_l(q) \tag{5.113}$$

which gives

$$D_l(q) = \frac{2\Omega_{ql}}{q_0^2 - \frac{2|g_{el}|^2\Omega_{ql}}{V(q)\kappa(q)} - \left[\Omega_{ql}^2 - \frac{2|g_{el}|^2\Omega_{ql}}{V(q)}\right] + i\delta} \tag{5.114}$$

In the last formula, Eq. 5.115, one can take into account that in the jellium model

$$\Omega_{ql}^2 = \frac{2|g_{el}|^2\Omega_{ql}}{V(q)} \tag{5.115}$$

which allows us to simplify $D_l(q)$ as

$$D_l(q) = \frac{2\Omega_{ql}}{q_0^2 - \frac{\Omega_{ql}^2}{\kappa(q)} + i\delta} \tag{5.116}$$

The dressed phonon frequencies are obtained from the poles of D

$$\omega_{ql}^2 = \frac{\Omega_{ql}^2}{\bar{\kappa}\left(\mathbf{q}, \omega_{ql}\right)} \tag{5.117}$$

In the last equation, $\bar{\kappa}$ is the analytic continuation of κ along the q_0 axis. Equation 5.118 suggests that the dressed frequencies are determined at the expense of reducing the effective force between ions or with the electronic dielectric constant. Since Ω_{ql}^2 is proportional to the ionic charge squared, Eq. 5.118 is the expected result.

When calculating the real part of the dressed phonon frequency ω_{ql}, it is sufficient to use a simpler static dielectric constant $\kappa\,(q, 0)$ instead of a more complicated function $\kappa\,(q, \omega)$ because the typical energies of electrons and phonons differ from each other by a small dimensionless parameter $\sim\sqrt{m/M} \approx 10^{-3}$. For the long wavelengths one obtains

$$\omega_{ql}^2 = \frac{\Omega_{ql}^2}{1 + \frac{k_s^2}{q^2}} = \frac{mZ}{3M} v_{\mathrm{F}}^2 q^2 \tag{5.118}$$

or otherwise

$$\omega_{ql} = \sqrt{\frac{mZ}{3M}} v_{\mathrm{F}} q \tag{5.119}$$

From Eq. 5.120 it follows that the dressed longitudinal phonons correspond to sound-like waves with acoustic dispersion law and sound velocity $v_{\mathrm{F}}\sqrt{mZ/(3M)}$, where v_{F} is the Fermi velocity. That jellium result has rather an illustrative significance since the situation in real metals is much more complicated. The first difference is that the bare phonon frequencies Ω_{ql} and the bare electron–phonon matrix elements depend on the crystallographic orientations. Besides, one also should take into account the Umklapp processes which have been disregarded in the simplified jellium model. Although the jellium model is oversimplified as compared to reality it, nevertheless, serves as a nice illustration of various computational methods. In particular, the screening effect diminishes the bare ion frequencies to sound-wave modes, (i.e., $\omega_q \propto q$) which indicates a correct tendency when considering the electron–phonon coupling effects.

Figure 5.15 Random phase approximation for the screened interaction between electrons and longitudinal phonons.

Let us consider the coupling between the dressed phonons and the electrons. On the one hand, the ion–ion interaction is screened, whereas the electron–ion interactions and thus electron–phonon interactions are not screened. We anticipate the answer for the screened electron–phonon matrix element to be of the kind

$$\bar{g}_{ql} \simeq \frac{g_{ql}}{\kappa\,(q)} \tag{5.120}$$

The above formula in Eq. 5.121 means that the bare matrix element must be divided by the dielectric function of the conduction electrons which depends on the wave number and frequency. The intuitive Eq. 5.121 is readily obtained by observing that the screened interaction is given by the series shown in Fig. 5.15, which represents the expansion of Eq. 5.121 in powers of the irreducible polarizability. When using the dressed phonon Green function together with the screened electron–phonon matrix element while calculating dynamical quantities, one should pay caution for avoiding of double count the vacuum polarization processes. Let us include longitudinal phonons along with the Coulomb interaction. Hence, we can use the former expression, whereas a one-phonon process gives

$$\Sigma^{\text{ph}}\,(p) = i \int G_0\,(p+q)\,\{\bar{g}_{ql}\}^2 D_l\,(-q)\,\frac{d^4q}{(2\pi)^4} \tag{5.121}$$

where $\{\bar{g}_{ql}\}^2 = \bar{g}_{ql}\bar{g}_{-ql}$. The total self-energy within this approximation is

$$\Sigma^{\text{ph}}\,(p) = i \int G_0\,(p+q)\left[V_c\,(q) + \{\bar{g}_{ql}\}^2\,D_l\,(-q)\right]\frac{d^4q}{(2\pi)^4} \tag{5.122}$$

A simple result is obtained when using the relation $2\,|g_{el}|^2\,/\,\Omega_{ql} = V\,(q)$, which holds for jellium. It gives

$$\Sigma\,(p) = i \int G_0\,(p+q)\,V_c\,(q)\left[\frac{q_0^2}{q_0^2 - \frac{\Omega_{ql}^2}{\kappa(q)} + i\delta}\right]\frac{d^4q}{(2\pi)^4} \tag{5.123}$$

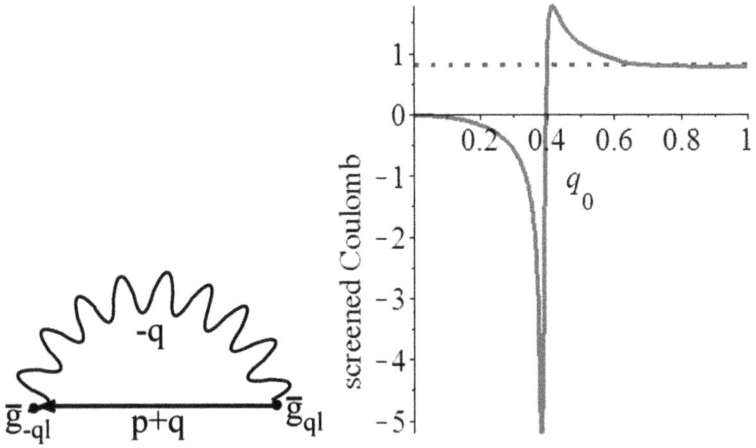

Figure 5.16 Left: The lowest second-order dressed phonon contribution to the electron self-energy. Right: The screened Coulomb interaction which contributes to the effective interelectron interaction Re$V(q)$ as a function of the energy transfer q_0 for a fixed momentum transfer $\mathbf{q}$. The resonance takes place at the dressed phonon frequency ω_q, which corresponds to the ionic overscreening of the bare Coulomb interaction for $q_0 < \omega$ and underscreening for $q_0 > \omega$. For high-frequency $q_0 >> \omega$, the ions do not respond and Re$V(q)$ approaches the bare Coulomb interaction reduced by the electronic dielectric function $\kappa(\mathbf{q}, q_0)$.

or

$$\Sigma(p) = i \int G_0(p+q) \left[\frac{V(q)}{1 + V(q) P(q) - \frac{\Omega_{ql}^2}{q_0^2} + i\delta} \right] \frac{d^4 q}{(2\pi)^4} \tag{5.124}$$

In the last Eq. 5.125 the denominator represents itself as a total dynamical dielectric constant. Here, the high-frequency expression $-\Omega_{ql}^2/q_0^2$ describes the contribution of ionic polarizability. The effective interaction between the screened Coulomb interaction and the exchange of a dressed phonon characterizes electronic and ionic polarizabilities of the system.

Function $V(q)$ is computed versus the energy transfer q_0 for a fixed momentum transfer $\mathbf{q}$. The plot on the right in Fig. 5.16 is shown for the RPA model of the "jellium" model of metal. One can notice the resonance which occurs at the dressed phonon frequency ω_q illustrating the ionic overscreening effect of the bare Coulomb interaction for $q_0 < \omega_q$ and underscreening for $q_0 < \omega_q$. In the high-frequency limit $q_0 >> \omega_q$ the ions do not respond and $V(q)$ approaches the bare Coulomb interaction reduced by the electronic dielectric function $\kappa(q)$.

The effective screened potential between conduction electrons is the bare Coulomb potential $V(q)$ divided by the total dielectric function of the medium $\kappa(q)$. From Eq. 5.124 it follows that at $|q_0| < |\omega_q|$ the effective potential is attractive which corresponds to an overscreening of the Coulomb repulsion between electrons by the vibrating ions. In the opposite case $|q_0| > |\omega_q|$ the effective potential is underscreened by the ions vibrating out of phase with the electrons. In the latter case the effective repulsion is stronger than the bare interaction. The underscreening represents just a dielectric anomaly which typically occurs when describing the dielectrics. When $|q_0| > |\omega_q|$, the bare potential is screened only by the conduction electrons, whereas the ionic polarization is negligible. Using the term "interaction" for the function $\mathrm{Re}V(q)$ is not fully adequate since the screened Coulomb interaction involves the retardation effects and is not Hermitian in general.

In graphene, a long-wavelength acoustic phonon interacts with electrons via the effective deformation potential

$$V_1 = g_1 \left(u_{xx} + u_{yy}\right) \tag{5.125}$$

where the interaction constant for graphene is evaluated numerically as $g_1 = 30$ eV and

$$u_{xx} = \frac{\partial u_x}{\partial x} + \frac{u_z}{r_\mathrm{T}}; u_{yy} = \frac{\partial u_y}{\partial y}; 2u_{xy} = \frac{\partial u_x}{\partial y} + \frac{\partial u_y}{\partial x} \tag{5.126}$$

The deformation potential V_1 constitutes a diagonal term of the matrix Hamiltonian. Another part of the electron–phonon interaction comes from the change in distance between the adjacent carbon atoms which affect the transfer integral. It gives the terms

$$\mathbf{H}'_K = \begin{pmatrix} 0 & V_2 \\ V_2^* & 0 \end{pmatrix}; \mathbf{H}'_{K'} = \begin{pmatrix} 0 & -V_2^* \\ -V_2 & 0 \end{pmatrix} \tag{5.127}$$

where

$$V_2 = g_2 e^{3i\eta} \left(u_{xx} - u_{yy} + 2iu_{xy}\right) \tag{5.128}$$

and also

$$g_2 = \frac{3\lambda_1\lambda_2}{4}\gamma_0 \tag{5.129}$$

Taking $\lambda_2 \simeq 2$ for s and p electrons, and $\lambda_1 \simeq 1/3$ one gets $g_2 \simeq \gamma_0/2$, or $g_2 \simeq 1.5$ eV. The spin part of the electron–phonon interaction matrix element is

$$\begin{aligned} M_{ep} &= \left\langle s,\ n,\ -k \middle| \begin{pmatrix} V_1 & V_2 \\ V_2^* & V_1 \end{pmatrix} \middle| s,\ n,\ +k \right\rangle \\ &= \kappa_{\nu\phi}\left(n\right) \frac{\sigma_{\nu\phi}^n}{\left|\sigma_{\nu\phi}^n\right|^2} V_1 + s\frac{\sigma_{\nu\phi}^n}{\left|\sigma_{\nu\phi}^n\right|} \mathrm{Re} V_2 \end{aligned} \tag{5.130}$$

where

$$\begin{aligned} \sigma_{\nu\phi}^n &= \kappa_{\nu\phi}\left(n\right) - ik \\ \kappa_{\nu\phi}\left(n\right) &= \frac{2}{d_{\mathrm{T}}}\left(n + \phi - \frac{\nu}{3}\right) \end{aligned} \tag{5.131}$$

In metallic nanotubes, the lower band is characterized by a linear dispersion law $\kappa_{\nu\phi}\left(n\right) = 0$. Then, only the 2nd term in Eq. 5.131 is relevant. The phonon drag contribution into the thermopower is

$$S_{\mathbf{drag}} = \frac{2\left|e\right|\, k_{\mathrm{B}}}{\sigma} \sum_{kq} \frac{\partial N_q^0}{\partial T} \alpha_{kq} \left(v_k\tau_k - v_{k+q}\tau_{k+q}\right)\cdot s_{\mathrm{g}} \tag{5.132}$$

where N_q^0 is the Bose–Einstein function, τ_k is the electron(-phonon) relaxation time, v_k is the electron velocity, s_{g} is the phonon group velocity. One can notice that the main contribution into the phonon drag comes from the back scattering when $v_{k+q} = -v_k$. In metallic nanotubes such back scattering processes are prohibited since they violate the conservation of pseudospin. They are however partially permitted in semiconducting nanotubes where the states with the pseudospin "up" and "down" are hybridized. The corresponding small parameter can be obtained similarly as for the electron–impurity scattering. The electron–phonon scattering time is computed as

$$\tau_{i\to f}^{-1} = \frac{2\pi}{\hbar}\sum_{f,\,q} \alpha_q^2 \left|\left\langle \psi_f \left| e^{\pm i\mathbf{q}\cdot\mathbf{r}} \right| \psi_{\mathrm{i}} \right\rangle\right|^2 \delta\left(E_f - E_{\mathrm{i}} \pm \omega_q\right) \left[N_q + \begin{Bmatrix} 1 \\ 0 \end{Bmatrix}\right] \tag{5.133}$$

where α_q^2 is the Fröhlich interaction strength.

One introduces

$$\kappa_{\nu n} = \frac{2}{d_{\mathrm{T}}}\left(n - \frac{\nu}{3}\right) \tag{5.134}$$

$$\varepsilon_{snk} = s\varepsilon_{nk} = s v_{\mathrm{F}}\sqrt{\kappa_{\nu n}^2 + k^2} \tag{5.135}$$

$$|k| = \sqrt{\left(\frac{\varepsilon_{snk}}{v_{\mathrm{F}}}\right)^2 - \kappa_{\nu n}^2} \tag{5.136}$$

$$\alpha_q \left\langle \psi_k \left| e^{\pm i\mathbf{q}\cdot\mathbf{r}} \right| \psi_{k'} \right\rangle = \kappa \frac{\kappa - ik}{\kappa^2 + k^2} g_1 \left(u_{xx} + u_{yy}\right) \tag{5.137}$$

$$\alpha_q^2 \left|\left\langle \psi_k \left| e^{\pm i\mathbf{q}\cdot\mathbf{r}} \right| \psi_{k'} \right\rangle\right|^2 = \frac{\Delta^2}{\varepsilon^2} g_1^2 \left\langle u_{xx} + u_{yy}\right\rangle^2 \tag{5.138}$$

$$\omega_q = s_{\mathrm{g}} q \tag{5.139}$$

where $g_1 \simeq 30\,\mathrm{eV}$ and for simplicity we set $\kappa_{\nu\phi}(n) \to \kappa$. Then, it gives

$$\begin{aligned} \tau_k^{-1} &= \frac{2\pi}{\hbar}\sum_{k'} \alpha_{k'-k}^2 \left|\left\langle \psi_k \left| e^{\pm i(k'-k)\cdot r} \right| \psi_{k'} \right\rangle\right|^2 \delta\left(E_k - E_{k'} \pm \omega_{k'-k}\right) \\ &\quad \left[N_{k'-k} + \begin{Bmatrix} 1 \\ 0 \end{Bmatrix}\right] \\ &= \frac{2\pi}{\hbar}\frac{g}{2\pi\hbar^2 v_{\mathrm{F}}^2} g_1^2 \left\langle u_{xx} + u_{yy}\right\rangle^2 \frac{\Delta^2}{\varepsilon^2}\int d\varepsilon' \left|\varepsilon'\right| \delta\left(\varepsilon - \varepsilon' \pm \omega_{\varepsilon'-\varepsilon}\right) \\ &\quad \left[N_{\varepsilon'-\varepsilon} + \begin{Bmatrix} 1 \\ 0 \end{Bmatrix}\right] \\ &= \frac{2\pi}{\hbar}\frac{g}{2\pi\hbar^2 v_{\mathrm{F}}^2} g_1^2 \left\langle u_{xx} + u_{yy}\right\rangle^2 \frac{\Delta^2}{\varepsilon^2}\sum_i \left|\varepsilon_i'\right| \left[N_{\varepsilon_i'-\varepsilon} + \begin{Bmatrix} 1 \\ 0 \end{Bmatrix}\right] \end{aligned} \tag{5.140}$$

where

$$\begin{aligned} \omega_{\varepsilon'-\varepsilon} &\to \frac{s_{\mathrm{g}}}{v_{\mathrm{F}}}\cdot\left(\sqrt{\varepsilon'^2 - \Delta^2} - \sqrt{\varepsilon^2 - \Delta^2}\right) \\ k_\varepsilon &\to \frac{1}{v_{\mathrm{F}}}\sqrt{\varepsilon^2 - \Delta^2}; \Delta = v_{\mathrm{F}}\kappa \end{aligned} \tag{5.141}$$

$$\varepsilon - \varepsilon' \pm \frac{s_g}{v_F} \cdot \left(\sqrt{\varepsilon'^2 - \Delta^2} - \sqrt{\varepsilon^2 - \Delta^2} \right) = 0$$

$$\left(B - \frac{v_F}{s_g} x \right)^2 = x^2 - \Delta^2 \quad (5.142)$$

$$\varepsilon_1' = \frac{s_g}{v_F} \frac{B^2 + \Delta^2}{2B} = \frac{s_g}{v_F} \frac{\left(\frac{v_F}{s_g}\varepsilon + \sqrt{\varepsilon^2 - \Delta^2} \right)^2 + \Delta^2}{2\left(\frac{v_F}{s_g}\varepsilon + \sqrt{\varepsilon^2 - \Delta^2} \right)}$$
$$= \frac{1}{2} \left(\varepsilon - \left(s_g / v_F \right) \sqrt{\varepsilon^2 - \Delta^2} \right)$$
$$\times \frac{\varepsilon^2 + 2\varepsilon \left(s_g / v_F \right) \sqrt{\varepsilon^2 - \Delta^2} + \varepsilon^2 \left(s_g^2 / v_F^2 \right)}{\varepsilon^2 + \left(s_g^2 / v_F^2 \right) \left(\Delta^2 - \varepsilon^2 \right)} \quad (5.143)$$

and also

$$\varepsilon_{2,3}' = s_g \frac{B v_F \pm \sqrt{B^2 s_g^2 - \Delta^2 v_F^2 + \Delta^2 s_g^2}}{v_F^2 - s_g^2}$$
$$= \frac{s_g}{v_F} \frac{B \pm \sqrt{\left(B^2 + \Delta^2 \right) \left(s_g^2 / v_F^2 \right) - \Delta^2}}{1 - s_g^2 / v_F^2}$$
$$= \frac{1}{1 - s_g^2 / v_F^2} \left(\varepsilon + \frac{s_g}{v_F} \sqrt{\varepsilon^2 - \Delta^2} \right.$$
$$\left. + \frac{s_g}{v_F} \sqrt{\varepsilon^2 - \Delta^2 + 2\varepsilon \frac{s_g}{v_F} \sqrt{\varepsilon^2 - \Delta^2} + \varepsilon^2 \left(\frac{s_g}{v_F} \right)^2} \right) \quad (5.144)$$

where

$$B = \frac{v_F}{s_g} \varepsilon + \sqrt{\varepsilon^2 - \Delta^2} \quad (5.145)$$

According to Perebeinos et al. (2005a, 2005b, 2005c), the strongest coupling of electrons is with the LO phonons, which give scattering roughly 2 orders of magnitude stronger than the acoustic modes. The KLO mode corresponds to interband scattering.

In Fig. 5.17d we show the coherence length, defined as $L_k = v_k \tau_k$, where v_k is the band velocity. It characterizes the spatial scale where the electron wavefunction preserves its coherence. We neglect phonon renormalization of the electron velocity v_k, which is only

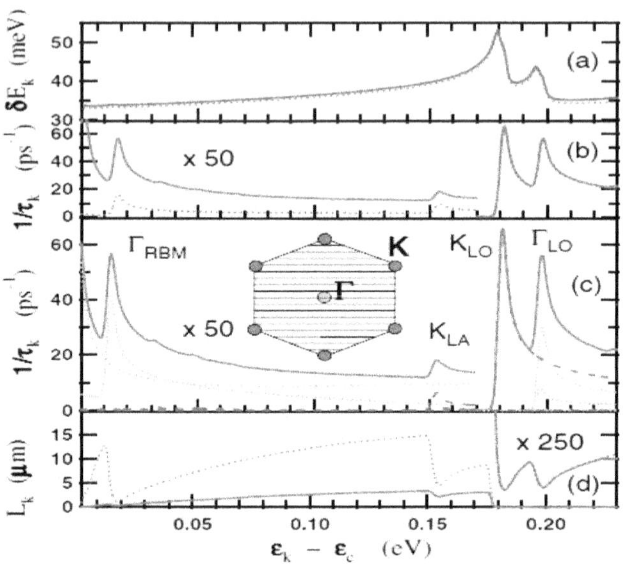

Figure 5.17 The electron–phonon scattering time along different crystallographic directions. The lower panel shows the energy dependence of coherence length on the electron–phonon scattering. Adapted from Perebeinos et al. (2005a, 2005b, 2005c) © APS.

a few percent. A stronger renormalization of the electron velocity occurs near the resonance with the optical phonons KLO and LO. We emphasize that the coherence length is strongly energy dependent, with a broad maximum located between the breathing mode and the optical phonons. The coherence length is very sensitive to temperature as well, but even at room temperature the coherence length can exceed one micron, consistent with the lengths inferred from experiment. The charge carriers can be injected into nanotubes well above the band edge, suggesting the possibility of a ballistic or even quantum-coherent scenario over a range of length scales. At higher energy, the carriers have a temperature-independent coherence length of around 20–40 nm, depending on energy. This is consistent with the lengths inferred from experiments for metallic nanotubes. Practically, it means that contribution of the acoustic phonons into the phonon drag involves a very small fraction of electrons with energies ε within a very narrow range in the vicinity of

$\varepsilon \sim \varepsilon_c + 0.175$ eV. One can avoid that phonon drag influence by setting the bias voltage V_{SD} outside the mentioned energy range.

5.7 Susceptibility of Graphene

In a pristine suspended graphene, one gets the following expression for the electron susceptibility

$$\chi(\omega) = -\sum_k \frac{|d_{\mathrm{cv}}|^2}{a^2}\left(f_{\mathrm{v},k} - f_{\mathrm{c},k}\right) \times \left[\frac{1}{\varepsilon_{\mathrm{v},k} - \varepsilon_{\mathrm{c},k} + \omega + i\gamma} - \frac{1}{\varepsilon_{\mathrm{c},k} - \varepsilon_{\mathrm{v},k} + \omega + i\gamma}\right] \quad (5.146)$$

For a finite $\mathbf{q} \neq 0$ (i.e., indirect interband transitions due to absorption/emission of a phonon) one can replace the summation by an integral as

$$\chi(q,\omega) = \sum_{s,s'=\pm 1} \int \frac{d^2k}{(2\pi)^2} \frac{f\left(E_{s,k+q}\right) - f\left(E_{s',k}\right)}{\omega - E_{s,k+q} + E_{s',k} + i\eta} \quad (5.147)$$

The computational results are shown in Fig. 5.18. The photons of external electromagnetic field excite electrons in the conductive band, while the holes are excited in the valence band (direct interband transitions). Then, one sets $\mathbf{q} = 0$, which gives

$$\chi(\omega) = -\sum_{s,s'=\pm 1} \int \frac{d^2k}{(2\pi)^2} \frac{|d_{\mathrm{cv}}|^2}{L^3} \frac{f\left(E_{s',\mathrm{v}}\right) - f\left(E_{s,\mathrm{c}}\right)}{E_{s',\mathrm{v}} - E_{s,\mathrm{c}} + \omega + i\eta} \quad (5.148)$$

Here $E_{s,\mathrm{v}} = \pm v_{\mathrm{F}}\,|\mathbf{q}|$ and the density of states is

$$N_{\mathrm{G}}(E) = \frac{3\sqrt{3}a^2}{\pi v_{\mathrm{F}}^2}\,|E| \quad (5.149)$$

Besides, if $ss' = -1$, then $E_{s',\mathrm{v}} - E_{s,\mathrm{c}} = \pm 2E$ and

$$\chi(\omega) = -\frac{6\sqrt{3}|d_{\mathrm{cv}}|^2}{\pi v_{\mathrm{F}}^2} \int_0^{v_{\mathrm{F}}\Lambda} dE\,E\,\left(f(-E) - f(E)\right) \times \left(\frac{1}{\omega + 2E + i\eta} - \frac{1}{\omega - 2E + i\eta}\right) \quad (5.150)$$

where $\Lambda \approx 1/a$ is the cutoff momentum. At zeroth temperature $T = 0$ one sets $f(E) = \theta(\mu - E)$ and for electron doping

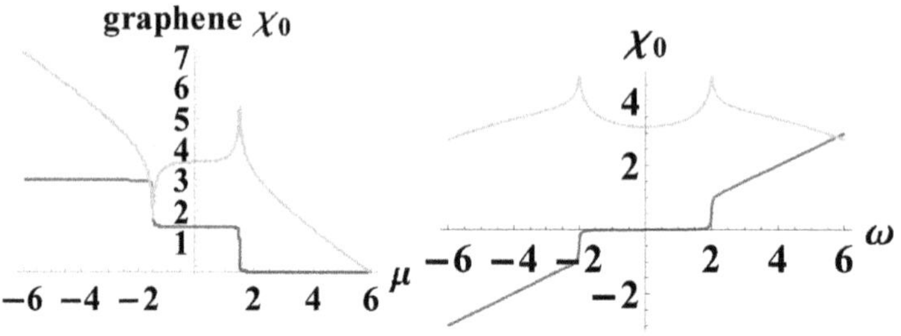

Figure 5.18 Left: 2D graphene $\chi(\omega)$ for $\Gamma = 0.2$, $\Lambda = 6$, and $\omega = 1$. Right: 2D graphene $\chi(\omega)$ for $\Gamma = 0.2$, $\Lambda = 6$, and $\mu = 1$.

when the electrochemical potential is positive, $\mu > 0$, one also sets $f(-E) = 1$. For the hole doping the electrochemical potential is negative, $\mu < 0$. The above simplifications allow us to obtain the following formula

$$\chi(\omega) = -\frac{6\sqrt{3}\,|d_{\text{cv}}|^2}{\pi v_F^2}\int_{\mu}^{v_F\Lambda}\left(\frac{1}{\omega+2E+i\eta}-\frac{1}{\omega-2E+i\eta}\right)EdE \tag{5.151}$$

where we have used that

$$\frac{1}{\omega+i\eta+2E}-\frac{1}{\omega+i\eta-2E}=\frac{4E}{4E^2-(\omega+i\eta)^2} \tag{5.152}$$

where $\zeta^2 = \left(\eta^2 - 2i\eta\omega - \omega^2\right)/4$ and also

$$f(-E)-f(E) = 1-\theta(\mu-E) = \begin{cases} 0 \text{ for } E < \mu \\ 1 \text{ for } E > \mu \end{cases} \tag{5.153}$$

One also obtains the following analytical expressions for the electron susceptibility

$$\chi(\omega,\mu) = -\frac{i}{\pi}\frac{3\sqrt{3}\,|d_{\text{cv}}|^2}{2\pi v_F^2} \times\left[\Lambda-\mu+\frac{\omega+i\Gamma}{2}\left(\tanh^{-1}\left(\frac{2\mu}{\omega+i\Gamma}\right) -\tanh^{-1}\left(\frac{2\Lambda}{\omega+i\Gamma}\right)\right)\right] \tag{5.154}$$

The last expression can be rewritten in another form as

$$\chi(\omega,\mu) = -\frac{6\sqrt{3}\,|d_{\text{cv}}|^2}{\pi v_F^2} \times\left[v_F\Lambda-\mu+\frac{\omega}{4}\left(\ln\left|\frac{\omega+2\mu}{\omega-2\mu}\right|+i\pi\theta(\omega-2\mu)\right)\right] \tag{5.155}$$

5.8 Graphene $n(\mu)$

The concentration of electric charge carriers in graphene is an important characteristic when studying the gated structure. One computes the number of excitations in the 2D graphene as

$$N = 2\sum_{\mathbf{k}} f_k \to -2 \sum_{s,\, s'=\pm 1} \int \frac{d^2\mathbf{k}}{(2\pi)^2} \left(f\left(E_{s',\,\mathrm{v}}\right) + f\left(E_{s,\mathrm{c}}\right)\right)$$
$$= \int_0^\infty d\varepsilon \rho_G\left(\varepsilon\right)\left(1 - f\left(-\varepsilon\right) + f\left(\varepsilon\right)\right)$$
$$= \frac{12\sqrt{3}a^2}{\pi v_\mathrm{F}^2} \int_0^\infty d\varepsilon \cdot \varepsilon \cdot f\left(\varepsilon\right) \tag{5.156}$$

Here, $E_{s,\,\mathrm{v}} = \pm v_\mathrm{F}\,|\mathbf{q}|$ and the density of electron states is

$$\rho_\mathrm{G}\left(\varepsilon\right) = \frac{3\sqrt{3}a^2}{\pi v_\mathrm{F}^2}\,|\varepsilon| \tag{5.157}$$

We used $ss' = -1$ which gives $E_{s',\,\mathrm{v}} - E_{s,\,\mathrm{c}} = \pm 2\varepsilon$ and

$$n = \frac{N}{L^2} = -\frac{12\sqrt{3}}{\pi v_\mathrm{F}^2} \int_0^{v_\mathrm{F}\Lambda} \varepsilon f\left(\varepsilon\right) d\varepsilon$$
$$= -\frac{12\sqrt{3}}{\pi v_\mathrm{F}^2} \int_0^{v_\mathrm{F}\Lambda} \varepsilon \frac{1}{e^{\beta(\varepsilon-\mu)}+1} d\varepsilon$$
$$= -\frac{12\sqrt{3}}{\pi v_\mathrm{F}^2 \beta^2} \int_0^{\beta v_\mathrm{F}\Lambda} x \frac{1}{e^{(x-\beta\mu)}+1} dx \tag{5.158}$$

where $x = \beta\varepsilon$

$$f\left(\varepsilon\right) = \frac{1}{e^{\beta(\varepsilon-\mu)}+1} \tag{5.159}$$

Then, one obtains the dependence of the electric charge carrier concentration versus the electrochemical potential μ (see also Fig. 5.19) in the form

$$n\left(\mu\right) = -\left(12\sqrt{3}T^2/\pi v_\mathrm{F}^2\right)\left[Li_2\left(-e^{-\frac{\mu}{T}}\right)\right.$$
$$\left. - Li_2\left(-e^{\frac{\Lambda-\mu}{T}}\right) + \Lambda\left(\Lambda - 2T\log\left(e^{\frac{\Lambda-\mu}{T}}+1\right)\right)/T^2\right] \tag{5.160}$$

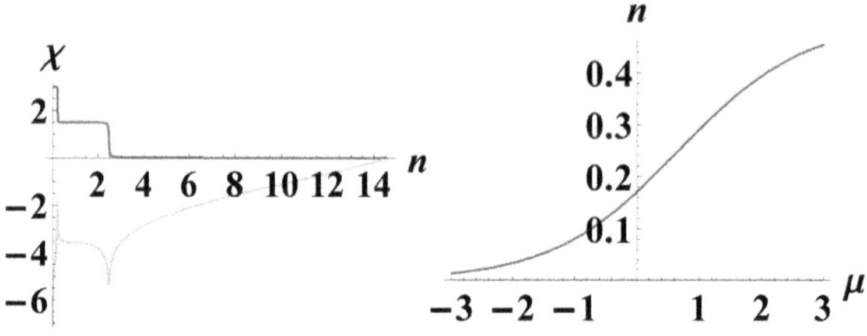

Figure 5.19 Left: $\chi(n)$ for $T = 1$, $\Lambda = 6$, $\Gamma = 0.02$, and $\omega = 1$. Right: Electron concentration versus chemical potential.

5.9 Dielectric Function of Graphene and CNTs

Screening of Coulomb interaction in graphene and CNTs is responsible for establishing the interaction strength between the charge carriers and, for example, charged lattice defects. In particular, the static screening of electric charge actually determines the contribution of elastic scattering into the electric conductivity and thermoelectric transport coefficients. The time-dependent dynamic screening actually shapes the strength of inelastic scattering of electrons on the lattice oscillations (phonons), and also on the charge and spin density oscillations.

There are many different approaches for computing the dielectric function. One of them is a traditional method with using the so-called RPA. In this approach one assumes that the quantum mechanical phases of the electron wavefunction mutually cancel each other due to a destructive interference. The 2D dielectric screening function is obtained in RPA approximation which gives

$$\varepsilon(q) = 1 + v_c(q)\,\Pi(q) \tag{5.161}$$

where $\Pi(q)$ is the graphene irreducible polarizability function and $v_c(q)$ is the Coulomb interaction. The bubble diagram gives

$$\Pi(q,\omega) = -\frac{g}{2A}\sum_{\mathbf{k}ss'} \frac{f_{\mathbf{k}s} - f_{\mathbf{k}'s'}}{\omega + \varepsilon_{sk} - \varepsilon_{sk'} + i\eta}\left(1 + ss'\cos\theta\right) \tag{5.162}$$

where $f_{ks} = 1/(\exp\beta(\varepsilon_{sk} - \mu) + 1)$. The summation over s, s' gives

$$\Pi(q,\omega) = -\frac{g}{2A}\sum_{\mathbf{k}}\left\{\frac{f_{k+} - f_{k'+}}{\omega + \varepsilon_{k+} - \varepsilon_{k'+} + i\eta} + \frac{f_{k-} - f_{k'-}}{\omega + \varepsilon_{s-} - \varepsilon_{s-} + i\eta}\right\}(1 + \cos\theta)$$
$$-\frac{g}{2A}\sum_{\mathbf{k}}\left\{\frac{f_{k+} - f_{k'-}}{\omega + \varepsilon_{k+} - \varepsilon_{k'-} + i\eta} + \frac{f_{k-} - f_{k'+}}{\omega + \varepsilon_{k-} - \varepsilon_{k'+} + i\eta}\right\}(1 - \cos\theta) \quad (5.163)$$

where $\mathbf{k}' = \mathbf{k} + \mathbf{q}$. The cos theorem gives $q^2 = k^2 + (k')^2 - 2kk'\cos\theta$ or

$$|\mathbf{k}'| = k' = k\cos\theta \pm \sqrt{q^2 + k^2\frac{\cos 2\theta - 1}{2}} \quad (5.164)$$

For graphene and for metallic CNTs the excitation energy reads

$$\varepsilon_{sk} = s v_{\mathrm{F}}|\mathbf{k}| \quad (5.165)$$

where $s = \pm 1$. The density of electron states in pristine graphene is

$$D(\varepsilon) = \frac{g}{2\pi\hbar^2 v_{\mathrm{F}}^2}|\varepsilon| \quad (5.166)$$

where $g = g_{\mathrm{s}}g_{\mathrm{v}}$ is the total degeneracy, $g_{\mathrm{s}} = 2$ and $g_{\mathrm{v}} = 2$ being the spin and valley degeneracy, respectively. One introduces

$$\varepsilon = v_{\mathrm{F}}k \text{ and } \Omega = v_{\mathrm{F}}q, \quad (5.167)$$

which gives the following expressions

$$\varepsilon_{k+} \to \varepsilon;\ \varepsilon_{k'+} \to \varepsilon\cos\theta \pm \sqrt{\Omega^2 + \varepsilon^2\frac{\cos 2\theta - 1}{2}}$$
$$\varepsilon_{k-} \to -\varepsilon;\ \varepsilon_{k'-} \to -\varepsilon\cos\theta \pm \sqrt{\Omega^2 + \varepsilon^2\frac{\cos 2\theta - 1}{2}} \quad (5.168)$$

Components of the polarization operator are expressed as

$$\Pi^{++}(q,\omega) = -\frac{g}{2A}\sum_{k}\left\{\frac{f_{k+} - f_{k'+}}{\omega + \varepsilon_{k+} - \varepsilon_{k'+} + i\eta}\right\}(1 + \cos\theta)$$
$$= -\frac{g}{2A}\frac{g}{2\pi\hbar^2 v_{\mathrm{F}}^2}\int d\theta\int_{\varepsilon}^{\varepsilon_2} d\varepsilon\,|\varepsilon|$$
$$\times\frac{1 + \cos\theta}{\omega + \varepsilon - \varepsilon\cos\theta \mp \sqrt{\Omega^2 + \varepsilon^2(\cos 2\theta - 1)/2} + i\eta}$$
$$\times\left(f_{\varepsilon} - f_{\varepsilon\cos\theta \pm \sqrt{\Omega^2 + \varepsilon^2(\cos 2\theta - 1)/2}}\right) \quad (5.169)$$

$$\Pi^{--}(q,\omega) = -\frac{g}{2A}\sum_k \left\{\frac{f_{k-} - f_{k'-}}{\omega + \varepsilon_{k-} - \varepsilon_{k'-} + i\eta}\right\}(1+\cos\theta)$$
$$= -\frac{g}{2A}\frac{g}{2\pi \mathrm{h}^2 v_F^2}\int d\theta \int d\varepsilon\, |\varepsilon|$$
$$\times \frac{1+\cos\theta}{\omega - \varepsilon + \varepsilon\cos\theta \mp \sqrt{\Omega^2 + \varepsilon^2(\cos 2\theta - 1)/2} + i\eta}$$
$$\times \left(f_{-\varepsilon} - f_{-\varepsilon\cos\theta \pm \sqrt{\Omega^2+\varepsilon^2(\cos 2\theta - 1)/2}}\right) \quad (5.170)$$

$$\Pi^{+-}(q,\omega) = -\frac{g}{2A}\sum_k \left\{\frac{f_{k+} - f_{k'-}}{\omega + \varepsilon_{k+} - \varepsilon_{k'-} + i\eta}\right\}(1-\cos\theta)$$
$$= -\frac{g}{2A}\frac{g}{2\pi \hbar^2 v_F^2}\int d\theta \int d\varepsilon\, |\varepsilon|$$
$$\times \frac{1-\cos\theta}{\omega + \varepsilon + \varepsilon\cos\theta \mp \sqrt{\Omega^2 + \varepsilon^2(\cos 2\theta - 1)/2} + i\eta}$$
$$\times \left(f_{\varepsilon} - f_{-\varepsilon\cos\theta \pm \sqrt{\Omega^2+\varepsilon^2(\cos 2\theta - 1)/2}}\right) \quad (5.171)$$

$$\Pi^{-+}(q,\omega) = -\frac{g}{2A}\sum_k \left\{\frac{f_{k-} - f_{k'+}}{\omega + \varepsilon_{k-} - \varepsilon_{k'+} + i\eta}\right\}(1-\cos\theta)$$
$$= -\frac{g}{2A}\frac{g}{2\pi \hbar^2 v_F^2}\int d\theta \int d\varepsilon\, |\varepsilon|$$
$$\times \frac{1-\cos\theta}{\omega - \varepsilon - \varepsilon\cos\theta \mp \sqrt{\Omega^2 + \varepsilon^2(\cos 2\theta - 1)/2} + i\eta}$$
$$\times \left(f_{-\varepsilon} - f_{+\varepsilon\cos\theta \pm \sqrt{\Omega^2+\varepsilon^2(\cos 2\theta - 1)/2}}\right) \quad (5.172)$$

One greatly simplifies calculations at zero temperature $T = 0$ when using the expressions below:

$$\int d\varepsilon\,(f_\varepsilon - f_{\varepsilon'}) \to \int_{-\infty}^{0} d\varepsilon - \int_{-\infty}^{\varepsilon_2} d\varepsilon \to \begin{cases} \int_{\varepsilon_2}^{0} d\varepsilon \text{ if } \varepsilon_2 < 0 \\ -\int_{0}^{\varepsilon_2} d\varepsilon \text{ if } \varepsilon_2 > 0 \end{cases} \quad (5.173)$$

$$\int d\varepsilon\,(f_\varepsilon - f_{\varepsilon'}) \to \int_{-\infty}^{0} d\varepsilon - \int_{-\infty}^{\varepsilon_2} d\varepsilon \to \begin{cases} \int_{\varepsilon_2}^{0} d\varepsilon \text{ if } \varepsilon_2 < 0 \\ -\int_{0}^{\varepsilon_2} d\varepsilon \text{ if } \varepsilon_2 > 0 \end{cases} \quad (5.174)$$

$$\int d\varepsilon\,(f_{-\varepsilon} - f_{-\varepsilon'}) \to \int_0^\infty d\varepsilon - \int_{\varepsilon_2}^\infty d\varepsilon \to \begin{cases} \int_0^{\varepsilon_2} d\varepsilon \text{ if } \varepsilon_2 > 0 \\ -\int_{\varepsilon_2}^0 d\varepsilon \text{ if } \varepsilon_2 < 0 \end{cases} \tag{5.175}$$

$$\int d\varepsilon\,(f_{\varepsilon} - f_{-\varepsilon'}) \to \int_{-\infty}^0 d\varepsilon - \int_{\varepsilon_2}^\infty d\varepsilon \tag{5.176}$$

$$\int d\varepsilon\,(f_{-\varepsilon} - f_{\varepsilon'}) \to \int_0^\infty d\varepsilon - \int_{-\infty}^{\varepsilon_2} d\varepsilon \tag{5.177}$$

where

$$\varepsilon_2 = \frac{\Omega}{\sqrt{\cos^2\theta - (\cos 2\theta - 1)/2}} \tag{5.178}$$

It gives

$$\begin{aligned} \Pi_0^{++}(q,\,\omega) &= -\frac{g^2}{4A\pi\hbar^2 v_F^2}\int d\theta \int d\varepsilon\,|\varepsilon| \\ &\times \frac{(1+\cos\theta)}{\omega + \varepsilon - \varepsilon\cos\theta \mp \sqrt{\Omega^2 + \varepsilon^2(\cos 2\theta - 1)/2} + i\eta} \\ &\times \left(\theta[-\varepsilon] - \theta\left[-\varepsilon\cos\theta \mp \sqrt{\Omega^2 + \varepsilon^2(\cos 2\theta - 1)/2}\right]\right) \\ &= \frac{g^2}{4A\pi\hbar^2 v_F^2}\int d\theta \frac{1+\cos\theta}{1-\cos\theta}\int_0^{\varepsilon_2} d\varepsilon \\ &\times \frac{\varepsilon}{\frac{\omega+i\eta}{1-\cos\theta} + \varepsilon \mp \frac{\sqrt{(\cos 2\theta - 1)/2}}{1-\cos\theta}\sqrt{\frac{\Omega^2}{(\cos 2\theta-1)/2} + \varepsilon^2}} \end{aligned} \tag{5.179}$$

A concise form for the above expressions is obtained by introducing auxiliary functions

$$\Xi_0(a,\,b,\,c,\,\omega,\,\Omega) = \int_a^b d\varepsilon \frac{\varepsilon}{\omega + \varepsilon + c\sqrt{\Omega^2 + \varepsilon^2}} \tag{5.180}$$

and

$$\Xi_1(a,\,b,\,\omega,\,\Omega) = \int_a^b d\varepsilon \frac{\varepsilon}{\omega + \varepsilon + \sqrt{\Omega^2 + \varepsilon^2}} \tag{5.181}$$

Then, one writes

$$\Pi_0^{++}(q,\ \omega) = -\frac{g^2}{4A\pi\hbar^2 v_{\mathrm{F}}^2}\int d\theta \frac{1+\cos\theta}{1-\cos\theta}\int_0^{\varepsilon_2} d\varepsilon \times \frac{\varepsilon}{\frac{\omega+i\eta}{1-\cos\theta}+\varepsilon\mp\frac{\sqrt{(\cos 2\theta-1)/2}}{1-\cos\theta}\sqrt{\frac{2\Omega^2}{\cos 2\theta-1}+\varepsilon^2}} \tag{5.182}$$

$$\Pi^{--}(q,\ \omega) = \frac{g^2}{4A\pi\hbar^2 v_F^2}\int d\theta \frac{1+\cos\theta}{1-\cos\theta}\int_0^{\varepsilon_2} d\varepsilon \times \frac{\varepsilon}{-\frac{\omega-i\eta}{1-\cos\theta}+\varepsilon\pm\frac{\sqrt{(\cos 2\theta-1)/2}}{1-\cos\theta}\sqrt{\frac{2\Omega^2}{\cos 2\theta-1}+\varepsilon^2}} \tag{5.183}$$

$$\Pi^{+-}(q,\ \omega) = -\frac{g^2}{4A\pi\hbar^2 v_F^2}\int d\theta \frac{1-\cos\theta}{1+\cos\theta}\left(\int_{-\infty}^{0} d\varepsilon - \int_{\varepsilon_2}^{\infty} d\varepsilon\right) \times \frac{|\varepsilon|}{\frac{\omega+i\eta}{1+\cos\theta}+\varepsilon\mp\frac{\sqrt{(\cos 2\theta-1)/2}}{1+\cos\theta}\sqrt{\frac{2\Omega^2}{\cos 2\theta-1}+\varepsilon^2}} \tag{5.184}$$

$$\Pi^{-+}(q,\ \omega) = \frac{g^2}{4A\pi\hbar^2 v_F^2}\int d\theta \frac{1-\cos\theta}{1+\cos\theta}\left(\int_{0}^{\infty} d\varepsilon - \int_{-\infty}^{\varepsilon_2} d\varepsilon\right) \times \frac{|\varepsilon|}{-\frac{\omega-i\eta}{1+\cos\theta}+\varepsilon\pm\frac{\sqrt{(\cos 2\theta-1)/2}}{1+\cos\theta}\sqrt{\frac{2\Omega^2}{\cos 2\theta-1}+\varepsilon^2}} \tag{5.185}$$

which is reduced to

$$\Pi_0^{++}(q,\ \omega) = \frac{g^2}{4A\pi\hbar^2 v_F^2}\int d\theta \frac{1+\cos\theta}{1-\cos\theta}\Xi_0\left(0,\ \alpha_1,\ \mp\alpha_2,\ \alpha_3,\ \alpha_4\right) \tag{5.186}$$

where

$$\alpha_1 = \frac{\Omega}{\sqrt{\cos^2\theta-(\cos 2\theta-1)/2}}$$
$$\alpha_2 = \frac{\sqrt{(\cos 2\theta-1)/2}}{1-\cos\theta}$$
$$\alpha_3 = \frac{\omega+i\eta}{1-\cos\theta}$$
$$\alpha_4 = \frac{2\Omega^2}{\cos 2\theta-1}. \tag{5.187}$$

$$\Pi^{--}(q,\omega) = \frac{g^2}{4A\pi\hbar^2 v_F^2}\int d\theta \frac{1+\cos\theta}{1-\cos\theta}\Xi_0\left(0,\ \alpha_1,\ \pm\alpha_2,\ -\alpha_3^*,\ \alpha_4\right) \tag{5.188}$$

$$\Pi^{+-}(q,\omega) = -\frac{g^2}{4A\pi\hbar^2 v_F^2}\int d\theta \frac{1-\cos\theta}{1+\cos\theta}[\Xi_0(-\infty,\ 0,\ \mp\alpha_2,\ \alpha_3,\ \alpha_4) - \Xi_0(\alpha_1,\ \infty,\ \mp\alpha_2,\ \alpha_3,\ \alpha_4)] \tag{5.189}$$

$$\Pi^{-+}(q,\omega) = \frac{g^2}{4A\pi\hbar^2 v_F^2}\int d\theta \frac{1-\cos\theta}{1+\cos\theta}[\Xi_0\left(0,\ \infty,\ \pm\alpha_2,\ -\alpha_3^*,\ \alpha_4\right) - \Xi_0\left(-\infty,\ \alpha_1,\ \pm\alpha_2,\ -\alpha_3^*,\ \alpha_4\right)] \tag{5.190}$$

5.10 Electron–Impurity Scattering Time in Graphene

The electron–impurity scattering can be incorporated using the Boltzmann transport equation as

$$\left(\frac{\partial f_{sk}}{\partial t}\right)_c = \frac{\partial \mathbf{k}}{\partial t}\cdot\frac{\partial f(\varepsilon_{sk})}{\partial k} = -e\mathbf{E}\cdot\mathbf{v}_{s\mathbf{k}}\frac{\partial f}{\partial \varepsilon_{sk}} = -\int \frac{d^2k}{(2\pi)^2}\left(g_{s\mathbf{k}} - g_{s\mathbf{k}'}\right) W_{s\mathbf{k},\,s\mathbf{k}'} \tag{5.191}$$

where we have assumed that the electric field $\mathbf{E}$ is small and $f_{sk} = f(\varepsilon_{sk}) + g_{s\mathbf{k}}$. The quantum mechanical scattering probability is

$$W_{s\mathbf{k},\,s\mathbf{k}'} = \frac{2\pi}{\hbar}n_{\rm i}\left|\left\langle V_{s\mathbf{k},\,s\mathbf{k}'}\right\rangle\right|^2 \delta\left(\varepsilon_{sk} - \varepsilon_{sk'}\right) \tag{5.192}$$

Since the elastic scattering is considered, the interband processes are neglected. Then, from Eqs. 5.192 and 5.193 one merely obtains

$$g_{s\mathbf{k}} = -\frac{\tau(\varepsilon_{sk})}{\hbar}e\mathbf{E}\cdot\mathbf{v}_{s\mathbf{k}}\left(\frac{\partial f_{sk}}{\partial \varepsilon_{sk}}\right) \tag{5.193}$$

where

$$\frac{1}{\tau(\varepsilon_{sk})} = \frac{2\pi n_i}{\hbar}\int \frac{d^2k}{(2\pi)^2}\left|\left\langle V_{s\mathbf{k},\,s\mathbf{k}'}\right\rangle\right|^2 (1-\cos\theta_{kk'})\,\delta\left(\varepsilon_{sk} - \varepsilon_{sk'}\right) \tag{5.194}$$

The electrical current density is obtained as

$$j = g\int \frac{d^2k}{(2\pi)^2} e\mathbf{v}_{s\mathbf{k}} f_{sk} \tag{5.195}$$

which gives the conductivity as

$$\sigma = \frac{e^2 v_F^2}{2} \int d\varepsilon D(\varepsilon) \tau(\varepsilon) \left(-\frac{\partial f}{\partial \varepsilon}\right) \tag{5.196}$$

and which at $T = 0$ gives the conventional formula

$$\sigma = \frac{e^2 v_F^2}{2} D(\varepsilon_F) \tau(\varepsilon_F) \tag{5.197}$$

One computes the matrix element of the scattering potential of randomly distributed screened impurity charge centers as

$$\left|\langle V_{s\mathbf{k}, s\mathbf{k}'}\rangle\right|^2 = \left|\frac{v_i(q)}{\varepsilon(q)}\right|^2 \frac{1 + \cos\theta}{2} \tag{5.198}$$

where $q = |\mathbf{k} - \mathbf{k}'|$, $\theta = \theta_{kk'}$, and

$$v_i(q) = \frac{2\pi e^2}{\kappa q} \exp(-qd) \tag{5.199}$$

where κ is the background lattice dielectric constant. The factor $(1 + \cos\theta)/2$ takes into account the sublattice symmetry (overlap of wavefunctions), $\varepsilon(q)$ is the 2D dielectric screening function

$$\varepsilon(q) = 1 + v_c(q)\, \Pi(q) \tag{5.200}$$

where $\Pi(q)$ is the graphene irreducible polarizability function and $v_c(q)$ is the Coulomb interaction. Then, one gets

$$\begin{aligned} \frac{1}{\tau(\varepsilon_{sk})} &= n_i \frac{\pi}{\hbar} \int \frac{d^2k}{(2\pi)^2} \left|\frac{v_i(q)}{\varepsilon(q)}\right|^2 (1 + \cos\theta)(1 - \cos\theta)\, \delta(\varepsilon_{sk} - \varepsilon_{sk'}) \\ &= n_i \frac{\pi}{\hbar} \int \frac{d^2k}{(2\pi)^2} \left|\frac{v_i(q)}{\varepsilon(q)}\right|^2 \frac{1 - \cos 2\theta}{2} \delta(\varepsilon_{sk} - \varepsilon_{sk'}) \end{aligned} \tag{5.201}$$

From the plot in Fig. 5.20 one can see that in contrast to a conventional 2D material the factor $\Phi(\theta) = 1 - \cos 2\theta$ ensures that both the forward ($\theta = 0$) and the backward scattering ($\theta = \pi$) are suppressed. In the last equation

$$\begin{aligned} \varepsilon_{k+} &\to \varepsilon;\ \varepsilon_{k'+} \to \varepsilon \cos\theta \pm \sqrt{\Omega^2 + \varepsilon^2 \frac{\cos 2\theta - 1}{2}} \\ \varepsilon_{k-} &\to -\varepsilon;\ \varepsilon_{k'-} \to -\varepsilon \cos\theta \pm \sqrt{\Omega^2 + \varepsilon^2 \frac{\cos 2\theta - 1}{2}} \end{aligned} \tag{5.202}$$

$$\int \frac{d^2k}{(2\pi)^2} \to \frac{g}{2\pi \hbar^2 v_F^2} \int d\theta \int_{\varepsilon}^{\varepsilon_2} d\varepsilon\, |\varepsilon| \tag{5.203}$$

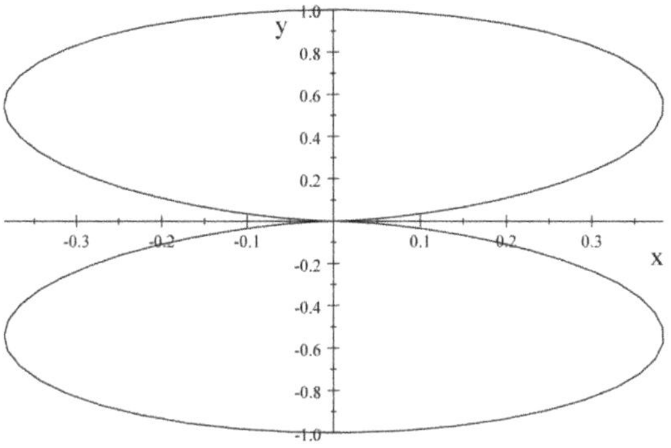

Figure 5.20 The angular dependence of the scattering factor $\Phi(\theta)$.

The cos theorem reads $q^2 = k^2 + (k')^2 - 2kk' \cos\theta$ or

$$|\mathbf{k}'| = k' = k\cos\theta \pm \sqrt{q^2 + k^2\frac{\cos 2\theta - 1}{2}}$$

which gives

$$\Pi(q, \omega) = -\frac{g}{2A}\sum_{\mathbf{k}ss'} \frac{f_{\mathbf{k}s} - f_{\mathbf{k}'s'}}{\omega + \varepsilon_{sk} - \varepsilon_{sk'} + i\eta}\left(1 + ss'\cos\theta\right) \tag{5.204}$$

The Dirac function in Eq. 5.202 contributes at

$$\varepsilon_{1,2} = \pm\frac{\Omega}{\sqrt{(1-\cos\theta)^2 - (\cos 2\theta - 1)/2}} \tag{5.205}$$

The angular factor for graphene is

$$\Omega\int_0^{\pi} d\theta \frac{1}{\sqrt{(1-\cos\theta)^2 - (\cos 2\theta - 1)/2}}\frac{1-\cos 2\theta}{2} = \frac{4}{3}\Omega \tag{5.206}$$

The magnetic length is

$$l = \sqrt{\frac{c\hbar}{eB_{\mathrm{fic}}}} = \sqrt{\frac{\hbar}{e\varphi_{\mathrm{fic}}}\pi d_{\mathrm{T}} L} = \sqrt{\frac{3d_{\mathrm{T}}L}{\nu}}, \tag{5.207}$$

which gives

$$\frac{1}{l^2} = \frac{\nu}{3d_{\mathrm{T}}L} \tag{5.208}$$

where for the sake of convenience we have introduced a fictitious magnetic field oriented perpendicular to the tube axis as

$B_{\text{fic}} = \varphi_{\text{fic}}/A$. In the last formula A is the effective area of the graphene stripe, and we have introduced a fictitious flux $\varphi_{\text{fic}} = -(\nu/3)\,\varphi_0$. The wavefunctions in a finite magnetic field B_{fic} are given by

$$\mathbf{F}_{sk}^{K} = \frac{1}{\sqrt{2L}}\begin{pmatrix} -is\,(k/|k|)\,F_-(x) \\ F_+(x) \end{pmatrix} \exp(iky) \tag{5.209}$$

where

$$F_\pm(x) = \frac{1}{\sqrt{\pi d_{\text{T}} I_0(2\zeta)}} \exp\left(\pm\zeta\cos\frac{2x}{d_{\text{T}}}\right) \tag{5.210}$$

and

$$\zeta = \frac{d_{\text{T}}^2}{4l^2} = \frac{\nu}{12}\frac{d_{\text{T}}}{L} \tag{5.211}$$

Here d_{T} and L are the tube diameter and length, $s = \pm 1$ for the conductance (valence) band, $I_0(2\zeta)$ is the modified Bessel function of the first kind.

In the K point vicinity, which corresponds to the vanishing of electron energy ($\varepsilon = 0$), there are two right-going channels K+ and K′+, and two left going channels K– and K′−. The electron–impurity matrix elements in the absence of a magnetic field are calculated as

$$V_{K\pm K+} = V_{K'\pm K'+} = \frac{1}{2}\left(\pm u_{\text{A}} + u_{\text{B}}\right) \tag{5.212}$$

$$V_{K\pm K'+} = V^{*}_{K'\pm K+} = \frac{1}{2}\left(\mp e^{i\eta}u'_{\text{A}} - \omega^{-1}e^{-i\eta}u'_{\text{B}}\right) \tag{5.213}$$

In the semiconducting nanotubes, the present symmetry is destroyed by the fictitious flux $\varphi = -(\nu/3)\,\varphi_0$ for the K point arising from the boundary conditions and giving rise to a nonzero gap. As a result, a finite back scattering is present and there is no perfectly conducting channel in semiconducting nanotubes. The real time reversal symmetry changes the wavefunction at the K point into one at the K′ point because they are mutually complex conjugated. Therefore, they cannot give any information on scattering within the K point or the K′ point. In a finite fictitious magnetic field $B_{\text{eff}} = \varphi_{\text{eff}}/A$ the electron–impurity matrix elements become as follows

$$V_{K\pm K+} = V_{K'\pm K'+} = \frac{1}{2A}\left(\pm u_{\text{A}}F_-(x_0)^2 + u_{\text{B}}F_+(x_0)^2\right) \tag{5.214}$$

$$V_{K'\pm K+} = \frac{1}{2A}\left(\mp e^{-i\eta}u'^{*}_{\text{A}} - \omega e^{i\eta}u'^{*}_{\text{B}}\right) F_+(x_0)\,F_-(x_0) \tag{5.215}$$

$$V_{K'\pm K'+} = \frac{1}{2A}\left(\pm u_A F_+ (x_0)^2 + u_B F_- (x_0)^2\right) \quad (5.216)$$

$$V_{K\pm K'+} = \frac{1}{2A}\left(\mp e^{i\eta} u'_A - \omega^{-1} e^{-i\eta} u'_B\right) F_+ (x_0)\ F_- (x_0) \quad (5.217)$$

5.11 Scattering of Phonons in Few-Layer Graphene

The phonon fraction of heat conductance experiences peculiar transformations during the 2D/3D transition. Those transitions are examined using graphene samples with different number of atomic layers. The experimental results obtained using a few layer graphene samples shown in Fig. 5.21 indicate that the heat conductance decreases as the number of layers increases. The saturation corresponding to the bulk graphene occurs as the number of layers exceeds four. That tendency is understood when considering the Umklapp scattering of phonons (Ghosh et al. 2010). In thin graphene films, consisting of just a few atomic layers, the phonon scattering on the upper and lower layers is rather weak because of the absence of the phonon momentum component perpendicular to layers. When the number of atomic layers increases, the phonon momentum acquires a perpendicular component across the layers. As a result, the phonon dispersion is also altered. It causes more initial and final states to become available for scattering in the phase space. The boundary scattering becomes stronger in thicker graphene films leading to an overall reduction of the thermal conductance. Besides, in real samples, it is hard to keep the same thickness around the whole area of the graphene sample. Consequently, the thermal conductance can be reduced far below its value in graphite. However, for thicker films, the thermal conductance acquires its value as in the bulk graphite. Variations of the thermal conductance versus the number of graphene layers is well described by the Fermi–Pasta–Ulam Hamiltonians for the crystal lattices (Saito and Dhar 2010) if the number of graphene layers n changes between one and four (Ghosh et al. 2010). For thicker multilayers with n = 1–8, a good agreement of the thermal conductance with the experiment has been achieved using molecular dynamics (MD) calculations for graphene nanoribbons. A plausible assumption

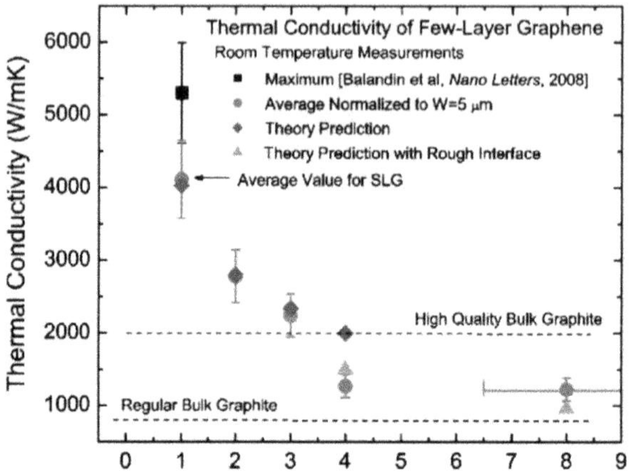

Figure 5.21 Measured thermal conductivity as a function of the number of atomic planes in few-layer graphene (FLG). The dashed straight lines indicate the range of bulk graphite thermal conductivities. The blue diamonds were obtained from the first-principles theory of thermal conduction in FLG on the basis of the actual phonon dispersion and accounting for all allowed three-phonon Umklapp scattering channels. The green triangles are Callaway–Klemens model calculations, which include extrinsic effects characteristic for thicker films. The figure is after Ghosh et al. (2010) reproduced with permission from the Nature Publishing Group.

when solving the Boltzmann's transport equation is that intraplane atomic interaction is described by the Tersoff potential, whereas the interplane interaction is mimicked by the Lennard–Jones potential (Lindsay and Broido 2011, Lindsay, Broido, and Mingo 2011). The theory predicts that the heat conductance drops sharply when going from a monolayer graphene to bilayer, whereas a slower decrease occurs when n exceeds two. The effect of the layer's top and bottom boundaries is explained encasing a few layer graphene between, for example, two layers of dielectrics. It results in a quite different thermal conductance dependence on the number of layers due to contribution of the acoustic phonon scattering. According to experiments (Jang et al. 2010) performed at $T = 310$ K an encased single-layer graphene is characterized by $K \approx 160$ W/m K. For graphite films with the thickness of 8 nm the heat conductivity increases to $\sim$1000 W/m K. The effect of encasing becomes stronger at low temperature where suppression of K follows the law

$K \propto T^{\beta}$ where $1.5 < \beta < 2$ (Jang et al. 2010). The major suppression factors are the disorder penetration through graphene and the rough boundary scattering both of which considerably limit the thermal conductance of the encased few layer graphene samples.

Problems

5-1. Explain why the conservation of a one-half pseudospin eliminates electron back scattering on impurities and phonons.
5-2. Discuss why the conservation of a one-half pseudospin is less pronounced in semiconducting carbon nanotubes (CNTs) than in metallic nanotubes. Explain also how the difference in pseudospin conservation affects electron back scattering with impurities and phonons in the two relevant materials.
5-3. Explain why the phonon drag effect is minimized in pristine graphene and metallic CNTs, while it remains finite in graphene stripes and in semiconducting nanotubes.
5-4. What are the necessary conditions for the plasma mode to exist?
5-5. Explain why phonon frequencies in conducting materials are renormalized.
5-6. For what reason does the electron–electron interaction become dependent on energy?
5-7. Explain why phonon scattering in graphene depends on the number of atomic layers.

References

Ando, T. (2005). Theory of electronic states and transport in carbon nanotubes. *J Phys Soc Jpn*, **74**(3): 777–817.

Bailyn, M. (1967). Phonon-drag part of thermoelectric power in metals. *Phys Rev*, **157**(3): 480.

Bohm, D., and D. Pines (1953). A collective description of electron interactions. 3. Coulomb interactions in a degenerate electron gas. *Phys Rev*, **92**(3): 609–625.

Gellmann, M., and K. A. Brueckner (1957). Correlation energy of an electron gas at high density. *Phys Rev*, **106**(2): 364–368.

Ghosh, S., W. Z. Bao, D. L. Nika, S. Subrina, E. P. Pokatilov, C. N. Lau, and A. A. Balandin (2010). Dimensional crossover of thermal transport in few-layer graphene. *Nat Mater*, **9**(7): 555–558.

Hubbard, J. (1957). The description of collective motions in terms of many-body perturbation theory. *Proc R Soc London, A*, **240**(1223): 539–560.

Jang, W. Y., Z. Chen, W. Z. Bao, C. N. Lau, and C. Dames (2010). Thickness-dependent thermal conductivity of encased graphene and ultrathin graphite. *Nano Lett*, **10**(10): 3909–3913.

Langer, J. S., and S. H. Vosko (1959). The shielding of a fixed charge in a high-density electron gas. *J Phys Chem Solids*, **12**(2): 196–205.

Lindhard, J. (1954). On the properties of a gas of charged particles. *Mat Fys Medd Dan Vid*, **28**(8): 1–57.

Lindsay, L., and D. A. Broido (2011). Enhanced thermal conductivity and isotope effect in single-layer hexagonal boron nitride. *Phys Rev B*, **84**(15).

Lindsay, L., D. A. Broido, and N. Mingo (2011). Flexural phonons and thermal transport in multilayer graphene and graphite. *Phys Rev B*, **83**(23).

Perebeinos, V., J. Tersoff, and P. Avouris (2005a). Effect of exciton-phonon coupling in the calculated optical absorption of carbon nanotubes. *Phys Rev Lett*, **94**(2).

Perebeinos, V., J. Tersoff, and P. Avouris (2005b). Electron-phonon interaction and transport in semiconducting carbon nanotubes. *Phys Rev Lett*, **94**(8).

Perebeinos, V., J. Tersoff, and P. Avouris (2005c). Radiative lifetime of excitons in carbon nanotubes. *Nano Lett*, **5**(12): 2495–2499.

Pines, D. (1957). Collective losses in solids. *Usp Fiz Nauk*, **62**(4): 399–425.

Rowland, T. J. (1962). Knight shift in silver base solid solutions. *Phys Rev*, **125**(2): 459.

Saito, K., and A. Dhar (2010). Heat conduction in a three dimensional anharmonic crystal. *Phys Rev Lett*, **104**(4).

Shafranjuk, S. E. (2009). Reversible heat flow through the carbon tube junction. *EPL*, **87**(5).

Tsaousidou, M., and K. Papagelis (2007). Phonon-drag thermopower of a ballistic semiconducting single-wall carbon nanotube. *AIP Conf Proc*, **893**: 1045–1046.

Chapter 6

Many-Body Effects and Excitations in Graphene

In the previous chapters, our consideration has been based on a model of noninteracting chiral fermions with a one-half pseudospin and the absence of an energy gap in graphene. Strictly speaking, the validity of the model is limited within the conic approximation. Since the real electron dispersion always deviates from linear form, the electron–hole binding becomes possible. In particular, the trigonal warping of the electron energy spectrum imposes additional selection rules for the electron–electron scattering in the neutrality point. Such selection rules represent the conditions for the electron–hole binding in graphene. Thus, the appearance of excitons actually is related with deviation of the electron dispersion from the conic shape.

6.1 Electron–Electron Coulomb Interaction

The electron–electron interaction in carbon nanotubes (CNTs) was calculated in terms of a screened Hartree–Fock approximation in the k–p scheme by Ando (1997). The k–p approximation also allows

Graphene: Fundamentals, Devices, and Applications
Serhii Shafraniuk

ISBN 978-981-4613-47-7 (Hardcover), 978-981-4613-48-4 (eBook)
www.panstanford.com

us to obtain the exciton-level energies. An alternative approach is based on the full random phase approximation (RPA) allowing us to compute (Sakai et al. 2004) the Coulomb potential between two electrons (see also Chapter 5 for more details). First, we consider the CNTs where a finite energy gap in the electron excitation spectrum originates from quantization in the perpendicular direction to the CNT axis. For the CNTs, when considering a cylindrical surface at $\mathbf{r} = (x, y)$ and $\mathbf{r}' = (x', y')$ one obtains (Wang et al. 1992, Lin and Shung 1993a, 1993b, Sato et al. 1993, Linchung and Rajagopal 1994a, 1994b) the potential of electron–electron interaction in the form

$$v\left(x - x', y - y'\right) = \frac{2e^2}{\kappa A}\sum_q e^{iq(y-y')} K_0\left(\frac{L|q|}{2\pi}\left|2\sin\frac{\pi\left(x - x'\right)}{L}\right|\right) \tag{6.1}$$

where $K_n(x)$ is the modified nth order Bessel function of the second kind. Using the effective-mass approximation, the potential (6.1) enters only the diagonal part of the matrix k–p Hamiltonian. Then, for the Coulomb matrix elements one obtains

$$\begin{aligned}&\left\langle \alpha, k, K; \beta', k' + q, K\right| v \left|\beta, k + q, K; \alpha', k', K\right\rangle \\ &= \frac{1}{A}\delta_{n-m,n'-m'}\left(\mathbf{F}^{K*}_{\alpha k}\cdot\mathbf{F}^{K}_{\beta k+q}\right)\left(\mathbf{F}^{K*}_{\beta' k'+q}\cdot\mathbf{F}^{K}_{\alpha' k'}\right) v_{n-m}(q)\end{aligned} \tag{6.2}$$

where

$$v_{n-m}(q) = \frac{2e^2}{\kappa} I_{|n-m|}\left(\frac{L|q|}{2\pi}\right) K_{|n-m|}\left(\frac{L|q|}{2\pi}\right) \tag{6.3}$$

In the above formulas, $\alpha = (s, n)$, $\beta = (s', m)$, and $I_n(x)$ is the modified Bessel function of the first kind. One can omit the matrix elements corresponding to the scattering between the K and K$'$ points, because a large momentum transfer is involved in the corresponding processes. In the RPA, the Coulomb interaction is screened by conducting electrons. The screening is described by introducing of the dynamical dielectric function $\varepsilon_{n-m}(q, m)$, given by

$$\varepsilon_{n-m}(q, m) = 1 + v_{n-m}(q)\, P_{n-m}(q, \omega) \tag{6.4}$$

where the polarization function $P_{n-m}(q, \omega)$ is expressed as

$$P_{n-m}(q, \omega) = -\frac{2}{A} \sum_{G=K,K'} \sum_{n',m'} \sum_{k'} \delta_{n-m,n'-m'} \times \left| \mathbf{F}^{G*}_{\alpha k} \cdot \mathbf{F}^{G}_{\beta k+q} \right|^2 g_0\left(\varepsilon^G_{+,m',k'}\right) g_0\left(\varepsilon^G_{-,n',-k'+q}\right) \times \left[\frac{1}{\omega - \varepsilon^G_{+,m',k'} + \varepsilon^G_{-,n',-k'+q} + i\delta} - \frac{1}{\omega + \varepsilon^G_{+,m',k'} - \varepsilon^G_{-,n',-k'+q} - i\delta} \right] \quad (6.5)$$

with $\delta \to 0$ being a positive infinitesimal, where G spans the K and K′ points.

6.2 Electron Self-Energy

In the dynamical RPA, one obtains the electron self-energy via the effective interaction

$$\tilde{\Sigma}^K_\alpha(k, \omega) = \frac{i}{2\pi A} \int d\omega' \sum_{\beta q} G^K_\beta\left(k + q, \omega + \omega'\right) \left| \mathbf{F}^{K*}_{\alpha k} \cdot \mathbf{F}^{K}_{\beta k+q} \right|^2 \times \frac{v_{n-m}(q)}{\varepsilon_{n-m}(q, \omega')} e^{i\delta\omega'} g_0\left(\varepsilon^K_{\beta,k+q}\right) \quad (6.6)$$

Here, the noninteracting Green's function $G^K_\beta(k, \omega)$ is written as

$$G^K_\alpha(q, \omega) = \frac{1}{\omega - \varepsilon^K_{\alpha,q} + i s_\alpha \delta} \quad (6.7)$$

Besides, in Eq. 6.6, $g_0(\varepsilon)$ is the cutoff function given by

$$g_0 = \frac{\varepsilon_c^{\alpha_c}}{|\varepsilon|^{\alpha_c} + \varepsilon_c^{\alpha_c}} \quad (6.8)$$

which depends on the two parameters α_c and ε_c. The cutoff function does not depend on these parameters, as long as ε_c is sufficiently large and α_c is not too large. Actually, by applying an appropriate procedure, the cutoff function Eq. 6.8 can be excluded. There is a particle–hole symmetry in the original k–p Hamiltonian in respect to zeroth energy $\varepsilon = 0$, and thus, in addition to a constant energy shift common to all bands, there is a symmetry about $\varepsilon = 0$ for the quasiparticle energies, too. Therefore, if one subtracts the constant shift of energy, the electron self-energy can be redefined as

$$\Sigma^{K}_{\pm,n}(k, \pm\omega) \equiv \pm\frac{1}{2}\left[\tilde{\Sigma}^{K}_{+,n}(k, \omega) - \tilde{\Sigma}^{K}_{-,n}(k, -\omega)\right] \qquad (6.9)$$

The upper sign in the right-hand side of Eq. 6.9 corresponds to $\Sigma^{K}_{+,n}(k, \omega)$, whereas the lower sign is related to $\Sigma^{K}_{-,n}(k, \omega)$. The procedure yields the self-energy which is particle–hole symmetric, $\Sigma^{K}_{+,n}(k, \omega) = \Sigma^{K}_{-,n}(k, -\omega)$.

If for a moment one disregards the frequency dependence of the dielectric function in Eq. 6.6, then $\varepsilon_{n-m}(q, \omega)$ must be replaced by $\varepsilon_{n-m}(q, 0)$. In the latter case, the self-energy corresponds to the screened Hartree–Fock approximation, which can be also regarded as a static RPA. The Hartree–Fock limit corresponds to the setting $\varepsilon_{n-m}(q, \omega) = 1$. Significance of the electron–electron interaction effects is deduced comparing different approximations.

6.3 Quasiparticle Excitation Energy

The above approach also allows computing of the contribution of many-body effects into the quasiparticle excitation energy $E^{K}_{s,n,k}$ which in the vicinity of K point becomes

$$E^{K}_{\pm,n,k} = \varepsilon^{K}_{\pm,n,k} + \Sigma^{K}_{\pm,n}\left(k, \varepsilon^{K}_{\pm,n,k}\right) \qquad (6.10)$$

The energy $E^{K}_{s,n,k}$ is determined from the above equation after replacing $\Sigma^{K}_{\pm,n}(k, \varepsilon^{K}_{\pm,n,k})$ with $\Sigma^{K}_{\pm,n}(k, E^{K}_{\pm,n,k})$. However, as soon as the electron self-energy is calculated only in the lowest order, the former Eq. 6.9 gives more precise results (Dubois 1959a, 1959b, Rice 1965). The single-particle energy Eq. 6.10 determines the band gaps $\varepsilon^{K}_{G}(n) = E^{K}_{+,n,0} - E^{K}_{-,n,0}$ and $\varepsilon^{K'}_{G}(n) = E^{K'}_{+,n,0} - E^{K'}_{-,n,0}$ and also yields the effective mass $m^{K*}_{\pm,n}$ and $m^{K}_{\pm,n}$ at the band edges. The ratio of the effective Coulomb energy $e^2/\kappa L$ to the typical kinetic energy $2\pi\gamma/L$, that is, $(e^2/\kappa L)^2/(2\pi\gamma/L)$, specifies the effective strength of the Coulomb interaction. The latter characteristic does not depend on the circumference length L of the CNT, for $\gamma_0 \approx 3$ eV and $a = 0.246$ nm, and thus one obtains $(e^2/\kappa L)^2/(2\pi\gamma/L) \approx 0.35/\kappa$. The parameter of electron–electron interaction is independent on the diameter of CNT if κ is independent of L. Therefore, for a fixed cutoff, the binding energy of exciton as well as the band gap are both inversely proportional to the diameter $d = L/\pi$. One frequently obtains a converging result, despite the divergence of the cutoff energy. It happens, for example, for a spontaneous in-plane Kekule and out-of-plane lattice distortions and for the acoustic-phonon distortion (Viet et al. 1994). However, for the strong short-range scatterers, such as lattice vacancies and Stone–Wales defects, respectively, and for localized eigenstates in nanotube caps (Yaguchi and Ando 2002), the results are quite sensitive to the choice of the cutoff. The divergences could be avoided by selecting the cutoff as $\varepsilon_c \approx 3\gamma_0$ which corresponds to the half of the π band width of the 2D graphite. It gives $\varepsilon_c(2\pi\gamma/L)^{-1} \sim (\sqrt{3}/\pi)(L/a)$. Within this consideration, the cutoff function causes tiny logarithmic correction to the band gap $\propto \ln[\varepsilon_c(2\pi\gamma/L)^{-1}] \sim (2\pi\gamma/L)\ln(d/a)$ which complements the other term with $2\pi\gamma/L \propto 1/d$. That happens because the energy band gap, which is proportional to the inverse of diameter, involves another term with a weak but finite logarithmic dependence versus the diameter of the CNT.

6.4 Computational Results

In Fig. 6.1 (left) we plot the numerical results of the first and second band gaps in the energy excitation spectrum of a semiconducting CNT and the first energy gap of a metallic CNT versus the effective strength of the Coulomb interaction. The cutoff here is chosen as $\varepsilon_c(2\pi\gamma/L)^{-1} = 5$. In the vicinities of K and K′ points and in the lowest energy band, Eq. 6.9 corresponds to $n = 0$, whereas $n = 1$ or $n = -1$ are related to the second band. In the range of Coulomb interaction parameter corresponding to Fig. 6.1, the static RPA gives the same results as the full dynamical approximation does. It suggests that in this range of parameters, the static approximation works good. A discrepancy increases for metallic nanotubes, although it can still remain small. Even in the regime of weak coupling, the Coulomb interaction causes the band gaps to be increased considerably. For example, if $(e^2/\kappa L)^2/(2\pi\gamma/L) = 1$, the first band gap, taking into account the electron–electron

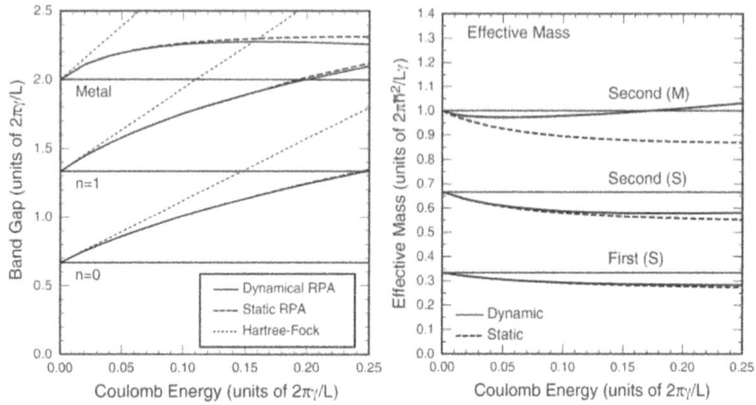

Figure 6.1 Left: The numeric results for the first and second band gaps in semiconducting nanotubes and first gap in metallic nanotubes for $\varepsilon_c\,(2\pi\gamma/L)^{-1} = 5$. Right: The effective electron mass calculated for the bottom of the first and second band gaps in semiconducting nanotubes and that of the first gap in metallic nanotubes for $\varepsilon_c\,(2\pi\gamma/L)^{-1} = 5$. Adapted from Ando (2005).

interaction, is about 1.5 times larger than without interaction. For semiconducting nanotubes, the Hartree–Fock approximation remains valid when $(e^2/\kappa L)^2/(2\pi\gamma/L) < 0.05$, whereas in metallic nanotubes it fails.

The numeric results for the band-edge effective mass are shown in Fig. 6.1 (right). The effective electron mass is only slightly diminished in the semiconducting nanotubes due to the electron–electron interaction. The dynamic and static RPAs both yield practically the coinciding results. On the other hand, the discrepancy is bigger in metallic nanotubes, because the dynamical effects are more important due to availability of metallic electrons. The effective mass changes less than 10% and thus they can be disregarded. The electron–electron interaction causes essentially different impacts on the band gap and the effective mass. For such reasons the influence of interaction cannot be accounted for as a mere renormalization of the single band parameter γ. If the interaction is turned off, the first and second band gaps become $4\pi\gamma/3L$ and $8\pi\gamma/3L$, respectively, and the corresponding effective masses are given by $2\pi\hbar^2/3L\gamma$ and $4\pi\hbar^2/3L\gamma$, respectively.

Therefore, renormalization of γ should give the result that the ratio between the first and second gaps should remain the same and that the effective mass should also be reduced by the same amount as that of the band-gap enhancement.

6.5 Excitons

In an ideal 1D electron–hole system, the exciton binding energy diverges (Elliott and Loudon 1959, Ogawa and Takagahara 1991a, 1991b). One consequence is that the exciton binding becomes very important, because it modifies the absorption spectra in CNTs and graphene stripes drastically. The exciton energy levels and relevant optical spectra have been computed using a mere k–p approach and the screened Hartree–Fock approximation (Ando 1997). Due to a small cutoff, the mentioned approach gives smaller absolute values of the band gaps. Exciton states are described with the many-body wavefunction

$$|u\rangle = \sum_n \sum_k \psi_n (k) c^{\dagger}_{+,n,k} c_{-,n,k} |g\rangle \tag{6.11}$$

where $|g\rangle$ is the ground-state wavefunction, and $c^{\dagger}_{\pm,n,k}$ $(c_{\mp,n,k})$ is the creation (annihilation) operator. The equation of motion for $\psi_n (k)$ is formulated in the form

$$\begin{aligned} \varepsilon_u \psi_n (k) = & \left(2\gamma\sqrt{\kappa_{\nu\varphi} (n)^2 + k^2} + \Delta\varepsilon_{nk}\right) \psi_n (k) \\ & - \sum_m \int \frac{dq}{2\pi} \frac{2e^2}{\kappa\varepsilon_{|n-m|} (q)} I_{|n-m|} \left(\frac{L|q|}{2\pi}\right) K_{|n-m|} \left(\frac{L|q|}{2\pi}\right) \\ & \times \frac{1}{2} \left(1 + \frac{\kappa_{\nu\varphi} (n) \kappa_{\nu\varphi} (m) + k(k+q)}{\sqrt{\kappa_{\nu\varphi} (n)^2 + k^2}\sqrt{\kappa_{\nu\varphi} (m)^2 + (k+q)^2}}\right) \psi_m (k+q) \end{aligned} \tag{6.12}$$

where, $\Delta\varepsilon_{nk}$ is the difference of the self-energies of the conduction and valence band. Using the screened Hartree–Fock approximation, one obtains

$$\Delta\varepsilon_{nk} = \sum_m \sum_q \frac{2e^2}{A\kappa\varepsilon_{|n-m|}(q)} I_{|n-m|}\left(\frac{L|q|}{2\pi}\right) K_{|n-m|}\left(\frac{L|q|}{2\pi}\right)$$
$$\times \frac{\kappa_{\nu\varphi}(n)\,\kappa_{\nu\varphi}(m) + k(k+q)}{\sqrt{\kappa_{\nu\varphi}(n)^2 + k^2}\sqrt{\kappa_{\nu\varphi}(m)^2 + (k+q)^2}} g_0$$
$$\times \left(\gamma\sqrt{\kappa_{\nu\varphi}(m)^2 + (k+q)^2}\right) \tag{6.13}$$

For the sake of convenience, to characterize the optical absorption, it is instructive to introduce the dynamical conductivity $\sigma_{yy}(\omega)$ which at the K point is

$$\sigma_{yy}(\omega) = \frac{\hbar e^2}{AL} \sum_u \frac{-2i\hbar\omega\left|\langle u|\, v_y\, |g\rangle\right|^2}{\varepsilon_u\left[\varepsilon_u^2 - (\hbar\omega)^2 - 2i\hbar^2\omega/\tau\right]} \tag{6.14}$$

where we have introduced the broadening in terms of the phenomenological relaxation time τ. Furthermore, we use

$$\langle u|\, v_y\, |g\rangle = i\frac{\gamma}{\hbar} \sum_n \sum_k \frac{\kappa_{\nu\varphi}(n)}{\sqrt{\kappa_{\nu\varphi}(n)^2 + k^2}} \psi_n^*(k) \tag{6.15}$$

As it follows from Eq. 6.13, the intensity of absorption is proportional to $\left|\langle u|\, v_y\, |g\rangle\right|^2/\varepsilon_u$. Similarly to the case of the free-electron systems, it is instructive to introduce dimensionless oscillator strength

$$f_u = 2m^* \frac{\left|\langle u|\, v_y\, |g\rangle\right|^2}{\varepsilon_u} \tag{6.16}$$

where m^* corresponds to the effective mass at the bottom of the lowest conduction band

$$m^* = \frac{2\pi\hbar^2}{3\gamma L} \tag{6.17}$$

For the lowest exciton one can use $f_u \approx 1$. In the systems of free noninteractive electrons, the oscillator strength conforms the f sum rule

$$\sum_u f_u = 1 \tag{6.18}$$

However, the sum rule, Eq. 6.18, is not satisfied for the case of interacting electrons. The rule, Eq. 6.18, is broken because the

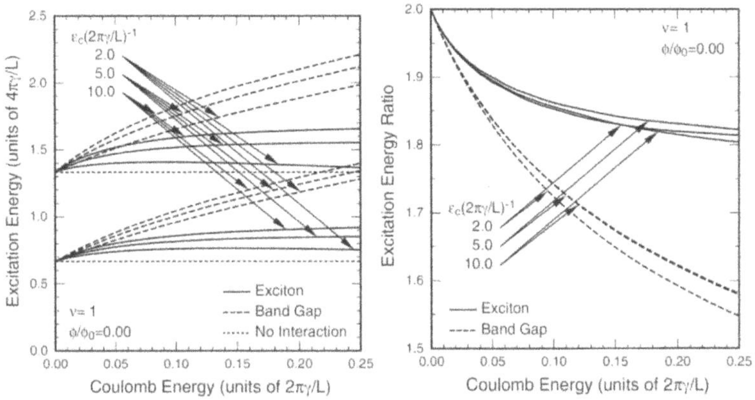

Figure 6.2 Left: The electron energy band gaps (dashed lines) and exciton absorption energies (solid lines) computed for the first and second bands versus the effective Coulomb energy $(e^2/\kappa L)^2/(2\pi\gamma/L)$ in the absence of flux for various cutoff energies. Right: The ratio of the absorption energies (solid lines) and band gaps (dashed lines) for the first and second energy bands as a function of the effective Coulomb energy $(e^2/\kappa L)^2/(2\pi\gamma/L)$. Adapted from Ando (2005).

electron propagation is described by Weyl's equation for massless neutrinos instead of the Schrödinger equation, which is valid for conventional spinless particles.

In Fig. 6.2 we show the lowest exciton absorption energies associated with the fundamental and second gaps and the band gaps versus the effective Coulomb energy $(e^2/\kappa L)^2/(2\pi\gamma/L)$ (Ando 2004). When the interaction strength increases, one can notice an increase of energy band gaps. The exciton binding energy becomes larger as the interaction grows, but is still below the band-gap enhancement. It causes the absorption energies to exceed the energy band gap of a noninteracting system. Furthermore, the influence of interaction on the band gap and on the exciton binding energy is more significant for the second band than for the first band. In Fig. 6.2 (right) we plot the ratio of excitation energies for the first and second gap versus the effective Coulomb energy. When the Coulomb energy increases, the ratio of absorption energies plunges from 2 to ~1.8. On the contrary, the band gap decreases much faster. A remarkable circumstance is that the energies ratio is practically

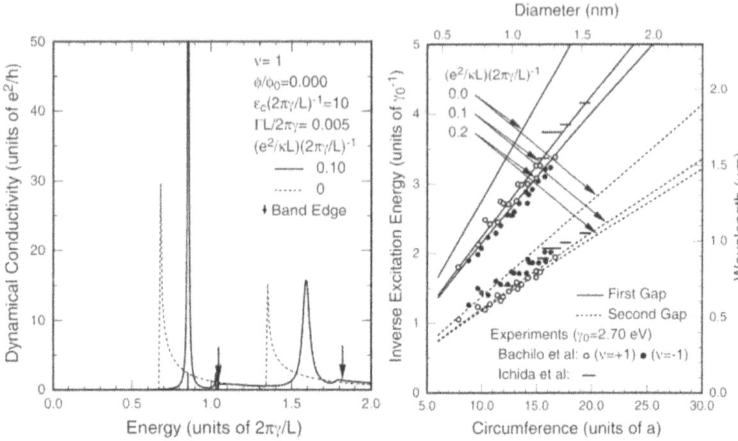

Figure 6.3 Left: Calculated optical absorption spectra of a semiconducting nanotube in the presence of AB magnetic flux. The solid lines represent spectra in the presence of interaction $(e^2/\kappa L)^2/(2\pi\gamma/L)$ and the dotted lines represent those in the absence of interaction. Right: The inverse of the absorption energies obtained by an interpolation as a function of the circumference and the diameter. The experimental data are adapted from Ichida et al. (1999), Ichida et al. (2002), and Bachilo et al. (2002) and are plotted using $\gamma_0 = 2.7$ eV.

independent versus the cutoff parameter. The mentioned ratio ~ 1.8 well explains the available experimental except for the limit when the interaction parameter becomes small.

The absorption spectrum computed for $\varepsilon_c\,(2\pi\gamma/L)^{-1} = 10$ and $(e^2/\kappa L)^2/(2\pi\gamma/L) = 0.1$ are shown in Fig. 6.3 (left). The exciton levels designated by vertical straight lines and the band gaps are signified by downward arrows. The intensity of interband transitions almost vanishes, whereas most of the optical intensity is transferred to the exciton bound states.

Such the tendency emerges from the 1D nature of the graphene stripes and CNTs. Similar behavior takes place in a variety of 1D systems (Ogawa and Takagahara 1991a, 1991b). Kataura et al. (1999) studied the optical absorption spectra of thin film samples of SWCNTs. Visibility of the excitonic effects is revealed from detailed comparison between the experimentally observed spectrum and theoretical predictions based on a simple tight-binding model

(Ichida et al. 1999, Ichida et al. 2002, Bachilo et al. 2002). Other activities (Bachilo et al. 2002, Hagen and Hertel 2003) have addressed the optical absorption and photoluminescence of individual nanotubes. Zaric et al. (2004) observed splitting of the absorption and emission peaks due to AB effect associated with magnetic flux passing through the cross section. According to Ando (2004), the electron–electron interactions leads to a slight enhancement of the AB splitting which originates from a cumulative effect of magnetic flux and of the shift due to strain. In the k–p scheme, the only remaining parameter is γ or γ_0 which are related to each other as $\gamma = \sqrt{3}a\gamma_0/2$. In Fig. 6.3 (right) we compare the experiments by Ichida et al. (1999), Ichida et al. (2002), and Bachilo et al. (2002) with computed absorption energies for $\gamma_0 =$ 2.7 eV. One can see that there is a good agreement between the experiments on the one hand and the overall dependence on the circumference and the diameter, on the other hand. Experimental data contain the dependence versus chirality as well as the diameter, in particular, for narrow nanotubes and graphene stripes. In the low-diameter nanotubes and in narrow graphene stripes there is an influence from a variety of higher-order effects like the trigonal warping and a finite curvature. The higher-order effects might cause, for example, the opposite energy shifts of the first and second gaps. Besides, it might also alter the signature between semiconducting nanotubes with $\nu = +1$ and –1. Such effects had been exploited for studying of the structure of individual SWCNTs. It was accomplished by the combination of absorption which was caused by transitions corresponding to the second gap on the one hand with the photoluminescence due to transitions related to the first gap, on the other hand (Bachilo et al. 2002, Hagen and Hertel 2003). Nevertheless, a strong presence of excitons and of the electron–electron interaction complicates an accurate estimation of these effects even in the k–p scheme. In the extremely narrow graphene stripes and CNTs one should appropriately consider mixing between the π and σ bands. The last effect can cause anisotropy, which is inherent when forming the cylindrical structure. An appropriate generalization of the tight-binding approach provides only a partial explanation of the aforementioned effects (Samsonidze et al. 2004).

6.6 Wannier–Mott Excitons

In this subsection we use an envelope approximation to illustrate the formation of the Wannier–Mott excitons near the conic point in a neutral graphene (Mahmoodian and Entin 2013). Here, we use a simple analytical model (Mahmoodian and Entin 2013) which illustrates the exciton formation with involvement of electron–hole bound states. We consider an ideal pristine graphene with no energy gap in its electron energy spectrum. Thus, we study the formation of an electron–hole bound state with the energy lying in the free-electron–hole continuum. In conventional semiconductors, such exciton state requires an energy gap to be present in the electron excitation spectrum of the material. However, the conic shape of spectrum in graphene makes a difference with a regular semiconductor where the electron dispersion law is determined by quadratic energy spectrum of particles $p^2/2m$. More specifically, in a quasiclassical approximation, an attraction between two particles characterized by the quadratic dispersion law causes their oscillatory motion which is confined inside their mutual well. On the other hand, the two particles behave differently if their dispersion law has a conic shape: Their moments remain constant and parallel to each other due to equality of velocities, whereas they maintain a constant separating distance between the two particles. Indeed, such an idealized picture can occur only for an infinite pristine graphene sheet and almost never occurs in the realistic samples. A more realistic picture of the quasiparticle's propagation should include nonlinear corrections to the dispersion law. In that case there appears a finite electron effective mass which is computed by expanding of the kinetic energy of the pair in the series over the relative momentum. The mass of a particle corresponds to the second derivative of the electron energy with respect to the relative momentum. As soon as the mass becomes finite, the particles cannot maintain the constant spatial separation any more. Thus, the propagation of particles with a finite mass is accompanied by variation of the electron–hole distance. If the mass is positive, the electron–hole pair forms a bound state. The depicted behavior retains validity also using the purely quantum language. Because the sign of the mass depends versus the angle of the exciton momentum, the interparticle binding occurs only in certain sectors of the momentum space, whereas the bound states are prohibited in the other sectors belonging to the same space.

Here, we assume that the binding energy is small as compared with the kinetic energy. It happens in graphene when the charge carriers interact weakly with each other because the interaction constant is small, $g = e^2/v\chi \ll 1$, where e is an electron charge, χ is the half-sum of dielectric constants of surrounding media, $\hbar = 1$. It allows implementing of the quasiclassical envelope function approximation and ignores the many-body effects which will be described in the second half of this chapter.

6.7 Excitonic States

We describe the exciton states in terms of the tight-binding electron Hamiltonian

$$H_{\mathrm{e}}(p) = \begin{pmatrix} 0 & \Omega_p \\ \Omega_p^* & 0 \end{pmatrix}$$
$$\Omega_p = -\gamma \sum_i e^{-i\mathbf{pl}_i} \qquad (6.19)$$

Here, $\gamma = 2v/a\sqrt{3}$, $l_1 = (0, 1)a/\sqrt{3}$, $l_2 = (-3, -\sqrt{3})a/6$, and $l_3 = (3, -\sqrt{3})a/6$, and we use primitive translation vectors $\mathbf{a} = a(1, 0)$ and $\mathbf{b} = a(-1/2, \sqrt{3}/2)$, where $a = 0.246$ nm is the lattice constant. The basis vectors of primitive reciprocal lattice are $\mathbf{a}^* = (2\pi/a)(1, 1/\sqrt{3})$ and $\mathbf{b}^* = (2\pi/a)(0, 2/\sqrt{3})$. One finds the energy eigenvalues of Hamiltonian Eq. 6.19 as

$$\varepsilon_{\pm}(\mathbf{p}) = \pm\sqrt{1 + 4\cos\frac{ap_x}{2}\cos\frac{\sqrt{3}ap_y}{2} + 4\cos\frac{ap_x}{2}} \qquad (6.20)$$

Here we focus on the exciton states which are formed within a narrow region of the momentum space in the vicinity of the K and K′ points, where $\mathbf{p} = \pm\mathbf{K}$, $\mathbf{K} = (1, 0)4\pi/3a$. In a pristine graphene sheet with infinite dimensions, the single-electron excitation spectrum has a conic shape $\varepsilon\pm(\mathbf{p}) = \pm v|\mathbf{p} - \mathbf{K}|$ and $\varepsilon\pm(\mathbf{p}) = \pm v|\mathbf{p} + \mathbf{K}|$, respectively. In the quasiclassical envelope wavefunction approximation, the two-particle excitonic Hamiltonian is

$$H_{\text{ex}} = H_{\text{e}}(\mathbf{p}_{\text{e}}) \otimes I_{\text{h}} - I_{\text{e}} \otimes H_{\text{h}}(\mathbf{p}_{\text{h}}) + I_{\text{e}} \otimes I_{\text{h}} V(\mathbf{r}_{\text{e}} - \mathbf{r}_{\text{h}}) \qquad (6.21)$$

Where $H_{\text{h}}(\mathbf{p}) = -H_{\text{e}}(\mathbf{p})$ is the hole Hamiltonian, $\mathbf{r}_{\text{e,h}}$ and $\mathbf{p}_{\text{e,h}}$ are the coordinates and momentums of electrons and holes. The electron and hole Hamiltonians $H_{\text{e,(h)}}\left(\mathbf{p}_{\text{e,(h)}}\right)$ refer to the electron and the hole subspaces of quantum numbers. The interaction between the electrons and holes is described in terms of the potential of $V(\mathbf{r}) = -e^2/\chi r$ is. The above Hamiltonian, Eq. 6.21, has a fourfold set of eigenstates. Every eigenstate belongs to the ordinary exciton, namely, a pair with "an electron in the conduction band and a hole in the valence band."

We diagonalize $H_e(\mathbf{p}_e)$ and $H_h(\mathbf{p}_h)$ by performing a unitary transformation $U = U_{\text{e}}(\mathbf{p}_{\text{e}})U_{\text{h}}(\mathbf{p}_{\text{h}})$ of the Hamiltonian, Eq. 6.21. The binding energy is assumed to be small which means that the exciton wavefunction is smooth because its spatial derivatives are small, whereas it spreads over a large extend. Besides, we can disregard the commutators of U and $V(\mathbf{r}_{\text{e}} - \mathbf{r}_{\text{h}})$. Within the listed assumptions we get an effective simplified Hamiltonian

$$H_{\text{ex}} = \varepsilon_{+}(\mathbf{p}_{\text{e}}) + \varepsilon_{+}(\mathbf{p}_{\text{h}}) + V(\mathbf{r}_{\text{e}} - \mathbf{r}_{\text{h}}) \qquad (6.22)$$

The electron and hole momentums $\mathbf{p}_{\text{e}}$ and $\mathbf{p}_{\text{h}}$ are both expressed via the momentum of the pair $\mathbf{q} = \mathbf{p}_{\text{e}} + \mathbf{p}_{\text{h}}$ and the relative momentum $\mathbf{p} = \mathbf{p}_{\text{e}} - \mathbf{p}_{\text{h}}$. The momentums can be either situated (*i*) in the K point vicinity or (*ii*) at the opposite conic points. Case (*i*) corresponds to the direct exciton with momentum $\mathbf{q} \to k \ll K$. For case (*ii*), one gets the indirect exciton with $\mathbf{q} = 2\mathbf{K} + \mathbf{k}$ and $k \ll K$, which is related to the opposite conic points (see Fig. 6.4). The Hamiltonian for describing of the latter case (ii) is obtained by expanding Eq. 6.22 over the small parameter $p/k \ll 1$. Then, the model Hamiltonian for the indirect exciton reads

$$H_{\text{ex}} = 2\varepsilon_{+}\left(\frac{\mathbf{q}}{2}\right) + \frac{p_i p_j}{2}\left(\frac{1}{m}\right)_{ij} - \frac{e^2}{\chi r} \qquad (6.23)$$

where $\mathbf{r} = \mathbf{r}_{\text{e}} - \mathbf{r}_{\text{h}}$. In Eq. 6.23, we have introduced the tensor of inverse masses $(m^{-1})_{i,j} = \nabla_i \nabla_j \varepsilon_{+}(\mathbf{q}/2)/2$ whose components are

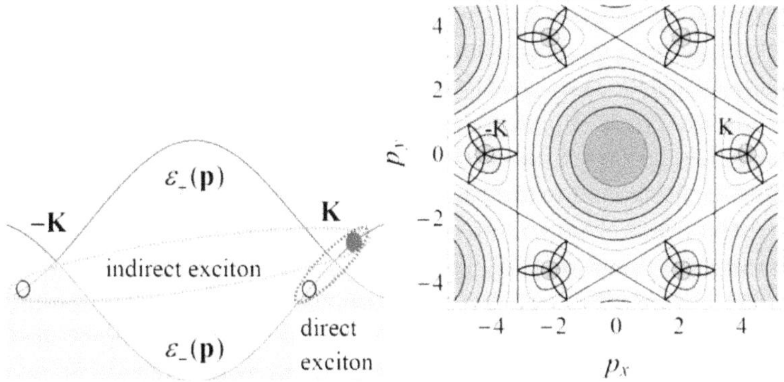

Figure 6.4 Left: Direct and indirect exciton formation in terms of a single-particle spectrum. Right: Contour plot of the single-electron spectrum. The electron energy varies from 0 at the K points to 3 in the center. The regions where the indirect excitons are formed are bound by trefoils. Adapted from Mahmoodian and Entin (2013).

$$\left(\frac{1}{m}\right)_{x,x} = \frac{1}{\varepsilon_+^3}\left(2 - 8c_x^4 - c_x c_y - 12c_x^3 c_y - 2c_y^2 - 2c_x^2\left(2 + c_y^2\right)\right)$$
$$\left(\frac{1}{m}\right)_{x,y} = \frac{\sqrt{3}}{\varepsilon_+^3}\left(1 + 2c_x c_y\right) s_x s_y$$
$$\left(\frac{1}{m}\right)_{y,y} = -\frac{3}{\varepsilon_+^3} c_x\left(2c_x + c_y\right)\left(1 + 2c_x c_y\right) \quad (6.24)$$

In the last formula, Eq. 6.24, we have introduced the following notations $c_x = \cos(q_x a/4)$, $c_y = \cos(\sqrt{3}q_y a/4)$, $s_x = \sin(q_x a/4)$, and $s_y = \sin(\sqrt{3}q_y a/4)$. Assuming a conic shape of the electron spectrum and by setting $m = k/v$, $M = \infty$, the Hamiltonian, Eq. 6.23, at small $k \ll K$ is simplified as

$$H_{ex} = E_{\text{kin}} + \frac{p_1^2}{2M} + \frac{p_2^2}{2m} - \frac{e^2}{\chi\sqrt{x_1^2 + x_2^2}} \quad (6.25)$$

Here, we use the coordinate system with the basis vectors $\mathbf{e}_1 \equiv \mathbf{k}/k$ and $\mathbf{e}_2 \perp \mathbf{e}_1$, the kinetic energy of a free pair is $E_{\text{kin}} = vk - (\mu k^2/2)$ $\cos 3\varphi_{\text{k}}$, and $\mu = va/(4\sqrt{3})$ is a warping parameter. To obtain finite M, one should go beyond the conic approximation. When moments of the indirect exciton are close to $\nu_{2\text{K}}$, $(\nu = \pm 1)$, one obtains the expression effective mass M as $1/M = \nu\mu \cos 3\varphi_{\text{k}}$ where φ_{k} is an angle between **k** and **K**. It suggests that M originates from trigonal contributions and is parametrically large because $M/m = \eta = v/(\nu\mu k\cos 3\varphi_k) \gg 1$. The sign of M is determined by the factor ν cos $3\varphi_k$. Thus, the binding condition for the electron–hole pairs is ν cos $3\varphi_k > 0$, whereas excitons are not formed if ν cos $3\varphi_k < 0$. Similarly, one can also examine the pair from the same valley which suggests that the trigonal contributions to $1/M$ vanish in the latter case. In the higher orders one obtains negative contribution into the mass $1/M = -kva^2(7 - \cos 6\varphi_k)/32$. Therefore, the binding between the electrons and holes is forbidden if they belong to the same valley. Using the condition that $p_i p_j (1/m)_{i,j}/2 > 0$, representing the kinetic energy which must be positively defined. We also check whether the indirect excitons might occur far away from the conic points or not. The latter condition can be reformulated as a requirement that the both eigenvalues of the mass tensor $(1/m)_{i,j}$ must be positive. From Eq. 6.24 we find

$$\begin{aligned}
&\left(2c_x^2 + c_x c_y + s_x s_y\right) < 0 \wedge \left(2c_x^2 + c_x c_y - s_x s_y\right) < 0 \vee \\
&\left(2c_x^2 + c_x c_y + s_x s_y\right) < 0 \wedge \left(1 + 2c_x c_y\right) < 0 \vee \\
&\left(2c_x^2 + c_x c_y - s_x s_y\right) < 0 \wedge \left(1 + c_x c_y\right) < 0
\end{aligned} \quad (6.26)$$

Figure 6.4 (left) depicts the region of exciton formation in the $\mathbf{p}_e$ space. The region involves just a small part of the Brillouin zone. The units for the momentum are $1/a$.

Coulomb states can be studied with the help of the Hamiltonian, Eq. 6.23, by exploiting a strong anisotropy of the energy spectrum. The free pair is characterized by the kinetic energy E_{kin} which represents a fraction of the total exciton energy

$$E_{nN} = E_{\text{kin}} - \varepsilon_{nN} \quad (6.27)$$

Here, the positive quantity $\varepsilon_{nN} > 0$ is the binding energy and the quantum numbers n and N characterize the exciton state. There is a similarity between our case and the case of molecular states where m and M represent electron and ion masses, correspondingly, while the ratio of masses is also large, $\eta = M/m \gg 1$. It allows us to use here a solution which is well known from the problem of the molecular levels. The scenario is as follows. Initially, we eliminate the term $p_1^2/2M$ and fix the ion coordinate x_1. Then, we establish the energy terms $-\zeta_n(x_1)$, which now pose as the potential energy of ions. Besides, we factorize the total wavefunction $\Psi_{nN}(x_1, x_2)$ by representing it as a product of electron and ion wavefunctions $\Psi_{nN}(x_1, x_2) = \psi_n(x_1;x_2)\ \Psi_{nN}(x_1)$. The Schrödinger equation for ions

$$\left(\frac{p_1^2}{2M} - \zeta_n(x_1) + \varepsilon_{nN}\right)\Psi_{nM}(x_1) = 0 \tag{6.28}$$

allows us to identify the binding energy ε_{nN} in the Nth ion state since the nth electron term $\zeta_n(x_1)$ is determined by the above Eq. 6.28. Energies of the lowest levels are roughly equal to the Bohr energy $\varepsilon_B = me^4/\chi^2$. At small x_2 ($x_2 \ll x_1$) we approximate the energy terms using a simplified form of the electron–hole interaction potential as $V(\mathbf{r}) \simeq -e^2/\chi(|x_1|+|x_2|)$, since it has the same asymptotic as the actual potential. Such a simplifying trick allows us to rewrite the Schrödinger equation in another form which is already solvable analytically

$$\frac{1}{2m}\frac{\partial^2\psi_n(x_2;x_1)}{\partial x_2^2} + \frac{e^2}{\chi(|x_1|+|x_2|)}\psi_n(x_2;x_1) = \zeta_n(x_1)\,\psi_n(x_2;x_1) \tag{6.29}$$

The boundary condition at $|x_2| \to \infty$ ($\zeta_n(x_1) > 0$) assumes that the wavefunction vanishes which corresponds to the solution of Eq. 6.29 in the form

$$\psi_n(x_2;x_1) = C_{\pm}e^{-P_n(x_1)(|x_1|+|x_2|)}(|x_1|+|x_2|)\,U \times \left[1 - \frac{1}{P_n(x_1)\,a_B}, 2, 2P_n(x_1)(|x_1|+|x_2|)\right] \tag{6.30}$$

The quasiclassical solution of the simplified Eq. 6.29 corresponds to the molecular energy terms at $x_1 \ll a_B$

$$\begin{aligned}
\zeta_0(x_1) &= 2\varepsilon_B \log^2\frac{a_B}{|x_1|}\\
\zeta_n^{\text{even}}(x_1) &= \frac{\varepsilon_B}{2n^2} - \frac{\varepsilon_B}{n^3\log^2\frac{a_B}{|x_1|}}\\
\zeta_n^{\text{odd}}(x_1) &= \frac{\varepsilon_B}{2n^2} - \frac{2\varepsilon_B}{n^3}\frac{|x_1|}{a_B}
\end{aligned} \tag{6.31}$$

where $n \geq 1$. Using the above energy terms in Eq. 6.28, we find the energy levels in the quasiclassical approximation as

$$\varepsilon_{00} \approx \frac{\varepsilon_B}{2}\log^2\eta + \ldots$$
$$\varepsilon_{nN}^{\text{even}} \approx \frac{\varepsilon_B}{2n^2} - \frac{\varepsilon_B}{n^3}\frac{2}{\eta\log\frac{2\eta}{\pi^2 n^3\left(N+\frac{1}{4}\right)}} + \ldots$$
$$\varepsilon_{nN}^{\text{odd}} \approx \frac{\varepsilon_B}{2n^2} - \frac{\varepsilon_B}{n^2}\left[\frac{3\pi}{4\eta^2}\left(N+\frac{3}{4}\right)\right]^{\frac{2}{3}} + \ldots \quad (6.32)$$

The interlevel spacing for ions $\varepsilon_{n,N+1} - \varepsilon_{n,N}$ is much narrower than the interlevel spacing for the electron levels $\varepsilon_{n+1,N} - \varepsilon_{n,N}$, with the smallness parameter $1/\eta$. Using the variational method in respect to the Hamiltonian, Eq. 6.23, with total variational wavefunctions

$$\Phi_1 = Be^{-\rho^2}$$
$$\Phi_2 = Be^{-\rho}$$
$$\rho^2 = \frac{x_1^2}{a_1^2} + \frac{x_2^2}{a_2^2} \quad (6.33)$$

one obtains

$$\varepsilon_{00} = \alpha\varepsilon_{\text{B}}\log^2\eta \quad (6.34)$$

where α is equal to $1/\pi$ and $4/\pi^2$, for Φ_1 and Φ_2, respectively. The magnitudes of α are close to 1/2 which is obtained from an exactly solvable model (see Eq. 6.30). For both models one gets the same $B = (2/\pi a_1 a_2)^{1/2}$. The binding energy ε_{00} of the lowest level is angular dependent, while for the other levels the dependence versus angle φ_{k} is rather weak. There are singular corrections for the higher energy levels, which depend on the ratio of masses η and which are different for odd and even states. In Fig. 6.5 (left) we plot the dependencies of ε_{00} on the exciton momentum for $\chi = 6.5$ obtained numerically. In Fig. 6.5 (right) we plot the total exciton energy along the medial line of the exciton formation region. The total exciton energy E_{nN} is always positive and belongs to the electron–hole continuum. The exciton slowly decays at the expense of weak scattering processes, although the momentum is conserved. In accordance with the total momentum of the indirect exciton it has a twofold degeneracy corresponding to an opposite symmetry in the k space.

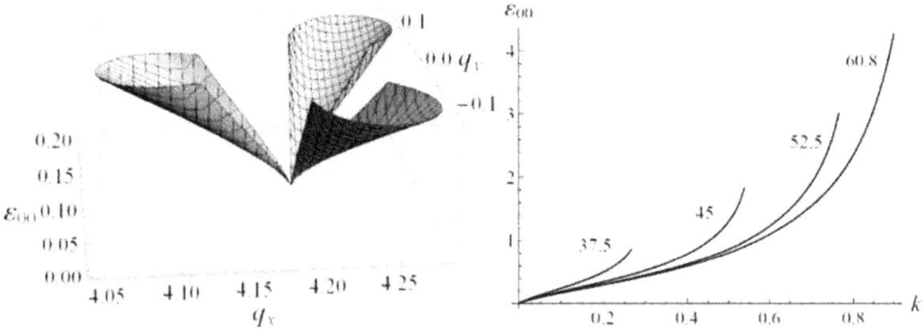

Figure 6.5 Left: A 3D plot of binding energy (in eV) of the exciton's ground state in graphene. The electron wavevector components q_x and q_y are in units of reciprocal lattice constants. The indirect exciton exists in the regions shown in Fig. 6.4. Right: Radial sections of the top figure at fixed angles in degrees. Adapted from Mahmoodian and Entin (2013).

6.8 Experimental Observation of Excitons in Graphene

The indirect excitons can be observed in graphene using of optical methods. Measuring the light emission in a semiconductor with the energy gap one can distinguish the excitons contribution. In contrast to conventional semiconductors with energy gap, the absence of gap in the electron excitation spectrum of pristine graphene causes the exciton energies to be distributed in the large range between zero and several tenths of electron volts. Since the indirect exciton has a large momentum, it blocks the recombination as well as the direct optical excitation. The good news is that the decay and the recombination of excitons occur pretty slowly since the electron recombination and the intervalley relaxation are quite weak. The slow decay improves the exciton stability. Another issue, when generating excitons, is a necessity to use photons with long wavelength (short wavevector). It assumes an implementation of very low photon energy. The latter complication is solved using absorption which is either phonon or impurity assisted. In order to facilitate the absorption of the light, one can also engineer the graphene samples with an artificial grating or rippled graphene sheets aimed to ensure an additional in-plane momentum for the electrons.

The frequency $\Omega_i(\mathbf{K})$ of the phonon branch i determines the threshold of the exciton phonon-assisted indirect optical absorption.

The relevant contribution into the light absorption close to the threshold $\Omega_i(\mathbf{K})$ is related to the exciton density of states which is proportional to $(\omega - \Omega_i + E_{nN})\theta(\omega - \Omega_i + E_{nN})$, where ω is the frequency of light. The behavior of the threshold is compared with the behavior of free-pair excitation whose threshold is smoother for $n \propto (\omega - \Omega_i)3\theta(\omega - \Omega_i)$. The indirect exciton is pronounced as a maximum in the second derivative of absorption in ω when $\omega = \Omega_i - E_{nN}$. Figure 6.4 illustrates the momentum dependence of $\ln(|F_S|^2)$ for the variational wavefunction Φ_2.

6.9 Electron Scattering on Indirect Excitons

The indirect excitons are observed in experiments utilizing specifics of the monoatomic layer graphene. One example is the forward or backward scattering of vacuum electrons on graphene sheet. In the experiment one measures dependence of the energy loss maximum versus the scattering angle. The intensity of scattering contains contribution from the Bose excitation, for example, from the excitons and also it depends on their spectrum. The spectrum of excitons is consistent with the incident electron energy about 10 eV, which is within the operation range of high-resolution electron energy loss spectroscopy (HREELS) (Lu et al. 2009). The relevant HREELS experiments (Lu et al. 2009) were conducted focusing on the continuum spectrum of electron–hole pair in graphene and on plasmarons (Koch et al. 2010). The exciton transitions are also identified by measuring the angular in-plane dependence of losses. When an external electron scatters between the states with momentums $\mathbf{P}$ and $\mathbf{P}'$ on free graphene, the transition amplitude is equal to $\mathcal{A}(\mathbf{P} - \mathbf{P})$. The amplitude $\mathcal{A}(\mathbf{Q})$, where $\mathbf{Q} = \mathbf{P} - \mathbf{P}'$ is a product of the Coulomb factor e^2/Q^2 and matrix element of $e^{i\mathbf{Q}\mathbf{r}}$ between the two electron states in graphene. In the vicinity of cones (i.e., near the K and K′ points), one finds

$$\mathcal{A}(\mathbf{Q}) = \frac{e^2}{Q^2} \int d^3 r u^*_{\mathrm{e;K}}(\mathbf{r})\, u_{\mathrm{h;K'}}(\mathbf{r})\, e^{i(\mathbf{Q}-\mathrm{K})\mathbf{r}} \tag{6.35}$$

where K is the smallest vector connecting the cone zero-energy points K and K′, $u_{\mathrm{e;K^*}}(\mathbf{r})$ and $u_{\mathrm{h;K'}}(\mathbf{r})$ are the Bloch amplitudes of

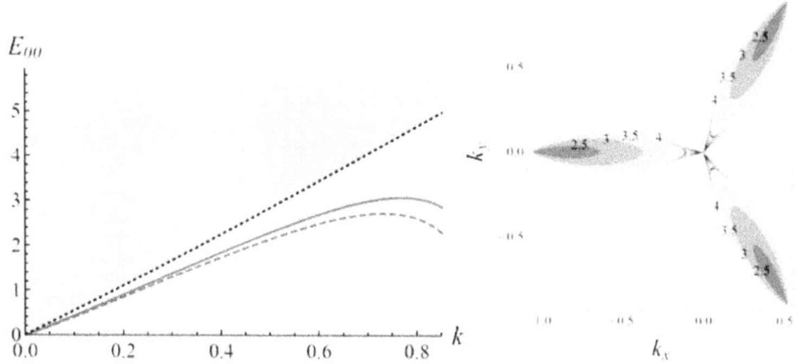

Figure 6.6 Left: The total exciton energy E_{00} versus κ along a fixed angle of 60.8 degrees. The data correspond to variational wavefunctions Φ_1 (solid) and Φ_2 (dashed). The filled area represents the free-electron and hole continuum. Right: The map of $\ln(|F_S|^2)$ for the variational wavefunction Φ_2. In addition, the functional dependence on the small momentum is given by the Sommerfeld factor. For the variational wave functions Φ_1 and Φ_2, we have $F_S = \text{B}$. Adapted from Mahmoodian and Entin (2013).

the electron/hole wavefunctions, respectively. The 2D momentum $\mathbf{Q}$ lies in-plane and has the values $\mathbf{K}+n\mathbf{a}^* + m\mathbf{b}^*$ which are adjusted by the reciprocal lattice vectors. The amplitude of exciton wavefunction is obtained as a product of the electron–hole pairing amplitude $\mathcal{A}(\mathbf{Q})$ and the Sommerfeld factor $F_S(\mathbf{k})$. The probability of inelastic diffraction is determined by the transition probability

$$W_{P\to P'} = 2\pi \left|F_S(\mathbf{k})\right|^2 \sum_{nm} \left|\mathcal{A}(\mathbf{Q})\right|^2 \delta\left(\frac{P'^2 - P^2}{2m_0} + E_{00}\right) \qquad (6.36)$$

where the sum is performed over the reciprocal lattice vectors and m_0 is the bare electron mass. The Sommerfeld factor $|F_S(\mathbf{k})|^2$, which enters the above formula, Eq. 6.36, contains a functional dependence on the small momentum $\mathbf{k}$, which is shown in Fig. 6.6 (right) for the variational wavefunction Φ_2. The geometry of the suggested experiment is depicted in Fig. 6.7. It is organized as follows. An electron beam is directed to the graphene plane. The momentum of the beam is $\mathbf{P} = (\mathbf{K}, P_z)$, $P_z^2/2m_0 >> E_{00}$. In the case when $4\pi/\sqrt{3}a > P > 4\pi/3a$, or $a^* > P > K$, only the diffraction peak at $n = m = 0$ is allowed. An elastic peak at $\mathbf{P}' = \mathbf{P}$ also occurs, and the scattered

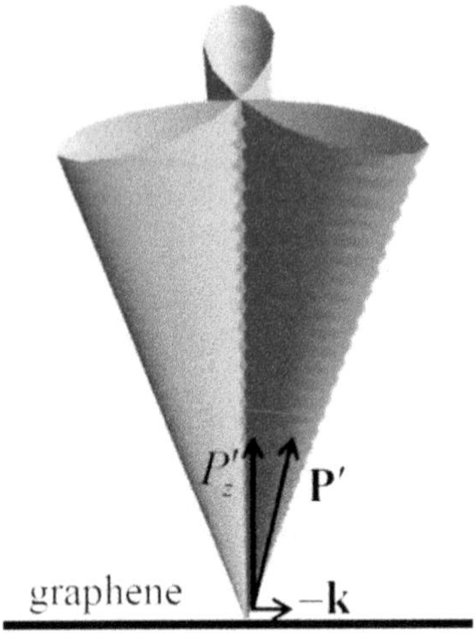

Figure 6.7 Sketch of proposed experiment for measurements of exciton spectrum. Adapted from Mahmoodian and Entin (2013).

electron momentum $(-\mathbf{k}, P'_z)$, where

$$P'_z = \sqrt{P_z^2 + K^2 - 2m_0 E_{00} - k^2} \tag{6.37}$$

is close to the normal. The exciton spectrum is extracted from the measurement of the normal component of the momentum $\mathbf{P}'_z$ in the diffraction peak as a function of the in-plane momentum $-\mathbf{k}$. Indirect excitons are detected when the in-plane incident momentum coincides with $\mathbf{K}$. The direction of the in-plane projection of $\mathbf{P}$ along $\mathbf{K}$ or $-\mathbf{K}$ allows us to choose one of the two degenerate exciton states under the condition that it cause the trigonal symmetry in the $\mathbf{P}'$ space.

A small magnitude of the interaction constant g represents the basic assumption in our description of the indirect excitons which are formed near the conic points. According to the above condition, the binding energy is small comparing to the kinetic energy of the exciton.

Another assumption is that the value of the electron scattering rate on impurities or phonons is low. It ensures stability of the exciton states, whereas the scattering destroys the moving excitons. The Auger process can also contribute to the seafaring at higher energies. It becomes significant in the regions of momentum space which coincides with the regions where exciton are formed. The electron–hole scattering rate is estimated as $\nu_{\text{e-h}} \propto g^2vk$, which must be compared against the exciton binding energy $g^2vk \ln^2\eta$. On one hand, the parameter $\ln^2\eta$ is large, which causes low damping. On the other hand, the binding energy becomes comparable with the width of excited exciton states. Indeed, the exciton states cannot be observable in the latter case.

An exciton in graphene can also be destroyed due to screening by the finite concentration of equilibrium charged carriers n_{ch}. In regular semiconductors with a finite energy gap there are slow excitons only. Quasistatic coupling between the electrons and holes, which compose the excitons, is screened due to the equilibrium electrons and holes. The moving excitons in graphene only partially are accompanied by the charge carriers. If the screening radius of the moving electron-hole pair is larger than the exciton radius, the binding energy remains steady. Then, one can write the condition for exciton existence as $n_{\text{ch}} < \alpha k^2 \log^2\eta$.

Besides, the above model cannot describe renormalizing of the electron spectrum like, for example, the exciton insulator. Such a renormalization is not included into the model because we have assumed that the interaction and trigonal corrections are small.

The indirect excitons considered in the above example emerge in the vicinity of conic K and K′ points. It differs from the exciton states emerging at the saddle point, the energies of which are about 5 eV (Katsnelson 2006, Yang et al. 2009, Mak et al. 2011, Yang 2011). There is no energy gap in the excitons spectrum. Although excitons with small wavevectors can be destroyed by means of the many-body interaction in concurrence with warping, such excitons might nevertheless survive, if the interaction constant remains small enough.

6.10 Tomonaga–Luttinger Liquid

A remarkable feature of the electron excitation spectrum of pristine graphene and metallic CNTs is the linear dispersion in the vicinity of the Fermi level. For such a reason, the mentioned materials are expected to be an ideal 1D conductor. It is well known that the interacting electrons in the 1D system with a linear dispersion might exhibit Tomonaga–Luttinger liquid (TTL) behavior (Tomonaga 1950, Luttinger 1963, Kane and Fisher 1996, Yi and Kane 1996, Kane and Mele 1997, Bockrath et al. 1999, Ishii et al. 2003). Such an unconventional electronic liquid state is characterized by the charge–spin separation, interaction-dependent power laws for transport quantities, and nonexistence of quasiparticle states. For the linear band with the lowest $n = 0$, each term of the perturbation series for the electron self-energy becomes diverging. Technically, for a linear band, the divergence originates from the fact that the electron energy is conserved automatically if the momentum is conserved. The low-energy electronic excitation properties of the TTL are exactly described in terms of the bosonization method (Tomonaga 1950, Luttinger 1963, Kane and Fisher 1996, Yi and Kane 1996, Kane and Mele 1997, Bockrath et al. 1999, Ishii et al. 2003). There was a lot of theoretical and experimental activity devoted to the TLL properties of CNTs and narrow graphene stripes (Snow and Perkins 2005, Snow, Perkins, and Houser 2005, Snow et al. 2005, Shafranjuk 2007, Shafranjuk 2008, Shafranjuk 2009, Rinzan et al. 2012, Yang et al. 2012, Yang et al. 2013). An effective low-energy theory has been focused on various characteristics like the energy gap at the Fermi level, tunneling conductance of the CNT–metallic contact, etc. The authors of a few experiments have noticed signatures of many-body effects. In particular, some experiments studied the conductance of bundles (ropes) of SWCNTs versus temperature and voltage. Metal contacts to nanotubes were either deposited over the top (so-called end contact) or the nanotubes were placed on the top of metal leads (bulk contact). The obtained experimental data conform the differential conductance having power law dependence versus temperature and the applied bias. The mentioned experiments have found a remarkable consistence between the obtained different values of the power for two samples and theoretical predictions for the TTL. Another direct evidence of the TLL behavior had been obtained in the photoemission experiments which have found the power law dependence of the density of states of the linear bands. The TLL features of narrow graphene stripes and CNTs can be described within the same approximation approach. It allows computing of the self-energy for the linear bands with $n = 0$, and yields a gapless linear band with a renormalized velocity. Although each term of perturbation series for the self-energy diverges, the RPA gives a finite result because the divergent polarization operators mutually cancel each other. That fact is inconsistent with the claim that only charge density and spin density excitations are stable, whereas the quasiparticle excitations are not clearly defined. Such a contradiction follows from the method of defining the quasiparticle energy via the self-energy. In the RPA, the imaginary part of the Green's function (the spectral function) shows double sharp peaks when computed for a system where the metallic linear bands are present only. The splitting of the peak into the presumable charge density and spin density excitations qualitatively agrees with the TTL scenario, which is discussed below. The mentioned singular behavior does not occur in the polarization function and in the perturbation expansion when parabolic bands are taking place in graphene stripes and in semiconducting and metallic CNTs. In the latter case, the self-energy and therefore quasiparticle states serve to provide an adequate description of the low-energy excitations.

6.11 Probing of the Intrinsic State of a 1D Quantum Well with Photon-Assisted Tunneling

Here, we describe a method which allows us to probe the TLL state using the photon-assisted tunneling (PAT) through an SWCNT quantum well (QW). The elementary TLL excitations inside the QW are density ($\rho_\pm$) and spin ($\sigma_\pm$) bosons. The bosons populate the quantized energy levels $\varepsilon_n^{\rho+} = \Delta n/g$ and $\varepsilon_n^{\rho-(\sigma\pm)} = \Delta n$ where $\Delta = hv_{\mathrm{F}}/L$ is the interlevel spacing, n is an integer number, L is the nanotube length, g. is the TLL parameter. Since the external electromagnetic field (EF) acts on the ρ_+ bosons only, whereas the neutral ρ_- and $\sigma_\pm$ bosons remain unaffected, the PAT spectroscopy is able to identify the ρ_+ levels in the QW setup. The spin $\varepsilon_n^{\sigma+}$ boson levels in the same QW are recognized from Zeeman splitting when applying a DC magnetic field $H \neq 0$. Basic TLL parameters are readily extracted from the differential conductivity curves.

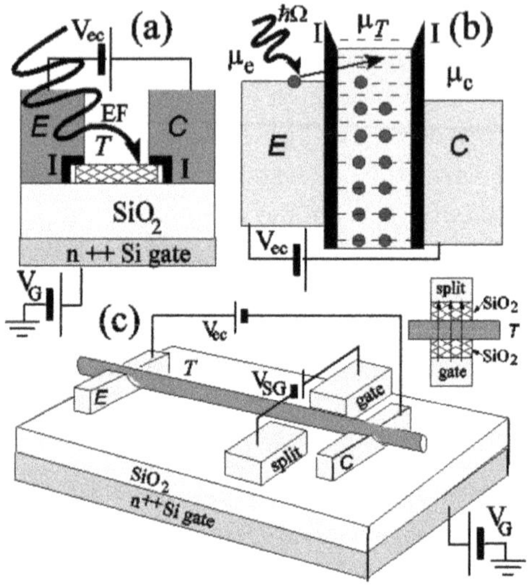

Figure 6.8 (a) A quantum well (QW) composed of the 1D section T with attached emitter (E) and collector (C) electrodes. The potential barriers are shown in black at the E/T and T/C interfaces. (b) Energy diagram of the PAT process in the QW. (c) The split-gate configuration of the QW. The right side inset shows how the electric field is applied to the T section. Adapted from Shafranjuk (2008).

One-dimensional QWs are very promising for scientific research and various practical applications (Snow and Perkins 2005, Snow, Perkins, and Houser 2005, Snow et al. 2005, Shafranjuk 2007, Shafranjuk 2008, Shafranjuk 2009, Rinzan et al. 2012, Yang et al. 2012, Yang et al. 2013). One spectacular example is the junction formed by an SWCNT T with emitter E and collector C electrodes attached to its ends (see the sketch in Fig. 6.8). Depending on the nanotube properties, that setup (see Fig. 6.8) corresponds to various condensed matter systems. Remarkable properties (Snow and Perkins 2005, Snow, Perkins, and Houser 2005, Snow et al. 2005, Shafranjuk 2007, Shafranjuk 2008, Shafranjuk 2009, Rinzan et al. 2012, Yang et al. 2012, Yang et al. 2013) of the carbon tubes emerge from their intrinsic structure (Dresselhaus et al. 2002a, 2002b, Saito et al. 2002). A single-wall carbon nanotube (SWCNT) $[n, m]$ is a rolled up atomic honeycomb monolayer formed with two sublattices A and B. The integer indices n and m (assuming that $n \geq m \geq 0$) of the rollup vector $\mathbf{R} = n\mathbf{R}_1 + m\mathbf{R}_2$ actually determine the electronic band structure of the tube. In particular, if $n - m = 3k$ (k being an integer) the nanotube is metallic, while it is semiconducting or insulating otherwise (Mintmire and White 1998). The charge carriers in metallic tubes conform to the linear dispersion law $\varepsilon_k = \pm v_{\mathbf{F}} |\mathbf{k}|$ [where "$\pm$" corresponds to the electrons (holes), and $v_{\mathbf{F}}$ is the Fermi velocity].

A lot of discussions are devoted to the intrinsic state of metallic nanotubes where the TTL state may presumably occur (Tomonaga 1950, Luttinger 1963, Kane and Fisher 1996, Yi and Kane 1996, Kane and Mele 1997, Bockrath et al. 1999, Ishii et al. 2003). In contrast to semiconducting nanotubes, where a general consensus is achieved (Dresselhaus et al. 2002a, 2002b), unconventional features of the metallic CNTs are not well understood yet. Much attention (Tomonaga 1950, Luttinger 1963, Kane and Fisher 1996, Yi and Kane 1996, Kane and Mele 1997, Bockrath et al. 1999, Ishii et al. 2003) is paid to the strong correlation effects, and to the 1D transport of the electric charge carriers. Along with the TLL state in metallic nanotubes (Tomonaga 1950, Luttinger 1963, Kane and Fisher 1996, Yi and Kane 1996, Kane and Mele 1997, Bockrath et al. 1999, Ishii et al. 2003) under the current elaboration there are models operating with noninteracting electrons, while other models exploit coupling of the nanotube to the external environment (Nazarov and Glazman 2003). Although there are indications of the TLL state in the shot noise (Recher et al. 2006) and in angle integrated photoemission measurements (Ishii et al. 2003), present experimental evidences are still indirect (Tarkiainen et al. 2001). Therefore, more efforts to clearly identify the intrinsic state of the 1D QWs formed of metallic carbon tubes are required. A typical QW setup (Tselev et al. 2009, Rinzan et al. 2012, Yang et al. 2012, Yang et al. 2013) is sketched in Fig. 6.9, where the 1D section is denoted as T. The bias voltage $V_{\mathbf{ec}}$ drops between the emitter (E) and collector (C) electrodes, while the gate voltage $V_{\mathbf{G}}$ is applied to the $n++$Si substrate, as shown in Fig. 6.8. The electrochemical potentials in E, T, and C are denoted as $\mu_{\mathrm{e,T,c}}$. The E and C electrodes are separated from the metallic nanotube section T by the interface barriers I shown in black in Fig. 6.8a,b. The potential barriers emerge from differences between the Fermi velocities in the adjacent electrodes.

In this section we suggest a method which identifies the quantized levels of charge and spin excitations in the TTL state

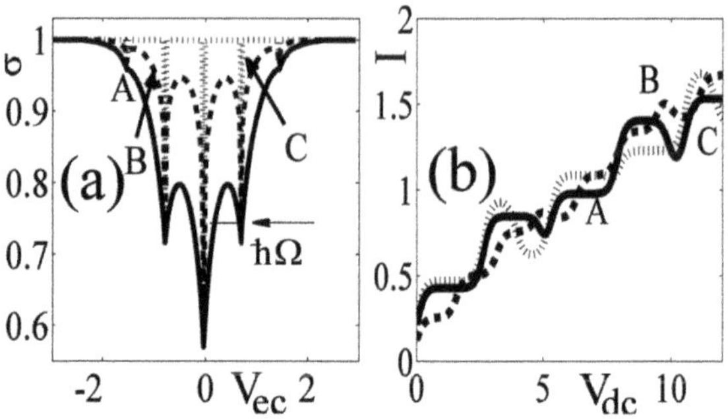

Figure 6.9 (a) The single-electron density of states inside the T section of a QW for the Luttinger parameter $g = 1$ (curve A), 0.4 (curve B), and 0.2 (curve C). (b) Contour E–g plot of the single-electron density of states inside the T section. Adapted from Shafranjuk (2008).

inside the 1D QW shown in Fig. 6.8a–c. Our method exploits the fact that the EF interacts with the charge excitations only, while the neutral particles remain unaffected. When a tunneling electron with energy ε absorbs n photons of the external EF, the intrinsic structure of the TLL state in T is pronounced in the multiphoton tunneling process probability. PAT influences probability of the single-electron tunneling (SET) which helps to elucidate the intrinsic state of the tube. In this Section we address QWs with long and short T sections. When the nanotube is long, the interlevel spacing $\Delta = hv_F/L$ (v_F is the Fermi velocity, L is the nanotube length) is small. Therefore, quantization of the electron motion inside T is negligible. In that case the local single-electron density of states $\mathcal{N}(\varepsilon)$ inside T has a dip at zero energy $\varepsilon = 0$. We will see that such a dip is clearly visible in photon-assisted and in SET characteristics which helps to identify the TLL state. In the opposite limit when the nanotube is short, the ballistic motion of the charged and neutral excitations inside T is quantized. During the tunneling, an electron splits into four $\rho_\pm$, $\sigma_\pm$ bosons (two density and two spin). The bosons populate quantized levels with different energies $\varepsilon_{\rho+} \neq \varepsilon_{\rho-(\sigma\pm)}$. The charge boson energy levels are detected with PAT (Shafranjuk 2008). We will see that the tunneling mechanism is sensitive to the emitter–collector $V_{\mathbf{ec}}$ and the gate $V_{\mathbf{G}}$ voltages. Therefore, the TLL properties are pronounced in the differential conductivity curves of the 1D

QWs. In the same setup, the spin boson levels are fingered from Zeeman splitting $\propto \mu_{\mathbf{B}}H$ when applying a finite DC magnetic field $H \neq 0$. The quantization of both the charge and spin excitations is proclaimed in the differential conductivity curves of the QW.

6.12 Photon-Assisted Tunneling into the TLL State

Here we address low-transparency double-barrier SWCNT junctions, assuming that the tunneling events across the E/T and T/C barriers are not phase-correlated with each other. When the external AC electric field vector is directed along the nanotube axis, it induces an AC bias voltage $V^{(1)}\cos\Omega t$ with the field frequency Ω across the whole double-barrier junction. The AC bias voltage effectively drops on the interface barriers I, whose partial resistance is assumed to be much higher than the resistance of T. Since typical length of the CNT junction (Tselev et al. 2009, Rinzan et al. 2012, Yang et al. 2012, Yang et al. 2013) is $L \approx 200$ nm –0.5 μm, the AC field wavelength of interest is 1 mm $\geq \lambda_{\mathbf{EF}} \geq$ 0.5 μm. This corresponds to the THz wave region. We describe the tunneling between the TLL state in T and the free-electron states in the emitter (collector) denoted as E(C) electrodes using microscopic methods (Keldysh 1965). For the sake of simplicity, we do not consider here the ratchet effect (Feldman et al. 2005, Feldman et al. 2006), which comes either from an asymmetric scattering potential, or from a nonlinearity of the electronic dispersion. Using methods of Keldysh (1965) one finds (see derivation details in Section 6.15) the time-averaged electric current through the quantum dot in the form

$$I = \Gamma_{\mathbf{n}}\frac{2e}{h}\int d\varepsilon \sum_{nm} \zeta_n \mathrm{Im}\mathcal{K}(\varepsilon)$$
$$\Big[J_m^2(\alpha_e)\left[f_+^e + G_\varepsilon\left(f_-^e - f_+^e\right)\right]$$
$$- J_m^2(\alpha_c)\left[f_+^c + \left(f_-^c - f_+^c\right)G_\varepsilon\right]\Big] \tag{6.38}$$

where K is the full electron correlator and $f_\pm^{\mathbf{e(c)}}$ are the electron distribution functions in the emitter (collector) denoted as E(C) electrodes for which we have used short notations $f\left(\varepsilon_{e(c)}^{m\pm}\right) \rightarrow f_{\pm}^{e(c)}$

$$\varepsilon_{e(c)}^{m+} = \varepsilon + E + \hbar m\Omega + \Delta U_n^{e,\,c}$$
$$\varepsilon_{e(c)}^{m-} = \varepsilon + E + \hbar m\Omega + \Delta U_{n-1}^{e,\,c} \tag{6.39}$$

In Eqs. 6.38 and 6.39 E is the energy of the occupied level in the well relative to the conductance band edge in the emitter at the zeroth DC bias voltage $V_{\mathbf{ec}} = 0$. In Eq. 6.38 $\Gamma^{c(e)} = h \cdot (4\pi e^2 \nu_{e(c)} R_{e(c)})^{-1}$, $\nu_{e(c)}$ is the electron density of states inside the E(C) electrodes, $R_{e(c)}$ is the tunnel resistance between the E(C) electrodes and the T section, $\Gamma_n = \Gamma^e\Gamma^c/(\Gamma^e + \Gamma^c)$, ζ_n is the probability to find n electrons inside the well determined by a master equation (see Section 6.15), and the integration is performed over the electron energy ε in the well. The Bessel function $J_m\,(\alpha_{e,\,c})$ of order m (m is the number of emitted [absorbed] photons) in Eq. 6.38 depends on the AC bias parameter $\alpha_{e,\,c} = eV_{e,\,c}^{(1)}/\hbar\Omega$, where $eV_{\mathbf{ec}}^{(1)}$ is the AC bias amplitude on the E(C) barriers and ε is the electron energy in T. The corresponding changes of emitter and collector electrostatic energy $\Delta U_n^{e,\,c}$ depend on the number n of electrons in the well as

$$\Delta U_n^e = \delta\left(n + \frac{1}{2}\right) - \eta e V_{\mathbf{ec}}$$
$$\Delta U_n^c = \delta\left(n + \frac{1}{2}\right) + (1 - \eta)\, e V_{\mathbf{ec}} \tag{6.40}$$

where $\delta = e^2/C$, $C = C_{\mathrm{e}} + C_{\mathrm{c}}$ is the net capacitance, $C_{\mathrm{e(c)}}$ is the emitter (collector) capacitance, η is the fraction of the net DC bias voltage V_{ec} so that ηV_{ec} drops between the emitter and the CNT. The electron distribution function G_ε inside the nanotube entering Eq. 6.38 must in general be obtained from the corresponding quantum kinetic equation (Keldysh 1965). However, for the sake of simplicity we will follow a procedure suggested by Averin et al. (1991). Namely, we approximate G_ε by a Fermi–Dirac distribution but with a finite chemical potential $\mu_{\mathrm{T}} \neq 0$ in the form

$$\mathrm{G}_\varepsilon \rightarrow \mathrm{G}_0(\varepsilon) = \frac{1}{\exp\left[(\varepsilon - \mu_{\mathrm{T}})\,/\,T\right] + 1} \tag{6.41}$$

A similar quasiequilibrium approximation was used describe the nonequilibrium superconductors (Elesin and Kopayev 1981). The chemical potential μ_T of electrons in T entering Eq. 6.41 is defined by the expression for the mean number of electrons

$$\langle n \rangle = \frac{2}{h} \frac{1}{\Gamma^e + \Gamma^c} \int d\varepsilon \sum_{nm} \zeta_n \mathrm{Im}\mathcal{K}(\varepsilon)$$

$$\Big[\Gamma^e J_m^2(\alpha_e) \left[1 - 2f_-^e + 2\left(f_-^e - f_+^e\right) \mathrm{G}_\varepsilon\right]$$

$$+ \Gamma^c J_m^2(\alpha_c) \left[1 - 2f_-^c + 2\left(f_-^c - f_+^c\right) \mathrm{G}_\varepsilon\right] \Big]$$

$$+ \langle n_\mathbf{G} \rangle \tag{6.42}$$

where $\langle n_\mathbf{G} \rangle$ is the number of extra electrons induced by the gate voltage $V_\mathbf{G} \neq 0$ applied, as shown in Fig. 6.8. When the AC bias is off ($\alpha_{\mathbf{e},\,\mathbf{c}} \to 0$), Eq. 6.38 yields a well-known formula for electric conductivity of a double-barrier low-transparency tunneling junction (Averin et al. 1991). In equilibrium and in absence of SET one sets $\mu_\mathbf{T} \equiv 0$, and terms containing the distribution function $\mathrm{G}_0(\varepsilon)$ in Eqs. 6.38, 6.42 disappear. In the latter case one also uses $f_\pm^{e(c)}(\varepsilon) = f_0(\varepsilon) = 1/(\exp[\varepsilon/T] + 1)$, where T is the temperature. Then, one simply gets

$$\sigma = (2e^2/h)\Gamma_n \overline{M}(eV_\mathbf{ec}), \tag{6.43}$$

where $\overline{M}(eV_\mathbf{ec}) = \int_{-\infty}^{\infty} d\varepsilon N(\varepsilon) \left(-\partial f_0(\varepsilon - eV_\mathbf{ec})/(\partial\varepsilon)\right)$ and $\mathbf{N}(\varepsilon)$ is the single-electron density of states. The comb-shaped free-electron density of states in the well is

$$\mathcal{N}(\varepsilon) = \frac{1}{\pi} \mathbf{Im} \sum_m \frac{1}{\varepsilon - \Delta m - i\gamma_n} \tag{6.44}$$

where Δ is the level spacing, γ_n is the quantized-level width (in the case of interest $\gamma_n << \Delta$), and h is the Plank constant.

6.13 TLL Tunneling Density of States of a Long Quantum Well

If the metallic nanotubesection T is long, $L > v_\mathbf{F}\tau_\mathbf{T}$ (where $\tau_\mathbf{T} = \hbar/\Gamma_n$ is the net tunneling time), the quantization inside T is negligible. The level separation for typical CNT junctions (Tselev et al. 2009, Rinzan et al. 2012, Yang et al. 2012, Yang et al. 2013) becomes indistinguishable when $L \geq 3$ μm. Strong electron correlations drive the electron system into the TLL state (Tomonaga 1950, Luttinger 1963, Kane and Fisher 1996, Yi and Kane 1996, Kane and Mele 1997, Bockrath et al. 1999, Ishii et al. 2003). According to Eq. 6.38, the electric current is expressed via the single-electron correlator K, which is related to the spectral density $A_{\kappa}(\varepsilon)$ of the right-moving ($\kappa = 1$) fermions as

$$\mathcal{N}(\varepsilon) = \frac{1}{\pi}\mathrm{Im}\sum_{m}\frac{1}{\varepsilon - \Delta m - i\gamma_n} \qquad (6.45)$$

At zero temperature $T = 0$, following Emery and Mukamel (1979), one finds

$$\begin{aligned} K(\varepsilon) = &-\sqrt{2/\pi i}|\varepsilon|^{2\gamma}\sin(\pi\gamma) \\ &\times[2(-1)^{\gamma}\gamma\Gamma(-2\gamma-1)|\varepsilon| + e^{i\pi\gamma}\mathrm{sign}(\varepsilon) \\ &(\Gamma(-2\gamma) + g^2(r^2/v_{\mathrm{F}}^2)\times\gamma(2\gamma+1)\Gamma(-2(\gamma+1))\varepsilon^2)] \end{aligned} \qquad (6.46)$$

where $\Gamma(x)$ is the gamma function of x, r is the cutoff parameter, $\gamma = \left(g^{-1} + g - 2\right)/8$, and g is the Luttinger liquid parameter. In the limit $g \to 1$, the expression for $A_{\kappa}(\varepsilon)$ transforms to the free electron spectral density $A_{\kappa}^{(0)}(\varepsilon) = \mathrm{sign}\varepsilon/(\sqrt{2\pi}v_{\mathrm{F}})$. Properties of the TLL state are sensitive to the Luttinger parameter g. The single-electron density of states $\mathcal{N}(E)$ of a 1D QW with a long T section is shown for different values of g in Fig. 6.10. The Luttinger parameter g can be controlled either by the gate voltage V_{G} or by the split-gate voltage V_{SG}, as shown in Fig. 6.8 (see also the right panel inset in the same figure). Since the gate voltage V_{G} affects the charge density q on T, $q = CV_{\mathrm{G}}$ (where the capacitance $C = 2\pi\varepsilon_0 L/\cosh^{-1}(2h_T/d)$, ε_0 is the vacuum permittivity, d is the nanotube diameter, and h_T is the distance from the nanotube to substrate) it allows changing of g. An altering of V_{G} renormalizes $g \to g + \beta_{\mathrm{G}}V_{\mathrm{G}}$ due to changes in the dielectric function $\varepsilon(k, \varepsilon)$ and in the Coulomb screening. According to Kane and Fisher (1996), Yi and Kane (1996), and Kane and Mele (1997), the Luttinger parameter g for a CNT depends on the electrostatic energy U_n as

$$g \simeq \frac{1}{\sqrt{1 + 2U_n/\Delta}} \qquad (6.47)$$

where $\Delta = hv_{\mathrm{F}}/L$ is the energy-level spacing, while the change of U_n is determined by Eq. 6.40. A simple evaluation from the band

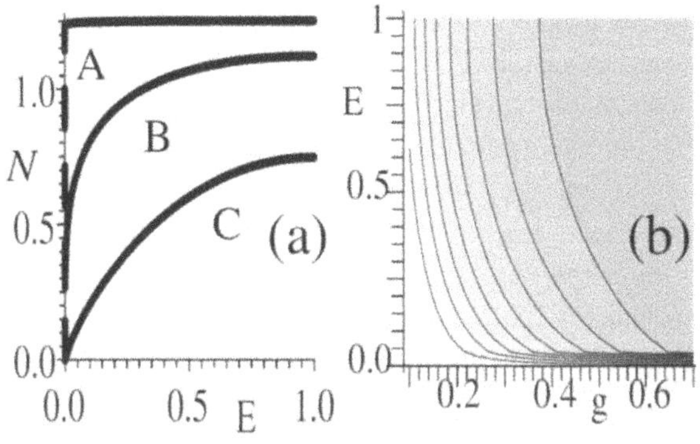

Figure 6.10 (a) Splitting of the zero-bias TLL dip in differential tunneling conductivity $\sigma(V_{\mathrm{dc}})$ (in units of $e^2 v_{\mathrm{F}} N(0)$) in a long CNT junction due to photon-assisted tunneling. Spacing between the zero dip and adjacent satellite dips is $\hbar\Omega$. (b) The Coulomb staircase in the PASET current–voltage characteristics $I(V_{\mathrm{dc}})$ versus reduced voltage V_{dc} (see text) of the CNT junction in the TLL state with $g = 0.4$ under the influence of the AC bias field with the amplitude $eV^{(1)} = 3.4$ (in units of $\delta = e^2/C$) and for a symmetric junction ($\eta = 0.5$). Curve A corresponds to $\Omega = 4.7$, curve B to $\Omega = 3.7$, and curve C to $\Omega = 6.7$. Adapted from Shafranjuk (2008).

structure calculations (Gunlycke, Jefferson et al. 2006, Gunlycke, Lambert et al. 2006) gives $\beta_{\mathrm{G}} = 0.005$–$0.03$ depending on directions of the rollup vector **R**. The $\mathcal{N}(E)$ shape is also controlled with V_{SG} utilizing the split-gate configuration (Gunlycke, Jefferson et al. 2006, Gunlycke, Lambert et al. 2006), as shown in Fig. 6.8c. In that setup, the electric field is perpendicular to the nanotube axis, as shown in the right inset of Fig. 6.8c. The split-gate setup allows us to drive the nanotube electron state from the semiconducting to the metallic one. The transversal electric field induces a finite dipole momentum directed perpendicular to the tube, which renormalizes g as well. The corresponding alteration of the Luttinger parameter g is evaluated. This gives $g \to g \cdot J_0^{-1}(V_{\mathrm{SG}} d/\hbar v_{\mathrm{F}})$. For a narrow-gap semiconducting T and for typical parameters of the nanotube QW, the split-gate-induced change is $g \to g + \beta_{\mathrm{SG}} V_{\mathrm{SG}}$ where $\beta_{\mathrm{SG}} = 0.01-0.05$ for different rollup vectors. If the transversal electric field V_{SG}/d inside T is sufficiently strong, one induces a semiconductor–metal transition. The electronic properties of the nanotube then switch from a 1D narrow gap semiconductor to the TLL.

The time-averaged conductance $\sigma(V_{\mathbf{ec}})$ of the long CNT junction exposed to an external EF is computed using Eqs. 6.38, 6.40, 6.42, and 6.43. We calculate $\sigma(V_{\mathbf{ec}})$ and the electric current $I(V_{\mathbf{dc}})$ [$V_{\mathbf{dc}} = (V_{\mathbf{ec}} - V_t)\eta$ is the reduced voltage, $V_t = (E - \varepsilon_{\mathbf{F}})/e\eta$ is the SET threshold voltage] for the two cases of interest. One limit corresponds to $\delta << T$ when SET is not essential ($\delta = e^2/C$, $C = C_{\mathrm{e}} + C_{\mathrm{c}}$ is the net capacitance of the double-barrier junction, $C_{\mathrm{e(c)}}$ is the emitter (collector) capacitance). Then, Eq. 6.38 for the tunneling current through the QW reads as

$$I(V_{ec}) = \frac{2e}{h}\Gamma_n \sum_m \int d\varepsilon \mathrm{Im}\mathcal{K}(\varepsilon, \Omega) \times [J_m^2(\alpha_e) f(\varepsilon_m^e) - J_m^2(\alpha_c) f(\varepsilon_m^c)] \quad (6.48)$$

where now

$$\varepsilon_m^e = \varepsilon_{\mathbf{ec}} - \eta e V_{\mathbf{ec}} + \hbar m \Omega$$
$$\varepsilon_m^c = \varepsilon_{\mathbf{ec}} + (1 - \eta) e V_{\mathbf{ec}} + \hbar m \Omega \quad (6.49)$$

The above Eqs. 6.38 and 6.40 are completed by Eqs. 6.42 and 6.43 to compute the DC differential tunneling conductance $\sigma(V_{\mathrm{ec}})$. The results are shown in Fig. 6.9a where we plot $\sigma(V_{\mathbf{ec}})$ of a long CNT junction in conditions of the PAT for $\Omega = 0.75$, $eV^{(1)} = 0.65$ and for three distinct values of the Luttinger liquid parameters $g = 0.2$ (curve A), $g = 0.4$ (curve B), and $g = 0.93$ (curve C). One can notice that the zero-bias dip in $\sigma(V_{\mathbf{ec}})$, which was positioned at $V_{\mathbf{ec}} = 0$ when the AC field was off ($eV^{(1)} = 0$) splits into additional satellite peaks spaced by $\hbar\Omega$.

Another limit corresponds to the SET which occurs if the condition

$$eV_{\mathbf{ec}}/2 - E - \hbar m\Omega - \delta \times (n + 1/2) \le \mu_{\mathbf{T}}(n) \le eV_{\mathbf{ec}}/2 - E - \hbar m\Omega - \delta \times (n - 1/2) \quad (6.50)$$

is fulfilled. The condition Eq. 6.50 can be independently accomplished by adjusting Ω, $V_{\mathbf{ec}}$, and $V_{\mathbf{G}}$ (which alters n). The zero-temperature conductivity for a symmetric junction then takes the form

$$\sigma(V_{\mathbf{ec}}) = \frac{\partial I(V_{\mathbf{ec}})}{\partial V_{\mathbf{ec}}} = \frac{2e}{h}\Gamma_{\mathrm{ec}} \sum_m J_m^2(\alpha) \zeta_n \times [\mathrm{Im}\mathcal{K}(E_+^m(n)) + \mathrm{Im}\mathcal{K}(E_{n-}^m) \times \theta(\mu_{\mathbf{T}}(n) - E_{n-}^m) - \mathrm{Im}\mathcal{K}(E_{n+}^m) \times \theta(\mu_{\mathbf{T}}(n) - E_{n+}^m)] \quad (6.51)$$

where ζ_n is the probability to find n electrons inside the well and E is the occupied-level energy in the QW relative to the conductance band edge in the emitter in the absence of the bias voltage. Equation 6.51 can be rewritten in shorter form as

$$\sigma\left(V_{\mathbf{ec}}\right)=\frac{2e}{h}\Gamma_{\mathrm{ec}}\sum_{m}J_{m}^{2}\left(\alpha\right)\zeta_{n}\times\mathbf{A}\left(n,\ m\right)\tag{6.52}$$

where

$$\mathbf{A}\left(n,\ m\right)=\begin{cases}\mathrm{Im}\mathcal{K}\left(E_{n+}^{m}\right) & \text{if}\,\mu_{\mathbf{T}}\left(n\right)<E_{n+}^{m}\\ 0 & \text{if}\,E_{n+}^{m}<\mu_{\mathbf{T}}\left(n\right)<E_{n-}^{m}\\ \mathrm{Im}\mathcal{K}\left(E_{n-}^{m}\right) & \text{if}\,\mu_{\mathbf{T}}\left(n\right)>E_{n-}^{m}\end{cases}\tag{6.53}$$

since one always gets $\mathbf{E}_{n+}^{m}<\mathbf{E}_{n-}^{m}$. For a consistent description of SET, Eqs. 6.38 and 6.40 must be complemented with an equation for $\mu_{\mathbf{T}}\left(n\right)$. For a symmetric junction ($\alpha_e=\alpha_c$) one gets

$$\langle n_G\rangle+\langle n\rangle=\frac{4}{h}\times\sum_{n,\,m}J_{m}^{2}\left(\alpha\right)\zeta_{n}\int d\varepsilon\;\mathrm{Im}\mathcal{K}\left(\varepsilon\right)\left[1-f_{-}^{\mathrm{ec}}+\left(f_{-}^{\mathrm{ec}}-f_{+}^{\mathrm{ec}}\right)G_{\varepsilon}\right]\tag{6.54}$$

where

$$f_{\pm}^{\mathrm{ec}}=f\left(\varepsilon+E+\hbar m\Omega+\delta\left(n\pm\frac{1}{2}\right)-eV_{\mathbf{ec}}/2\right)+f\left(\varepsilon+E+\hbar m\Omega+\delta\left(n\pm\frac{1}{2}\right)+eV_{\mathbf{ec}}/2\right)\tag{6.55}$$

and $\mathbf{G}_{\varepsilon}$ is approximated by Eq. 6.41 and $n_{\mathbf{G}}$ is the additional electron density induced by a finite gate voltage $V_{\mathbf{G}}\neq 0$. The condition which determines the $(k,\,m)$th (k being the integer SET index, m being the integer PAT index) the vertical step in the $I\left(V_{\mathrm{ec}}\right)$ [or a sharp peak in $\sigma\left(V_{\mathrm{ec}}\right)$] is

$$e\eta\left(V_{k-1,\,k}^{m}-V_{t}\right)=\delta\left(k-1\right)+\varepsilon+m\hbar\Omega\tag{6.56}$$

where k is integer. At $\alpha_{e,\,c}=0$ one gets

$$V_{k-1,\,k}=V_{t}+\frac{\delta\left(k-1\right)+\varepsilon}{e\eta}\tag{6.57}$$

The spacing between two adjacent steps at $\alpha_{e,\,c}=0$ is

$$V_{k,\,k+1}-V_{k-1,\,k}=\frac{\delta+\varepsilon_{k+1}-\varepsilon_{k}}{e\eta}\tag{6.58}$$

When the external AC field is finite ($\alpha_{e,c} \neq 0$), the steps split additionally by $\pm\hbar\Omega$.

The current–voltage characteristics $I(V_{\mathbf{dc}})$ [where $V_{\mathbf{dc}} = (V_{\mathbf{ec}} - V_t)\eta$] in condition of the photon-assisted single-electron tunneling (PASET) across the QW in the TTL state are shown in Fig. 6.9b. According to Averin et al. (1991), the equilibrium shape of the $I(V_{\mathbf{dc}})$ curves (quoted as Coulomb staircase) is extremely sensitive to the double-barrier junction's parameters such as barrier transparencies, capacitance, symmetry, purity of the CNT section, and the energy-level spacing. The PAT induced by the external EF introduces additional features in those curves. We have computed PASET curves for a QW with a long T section where SET takes place. The external $\hat{x}$-polarized EF induces an AC bias voltage across the junction as $V^{(1)} \cos \Omega t$. The most remarkable elements of the $I(V_{\mathbf{dc}})$ curves A–C in Fig. 6.9b are local dips which originate from an interference between the zero-energy TLL anomaly pronounced in equilibrium at $\varepsilon = 0$ (see Fig. 6.10a) and the PASET processes. The Coulomb staircase curve A in Fig. 6.9b corresponds to $\Omega = 4.7$, curve B to $\Omega = 3.7$ and curve C to $\Omega = 6.7$ computed for $g = 0.4$.

6.14 Identifying the Charge and the Spin Boson Energy Levels

In the opposite limit when the T section is short, the quantized energy levels are well resolved, since the condition $\Gamma^{e,c} << \Delta$ is observed (here $\Gamma^{e,c}$ denotes the E $\Leftrightarrow$ T and T $\Leftrightarrow$ C electron tunneling rates, $\Delta = hv_F/L$ is the interlevel spacing inside the middle section T). In this section we neglect the SET contribution (Averin et al. 1991) (Coulomb blockade phenomena). That is justified when the temperature T is not too low, $T >> \Gamma^{e,c}$. In that limit we use Eq. 6.38 again but with a different Im$\mathcal{K}(\varepsilon)$ which now acquires a comb-like shape. Due to the spin–charge separation in the TLL there are two sets of quantized energy levels in a low-transparency QW with a short T section. For the QW transparency $\tilde{T} = 0.3$ (where $\tilde{T} = 4\pi\Gamma_n L_n/(\hbar v_F)$, $L_n = L_e + L_c$, L_e and L_c are the E and C thicknesses, respectively, and $v_F = 8.1 \times 10^5$ m/s) one gets $\Gamma_n \approx 0.3$ meV. For the nanotube length $L = 3$ μm one obtains spacing between the quantized levels as $\Delta = 1$ meV. The photon-assisted processes cause an additional splitting ˜0.6 meV which corresponds to the AC bias frequency $\Omega \approx 1$ THz. Following Kane and Fisher (1996, Yi and Kane (1996, and Kane and Mele (1997), one defines the transmission coefficient as $\tilde{T}(E) = |i\hbar G^R(L, E)|^2$. We assume that coupling of the SWCNT segment T to the external E and C electrodes is weak. In this approximation we compute the local-electron density of states $\mathbf{N}(\varepsilon)$ implementing boundary conditions (Kane and Fisher 1996, Yi and Kane 1996, Kane and Mele 1997) for the electron wavefunction inside a short CNT section T. Then, the quantized energy levels are firmly separated from each other and resolved. The retarded single-electron Green function is $G^R(L, t) = \Pi_a \mathbf{G}_a^R(L, t)$,whose Fourier transform has a comb-like shape

$$G_n^R(L, \varepsilon) = i\sqrt{\frac{2}{\pi}} \sum_a \frac{4^{g_a^-} \sin^{2g_a^-}(\pi\lambda/L)}{\varepsilon_a}$$
$$\times \sum_n \Theta_n (-1)^{-2g_a^+ + n} \cdot \frac{\Gamma(1 - 2g_a^+ + n)\,\Gamma(\varepsilon/\varepsilon_a)}{\Gamma(1 - 2g_a^+ + n + \varepsilon/\varepsilon_a)} \qquad (6.59)$$

where $a = (\rho_\pm, \sigma_\pm)$ is the TLL boson index, that is, $\varepsilon_n^{\rho+} = \Delta/g$, while $\varepsilon_n^{\rho-(\sigma\pm)} = \Delta$, Θ_n is the coordinate-dependent coefficient inside the nanotube, and the length parameter $\lambda << L$ effectively incorporates influence of the interface barriers (Kane and Fisher 1996, Yi and Kane 1996, Kane and Mele 1997), n is the quantization index, and

$$g_a^\pm = \frac{1}{16}\left(\frac{1}{g_a} \pm g_a\right). \qquad (6.60)$$

The parameters Θ_n and λ are determined by the integer charge and by the sum of phase shifts at the interfaces. The charge ρ_+ bosons populate the energy levels

$$\varepsilon_n^{\rho+} = \frac{hv_F n}{Lg} = \frac{n\Delta}{g} \qquad (6.61)$$

(where n is integer number), while three other neutral ρ_- and $\sigma_\pm$ boson energy levels have conventional values

$$\varepsilon_n^{\rho-(\sigma\pm)} = \frac{hv_F n}{L} = n \cdot \Delta \qquad (6.62)$$

The tunneling differential conductivity $\sigma(V_{\mathbf{ec}})$ of a "clean" (i.e., without impurities on T) sample is a combination of two combs with

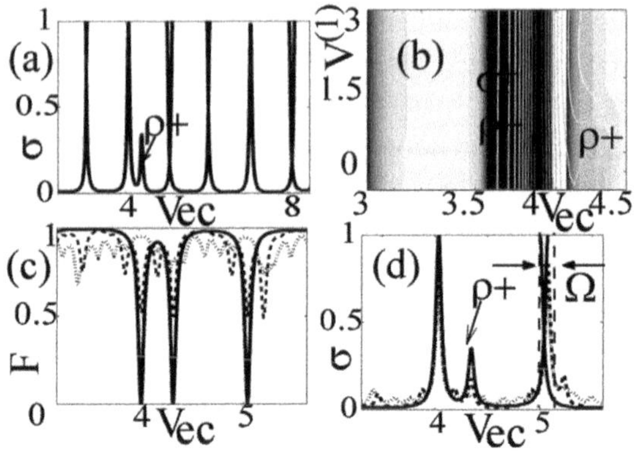

Figure 6.11 (a) The steady-state tunneling differential conductivity $\sigma\,(V_{\mathrm{dc}})$ (in units of $e^2 v_{\mathrm{F}} N\,(0)$) in the TLL state. The peak at $V_{dc} = 4.3$ (in units of Δ/e) corresponds to the ρ_+ boson. (b) The contour plot $\sigma\left(V_{\mathrm{dc}}, V^{(1)}\right)$ (V_{dc} and $V^{(1)}$ are the DC and AC bias amplitudes in units of Δ/e). (c) The Fano factor $F\,(\varepsilon)$ of the QW in conditions of PAT. (d) $\sigma\,(V_{\mathrm{dc}})$ for different AC field frequencies Ω. Adapted from Shafranjuk (2008).

different periods shown in Fig. 6.11a for $g = 0.23$. One of the combs corresponds to the ρ_+ boson, while another comb is related to the three remaining neutral (ρ_-, σ_+, σ_-) bosons. In the steady state, when the EF is off (i.e., $\alpha_{\mathbf{ec}} \equiv 0$), during the tunneling, for example, from E to T, an electron splits into four bosons as

$$e \to \rho_+ + \rho_- + \sigma_+ + \sigma_- \tag{6.63}$$

assuming the energy conservation as

$$\begin{aligned} E + \eta e V_{\mathbf{ec}} &= \varepsilon_n^{\rho+} + \varepsilon_n^{\rho-} + \varepsilon_n^{\sigma+} + \varepsilon_n^{\sigma-} \\ &= n\left(3 + \frac{1}{g}\right)\Delta \end{aligned} \tag{6.64}$$

(n being the integer number). It corresponds to a resonance tunneling through the quantized TLL states tuned by $V_{\mathbf{ec}}$. However, if the EF is on ($\alpha_{\mathbf{e,\,c}} \neq 0$), the resonance tunneling condition changes. That happens because the AC field acts on the charge ρ_+ bosons only, which absorb the EF photons during the PAT processes. The photons do not excite the neutral ρ_- and $\sigma_\pm$ bosons since they do not interact with the external EF. The AC field-modified resonance condition depends on both $V_{\mathbf{ec}}$ and Ω simultaneously

$$E + \eta e V_{\mathbf{ec}} = \left(\varepsilon_n^{\rho+} + m\hbar\Omega\right) + \varepsilon_n^{\rho-} + \varepsilon_n^{\sigma+} + \varepsilon_n^{\sigma-}$$
$$= n\left(3 + \frac{1}{g}\right)\Delta + m\hbar\Omega \qquad (6.65)$$

where n and m are integer numbers. The external EF splits the conductivity peaks selectively. Because $V_{\mathbf{ec}}$ and Ω are bound with the condition Eq. 6.65, it imposes a constraint on the net PAT resonant current through the QW. Using Eq. 6.65 one immediately extracts g and Δ from the DC PAT current–voltage characteristics. More specifically, the values of g and Δ are deduced from periods of the two steady state combs $\sigma(V_{\text{ec}})$ shown in Fig. 6.11a. This is illustrated by the PAT differential tunneling conductivity $\sigma(V_{\mathbf{ec}})$ for $\alpha_{\mathbf{e,\,c}} \neq 0$ shown in Fig. 6.11b,d. Figure 6.11b is the contour plot $\sigma(V_{\mathbf{ec}},\ V_{e,\,c}^{(1)})$ ($V_{\mathbf{ec}}$ and the AC bias amplitude $V_{e,\,c}^{(1)}$ being in units of Δ/e). The same quantity $\sigma(V_{\mathbf{ec}})$ for the fixed $\Omega = 0.85$ and $eV_{\mathbf{e,\,c}}^{(1)} = 0.01$ (solid curve), $eV_{\mathbf{e,\,c}}^{(1)} = 1$ (dashed curve), and $eV_{\mathbf{e,\,c}}^{(1)} = 5$ (dotted curve) is presented in Fig. 6.11d. In an experiment one obtains series of peaks in the differential conductivity $\sigma(V_{\mathbf{ec}}) = \partial I/\partial V_{\mathbf{ec}}$ curves for the steady state ($\alpha_{\mathbf{e,\,c}} \equiv 0$). When the AC field is on ($\alpha_{\mathbf{e,\,c}} \neq 0$) one also gets the satellite PAT peaks with an additional spacing

$$\Delta \to \Delta \pm \hbar m\Omega \qquad (6.66)$$

As shown in Fig. 6.11c,d. Then one determines the ratio $r_1 = A_1^{\rho+}/A_0^{\rho+}$ where $A_{0(1)}^{\rho+}$ is the ρ_+ boson peak height, which corresponds to the number of emitted (absorbed) photons $m = 0,\ 1$. The ratio r_1 allows us to extract the actual AC field amplitude $V^{(1)}$ acting on the junction. The splitting of the charged ρ_+ boson peaks with the AC field helps to identify the TTL state. The method is illustrated further in Fig. 6.12 (left) where we show a single peak in $\mathbf{N}(\varepsilon)$ corresponding to a quantized free electron energy level, as shown in Fig. 6.12 (left). For noninteracting electrons ($g = 1$) the same single-level splits either by an AC field due to the PAT phenomena with spacing $\propto m\hbar\Omega$ (m being integer) or by a DC magnetic field with the Zeeman spacing $\propto \mu_{\mathbf{B}}H$. The situation is remarkably different in the Luttinger liquid state when $g \neq 1$ and the charge ρ_+ and spin σ_+ levels have distinct energies $\varepsilon_n^{\rho+} \neq \varepsilon_n^{\sigma+}$. Then, one easily identifies the charge and spin levels merely by applying the AC field and DC

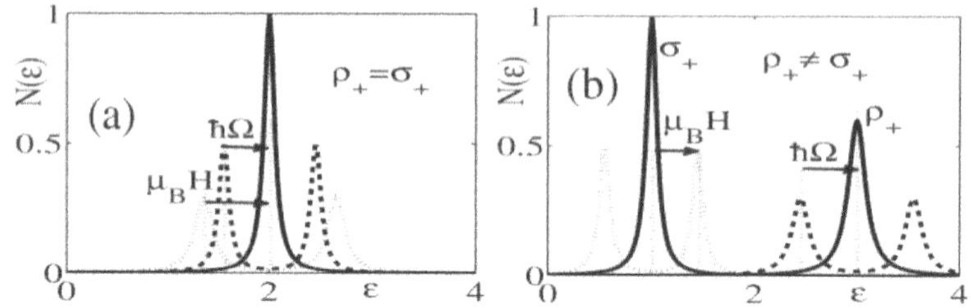

Figure 6.12 Splitting of quantized levels due to photon-assisted tunneling and the Zeeman effect, as pronounced in the single-electron density of states $N(\varepsilon)$ of a short CNT junction. Left: A free-electron quantized level (solid curve), for which the charge ρ_+ and spin σ_+ bosons coincide, splits when the AC electromagnetic field (with spacing $\propto \hbar\Omega$) and a DC magnetic field (with spacing $\propto \mu_B H$) are applied simultaneously. Right: The quantized levels of spin σ_+ and charge ρ_+ bosons have different energies $\varepsilon_n^{\rho+} \neq \varepsilon_n^{\sigma+}$ in the TLL state. The AC bias splits the charge boson levels (dashed curve on the right), while the DC magnetic field splits the spin boson levels (dotted curve on the left) only. The charge boson localized energy level $\varepsilon_n^{\rho+}$ splits in the two satellite peaks with spacing $2\hbar\Omega$. Although the AC field has no influence on the neutral spin bosons, the spin level $\varepsilon_n^{\sigma+}$ splits in two sublevels spaced with $\Delta_Z \propto \mu_B H$ (both in units of Δ) due to the Zeeman effect when a DC magnetic field $H \neq 0$ is applied. Adapted from Shafranjuk (2008).

magnetic field to the same QW. If a level splits with spacing $\propto m\hbar\Omega$ by the AC field only (showing no response to the DC field) then it certainly is a ρ_+ charge boson level ($g \neq 1$). If it splits by the DC magnetic field (Kane and Fisher 1996) with the Zeeman spacing $\propto \mu_B H$ showing no response to the AC field, then it must be associated with the spin bosons σ_+. However, if both AC and DC magnetic fields split the same level, the level belongs to the noninteractive electrons ($g = 1$), as stated above. Thus, one perceives the charge and spin bosons in experiments when applying AC EF in combination with the DC magnetic field to a CNT junction. An important requirement to the experimental metallic CNT QW samples is that they must be clean. An electron–impurity scattering in real samples leads to a formation of additional pairs of combs with different periods. Then, an identification of the TLL state becomes possible with a mere generalization of the method described above. Ratio of the noise power to the mean current (Fano factor) is computed as

$$F = \frac{\sum_n T_n(1 - T_n)}{\sum_n T_n} \tag{6.67}$$

where the summation is performed over the conducting channels. The result is shown in Fig. 6.11c. One can see that at the energies of quantized levels the noise is much lower than the Poisson noise of a conventional tunnel junction where $F = 1$. Remarkably, the multiphoton absorption is pronounced in the noise spectra as well. Thus, Fig. 6.11c suggests a method of the noise spectroscopy for studying the PAT into the TLL state.

The phenomena considered in this Section originate from a specific physics of the charge and spin carriers behaving like a blend of four noninteracting bosons. The TTL state occurs inside the 1D QW formed by a metallic SWCNT. The TLL state is tested applying an external AC EF and of a DC magnetic field simultaneously. The AC field splits the charge boson energy levels due to the PAT, while the DC magnetic field splits the spin boson levels due to the Zeeman effect. This allows us to identify the quantized energy levels associated with the charge and spin bosons forming the TLL state in relevant experiments. Besides, one also determines the quantized-level spacing Δ and the TLL parameter g. The unconventional electronic and photonic properties of the metallic CNT QW can be utilized in various nanodevice applications, including THz field sensors and nanoemitters.

6.15 Useful Relationships

Here we derive analytical expressions for the time-averaged electric current through the TLL QW in conditions of PASET. The external EF is applied to the nanotube junction, as shown in Fig. 6.8. Our model describes a low-transparency junction whose average conductance is small, $G << R_Q^{-1}$ where

$$R_Q = \frac{\pi \hbar}{2e^2} \simeq 6.5\text{k}\Omega \tag{6.68}$$

Since coupling to the electrodes is weak, the emitter–tube (E $\Leftrightarrow$ T) and the tube–collector (T $\Leftrightarrow$ C) tunneling processes are assumed as being not phase correlated. The SET dynamics in that approximation is well described by a simple master equation (Averin et al. 1991). The external EF with frequency Ω induces an AC bias voltage $V^{(1)}\cos\Omega t$ with amplitude $V^{(1)}$ across the junction. The AC voltage modulates phases of the tunneling electrons as

$$\varphi^{e,\,c}(t) = \frac{eV^{(1)}_{e,\,c}}{\hbar}\int_t \cos\Omega t' dt' \quad (6.69)$$

where $V^{(1)}_e = \eta V^{(1)}$, $V^{(1)}_c = (1-\eta)\ V^{(1)}$ are corresponding fractions of the AC voltage drop on the emitter and collector, η is the fraction of the net AC bias voltage $V^{(1)}$, so $\eta V^{(1)}$ drops between the emitter and CNT. The time-averaged SET electric current through the double barrier junction is expressed via partial tunneling rates $w^{e,\,c}_{+}$ and $w^{e,\,c}_{-}$ from emitter (collector) to the energy levels inside the well (Averin et al. 1991)

$$w^{e,\,c}_{+}(n,\ m) = J^2_m(\alpha_{e,\,c})\int d\varepsilon\Gamma^{e,\,c}_{\varepsilon}\left(\varepsilon^{m+}_{e(c)}\right) \times f\left(\varepsilon^{m+}_{e(c)}\right)G^{-}_m(\varepsilon,\ \Omega)$$

$$w^{e,\,c}_{-}(n,\ m) = -J^2_m(\alpha_{e,\,c})\int d\varepsilon\Gamma^{e,\,c}_{\varepsilon}\left(\varepsilon^{m-}_{e(c)}\right) \times\left[1 - f\left(\varepsilon^{m-}_{e(c)}\right)\right]G^{+}_m(\varepsilon_k,\ \Omega) \quad (6.70)$$

$$w^{\pm}_{nm} = w^{e}_{\pm}(n,\ m) + w^{c}_{\pm}(n,\ m) \quad (6.71)$$

where the integration is performed over the electron energy ε in the well, E is the energy of the occupied level in the well relative to the conductance band edge in the emitter at the zeroth DC bias voltage $V_{\mathbf{e(c)}} = 0$, m is the number of emitted (absorbed) photons, and the electron energy arguments are

$$\varepsilon^{m+}_{e(c)} = \varepsilon + E + \hbar m\Omega + \Delta U^{e,\,c}_n$$
$$\varepsilon^{m-}_{e(c)} = \varepsilon + E + \hbar m\Omega + \Delta U^{e,\,c}_{n-1} \quad (6.72)$$

The Bessel function $J_m(\alpha_{e,\,c})$ of order m in Eq. 6.71 depends on the AC bias parameter $\alpha_{e,\,c} = eV^{(1)}_{e,\,c}/\hbar\Omega$. Corresponding changes $\Delta U^{e,\,c}_n$ of the emitter and collector electrostatic energy depend on the number n of electrons in the well as

$$\Delta U^{e}_n = \delta\left(n + \frac{1}{2}\right) - \eta eV_{\mathbf{ec}}$$

$$\Delta U^{c}_n = \delta\left(n + \frac{1}{2}\right) + (1-\eta)\ eV_{\mathbf{ec}} \quad (6.73)$$

where $\delta = e^2/C$, $C = C_e + C_c$ is the net capacitance, $C_{e(c)}$ is the emitter (collector) capacitance. In Eq. 6.71, the electron Keldysh–Green function (Keldysh 1965) $G^{\pm}$ in the well is defined in the $\{\mathbf{r},\ t\}$ presentation as

$$G^{\pm}\left(\mathbf{r},\ \mathbf{r}',\ t,\ t'\right) = \pm i\left\langle \psi\left(\mathbf{r},\ t_{\pm}\right)\psi^{\dagger}\left(\mathbf{r}',\ t'_{\mp}\right)\right\rangle \tag{6.74}$$

where $\mathbf{r}, \mathbf{r}'$ are electron coordinates and $t_{\pm}, t'_{\mp}$ are the time moments assigned to points lying either on the positive (+) or on the negative (-) branch of the contour c circled around the time axis $-\infty < t < \infty$, $\langle\ldots\rangle$ means averaging (Keldysh 1965) with the full Hamiltonian $\mathbf{H}$, which includes all the interactions in the system. Following Keldysh (1965) one can introduce the auxiliary right-moving ($\kappa = 1$) free-fermion Green–Keldysh function as

$$G_{10}^{\pm} = \frac{i\pi T/v_{\mathrm{F}}}{\sin h\left(\pi T\left(x - v_{\mathrm{F}}t\right)/v_{\mathrm{F}}\right)} \pm i\pi\delta\left(x - v_{\mathrm{F}}t\right) \tag{6.75}$$

and for the left-moving ($\kappa = 2$) free fermion

$$G_{20}^{\pm} = [G_{10}^{\pm}\{v_{\mathbf{F}} \to -v_{\mathbf{F}}\}]^{*} \tag{6.76}$$

The Fourier transforms of (G0) at $x = 0$ are

$$\begin{aligned} G_{10}^{+}(\varepsilon) &= i\mathbf{N}(\varepsilon)\left(1 - \mathbf{G}_0(\varepsilon)\right) \\ G_{10}^{-}(\varepsilon) &= -i\mathbf{N}(\varepsilon)\,\mathbf{G}_0(\varepsilon) \end{aligned} \tag{6.77}$$

where $\mathbf{N}(\varepsilon)$ is the single-electron density of states inside the well,

$$\mathbf{G}_0(\varepsilon) = \frac{1}{2}\left(1 - \tan h\left(\frac{\varepsilon}{2T}\right)\right) = \frac{1}{\exp\left(\frac{\varepsilon}{T}\right) + 1} \tag{6.78}$$

is the equilibrium electron distribution function in the well. The electron tunneling rate $\Gamma_k^{e,\,c}$ is modified by the external AC bias as

$$\begin{aligned} \Gamma^{\mathrm{e,c}}(t) &\to \Gamma^{\mathrm{e,c}}(t)\, e^{i\alpha_{\mathrm{e,c}}\cos\Omega t} \\ &= \int d\varepsilon'\Gamma^{\mathrm{e,c}}\left(\varepsilon'\right) e^{i\varepsilon' t/\hbar} e^{i\alpha_{\mathrm{e,c}}\cos\Omega t} \end{aligned} \tag{6.79}$$

The backward Fourier transform gives

$$\begin{aligned} \Gamma^{\mathrm{e,c}}(\varepsilon) &= \int dt \int d\varepsilon'\Gamma^{\mathrm{e,c}}\left(\varepsilon'\right) e^{i\alpha\cos\Omega t} e^{-i\varepsilon t/\hbar + i\varepsilon' t/\hbar} \\ &= \sum_m J_m^2\left(\alpha_{\mathrm{e,c}}\right) \int d\varepsilon' \int dt\Gamma^{\mathrm{e,c}}\left(\varepsilon'\right) e^{im\Omega t - i\varepsilon t/\hbar + i\varepsilon' t/\hbar} \\ &= \sum_m J_m^2\left(\alpha_{\mathrm{e,c}}\right) \int d\varepsilon'\Gamma^{\mathrm{e,c}}\left(\varepsilon'\right)\delta\left(m\hbar\Omega - \varepsilon + \varepsilon'\right) \end{aligned} \tag{6.80}$$

which indicates that photons of the external AC field shift energies of tunneling electrons by $m\hbar\Omega$. Then,

$$w_+(m) - w_-(m) = \int d\varepsilon J_m^2(\alpha_{e,c})\, \Gamma^{e,c}\left(\varepsilon^m_{k,e(c)}\right) \times \left(f\left(\varepsilon^m_{e(c)}\right) G^-_m(\varepsilon, \Omega) + \left[1 - f\left(\varepsilon^m_{e(c)}\right)\right] G^+_m(\varepsilon, \Omega)\right) \tag{6.81}$$

If the energy dependence of $\Gamma^{e,c}(\varepsilon^m_{k,e(c)})$ is negligible, one gets

$$\begin{aligned} & w^{e,c}_+(n, m) - w^{e,c}_-(n, m) \\ &= \Gamma^{e.c} J_m^2(\alpha_{e,c}) \int d\varepsilon \left(f^{e(c)}_+ G^- + \left[1 - f^{e(c)}_-\right] G^+\right) \\ &= \Gamma^{e.c} J_m^2(\alpha_{e,c}) \int d\varepsilon \left(f^{e(c)}_+ (K + i\mathrm{Re}\mathcal{K}) + \left(1 - f^{e(c)}_-\right)(K - i\mathrm{Re}\mathcal{K})\right) \\ &= \Gamma^{e.c} J_m^2(\alpha_{e,c}) \int d\varepsilon \left(K\left(1 + f^{e(c)}_+ - f^{e(c)}_-\right)\right. \\ &\quad \left. + i\mathrm{Re}\mathcal{K}\left(f^{e(c)}_+ + f^{e(c)}_- - 1\right)\right) \end{aligned} \tag{6.82}$$

where we have used the short notations $f(\varepsilon^{m\pm}_{e(c)}) \to f^{e(c)}_\pm$ and $G^\pm = \mathbf{K} \mp i\mathrm{Re}\mathcal{K}$. Here

$$\mathbf{K} = \frac{G^- + G^+}{2} \tag{6.83}$$

is the full electron correlator from which the retarded Green function is obtained as

$$G^R(x, t) = -2i\theta(t)\mathrm{Re}\mathcal{K}(x, t) \tag{6.84}$$

The time-averaged partial electric current I^e between the emitter and the QW takes the form

$$\begin{aligned} I^e &= e\sum_n \zeta_n \left(w^e_+(n, m) - w^e_-(n, m)\right) \\ &= e\Gamma^e \int d\varepsilon \sum_{nm} J_m^2(\alpha_{e,c})\, \zeta_n \left[K(\varepsilon) + K(\varepsilon)\left[f^{e(c)}_+ - f^{e(c)}_-\right]\right. \\ &\quad \left. + \mathrm{Im}\mathcal{K}(\varepsilon)\left(1 - f^{e(c)}_+ - f^{e(c)}_-\right)\right] \\ &= e\Gamma^e \sum_n \zeta_n n + e\Gamma^e \int d\varepsilon \sum_{nm} J_m^2(\alpha_{e,c})\, \zeta_n K(\varepsilon)\left[f^{e(c)}_+ - f^{e(c)}_-\right] \\ &\quad + e\Gamma^e \int d\varepsilon \sum_{nm} J_m^2(\alpha_{e,c})\, \zeta_n \mathrm{Im}\mathcal{K}(\varepsilon)\left(1 - f^{e(c)}_+ - f^{e(c)}_-\right) \end{aligned} \tag{6.85}$$

where the number of extra electrons in the well is

$$n = \frac{1}{2\pi i}\sum_k \mathbf{K}(\varepsilon) \tag{6.86}$$

The electron tunneling between the QW and the electrodes causes a time evolution of the probability ζ_n to find n electrons inside the well. The time dependence of $\zeta_n(t)$ satisfies the master equation

$$\dot{\zeta}_n = w_{n+1}^- \zeta_{n+1} + w_{n-1}^+ \zeta_{n-1} - \left(w_n^+ + w_n^-\right)\zeta_n \tag{6.87}$$

The collector part of the electric current inside the well follows from the equilibrium condition

$$\begin{aligned} 0 \equiv I^e - I^c = \frac{2}{h}\cdot e\Gamma^e \int d\varepsilon \sum_n \zeta_n \big[K(\varepsilon)\, J_m^2(\alpha_e)\left[f_+^e - f_-^e\right] \\ -\mathrm{Im}\mathcal{K}(\varepsilon)\, J_m^2(\alpha_e)\left[1 - f_+^e - f_-^e\right]\big] \\ -e\Gamma^c \int d\varepsilon \sum_n \zeta_n \big[\mathrm{Im}\mathcal{K}(\varepsilon)\, J_m^2(\alpha_c)\left[1 - f_+^c - f_-^c\right] \\ -\mathrm{K}(\varepsilon)\, J_m^2(\alpha_c)\left[f_+^c - f_-^c\right]\big] - e\left[\Gamma^e + \Gamma^c\right]\sum_n \zeta_n n \end{aligned} \tag{6.88}$$

The above equation gives

$$\begin{aligned} e\left[\Gamma^e + \Gamma^c\right]\sum_n \zeta_n n = \frac{2}{h}\cdot \int d\varepsilon \sum_n \zeta_n \big[K(\varepsilon)\left(e\Gamma^e J_m^2(\alpha_e)\left[f_+^e - f_-^e\right]\right. \\ \left. + e\Gamma^c J_m^2(\alpha_c)\left[f_+^c - f_-^c\right]\right) - \mathrm{Im}\mathcal{K}(\varepsilon)\left(e\Gamma^e J_m^2(\alpha_e)\right. \\ \left. \times\left[1 - f_+^e - f_-^e\right] + e\Gamma^c J_m^2(\alpha_c)\left[1 - f_+^c - f_-^c\right]\right)\big] \end{aligned} \tag{6.89}$$

The average number of electrons in the well is

$$\langle n\rangle = \sum_n \zeta_n n + n_{\mathbf{G}} \tag{6.90}$$

where $n = \sum_k /(2\pi i)$, and $\mathbf{K} = (G^- + G^+)/2$ is the full electron correlator, $G^\pm$ is the electron Keldysh-Green function, Eq. 6.75. The second term $n_{\mathbf{G}}$ is controlled by the gate voltage $V_{\mathbf{G}} \neq 0$ applied to the QW, as shown in Fig. 6.8. Then one gets the following expression for the average number of electrons in the well

$$\begin{aligned} \langle n\rangle = \frac{2}{h}\frac{1}{\Gamma^e + \Gamma^c}\cdot \int d\varepsilon \sum_{nm} \zeta_n \big[K(\varepsilon)\left(\Gamma^e J_m^2(\alpha_e)\left[f_+^e - f_-^e\right]\right. \\ \left. + \Gamma^c J_m^2(\alpha_c)\left[f_+^c - f_-^c\right]\right) + \mathrm{Im}\mathcal{K}(\varepsilon)\left(\Gamma^e J_m^2(\alpha_e)\left[1 - f_+^e - f_-^e\right]\right. \\ \left. + \Gamma^c J_m^2(\alpha_c)\left[1 - f_+^c - f_-^c\right]\right)\big] + n_G \end{aligned} \tag{6.91}$$

The equation (n) actually determines the chemical potential $\mu_{\mathrm{T}}(n)$ of electrons in the well. In absence of the SET one gets $f(\varepsilon_+^e) = f(\varepsilon_-^e)$ and the first term under $\int d\varepsilon$ in Eq. 6.91 vanishes. Then, one simply gets

$$\begin{aligned}\frac{I^e + I^c}{2} &= \frac{e\Gamma^e}{h}\int d\varepsilon \sum_n \zeta_n \left[K(\varepsilon)\, J_m^2(\alpha_e)\left[f_+^e - f_-^e\right]\right.\\ &\quad -\mathrm{Im}\mathcal{K}(\varepsilon)\, J_m^2(\alpha_e)\left[1 - f_+^e - f_-^e\right]\Big]\\ &\quad +e\Gamma^c\int d\varepsilon \sum_n \zeta_n \left[\mathrm{Im}\mathcal{K}(\varepsilon)\, J_m^2(\alpha_c)\left[1 - f_+^c - f_-^c\right]\right.\\ &\quad -K(\varepsilon)\, J_m^2(\alpha_c)\left[f_+^c - f_-^c\right]\Big] - e\left[\Gamma^e - \Gamma^c\right]\sum_n \zeta_n n\end{aligned} \tag{6.92}$$

If one also sets $n_{\mathrm{G}} = 0$ (i.e., when no gate voltage is applied $V_{\mathrm{G}} \equiv 0$) then

$$\begin{aligned}\frac{I^e + I^c}{2} &= \frac{e\Gamma^e}{h}\int d\varepsilon \sum_{k,n} \zeta_n \left[K(\varepsilon)\, J_m^2(\alpha_e)\left[f_+^e - f_-^e\right]\right.\\ &\quad -\mathrm{Im}\mathcal{K}(\varepsilon)\, J_m^2(\alpha_e)\left[1 - f_+^e - f_-^e\right]\Big]\\ &\quad +e\Gamma^c \sum_{k,n} \zeta_n \left[\mathrm{Im}\mathcal{K}(\varepsilon)\, J_m^2(\alpha_c)\left[1 - f_+^c - f_-^c\right]\right.\\ &\quad -K(\varepsilon)\, J_m^2(\alpha_c)\left[f_+^c - f_-^c\right]\Big]\\ &\quad -\frac{2e}{h}\frac{\Gamma^e - \Gamma^c}{\Gamma^e + \Gamma^c}\cdot\int d\varepsilon \sum_{nm} \zeta_n \Big[K(\varepsilon)\left(\Gamma^e J_m^2(\alpha_e)\left[f_+^e - f_-^e\right]\right.\\ &\quad +\Gamma^c J_m^2(\alpha_c)\left[f_+^c - f_-^c\right]\Big)\\ &\quad -\mathrm{Im}\mathcal{K}(\varepsilon)\left(e\Gamma^e J_m^2(\alpha_e)\left[1 - f_+^e - f_-^e\right]\right.\\ &\quad +e\Gamma^c J_m^2(\alpha_c)\left[1 - f_+^c - f_-^c\right]\Big)\Big]\end{aligned} \tag{6.93}$$

Using that

$$\Gamma_{\mathbf{n}} = \Gamma^{e(c)}\left(1 \mp \frac{\Gamma^e - \Gamma^c}{\Gamma^e + \Gamma^c}\right) = \frac{2\Gamma^e\Gamma^c}{\Gamma^e + \Gamma^c} \tag{6.94}$$

the DC PASET electric current, Eq. 6.93, across the QW reads

$$\begin{aligned}I = \frac{I^e + I^c}{2} &= \Gamma_n \frac{2e}{h}\int d\varepsilon \sum_{nm} \zeta_n \Big[K(\varepsilon)\left[J_m^2(\alpha_e)\left(f_+^e - f_-^e\right)\right.\\ &\quad -J_m^2(\alpha_c)\left(f_+^c - f_-^c\right)\Big]\\ &\quad +\mathrm{Im}\mathcal{K}(\varepsilon)\left[J_m^2(\alpha_e)\left(f_+^e + f_-^e\right) - J_m^2(\alpha_c)\left(f_+^c + f_-^c\right)\right]\Big]\end{aligned} \tag{6.95}$$

In equilibrium one makes use of the Fourier transform

$$\mathrm{K}_0(k) = \frac{i}{2}\tan h\left(\frac{\beta v_{\mathrm{F}}|k|}{2}\right) = \frac{i}{2}\left(1 - 2\mathrm{G}_0(\beta v_{\mathrm{F}}|k|)\right) \tag{6.96}$$

where $\mathrm{G}_0(\varepsilon) = 1/(\exp(\varepsilon/T) + 1)$, k is the fermion momentum. The noninteractive equilibrium right-moving fermion correlator $K_0(0, t)$ is

$$K_0(x, t) = \frac{i}{x - v_{\mathrm{F}}t + i\delta}\frac{\pi(x - v_{\mathrm{F}}t)/\beta v_{\mathrm{F}}}{\sin h\left(\pi(x - v_{\mathrm{F}}t)/\beta v_{\mathrm{F}}\right)} \tag{6.97}$$

where $\delta \to +0$. The electron distribution function G_ε is introduced by an ansatz (see, for example, Keldysh (1965)):

$$K(\varepsilon) = \mathrm{Im}\mathcal{K}(\varepsilon)\,(1 - 2\mathrm{G}_\varepsilon) \tag{6.98}$$

This gives the expression for the net time-averaged electric current across the double-barrier junction as follows:

$$I = \Gamma_n \frac{2e}{h}\int d\varepsilon \sum_{nm} \zeta_n \mathrm{Im}\mathcal{K}(\varepsilon)\left[J_m^2(\alpha_e)\left[f_+^e + G_\varepsilon\left(f_-^e - f_+^e\right)\right]\right.$$
$$\left. - J_m^2(\alpha_c)\left[f_+^c + \left(f_-^c - f_+^c\right)G_\varepsilon\right]\right] \tag{6.99}$$

From Eq. 6.99 in the limits $\alpha_{\mathrm{e,c}} = 0$ one easily recovers the expressions for tunneling current used before.

Problems

6-1. Discuss why Mott excitons require special conditions for appearing in graphene.

6-2. Explain how conductive electron screening changes the electron–electron Coulomb interaction and effective electron mass in carbon nanotubes (CNTs).

6-3. Explain why the binding of two particles causes no bound states if their dispersion law has a conic shape.

6-4. Explain why interparticle binding occurs only in certain sectors of the momentum space, whereas bound states are prohibited in the other sectors belonging to the same space.

6-5. Explain why for generating Mott excitons in graphene, photons with a short wavevector are required.

6-6. How does the use of phonon- or impurity-assisted absorption help to excite Mott excitons in graphene?

6-7. Discuss the role of screening with the charge carriers of moving excitons in graphene.

6-8. What are the signatures of the Tomonaga–Luttinger liquid (TTL) state in narrow graphene stripes and CNTs which have been observed in experiments?

6-9. Explain how the quantized levels of charge and spin excitations in the TTL state inside a 1D quantum well (QW) can be separated from each other and are identified in experiments.

References

Ando, T. (1997). Excitons in carbon nanotubes. *J Phys Soc Jpn*, **66**(4): 1066–1073.

Ando, T. (2004). Excitons in carbon nanotubes revisited: dependence on diameter, Aharonov-Bohm flux, and strain. *J Phys Soc Jpn*, **73**(12): 3351–3363.

Ando, T. (2005). Theory of electronic states and transport in carbon nanotubes. *J Phys Soc Jpn*, **74**(3): 777–817.

Averin, D. V., A. N. Korotkov, and K. K. Likharev (1991). Theory of Single-electron charging of quantum-wells and dots. *Phys Rev B*, **44**(12): 6199–6211.

Bachilo, S. M., M. S. Strano, C. Kittrell, R. H. Hauge, R. E. Smalley, and R. B. Weisman (2002). Structure-assigned optical spectra of single-walled carbon nanotubes. *Science*, **298**(5602): 2361–2366.

Bockrath, M., D. H. Cobden, J. Lu, A. G. Rinzler, R. E. Smalley, L. Balents, and P. L. McEuen (1999). Luttinger-liquid behaviour in carbon nanotubes. *Nature*, **397**(6720): 598–601.

Dresselhaus, M. S., G. Dresselhaus, A. Jorio, A. G. Souza, M. A. Pimenta, and R. Saito (2002a). Single nanotube Raman spectroscopy. *Acc Chem Res*, **35**(12): 1070–1078.

Dresselhaus, M. S., G. Dresselhaus, A. Jorio, A. G. Souza, and R. Saito (2002b). Raman spectroscopy on isolated single wall carbon nanotubes. *Carbon*, **40**(12): 2043–2061.

Dubois, D. F. (1959a). Electron interactions. 1. Field theory of a degenerate electron gas. *Ann Phys*, **7**(2): 174–237.

Dubois, D. F. (1959b). Electron interactions. 2. Properties of a dense electron gas. *Ann Phys*, **8**(1): 24–77.

Elesin, V. F., and Y. V. Kopayev (1981). Superconductors with excess quasi-particles. *Usp Fiz Nauk*, **133**(2): 259–307.

Elliott, R. J., and R. Loudon (1959). Theory of fine structure on the absorption edge in semiconductors. *J Phys Chem Solids*, **8**: 382–388.

Emery, V. J., and D. Mukamel (1979). Locking of the 2 charge-density waves in NbSe 3. *J Phys C- Solid State*, **12**(17): L677–L679.

Feldman, D. E., S. Scheidl, and V. M. Vinokur (2005). Rectification in Luttinger liquids. *Phys Rev Lett*, **94**(18).

Feldman, D. E., S. Scheidl, and V. M. Vinokur (2006). Ratchet effects in Luttinger liquids. *Theor Quantum Transport Met Hybrid Nanostruct*, **230**: 147–155.

Gunlycke, D., J. H. Jefferson, S. W. D. Bailey, C. J. Lambert, D. G. Pettifor, and G. A. D. Briggs (2006). Zener quantum dot spin filter in a carbon nanotube. *J Phys: Condens Mater*, **18**(21): S843–S849.

Gunlycke, D., C. J. Lambert, S. W. D. Bailey, D. G. Pettifor, G. A. D. Briggs, and J. H. Jefferson (2006). Bandgap modulation of narrow-gap carbon nanotubes in a transverse electric field. *Europhys Lett*, **73**(5): 759–764.

Hagen, A., and T. Hertel (2003). Quantitative analysis of optical spectra from individual single-wall carbon nanotubes. *Nano Lett*, **3**(3): 383–388.

Ichida, M., S. Mizuno, Y. Saito, H. Kataura, Y. Achiba, and A. Nakamura (2002). Coulomb effects on the fundamental optical transition in semiconducting single-walled carbon nanotubes: divergent behavior in the small-diameter limit. *Phys Rev B*, **65**(24).

Ichida, M., S. Mizuno, K. Tani, Y. Saito, and A. Nakamura (1999). Exciton effects of optical transitions in single-wall carbon nanotubes. *J Phys Soc Jpn*, **68**(10): 3131–3133.

Ishii, H., H. Kataura, H. Shiozawa, H. Yoshioka, H. Otsubo, Y. Takayama, T. Miyahara, S. Suzuki, Y. Achiba, M. Nakatake, T. Narimura, M. Higashiguchi, K. Shimada, H. Namatame, and M. Taniguchi (2003). Direct observation of Tomonaga-Luttinger-liquid state in carbon nanotubes at low temperatures. *Nature*, **426**(6966): 540–544.

Kane, C. L., and M. P. A. Fisher (1996). Thermal transport in a Luttinger liquid. *Phys Rev Lett*, **76**(17): 3192–3195.

Kane, C. L., and E. J. Mele (1997). Size, shape, and low energy electronic structure of carbon nanotubes. *Phys Rev Lett*, **78**(10): 1932–1935.

Kataura, H., Y. Kumazawa, Y. Maniwa, I. Umezu, S. Suzuki, Y. Ohtsuka, and Y. Achiba (1999). Optical properties of single-wall carbon nanotubes. *Synth Met*, **103**(1–3): 2555–2558.

Katsnelson, M. I. (2006). Nonlinear screening of charge impurities in graphene. *Phys Rev B*, **74**(20).

Keldysh, L. V. (1965). Diagram technique for nonequilibrium processes. *Sov Phys JETP: USSR*, **20**(4): 1018.

Koch, R. J., T. Haensel, S. I. U. Ahmed, T. Seyller, and J. A. Schaefer (2010). HREELS study of graphene formed on hexagonal silicon carbide. *Phys Status Solidi C*, **7**(2): 394–397.

Lin, M. F., and K. W. K. Shung (1993a). Elementary excitations in cylindrical tubules. *Phys Rev B*, **47**(11): 6617–6624.

Lin, M. F., and K. W. K. Shung (1993b). Magnetoplasmons and persistent currents in cylindrical tubules. *Phys Rev B*, **48**(8): 5567–5571.

Linchung, P. J., and A. K. Rajagopal (1994a). Electronic excitations in nanoscale systems with helical symmetry. *J Phys: Condens Mater*, **6**(20): 3697–3706.

Linchung, P. J., and A. K. Rajagopal (1994b). Magnetoplasma oscillations in nanoscale tubules with helical symmetry. *Phys Rev B*, **49**(12): 8454–8463.

Lu, J., K. P. Loh, H. Huang, W. Chen, and A. T. S. Wee (2009). Plasmon dispersion on epitaxial graphene studied using high-resolution electron energy-loss spectroscopy. *Phys Rev B*, **80**(11).

Luttinger, J. M. (1963). An exactly soluble model of a Many-Fermion system. *J Math Phys*, **4**(9): 1154.

Mahmoodian, M. M., and M. V. Entin (2013). Moving zero-gap Wannier-Mott excitons in graphene. *EPL*, **102**(3).

Mak, K. F., J. Shan, and T. F. Heinz (2011). Seeing many-body effects in single- and few-layer graphene: observation of two-dimensional saddle-point excitons. *Phys Rev Lett*, **106**(4).

Mintmire, J. W., and C. T. White (1998). Universal density of states for carbon nanotubes. *Phys Rev Lett*, **81**(12): 2506–2509.

Nazarov, Y. V., and L. I. Glazman (2003). Resonant tunneling of interacting electrons in a one-dimensional wire. *Phys Rev Lett*, **91**(12).

Ogawa, T., and T. Takagahara (1991a). Interband absorption-spectra and Sommerfeld factors of a one-dimensional electron-hole system. *Phys Rev B*, **43**(17): 14325–14328.

Ogawa, T., and T. Takagahara (1991b). Optical-absorption and sommerfeld factors of one-dimensional semiconductors: an exact treatment of excitonic effects. *Phys Rev B*, **44**(15): 8138–8156.

Recher, P., N. Y. Kim, and Y. Yamamoto (2006). Tomonaga-Luttinger liquid correlations and Fabry-Perot interference in conductance and finite-frequency shot noise in a single-walled carbon nanotube. *Phys Rev B*, **74**(23).

Rice, T. M. (1965). Effects of electron-electron interaction on properties of metals. *Ann Phys*, **31**(1): 100.

Rinzan, M., G. Jenkins, H. D. Drew, S. Shafranjuk, and P. Barbara (2012). Carbon nanotube quantum dots as highly sensitive terahertz-cooled spectrometers. *Nano Lett*, **12**(6): 3097–3100.

Saito, R., A. Jorio, A. G. Souza, G. Dresselhaus, M. S. Dresselhaus, and M. A. Pimenta (2002). Probing phonon dispersion relations of graphite by double resonance Raman scattering. *Phys Rev Lett*, **88**(2).

Sakai, H., H. Suzuura, and T. Ando (2004). Effective-mass approach to interaction effects on electronic structure in carbon nanotubes. *Phys E-Low-Dimens Syst Nanost*, **22**(1–3): 704–707.

Samsonidze, G. G., R. Saito, N. Kobayashi, A. Gruneis, J. Jiang, A. Jorio, S. G. Chou, G. Dresselhaus, and M. S. Dresselhaus (2004). Family behavior of the optical transition energies in single-wall carbon nanotubes of smaller diameters. *Appl Phys Lett*, **85**(23): 5703–5705.

Sato, O., Y. Tanaka, M. Kobayashi, and A. Hasegawa (1993). Exotic behavior of the dielectric function and the plasmons of an electron-gas on a tubule. *Phys Rev B*, **48**(3): 1947–1950.

Shafranjuk, S. E. (2007). Sensing an electromagnetic field with photon-assisted Fano resonance in a two-branch carbon nanotube junction. *Phys Rev B*, **76**(8).

Shafranjuk, S. E. (2008). Probing the intrinsic state of a one-dimensional quantum well with photon-assisted tunneling. *Phys Rev B*, **78**(23).

Shafranjuk, S. E. (2009). Reversible heat flow through the carbon tube junction. *EPL*, **87**(5).

Snow, E. S., and F. K. Perkins (2005). Capacitance and conductance of single-walled carbon nanotubes in the presence of chemical vapors. *Nano Lett*, **5**(12): 2414–2417.

Snow, E. S., F. K. Perkins, and E. J. Houser (2005). Chemical detection using single-walled carbon nanotubes. *Abstr Pap Am Chem Soc*, **230**: U312–U312.

Snow, E. S., F. K. Perkins, E. J. Houser, S. C. Badescu, and T. L. Reinecke (2005). Chemical detection with a single-walled carbon nanotube capacitor. *Science*, **307**(5717): 1942–1945.

Tarkiainen, R., M. Ahlskog, J. Penttila, L. Roschier, P. Hakonen, M. Paalanen, and E. Sonin (2001). Multiwalled carbon nanotube: Luttinger versus Fermi liquid. *Phys Rev B*, **64**(19).

Tomonaga, S. (1950). Remarks on Blochs method of sound waves applied to Many-Fermion problems. *Prog Theor Phys*, **5**(4): 544–569.

Tselev, A., Y. F. Yang, J. Zhang, P. Barbara, and S. E. Shafranjuk (2009). Carbon nanotubes as nanoscale probes of the superconducting proximity effect in Pd-Nb junctions. *Phys Rev B*, **80**(5).

Viet, N. A., H. Ajiki, and T. Ando (1994). Lattice instability in metallic carbon nanotubes. *J Phys Soc Jpn*, **63**(8): 3036–3047.

Wang, L., P. S. Davids, A. Saxena, and A. R. Bishop (1992). Quasi-particle energy-spectra and magnetic response of certain curved graphitic geometries. *Phys Rev B*, **46**(11): 7175–7178.

Yaguchi, T., and T. Ando (2002). Electronic states in capped carbon nanotubes [**70**, 1327, (2001)]. *J Phys Soc Jpn*, **71**(11): 2824–2824.

Yang, L. (2011). Excitons in intrinsic and bilayer graphene. *Phys Rev B*, **83**(8).

Yang, L., J. Deslippe, C. H. Park, M. L. Cohen, and S. G. Louie (2009). Excitonic effects on the optical response of graphene and bilayer graphene. *Phys Rev Lett*, **103**(18).

Yang, Y., G. Fedorov, P. Barbara, S. E. Shafranjuk, B. K. Cooper, R. M. Lewis, and C. J. Lobb (2013). Coherent nonlocal transport in quantum wires with strongly coupled electrodes. *Phys Rev B*, **87**(4).

Yang, Y. F., G. Fedorov, J. Zhang, A. Tselev, S. Shafranjuk, and P. Barbara (2012). The search for superconductivity at van Hove singularities in carbon nanotubes. *Supercond Sci Tech*, **25**(12).

Yi, H. M., and C. L. Kane (1996). Coulomb blockade in a quantum dot coupled strongly to a lead. *Phys Rev B*, **53**(19): 12956–12966.

Zaric, S., G. N. Ostojic, J. Kono, J. Shaver, V. C. Moore, M. S. Strano, R. H. Hauge, R. E. Smalley, and X. Wei (2004). Optical signatures of the Aharonov-Bohm phase in single-walled carbon nanotubes. *Science*, **304**(5674): 1129–1131.

Chapter 7

Andreev Reflection at the Graphene/Metal Interface

In this chapter we consider properties of graphene junctions with superconducting electrodes. Besides, we also discuss the experimental search for intrinsic superconductivity in carbon nanotubes and graphene stripes.

7.1 The Graphene/Superconductor Interface

The study of interaction between a superconductor and graphene is interesting because the two distinct materials represent different types of intrinsic coherence. An interest in mechanism of interaction between graphene and superconducting materials is motivated by the following. On the one hand, inside the superconductor, the superfluid condensate of Cooper pair is responsible for numerous unconventional phenomena like, for example, the Meissner effect, Abrikosov vortices, Josephson current, etc. A key microscopic process, which causes the superconducting state, is represented by Cooper coupling between an electron state, say, with spin "$\uparrow$" and its time-reversed state, which is a hole with spin "$\downarrow$." On the other hand,

Graphene: Fundamentals, Devices, and Applications
Serhii Shafraniuk

ISBN 978-981-4613-47-7 (Hardcover), 978-981-4613-48-4 (eBook)
www.panstanford.com

in graphene, the intrinsic coherence emerges from conservation of the one-half pseudospin. Because the pseudospin is preserved, the electrons and holes acquire an extra robustness which diminishes the electron scattering on atomic defects and lattice oscillations. For such reasons, graphene is regarded as intrinsically "purified," since the electron–impurity and electron–phonon scatterings are substantially reduced due to the pseudospin conservation. Besides, graphene (G) is attractive as a barrier material for Josephson junctions due to the high mobility of carriers along with an unsurpassed flexibility in controlling of transport properties using a variety of methods. In addition, such junctions offer an opportunity for physicists to explore relativistic superconductivity and unusual proximity effects (Beenakker 2006). Studying these effects and making useful devices is hampered, however, by the quality of the contacts between the graphene and metal banks (Liang et al. 2008, Khomyakov et al. 2009, Mueller et al. 2009, Avouris 2010). Due to the difference, the workfunctions of graphene and metals, Schottky-type barrier are formed at the interface, thereby significantly changing the transport properties of the metal/G devices.

7.2 Conversions between Electrons and Holes at the N/S Interface

When an electric current flows across the interface between a normal material N and a superconductor S, it is accompanied by a conversion between different types of the charge carriers. On the one hand, the electric charge carriers in the normal materials are quasiparticle excitations, electrons and holes. On the other hand, in a superconductor, the electric current occurs via superfluid condensate of Cooper pairs. To satisfy the electric current continuity equation, different charge carriers must be converted from one type to another. For the normal metal/superconducting (N/S) interface, the Andreev reflection (AR) is the mechanism of conversion of electron into hole excitations by the superconducting pair potential (Andreev 1964). In respect to AR, the electron excitation is a filled state at positive energy $\varepsilon > 0$ which is above the Fermi energy E_F, whereas the hole excitation is an empty state at $-\varepsilon < 0$ below E_F.

Because the excitation energy ε is essentially the same, AR is an elastic process. However, since the electric charges of electron (-e) and of the hole ($+e$) are opposite, an extra charge of $2e$ is missing in the conversion process. The lost charge is transferred to the superfluid condensate which is interpreted as a creation of an extra Cooper pair superconductor. If the electron energy E is less than the superconducting gap Δ, a finite electric current can take place only when a Cooper pair is created inside S. To fulfil the electric charge conservation at the N/S interface, an incident electron which had been transmitted through the N/S interface must pick up another electron inside S to create a Cooper pair. Thus, the hole which is created during AR is an empty state which is left behind the electron when it is paired with an incident electron to form a Cooper pair. The total momentum of Cooper pair is equal to zero and, therefore, the electron and hole are taken from opposite corners $\pm$ K of the Brillouin zone. It means that the valleys are switched due to AR in graphene (Beenakker 2006). The above scenario corresponds to the conventional s-wave pairing symmetry of the superfluid condensate. There is an analogy between the switching of valleys by AR due to s-wave pairing in the superconductor and the switching of spin bands due to singlet pairing. One can detect that phenomena experimentally by producing a spin polarization in the normal metal (Dejong and Beenakker 1995). A similar scenario can also be used in graphene where the switching the valleys can be observed by producing a valley polarization. In a conventional case, the electron and hole both belong to the conduction band. That corresponds to electron energy $\varepsilon < E_F$ and creates the intraband AR. When $\varepsilon > E_F$, then, in contrast to the former case, the hole belongs to the valence band. In a pristine infinite undoped graphene sheets, at $E_F = 0$, AR is interband at all excitation energies. Such interband process is not possible in regular metals, with no excitation gap Δ separating the conduction and valence bands.

7.3 The BTK Model of Andreev Reflection

Below, we briefly delineate a simple, popular, and comprehensive model of AR, which is regarded as the Blonder-Tinkham-Klapwijk

(BTK) model (Blonder et al. 1982). The process occurs at the interface separating a conventional normal material N and superconductor S. The BTK model is formulated in terms of 1D Bogoliubov equations written in the form

$$i\hbar\frac{\partial f}{\partial t}=\left[-\frac{\hbar^2\nabla^2}{2m}-\mu(x)+V(x)\right]f(x,t)+\Delta(x)g(x,t) \quad (7.1)$$

$$\hbar\frac{\partial g}{\partial t}=-\left[-\frac{\hbar^2\nabla^2}{2m}-\mu(x)+V(x)\right]g(x,t)+\Delta(x)f(x,t) \quad (7.2)$$

In Eqs. 7.1 and 7.2, $\Delta(x)$ is the energy gap and $\mu(x)$ is the chemical potential. In the normal metal $\Delta(x) = 0$ and Eq. 7.1 represents the Schrödinger equation for electrons, whereas Eq. 7.2 is the time-reversed Schrödinger equation for electrons. Here, we exploit a close similarity between the properties of an electron satisfying the time-reversed Schrödinger equation and a hole. Therefore, in this chapter, we define the hole as a time-reversed electron. Since inside the superconductor the pairing potential is finite, there is a finite BCS energy gap $\Delta \neq 0$. Therefore, according to Eqs. 7.1 and 7.2, the electron and hole wavefunctions couple to each other. To derive the electron excitation spectrum for a homogeneous superconductor, we set the functions $\Delta(x)$, $\mu(x)$, and $V(x)$ in Eqs. 7.1 and 7.2 as coordinate-independent constants. Using the plane wares $f = \tilde{u}e^{ikx-iEt/\hbar}$ and $g = \tilde{v}e^{ikx-iEt/\hbar}$ as trial wavefunctions we immediately get

$$E\tilde{u}=\left[\frac{\hbar^2k^2}{2m}-\mu\right]\tilde{u}+\Delta\tilde{v} \quad (7.3)$$

$$E\tilde{v}=-\left[\frac{\hbar^2k^2}{2m}-\mu\right]\tilde{v}+\Delta\tilde{u} \quad (7.4)$$

The solution of Eqs. 7.3 and 7.4 in respect to E gives

$$E^2=\left[\frac{\hbar^2k^2}{2m}-\mu\right]^2+\Delta^2 \quad (7.5)$$

Thus, an energy gap appears in the E vs k relation, just as in the BCS theory. There are two roots for E. For excitations above the ground state (7.5) we obtain

$$\tilde{u}^2=\frac{1}{2}\left[1\pm\frac{\sqrt{E^2-\Delta^2}}{E}\right]=1-\tilde{v}^2 \quad (7.6)$$

$$\hbar k^{\pm} = \sqrt{2m}\sqrt{\mu \pm \sqrt{E^2 - \Delta^2}}. \qquad (7.7)$$

We also define

$$1 - v_0^2 = u_0^2 = \frac{1}{2}\left[1 + \frac{\sqrt{E^2 - \Delta^2}}{E}\right]. \qquad (7.8)$$

For the energies below the gap, $|E| < \Delta$, $\tilde{u}$ and $\tilde{v}$ are complex conjugated. It is different from the BCS convention where u_k and v_k are not defined for $E < \Delta$.In the spinor notation

$$\psi = \begin{bmatrix} f(x) \\ g(x) \end{bmatrix}. \qquad (7.9)$$

We distinguish four types of quasiparticle waves. For certain energy, the spinors are written as

$$\psi_{\pm k^+} = \begin{bmatrix} u_0 \\ v_0 \end{bmatrix} e^{\pm ik^+ x} \qquad (7.10)$$

and

$$\psi_{\pm k^-} = \begin{bmatrix} v_0 \\ u_0 \end{bmatrix} e^{\pm ik^- x}. \qquad (7.11)$$

The above spinor functions, Eqs. 7.10 and 7.11, are the space representations of the quasiparticle creation and annihilation operators $\gamma_{k0}^* = u_k c_{k\uparrow}^* - v_k c_{-k\downarrow}$, since $c_{k\uparrow}^*$ produces the wave e^{ikx}, while $c_{-k\downarrow}$ destroys an electron at $-k$. Thus, the system acquires a net positive momentum emerging from the creation of a hole wave e^{ikx}.

Another interesting property, which directly follows from the quasiparticle dispersion law, Eq. 7.5, is that the group velocity $v_g = dE/d(\hbar k)$ goes to zero at the gap edges $E = \pm\Delta$, even while the phase velocity remains close to the normal state value. The quasiparticle current contains contributions from both the electrons and holes with the same sign. Let us consider an electron $f \propto e^{ik^+ x}$ which propagates in the positive direction and carries a positive current. The time-reversed wave $g \propto e^{ik^+ x}$, a hole, propagates in the negative direction, but it also carries a positive current because it has a positive electric charge. On the one hand, in the terms of our ballistic model, both the electron and hole propagate with the Fermi velocity in opposite directions inside the normal metal N. On the other hand, inside the bulk of superconductor, the electron current is carried by Cooper pairs which belong to the superfluid condensate. Now the question is how the electric current is transmitted through the N/S interface? And what is the mechanism of conversion between the normal current in N and the supercurrent in S? Since the Bogoliubov equations, Eqs. 7.1 and 7.2, are valid for both, the normal metal and the superconductor, we would like to compute the electric current by using the conventional Landauer formula

$$I_{\text{NS}} = \int_{-\infty}^{\infty} M(E)\,[f(E - eV_{\text{SN}}) - f(E)]\,dE \qquad (7.12)$$

where $f(E)$ is the electron distribution function which coincides with the Fermi function in equilibrium, $f_0(E) = 1/\left(e^{(E-\mu)/k_{\text{B}}T} + 1\right)$, $M(E)$ is the number of quantum channels per energy interval

$$M(E) = M_0\,[1 + A(E) - B(E)] \qquad (7.13)$$

where $A(E)$ and $B(E)$ are coefficients of the so-called Andreev and conventional reflection, respectively, M_0 is the value of $M(E)$ in the bulk of N where $A(E) = B(E) = 0$. For instance, for a regular plane N/S junctions one uses the well-known expression (Blonder et al. 1982)

$$M_0 = 2N(0)\,ev_{\text{F}}A \qquad (7.14)$$

where A is the junction's area and $N(0)$ is the density of electron states in the normal metal. The coefficients where $A(E)$ and $B(E)$ are computed from solutions of Eqs. 7.1 and 7.2 complemented by appropriate boundary conditions at the N/S interface. The boundary conditions at the N/S interface for steady-state plane-wave solutions are as follows:

(i) Continuity of the electron wavefunction ψ at $x = 0$, so $\psi_{\text{S}}(0) = \psi_{\text{N}}(0) \equiv \psi(0)$.
(ii) $(\hbar/2m)\left[\psi_{\text{S}}' - \psi_{\text{N}}'\right] = H\psi(0)$, the derivative boundary condition appropriate for δ functions.
(iii) Incoming (incident), reflected, and transmitted wave directions are defined by their group velocities. We assume that the incoming electron produces only outgoing particles.

The trial wavefunctions are written as plane waves propagating in-perpendicular to the N/S interface

$$\psi_{\text{inc}} = \begin{bmatrix} 1 \\ 0 \end{bmatrix} e^{iq^+x}$$

$$\hbar q^{\pm} = \sqrt{2m}\sqrt{\mu \pm E} \tag{7.15}$$

$$\psi_{\text{refl}} = a \begin{bmatrix} 0 \\ 1 \end{bmatrix} e^{iq^-x} + b \begin{bmatrix} 1 \\ 0 \end{bmatrix} e^{-iq^+x} \tag{7.16}$$

$$\psi_{\text{trans}} = c \begin{bmatrix} u_0 \\ v_0 \end{bmatrix} e^{ik^+x} + d \begin{bmatrix} v_0 \\ u_0 \end{bmatrix} e^{-ik^-x} \tag{7.17}$$

The boundary conditions (i)–(iii) yield a system of linear algebraic equations the solution of which is

$$a = \frac{u_0 v_0}{\gamma} \tag{7.18}$$

$$b = -\frac{\left(u_0^2 - v_0^2\right)\left(Z^2 + iZ\right)}{\gamma} \tag{7.19}$$

$$c = \frac{u_0\,(1 - iZ)}{\gamma} \tag{7.20}$$

$$d = \frac{i v_0 Z}{\gamma} \tag{7.21}$$

where the interface barrier strength is defined as

$$Z = \frac{mH}{\hbar^2 k_{\text{F}}} = \frac{H}{\hbar v_{\text{F}}} \tag{7.22}$$

and

$$\gamma = u_0^2 + \left(u_0^2 - v_0^2\right) Z^2. \tag{7.23}$$

In the above formulas, for simplicity, we set $k^+ = k^- = q^+ = q^- = k_{\text{F}}$, whenever that substitution does not lead to a qualitative change. If the interface transparency is ideal, that is, in the absence of a barrier (i.e., $Z = 0$), then $b = d = 0$. It means that there is no branch crossing because all the reflections become just ARs.

The coefficients $A(E)$ and $B(E)$ entering the Landauer formula for the electric current, Eqs. 7.12 and 7.13, actually determine the corresponding probability currents, for the particles, measured in units of the Fermi velocity v_F. In particular, one defines $A = a^*a$, and $B = b^*b$. Since the plane-wave currents are spatially homogeneous, there is no need to specify the position at which they are evaluated. Nevertheless, if $E < \Delta$, while the imaginary parts of k^+ and k^- in the superconductor remain small, there is an exponential decay of A on the length scale λ

$$\lambda = \frac{\hbar v_{\mathrm{F}}}{2\Delta}\sqrt{1-\left(\frac{E}{\Delta}\right)^2} \tag{7.24}$$

Although the length diverges right at the gap edge, the characteristic length, nevertheless, is $\hbar v_{\mathrm{F}}/2\Delta = 1.22\xi\,(T)$. Thus, in order of magnitude terms, one can say that the particles penetrate a depth $\sim \xi\,(T)$ before the current is converted to a supercurrent carried by the condensate.

7.4 Experimental Study of Andreev Reflection in Graphene

The fabrication and the experimental study of graphene junctions have recently become a subject of sustainable experimental research activity. The graphene/metal junctions have been typically fabricated by means of e-beam lithography using graphene flakes. One example is represented by Nb(/Ti)–G–(Ti/)Nb junctions (Li et al. 2013) characterized at temperatures above 1.8 K where the Josephson current is not observed, but the differential conductivity revealed features below the critical temperature of Nb. An overall metallic conductivity of those junctions, in spite of a high junctions resistance, is presumably associated with the formation of potential barriers at the Nb(/Ti)–G interfaces. To explain the unconventional properties we consider a theoretical model which involves two very different graphene "access" lengths (Kaufman et al. 2000). The shorter length characterizes ordinary tunneling between the 3D Nb(/Ti) electrode and 2D graphene, while the second, much longer length, is related to AR (Andreev 1964) at the 3D/2D Nb(/Ti)–G interface. Associated transmission factors are small in the first

Table 7.1 Summary of device parameters. Adapted from Nevirkovets (2014)

Device number	Material of leads (in parentheses) (thickness in nm)	Spacing between the leads (nm)	Graphene thickness (nm)	Device resistance at 5 K and 5 mV (Ω)
G1	Ti(2)/Nb(40)	430	148	224
G2	Nb(40)	640	19	-
G3	Ti(2)/Nb(40)	440	9	369
G4	Ti(2)/Nb(40)	170	8	600

case and much larger in the second, which explains the apparent contradiction of the observed behaviors.

Here, we describe the transport properties of Nb–G–Nb junctions fabricated on graphene flakes which accommodate four devices in total. Graphene flakes were deposited onto oxidized Si substrates by mechanical exfoliation of highly oriented pyrolytic graphite (HOPG). Using the e-beam lithography, a PMMA mask was patterned on the graphene flakes. Then, 2 nm of Ti was deposited, followed by 40 nm of Nb, to form devices G1 and G3; in devices G2 and G4, 40 nm Nb film was deposited directly onto the graphene. Prior to deposition of the Ti and Nb layers, ion milling was used to remove about 4 nm of the surface layer for cleaning in device G1 with thicker graphene flake; no ion milling was used for the rest of devices (G2–G4) with thinner flakes. The thickness of flakes was measured using AFM. The device parameters are summarized in Table 7.1 above. Figure 7.1 shows an SEM image of a typical device structure.

Current–voltage characteristics ($I-V$ curves) of the devices were measured out in a PPMS Quantum Design cryostat at temperatures down to 1.8 K using the two-probe method. Due to the latter, the device resistance (see Table 7.1) contains a 25 Ω contribution from the wires. Measurable characteristics were obtained for devices G1, G3, and G4; resistance of the device G2 was too high to be measured with our technique. The measured characteristics of different devices were similar, displaying a nonlinearity of the $I-V$ curve which was most pronounced for the device G1. The current–voltage curves of this device, taken at various temperatures, are shown in Fig. 7.2a. Considerable change of the junction resistance

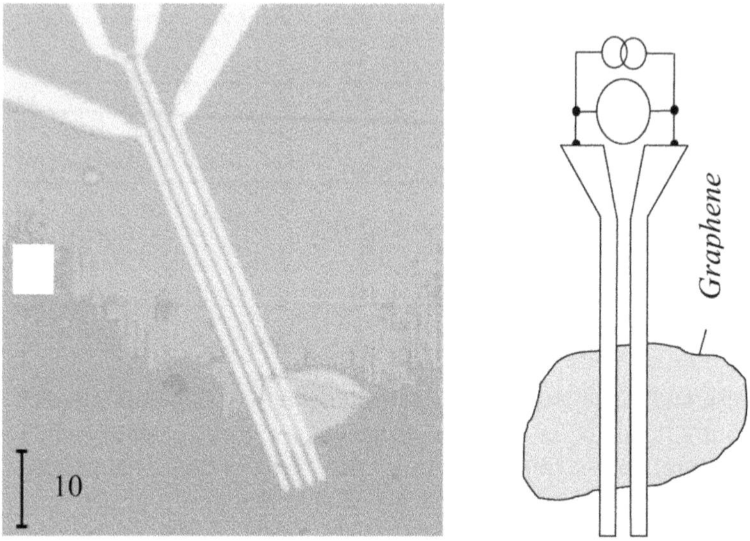

Figure 7.1 Left: SEM image of the device G1 made on a 148 nm thick graphene flake. Right: Schematic of the I–V curve measurement. Adapted from Nevirkovets (2014).

starting from about 7.0 K indicates beginning of transition of the Nb film into a resistive state (the critical temperature, T_c, is reduced for a 40 nm thick Nb film as compared to usual $T_c \approx 9.0$ K for our thicker films).

At the temperatures of the experiment, Josephson current was not observed, but the differential conductivity (see Fig. 7.2b) revealed features below the critical temperature of the superconducting transition T_c of Nb. The junctions show an overall metallic-like conductivity (initial portion is concave down), in spite of a high junction resistance. Such behavior is presumably associated with formation of the potential barriers at the Nb/Ti–G interfaces. In samples G3 and G4 the nonlinearity is weaker, and the junction resistance is higher, as follows from Table 7.1. Below we consider properties of the sample G1 in more detail.

Geometrical dimensions of the sample are determined by AFM (Fig. 7.1a): $L = 430$ nm (L is the spacing between the Nb/Ti leads), $W = 10$ μm, and the thickness of the flake of 148 nm.

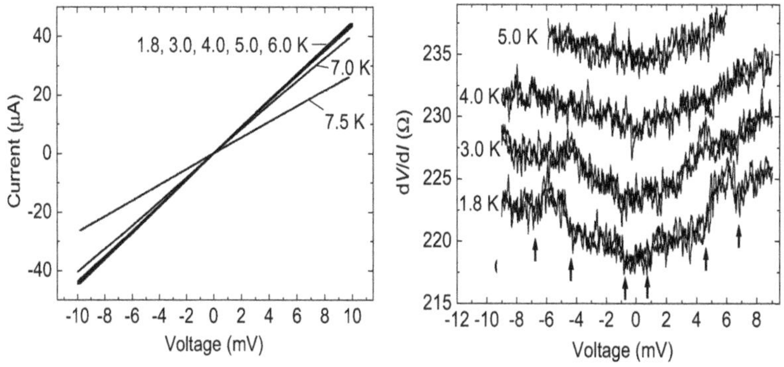

Figure 7.2 Left: I–V curves of a Nb/Ti-graphene-Ti/Nb device (G1) at various temperatures from 1.8 to 7.5 K. Right: Numerically taken derivatives dV/dI (V) for the I–V curves measured at 1.8, 3.0, 4.0, and 5.0 K. Adapted from Nevirkovets (2014).

Given this thickness, the electric properties of the flake should be regarded as those of the graphite. Then, assuming that the resistivity of graphite is about $9\times10^{-6}\,\Omega{\cdot}\text{m}$, and taking into account its temperature dependence, we estimate that the resistance of our junction should be about 6 Ω; in fact, it is 224 Ω at low temperatures. Excluding the contribution of 25 Ω from the wires and 6 Ω from the graphene flake, we obtain the resistance of 193 Ω, which is probably originating from the interfaces between the Ti/Nb and the graphene.

Assuming that the two interfaces are identical, with an average area of $A = 1\ \mu\text{m} \times 10\ \mu\text{m}$, we obtain the specific tunneling resistance ($R \times A$) of the interface to be of the order of $10^{-5}\ \Omega \cdot \text{cm}^2$, indicating a rather strong barrier. It is known that such a barrier appears at the metal–graphene interface due to difference between electronic concentrations and work functions (Lee et al. 2008, Avouris 2010).

At lower temperatures, the features are observed in the $dV/dI(V)$ dependences (marked by arrows in Fig. 7.2b). The conductance peaks at a voltage of about 1 mV can be attributed to manifestation of the Nb superconducting energy gap. Indeed, using the Bardeen–Cooper–Schrieffer (BCS) relation $2\Delta/k_\text{B}T_\text{c} = 3.52$ (where Δ is the superconducting energy gap, k_B is the Boltzmann constant, $k_\text{B} =$

8.62×10^{-5} eV/K), with $T_c \approx 7$ K, as deduced above, we obtain an estimated maximum value $\Delta \approx 1$ meV.

The conductivity anomalies at higher voltages ($\sim$4 and 7 mV) are unusual. Similar anomaly (as well as metallic junction type) was observed by authors (Choi et al. 2010) for devices claimed to be made of monolayer graphene. In case if the tunnel barrier is not too strong, one may expect manifestation of the features in the conductivity at $V < \Delta/e$, associated with ARs (Andreev 1964) at the superconductor/graphene interface. (Ossipov et al. 2007, Titov et al. 2007) The nature of peaks at $V > \Delta/e$ (Fig. 7.2b) is not clear and requires further experimental material and analysis.

The device conductivity has maximum at zero voltage (i.e., it is of metallic type). Metallic type of conductivity takes place in junctions with high-conductivity channels. Also, conductivity may continuously increase with voltage if the barrier is not rectangular but its width decreases with energy; it is suggested that the metal–graphene interface barrier has essentially a triangular shape (Avouris 2010, Chiu et al. 2010, Choi et al. 2010). The barrier is probably also asymmetric, as follows from the asymmetry of the $dV/dI(V)$ dependences with respect to zero voltage (see Fig. 7.2b). However, if nonrectangular barrier was the only reason for increase in conductivity, then the conductivity would not have an inflection point, as it seems to be in our case and in (Avouris 2010, Chiu et al. 2010, Choi et al. 2010). Therefore, we have to look for another mechanism of such a behavior.

7.5 Interpretation of Andreev Reflection Experiments in Graphene-Based Junctions

Here, we focus on the explanation of metallic-like conductivity of the graphene-based junctions, which is an interesting and unusual feature considering a rather strong barrier between the metal and graphene. Qualitatively, the experimental data can be described by the orthodox BTK model (Blonder et al. 1982) appropriately modified to take into account the present nanodevice geometry. First, we analyze the junction resistance in more detail. In general, there are three contributions into the junction resistance: (i) the

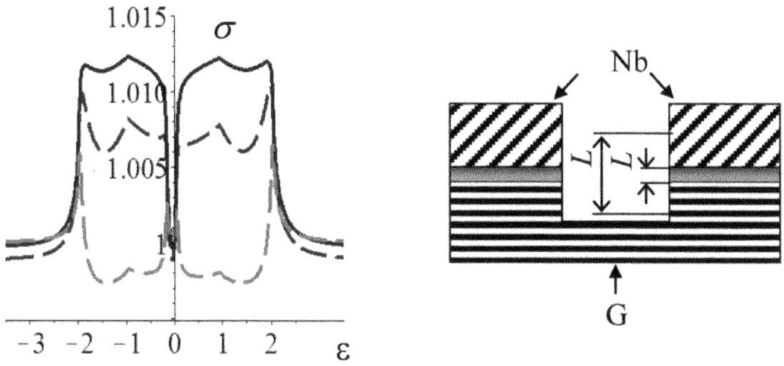

Figure 7.3 Left: The excess conductivity σ of just one SIS' junction normalized to $\sigma_0 = 1/(193\ \Omega)$ versus bias voltage eV (in units of Δ_{Nb}) due to Andreev reflection for $T_2 = 0.9$, 0.6, and 0.1 (blue, red, and green curves, respectively). Right: Setup of the Nb/G junction, where G is graphene.

Schottky barrier resistance due to the difference of work functions, (ii) a change of the number of conducting quantum channels during the tunneling from 3D metallic electrode into the 2D graphene, and (iii) the own resistance of the graphene sheet. The contribution (iii) is estimated to be 6 Ω for the device G1, as discussed above. A separation between the contributions (i) and (ii) due to the interface resistance of 193 Ω can be made from our experimental data from the ratio of the excess zero-voltage conductance measured at a very low temperature to the normal state conductance. We estimate this ratio by comparing zero-voltage differential resistance values at 1.8 K (the lowest temperature accessible in this experiment) and 5.0 K. The choice of the curve for 5.0 K is dictated by the fact that, at higher temperatures, an increasing overall shift of the differential resistance curve appears, indicating that some regions of the Nb leads become resistive below an estimated T_c value of 7 K; this makes the curves for higher temperatures unsuitable for the estimation. Then, from curves in Fig. 7.2b we obtain the excessive resistance for the 5.0 K curve to be 2.7 Ω, which means that the excessive zero-voltage conductance due to the contribution (ii) considered above is about 1.4% of the interface conductance. We will use this value in the modelling described below.

An important distinction between the assumptions of BTK model (Blonder et al. 1982) and our consideration of the real device

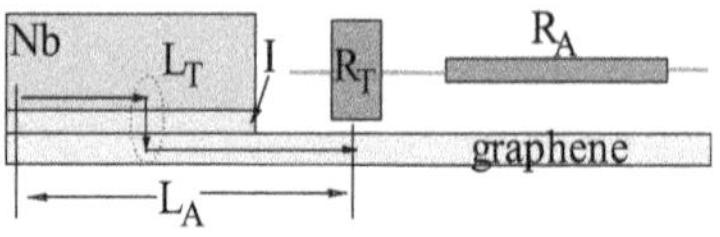

Figure 7.4 Schematics of the metal/graphene junction. There are two different spatial scales, LA and LT, with LA >>LT. The interface barrier I is formed due to imperfections and the Schottky effect. The electric resistances RA and RT in series mimic the resistances of the interface and the graphene region, respectively.

sketched in Fig. 7.4 is the change of electron state dimensionality 3D $\rightarrow$ 2D while tunneling between the Nb/Ti electrode (S electrode) and the graphene (G). The number of quantum channels in graphene is finite. Therefore, the tunneling probability from Nb/Ti into G is restricted by the limited amount of conducting channels. Another distinction between the BTK model and our setup stems from a specific electron momentum conservation in our geometry shown in Fig. 7.4. On the one hand, only the electrons with momentum $p_{\perp}$ perpendicular to the interface contribute into the conventional tunneling (CT) between the Nb/Ti electrode and graphene. On the other hand, only the electrons, the momentum $p_{||}$ of which is parallel to the interface, actually contribute into AR. The aforementioned distinction between the BTK model on the one hand and the real device on the other hand allows us to separate the CT and AR processes from each other in the latter system. The two processes, CT and AR, are characterized by two very different times τ_{T} and τ_{A}, respectively. For these reasons, the electric charge transfer through the Nb/graphene/Nb junction occurs actually in two time stages which are related to the two very different time scales τ_{T} and τ_{A}. During the initial CT stage happening during the short time τ_{T}, the electrons tunnel from 3D Nb/Ti electrode into the 2D graphene. The CT goes in-perpendicular to the interface, which means that an electron spends a very short time $\tau_{\mathrm{T}} \approx D_{\mathrm{B}}/v_{\perp}$ near the barrier before the tunneling. Here D_B is the interface barrier thickness, $v_{\perp} \approx v_{\mathrm{F}}$, v_{F} is the Fermi velocity in Ti. The AR process happens during much longer time $\tau_{\mathrm{A}} >> \tau_{\mathrm{T}}$, where $\tau_{\mathrm{A}} = \xi_{\mathrm{Nb}}/v_{\mathrm{F}}$, ξ_{Nb} and Δ_{Nb} are the superconducting coherence length and the energy gap in Nb, respectively; $\xi_{\mathrm{Nb}} = \hbar v_{\mathrm{F}}/\pi\Delta_{\mathrm{Nb}}$ in the "clean" limit.

A theoretical model is proposed to explain an experimental finding that the considerable value of the obtained junction resistance ($R_0 = 193\ \Omega$) comes into a contradiction to the metallic-like shape of the $\sigma(V)$ curves at $T < T_{c,Nb}$, as seen in Fig. 7.2b. On the one hand, the low value of $\sigma(V)$ suggests that there is a sufficient interface resistance. On the other hand the metallic-like shape of $\sigma(V)$ assumes that the interface resistance must be very low, which corresponds to a necessary condition for AR. The controversy can be resolved using a simple model which involves two very different graphene "access" lengths (Kaufman et al. 2000), $L_{\Delta 1}$ and $L_{\Delta 2}$, which are related as $L_{\Delta 1} << L_{\Delta 2}$. The short length $L_{\Delta 1} \approx a$ (here a is the atomic lattice constant) characterizes the regular tunneling of electrons between the 3D Ti/Nb electrode and 2D graphene perpendicular to the Nb/Ti/graphene interface ($p_{\perp} \neq 0$). The much longer $L_{\Delta 2} \approx \xi_{Nb}$ (where ξ_{Nb} is the superconducting coherence length of Nb) is related to AR at the 3D/2D Ti/Nb/graphene interface occurring parallel to the Nb/Ti/graphene interface ($p_{||} \neq 0$). Then, the coupling between the 3D contact metal and the 2D graphene can be effectively described by introducing two different contact transmission factors T_1 and T_2, for which $T_{1,2} < 1$. Because $L_{\Delta 2} \approx \xi_{Nb} >> a$, the sufficient number of 2D conducting channels ensures a good AR. For this case, the 3D/2D transmission factor T_2 for AR is sufficiently large ($T_2 \approx 1$). On the contrary, the factor T_1 describing regular tunneling is determined by a very short, $L_{\Delta 1} \approx a$, which corresponds to just a few conducting 2D channels and gives $T_1 << T_2$. Due to the lack of sufficient scattering along the short path $\sim L_{\Delta 1}$, the 3D/2D interface resistance gains considerable 193 Ω. The CT, acting during the first stage, actually restricts AR to just a small fraction of electrons coming from Nb to graphene. The next stage is dominated by AR which takes place on a much longer time scale τ_A. This AR process involves only the electrons whose momentum is parallel to the barrier component, that is, $p_{||} \neq 0$. In the latter case, since the barrier length $L_B >> D_B$, the electrons spend much longer time L_B/v_F near the barrier before being Andreev reflected. Because $L_B/v_F >> \tau_T$, the prolonged stay of electrons near the Nb/Ti-G barrier strongly increases the probability of AR, T_2, as compared to the CT probability, T_1, for electrons with $p_{||} \neq 0$. To sum up, all the electrons with $p_{\perp} \neq 0$ contribute into CT although its probability could be small due to the presence of a finite interface barrier. On the other hand, only a small fraction of electrons with $p_{||} \neq 0$ contribute into AR from the Nb/Ti–G interface, although the AR process probability is high. The associated transmission factors are small in the first case and much larger in the second, which explains the apparent contradiction of the observed behaviors. The calculations had been performed using solutions of the Dirac equation for graphene and the S-matrix technique extended to include superconducting correlations (Beenakker 2008, Beenakker et al. 2008).

The transmission through the graphene/superconducting metal (G/S) interface occurs in the z direction, while AR proceeds in the x direction. That broken line trajectory is different from the straight line trajectory considered in the BTK paper (Blonder et al. 1982). The *major distinction* between this geometry and the original BTK model is that there is no conventional reflection in our geometry: When an electron moving inside the graphene sheet approaches the SIC interface, it either penetrates through it with a certain probability $T = |t|^2$ (t is the transmission amplitude) or it continues to move ahead inside the same graphene sheet passing the SIC contact area without reflection. AR there happens inside Nb on the longer scale $\sim\xi_{\mathrm{BCS}}$. It occurs only for electrons which have crossed the SIC interface. The other electrons do not contribute into the AR process. Following Andreev (1964), AR in graphene is to be taken into account for the electron/hole chirality (Beenakker 2008).

The transport properties of the graphene junction are described using the scattering matrix approach. The obtained theoretical results for the excessive conductance are shown in Fig. 7.5. We restrict our attention to the electron states which are located in the vicinity of K(K′) point. The envelope wavefunctions of incoming $\breve{\Psi}_{\mathrm{i}}$ and outgoing $\breve{\Psi}_{\mathrm{o}}$ electrons are connected by the S-matrix as $\breve{\Psi}_{\mathrm{o}} = \breve{S}\breve{\Psi}_{\mathrm{i}}$. The $\breve{\Psi}_{\mathrm{i(o)}}$ states are represented by the spinors

$$\breve{\Psi}_{\mathrm{o(i)}} = \begin{pmatrix} \hat{\psi}_{\mathrm{L}}^{\mathrm{o}} \\ \hat{\psi}_{\mathrm{R}}^{o} \end{pmatrix};\ \hat{\psi}_{\mathrm{L(R)}}^{\mathrm{o(i)}} = \begin{pmatrix} u_{\mathrm{L(R)}}^{\mathrm{o(i)}} \\ v_{\mathrm{L(R)}}^{\mathrm{o(i)}} \end{pmatrix} \tag{7.25}$$

where $\hat{\psi}_{\mathrm{L(R)}}^{\mathrm{o(i)}}$ are the Nambu spinors composed of the electron ($u_{\mathrm{L(R)}}^{\mathrm{o(i)}}$) and of its time-reversed hole ($v_{\mathrm{L(R)}}^{\mathrm{o(i)}}$) states. The wavefunctions $u_{\mathrm{L(R)}}^{\mathrm{o(i)}}$

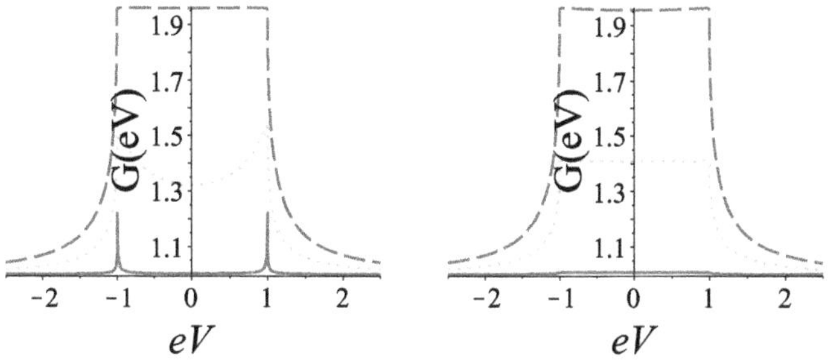

Figure 7.5 Excessive normalized conductance $G_S(eV)/G_N$ versus the source–drain bias voltage due to conventional Andreev reflection for three different interface barrier strengths: $t_1 = 0.3$ (blue dash), 0.8 (green dot), and 0.95 (red solid). Left: Decrease of $G_S(eV)/G_N$ at $|eV| < D$ is caused by the regular reflection of electrons. Right: Conductance with Andreev reflection without regular reflection.

and $v_{L(R)}^{o(i)}$ describe incoming and outgoing electrons and holes from the left (L) and right (R) sides of the scatterer which constitutes the $SIC_{1,2}$ interface. As compared to spinless electron states in conventional conductors, the electron and holes in graphene are characterized by additional quantum numbers which are two one-half pseudospins. Therefore, an electron state $u_{L(R)}^{o(i)}$ is represented by a 4D vector $u = (\phi_A, \phi_B, -\phi'_B, \phi'_A)$ where indices A and B denote two different graphene sublattices, while the prime indicates the different valley. The corresponding hole state $v_{L(R)}^{o(i)}$ is represented as $v = Tu = (\phi_A'^*, \phi_B'^*, -\phi_B^*, \phi_A^*)$ where T is the time reversal operator (Beenakker 2008, Beenakker et al. 2008). The Cooper coupling between u and v is determined from the Dirac–Bogoliubov–de Gennes equation

$$\begin{pmatrix} H - E_F & \Delta(\sigma_0 \otimes \tau_0) \\ \Delta^*(\sigma_0 \otimes \tau_0) & -(H - E_F) \end{pmatrix} \begin{pmatrix} u \\ v \end{pmatrix} = \varepsilon \begin{pmatrix} u \\ v \end{pmatrix} \tag{7.26}$$

where $H = v(\mathbf{p} \cdot \sigma) \otimes \tau_0 + U(r)\sigma_0 \otimes \tau_0$, Δ is the superconducting pair potential which couples u and its time-reversed state v, and $\otimes$ is the Kronecker product (see also Eq. 2.89). Given an $m \times n$ matrix A and a $p \times q$ matrix B, their Kronecker product $C = A \otimes B$, also called their matrix direct product, is an $(mp) \times (nq)$ matrix with elements defined by $c_{\alpha\beta} = a_{ij} b_{kl}$ where $\alpha = p(i-1)+k$ and $\beta = q(j-1)+l$.

The S-matrix has the following structure:

$$\check{S} = \begin{pmatrix} \hat{r} & \hat{t} \\ \hat{t}^{\dagger} & \hat{r} \end{pmatrix}, \hat{r} = \begin{pmatrix} r & r_{\mathrm{A}}^{*} \\ r_{\mathrm{A}} & r^{*} \end{pmatrix}, \hat{t} = \begin{pmatrix} t & t_{\mathrm{A}}^{*} \\ t_{\mathrm{A}} & t^{*} \end{pmatrix} \tag{7.27}$$

where one might notice that besides the diagonal elements which correspond to conventional reflection (r) and transmission (t) amplitudes there are also nondiagonal elements r^{A} and t^{A}. The nondiagonal elements describe the AR processes preserving the time reversal symmetry. The reflection (r) and transmission (t) amplitudes are 4×4 matrices since they also account for the one-half pseudospin flips.

Because of more complex structure of electron and hole states in graphene, AR at the superconducting metal/graphene interface is, in general, more complex than in conventional materials. Following Andreev (1964), Beenakker (2008), and Beenakker et al. (2008), in the former system, one can observe not only conventional Andreev retroreflection which takes place inside the same (conduction) band but also a specular AR which occurs at the expense of interband processes. Another distinguished feature of our setup is that the electron trajectory has a broken line shape: An electron which moves inside the graphene sheet approaches the G/S interface parallel to the interface and then it tunnels between the superconducting metal and graphene perpendicular to the G/S interface. After the tunneling, for example, from G to S, it is gradually converted into the Cooper pair and a retroreflected hole. Besides, CT and AR occur on different scales $\sim a$ for the former and $\sim \xi_{\mathrm{BCS}}$ for the latter. This situation is quite different from the scenario considered in (Blonder et al. 1982) where the geometry was actually 1D. Because in our case the trajectory is not along the straight line, the two different scales serve to separate the microscopic tunneling from AR in space. Therefore, we treat them as occurring in a sequence and as being described by two S-matrices S_{T} and S_{A}. Then two consequent scattering processes are described by a composition

$$S_{\mathrm{tot}} = S_{\mathrm{T}} \tilde{\otimes} S_{\mathrm{A}} \tag{7.28}$$

where S_{T} describes pure tunneling on a "short" scale $\sim a$ of the atomic lattice constant a perpendicular to the S/G interface, while S_{A} corresponds to a pure AR happening on a "long" scale $\sim \xi_{\mathrm{BCS}}$ inside

S. The matrix elements are reflection and transmission amplitudes r, r', and t, t'

$$
\begin{aligned}
t_{\text{tot}} &= t_2 \left(I - r_1' r_2\right)^{-1} t_1 \\
r_{\text{tot}} &= r_1 + t_1' r_2 \left(I - r_1' r_2\right)^{-1} t_1 \\
t_{\text{tot}}' &= t_1' \left(I + r_2 \left(I - r_1' r_2\right)^{-1} r_1'\right) t_2' \\
r_{\text{tot}}' &= r_2' + t_2 \left(I - r_1' r_2\right)^{-1} r_1' t_2'
\end{aligned}
\tag{7.29}
$$

Every partial S-matrix S_i (here $i = T, A$) connects the incoming and outgoing states at each ith scatterer. Reflection and transmission amplitudes in Eq. 7.29 are represented by matrices 8×8 (because for each u–v couple there are two $\pm$ orientations for the two one-half pseudospins).

For certainty, we also assume that the electron transport is coherent, that is, the CT is phase-correlated with AR. Then, we combine the successive sections coherently. Under an opposite assumption of incoherent transmission, one should use the scattering probabilities in the place of scattering amplitudes (Datta and Mclennan 1990, Datta 1992).

For the sake of simplicity, we begin with conventional AR when an incident electron is converted into a Cooper pair and a retroreflected hole. Another simplification is that we use the BTK model of the interface barrier in the form of Dirac δ function. Under the listed assumptions, the S-matrix of a pure tunneling amplitude through the interface barrier is approximated by

$$
S_{\text{T}} = \begin{pmatrix} \hat{r}_1 & \hat{t}_1 \\ \hat{t}_1^{\dagger} & \hat{r}_1' \end{pmatrix}, \hat{r} = \begin{pmatrix} r_1 & 0 \\ 0 & r_1' \end{pmatrix}, \hat{t} = \begin{pmatrix} t_1 & 0 \\ 0 & t_1' \end{pmatrix} \tag{7.30}
$$

where we have completely neglected the AR processes. On the other hand, the S-matrix of a pure AR is

$$
\breve{S} = \begin{pmatrix} \hat{r}_2 & \hat{t}_2 \\ \hat{t}_2^{\dagger} & \hat{r}_2 \end{pmatrix}, \hat{r}_2 = \begin{pmatrix} 0 & r_{\text{A}}^* \\ r_{\text{A}} & 0 \end{pmatrix}, \hat{t}_2 = \begin{pmatrix} t_2 & 0 \\ 0 & t_2' \end{pmatrix} \tag{7.31}
$$

where we omitted the conventional reflection and tunneling amplitudes. The reflection and transmission amplitudes $r_1(r_1')$, $t_1(t_1')$, and $t_2(t_2')$ entering the above Eqs. 7.30 and 7.31 connect incoming and outgoing states $u = (\phi_{\text{A}}, \phi_{\text{B}}, -\phi_{\text{B}}', \phi_{\text{A}}')$ and $v = (\phi_{\text{A}}'^*, \phi_{\text{B}}'^*, -\phi_{\text{B}}^*, \phi_{\text{A}}^*)$; therefore, they are matrices 4×4. The CT preserves the particle's chirality, hence, one can set $t_i = t \cdot \hat{1}$ where $\hat{1}$ is the 4×4 unit matrix. Conventional AR preserves the time invariance and couples the electron state u and its time-reversed hole state v with the opposite momentum (i.e., the corresponding electron and hole are located in K and K′ points). Thus, one sets $r_A^i = r_A \cdot \mathrm{T} = -r_A(\tau_2 \otimes \sigma_2)\mathrm{C}$ where T and C are the time reversal and the complex conjugation operators.

Further simplifying assumptions are as follows. Our experimental data are well understood when considering a conventional AR as in the BTK model. As an illustrative example we first assume that no proximity superconducting gap is induced in graphene which is purely in a normal state. Then, we use

$$r = Z\frac{(2i - Z)}{D_A}\left(u_0^2 - v_0^2\right); r_A = 4\frac{v_0\, u_0\, e^{i\varphi}}{D_A};$$
$$t = 2\frac{u_0\,(2 + iZ)}{D_A}e^{i\frac{\varphi}{2}}; t_A = -\frac{2iv_0\, Z}{D_A}e^{i\frac{\varphi}{2}} \tag{7.32}$$

where $D_A = 4u_0^2 + Z^2\left(u_0^2 - v_0^2\right)$, $u_0 = \sqrt{1 + \xi_\varepsilon/\varepsilon}/\sqrt{2}$, $v_0 = \sqrt{1 - \xi_\varepsilon/\varepsilon}/\sqrt{2}$. One also obtains $R = |r|^2 = Z^2/(Z^2 + 4)$; $T = |t|^2 = 1 - R = 1/((Z/2)^2 + 1)$, which, for example, for $T = 10^{-6}$ gives $Z = 2 \times 10^2$. The interface barrier strength Z is expressed via the interface barrier transparency T as $Z = 2\sqrt{(1 - T)/T}$. In the simplest approximation of an ideal interface barrier $Z = 0$ one writes

$$t_2 = \frac{1}{u_0}e^{i\varphi/2}; r_A = \frac{v_0}{u_0}e^{i\varphi/2} \tag{7.33}$$

The amplitude of composite AR is then obtained as

$$r_A^{1-2} = \frac{e^{i\phi}t_1^2 r_A^*}{1 - r_1^2\,|r_A|^2} \simeq t_1^2 r_A^* << 1 \tag{7.34}$$

while the composite conventional reflection amplitude is

$$r = \frac{r_1\left(-t_1^2\,|r_A|^2 + r_1^2\,|r_A|^2 - 1\right)}{r_1^2\,|r_A|^2 - 1} = r_1\frac{1 + \left(t_1^2 - r_1^2\right)\,|r_A|^2}{1 - r_1^2\,|r_A|^2} \simeq r_1 \tag{7.35}$$

The regular transmission amplitude is

$$t = \frac{t_1 t_2}{1 - r_1^2\,|r_A|^2} \tag{7.36}$$

The above formulas allow a simple illustrative explanation of the obtained experimental data. The corresponding excessive conductance is shown in Fig. 7.5.

7.6 Amplitude of the Composite Andreev Reflection

Electric conductance σ of the SIS' contact is shown in Fig. 7.3 for three different interface transmission factors $T_2 = 0.9, 0.6$, and 0.1 of AR. An excessive conductance σ is visible at voltages

$$-\frac{\Delta_{\text{Nb}} + \Delta_{\text{G}}}{e} < V < \frac{\Delta_{\text{Nb}} + \Delta_{\text{G}}}{e} \quad (7.37)$$

The mentioned maximum of conductance is occurring when $1 > T_2 > 0.5$. Excessive conductance, which is observed in our experiment, originates from AR. The process occurs in the bias voltage interval related with four SIS' junctions connected in series. Two of the junctions are formed at the Nb/graphene 3D/2D interfaces, whereas two other SIS' junctions are formed inside the graphene layer due to presence of a depleted sublayer inside of it, as shown on the lower panel in Fig. 7.3. Thus, the four SIS' junctions connected in sequence provide the bias voltage interval

$$-4\frac{\Delta_{\text{Nb}} + \Delta_{\text{G}}}{e} < V < 4\frac{\Delta_{\text{Nb}} + \Delta_{\text{G}}}{e} \quad (7.38)$$

where $(\Delta_{\text{Nb}} + \Delta_{\text{G}})/e = 1.9$ mV for the excess Andreev conductance to happen. Similar phenomena were reported recently for the Nb/Pd/CNT/Pd/Nb junctions (Zhang et al. 2006, Tselev et al. 2009, Yang et al. 2012).

It should be noted that the interfacial phenomena between the superconductor and graphene (and graphite as well) are not well studied, both experimentally and theoretically. Clearly, more experimental work is needed to study this system.

7.7 Andreev Specular Reflection versus Retroflection

According to the original quasiclassical model suggested by (Andreev 1964) the conversion of electron into a Cooper hole at the N/S interface is accompanied by a retroreflection of a hole. It means that the Andreev-reflected hole propagates in the reverse direction along the path of the incident electron as sketched in the left panel of Fig. 7.6. Besides, all components of the hole's velocity flip their sign. The story might be remarkably different for the undoped graphene

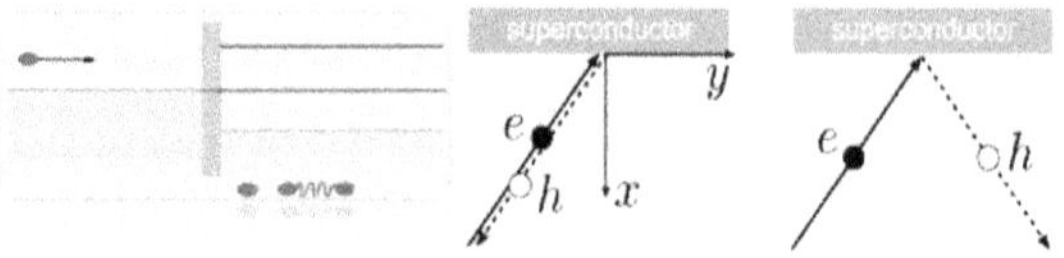

Figure 7.6 Left: Schematics of the energy diagram of Andreev reflection at the N/S interface, where blue and red circles denote electrons and holes, respectively. Middle: Sketch of the Andreev retroreflection at the N/S interface between a normal metal and the superconductor. Arrows indicate direction of the velocity, and solid or dashed lines distinguish whether the particle is a negatively charged electron (e) or a positively charged hole (h). Right: Specular Andreev reflection at the interface between undoped graphene and a superconductor (Beenakker 2006, Beenakker 2008).

where AR might be specular as illustrated in the right panel. In the latter case, only the component perpendicular to the interface changes sign (Beenakker 2006, Beenakker 2008). Here we discuss interesting peculiarities of the intraband and interband ARs. The difference between the two aforementioned scenarios is understood from the dispersion relation in graphene.

We will see why the intraband AR leads to retroreflection, while interband AR causes specular reflection. To model the intraband processes, one can use the linear dispersion relation for the elections in graphene in terms of the excitation energy $\varepsilon = |E - E_{\mathrm{F}}|$ as

$$\varepsilon = \left| E_{\mathrm{F}} \pm \hbar v_{\mathrm{F}} \sqrt{\delta k_x^2 + \delta k_x^2} \right| \tag{7.39}$$

Here the sign $\pm$ indicates that the excitation is being attributed either to the conduction or to the valence band. We select the coordinates where the interface with the superconductor S is positioned at $x = 0$ and the electron comes to the interface from $x>0$. The y component of the electron momentum δk_y which is parallel to the N/S interface and ε both are conserved during the reflection. Thus, the reflected wave is a superposition of the four waves characterized by the δk_x values that solve Eq. 7.39 at given δk_y and ε. The expectation value of the electron velocity in the x direction v_x is obtained by computing the derivative $\hbar^{-1} d\varepsilon/d\,(\delta k_x)$. Thus, the reflected wave is composed by the two waves with δk_{x} values characterized by a positive slope. Those two δk_{x} values correspond either to an electron or to a hole excitation. There are

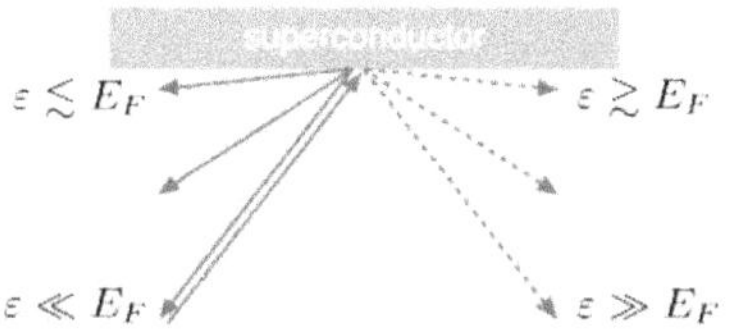

Figure 7.7 Various possibilities for Andreev reflection versus excitation energies ε relative to the Fermi energy E_F, at a fixed angle of incidence. The trajectories of an incident electron are shown as red, and the Andreev-reflected hole are blue. The hole belongs to the conduction band for energies $\varepsilon < E_F$, as shown by solid lines, while it is in the valence band for energies $\varepsilon > E_F$. the hole is in the (dashed lines). When the energy ε increases, the reflected trajectories rotate clockwise. They jump by 180° as soon as $\varepsilon < E_F$ (Beenakker 2006, Beenakker 2008).

two possibilities for a reflected hole to be ascribed either to an empty state in the conduction band (for $\varepsilon < E_F$) or an empty state in the valence band ($\varepsilon > E_F$). Those two possibilities are sketched in Fig. 7.7 (see also the energy diagram in Fig. 7.8). If the reflected hole belongs to the conduction band, its wavevector is directed opposite to the velocity. Therefore, v_y changes sign as well as v_x does, which is regarded as the retroreflection. A different story occurs if the reflected hole belongs to the valence band. In the latter case, the parallel component of the velocity v_y remains the same whereas the normal component v_x flips the sign which is now quoted as specular reflection (see the diagram in Fig. 7.8). In other words, the hole moves in the same direction as is pointed by its wavevector. One initializes the transition between the two different regimes of retroreflection and specular reflection. When the electron energy ε increases beyond E_F, the reflection angle α_{out} in respect to the normal increases above the angle of incidence α_{in}. However, when $\varepsilon < E_F$, the reflection angle switches from +90° to −90° and finally approaches α_{in} as soon as $\varepsilon > E_F$, as schematically shown in Fig. 7.7.

A remarkable feature of specular AR is that it causes charge-neutral propagating modes along an undoped graphene channel with superconducting boundaries (Titov et al. 2007), as shown in Fig. 7.10. The consequence of specular AR makes it different from conventional Andreev retroreflection which, instead, creates

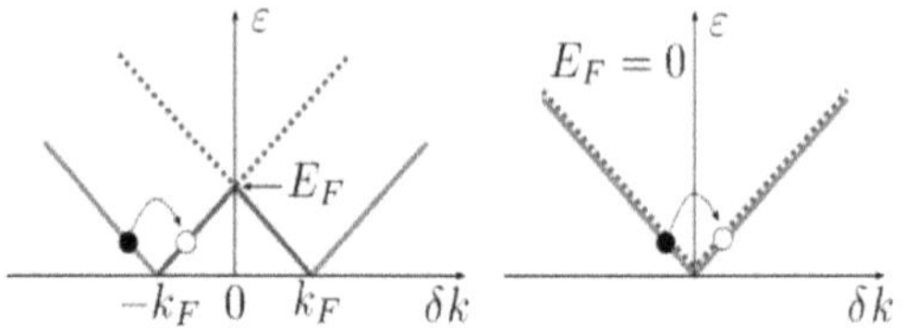

Figure 7.8 The difference between conventional Andreev reflection (left panel) and the specular Andreev reflection in graphene shown in the right panel. Conventional Andreev reflection happens when the electron and the hole both lie in the conduction band. If a conduction band electron is converted into a valence band hole, the Andreev reflection is specular, as shown in the right panel. The electron branch of the excitation spectrum is shown by red lines (it corresponds to filled states above the Fermi level, from one valley). The hole branch of excitations is shown by blue lines (it is related to empty states below the Fermi level, from the other valley). The sketch corresponds to two values of the Fermi energy $E_F = \hbar v k_F$, when the incidence is normal, ($\delta k_y = 0, \delta k_x = 0$). The conduction and valence bands are marked by solid and dotted lines, respectively. The arrows show the electron–hole conversion upon reflection at a superconductor (Beenakker 2006, Beenakker 2008).

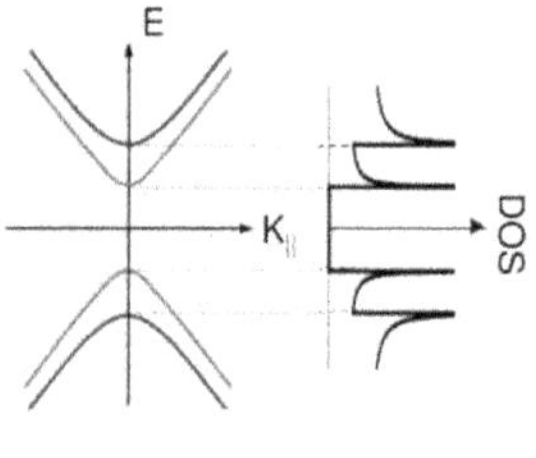

Figure 7.9 Van Hove singularities in the electron density of states.

bound states known as Andreev levels (Andreev 1964, Gogadze and Kulik 1970, Kulik 1970). The propagating "Andreev modes" may be used to carry a charge-neutral spin current along the channel (Greenbaum et al. 2007). That ability emerges from a phase-sensitive contribution of the "Andreev modes" to the thermal conductance along the graphene channel. The mentioned phase-coherent phenomena might serve as a firm signature of specular AR at the graphene/superconducting interface.

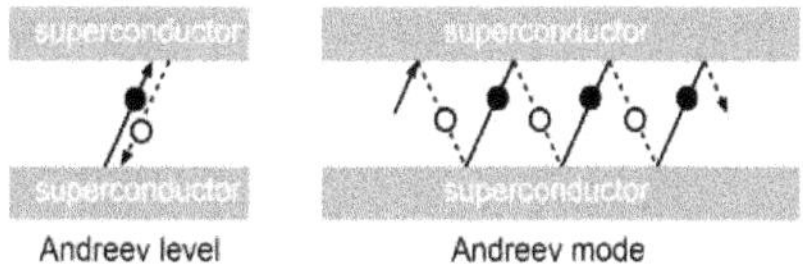

Figure 7.10 Interplay between the retroreflection and the specular Andreev reflection in a graphene channel with superconducting boundaries. The transition between the two different types of Andreev reflection transforms the localized level (left) to a propagating mode (right). The propagating mode affects the thermal transport along the channel but causes no contribution to the electrical transport (Titov et al. 2007).

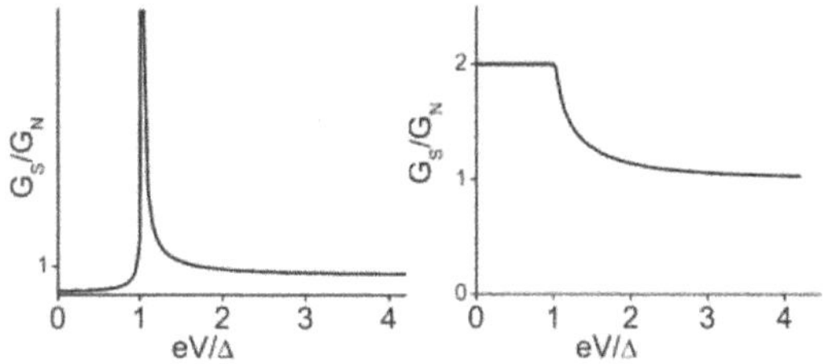

Figure 7.11 Normalized conductance as a function of the voltage applied across the interface calculated using the BTK model (Blonder et al. 1981) at zero temperature for high (a) and low (b) interface transparency. $Z = 0$ in (a), and $Z = 1.2$ in (b), where Z is the interface barrier height defined by Blonder et al. 1982, Klapwijk et al. (1982).

7.8 Van Hove Singularities and Superconductivity in Carbon Nanotubes and Graphene Stripes

A possibility of intrinsic superconductivity has been studied experimentally in carbon nanotube (CNT) arrays, ropes, in multiwalled CNTs (CNTs nested into each other), and in films of boron-doped CNTs. The obtained results show a wide range of superconducting critical temperatures, from 500 mK to 12 K. A probable reason for the large spread of critical temperature is that in samples containing multiple nanotubes there might be tubes with different chirality which have distinct electronic properties. In this section, we address samples fabricated of isolated single-walled CNTs to distinguish the relevance of superconductivity to their density of states. In experiments, one measures anomalous transport features which suggest appearance of a hypothetic superconductivity in the nanotubes as soon as the gate voltage shifts the Fermi energy into Van Hove singularities (VHSs) of the electronic density of states whose energy diagram is shown schematically in Fig. 7.9. Then, the electron transport is influenced by the proximity effect at the interface between the normal electrode and the superconducting nanotube whereas the superconducting properties can be tuned by chemical doping or by applying an electric field. The electron transport properties are sensitive to positions of the VHSs. The VHSs indicate the onset of new subbands in the density of states of CNTs. The influence of VHS on the conductance of nanotubes is not well understood yet, because they are obscured in electrical transport measurements since VHSs are not typically pronounced as notable features. In this section we describe experimental measurements of conductance of CNTs with clearly revealed VHSs by means of interference patterns of the electronic wavefunctions. The data indicate clear presence of both, a sharp increase of quantum capacitance, and a sharp reduction of electronic energy level spacing, which is consistent with an upsurge of density of states. Below critical temperature of about 30 K, at the positions of VHSs on the $I-V$ curves, there is an anomalous increase of metal/nanotube interface conductance posing as a sharp reduction of electronic energy level spacing, consistent with an upsurge of density of states.

A single-walled CNT represents an example of a 1D conductor which consists of a layer of carbon atoms wrapped in the shape of a cylinder, with diameter of a few nanometers and length of several micrometers. Depending on the chirality, CNTs can be either semiconducting or metallic. However, a possibility for an occurrence of superconductivity in this system with a very small number of conducting channels must also be explored. It was the purpose of several experimental works that have been done on multiple and on single nanotubes. Besides, we shall discuss advantages and disadvantages of these approaches as well as the general challenges that are involved in these experiments. In the following sections, we will describe the physics of superconducting proximity effect in CNT devices. We begin with a simple CNT junction which involves CNT/superconductor interfaces. The superconductivity is induced from a superconducting electrode into a CNT that is not superconducting. The transport properties of such system are similar as in the setup where a CNT is attached to normal electrodes, if Cooper pairs are present in the CNT due to intrinsic superconductivity. The obtained physical picture also serves to interpret transport anomalies in the CNT devices at certain values of gate voltage. Thus, we model the superconductivity arising in CNTs as soon as the gate voltage shifts from the Fermi energy into VHSs of the density of states which pop up at the onset of each 1D subband.

7.8.1 *Superconductivity in Carbon Nanotubes: Experimental Challenges*

CNTs with their own intrinsic superconductivity have been a subject of experiments involving electrical transport or magnetic measurements. Most of observed anomalies pointing to intrinsic superconductivity had been detected on samples made of multiple tubes, including multiwalled CNTs (Takesue et al. 2006), arrays or film of single-walled nanotubes (Tang et al. 2001, Murata et al. 2008, Matsudaira et al. 2010), and suspended ropes of single-walled CNTs. The range of reported critical temperatures for CNT samples is quite large spreading between 0.5 K to 15 K (Tang et al. 2001, Murata et al. 2008, Matsudaira et al. 2010). One can understand the large spread of critical temperatures by taking into account the problems of controlling the structure of CNTs. Furthermore, most samples were made of a mixture of CNTs with different diameter and chirality. Because of variations of fundamental physical properties like density of states, and vibrational modes, different nanotubes will have different electronic properties. Consequently, it causes wide variations of critical temperatures. The aforementioned ambiguity has motivated a more systematic search for superconductivity in CNTs. One option is to look for superconducting anomalies in one isolated single-walled CNT. It provides an opportunity of focusing the attention on an individual CNT and obtaining the density of states, as well as of lattice vibrational modes. There are two serious obstacles which emerge on that path. One issue is about thermal and quantum fluctuations which significantly broaden the phase transition and weaken the superconducting properties

(Bezryadin et al. 2000) as soon as the sample size shrinks to the nanowire dimensions. Another problem emerges when attaching normal metal electrodes to each end of the graphene stripe or nanotube to supply an electrical current through it. In the latter case, we have to specify how the current is propagating through the electrode/CNT interface if there are Cooper pairs in nanotubes and electrons–hole pairs in the electrodes. We analyze the phenomenon using classic picture as a starting point. The simplest setup is represented by CNT in the normal state which is connected to a superconducting electrode (Zhang et al. 2006, Tselev et al. 2009). It allows interpreting of the proximity effect in terms of AR. Then, the same method will be extended to a superconducting nanotube connected to normal electrodes, but already assuming that Cooper pairs reside in the nanotube whereas the normal excitations (i.e., electrons and holes) are located in the metal electrodes.

7.8.2 *Andreev Reflection and CNT Point Contact Spectroscopy*

AR is an elementary quantum mechanical process which is responsible for conversion of quasiparticle excitations into the Cooper pairs. Therefore, it governs the electric current flow through the interface between the superconductor and normal metal. If the electron energy is less than the superconducting energy gap, the conventional transmission of conducting electrons is prohibited because there are no states available in the superconductor. A finite electron current can flow only if an incident electron is getting paired with another electron having opposite momentum and spin, to form a Cooper pair in the superconductor. The described scenario is also represented like an incident electron being reflected back as a hole, a phenomenon which is named AR (Andreev 1964). All the quasiparticles involved in AR, that is, the electrons, holes, and Cooper pairs participate in a phase-coherent process whose spatial scale is of order of the BCS coherence length $\xi \simeq \hbar v_F/\pi\Delta$ (here Δ is the superconducting energy gap). The phase coherence is extended toward both sides of the normal metal/superconductor interface. Due to conservation of momentum, spin, and electric charge, a retroreflected hole propagates backward along the same

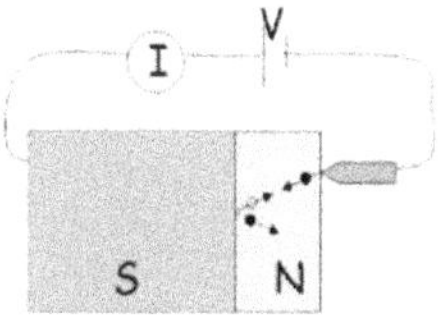

Figure 7.12 The electrodes shaped as narrow points can effectively measure the current of retroreflected holes. Electrons undergoing normal reflection at the S/N interface will be deflected away from the point contact and will not contribute to the current. Electrons and holes are represented as black and empty circles, respectively (Tselev et al. 2009).

trajectory as the incoming electron does. During the AR process, an incident electron with energy $\varepsilon < \Delta$ and the electric charge $-e$ induces a Cooper pair having a twice larger charge $-2e$ which physically means that the electric current becomes twice larger for those electron energies. According to Section 7.3, it corresponds to a twofold conductance increase when the bias voltage $-\Delta/e < V < \Delta/e$ when compared to the conductance of the normal state G_{N} (the conductance measured through the same interface when the superconducting material is in the normal state). In experiments, the excess current due to AR is readily measured using narrow tip-shaped electrodes which form point contacts, as sketched in Fig. 7.12. Ideally, the point contact diameter is required to be much smaller compared to the thickness of the normal layer. It allows us to achieve focusing of the retroreflected holes into the point contact, whereas the electrons undergoing normal reflection will be deflected away and will not contribute to the current. The smallest point contacts are obtained using the CNTs corresponding to ideal geometry used for probing the superconducting proximity effect with normal layers as thin as a few nanometers (Tselev et al. 2009).

7.8.3 *Proximity Effect at the Carbon Nanotube/Superconductor Interface*

AR can be maximized at the highly transparent interfaces. Therefore, finding a superconducting metal which makes a good contact with CNTs represents a critical issue for conducting the corresponding experiment. On the one hand, it is quite challenging to get a

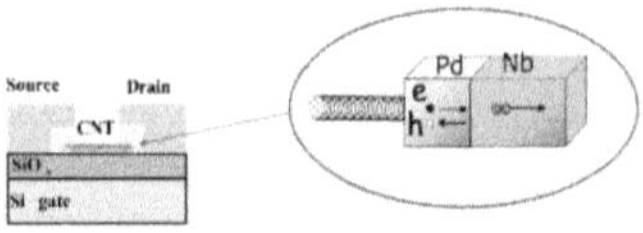

Figure 7.13 Schematics of a carbon nanotube field-effect transistor with superconducting electrodes. A thin layer of Pd is used for high contact transparency with the carbon nanotube. Adapted from Tselev et al. (2009).

superconducting material for the electrodes with a good contact with the CNT. On the other hand, there are several normal metals, for example, palladium, yielding highly transparent interfaces with CNTs (Javey et al. 2003, Mann et al. 2003). It represents a good opportunity for the fabrication of proximity effect nanodevices where a thin layer of one of those materials is placed between the superconductor and the nanotube (Kasumov et al. 1999, Jarillo-Herrero et al. 2006, Cleuziou et al. 2006). An example of such a proximity effect device is sketched in the enlargement of Fig. 7.13, where a contact at one end of the nanotube with the specific materials is utilized. The samples used in the discussed experiments are summarized in Table 1. All the devices are characterized by a highly transparent interface which is located between the metal which is Pd and the CNT. Applying the bias voltage between the source and drain electrodes, the whole CNT device performs as a nanoscale point contact which distinguishes the excess conductance at the Pd/Nb interface (Zhang et al. 2006, Tselev et al. 2009). The CNTs have many additional unique features which can be exploited for obtaining additional benefits to extend capabilities of the point contact spectroscopy. One notable feature, as compared to traditional point contacts, is that the normal state conductance G_N of CNTs can be controlled merely applying the gate voltage to the conducting (doped) silicon substrate. If the CNT is semiconducting, one can switch the conductance on and off merely varying the gate voltage and shifting the nanotube Fermi energy from the gap to the valence band ($V_G < 0$), or otherwise from the gap to the conduction band ($V_G > 0$) (Tans et al. 1998), as shown in Fig. 7.14. It corresponds to operating of the CNTs as field-effect transistors (FETs). The operation was accomplished by many research groups (Anatram and L'eonard 2006). In particular, at low temperatures, one might

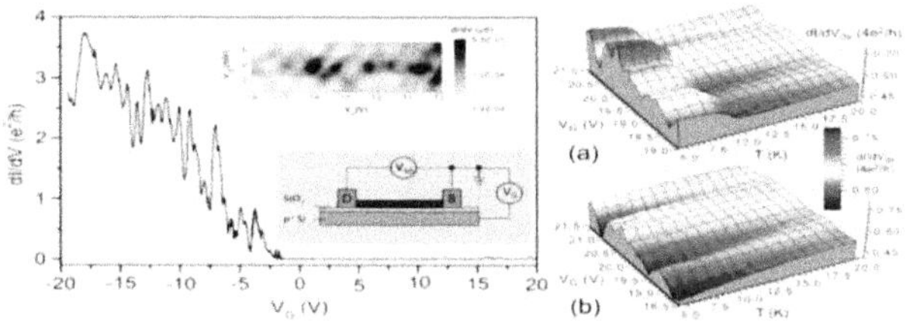

Figure 7.14 Left: Zero-bias differential conductance versus gate voltage (sample 4). The lower inset shows a schematic of a carbon nanotube field-effect transistor. The 2D plot is a gray-scale map of differential conductance as a function of gate voltage and source–drain bias voltage. Right: Differential conductance as a function of gate voltage and temperature (sample 1). In (b) superconductivity in the Nb electrode is suppressed by a magnetic field (**B** = 3 T) applied in a direction perpendicular to the substrate. Adapted from Tselev et al. (2009).

observe interference effects of the electronic wavefunctions which emerge as irregular oscillations of conductance versus the gate voltage and the source–drain voltage (see Fig. 7.14 and the energy diagram in Fig. 7.15). Such interference effects result from scattering of electrons either by defects in the nanotube (Kong et al. 2001) or by the contacts (Liang et al. 2008). A typical interference pattern is shown in Fig. 7.14 while the energy diagram is depicted in Fig. 7.15. When the electron propagation inside the nanotubes is ballistic it behaves as a Fabry–Perot resonator for electrons (Liang et al. 2008). The period of inference pattern is deduced from the resonant condition that the length L of the nanotube section enclosed between the source and drain electrodes matches an integer number of wavelengths. Thus, the characteristic energy scale actually determines the period of the Fabry–Perot interference pattern which varies with the nanotube length L as $\Delta E_{\mathrm{FP}} = hv_{\mathrm{F}}/L$, where v_{F} is the Fermi velocity in the nanotube (Liang et al. 2008). In Fig. 7.14 (right) we plot the experimental data of conductance versus the gate voltage and temperature, in a region of gate voltage where the FET is conducting. The aforementioned electron interference is pronounced as irregular oscillations of electric conductance versus the gate voltage and also depend on temperature.

If the temperature of the sample is below the superconducting transition of the Nb electrode, which is approximately 9 K, one can

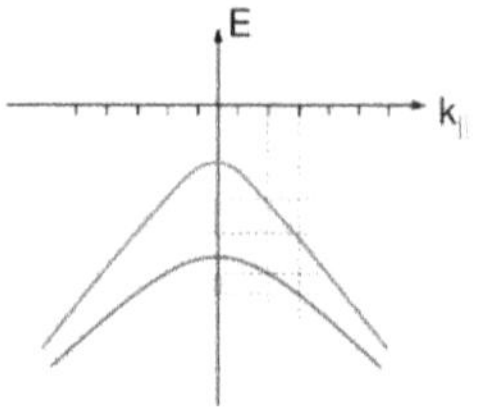

Figure 7.15 The same wavevector interval $\Delta k_{\parallel}$ corresponds to two different energy intervals ΔE_1 and ΔE_2 in the two VHS subbands, respectively. Adapted from Tselev et al. (2009).

notice that the electric conductance increases for all the values of gate voltage (see Fig. 7.14, right, a), suggesting that conductance of the point contact in the superconducting state, G_S, always exceeds the corresponding normal state value of conductance G_N. In Fig. 7.14 (right, b) we show the normal conductance G_N, measured with a suppressing of superconductivity in the Nb electrodes by a magnetic field in the temperature range of our experiment. At certain values of gate voltage and temperature, the conductance is measured versus the source–drain voltage, V_{ds}. Typical curves of conductance G_S and G_N versus V_{ds}, which have been measured at certain values of temperature and gate voltage are shown in Fig. 7.16. Although, at zero bias, the curve G_N versus V_{ds} shows either a peak or a dip, the ratio G_S/G_N always indicates a zero-bias conductance (ZBC) peak (see inset in Fig. 7.16). Such behavior is understood in terms of simple BTK model formulated in Section 7.3 which we have adapted to our devices—see Zhang et al. (2006) and Tselev et al. (2009). One should take into account that the superconductivity in the normal Pd layer was induced due to the proximity effect. It allows modeling the contact roughly as a single interface between the CNT and a superconducting electrode. The experimental data, however, suggest that the proximity effect must be accounted for properly by considering the two interfaces. The sufficient model includes both the interfaces, CNT/Pd and Pd/Nb as shown in Fig. 7.4 characterized by corresponding interface transparencies T_1 and T_2. Thus, the CNT actually acts as a point contact and detects excess current due to AR. AR occurs at the Pd/Nb interface. The outlined model is utilized for fitting the experimental data (see inset in Fig. 7.17) which permits

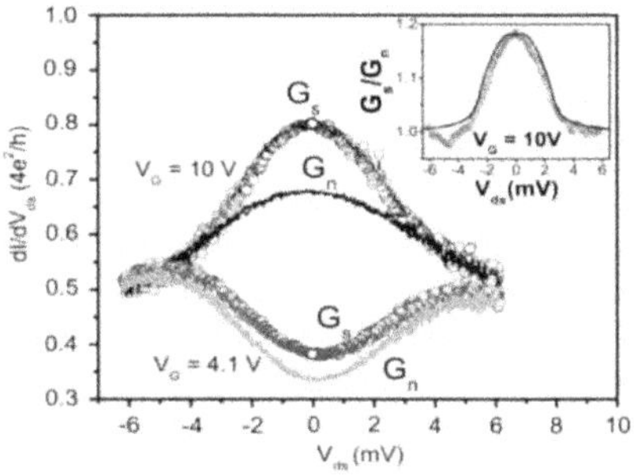

Figure 7.16 Differential conductance as a function of gate voltage and temperature (sample 1). From Tselev et al. (2009). For the G_N curves, the superconductivity in the Nb electrode is suppressed by a magnetic field (B $=$ 3 T) applied in a direction perpendicular to the substrate. Inset: Ratio of superconducting state and normal state differential conductances ($G_S = G_N$). The line is a fit to the experimental data (dots), using the model described in the text and Tselev et al. (2009), with interface transparencies $T_1 = 0.85$ and $T_2 = 0.8$. Adapted from Zhang et al. (2006) and Tselev et al. (2009).

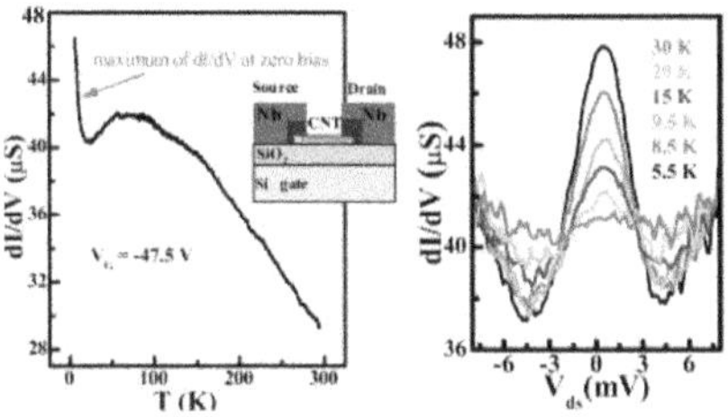

Figure 7.17 Left: Zero-bias differential conductance as a function of temperature (sample 2). The conductance upturn at low temperature only occurs in a narrow range ($\Delta V_G = 1$ V) of gate voltage around $V_G = -47.5$ V. Right: Temperature dependence of the zero-bias conductance peak measured at $V_G = -47.5$ V (sample 2). Adapted from Zhang et al. (2006) and Tselev et al. (2009).

an extracting of transparencies selectively for the two interfaces. We should emphasize that the CNT performs as a nanoscale point contact which is fully capable of detecting AR and of studying the proximity effect with layers of normal metal as thin as 5 nm.

7.8.4 *Peak of Zero-Bias Conductance as a Signature of Intrinsic Superconductivity in Carbon Nanotubes with Normal Electrodes*

As it follows from the former Section 7.8.3, the presence of the ZBC peak in the G_S/G_N curves versus V_{ds} for nanotubes with superconducting electrodes suggests a contribution of AR between Cooper pairs in the Nb electrodes and phase coherent electrons and holes in the Pd. The ZBC peak appears at all values of gate voltage and at temperatures below the critical temperature of the Nb electrodes. Besides, in a wider gate voltage range, −50 V $< V_G <$50 V, there are new interesting features emerging at large negative values and in a narrow interval ($\Delta V_G = 1$ V) of gate voltage. One feature is that the zero-bias differential conductance experiences dramatic upturn at low temperature. Furthermore, there is a ZBC peak, which is strongly temperature dependent. The mentioned features are illustrated in Fig. 7.18. In this case, the CNT section between the electrodes is quite long ($>$1 μm) for FET samples like shown in Fig. 7.18 whereas the changes of conductance due to Fabry–Perot resonances are tiny. Typically, as one changes the gate voltage, the conductance remains steady, and there is no upturn at low temperature. The dependence of the pattern versus bias voltage shows a dip at zero bias. When the temperature is reduced, the conductance dip at zero-bias voltage diminishes along with decreasing the temperature. The anomalous behavior represented in Fig. 7.18 only occurs in the narrow range of gate voltage, indicated above. A tricky point which follows from Fig. 7.18 is that the ZBC peak is not related to the superconductivity in the metal electrodes since it persists well above the critical temperature of Nb. Besides, similar anomalies are also taking place in samples made with Pd electrodes. It is illustrated in Fig. 7.19 where we show similar nonmonotonic temperature dependence for the conductance as a function of temperature for a sample with Pd electrodes. One can see that in addition to the upturn, there is a strongly temperature-dependent ZBC peak (Zhang et al. 2006). In principle, one can consider a few other physical mechanisms that might cause a ZBC peak, including Kondo effect, resonant scattering and Fano resonance. However, none of the listed

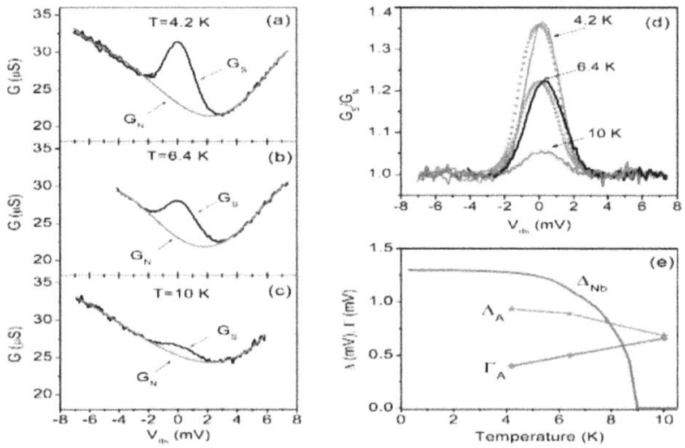

Figure 7.18 (a–c) Differential conductance versus the bias voltage at different temperatures (sample 3). The fitting of the high-bias data for G_N is shown by red curves. (d) Ratio G_S/G_N versus the source–drain bias voltage. The experimental data obtained from the curves in (a–c) are presented by solid lines. The fits to the Andreev reflection model are shown by dotted lines. (e) Temperature dependence of the fitting parameters Δ_A and Γ_A used in the model. For comparison we also plot the temperature dependence of a BCS superconductor with the critical temperature of Nb. Adapted from Zhang et al. (2006) and Tselev et al. (2009).

scenarios could successfully describe the transport anomalies in our samples (Zhang et al. 2006). The most probable explanation of the ZBC peak is the intrinsic superconductivity in the CNT which leads to AR at the interfaces between the nanotube and the Pd electrodes. An in-depth analysis of the AR experimental data is based on the measurement of ratio G_S/G_N versus V_{ds} at various temperatures. Similarly, as in the previous experiment, the conductance G depends on temperature and V_{ds}. It means that the G_N curves must be measured at various temperatures and V_{ds}. Using the superconducting Nb electrode to study AR, one applies a finite magnetic field higher than the Nb critical field to obtain background data. Thus, the superconductivity in the Nb electrode is suppressed at all temperatures (see Fig. 7.16). A more interesting situation takes place, when the superconductivity develops in CNTs. In the latter case, even applying a sufficiently large magnetic field 7 T is not enough for a full suppression of the hypothetic superconductivity which presumably occurs in the CNTs. According to the experiment

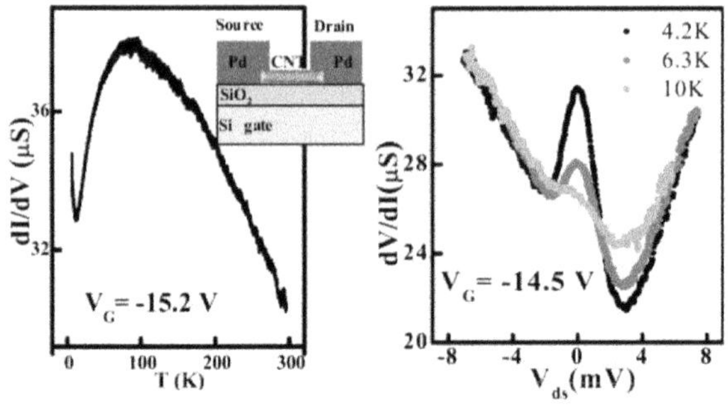

Figure 7.19 Left: Differential conductance as a function of temperature (sample 3). The conductance upturn at low temperature only occurs in a narrow range ($\Delta V_G = 1$ V) of gate voltage around ($V_G = -15.2$ V). Right: Temperature dependence of the zero-bias conductance peak measured at $V_G = -14.5$ V (sample 3). Adapted from Zhang et al. (2006) and Tselev et al. (2009).

(Zhang et al. 2006, Tselev et al. 2009), the ZBC peak is only slightly lowered. For the above reasons, a direct measurement of the curve G_N versus V_{ds} to obtain the ratio G_S/G_N versus V_{ds} represents a challenge. Instead, one can focus attention on the conductance at zero-bias, whereas the AR model is implemented to fit the ZBC versus temperature. Such a fit serves to provide the data about temperature dependence of the hypothetical superconducting gap in the nanotube. A more flexible and advanced approach that gives a better stringent test of the AR model is as follows. We use the high-bias data to extrapolate the curve G_N versus V_{ds} at a fixed temperature. The extrapolated curves G_N versus V_{ds} are then utilized to analyze the AR model.

Plots of G_N versus V_{ds} as well as extrapolated G_N versus V_{ds} at three different temperatures for the raw experimental data in Fig. 7.19 are shown in Fig. 7.18. One can see that the ZBC peak at 4.2 K in the ratio G_S/G_N versus V_{ds} curve shows that the magnitude of G_S is as much as 35% higher than G_N. The data were analyzed utilizing model similar to the one used in Zhang et al. (2006). In the model one assumes that resistance of sample is dominated by the contacts, therefore, one safely neglects contribution to the resistance emerging from thermal or quantum fluctuations (Bezryadin et al. 2000, Lau et al. 2001). For each interface, which is separating the metal electrode and CNT, the interface resistance (and corresponding conductance) is computed using a modified BTK model, with a temperature-dependent superconducting energy gap $\Delta_A(T)$. Another important parameter is the lifetime broadening $\Gamma_A(T)$ due to the inelastic scattering processes. The latter quantity is modeled by adding an imaginary part to the electron energy as $E_A = E + i\Gamma_A(T)$ (Dynes et al. 1984, Daghero et al. 2011). The quantities $\Delta_A(T)$ and $\Gamma_A(T)$ are then obtained from fitting the model to the experimental curves, (see Fig. 7.18d). The described procedure provides the temperature dependence of $\Delta_A(T)$ and $\Gamma_A(T)$ which is represented in Fig. 7.18e. The temperature dependence of $\Delta_A(T)$ is remarkably different from the conventional BCS curve (we also show the temperature dependence of the Nb gap for comparison). Such a nontrivial deviation from the BCS dependence behavior was experimentally observed by Tang et al. (2001) an ascribed to the 1D nature of CNTs. The values of the broadening parameter $\Gamma_A(T)$ and temperature dependence $\Delta_A(T)$ are consistent with electron–phonon inelastic scattering.

7.8.5 *Probe of the Density of States and Its Van Hove Singularities*

There are several reports of superconductivity occurring in multiple nanotubes, with critical temperatures ranging between 0.5 K (Kociak et al. 2001) and 15 K (Tang et al. 2001). However, in most of the known experiments the authors had not utilized an important advantage of the CNT FETs. However, they omitted a capability of CNT FETs to adjust the Fermi level. It allows to match the electron spectral singularities when the gate voltage is applied. Recently, such experiments have been reported by Zhang et al. (2006) and Tselev et al. (2009). One novelty in the experiments (Zhang et al. 2006, Tselev et al. 2009) is that the CNT FET devices have been fabricated with isolated single-walled CNTs. Another advantage is that a zero-bias anomaly (ZBA) occurs only in a narrow range and at very large negative values of gate voltage. It allows the gate voltage to shift the Fermi energy of the electrons in the nanotube and, for sufficiently large gate voltages, the Fermi energy will be shifted to values of energy corresponding to VHSs in the electronic density of states of the nanotube. The idea of the experiment is related to a well-known study of A-15 superconductors (Izyumov and Kurmaev 1974). Izyumov and Kurmaev (1974) noticed the correlation between the density of states and the critical temperature of superconductors. Besides, the VHSs were then proposed as an explanation for the properties of high-TC cuprates (Dzyaloshinskii 1987a, 1987b). In CNTs, the Fermi energy can be adjusted into VHSs either by chemical doping (Murata et al. 2008, Matsudaira et al. 2010) or by electrostatic doping with a gate voltage applied (Zhang et al. 2006). In the experiments (Zhang et al. 2006, Tselev et al. 2009), the authors used the electrostatic doping. Owing to sharp singularities in the electron density of states, the 1D nanotube becomes superconducting only in a narrow range of gate voltage. According to Mintmire and White (1998a, 1998b), the diameter D of single-walled CNTs uniquely determines energy spacing of the VHSs. For semiconducting nanotubes, the lowest singularity whose energy lies right below the Fermi energy E_{F0} of the undoped nanotube, corresponds to the upper boundary of the valence band and it is spaced from E_{F0} by

$$\Delta E = \frac{|V_{pp\pi}|\, d}{D} = \frac{E_{\mathrm{G}}}{2} \tag{7.40}$$

where E_{G} is the energy gap, $V_{pp\pi}$ is the nearest-neighbor interaction and d is the carbon–carbon bond distance. In the experiments (Zhang et al. 2006, Tselev et al. 2009), where the CNT is part of the FET setup, the lowest singularity can approximately coincide with the threshold voltage in the conductance versus V_{G} curves. The zero-bias anomalies at the threshold were not been measured, probably because of the presence of the Schottky barriers at the Pd/CNT interfaces which causes the low transparency.

In principle, the shape of the Schottky barrier depends on the band bending when the Fermi energy is moved away from the threshold by the applied gate voltage. The process results in an increase of interface transparency.

The spacing between the second singularity and the lowest singularity (at the threshold gate voltage) is also equal to ΔE in Eq. 7.40. An expected result is that the Fermi energy is adjusted to the next VHS when the gate voltage is changed from the threshold voltage by an amount $\Delta V_{\mathrm{G}} = \Delta E/(\alpha \cdot e)$ where α is the gate efficiency factor. Then, one can verify whether the location of VHSs of the density of states coincides with the values of gate voltages corresponding to the occurrence of zero-bias anomalies. Zhang et al. (2006) and Tselev et al. (2009) determined the diameters of three nanotube samples indicating zero-bias anomalies. It allowed to estimate the energy spacing between the lowest two VHSs in the density of states for each of them (Zhang et al. 2006). A quantitative comparison with the experimental data is made for $V_{pp\pi} = 2.5$ eV (Odom et al. 1998) and $d = 0.14$ nm. Another important quantity is the gate efficiency α which corresponds to the gate voltage necessary to shift the Fermi energy by ΔE. The last procedure requires experimental obtaining of clear periodic patterns of conductance versus V_{ds} and V_{G} for every device which unfortunately is not always possible. For Fabry–Perot patterns, the efficiency α can be attained from the ratio of voltages $V_{\text{c1}}/\Delta V_{\text{g1}}$ presented in Fig. 7.18. The data were obtained by Zhang et al. (2006) where the samples were made by photolithography (Tselev et al. 2004) and the nanotube section between the source and drain electrodes was too long ($L > 1200$ nm) to clearly resolve Fabry–Perot patterns at the lowest temperatures. Nevertheless, we found that the gate voltage corresponding to the zero-bias anomalies was in good agreement with the expected location of VHSs by assuming $\alpha = 0.75\%$ (Zhang et al. 2006). Another experiment (Tselev et al. 2009) was performed on recently fabricated CNT FETs by using e-beam lithography and obtained devices of different lengths from sections of the same semiconducting CNT. In a typical device there were two different sections, A and B, with lengths of 400 nm and 1200 nm, respectively. On the one hand, both sections belong to the same CNT, which suggests that any features related to VHSs will have the same energy spacing. On the other hand, features related to the CNT length, such as Fabry–Perot interference patterns, will have very different characteristic energies. At the temperature $T = 4$ K of the experiment the energy spacing of peaks in Fabry–Perot patterns for device A was indeed sufficiently large to be clearly measured. The obtained gate efficiencies α ranges between 0.7% and 1% (Yang et al. 2012), which is consistent with the value assumed earlier by Zhang et al. (2006). The shift of gate voltage between the threshold voltage

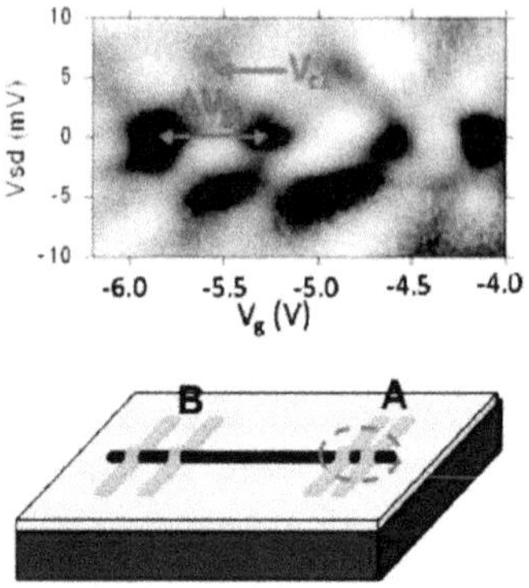

Figure 7.20 Top: 2D plot of differential conductance as a function of source–drain bias and gate voltage for device A (sample 4) at $T \approx 5$ K. The gate efficiency can be extracted from the ratio of voltages indicated in the figure, $\alpha = V_{c1}/\Delta V_{g1}$. Bottom: Schematics of the sample showing devices A (sample 4) and B (sample 5). Adapted from Zhang et al. (2006) and Tselev et al. (2009).

corresponding to the next VHS is estimated from the measured gate efficiency and CNT diameter, which gives $\Delta V_G = \Delta E / (\alpha \cdot e)$.

For the CNT, whose diameter is $D = 3.0$–3.5 nm, it gives $\Delta V_G = 11$ V. As one can see from Fig. 7.20, the Fabry–Perot pattern contains a remarkably bright high conductance region when the gate voltage is shifted about 10 V from the threshold. The differential conductance versus the source–drain bias voltage in the region has a broad peak at zero bias as marked by blue curves in the curves extracted from the 2D plots in the top part of Fig. 7.20. In the actual experiment, the peak appears to be much broader and becomes more pronounced than the typical Fabry–Perot oscillation peaks, shown by the black curves on the same plot for comparison. As it follows from the experimentally measured temperature dependence of the ZBC peak it survives up to temperatures of about 30 K. On the contrary, the Fabry–Perot peaks disappear at temperatures above 12 K. Section B also shows a broad ZBC peak, as is seen in the bottom part of Fig. 7.20. The characteristic Fabry–Perot energy scale becomes too small for the longer sample, since at the temperature of experiment $T = 4$ K no Fabry–Perot pattern can be resolved. Nevertheless, most of conductance curves versus the source–drain

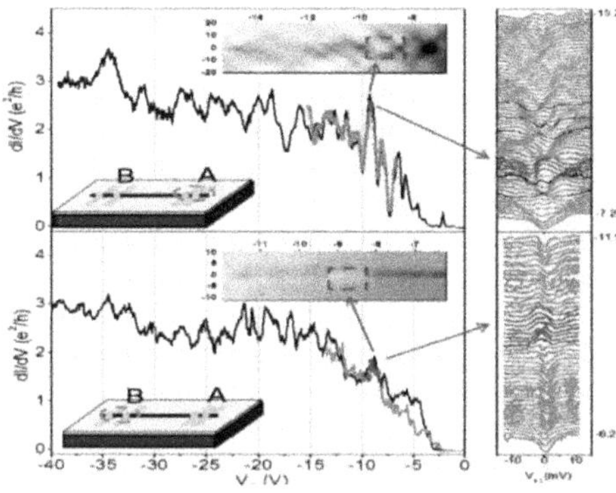

Figure 7.21 Zero-bias differential conductance as a function of gate voltage for device A (top) and device B (bottom). Insets: 2D plots of differential conductance as a function of source–drain bias and gate voltage for devices A (top) and B (bottom). The red dots superimposed to the black curves are data extracted from the 2D plot corresponding to zero source–drain bias. The blue dots correspond to values of gate voltage where the ZBA occurs. Right: Curves of differential conductance as a function of source–drain bias from the corresponding 2D plot on the left. Curves corresponding to different values of gate voltage are shifted vertically from $V_G = -7.2$ V (-6.2 V) to $V_G = -15.2$ V (-11.2 V). All measurements are at $T \approx 5$ K. Adapted from Zhang et al. (2006) and Tselev et al. (2009).

bias exhibit a pronounced dip at zero bias as seen from red curves in the lower right panel of Fig. 7.21. Otherwise, a broad peak is seen for a small gate voltage range, as seen from blue curves in lower right panel of Fig. 7.21. As follows from the experimental temperature dependence of the ZBA, it persists up to temperatures of $\sim$25 K. At the value of gate voltage where the ZBA occurs, the spacing ΔV_G between ZBA and the threshold may be slightly shifted from the actual edge of the valence band. It happens due to the presence of Schottky barriers which impede the electric current flow when the Fermi energy is aligned with the edge of the valence band. It is consistent with appearance of the Coulomb blockade peaks emerging at lower temperature (50 mK) in the region closer to and slightly above $V_G = 0$ (Yang et al. 2012). Because both devices are fabricated on the same nanotube, the spacing between the values of gate voltage where the ZBA occurs on the one hand and the threshold for the devices A and B (see Fig. 7.22) on the other

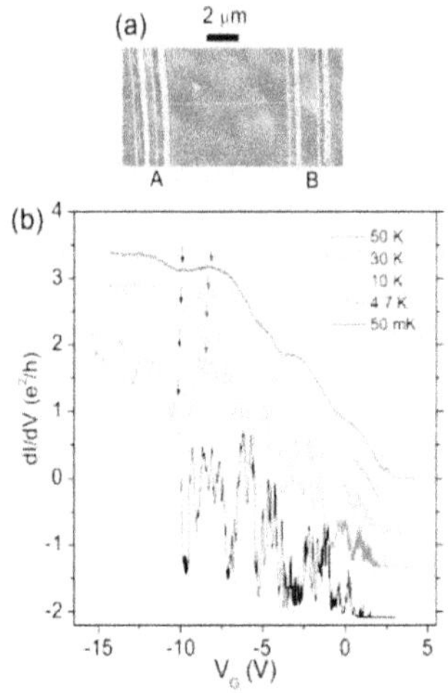

Figure 7.22 (a) AFM image of the sample showing the short ($L = 400$ nm) device A on the left and the long ($L = 1200$ nm) device B on the right. (b) Zero-bias differential conductance as a function of back-gate voltage for device A at different temperatures. Adapted from Zhang et al. (2006) and Tselev et al. (2009).

hand are remarkably similar to each other. The mentioned spacing is also consistent with the VHS spacing value. The aforementioned experimental data had been obtained for the samples A and B fabricated with a large diameter nanotube. The data confirm that either the tube is a single-walled nanotube, or it is a double-wall CNT with only the outer shell contributing to the conductance. Such a conclusion follows from the fact that both devices A and B are completely turned off above the threshold gate voltage, ruling out the possible existence of an inner metallic tube. In the other case, when an inner semiconducting tube is present, its threshold and VHSs can be accessed at a much higher gate voltage. It happens because the diameter is small, and it should not affect the ZBA corresponding to the second VHS of the outer nanotube. The single-wall CNTs with diameter as large as 4.3 nm were fabricated by chemical vapor deposition (CVD) (Cheung et al. 2002, Yang et al. 2003).

The described experiment utilizing isolated CNTs, demonstrates an occurrence of anomalous ZBC peaks at the values of gate voltage corresponding to VHSs of the electron density of states. Analyzing various physical mechanisms it can be concluded that the most likely explanation for this ZBA involves superconductivity in the CNT causing AR at the carbon-nanotube/electrode interface. Other mechanisms that are not related to superconductivity and are known to cause ZBC peak in CNTs provide no consistent interpretation of the obtained experimental data. None of other effects besides AR could explain the experimental data (Zhang et al. 2006, Tselev et al. 2009). The temperature and bias dependence of the ZBA can be nicely fit by the AR model, including broadening due to electron–phonon scattering. From the fits, one can obtain the temperature dependence of the hypothetical superconducting gap and of the broadening parameter.

7.9 Theoretical Model

Below we theoretically describe the electron transport properties of the Pd/CNT/Pd junction. Here, we assume that the CNT becomes superconducting when the gate voltage shifts the Fermi level to match the VHS energy. Then, the electron transport through the Pd/CNT and CNT/Pd interfaces is affected by AR (Andreev 1964). We implement a combined approach which involves the scattering matrix technique (Datta 1992) and the BTK model (Blonder et al. 1982). Following Beenakker (2006, 2008) and Beenakker et al. (2008), AR (Andreev 1964) in CNT is influenced by the electron/hole chirality (Ando 2005). It causes a more complex structure of electron and hole states in CNT as compared to conventional metals. In turn, the particle chirality (Ando 2005) introduces new features into AR (Andreev 1964) at the metal/CNT interface (Beenakker 2006, Beenakker 2008, Beenakker et al. 2008).

AR occurs at the N/S interfaces where the superconducting energy gap Δ diminishes from its bulk value Δ_0 in S (which is CNT in our experimental setup) to zero in N (which is Pd). When the gate voltage V_G adjusts the Fermi level to match the VHS singularity in the density of electron states in CNT, it substantially enlarges the Cooper coupling constant $\lambda = N(0)\ V$ ($N(0)$ is the electron density of states at the Fermi level and V is the potential of electron–electron

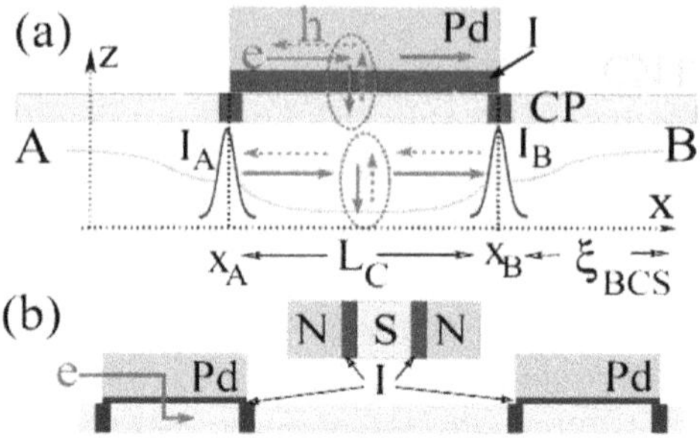

Figure 7.23 Model of the "reflectionless" Andreev reflection (AR) in the Pd/CNT junction. (a) Upper panel: The Pd/CNT contact with the normal metal Pd electrode. The single-walled carbon nanotube (CNT) splits in two parts, $T \rightarrow T' + T''$, where T' is a CNT section immediately under Pd and T'' is the open CNT section outside the contact. We assume that Pd, T', and T'' are separated from each other by the potential barriers I_A and I_B. Thus, each Pd/CNT section of the whole junction is represented as a multisectional N-I-T′-I-T″ junction. The electrons (blue solid arrows) and holes (red dash arrows) experience multiple Andreev and conventional reflections inside the contact region of the N-I-T′-I-T″ junction before being transmitted between the T′ and T″ regions. Lower panel: The multiple electron and hole reflections inside the contact area between the two potential barriers I_A and I_B located at $x = x_{A,B}$. (b) The double-barrier Pd/CNT/Pd junction formed as a sequence of the two sections shown above in (a). The whole Pd/CNT/Pd structure can be effectively represented as an N-I-S-I-N junction sketched in the upper panel.

attractive interaction). In turn, the superconducting energy gap Δ and the critical temperature T_c are both increasing.

The above scenario can presumably cause the development of own superconductivity inside the CNT (Tselev et al. 2009, Yang et al. 2012, Yang et al. 2013). Hypothetical superconductivity is detected in the experiment by measuring excessive conductance of the Pd/CNT/Pd junction which acquires contribution from AR at the N/S interface. Geometries of the Pd/CNT and Pd/CNT/Pd contacts and the electron/hole trajectories are sketched in Fig. 7.23. In Fig. 7.23, the Pd electrode is shown by green, the CNT is blue, the electrons are shown as dark blue arrows, the holes as red dash arrows, and Cooper pairs are thick yellow arrows. We assume that the electron transmission trajectory through the whole N-I-S-I-N junction is represented by a broken line. The CNT section itself has a complex structure; it is composed of two different pieces, (a) the CNT section located immediately under the Pd electrode (T′-section) and (b) the uncovered CNT section outside the Pd electrode (T′-section). Distinct fragments of the whole electron trajectory through the whole junction involve different types of the electron and hole transport.

For the sake of convenience we split the whole electron trajectory into several pieces: (i) the 3D electron propagation in the normal Pd, (ii) the tunneling through the Pd/CNT interface barrier I which emerges because the difference of workfunctions in Pd and in CNT, (iii) the 1D electron propagation in the T′-section, (iv) the 1D electron motion in the T′-section, and (v) the transmission through the T′/T″ interface barrier between the T′- and T″-sections. The barrier at the T′/T″-interface is formed during the Pd electrode deposition at the top of CNT. There are two different types of transmission processes contributing into (ii) and (v) which are the CT and AR. Both processes, CT and AR, proceed in the transversal z direction for the (ii) trajectory fragment and in x direction for the (v) trajectory fragment. The hypothetic superconductivity taking place inside CNT is suppressed in the T′-section at $x_{\mathbf{A}} < x < x_{\mathbf{B}}$ (x is the coordinate along the CNT axis) due to the proximity effect. Thus, the energy gap Δ' in the T′-region is diminished as compared to its value Δ'' in the uncovered CNT section T″, that is, $\Delta' < \Delta''$.

As it follows from the experimental data, the ratio $\delta G/G_{\mathrm{N}}$ of excessive conductance $\delta G = G_{\mathrm{S}} - G_{\mathrm{N}}$ to the normal state conductance G_{N} for our Pd/CNT/Pd junction exceeds 1. The measured excessive value $\delta G/G_{\mathrm{N}} > 1$ cannot be understood within a simple AR model (Blonder et al. 1982) which always gives $\delta G/G_{\mathrm{N}} < 1$. However, the seeming discrepancy between our experiment and the theory (Blonder et al. 1982) can be reconciled taking into account the so-called reflectionless AR (Vanwees et al. 1992, Popinciuc et al. 2012). We take into account the reflectionless AR occurring when an electron spends sufficient time in the N/S interface vicinity. An important outcome of the reflectionless AR is that the excessive conductance δG is strongly increased compared with a single AR effect. The reflectionless AR contributes when the corresponding dwell time, $\tau_{\mathbf{dw}}$, is much longer than the duration of an individual AR process, $\tau_{\mathbf{AR}}$, as is the case for the electron residing in region

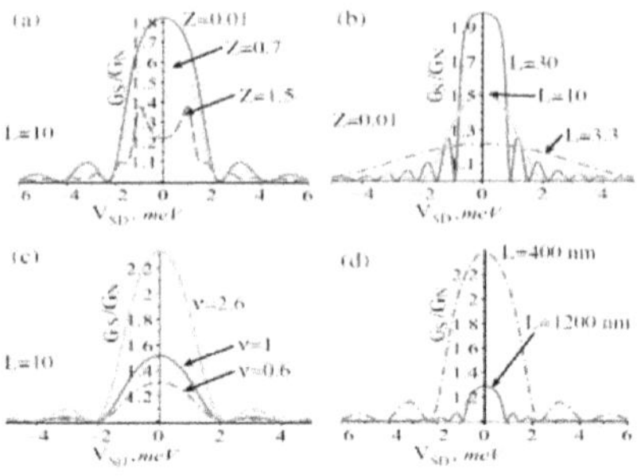

Figure 7.24 Theoretical curves of normalized conductance vs. source–drain bias for the N-I-S-I-N structure in Fig. 7.1, calculated for different values of the parameters introduced in the model. (a) The Rowell–McMillan (RM) oscillations of the normalized conductance, with different strength Z of the interface barriers shown for the short section with $L = 10$ (in units of BCS coherence length ξ). The corresponding values are $Z = 15$ (dashed blue curve), $Z = 07$ (dotted green curve), and $Z = 001$ (solid red curve). (b) A similar plot for fixed $Z = 001$ and different $L = 33$, 10, and 30. (c) A similar plot for fixed $L = 10$ and different $\nu = \tau_{\mathrm{dw}}\Delta/\hbar = 06$, 1, and 26. (d) The solid curve is $G(V_{\mathrm{SD}})$ of a "long" junction with the middle CNT section length $L = 1200$ nm and $\nu = 06$ whereas the dashed blue curve is $G(V_{\mathrm{SD}})$ for the short junction with $L = 400$ nm and $\nu = 26$.

T′. Then, an electron can experience multiple ARs when traveling in the vicinity of the CNT/Pd interface. Furthermore, the dwell time $\tau_{\mathbf{dw}}$ is energy dependent. In our theoretical model, we assume that the prolonged dwell time $\tau_{\mathbf{dw}}$ in the region T′ is caused by multiple reflections of electrons back and forth from the barriers $\mathrm{I_A}$ and $\mathrm{I_B}$ (see Fig. 7.24). In this model, the energy dependence of the time $\tau_{\mathbf{dw}}$ naturally originates from the energy dependence of the transparencies of the barriers $\mathrm{I_A}$ and $\mathrm{I_B}$. At low energies, $\varepsilon \simeq 0$, the barriers $\mathrm{I_A}$ and $\mathrm{I_B}$ are thicker and therefore less transparent, which corresponds to a longer dwell time $\tau_{\mathbf{dw}} >> \tau_{\mathbf{AR}}$. The barriers $\mathrm{I_A}$ and $\mathrm{I_B}$ are getting thinner and more transparent as the electron energy ε increases. It makes the dwell time $\tau_{\mathbf{dw}}$ shorter, $\tau_{\mathbf{dw}} \simeq \tau_{\mathbf{AR}}$ because the electrons experience less bounces back and forth within the CNT region under metal. Therefore, probability of the reflectionless AR is higher at low energies, and a conductance peak appears around zero voltage (in our case within the voltage interval of about ~5 mV). The peak width is determined by energy dependence of the transparencies of the barriers I_A and I_B rather than by the Nb energy gap magnitude. Energy dependence of the barrier transparency is incorporated in our calculations.

The ratio of the open nanotube section length L_o to the section lengths $2L_m$ covered by metal actually determines whether an electron experiences a conventional AR or reflectionless AR. For the N-I-S-I-N junction with electrode length $L_m = 650$ nm and a short open CNT section $L_o = 400$ nm one gets the ratio $L_o/2L_m = 0.3$ whereas for the long section with $L_o = 1300$ nm the corresponding ratio is $L_o/2L_m = 0.92$. We take into account that the net transmission time τ_J from the left metal electrode through the N-I-S-I-N junction to the right metal electrode is determined by the junction eigenenergy. We evaluate the time τ_m which electron spends in the open CNT section $\tau_o \simeq \tau_J L_o/2L_m$ and in the vicinity of metal/CNT contact $\tau_m \simeq \tau_J(1 - L_o/2L_m)$. Besides, for our junction geometry and high interface transparency, we find $\tau_J \approx E_J \approx \Delta$ which almost coincides for the two cases with different L_o. It means that $\tau_m \simeq \tau_o$, that is, the electron spends almost equal time inside the open section and in the immediate contact when L_o is long. Then, AR occurs in a conventional fashion. The story is different for the short N-I-S-I-N junction. The N-I-S-I-N energy eigenvalue is similar for both junctions, short and long. This also means that τ_J is roughly the same for the long and the short junctions, and hence, τ_m appears to be very different for the two cases. Namely, for the short junction $\tau_m >> \tau_o$, that is, the electron spends much longer time in the immediate vicinity of the metal /nanotube contact, while its dwelling along the short open CNT section occurs much faster. It implies that the whole time τ_J which an electron spends while getting transmitted through the N-I-S-I-N junction for the short and long cases must be approximately the same. The long time τ_m in the contact area of the short N-I-S-I-N junction is a reason why the electron experiences several ARs during every transmission through the whole N-I-S-I-N junction. This is illustrated in Fig. 7.23 where the "reflectionless" AR is represented as a "zigzag" trajectory.

Another important contribution into conductance of the Pd/CNT/Pd junction emerges from Rowell–McMillan (RM) oscillations (Rowell and Mcmillan 1966). The RM oscillations originate from the interference between the electron and hole evanescent waves

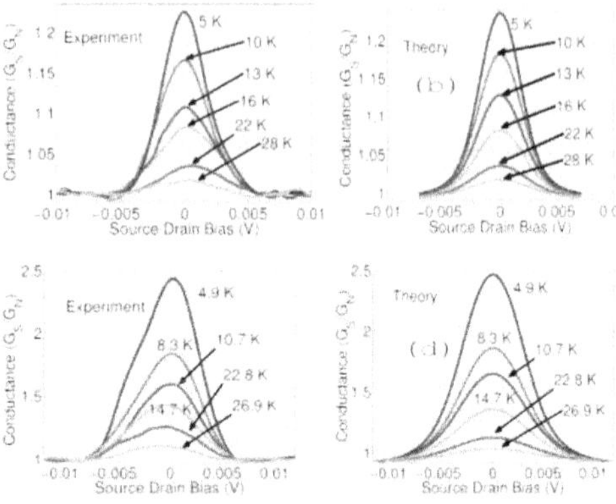

Figure 7.25 (a) Experimentally measured conductance normalized to its normal state value versus the source–drain bias voltage through the "long" Pd/CNT/Pd junction for temperatures $T = 4.9$ K, 8.3 K, 10.7 K, 14.7 K, 22.8 K, and 26.9 K. (b) Theoretically computed normalized conductance versus the source–drain bias voltage through the "long" Pd/CNT/Pd junction for the same temperatures. (c, d) Similar curves indicating a much bigger excess conductance for the "short" junction. Adapted from Zhang et al. (2006) and Tselev et al. (2009).

in the middle electrode of the complex Pd/CNT/Pd junction. The Pd/CNT/Pd geometry is effectively represented by the N-I-S-I-N junction sketched in the upper panel Fig. 7.23. The model calculations are performed using the methods (Blonder et al. 1982) generalized for the CNTs (Beenakker 2006, Beenakker 2008, Beenakker et al. 2008). Theoretical results for conductance $G(V_{\mathrm{SD}})$ of the Pd/CNT/Pd junction are represented in Fig. 7.25 where we show zero-bias AR peak fringed with the RM oscillations (Rowell and Mcmillan 1966). Zero-bias peak and RM oscillations in $G(V_{\mathrm{SD}})$ the Pd/CNT/Pd junction are influenced by interface barriers I, as illustrated in Fig. 7.25 where different $G(V_{\mathrm{SD}})$ curves correspond to the barrier strengths $Z = 1.5$ (dashed blue curve), $Z = 0.7$ (dotted green curve), and $Z = 0.01$ (solid red curve). As one can see from Fig. 7.25a, the zero-bias peak in $G(V_{\mathrm{SD}})$ at $V_{\mathrm{SD}} = 0$ splits in two smaller subpeaks when the barrier strength $Z \geq 0.5$. Besides, the interference of evanescent waves which creates RM also causes smoothing of the zero-bias peak of $G(V_{\mathrm{SD}} = 0)$. The RM period $\delta\varepsilon$ is determined by the CNT section length L and by the value of the

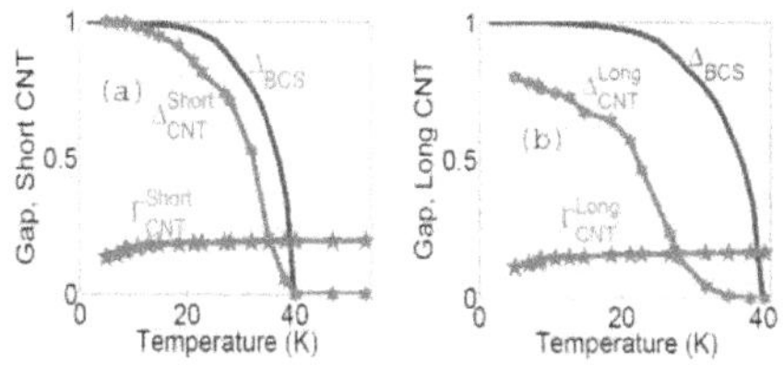

Figure 7.26 Temperature dependence of the hypothetical energy gap $\Delta_{\text{CNT}}(T)$ extracted from the Andreev reflection experiment (red curve). Blue curve shows the corresponding temperature dependence according to the BCS model. In the same plot we also show temperature dependence of the inelastic electron energy-scattering rate $\Gamma_{\text{CNT}}(T)$ extracted from the experimental data (green curve). Adapted from Zhang et al. (2006) and Tselev et al. (2009).

hypothetic CNT energy gap, Δ as it follows from the next Fig. 7.25b. In particular—see Rowell and Mcmillan (1966)—at a small $Z << 1$, one obtains the RM period as

$$\delta\varepsilon \approx 4\text{meV} \times \sqrt{1 - \frac{\Delta^2}{\varepsilon^2}} \tag{7.41}$$

In Fig. 7.25b we plot the computed conductance $G(V_{\text{SD}})$ of the Pd/CNT/Pd junction versus the source drain voltage V_{SD} with different length L of the middle CNT section. From Fig. 7.25b one notices that the RM period is increased for the junction with a short CNT section $L = 400$ nm (dashed curve). Solid curve in Fig. 7.25b with a lower oscillation period is the conductance of a junction with the long CNT section $L = 1200$ nm.

We compare the conductance curves $G(V_{\text{SD}})$ at different temperatures $T = 4.9$ K, 8.3 K, 10.7K, 14.7 K, 22.8 K, and 26.9 K obtained in the experiment (see Fig. 7.24a), and computed theoretically (presented in Fig. 7.24b). It confirms a good correspondence between the model of AR and the experiment.

Fitting the experimental $G(V_{\text{SD}})$ curves allows to extract the model parameters. The most interesting parameter is the temperature dependence of hypothetic CNT energy gap $\Delta(T)$ shown in Fig. 7.26 which was obtained fitting the experimental $G(V_{\text{SD}})$ curves measured at different temperatures. Another important parameter is the inelastic collision rate $\Gamma(T)$ shown for both junctions with the short (Fig. 7.26a) and long (Fig. 7.26b) CNT sections. It slightly differs from the classic temperature dependence of the BCS energy gap $\Delta_{\mathrm{BCS}}(T)$. Comparing Figs. 7.26a and 7.26b, one also concludes that both the magnitude of energy gap Δ_{CNT} and the critical temperature T_c^{CNT} of the presumable superconducting transition in the middle of CNT section are bigger for the junction with the short CNT section for which $\Delta_{\mathrm{CNT}}^{\mathrm{Short}}(T = 0) \simeq 5$ mV and $T_c^{\mathrm{CNT,\,Short}} \simeq 37$ K, respectively. Instead, for the junction with the longer CNT section one obtains lower values $\Delta_{\mathrm{CNT}}^{\mathrm{Long}}(T = 0) \simeq 3.8$ mV and $T_c^{\mathrm{CNT,\,Long}} \simeq$ 29 K instead.

Problems

7-1. Discuss a possible relationship between the intrinsic coherence of graphene and quantum coherence in conventional superconductors.

7-2. Which conservation laws determine the Andreev reflection (AR) from the N/S interface which separates the normal metal N and superconductor S?

7-3. Explain why, when the barrier strength Z is finite, AR occurs mostly at energies $E = \Delta$, whereas it is suppressed at lower electron energies $E < \Delta$.

7-4. Explain why the differential conductance of graphene/ metal junctions depends on the bias voltage and is asymmetric.

7-5. What are the most important distinctions between the classical BTK model and a real graphene/metal junction?

7-6. Explain why the excessive conductance due to AR appears only below the threshold voltage Δ/e.

7-7. Explain why the reflection (r) and transmission (t) amplitudes entering Eq. 7.27 for the graphene/superconductor interface are 4×4 matrices.

7-8. Explain why conventional AR happens when electrons and holes both lie in the conduction band, whereas specular AR occurs when they are in different, conduction and valence, bands.

7-9. Explain why reflectionless AR can cause an increase of the N/S interface conductance more than twice.

7-10. Explain why the period of Rowell–McMillan oscillations is energy dependent and why it changes versus the barrier strength.

References

Ando, T. (2005). Theory of electronic states and transport in carbon nanotubes. *J Phys Soc Jpn,* **74**(3): 777–817.

Andreev, A. F. (1964). The thermal conductivity of the intermediate state in superconductors. *Sov Phys JEPT: USSR,* **19**(5): 1228–1231.

Avouris, P. (2010). Graphene: electronic and photonic properties and devices. *Nano Lett,* **10**(11): 4285–4294.

Beenakker, C. W. J. (2006). Specular andreev Reflection in graphene. *Phys Rev Lett,* **97**(6).

Beenakker, C. W. J. (2008). Colloquium: Andreev reflection and Klein tunneling in graphene. *Rev Mod Phys,* **80**(4): 1337–1354.

Beenakker, C. W. J., A. R. Akhmerov, P. Recher, and J. Tworzydlo (2008). Correspondence between Andreev reflection and Klein tunneling in bipolar graphene. *Phys Rev B,* **77**(7).

Bezryadin, A., C. N. Lau, and M. Tinkham (2000). Quantum suppression of superconductivity in ultrathin nanowires. *Nature,* **404**(6781): 971–974.

Blonder, G. E., M. Tinkham, and T. M. Klapwijk (1982). Transition from metallic to tunneling regimes in superconducting micro-constrictions: excess current, charge imbalance, and super-current conversion. *Phys Rev B,* **25**(7): 4515–4532.

Cheung, C. L., A. Kurtz, H. Park, and C. M. Lieber (2002). Diameter-controlled synthesis of carbon nanotubes. *J Phys Chem B,* **106**(10): 2429–2433.

Chiu, H. Y., V. Perebeinos, Y. M. Lin, and P. Avouris (2010). Controllable p-n junction formation in mono layer graphene using electrostatic substrate engineering. *Nano Lett,* **10**(11): 4634–4639.

Choi, J. H., H. J. Lee, and Y. J. Doh (2010). Above-gap conductance anomaly studied in superconductor-graphene-superconductor Josephson junctions. *J Korean Phys Soc,* **57**(1): 149–155.

Datta, S. (1992). Exclusion-principle and the Landauer-Buttiker formalism. *Phys Rev B,* **45**(3): 1347–1362.

Datta, S., and M. J. Mclennan (1990). Quantum transport in ultrasmall electronic devices. *Rep Prog Phys,* **53**(8): 1003–1048.

Dejong, M. J. M., and C. W. J. Beenakker (1995). Andreev reflection in ferromagnet-superconductor junctions. *Phys Rev Lett,* **74**(9): 1657–1660.

Dzyaloshinskii, I. E. (1987a). Maximal increase of the superconducting transition-temperature due to the presence of Vanthoff singularities. *JETP Lett,* **46**(3): 118–121.

Dzyaloshinskii, I. E. (1987b). Superconducting transitions due to Van-Hove singularities in the electron spectrum. *Zh Eksp I Teoret Fiz* **93**(4): 1487–1498.

Gogadze, G. A., and I. O. Kulik (1970). Theory of dimensional (size) oscillation effects and geometrical resonances in normal metal films. *Sov Phy Solid State, USSR* **11**(8): 1762–&.

Greenbaum, D., S. Das, G. Schwiete, and P. G. Silvestrov (2007). Pure spin current in graphene normal-superconductor structures. *Phys Rev B,* **75**(19).

Izyumov, Y. A., and E. Z. Kurmaev (1974). Physical-properties and electronic-structure of superconducting compounds with beta tungsten structure. *Usp Fiz Nauk,* **113**(2): 193–238.

Kaufman, D., B. Dwir, A. Rudra, I. Utke, A. Palevski, and E. Kapon (2000). Direct evidence for quantum contact resistance effects in V-groove quantum wires. *Phys E* **7**(3–4): 756–759.

Khomyakov, P. A., G. Giovannetti, P. C. Rusu, G. Brocks, J. van den Brink, and P. J. Kelly (2009). First-principles study of the interaction and charge transfer between graphene and metals. *Phys Rev B,* **79**(19).

Klapwijk, T. M., G. E. Blonder, and M. Tinkham (1982). Explanation of sub-harmonic energy-gap structure in superconducting contacts. *Phys B, C* **109**(1–3): 1657–1664.

Kulik, I. O. (1970). Macroscopic quantization and proximity effect in S-N-S junctions. *Sov Phys JEPT: USSR,* **30**(5): 944–&.

Lee, E. J. H., K. Balasubramanian, R. T. Weitz, M. Burghard, and K. Kern (2008). Contact and edge effects in graphene devices. *Nat Nanotechnol,* **3**(8): 486–490.

Li, L., K. H. Seng, Z. X. Chen, H. K. Liu, I. P. Nevirkovets, and Z. P. Guo (2013). Synthesis of Mn3O4-anchored graphene sheet nanocomposites via a facile, fast microwave hydrothermal method and their supercapacitive behavior. *Electrochim Acta,* **87**: 801–808.

Liang, G. C., N. Neophytou, M. S. Lundstrom, and D. E. Nikonov (2008). Contact effects in graphene nanoribbon transistors. *Nano Lett,* **8**(7): 1819–1824.

Matsudaira, M., J. Haruyama, H. Sugiura, M. Tachibana, J. Reppert, A. Rao, T. Nishio, Y. Hasegawa, H. Sano, and Y. Iye (2010). Pressure-induced superconductivity and phonon frequency in paperlike thin films of boron-doped carbon nanotubes. *Phys Rev B,* **82**(4).

Mintmire, J. W., and C. T. White (1998a). First-principles band structures of armchair nanotubes. *Appl Phys A,* **67**(1): 65–69.

Mintmire, J. W., and C. T. White (1998b). Universal density of states for carbon nanotubes. *Phys Rev Lett,* **81**(12): 2506–2509.

Mueller, T., F. Xia, M. Freitag, J. Tsang, and P. Avouris (2009). Role of contacts in graphene transistors: a scanning photocurrent study. *Phys Rev B,* **79**(24).

Murata, N., J. Haruyama, J. Reppert, A. M. Rao, T. Koretsune, S. Saito, M. Matsudaira, and Y. Yagi (2008). Superconductivity in thin films of boron-doped carbon nanotubes. *Phys Rev Lett,* **101**(2).

Odom, T. W., J. L. Huang, P. Kim, and C. M. Lieber (1998). Atomic structure and electronic properties of single-walled carbon nanotubes. *Nature,* **391**(6662): 62–64.

Ossipov, A., M. Titov, and C. W. J. Beenakker (2007). Reentrance effect in a graphene n-p-n junction coupled to a superconductor. *Phys Rev B,* **75**(24).

Popinciuc, M., V. E. Calado, X. L. Liu, A. R. Akhmerov, T. M. Klapwijk, and L. M. K. Vandersypen (2012). Zero-bias conductance peak and Josephson effect in graphene-NbTiN junctions. *Phys Rev B,* **85**(20).

Rowell, J. M., and W. L. Mcmillan (1966). Electron interference in a normal metal induced by superconducting contacts. *Phys Rev Lett,* **16**(11): 453–&.

Takesue, I., J. Haruyama, N. Kobayashi, S. Chiashi, S. Maruyama, T. Sugai, and H. Shinohara (2006). Superconductivity in entirely end-bonded multiwalled carbon nanotubes. *Phys Rev Lett,* **96**(5).

Tang, Z. K., L. Y. Zhang, N. Wang, X. X. Zhang, G. H. Wen, G. D. Li, J. N. Wang, C. T. Chan, and P. Sheng (2001). Superconductivity in 4 angstrom single-walled carbon nanotubes. *Science,* **292**(5526): 2462–2465.

Titov, M., A. Ossipov, and C. W. J. Beenakker (2007). Excitation gap of a graphene channel with superconducting boundaries. *Phys Rev B,* **75**(4).

Tselev, A., Y. F. Yang, J. Zhang, P. Barbara, and S. E. Shafranjuk (2009). Carbon nanotubes as nanoscale probes of the superconducting proximity effect in Pd-Nb junctions. *Phys Rev B,* **80**(5).

Vanwees, B. J., P. Devries, P. Magnee, and T. M. Klapwijk (1992). Excess conductance of superconductor-semiconductor interfaces due to phase conjugation between electrons and holes. *Phys Rev Lett,* **69**(3): 510–513.

Yang, Q. H., S. Bai, J. L. Sauvajol, and J. B. Bai (2003). Large-diameter single-walled carbon nanotubes synthesized by chemical vapor deposition. *Adv Mater,* **15**(10): 792–+.

Yang, Y., G. Fedorov, P. Barbara, S. E. Shafranjuk, B. K. Cooper, R. M. Lewis, and C. J. Lobb (2013). Coherent nonlocal transport in quantum wires with strongly coupled electrodes. *Phys Rev B,* **87**(4).

Yang, Y. F., G. Fedorov, J. Zhang, A. Tselev, S. Shafranjuk, and P. Barbara (2012). The search for superconductivity at van Hove singularities in carbon nanotubes. *Supercond Sci Technol,* **25**(12).

Zhang, J., A. Tselev, Y. F. Yang, K. Hatton, P. Barbara, and S. Shafraniuk (2006). Zero-bias anomaly and possible superconductivity in single-walled carbon nanotubes. *Phys Rev B,* **74**(15).

Chapter 8

Nonequilibrium Effects in Graphene Devices

Graphene devices utilize quantum states in extensive nonequilibrium conditions caused with strong fields. Therefore, adequate understanding of the nonequilibrium quantum dynamics of nanoscale devices nowadays is regarded as an important problem. Graphene quantum well represents a new type of a device and can be constructed using the graphene field-effect transistors. Such a system functions in competitive situations while balancing the thermal relaxation and the external driving which emerge under extreme nonequilibrium conditions. The theoretical methods for understanding such systems, however, are still under development.

8.1 Relevance of Nonequilibrium Effects in a Graphene Junction

Nonequilibrium effects in graphene and carbon nanotube devices emerge in conditions when either an external field or a heat flow are applied to the sample (Shafranjuk 2008, Shafranjuk 2009, Shafranjuk 2011a, 2011b, Rinzan et al. 2012). If the energy supply from outside the system exceeds the dissipation and the energy

Graphene: Fundamentals, Devices, and Applications
Serhii Shafraniuk

ISBN 978-981-4613-47-7 (Hardcover), 978-981-4613-48-4 (eBook)
www.panstanford.com

escape from the system to outside, then the system state can deviate from equilibrium. At first, when energy is supplied into a system, it can be absorbed with a certain subsystem and only later can be redistributed among the other subsystems of the same system. In particular, if a graphene device is exposed to an external electromagnetic field (EF), the field acts directly on the electrically charged particles, which are the chiral fermions (Shafranjuk 2009, Shafranjuk 2011a, 2011b). Therefore, at the initial stage, the external field energy is absorbed by the chiral fermions. On the next stage, owing to the inelastic electron–phonon collisions, part of the absorbed energy is transferred from electrons to phonons, which carry it far away from the active region. In principle, depending on the system's geometry and the average temperature, the whole scenario might be more complicated due to secondary reabsorption of the excessive phonon energy by the chiral fermions again. In general, there are several most important micro- and macroscopic parameters which characterize the energy exchange between the system and the external environment. Some essential quantities might be evaluated theoretically, while others can be extracted from experiments. For instance, the electron density of states in pristine graphene is computed as

$$N(E) = 4\sum_{k} \delta(E - E_k) = 4\frac{WD}{2\pi}\frac{|E|}{\hbar^2 v^2} \tag{8.1}$$

Another important parameter is broadening of the electron energy levels. The broadening actually characterizes the rate of inelastic electron–phonon collisions Γ_{ep}. The value of Γ_{ep} is extracted from measurements of broadening of the electron spectral singularities, like quantized levels, or Van Hove singularities. The value of Γ_{ep} can be readily obtained from measurements of Raman spectra (see Chapter 4). Besides, from the experimental measurements of differential conductance (Zhang et al. 2006, Tselev et al. 2009, Yang et al. 2012) at different temperatures (see Chapter 7) one finds, for example, that at $T = 100$ K, the broadening achieves $\Gamma_{ep} = 2$ meV, which corresponds to the electron–phonon inelastic scattering time

$$\tau_{ep} = \frac{\hbar}{\Gamma_{ep}(100K)} = 3.3 \times 10^{-13}\text{s} \tag{8.2}$$

Instead, at $T = 10$ K one finds $\tau_{\text{ep}} = \hbar/\Gamma_{\text{ep}}(10K) \approx 3 \times 10^{-12}$ s. Another important parameter of a nonequilibrium system with tunneling injection is the rate of injection $\Gamma_E^{\mathbf{T}}$. For the geometry of a graphene ribbon with length D where nonequilibrium particles are injected from an attached metallic electrode, one obtains

$$\Gamma_E^{\mathbf{T}} = \frac{2\hbar}{\tau_{\mathbf{dw}}} = \frac{2\hbar}{D}v \simeq 10^{-3}\text{eV} \qquad (8.3)$$

In the last Eq. 8.3 we have introduced the electron dwell time $\tau_{\mathbf{dw}}$ inside the ribbon whose approximate value for $D = 1$ μm is $\tau_{\mathbf{dw}} = D/v \approx 10^{-12}$s. One can also evaluate the ratio $\tau_{\mathbf{ep}}/\tau_{\mathbf{dw}}$ when $T = 300$K and the electron–phonon relaxation time is $\tau_{\mathbf{ep}} = 3.3 \times 10^{-13}$s, which gives

$$\frac{\tau_{\mathbf{ep}}}{\tau_{\mathbf{dw}}} = \frac{\tau_{\text{ep}}\Gamma_E^{\mathbf{T}}}{\hbar} \simeq 0.16 \qquad (8.4)$$

A similar estimation at a lower temperature $T = 100$ K gives $\tau_{\mathbf{ep}}/\tau_{\mathbf{dw}} = \tau_{ep}\Gamma_E^{\mathbf{T}}/\hbar \approx 0.5$, whereas at $T = 10$ K one obtains the electron mean free path $v\tau_{\mathbf{ep}} = 2.7$ μm and the much higher ratio $\tau_{\mathbf{ep}}/\tau_{\mathbf{dw}} \simeq 5$, respectively. The last number means that the nonequilibrium effects are not essential at room temperatures. However, they might become already noticeable as the temperature decreases to $T = 100$ K, whereas at $T = 10$ K the nonequilibrium effects are quite essential.

The simplest experimental setup is the double-barrier junction where the two electrodes L and R are attached to the middle section T on both sides (Shafranjuk 2011a, 2011b). In many cases, electron states of the L and R electrodes are considered as stationary and equilibrium, because the L and R electrodes are connected to the external circuit. The AC field acts only on the T section, thus affecting the chiral transport inside it. If the dwell time $\tau_{\mathbf{D}}$ inside T is short, $\tau_{\mathbf{D}} = D/v << \min[\tau_{\mathbf{ep}}, \tau_{\mathbf{T}}, 2\pi/\omega]$ (where $\tau_{\mathbf{ep}}$ is the electron–phonon inelastic scattering time, $\tau_{\mathbf{T}}$ is the tunneling time between L, T, and R), no scattering events are taking place in T. Then one quotes the electron transmissions across the L/T and T/R interfaces as mutually phase-correlated events. Besides, we also assume that $\tau_{\mathbf{ep}} >> \tau_{\mathbf{T}}$, that is, the electrons arrive and escape the T section faster than the electron–phonon scatterings occur. In this limit of the coherent transmission, the electron distribution function n_ε inside

T remains close to equilibrium. Under the above assumptions, an electron inside T remains in the same state $|\kappa_m>$ during the dwell time $\tau_\mathbf{D}$. Since the time-inhomogeneous corrections are as small as $(\tau_\mathbf{D}/\tau_\mathbf{s})^2 << 1$ (where $\tau_\mathbf{s}^{-1} = \tau_\mathbf{ep}^{-1} + \tau_\mathrm{T}^{-1}$), it greatly simplifies calculation of the time-averaged electric current. The mentioned condition is accomplished when considering an AC field with a low frequency $\omega << \Delta$ ($\Delta = 2\hbar v/d_\mathrm{T}$ is the van Hove singularity spacing, d_T the nanotube diameter). In particular, for a grapheme stripe or a nanotube with $d_\mathrm{T} = 1$ nm, for example, one gets $\Delta = 1$ eV. We also take into account that for relevant electron energies $\varepsilon \sim 10^{-2}\Delta$ one typically gets $\tau_\mathbf{ep} \geq 10^{-12}$ s.

Let us consider an example of a setup where the Schottky barrier of height $U_0 = 0.5\Delta$ is located at the LIT/TIR interface. Then, for example, for a nanotube with diameter $d_\mathrm{T} = 1$ nm or a graphene stripe width $W = \pi \cdot 1$ nm, one estimates the tunneling time as $\tau_\mathrm{T} \geq 10^{-12}$ s. For the T section length $L_\mathrm{T} = 1$ μm, the corresponding dwell time is $\tau_\mathbf{D} = L_\mathrm{T}/v \simeq 10^{-12}$ s. When calculating the time-averaged electric current, the above conditions also guarantee that the nonequilibrium effects are small. It means that the electron distribution function can be approximated by $n_\varepsilon \approx 1/(\exp(\varepsilon/T) + 1)$.

8.2 Tunneling Rates for a Graphene Junction

A nonequilibrium state in a graphene device can originate either from applying external fields or from injecting nonequilibrium particles. For instance, one can induce the nonequilibrium effects by exposing a graphene sample with an EF or injection of nonequilibrium phonons and electrons from an external source attached to graphene. There were reports concerning the tunneling injection of electrons from an adjacent conducting electrode into the graphene sample (Lin et al. 2008, Chiu et al. 2010). Theoretically, the nonequilibrium electrons are characterized in terms of the nonequilibrium distribution function (Rinzan et al. 2012). The kinetic equation, which governs the evolution of nonequilibrium excitations, is obtained using a variety of methods (Keldysh 1965, Gorkov and Eliashbe.Gm 1968, Eliashbe.Gm 1970, Ivlev and

Eliashbe.Gm 1971). Along with the nonequilibrium parameters introduced in the former subsection, there is another important parameter which is the tunneling rate $\Gamma_E^{\mathbf{T}}$. A typical value of $\Gamma_E^{\mathbf{T}}$ for a nanoribbon with width $W = 10$ nm and length $D = 1000$ nm at electron energy $E = 1$ meV is $\Gamma_E^{\mathbf{T}} = 7 \times 10^{-2}$ meV. It means that one can neglect by the nonequilibrium effects in this particular case. At $T = 100$ K, when the electron–phonon relaxation time achieves the value of $\tau_{\mathbf{ep}} = 3.3 \times 10^{-13}$s, one gets $\tau_{\mathrm{ep}}\Gamma_E^{\mathbf{T}}/\hbar \simeq 0.1$, while at $T = 10$ K the last quantity becomes $\tau_{\mathrm{ep}}\Gamma_E^{\mathbf{T}}/\hbar \simeq 0.1$ instead. As the temperature is lowered to $T = 2$ K, one can even get a larger value $\tau_{\mathrm{ep}}\Gamma_E^{\mathbf{T}}/\hbar \simeq 10$ which suggests that the nonequilibrium effects become important (Rinzan et al. 2012).

The scattering approach has to be modified when an external field supplies its energy into the graphene device. The energy supply occurs owing to absorption of photons by the charge carriers. Example of a graphene field-effect transistor (FET) exposed to an external EF is presented in Fig. 8.1. Profile $U(x)$ of the chiral barrier which is spreading along the graphene stripe is defined applying the local gate voltage V_{G} by means of a doped Si gate electrode with length D, which is placed underneath. In the same Fig. 8.1, we also show schematics of the photon-assisted tunneling

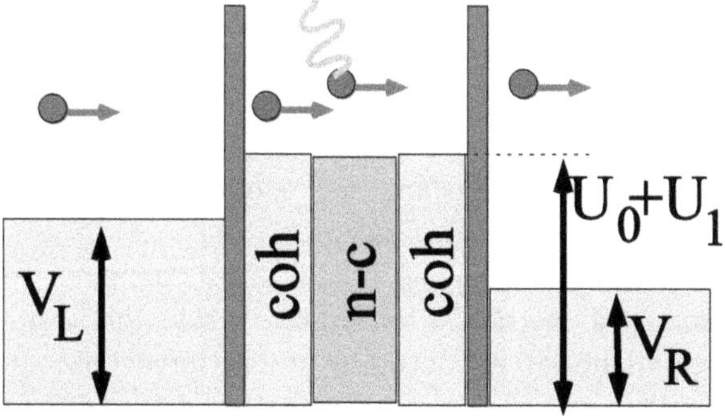

Figure 8.1 (a) Typical setup of a graphene field-effect transistor exposed to an external electromagnetic field (EF). The chiral barrier profile $U(x)$ is defined applying of the local gate voltage V_{G} via the doped Si gate electrode of length D placed underneath. (b) Schematics of photon-assisted tunneling through the gate-induced chiral barrier. Adapted from Shafranjuk (2011).

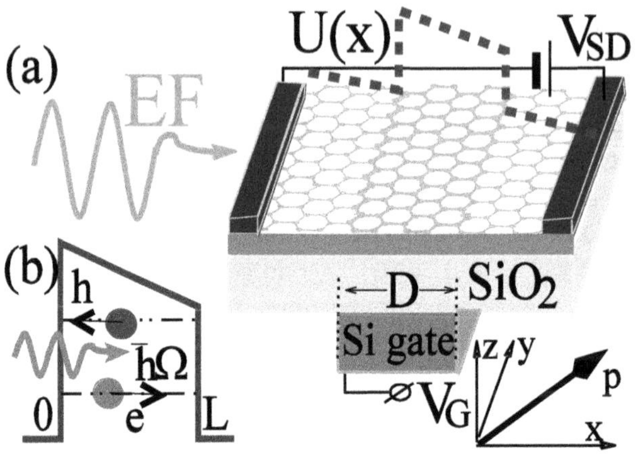

Figure 8.2 An example of a double-barrier junction where the electron propagation consists of coherent and noncoherent (nc) pieces.

through the gate-induced chiral barrier in the FET. Even the phase coherence is broken due to an energy absorption inside the FET; one can describe the nonequilibrium electron transport in the system theoretically. The task is accomplished combining the tunneling probability matrices instead of the S-matrices, which is valid for the former coherent transport only. On either side of chiral barrier, which is formed in the graphene FET, the phase coherence is preserved on a certain scale, in the left and right sections. Then, the corresponding probabilities of electron transmission are computed combining the S-matrices in the separate sections only. After that, one combines previously computed separate tunneling probabilities of the left and right sections and then computes the full tunneling probability through the whole junction.

In a more general case, one should use a combined approach which assumes a piece-wise coherence inside the system. The incoherence might be introduced by atomic impurities, structural defects, ripples and roughness of surfaces and interfaces, etc. An example of such geometry is a double-barrier junction shown in Fig. 8.2, where the electron propagation consists of coherent and noncoherent fragments. The piece-wise representation of the electron trajectories greatly simplifies computing of basic transport

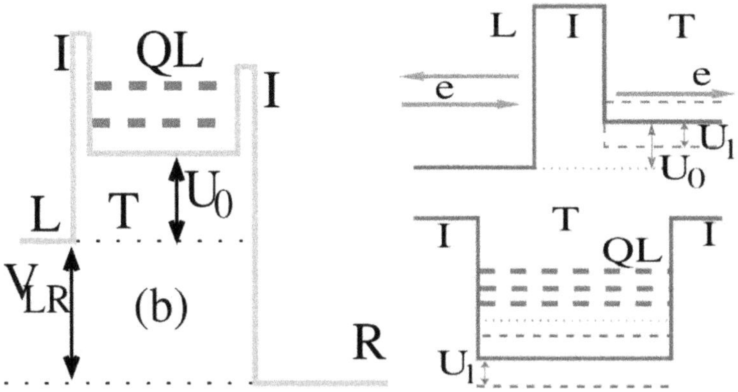

Figure 8.3 Left: Quantized levels (QLs) shown as red dashed lines are formed inside the biased LITIR junction. Right: The left LIT (upper panel) and the middle ITI (lower panel) cranks used in the S-matrix composition. Adapted from Shafranjuk (2011).

characteristics of the complex system. Using this approach, the whole electron trajectory is composed from separate coherent and noncoherent pieces (see, for example, Fig. 8.2). The separate pieces of trajectories with different types of transport are linked with each other by matching different coherent and noncoherent pieces. The electron propagation along coherent pieces of trajectories between the scatterers are described in terms of the S-matrices, while the incoherent connecting parts of electron propagation are determined via the previously computed probability matrices and the distribution functions.

The coherent approximation is illustrated with a simple example of a double-barrier LITIR junction (Shafranjuk 2009, Shafranjuk 2011a, 2011b) where left (L) and right (R) electrodes are attached to the middle section T whose energy diagram is shown in Fig. 8.3. The L and R electrodes are separated from T by finite potential barriers I. To compute the transmission coefficient $t_{\mathbf{LR}}(\varepsilon)$ through the whole junction, we consider it as a composition LITIR $\otimes$ LIT $\otimes$ ITI $\otimes$ TIR of three consequent sections (each section of the whole setup is a crank like the one sketched in Fig. 8.3). The whole three-crank composition $\hat{S}_{\mathbf{LR}} = \hat{S}_{\mathrm{L}} \otimes \hat{S}_{\mathrm{T}} \otimes \hat{S}_{\mathbf{R}}$ yields the net S-matrix $S_{\mathbf{LR}}$ of the whole junction.

The composition operator $\otimes$ acts, for example, as

$$\hat{S}_{\rm LT} = \hat{S}_{\rm L} \otimes \hat{S}_{\rm T} = \begin{pmatrix} r_{\rm L} & t_{\rm L} \\ t_{\rm L}^{c} & r_{\rm L}^{c} \end{pmatrix} \otimes \begin{pmatrix} r_{\rm T} & t_{\rm T} \\ t_{\rm T}^{c} & r_{\rm T}^{c} \end{pmatrix}$$

$$= \frac{1}{1 - r_{\rm T} r_{\rm L}^{c}} \begin{pmatrix} \zeta_1 & t_{\rm L} t_{\rm T} \\ t_{\rm L}^{c} t_{\rm T}^{c} & \zeta_2 \end{pmatrix} \tag{8.5}$$

where $\zeta_1 = (r_{\rm T} r_{\rm L}^{c} - 1)\, r_{\rm L} - r_{\rm T} t_{\rm L} t_{\rm L}^{c}$, $\zeta_2 = (r_{\rm T} r_{\rm L}^{c} - 1)\, r_{\rm L}^{c} - r_{\rm L}^{c} t_{\rm T} t_{\rm L}^{c}$, while L and T refer to the corresponding adjacent sections. The electron wavefunction in the middle T section satisfies the quasiclassical Dirac equation completed by corresponding time-dependent boundary conditions. In our system, the Dirac equation describes chiral particles with two pseudospins per each particle. It allows the partial transmission $t_{\rm L(R)}$ and $t_{\rm T}$ coefficients to be computed microscopically from solutions of the Dirac equation. In many practical cases, the electron transport through the LIT and TIR cranks (see Fig. 8.3, right) is a nonchiral one, while the transport through the middle ITI crank is chiral since the pseudospin is conserved.

Under the above assumptions, and using the composition rule, Eq. 8.5, one obtains the net transmission coefficient through the whole LITIR junction $t_{\rm LR}\,(t)$ as

$$t_{\rm LR}\,(\varepsilon) = [\hat{S}_{\rm LR}]_{12} = \frac{t_{\rm L} t_{\rm T} t_{\rm R}}{Z_{\rm LR}} \tag{8.6}$$

where

$$Z_{\rm LR} = (1 - r_{\rm L} r_{\rm T})\,(1 - r_{\rm R} r_{\rm T}) - r_{\rm L} r_{\rm R} t_{\rm T}^{2} \tag{8.7}$$

and $r_{\rm R(L)} = \sqrt{1 - t_{\rm R(L)}^{2}}$

8.3 Nonequilibrium Electric Current

When an external field supplies energy into the system, excessive energy is transferred from the field to the quasiparticle excitations. As a result, the distribution functions of the excitations, which in our case are chiral electrons and phonons, begin to deviate from the steady-state equilibrium value. Therefore, in conditions of external influence, computing of the transport characteristics must account for the nonequilibrium effects. The electric current consists of two terms:

$$I = 4ev \int d\varepsilon N(\varepsilon)\, \mathbf{Tr}\left(G^K_{\mathbf{R}}(0,\ D,\ \varepsilon) - G^K_{\mathbf{L}}(0,\ D,\ \varepsilon)\right)$$
$$= I_0 + I_{\mathbf{ne}} \tag{8.8}$$

The first term in Eq. 8.8 corresponds to the equilibrium electric current, whereas the second term is caused by the field-induced nonequilibrium effects. In Eq. 8.8, the electric current is expressed via the Keldysh–Green functions (Keldysh 1965) $G^K_{\mathbf{L(R)}}(0,\ D,\ \varepsilon)$ of the L and R electrodes. We will compute these functions in the following subsections.

8.4 The Green–Keldysh Function of Nonequilibrium Electrons

In the presence of external time-dependent field, a straightforward calculation of the Green–Keldysh function (Keldysh 1965) inside the chiral barrier gives

$$\begin{aligned}
G^K_{\mathbf{T}}\left(x_{\mathbf{D-0}},\ t;\ x'_{\mathbf{D-0}},\ t'\right) &= -i\left\langle\left[\psi_{\mathbf{T}\sigma}\left(x_{\mathbf{D-0}},\ t\right)\psi^{\dagger}_{\mathbf{T}\sigma'}\left(x'_{\mathbf{D-0}},\ t'\right)\right]_-\right\rangle \\
&= -i\sum_{k,\ k'}\left(\alpha_\kappa(t)\sum_m e^{i\zeta(t)}\chi_{n,\ \kappa_m}e^{i\kappa(t)\,D}\right. \\
&\quad\left.-\beta_{-\kappa}(t)\sum_m e^{i\zeta(t)}\chi_{n,-\kappa}e^{-i\kappa D}\right) \\
&\quad\times\left(\alpha_{\kappa'}(t)\sum_{m'} e^{i\zeta(t')}\chi_{n,\ \kappa'_m}e^{i\kappa' D}\right. \\
&\quad\left.-\beta_{-\kappa'}(t)\sum_{m'} e^{i\zeta(t')}\chi_{n,-\kappa'_m}e^{-i\kappa'_m D}\right)^* \\
&\quad\times e^{i[\zeta(t)-\zeta(t')]}\varpi_\kappa\varpi^{\dagger}_{\kappa'}e^{i\kappa x-i\kappa' x'}\left\langle\left[a_{\sigma\kappa},\ a^{\dagger}_{\sigma'\kappa'}\right]_-\right\rangle \\
&= -i\delta_{\sigma\sigma'}\sum_{m,\ m',\ \kappa,\ \kappa'}\Xi_{\kappa,\ \kappa'}e^{i[\zeta(t)-\zeta(t')]}e^{i\kappa(x-x')} \\
&\quad\times\left(2n^{\mathbf{T}}\left(\kappa,\ \kappa'\right)-1\right)
\end{aligned} \tag{8.9}$$

where $\kappa = \kappa(t)$, $\kappa' = \kappa(t')$, and we have introduced the auxiliary functions

$$\Xi_{\kappa,\kappa'} = \varpi_{\kappa}\varpi_{\kappa'}^{\dagger}\left(\alpha_{\kappa}\alpha_{\kappa'}^{*} + \beta_{-\kappa}\beta_{-\kappa'}^{*}\right) \tag{8.10}$$

and

$$\zeta(t) = \frac{1}{\hbar}\int_{-\infty}^{t} dt\,(E + U_0 + U_1(t)) \tag{8.11}$$

which take into account an influence of the AC field inside the chiral barrier. Besides, we have introduced the electron wavevectors as

$$\kappa(t) = \sqrt{(E + U_0 + U_1(t))^2/(\hbar v)^2 - q^2} \tag{8.12}$$

$$\pi = \sqrt{\kappa^2 + q^2} = \frac{|E + U_0 + U_1(t)|}{\hbar v} \tag{8.13}$$

and

$$\begin{aligned}\sigma(t) &= \mathbf{sign}\,(E + U_0 + U_1(t))\\ k &= \sqrt{E^2/(\hbar v)^2 - q^2},\\ p &= \sqrt{k^2 + q^2} = \frac{|E|}{\hbar v},\\ s &= \mathbf{sign}\,(E)\end{aligned} \tag{8.14}$$

In the above Eqs. 8.14, the coherence factor functions α_κ and $\beta_{-\kappa}$ are computed using appropriate nonstationary boundary conditions for a graphene junction. In the steady state, for a single symmetric chiral crank one gets

$$\alpha_{\kappa} = -\frac{2ks\pi\,(iq\sigma p - iqs\pi + ks\pi + \kappa\sigma p)}{D_{\mathbf{en}}}e^{-\frac{i\phi}{2}} \tag{8.15}$$

$$\beta_{-\kappa} = \frac{2ks\pi\,(iq\sigma p - iqs\pi + ks\pi - \kappa\sigma p)}{D_{\mathbf{en}}}e^{2i\kappa D - \frac{i\phi}{2}} \tag{8.16}$$

$$\begin{aligned}D_{\mathbf{en}} &= \left(e^{2i\kappa D} - 1\right)\left(k^2\pi^2 + q^2\left(\pi^2 - 2s\sigma p\pi\right)\right)\\ &\quad -2k\kappa s\sigma p\left(e^{2i\kappa D} + 1\right)\pi\end{aligned} \tag{8.17}$$

where

$$\begin{aligned}\pi &= \sqrt{\kappa^2 + q^2}\\ p &= \sqrt{k^2 + q^2}\end{aligned} \tag{8.18}$$

$$\varpi_{mn,\,k}\varpi^{\dagger}_{mn,\,k'} = \frac{1}{e^{i2yq_n}} \begin{pmatrix} a_n e^{2(iyq_n)} + b_n z_{n,\,k} \\ a_n z_{n,\,k} e^{2(iyq_n)} + b_n \\ a'_n z_{n,\,k} e^{2(iyq_n)} + b'_n \\ a'_n e^{2(iyq_n)} + b'_n z_{n,\,k} \end{pmatrix}$$

$$\times \left(a_n e^{2iyq_n} + b_n z_{n,\,k} b_n + a_n e^{2iyq_n} z_{n,\,k} b'_n + a'_n e^{2iyq_n} z_{n,\,k} b'_n z_{n,\,k} + a'_n e^{2iyq_n}\right), \quad (8.19)$$

and one uses the properties of the $\delta_{\kappa\kappa'}$ factor which gives

$$\kappa_m = \kappa'_{m'} = \kappa'_{m+l} \quad (8.20)$$

and

$$z_k = \pm \frac{k + iq}{\sqrt{k^2 + q^2}} \quad (8.21)$$

where $\pm$ signs apply to conductive (valence) bands. It satisfies the identity

$$z_k z_{-k} = -1 \quad (8.22)$$

Dependence of the above quantities on the AC frequency and amplitude enters solely via κ_m and π_m. In particular, one obtains

$$\psi_{\mathbf{T}\sigma}\left(x_{\mathbf{D-0}},\, t\right) = \alpha_\kappa\left(t\right) e^{i\zeta(t)} \chi_\kappa e^{i\kappa D} + \beta_{-\kappa}\left(t\right) e^{i\zeta(t)} \chi_{-\kappa} e^{-i\kappa D} \quad (8.23)$$

where we also neglect terms $\propto e^{\pm i2\kappa D}$. One can see that if $l \neq 0$, $G^K_{\mathbf{T}}$ is nonlocal in time. In the latter case, $G^K_{\mathbf{T}}$ depends not only on $t - t'$ but also on $t + t'$. The $\delta_{\kappa\kappa'}$ factor in Eq. 8.9 ensures the electron momentum conservation during the tunneling, thus providing that $\kappa\left(t\right) = \kappa\left(t'\right)$. We also introduce the Wigner function by

$$\left\langle \left[a_{\sigma\kappa},\, a^{\dagger}_{\sigma'\kappa'}\right]_{-} \right\rangle = 2n^{\mathbf{T}}\left(\kappa,\, \kappa'\right) - 1 \quad (8.24)$$

Equation 8.24 actually serves as a definition of the Wigner function $n^{\mathbf{T}}_{\kappa}\left(\kappa,\, \kappa'\right)$ whose time dependence is reflected in its dependence on two quasimoments κ and κ'. Curiously, in this formulation, the nonstationary effects enter via the two electron momenta inside the chiral barrier κ and κ'.

To sum up, we finally arrive at an expression for the time-dependent zero-order Keldysh–Green function of noninteracting electrons in the form

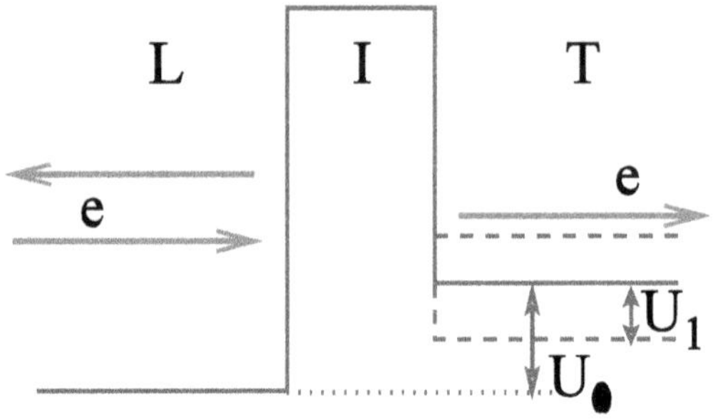

Figure 8.4 An example of a chiral barrier geometry where an AC field of frequency ω adjusts the electrochemical potential in the left electrode by the amount $U_1 \cos \omega t$.

$$G_{\mathbf{T}}^{K}\left(x_{\mathbf{D}-\mathbf{0}},\ t;\ x'_{\mathbf{D}-\mathbf{0}},\ t'\right) = -i\delta_{\sigma\sigma'} \sum_{\kappa} \Xi_{\kappa,\,\kappa'} e^{i[\varsigma(t)-\varsigma(t')]} e^{i\kappa(x-x')} \times \left(2n^{\mathbf{T}}\left(\kappa,\ \kappa'\right) - 1\right) \tag{8.25}$$

8.5 Homogeneous Approximation inside the Chiral Barrier

The Green–Keldysh function Eq. 8.25 of the electron system subjected to influence of an external field depends on two time variables t and t'. The dependence on the two time variables complicates obtaining of the handy kinetic equation which describes the behavior of nonequilibrium electrons. The mentioned issue is resolved using the homogeneous approximation which transforms the Green function Eq. 8.25 into quasihomogeneous form. It allows to obtain a Green function which is dependent just on a single time variable. If the chiral barrier (Fig. 8.4) is relatively short, that is, $L \leq 5\ \mu\text{m}$, and if the AC field is sufficiently weak, which is the case, for example, for amplitude $U_{01} \leq 3 \times 10^{-4}$ eV, Eq. 8.25 becomes drastically simplified. The distribution function $n^{\mathbf{T}}(\kappa,\ \kappa')$ entering Eq. 8.25 can be set as $n^{\mathbf{T}}(\kappa,\ \kappa') \simeq n^{\mathbf{T}}(\kappa)$. It allows to retain the dependence of $n^{\mathbf{T}}(\kappa)$ only on time difference t–t', while the dependence on $t + t'$ in $n^{\mathbf{T}}(\kappa)$ can be neglected. Such a trick corresponds to the use of the so-called homogeneous approximation which insures conservation of the electron momentum on the scale $\Delta x_{\mathbf{ac}}$. Furthermore, in accordance with the aforesaid, in Eq. 8.25, one can replace

$$\Xi_{\kappa,\,\kappa'} \rightarrow \Xi_{\kappa 0}$$
$$n_{\mathbf{B}}\left(\kappa,\ \kappa'\right) \rightarrow \delta_{\kappa,\,\kappa'} n_{\mathbf{B}}\left(\kappa,\ \kappa'\right) = n_{\mathbf{B}}\left(\kappa_0\right) \tag{8.26}$$

The shortest relevant spatial extent $\Delta x_{\mathbf{ac}}$ is then evaluated as

$$\Delta x_{\mathbf{ac}} \leq \min\left\{\lambda_{ac},\ \frac{\pi}{\kappa\left(t\right)-\kappa\left(t'\right)}\right\} \simeq \min\left\{\lambda_{ac},\ \frac{\pi\hbar v}{U_{01}}\right\} \tag{8.27}$$

where

$$\kappa = \kappa\left(t\right) = \sqrt{\left(E + U_0 + U_1\left(t\right)\right)^2 / \left(\hbar v\right)^2 - q^2} \tag{8.28}$$

and

$$\frac{1}{\kappa\left(t\right)-\kappa\left(t'\right)} = \frac{1}{\sqrt{\left(E + U_0 + U_1\left(t\right)\right)^2 / \left(\hbar v\right)^2 - q^2} - \sqrt{\left(E + U_0 + U_1\left(t'\right)\right)^2 / \left(\hbar v\right)^2 - q^2}} \tag{8.29}$$

The above Eqs. 8.28 and 8.29 actually mean that the shortest spatial inhomogeneity is determined rather by the AC field amplitude than by the field frequency ω for the chiral electrons propagating inside the monolayer graphene barrier. One can now evaluate the required contact dimensions for certain AC field parameters which justify the quasihomogeneous form.

Assuming that the external field is weak enough, and using the values $f = 1$ THz, $U_{01} = 0.1$ meV, we find the relevant spatial inhomogeneity as

$$\Delta x_{\mathbf{ac}} \leq \min\left\{17\,\mu\text{m}\right\}, \tag{8.30}$$

whereas for $f = 100$ THz, $U_{01} = 1$ meV

$$\Delta x_{\mathbf{ac}} \leq \min\left\{3\ \mu\text{m},\ 1.6\ \mu\text{m}\right\} \tag{8.31}$$

If the AC field amplitude is constant and if $D \leq \Delta x_{\mathbf{ac}}$, the homogeneous quasistationary approximation is justified, which simplifies Eq. 8.25 with setting $\kappa = \kappa'$ in the formula.

8.5.1 *Tunneling Time and Junction Transparency*

In the Landauer–Büttiker formula for the electric *conductance*

$$G = \frac{2e^2}{h} \int dET(E)\ M(E) \left(-\frac{\partial f}{\partial E}\right) \tag{8.32}$$

where $T(E)$ is an energy-dependent transmission probability and $M(E)$ is the number of conducting channels per energy interval. In the case of pristine *graphene* one merely uses

$$M(E) = \frac{2W|E|}{\pi \hbar v} \tag{8.33}$$

In an assumption of ballistic electron motion through the homogeneous sample without barriers we can also set $T \equiv 1$, which corresponds to an ideal transmission probability. It leads to the simple expression

$$G = \frac{2e^2}{h} \frac{2W|E_{\mathrm{F}}|}{\pi \hbar v} \tag{8.34}$$

Under an assumption of a constant amplitude of the AC field and considering the short junction with length $D \leq \Delta x_{\mathbf{ac}}$, one can apply the homogeneity conditions Eqs. 8.28 and 8.29. Then, in accordance with the quasistationary approximation, in Eq. 8.25, one merely sets $\kappa = \kappa'$, which gives

$$G_{\mathbf{B}}^{K}\left(x_{\mathbf{D-0}},\ t;\ x'_{\mathbf{D-0}},\ t'\right) = -i\delta_{\sigma\sigma'} \sum_{\kappa} \Xi_{\kappa_0} e^{i[\zeta(t)-\zeta(t')]} e^{i\kappa(x-x')} \times \left(2n_{\mathbf{B}}(\kappa_0) - 1\right) \tag{8.35}$$

In the above Eq. 8.35, $n_{\mathbf{B}}(\kappa_0)$ is the stationary nonequilibrium distribution function inside the chiral barrier T obtained in the homogeneous approximation. Besides, in Eq. 8.35

$$\Xi_{\kappa} = |\varpi_{\kappa}|^2 \left(|\alpha_{\kappa}|^2 + |\beta_{-\kappa}|^2\right), \tag{8.36}$$

and the Kronecker symbol $\delta_{\sigma,\sigma'}$ reflects our interest in the homogeneous solutions (on the scale $\sim D$ of the gated barrier region). In many practical considerations, it is frequently convenient to use the Fourier transform of the Green function

$$G_B^K(\kappa,\ \varepsilon,\ T) = \frac{1}{\pi}\int d\tau G_B^K(\kappa,\ \tau,\ T)\ e^{-i\tau\varepsilon} \tag{8.37}$$

where we have introduced new time variables

$$\begin{gathered}\tau = t - t' \text{ and } T = t + t' \\ t = \frac{T+\tau}{2};\ t' = \frac{T-\tau}{2}\end{gathered} \tag{8.38}$$

In the above representation, the Green function depends on two time arguments, τ and T.

8.6 Expressions for Advanced Green Functions

Using the homogeneous quasistationary approximation, one obtains

$$\begin{aligned}G_{0\mathbf{T}}^R(x,\ t;\ x',\ t') &= -i\theta(t-t')\left\langle\left[\psi_{\mathbf{T}\sigma}(x,\ t)\ \psi_{\mathbf{T}\sigma'}^\dagger(x',\ t')\right]_+\right\rangle \\ &= -i\theta(t-t')\sum_{m,\ m',\ \kappa,\ \kappa'}\left(\alpha_\kappa(t)\ e^{i\varsigma(t)}\chi_\kappa e^{i\kappa D}\right. \\ &\quad\left.-\beta_{-\kappa}(t)\ e^{i\varsigma(t)}\chi_{-\kappa}e^{-i\kappa D}\right) \\ &\quad\times\left(\alpha_{\kappa'}(t)\ e^{i\varsigma(t')}\chi_{\kappa'}e^{i\kappa' D}\right. \\ &\quad\left.-\beta_{-\kappa'}(t)\ e^{i\varsigma(t')}\chi_{-\kappa'}e^{-i\kappa' D}\right)^* \\ &\quad\times e^{i\varsigma(t)-i\varsigma(t')}\varpi_\kappa\varpi_{\kappa'}^\dagger e^{i\kappa x-i\kappa' x'}\left\langle\left[a_{\sigma k},\ a_{\sigma' k'}^\dagger\right]_+\right\rangle \\ &= -i\theta(t-t')\ \delta_{\sigma\sigma'}\sum_\kappa \Xi_\kappa e^{i\kappa(x-x')}e^{i\varsigma(t)-\varsigma(t')}\end{aligned} \tag{8.39}$$

where

$$\varsigma(t) = \frac{1}{\hbar}\int_{-\infty}^t dt\,(E + U_0 + U_{01}\cos\omega t) \tag{8.40}$$

$$\varsigma(t) - \varsigma(t') = \frac{1}{\hbar}(E+U_0)\ (t-t') + \frac{U_{01}}{\hbar\omega}(\sin\omega t - \sin\omega t') \tag{8.41}$$

$$
\begin{aligned}
e^{i[\zeta(t)-\zeta(t')]} &= \exp\left\{i\left[\frac{1}{\hbar}(E+U_0)(t-t') + \frac{U_{01}}{\hbar\omega}(\sin\omega t - \sin\omega t')\right]\right\} \\
&= \exp\left\{i\left[\frac{1}{\hbar}(E+U_0)(t-t')\right]\right\} \\
&\quad \exp\left\{i\frac{U_{01}}{\hbar\omega}(\sin\omega t - \sin\omega t')\right\} \\
&= \exp\left\{i\left[\frac{1}{\hbar}(E+U_0)(t-t')\right]\right\} \\
&\quad \exp\left\{i\frac{U_{01}}{\hbar\omega}\sin\omega t\right\}\exp\left\{-i\frac{U_{01}}{\hbar\omega}\sin\omega t'\right\} \\
&= e^{i(E+U_0)(t-t')/\hbar}\sum_{m,\,n} J_m\left(\frac{U_{01}}{\hbar\omega}\right) J_n\left(\frac{U_{01}}{\hbar\omega}\right) e^{im\omega t - in\omega t'} \\
&= e^{i(E+U_0)\tau/\hbar}\sum_{m,\,n} J_m\left(\frac{U_{01}}{\hbar\omega}\right) J_n \\
&\quad \times\left(\frac{U_{01}}{\hbar\omega}\right) e^{i(m+n)\omega\tau/2} e^{i(m-n)\omega T/2}
\end{aligned} \tag{8.42}
$$

The last formula is practically of no utility unless we admit that the system inside the chiral barrier is stationary and homogeneous. Then, the electron momentum is conserved inside the barrier, $\kappa = \kappa'$, which in general is not evident. If that is the case, then

$$
\begin{aligned}
G_{0\mathbf{B}}^{R(A)}(\kappa,\,\varepsilon,\,T) &= \frac{1}{\pi}\int d\tau G_{0\mathbf{B}}^{R(A)}(\tau,\,T)\,e^{-i\tau\varepsilon} \\
&= -\,i\delta_{\sigma\sigma'}\sum_{\kappa}\Xi_{\kappa}e^{i\kappa(x-x')}\sum_{m,\,n} J_m\left(\frac{U_{01}}{\hbar\omega}\right) J_n\left(\frac{U_{01}}{\hbar\omega}\right) \\
&\quad \times\int d\tau\theta(\tau)\,e^{-i\tau\varepsilon}e^{i(E_\kappa+U_0)\tau/\hbar}e^{i(m+n)\omega\tau/2}e^{i(m-n)\omega T/2}
\end{aligned} \tag{8.43}
$$

In the last Eq. 8.43, for a time-averaged Green function which is independent versus $T = t + t'$, one sets $m = n$, which gives

$$
\begin{aligned}
G_{0\mathbf{B}}^{R(A)}(\kappa,\,\varepsilon) &= -\,i\delta_{\sigma\sigma'}\sum_{\kappa}e^{i\kappa(x-x')}\sum_{n} J_n^2\left(\frac{U_{01}}{\hbar\omega}\right)\frac{1}{2\pi i} \\
&\quad \times\int d\tau\,\Xi_{\kappa}\int_{-\infty}^{\infty} d\Omega\frac{e^{i\Omega\tau}}{\Omega\pm i\delta}e^{i\tau[E_\kappa+U_0-\varepsilon+\hbar n\omega]/\hbar} \qquad (8.44) \\
&= \Xi_{\kappa}\sum_{n} J_n^2\left(\frac{U_{01}}{\hbar\omega}\right)\frac{1}{\varepsilon - E_\kappa - U_0 - \hbar n\omega \pm i\delta}
\end{aligned}
$$

$$G_{0\mathbf{B}}^{K}(\kappa,\ \varepsilon) = \overline{\Xi}_{\kappa} \sum_{n} J_n^2 \left(\frac{U_{01}}{\hbar\omega}\right) \delta(\varepsilon - E_{\kappa} - U_0 - \hbar n\omega) \left[2n_{\mathbf{B}}(\varepsilon) - 1\right] \tag{8.45}$$

where $\kappa = \kappa(\tau)$ and we introduced the time-averaged function

$$\overline{\Xi}_{\kappa} = \frac{\omega}{2\pi} \int d\tau\, \Xi_{\kappa(\tau)} \tag{8.46}$$

where $W_m = E + U_0 + m\hbar\Omega + \varepsilon$. If, for a moment, one sets $\Xi_{\kappa(\tau)} = 1$, then $\varsigma(\Omega) = \delta(\Omega)$, that is, the function Eq. 8.40 coincides with the Dirac δ function. Besides, one uses

$$\frac{1}{2\pi i} \int_{-\infty}^{\infty} d\Omega \frac{e^{i\Omega\tau}}{\Omega \pm i\delta} = \begin{cases} \theta(\tau) \\ -\theta(-\tau) \end{cases} \tag{8.47}$$

where the Heaviside step function is defined as

$$\theta(\tau) = \begin{cases} 0 & \text{at} \quad \tau < 0 \\ 1/2 & \text{at} \quad \tau = 0 \\ 1 & \text{at} \quad \tau > 0 \end{cases} \tag{8.48}$$

8.7 The δ Function Approximation

Here we exploit the time periodicity of the Ξ_{κ} function and, for convenience, we also introduce an auxiliary function

$$\zeta_{\kappa} = |\varpi_{\kappa 0}|^2 \left(|\alpha_{\kappa 0}|^2 + |\beta_{-\kappa 0}|^2\right) \tag{8.49}$$

where we have introduced the electron wavevector inside the chiral barrier as $\kappa_0 = \sqrt{(E + U_0)^2 / (\hbar v)^2 - q^2}$. Besides, taking into account our homogeneous quasistationary approximation, one gets

$$G_{0\mathbf{B}}^{R(A)}(\kappa,\ \varepsilon) = \zeta_{\kappa} \sum_{n} J_n^2 \left(\frac{U_{01}}{\hbar\omega}\right) \frac{1}{\varepsilon - E_{\kappa} - U_0 - \hbar n\omega \pm i\delta} \tag{8.50}$$

$$G_{0\mathbf{B}}^{K}(\kappa,\ \varepsilon) = \zeta_{\kappa} \sum_{n} J_n^2 \left(\frac{U_{01}}{\hbar\omega}\right) \delta(\varepsilon - E_{\kappa} - U_0 - \hbar n\omega) \left[2n_{\mathbf{B}}(\varepsilon) - 1\right] \tag{8.51}$$

$$G_{0\mathbf{R}}^{R(A)}(\kappa,\ \varepsilon) = \mathcal{T}_k |\varpi_k|^2 \frac{1}{\varepsilon - \varepsilon_k - eV_{\mathbf{R}} \pm i\delta} \tag{8.52}$$

$$\Sigma_{\mathbf{L(R)}}^{K} = \mathcal{T}_{\mathbf{L(R)}}^{\mathbf{2}} G_{\mathbf{T}}^{K}(\kappa,\ \varepsilon) \tag{8.53}$$

$$\Sigma_{\mathbf{T}}^{K} = \mathcal{T}_{\mathbf{L}}^{2} G_{\mathbf{L}}^{K}(k,\ \varepsilon - eV/2) + \mathcal{T}_{\mathbf{R}}^{2} G_{\mathbf{R}}^{K}(k,\ \varepsilon + eV/2) \tag{8.54}$$

$$\begin{aligned}
\Sigma_{\mathbf{T}}^{R(A)} &= \mathcal{T}_{\mathbf{L}}^{2} G_{\mathbf{L}}^{R(A)}(k,\ \varepsilon - eV/2) + \mathcal{T}_{\mathbf{R}}^{2} G_{\mathbf{R}}^{R(A)}(k,\ \varepsilon + eV/2) \\
&= \mathcal{T}_{\mathbf{L}}^{2} \sum_{k} T_k \frac{1}{\varepsilon - \varepsilon_k + eV/2 \pm i\delta} \\
&\quad + \mathcal{T}_{\mathbf{R}}^{2} \sum_{k} T_k \frac{1}{\varepsilon - \varepsilon_k - eV/2 \pm i\delta}
\end{aligned} \tag{8.55}$$

$$\begin{aligned}
\Gamma_{\mathcal{T}} &= i\left[\Sigma_{\mathcal{T}}^{R} - \Sigma_{\mathcal{T}}^{A}\right] \\
&= i\,\mathrm{Im}\left[\mathcal{T}_{\mathbf{L}}^{2} \sum_{k} T_k \left|\varpi_k\right|^2 \frac{1}{\varepsilon - \varepsilon_k + eV/2 \pm i\delta}\right. \\
&\quad \left. + \mathcal{T}_{\mathbf{R}}^{2} \sum_{k} T_k \left|\varpi_k\right|^2 \frac{1}{\varepsilon - \varepsilon_k - eV/2 \pm i\delta}\right] \\
&= \pi\left[\mathcal{T}_{\mathbf{L}}^{2} T_{E+eV/2} + \mathcal{T}_{\mathbf{R}}^{2} T_{E-eV/2}\right]
\end{aligned} \tag{8.56}$$

where we've utilized that

$$\begin{aligned}
G_{\mathbf{R}}^{K}\left(x,\ t;\ x',\ t'\right) &= -i\left\langle\left[\psi_{\mathbf{R}\sigma}(x,\ t)\ \psi_{\mathbf{R}\sigma'}^{\dagger}\left(x',\ t'\right)\right]_{-}\right\rangle \\
&= -i\sum_{k,\ k'} t_k(t)\ t_k^{*}\left(t'\right)\ e^{i\varepsilon_k t - i\varepsilon_{k'} t'} e^{iV_{\mathbf{R}}(t-t')} \\
&\quad \times \varpi_k \varpi_{k'}^{\dagger} e^{ik(x-D) - ik'(x'-D)} \left\langle\left[a_{\sigma k},\ a_{\sigma' k'}^{\dagger}\right]_{-}\right\rangle \\
&= -i\sum_{k,\ k'} t_{kn}(t)\ t_{kn}^{*}\left(t'\right)\ e^{i\varepsilon_k(t-t')} e^{iV_{\mathbf{R}}(t-t')} \\
&\quad \times \varpi_k \varpi_{k'}^{\dagger} e^{ik(x-x')} \delta_{\sigma\sigma'} \delta_{kk'}\left(2n_k^{\mathbf{R}} - 1\right) \\
&= -i\delta_{\sigma\sigma'} \sum_{k,\ k'} T_k e^{i(\varepsilon_k - eV_{\mathbf{R}})(t-t')} \varpi_k \varpi_{k'}^{\dagger} e^{ik(x-x')} \delta_{kk'} \\
&\quad \times \left(2n_k^{\mathbf{R}} - 1\right) \\
&= -i\delta_{\sigma\sigma'} \sum_{k} T_k \left|\varpi_k\right|^2 \left(2n_k^{\mathbf{R}} - 1\right)\ e^{i(\varepsilon_k - eV_{\mathbf{R}})(t-t')} e^{ik(x-x')}
\end{aligned} \tag{8.57}$$

8.8 Photon-Assisted Tunneling Current through the Chiral Barrier

Many nanodevices are constructed as arrays of tiny islands with attached electrodes. The electron concentrations inside the islands are controlled applying of electric potential to the local electrostatic gates. Furthermore, the nonequilibrium electrons are injected to the islands with the external electrodes. When an external AC field is applied, the electron tunneling becomes photon assisted. Electron tunneling injection, along with the absorption of photons of the AC field, serves as a source of energy supplied to each island from outside. Let us consider a tiny island with two electrodes connected to it, on the left (L) and on the right (R). The island is subjected to an external AC field which makes the tunneling to be photon assisted. In conditions of the photon-assisted tunneling through the graphene chiral barrier, one uses the modified Landauer–Büttiker formula for the electric current

$$\begin{aligned} I\left(V_{\mathbf{LR}}\right) &= 4ev\int d\varepsilon\, N\left(\varepsilon\right)\sum_{m,\,k}\mathrm{Tr}\left\{\chi_{\varepsilon}\chi_{\varepsilon}^{\dagger}\right\} \\ &\quad\times\left(T_k\left(2n_k^{\mathbf{R}}-1\right)\delta\left(\varepsilon-\varepsilon_k+eV_{\mathbf{R}}\right)\right. \\ &\quad\left.-T_k\left(2n_k^{\mathbf{L}}-1\right)\delta\left(\varepsilon-\varepsilon_k-eV_{\mathbf{L}}\right)\right) \\ &= 4ev\int d\varepsilon\, N\left(\varepsilon\right)\mathrm{Tr}\left\{\chi_{\varepsilon}\chi_{\varepsilon}^{\dagger}\right\}\,T_{\varepsilon}\left(n_{\varepsilon}-n_{\varepsilon-eV_{\mathbf{LR}}}\right) \end{aligned} \tag{8.58}$$

An extra factor $\mathrm{Tr}\{\chi_{\varepsilon}\chi_{\varepsilon}^{\dagger}\}$ in the last formula, Eq. 8.58, accounts for the effect of electron chirality. The right and left junction's electrodes L and R are strongly coupled to external wires. Therefore, in the above Eq. 8.58, the electron distribution functions $n_{\varepsilon}^{\mathbf{R}}$ and n_{ε}^{L} in the L and R electrodes are well approximated by the equilibrium Fermi distribution

$$n_{\varepsilon}^{\mathbf{R}} = n_{\varepsilon}^{\mathbf{L}} = n_{\varepsilon} = \frac{1}{\exp\left(\frac{\varepsilon}{T}\right)+1} \tag{8.59}$$

For a monoatomic infinite pristine graphene sheet, the electron density of states is written as

$$N\left(\varepsilon\right) = \frac{WL|\varepsilon|}{2\pi\hbar^2 v^2} \tag{8.60}$$

where W is the ribbon width and L is the chiral barrier length. If the sample has finite dimensions, the electron density of states $N\left(\varepsilon\right)$ also takes into account the singularities which originate from geometrical quantization and interaction with the substrate. An asymmetry of the electron–hole shoulders in our symmetric FET nanotube setup originates from a different energy dependence of the transmission coefficient $t_\varepsilon^{p,\,n}$ through the Schottky barrier for electron and holes. For a symmetric double-barrier junction, the Landauer–Büttiker formula reads

$$I_0 = 4ev \int d\varepsilon N(\varepsilon)\, \breve{T}(\varepsilon) \cdot [n_0(\varepsilon) - n_0(\varepsilon - eV_{LR})] \qquad (8.61)$$

$$I_{\mathbf{ne}} = 4ev \int d\varepsilon N(\varepsilon) \cdot \sum_m (\mathbf{A}_{\varepsilon - \hbar m\omega})^2 \cdot \breve{T}(\varepsilon - \hbar m\omega)$$
$$\times [n_{\mathbf{B}}(\varepsilon) - n_{\mathbf{B}}(\varepsilon - eV_{\mathbf{LR}})] \qquad (8.62)$$

where $\breve{T}(\varepsilon) = \mathrm{Tr}\{\chi_\varepsilon \chi_\varepsilon^\dagger\}\, T_\varepsilon$, $\mathbf{A}_\varepsilon = |\alpha_\varepsilon|^2 + |\beta_\varepsilon|^2$, α_ε and β_ε are given in the previous subsections, and $n_{\mathbf{B}}(\varepsilon)$ is the nonequilibrium distribution function on the central island. The additional term contributes in highly transparent junctions when the external AC field amplitude is sufficiently high.

8.9 Electron Self-Energy and Many-Body Effects

The electron Green–Keldysh function (Keldysh 1965) inside the central island T depends on the self-energy as

$$\left(i\frac{\partial}{\partial t} - W_\kappa - eU_1(t)\right) G_T^K = \Sigma_T^K \circ G_T^A + \Sigma_T^R \circ G_T^K \qquad (8.63)$$

where W_κ is the energy of a particle with wavevector κ, $U_1(t) = U_1 \cos\omega t$ is the AC bias of the electrochemical potential. In the nonstationary case, the product $A \circ B$ means

$$A \circ B_{t,\,t'} \to \int A(t,\, t_1)\, B(t_1,\, t')\, dt_1. \qquad (8.64)$$

The integral over time in Eq. 8.64 originates from the dependence on two time variables $\tau = t - t'$ and $T = t + t'$ (see also Eq. 8.25). The product, Eq. 8.64, can be simplified using the homogeneous approximation as discussed in Section 8.5. The right part in Eq. 8.63 contains the electron self-energies which account for the interaction of electrons with phonons, with the other electrons, and with the attached electrodes. Introducing the electron tunneling self-energy, one describes the effect of the nonequilibrium electron distribution on the central island induced by an electric current flowing through the quantum well.

Coupling of the central island to the L and R electrodes is included into the full Green function (Keldysh 1965), which acquires the form

$$G_{\mathbf{L}}^{K} \simeq G_{0\mathbf{L}}^{K} + G_{0\mathbf{L}}^{R}\Sigma_{\mathbf{L}}^{K}G_{\mathbf{L}}^{A} \qquad (8.65)$$

In the last Eq. 8.65, the transmission self-energy Σ_L^K describes coupling between the central island T and attached L and R electrodes, which takes the form

$$\begin{aligned}\Sigma_{\mathbf{L}}^{K} &= \mathbf{T}^{\mathbf{LT}}G_{\mathbf{T}}^{K}\left(x_{+0},\ x'_{+0}\right)\\ \Sigma_{\mathbf{T}}^{K} &= \mathbf{T}^{\mathbf{TL}}G_{\mathbf{L}}^{K}\left(x_{-0},\ x'_{-0}\right) + \mathbf{T}^{\mathbf{TR}}G_{\mathbf{R}}^{K}\left(x_{\mathbf{D}+0},\ x'_{\mathbf{D}+0}\right)\end{aligned} \qquad (8.66)$$

In the above Eq. 8.66, $\mathbf{T}^{\mathbf{ij}}$ are overlap integrals between two atomic sites located in adjacent electrodes.

8.10 Kinetic Equation for n_κ^T

Below we derive the kinetic equation for the electron distribution function inside the chiral barrier B. One generally computes the deviation $\delta n_\varepsilon^{\mathbf{B}} = n_\varepsilon^{\mathbf{B}} - n_\varepsilon^{(0)}$ of the electron distribution $n_\varepsilon^{\mathbf{B}}$ from its equilibrium steady-state value $n_\varepsilon^{(0)} = 1/\left(\exp\{\varepsilon/T\} + 1\right)$ utilizing the kinetic equation (Keldysh 1965) for the electron distribution function $n_\varepsilon^{\mathbf{B}}$ on the island. For the electric potential in the left (L) and right (R) electrodes one sets $V_{\mathbf{L(R)}} = \pm V/2$. Following Keldysh (1965), the kinetic equation is obtained using the above Eq. 8.63 complemented by a similar equation containing the time derivative $\partial/\partial t'$ instead of $\partial/\partial t$. The two equations are integrated over the energy variable ε and subtracted from one another. In the term containing $i\left(\partial/\partial t - \partial/\partial t'\right)$ one replaces

$$\frac{\partial}{\partial\left(t+t'\right)} \rightarrow \frac{\partial}{\partial T} \qquad (8.67)$$

$$\left(i\frac{\partial}{\partial t} - E - U_0 - \overset{\text{changes the wf phase}}{U_1(t)}\right) G_{\mathbf{T}}^{K} = \Sigma_{\mathbf{T}}^{K}G_{\mathbf{T}}^{A} + \Sigma_{\mathbf{T}}^{R}G_{\mathbf{T}}^{K} + \breve{I}_{\mathbf{ep}} + \breve{I}_{\mathbf{es}} \qquad (8.68)$$

In the collision integral, which is obtained using the right-hand side of Eq. 8.68, one gets

$$\Sigma_{\mathbf{T}}^{K}G_{\mathbf{T}}^{A} + \Sigma_{\mathbf{T}}^{R}G_{\mathbf{T}}^{K} + \breve{I}_{\mathbf{ep}} + \breve{I}_{\mathbf{es}} \qquad (8.69)$$

where

$$\Sigma_{\mathbf{T}}^{K} G_{\mathbf{T}}^{A} = \left(\mathcal{T}_{\mathbf{L}}^{\mathbf{2}} G_{\mathbf{L}}^{K}(k,\ \varepsilon - eV/2) + \mathcal{T}_{\mathbf{R}}^{\mathbf{2}} G_{\mathbf{R}}^{K}(k,\ \varepsilon + eV/2)\right) \cdot G_{\mathbf{T}}^{A}$$
$$= (-i)\ \mathcal{T}_{\mathbf{L}}^{\mathbf{2}} \breve{T}_{\varepsilon} \delta\left(\varepsilon - \varepsilon_{k} - eV/2\right)\left(2n_{k}^{\mathbf{L}} - 1\right)$$
$$\times \zeta_{\kappa} \sum_{n} J_{n}^{2}\left(\frac{U_{01}}{\hbar\omega}\right) \frac{1}{\varepsilon - W_{\kappa} - \hbar\omega m - i\delta}$$
$$+ (-i)\ \mathcal{T}_{\mathbf{R}}^{\mathbf{2}} \breve{T}_{\varepsilon} \delta\left(\varepsilon - \varepsilon_{k} + eV/2\right)\left(2n_{k}^{\mathbf{R}} - 1\right)$$
$$\times \zeta_{\kappa} \sum_{n} J_{n}^{2}\left(\frac{U_{01}}{\hbar\omega}\right) \frac{1}{\varepsilon - W_{\kappa} - \hbar\omega m - i\delta} \tag{8.70}$$

In Eq. 8.70, we have introduced the electron energy $W_{\kappa} = E + U_{0}$ shifted by the gate-induced potential U_{0}. After integrating over the electron energy variable ε one obtains

$$\int d\varepsilon \Sigma_{\mathbf{T}}^{K} G_{\mathbf{T}}^{A} = \zeta_{\kappa} \sum_{m} J_{m}^{2}\left(\frac{U_{01}}{\hbar\omega}\right) \breve{T}_{k,\kappa}\left[T_{\mathbf{L}}^{2}\left(2n_{k}^{\mathbf{L}} - 1\right)\right.$$
$$\times \frac{1}{\varepsilon_{k} + eV/2 - W_{\kappa} - \hbar\omega m - i\delta}$$
$$\left. + T_{\mathbf{R}}^{2}\left(2n_{k}^{R} - 1\right) \frac{1}{\varepsilon_{k} - eV/2 - W_{\kappa} - \hbar\omega m - i\delta}\right] \tag{8.71}$$

$$\Sigma_{\mathbf{T}}^{R} G_{\mathbf{T}}^{K} = \left[\mathbf{T}_{\mathbf{L}}^{\mathbf{2}} G_{\mathbf{L}}^{R}(k,\ \varepsilon - eV/2) + \mathbf{T}_{\mathbf{R}}^{\mathbf{2}} G_{\mathbf{R}}^{R}(k,\ \varepsilon + eV/2)\right] \cdot G_{\mathbf{T}}^{K}$$
$$= \left(\mathbf{T}_{\mathbf{R}}^{\mathbf{2}} \frac{1}{\varepsilon - \varepsilon_{k} + eV/2 + i\delta} + \mathbf{T}_{\mathbf{L}}^{\mathbf{2}} \frac{1}{\varepsilon - \varepsilon_{k} - eV/2 + i\delta}\right) \cdot \breve{T}_{\varepsilon} G_{\mathbf{T}}^{K} \tag{8.72}$$

Furthermore, one gets

$$\int d\varepsilon \Sigma_{\mathbf{T}}^{R} G_{\mathbf{T}}^{K} = -i\delta_{\sigma\sigma'} \frac{1}{\pi} \zeta_{\kappa} \sum_{m} J_{m}^{2}\left(\frac{U_{01}}{\hbar\omega}\right) \breve{T}_{k\kappa}\left(2n^{\mathbf{T}}(W_{\kappa}) - 1\right)$$
$$\left[\mathbf{T}_{\mathbf{R}}^{2} \frac{1}{W_{\kappa} + \hbar\omega m - \varepsilon_{k} + eV/2 + i\delta}\right.$$
$$\left. + \mathbf{T}_{\mathbf{L}}^{2} \frac{1}{W_{\kappa} + \hbar\omega m - \varepsilon_{k} - eV/2 + i\delta}\right] \tag{8.73}$$

where one uses that

$$G_{\mathbf{T}}^{K}(\kappa,\ \varepsilon,\ T) = -i\delta_{\sigma\sigma'} \frac{1}{\pi} \zeta_{\kappa} \sum_{m} J_{m}^{2}\left(\frac{U_{01}}{\hbar\omega}\right)$$
$$\times \left(2n^{\mathbf{T}}(W_{\kappa}) - 1\right) \delta\left(W_{\kappa} + \hbar\omega m - \varepsilon\right) \tag{8.74}$$

$$G_{0\mathbf{L(R)}}^{R(A)}(k,\ \varepsilon) = \breve{T}_\varepsilon \frac{1}{\varepsilon - \varepsilon_k \mp eV_{\mathbf{L(R)}} \pm i\delta} \tag{8.75}$$

$$G_{0\mathbf{L(R)}}^{K}(k,\ \varepsilon) = (-i)\ \breve{T}_\varepsilon \delta\left(\varepsilon - \varepsilon_k \mp eV_{\mathbf{L(R)}}\right)\left(2n_k^{\mathbf{L(R)}} - 1\right) \tag{8.76}$$

$$\Sigma_{\mathbf{L(R)}}^{K} = \mathcal{T}_{\mathbf{L(R)}}^{\mathbf{2}} G_{\mathbf{B}}^{K}(\kappa,\ \varepsilon) \tag{8.77}$$

$$\Sigma_{\mathbf{T}}^{R} = \mathcal{T}_{\mathbf{L}}^{\mathbf{2}} G_{\mathbf{L}}^{R}(k,\ \varepsilon - eV/2) + \mathcal{T}_{\mathbf{R}}^{\mathbf{2}} G_{\mathbf{R}}^{R}(k,\ \varepsilon + eV/2) \tag{8.78}$$

where $T^{\mathbf{L(R)}}$ are the partial transparencies of the left (right) interfaces, T_k is the net junction transparency, and $\breve{T}_{k,m} = T_k \varpi_{n,k} \varpi_{n,k'}^{\dagger}$ depends on the crystallographic orientation.

Collecting the terms in the above formulas, one gets

$$\begin{aligned}\int \left(\Sigma_{\mathbf{T}}^{K} G_{\mathbf{T}}^{A} + \Sigma_{\mathbf{T}}^{R} G_{\mathbf{T}}^{K}\right) d\varepsilon = \zeta_\kappa \sum_m J_m^2\left(\frac{U_{01}}{\hbar\omega}\right) \breve{T}_{k,\kappa} \Bigg[& T_{\mathbf{L}}^2 \left(2n_k^L - 1\right) \\ & \times \frac{1}{\varepsilon_k + eV/2 - W_\kappa - \hbar\omega m - i\delta} \\ & + \mathbf{T}_{\mathbf{R}}^2 (2n_k^R - 1) \frac{1}{\varepsilon_k - eV/2 - W_\kappa - \hbar\omega m - i\delta} \\ & - \left(2n^{\mathbf{T}}(W_\kappa) - 1\right) \\ & \times \Bigg[\mathbf{T}_{\mathbf{R}}^2 \frac{1}{-W_\kappa - \hbar\omega m + \varepsilon_k - eV/2 - i\delta} \\ & + \mathbf{T}_{\mathbf{L}}^2 \frac{1}{-W_\kappa - \hbar\omega m + \varepsilon_k + eV/2 - i\delta}\Bigg]\Bigg]\end{aligned} \tag{8.79}$$

The photon-assisted tunneling source of the nonequilibrium electron distribution inside the central island is obtained as

$$\begin{aligned}\int \left(\Sigma_{\mathbf{T}}^{K} G_{\mathbf{T}}^{A} + \Sigma_{\mathbf{T}}^{R} G_{\mathbf{T}}^{K}\right) d\varepsilon = & 2\zeta_\kappa \sum_m J_m^2\left(\frac{U_{01}}{\hbar\omega}\right) \breve{T}_{k,\kappa} \\ & \times \left[\mathbf{T}_{\mathbf{L}}^2 \frac{n_k^L - n_B\left(\kappa_m,\ \kappa'_{m+l}\right)}{\varepsilon_k + eV/2 - W_\kappa - \hbar\omega m - i\delta} + \mathbf{T}_{\mathbf{R}}^2 \frac{n_k^R - n_B\left(\kappa_m,\ \kappa'_{m+l}\right)}{\varepsilon_k - eV/2 - W_\kappa - \hbar\omega m - i\delta}\right]\end{aligned} \tag{8.80}$$

and again, $T^{\mathbf{L(R)}}$ are fractional transparencies of the left (right) interfaces, $T_{k,\kappa}^m$ is the net junction transparency,

$$\breve{T}_{k,\kappa} = T_{k,\kappa} \left|\varpi_k\right|^2, \tag{8.81}$$

and the product $\varpi_k \varpi_{k'}^{\dagger}$ depends on the crystallographic orientation. The tunneling source I_T takes the form

$$J_{\mathbf{T}} = Im \int \left(\Sigma_{\mathbf{T}}^{K} G_{\mathbf{T}}^{A} + \Sigma_{\mathbf{T}}^{R} G_{\mathbf{T}}^{K}\right) d\varepsilon = 2\pi \zeta_\kappa \sum_m J_m^2 \left(\frac{U_{01}}{\hbar\omega}\right) \times \left[T_{\mathbf{L}}^2 \left[n_k^L - n_B\left(W_\kappa\right)\right] \delta\left(\varepsilon_k + eV/2 - W_\kappa - \hbar\omega m\right) + T_{\mathbf{R}}^2 \left[n_k^R - n_B\left(W_\kappa\right)\right] \delta\left(\varepsilon_k - eV/2 - W_\kappa - \hbar\omega m\right)\right] \tag{8.82}$$

where

$$\zeta_\kappa = \left|\varpi_{\kappa 0}\right|^2 \left(\left|\alpha_{\kappa 0}\right|^2 + \left|\beta_{-\kappa 0}\right|^2\right), \tag{8.83}$$

which gives the time-averaged nonequilibrium photon-assisted tunneling source in the form

$$\begin{aligned} \mathbf{J}_T^0 &= 2\pi \zeta_\kappa \sum_m J_m^2 \left(\frac{U_{01}}{\hbar\omega}\right) \breve{T}_\kappa^{(m)} [\mathbf{T}_{\mathbf{L}}^2 \left[n_0\left(W_\kappa - eV_{\mathbf{LR}}/2\right)\right. \\ &\quad \left. - n_{\mathbf{B}}\left(W_\kappa + \hbar\omega m\right)\right] \\ &\quad + \mathcal{T}_{\mathbf{R}}^2 \left[n_0\left(W_\kappa + eV/2\right) - n_{\mathbf{B}}\left(W_\kappa + \hbar\omega m\right)\right]] \\ &= 2\pi \sum_m J_m^2 \left(\frac{U_{01}}{\hbar\omega}\right) \left[\Gamma_\kappa^{\mathbf{L}} \left[n_0\left(W_\kappa - eV_{\mathbf{LR}}\right) - n_{\mathbf{B}}\left(W_\kappa + \hbar\omega m\right)\right]\right. \\ &\quad \left. + \Gamma_\kappa^{\mathbf{R}} \left[n_0\left(W_{\kappa_m}\right) - n_{\mathbf{B}}\left(W_\kappa + \hbar\omega m\right)\right]\right] \end{aligned} \tag{8.84}$$

where the tunneling rates $\Gamma_\kappa^{\mathbf{L(R)}}$ between the left (right) electrodes and the central island are

$$\Gamma_\kappa^{\mathbf{L(R)}} = \zeta_\kappa \breve{T}_\kappa \mathbf{T}_{\mathbf{L(R)}}^{\mathbf{2}} = \mathbf{T}_{\mathbf{L(R)}}^{\mathbf{2}} T_{k,\kappa} \left|\varpi_k\right|^4 \cdot \left(\left|\alpha_\kappa\right|^2 + \left|\beta_{-\kappa}\right|^2\right) \tag{8.85}$$

In the above formulas, we have made the following simplifications: We have replaced the nonequilibrium electron distribution functions $n_k^{\mathbf{L}}$ in the electrodes with their equilibrium values, Eq. 8.59, which is consistent with imposing of the equilibrium boundary conditions. Besides, we have replaced the electron Wigner function $n_{\mathbf{B}}\left(\kappa_m, \kappa_{m+l}'\right)$ on the island by a nonequilibrium electron distribution function $n_{\mathbf{B}}\left(W_\kappa\right)$.

$$\alpha_\kappa = -\frac{2ks\pi\left(iq\sigma p - iqs\pi + ks\pi + \kappa\sigma p\right)}{D_{\mathbf{en}}} e^{-\frac{i\phi}{2}} \tag{8.86}$$

$$\beta_{-\kappa} = \frac{2ks\pi\left(iq\sigma p - iqs\pi + ks\pi - \kappa\sigma p\right)}{D_{\mathbf{en}}} e^{2i\kappa D - \frac{i\phi}{2}} \tag{8.87}$$

$$D_{\mathbf{en}} = \left(e^{2i\kappa D} - 1\right)\left(k^2\pi^2 + q^2\left(\pi^2 - 2s\sigma p\pi\right)\right) - 2k\kappa s\sigma p\left(e^{2i\kappa D} + 1\right)\pi \tag{8.88}$$

where we have used the following notations:

$$\pi = \sqrt{\kappa^2 + q^2},\ p = \sqrt{k^2 + q^2} \tag{8.89}$$

8.11 Symmetric Junction

For the case of a symmetric double-barrier junction with identical L and R electrodes and barriers (see Fig. 8.3, left), in Eq. 8.84 one sets $\Gamma_E^{\mathbf{L(R)}} = \Gamma_\kappa^{\mathbf{ext}}$, which gives the tunneling source entering the quantum kinetic equation in the form

$$\begin{aligned}\mathbf{J}_T^0 = 2\pi \sum_m J_m^2 \left(\frac{U_{01}}{\hbar\omega}\right) \Gamma_\kappa^{\mathbf{ext}} \big[& n_0 \left(W_{\kappa_m} - eV_{\mathbf{LR}}\right) \\ & + n_0 \left(W_{\kappa_m}\right) - 2n_{\mathbf{B}} \left(W_{\kappa_m} + \hbar\omega m\right) \big]\end{aligned} \tag{8.90}$$

The above source, Eq. 8.90, of tunneling injection constitutes the photon-assisted transmissions between the L and R electrodes on the one hand and the central island T on the other hand. The above Eq. 8.90 describes nonequilibrium pumping of the central island due to the tunneling from adjacent L and R electrode regions and due to the photon-assisted tunneling.

8.12 Nonequilibrium Contribution

The different signs of $\Gamma_\kappa^{\mathbf{ext}}$ entering Eq. 8.90 correspond to the *tunneling injection* (*extraction*). For the sake of convenience, one might introduce also a deviation from the equilibrium

$$\delta n_{\mathbf{B}} (E) = n_{\mathbf{B}} (E) - n_0 (E) \tag{8.91}$$

The nonequilibrium deviation $\delta n_B (E)$ from its steady-state value $n_0 (E) = 1/(\exp\{E/T\} + 1)$ is determined from the stationary kinetic equation on $\delta n^{\mathbf{T}}$

$$\begin{aligned}\frac{\partial}{\partial t}\delta n_{\mathrm{B}} \left(W_{\kappa_m}\right) \equiv 0 = 2\pi & \sum_m J_m^2 \left(\frac{U_{01}}{\hbar\omega}\right) \Gamma_\kappa^{\mathbf{ext}} \\ & \times \left[n_0 \left(W_{\kappa_m} - eV_{\mathbf{LR}}\right) + n_0 \left(W_{\kappa_m}\right) - 2n_{\mathbf{B}} \left(W_{\kappa_m} + \hbar\omega m\right)\right] \\ & - \hbar\frac{n_{\mathbf{B}} \left(W_{\kappa_m}\right) - n_0 \left(W_{\kappa_m}\right)}{\tau_{\mathbf{ep}}}\end{aligned} \tag{8.92}$$

In a simplest case of an unbiased junction which is exposed to an external AC field, we set $V_{\mathbf{LR}} = 0$ in Eq. 8.92. It gives an approximate solution in the form

$$\delta n_{\mathbf{B}}\left(W_{\kappa_m}\right) \simeq -\frac{4\pi\tau_{\mathbf{ep}}}{\hbar}\sum_m J_m^2\left(\frac{U_{01}}{\hbar\omega}\right)\Gamma_\kappa^{\mathbf{ext}} n_0\left(W_\kappa + \hbar\omega m\right) \quad (8.93)$$

The last result, Eq. 8.93, also suggests that the nonequilibrium effects are significant as soon as

$$J_1^2\left(\frac{U_{01}}{\hbar\omega}\right)\cdot\Gamma_\kappa^{\mathbf{ext}} >> \Gamma_{\mathbf{ep}} \quad (8.94)$$

The above condition Eq. 8.94 becomes satisfied either if the external AC field is sufficiently strong $U_{01}/\hbar\omega \simeq 1$, or otherwise if $\Gamma_\kappa^{\mathbf{ext}} >> \Gamma_{\mathbf{ep}}$. It happens when the rate $\Gamma_{\mathbf{ep}}$ of electron-phonon collisions is low, which is the case when temperatures are very low. For particular values of the AC field parameters, for example, $f = 1$ THz and $U_{01} = 10^{-4}$ eV one obtains

$$\frac{U_{01}}{\hbar\omega} = 0.15, \quad (8.95)$$

whereas for $f = 1$ THz and $U_{01} = 1$ meV one gets

$$\frac{U_{01}}{\hbar\omega} = 1.5 \quad (8.96)$$

where at temperature $T = 10$ K one uses $\Gamma_{\mathbf{ep}}^{\mathbf{10K}} \simeq 2\times10^{-4}$ eV.

8.13 Photon-Assisted Electric Current

The electric current can be decomposed into two terms:

$$\begin{aligned} I &= ev\sum_k\int d\varepsilon\left(G_{\mathbf{R}}^K(k,\varepsilon) - G_{\mathbf{L}}^K(k,\varepsilon)\right) \\ &= I_0 + I_{ne} \end{aligned} \quad (8.97)$$

We introduce the nonequilibrium electron distribution function on the central island $n_\varepsilon^{\mathbf{B}}$, which deviates from equilibrium immediately after an electron jumps into the central island. The electron Green functions entering the above Eq. 8.97 are computed as follows:

$$
\begin{aligned}
G_{\mathbf{L}}^{K} &\simeq G_{0\mathbf{L}}^{K} + G_{0\mathbf{L}}^{R}\Sigma_{\mathbf{L}}^{K}G_{0\mathbf{L}}^{A} = \\
&G_{0\mathbf{L}}^{K} + G_{0\mathbf{L}}^{R}(k,\ \varepsilon + eV/2)\cdot \mathbf{T}_{\mathbf{L}}^{2}G_{\mathbf{B}}^{K}(\kappa,\ \varepsilon)\ G_{0\mathbf{L}}^{A}(k,\ \varepsilon + eV/2) \\
&= \breve{T}_{\varepsilon}\left(2n_{k}^{\mathbf{L}} - 1\right)\delta\left(\varepsilon - \varepsilon_{k} + eV_{\mathbf{L}}\right) + \breve{T}_{\varepsilon}\frac{1}{\varepsilon - \varepsilon_{k} + eV_{\mathbf{L}} + i\delta}\mathbf{T}_{\mathbf{L}}^{2}\zeta_{\kappa} \\
&\times\sum_{n} J_{n}^{2}\left(\frac{U_{01}}{\hbar\omega}\right)\delta\left(\varepsilon - E_{\kappa} - U_{0} - \hbar n\omega\right)\left[2n_{\mathbf{B}}(\varepsilon) - 1\right] \\
&\times\breve{T}_{\varepsilon}\frac{1}{\varepsilon - \varepsilon_{k} + eV_{\mathbf{L}} - i\delta} \\
&= \breve{T}_{\varepsilon}\left(2n_{k}^{\mathbf{L}} - 1\right)\delta\left(\varepsilon - \varepsilon_{k} + eV_{\mathbf{L}}\right) + \breve{T}_{\varepsilon}^{2}\mathcal{T}_{\mathbf{L}}^{2}\zeta_{\kappa}\frac{1}{\left(\varepsilon - \varepsilon_{k} + eV_{\mathbf{L}}\right)^{2} + \delta^{2}} \\
&\times\sum_{n} J_{n}^{2}\left(\frac{U_{01}}{\hbar\omega}\right)\delta\left(\varepsilon - E_{\kappa} - U_{0} - \hbar n\omega\right)\left[2n_{\mathbf{B}}(\varepsilon) - 1\right] \qquad (8.98)
\end{aligned}
$$

$$
\begin{aligned}
G_{\mathbf{R}}^{K} &\simeq \breve{T}_{\varepsilon}\left(2n_{k}^{\mathbf{R}} - 1\right)\delta\left(\varepsilon - \varepsilon_{k} + eV_{\mathbf{R}}\right) + \breve{T}_{\varepsilon}^{2}\mathcal{T}_{\mathbf{R}}^{2}\zeta_{\kappa}\frac{1}{\left(\varepsilon - \varepsilon_{k} + eV_{\mathbf{R}}\right)^{2} + \delta^{2}} \\
&\times\sum_{n} J_{n}^{2}\left(\frac{U_{01}}{\hbar\omega}\right)\delta\left(\varepsilon - E_{\kappa} - U_{0} - \hbar n\omega\right)\left[2n_{\mathbf{B}}(\varepsilon) - 1\right] \qquad (8.99)
\end{aligned}
$$

where we utilize that

$$G_{0\mathbf{L}(\mathbf{R})}^{R(A)}(k,\ \varepsilon) = \breve{T}_{\varepsilon}\frac{1}{\varepsilon - \varepsilon_{k} + eV_{\mathbf{L}(\mathbf{R})} \pm i\delta} \qquad (8.100)$$

$$G_{0\mathbf{L}(\mathbf{R})}^{K}(k,\ \varepsilon) = \breve{T}_{\varepsilon}\delta\left(\varepsilon - \varepsilon_{k} + eV_{\mathbf{L}(\mathbf{R})}\right)\left(2n_{k}^{\mathbf{L}(\mathbf{R})} - 1\right) \qquad (8.101)$$

$$\Sigma_{\mathbf{L}(\mathbf{R})}^{K} = \mathbf{T}_{\mathbf{L}(\mathbf{R})}^{2}G_{\mathbf{B}}^{K}(\kappa,\ \varepsilon) \qquad (8.102)$$

$$
\begin{aligned}
\Sigma_{\mathbf{B}}^{R(A)} &= \mathbf{T}_{\mathbf{L}}^{2}G_{\mathbf{L}}^{R(A)}(k,\ \varepsilon + eV_{\mathbf{L}}) + \mathbf{T}_{\mathbf{R}}^{2}G_{\mathbf{R}}^{R(A)}(k,\ \varepsilon + eV_{\mathbf{R}}) \\
&= \mathbf{T}_{\mathbf{L}}^{2}\sum_{k} T_{k}\frac{1}{\varepsilon - \varepsilon_{k} + eV_{\mathbf{L}} \pm i\delta} + \mathbf{T}_{\mathbf{R}}^{2}\sum_{k} T_{k}\frac{1}{\varepsilon - \varepsilon_{k} + eV_{\mathbf{R}} \pm i\delta}
\end{aligned}
\qquad (8.103)
$$

$$\Sigma_{\mathbf{B}}^{K} = \mathbf{T}_{\mathbf{L}}^{2}G_{\mathbf{L}}^{K}(k,\ \varepsilon + eV_{\mathbf{L}}) + \mathbf{T}_{\mathbf{R}}^{2}G_{\mathbf{R}}^{K}(k,\ \varepsilon + eV_{\mathbf{R}}) \qquad (8.104)$$

$$G_{0\mathbf{B}}^{R(A)}(\kappa,\ \varepsilon) = \zeta_{\kappa}\sum_{n} J_{n}^{2}\left(\frac{U_{01}}{\hbar\omega}\right)\frac{1}{\varepsilon - E_{\kappa} - U_{0} - \hbar n\omega \pm i\delta} \qquad (8.105)$$

$$G_{0\mathbf{B}}^{K}(\kappa,\ \varepsilon) = \zeta_\kappa \sum_n J_n^2 \left(\frac{U_{01}}{\hbar\omega}\right) \delta\left(\varepsilon - E_\kappa - U_0 - \hbar n\omega\right) \left[2n_{\mathbf{B}}(\varepsilon) - 1\right] \tag{8.106}$$

We have introduced the following auxiliary functions

$$\Gamma_{\mathbf{T}} = i\left[\Sigma_{\mathbf{T}}^{R} - \Sigma_{\mathbf{T}}^{A}\right] = \pi\left[\mathcal{T}_{\mathbf{L}}^{2} T_{E+eV_{\mathrm{L}}} + \mathcal{T}_{\mathbf{R}}^{2} T_{E+eV_{\mathbf{R}}}\right] \tag{8.107}$$

$$\zeta_\kappa = \left(|\alpha_{\kappa 0}|^2 + |\beta_{-\kappa 0}|^2\right) |\varpi_{\kappa 0}|^2 \tag{8.108}$$

In the above formulas Eqs. 8.100 and 8.101, we also use a renormalized tunneling matrix element $\breve{T}_{k,\,\kappa}$, which is defined as

$$\breve{T}_{k,\,\kappa} = T_{k,\,\kappa} |\varpi_{k0}|^2, \tag{8.109}$$

whereas the electron tunneling rate between the electrodes and central region is

$$\Gamma_\kappa^{\mathbf{L(R)}} = \zeta_\kappa \breve{T}_\kappa \mathbf{T}_{\mathbf{L(R)}}^{2} = \mathbf{T}_{\mathbf{L(R)}}^{2} T_{k,\,\kappa} |\varpi_{k0}|^4 \cdot \left(|\alpha_{\kappa 0}|^2 + |\beta_{-\kappa 0}|^2\right) \tag{8.110}$$

For the sake of simplicity, in the above Eqs. 8.100–8.110 we have neglected the renormalizing of electric current due to the interelectrode coupling.

$$\begin{aligned} G_{\mathbf{R}}^{K} &\simeq G_{0\mathbf{R}}^{K} + G_{0\mathbf{R}}^{R} \Sigma_{\mathbf{R}}^{K} G_{0\mathbf{R}}^{A} = G_{0\mathbf{R}}^{K} \\ &+ G_{0\mathbf{R}}^{R}(k,\ \varepsilon - eV/2)\, T^{\mathbf{R}} G_{\mathbf{B}}^{K}(\varepsilon)\, G_{0\mathbf{R}}^{A}(k,\ \varepsilon - eV/2) \end{aligned} \tag{8.111}$$

The electric current has two terms

$$\begin{aligned} I &= ev \sum_k \int d\varepsilon \left(G_{\mathbf{R}}^{K}(k,\ \varepsilon) - G_{\mathbf{L}}^{K}(k,\ \varepsilon)\right) \\ &= I_0 + I_{\mathrm{ne}} \end{aligned} \tag{8.112}$$

where the first term in the second line corresponds to the regular expression, while the second term describes the influence of nonequilibrium effects.

8.14 Equilibrium Current

The conventional part of electric current is computed as

$$\begin{aligned}
I_0 =& ev \sum_k \int d\varepsilon [\breve{T}_\varepsilon \left(2n_k^{\mathbf{L}} - 1\right) \delta\left(\varepsilon - \varepsilon_k + eV_{\mathbf{L}}\right) \\
&- \breve{T}_\varepsilon \left(2n_k^{\mathbf{R}} - 1\right) \delta\left(\varepsilon - \varepsilon_k + eV_{\mathbf{R}}\right)] \\
=& ev \int d\varepsilon d\varepsilon_k N\left(\varepsilon_k\right) [\breve{T}_\varepsilon \left(2n_k^{\mathbf{L}} - 1\right) \delta\left(\varepsilon - \varepsilon_k + eV_{\mathbf{L}}\right) \\
&- \breve{T}_\varepsilon \left(2n_k^{\mathbf{R}} - 1\right) \delta\left(\varepsilon - \varepsilon_k + eV_{\mathbf{R}}\right)] \\
=& ev \int d\varepsilon \breve{T}_\varepsilon [N\left(\varepsilon + eV_{\mathbf{L}}\right) \left(2n\left(\varepsilon + eV_{\mathbf{L}}\right) - 1\right) \\
&- N\left(\varepsilon + eV_{\mathbf{R}}\right) \left(2n\left(\varepsilon + eV_{\mathbf{R}}\right) - 1\right)]
\end{aligned} \tag{8.113}$$

which can be simplified as

$$\begin{aligned}
I_0 =& 2ev \int d\varepsilon \breve{T}_\varepsilon [N\left(\varepsilon + eV_{\mathbf{L}}\right) n\left(\varepsilon + eV_{\mathbf{L}}\right) - N\left(\varepsilon + eV_{\mathbf{R}}\right) n\left(\varepsilon + eV_{\mathbf{R}}\right)] \\
&+ ev \int d\varepsilon \breve{T}_\varepsilon \left(N\left(\varepsilon + eV_{\mathbf{L}}\right) - N\left(\varepsilon + eV_{\mathbf{R}}\right)\right)
\end{aligned} \tag{8.114}$$

For a conventional conductor with a roughly constant electron density of states

$$N\left(\varepsilon\right) = N_0 = \text{const} \tag{8.115}$$

one arrives at a regular Landauer–Büttiker expression for the electric current, except the transparency $\breve{T}_{E_k}$, which depends on the photon-assisted coherent tunneling through the central island:

$$I_0 = 2evN_0 \int d\varepsilon \breve{T}_\varepsilon \left(n\left(\varepsilon + eV/2\right) - n\left(\varepsilon - eV/2\right)\right) \tag{8.116}$$

8.15 Gate Current

The nonequilibrium part of the electric current becomes finite only if the gate voltage acquires a finite value by inducing a finite shift of the electron electrochemical potential $U_0 \neq 0$. Then, the gate current is computed as

$$I_{\text{ne}} = ev \sum_k \int d\varepsilon \left(\delta G_{\mathbf{R}}^K (k, \varepsilon) - \delta G_{\mathbf{L}}^K (k, \varepsilon) \right)$$

$$= ev \sum_k \int d\varepsilon \left[\breve{T}_\varepsilon^2 T_{\mathbf{R}}^2 \zeta_\kappa \frac{1}{(\varepsilon - E_k + eV_{\mathbf{R}})^2 + \delta^2} \right.$$

$$\times \sum_n J_n^2 \left(\frac{U_{01}}{\hbar\omega} \right) \delta (\varepsilon - E_\kappa - U_0 - \hbar n\omega) \left[2n_{\mathbf{B}} (\varepsilon) - 1 \right]$$

$$- \breve{T}_\varepsilon^2 T_{\mathbf{L}}^2 \zeta_\kappa \frac{1}{(\varepsilon - E_k + eV_{\mathbf{L}})^2 + \delta^2}$$

$$\left. \times \sum_n J_n^2 \left(\frac{U_{01}}{\hbar\omega} \right) \delta (\varepsilon - E_\kappa - U_0 - \hbar n\omega) \left[2n_{\mathbf{B}} (\varepsilon) - 1 \right] \right] \tag{8.117}$$

where one sets $\mathcal{T}_{\mathbf{R}}^2 = \mathcal{T}_{\mathbf{L}}^2 = 1$. This gives

$$I_{\text{ne}} = ev \sum_k \int d\varepsilon \left(\delta G_{\mathbf{R}}^K (k, \varepsilon) - \delta G_{\mathbf{L}}^K (k, \varepsilon) \right) \tag{8.118}$$

or

$$I_{\text{ne}} = ev \sum_k \int d\varepsilon \breve{T}_\varepsilon^2 \zeta_\kappa \sum_n J_n^2 \left(\frac{U_{01}}{\hbar\omega} \right) \delta (\varepsilon - E_\kappa - U_0 - \hbar n\omega)$$

$$\times \left[1 - 2n_B (\varepsilon) \right] \left[\frac{1}{(\varepsilon - E_k - eV/2)^2 + \delta^2} - \frac{1}{(\varepsilon - E_k + eV/2)^2 + \delta^2} \right] \tag{8.119}$$

or in another form

$$I_{\text{ne}} = ev \sum_k \breve{T}_\varepsilon^2 \zeta_\kappa \sum_n J_n^2 \left(\frac{U_{01}}{\hbar\omega} \right) \left[1 - 2n_{\mathbf{B}} (E_\kappa + U_0 + \hbar n\omega) \right]$$

$$\times \left[\frac{1}{(U_0 + \hbar n\omega - eV/2)^2 + \delta^2} - \frac{1}{(U_0 + \hbar n\omega + eV/2)^2 + \delta^2} \right] \tag{8.120}$$

For a symmetric junction gives

$$I_0 = 2evN_0 \int d\varepsilon \breve{T}_\varepsilon \left(n (\varepsilon + eV/2) - n (\varepsilon - eV/2) \right) \tag{8.121}$$

$$I_{\text{ne}} = ev\Upsilon (U_0, \omega, V) \sum_k \zeta_\kappa \breve{T}_k^2 \left[1 - 2n_{\mathbf{B}} (E_\kappa + U_0 + \hbar n\omega) \right] \tag{8.122}$$

where $n_{\mathbf{B}} (w_\kappa)$ is the electron distribution function on the central island, $T_{w_\kappa}^{\mathbf{s}} = T^{\mathbf{L}} = T^{\mathbf{R}}$. One can see that when $V = 0$ or $U_0 = 0$

$$\Upsilon\left(U_0,\ \omega,\ V\right) = 0 \tag{8.123}$$

which reflects the gate-induced origin of $I_{\mathbf{as}}$. In the last case, Υ vanishes due to summation over positive and negative n. The additional term contributes in highly transparent junctions when the external AC field amplitude is sufficiently high. We have used the identity

$$1 - 2\frac{1}{e^x + 1} = \frac{e^x - 1}{e^x + 1} = \tan h\frac{x}{2} \tag{8.124}$$

Then, if $n_{\mathbf{B}}\left(E_k\right)$ is close to equilibrium, one gets

$$I = I_0 + I_{\text{ne}} \tag{8.125}$$

$$I_0 = 2evN_0 \int dk\breve{T}_k\left(n_0\left(E_k\right) - n_0\left(E_k - eV\right)\right) \tag{8.126}$$

$$I_{\text{ne}} = ev\Upsilon\left(U_0,\ \omega,\ V\right)\sum_k \zeta_\kappa \breve{T}_k^2 \tanh\left(\frac{E_\kappa + U_0 + \hbar n\omega}{2T}\right) \tag{8.127}$$

In the above formulas we have introduced the coherence factor

$$\zeta_\kappa = \left(\left|\alpha_{\kappa 0}\right|^2 + \left|\beta_{-\kappa 0}\right|^2\right)\left|\varpi_{\kappa 0}\right|^2 \tag{8.128}$$

and the renormalized tunneling matrix element

$$\breve{T}_{k,\,\kappa} = T_{k,\,\kappa}\left|\varpi_{k0}\right|^2 \tag{8.129}$$

The corresponding wavevectors in Eqs. 8.128 and 8.129 are

$$\kappa_0 = \sqrt{\left(\frac{E + U_0}{\hbar v}\right)^2 - q^2} \tag{8.130}$$

$$\pi = \sqrt{\kappa} = \frac{\left|E + U_0\right|}{\hbar v} \tag{8.131}$$

$$\sigma_E = \text{sign}\left(E + U_0\right) \tag{8.132}$$

$$k = \sqrt{\left(\frac{E}{\hbar v}\right)^2 - q^2}$$

$$p = \sqrt{k^2 + q^2} = \left|\frac{E}{\hbar v}\right|$$

$$s = \mathbf{sign}\left(E\right) \tag{8.133}$$

The functions α_κ and $\beta_{-\kappa}$ are computed using appropriate nonstationary boundary conditions for a graphene junction. For a single symmetric chiral crank in the steady state one gets

$$\alpha_\kappa = -2ks\pi\frac{iq\sigma p - iqs\pi + ks\pi + \kappa\sigma p}{D_{\mathbf{en}}} \tag{8.134}$$

$$\beta_{-\kappa} = \frac{2ks\pi\left(iq\sigma p - iqs\pi + ks\pi - \kappa\sigma p\right)}{D_{\mathbf{en}}} \tag{8.135}$$

$$D_{\mathbf{en}} = \left(e^{2i\kappa D} - 1\right)\left(k^2\pi^2 + q^2\left(\pi^2 - 2s\sigma p\pi\right)\right)$$
$$-2k\kappa s\sigma p\left(e^{2i\kappa D} + 1\right)\pi \qquad (8.136)$$

where $\pi = \sqrt{\kappa^2 + q^2}$ and $p = \sqrt{k^2 + q^2}$.

8.16 Excessive Regular Current

There is a fraction of electric current $I_0^{(2)}$ which does not depend on $n_0(E_k)$:

$$I_0^{(2)} = ev \int_{-D_c/2}^{D_c/2} d\varepsilon \breve{T}_\varepsilon \left(N\left(\varepsilon + eV_{\mathbf{R}}\right) - N\left(\varepsilon + eV_{\mathbf{L}}\right)\right) \qquad (8.137)$$

where $V_{\mathbf{R(L)}} = \pm V/2$ and D_c is the conducting bandwidth, and the energy-dependent density of states in pristine graphene is

$$N(E) = 4\frac{WD}{2\pi}\frac{|E|}{\hbar^2 v^2} \qquad (8.138)$$

which allows the current to be rewritten as

$$I_0^{(2)} = 4\frac{WD}{2\pi}\frac{e}{\hbar^2 v}\int_{-D_c/2}^{D_c/2} d\varepsilon \breve{T}_\varepsilon \left(|E + eV/2| - |E - eV/2|\right) \quad (8.139)$$

The last integral vanishes if $\breve{T}_\varepsilon$ is even versus ε.

8.17 Absorbed Power

When a quantum well is exposed to an external AC field, the tunneling electrons absorb/emit energy from/to the field. The absorbed power is

$$P_{\mathrm{ac}} = \sum_{\kappa, m} W_\kappa \left[n_T\left(W_\kappa\right) - n_0\left(W_\kappa\right)\right] \qquad (8.140)$$

where $n_T(W_\kappa)$ is the nonequilibrium distribution function for an electron excitation with pseudoenergy W_κ and $n_0(W_\kappa) = 1/(\exp(W_\kappa/T) + 1)$ is the Fermi distribution function. We assume that the power W_{ac} pumped into the system is enough to deviate the electron distribution function from equilibrium. We evaluate the power W_{ac} supplied by the AC field of the given amplitude into the barrier region and compare it to the power W_{out} taken away due to quasiparticle diffusion into the electrodes and due to escape of phonons into the electrodes and substrate. Namely, we assume that the diffusion of quasiparticle excitations from the barrier region into the electrodes and the phonons created in the electron–phonon collisions escaping into the electrodes and substrate drive the energy away from the system faster than the AC field supplies it. It means that the electron distribution function is nearly equilibrium in the barrier region and is surely equilibrium in the electrodes. Besides it means that the Green functions in the electrodes must be time independent. The Green function inside the barrier is time dependent. We compute the time-averaged electric current depending on the time-averaged Green function which coincides with an equilibrium Green function.

The general interest is directed toward the bottom part of the THz domain corresponding to the frequency range 0.5–5 THz (for which $2\pi f t_{\mathrm{dwell}} < 1$ or even $2\pi f t_{\mathrm{dwell}} << 1$ if only $D < 1$ μm). In that case the AC field wavelength is comparable to or longer than D. Considering presently available capabilities of nanolithography, one in principle may even fabricate junctions with much shorter $D \approx 50$ nm, which are suitable, for example, for AC field frequencies up to $f = 20$ THz.

The topic skin depth of metal electrodes, for example, aluminum at $f = 1$ THz is

$$\delta = \frac{1}{\sqrt{\pi \mu_0}} \sqrt{\frac{\rho}{\mu_r f}} = 84\,\mathrm{nm}, \tag{8.141}$$

while for the light with $\lambda = 0.5$ μm and $f = c/\lambda = 6 \times 10^{14} \mathrm{s}^{-1}$ one obtains

$$\delta = 3.4\,\mathrm{nm} \tag{8.142}$$

8.18 Jarzynski Equality for Quantum Systems

Theoretical description of nonequilibrium effects on nanoscale turns to be trickier when the number of particles in the small system becomes finite. It is becoming an increasing issue when modeling a nanodevice the components of which are fabricated, for example, from single molecules. A conventional technique fails to work properly in small systems due to an ambiguity of the statistical averaging procedure. Fortunately, the issue can be solved using other approaches which have been developed recently, describing the nonequilibrium properties of small systems.

Here we briefly discuss a quantum version of the Jarzynski equality which is also valid for a small quantum system subjected to a thermal heat bath and undergoing a nonequilibrium switching process (Yukawa et al. 1997a, 1997b, Yukawa 2000). The equality establishes a correspondence between the ensemble averages of exponentiated work with the Helmholtz free-energy difference. For illustration, we will consider an open quantum system represented by a one-half spin system subjected to alternating magnetic field and interacting with a thermal heat bath. The obtained solution confirms that the quantum version of the Jarzynski equality works very well. In this section, we discuss the quantum Jarzynski equality, which remains valid under nonequilibrium conditions when a transition process involves the thermal relaxation and an energy supply from external sources. The mentioned equality represents a quantum analogue (Yukawa 2000) of the original Jarzynski equality for classic systems (Jarzynski 1997a, 1997b, 1997c). There are many important issues relevant to the quantum nonequilibrium thermodynamics when implementing such equality. When a classical system is coupled to an isothermal bath characterized by an inverse temperature β, one needs to wait a sufficiently long relaxation time until the system achieves a state of thermal equilibrium. A description of the transition process is facilitated if one assumes that the system depends on a macroscopic parameter λ. If one alters λ infinitely slowly between 0 and 1, the system switches into another thermal equilibrium state corresponding to $\lambda = 1$, which is distinct from an equilibrium state with $\lambda = 0$. Making the change of λ sufficiently slow, that is, when selecting a much longer switching time t_s, one keeps the system remaining in a quasistatic equilibrium during a switching process. Then, the Helmholtz free-energy difference ΔF between the initial state and the final state is equal to the total work $W(t_s = \infty)$ performed on the system by the outside through the parameter λ

$$W\left(t_s = \infty\right) = \Delta F \equiv F_0 - F_1 \tag{8.143}$$

where F_λ is the Helmholtz free energy of the thermal equilibrium state corresponding to certain value of λ. However, the work exceeds the free-energy difference

$$W(t_s) \geq \Delta F \tag{8.144}$$

if the switching happens much faster, that is, when t_s is finite. The above inequality, Eq. 8.144, represents the least-work principle. Due to the thermodynamic stability during a rapid switching process, an excessive work is spent. Since the system interacts with the isothermal heat bath, the extra work which was performed during the switching process eventually is transferred into the heat bath. Instead of the inequality Eq. 8.144 one can use an exact equality which had been suggested by Jarzynski in 1997 (Jarzynski 1997a, 1997b, 1997c). Originally, the Jarzynski equality was formulated for classic systems and was written as

$$\langle \exp\{-\beta W(t_s)\}\rangle_{\text{path}} = \exp(-\beta \Delta F) \tag{8.145}$$

The brackets $\langle \ldots \rangle_{\text{path}}$ in the above formula, Eq. 8.145, indicate that during the switching process, the ensemble average is taken along a single path. The switching process starts from a thermal equilibrium ensemble, which is an initial state characterized by certain β and $\lambda = 0$. Because one can use an arbitrary dynamics of the system during the process, it can be merely regarded as thermal relaxation.

The Jarzynski equality is closely related to the least-work principle; therefore it is general. According to Jarzynski, the equality remains valid for different dynamics, including Monte Carlo dynamics, Nos'e–Hoover dynamics, Langevin dynamics, and even Hamiltonian dynamics without a thermal heat bath. (Jarzynski 1997a, 1997b, 1997c) In particular, in the case of Monte Carlo dynamics, a single path is related to a certain sampling series. Here we would like to address the Jarzynski equality for a quantum system. Let us define several notations and quantities. In the quantum case, the partition function Z_λ and the Helmholtz free energy F_λ are written in the form

$$\begin{aligned} Z_{\lambda(t)} &= Tr\exp\left\{-\beta H_{\lambda(t)}\right\} \\ F_{\lambda(t)} &= -\beta^{-1}\ln Z_{\lambda(t)}(\beta) \end{aligned} \tag{8.146}$$

where $H_\lambda(t)$ and $\lambda(t)$ are the Hamiltonian of the system and a switching parameter operated from the outside at time t, respectively. We also introduce the operator of work $\hat{W}(t)$ performed while switching the system during a time interval $[0, t]$ as

$$\hat{W}(t) \equiv \int_0^t ds \frac{\partial \lambda(s)}{\partial s} \frac{\partial H_{\lambda(s)}}{\partial \lambda} \tag{8.147}$$

The above definition remains valid when λ is a macroscopic parameter; otherwise the intuitive view of work is not evident. Therefore, let us focus on the situation when λ represents a macroscopic parameter. The total average work $\langle \hat{W}(t) \rangle$ performed on the system during the switching process can be defined implementing the density matrix of the system $\rho(t)$ as

$$\langle \hat{W}(t) \rangle \equiv Tr \left\{ \int_0^t ds \frac{\partial \lambda(s)}{\partial s} \frac{\partial H_{\lambda(s)}}{\partial \lambda} \rho(s) \right\} \tag{8.148}$$

One can check that the least-work principle

$$\langle \hat{W}(t) \rangle \geq \Delta F \tag{8.149}$$

is preserved for any arbitrary switching processes. If the switching is quasistatic, the equality remains valid as it happens in the classical limit $\hbar \to 0$. Implementing the density matrix approach, the classical path average can be extended to the quantum case $\hbar \neq 0$. The averaging with the density matrix is used because it is unclear how the thermal relaxation of an ensemble occurs in the quantum dynamics. A path average of the exponentiated work is similar as it stands on the left-hand side of the classical Jarzynski equality and, thus, is expressed in the quantum language as

$$\begin{aligned} \langle \exp\{-\beta W(t)\}\rangle_{\text{path}} &\approx \overline{\exp\left\{-\beta \hat{W}(t)\right\}} \\ &\equiv \mathrm{Tr}\left\{ \lim_{N\to\infty} \tilde{\mathrm{T}} \prod_{n=0}^{N-1} \left\{ \tilde{P}^{\delta t}_{\lambda(t_n+1)} e^{-\beta H_{\lambda(t_{n+1})}} e^{+\beta H_{\lambda(t_n)}} \right\} \times \frac{e^{-\beta H_{\lambda(0)}}}{Z_{\lambda(0)}} \right\} \end{aligned} \tag{8.150}$$

where $\delta t = t/N$ and $t_n = n\delta t$, $\tilde{\mathrm{T}}$ shows that the time ordering is conducted in the direction from the right to the left. In Eq. 8.150, we have used the factor $e^{-\beta H_{\lambda(t_{n+1})}} e^{+\beta H_{\lambda(t_n)}}$, which represents the exponentiated infinitesimal work, and $\tilde{P}^{\delta t}_{\lambda(t_n+1)}$ is the time evolution infinitesimal superoperator for a small time interval δt at time t:

$$\tilde{P}^{\delta t}_{\lambda(t)} = e^{L(t)\delta t} \tag{8.151}$$

In Eq. 8.151, we have introduced the Liouville superoperator $L(t)$, which governs the time evolution of the density matrix

$$\frac{\partial \rho(t)}{\partial t} = L(t)\,\rho(t) \tag{8.152}$$

Equation 8.150 actually is an analogue of the classical path average, while the least property will be discussed later. The average of exponentiated work reads

$$\begin{aligned}\langle \exp\{-\beta W(t)\}\rangle_{\text{path}} = {} & \lim_{N\to\infty} \int dx_N \prod_{n=0}^{N-1} \\ & \times \left\{ \int dx_n \tilde{P}^{\delta t}_{\lambda(t_n+1)}\left(x_{n+1}|x_n\right) e^{-\beta\delta w(x_n)} \right\} \\ & \times \frac{e^{-\beta H_{\lambda(t_0)}(x_0)}}{Z_{\lambda(0)}} \end{aligned} \tag{8.153}$$

The integration in the above Eq. 8.153 is performed over the canonical variables x_n, which describes the classical system. The infinitesimal work, $\delta w(x_n)$ is introduced as $H_{\lambda(t_{n+1})}(x_n) - H_{\lambda(t_n)}(x_n) \cdot \tilde{P}^{\delta t}_{\lambda(t_n+1)}(x_{n+1}|x_n)$, which is an infinitesimal time evolution operator such as a transition probability for the Monte Carlo dynamics. The suggested approach disregards details of quantum dynamics which are not important here. Therefore, we quote Eq. 8.150 as a direct generalizing classical approach. The averaging with the density matrix introduced in Eq. 8.150 represents the quantum analogue of the classical path-averaged exponentiated work. Now we would like to make sure that the obtained formulas immediately correspond to the difference of exponentiated Helmholtz free energy. At this point, we implement the basic feature of the Liouville superoperator. Namely, the Liouville superoperator vanishes in thermal equilibrium which is characterized by the density matrix $\propto \exp\left(-\beta H_{\lambda(t)}\right)$. In the latter case, the system constantly interacts with the thermal heat bath which in general is a nonstationary process. The equilibrium thermal distribution $\exp\left(-\beta H_{\lambda(t)}\right)$ is a singular solution of the dynamics. Besides, when the thermal distribution is equilibrium, a fraction of the exponentiated infinitesimal work does not evolve.

$$\tilde{P}^{\delta t}_{\lambda(t_n+1)} e^{-\beta H_{\lambda(t_{n+1})}} = e^{-\beta H_{\lambda(t_{n+1})}} \tag{8.154}$$

It gives

$$\begin{aligned}
&\mathrm{Tr}\left\{\lim_{N\to\infty} \tilde{\mathrm{T}} \prod_{n=0}^{N-1} \left\{\tilde{P}^{\delta t}_{\lambda(t_n+1)}\, e^{-\beta H_{\lambda(t_{n+1})}} e^{+\beta H_{\lambda(t_n)}}\right\} \times \frac{e^{-\beta H_{\lambda(0)}}}{Z_{\lambda(0)}}\right\} \\
&= \frac{1}{Z_{\lambda(0)}} \mathrm{Tr}\left\{\lim_{N\to\infty} \tilde{P}^{\delta t}_{\lambda(t_N)}\, e^{-\beta H_{\lambda(t_N)}} e^{+\beta H_{\lambda(t_{N-1})}} \ldots \right. \\
&\left. \tilde{P}^{\delta t}_{\lambda(t_1)}\, e^{-\beta H_{\lambda(t_1)}} e^{+\beta H_{\lambda(t_0)}} e^{-\beta H_{\lambda(t_0)}} \right\} \\
&= \frac{1}{Z_{\lambda(0)}} \mathrm{Tr}\left\{\lim_{N\to\infty} \tilde{P}^{\delta t}_{\lambda(t_N)}\, e^{-\beta H_{\lambda(t_N)}} e^{+\beta H_{\lambda(t_{N-1})}} \ldots \right. \\
&\left. \tilde{P}^{\delta t}_{\lambda(t_2)}\, e^{-\beta H_{\lambda(t_2)}} e^{+\beta H_{\lambda(t_1)}} e^{-\beta H_{\lambda(t_1)}} \right\} \\
&= \frac{1}{Z_{\lambda(0)}} \mathrm{Tr}\left\{e^{-\beta H_{\lambda(t)}}\right\}
\end{aligned} \tag{8.155}$$

which is equivalent to the equality

$$\overline{\exp\left\{-\beta \hat{W}(t)\right\}} = \frac{Z_{\lambda(t)}}{Z_{\lambda(0)}} \tag{8.156}$$

The above formula actually is the quantum analogue of the classical Jarzynski equality which we have extended to the quantum case. We emphasize that the quantum Jarzynski equality, Eq. 8.156, remains true for both the case of a pure quantum dynamics and the classical dynamics. In the limit of pure dynamics, if one applies the superoperator to an arbitrary operator A, $\tilde{P}^{\delta t}_{\lambda(t_n+1)} A$, the result is $e^{-iH_{\lambda(t_{n+1})}\delta t/\hbar} A e^{iH_{\lambda(t_{n+1})}\delta t/\hbar}$. For such reasons, any arbitrary infinitesimal time evolution superoperators are actually the identity operators. Therefore, one arrives at the exponentiated free-energy difference once again.

Amid the Jarzynski equality, Eq. 8.145, for the classical systems, a similar equality, Eq. 8.156, for the quantum case is formulated in terms of operators. For the same reason, the transition process is characterized by introducing the operators but not the variables. The presence of the operators instead of the variables complicates the decomposition of the time variable. One can see that from the following example. For instance, one could use a slightly different definition of the infinitesimal exponentiated work by setting it either as $e^{+\beta H_{\lambda(t_n)}} e^{-\beta H_{\lambda(t_{n+1})}}$ or $e^{-\beta\left(H_{\lambda(t_{n+1})} - H_{\lambda(t_n)}\right)}$. However, the latter definition leads to errors at every step. Them, the error which is accumulated during the whole transition process is estimated as

$$\frac{1}{Z_{\lambda(0)}}\mathrm{Tr}\left\{\lim_{N\to\infty}\left\{\sum_{l=0}^{N-1}\left(\tilde{\mathrm{T}}\prod_{n=l+1}^{N-1}{}^{*}\tilde{P}^{\delta t}_{\lambda(t_{n+1})}\, e^{-\beta H_{\lambda(t_{n+1})}} e^{+\beta H_{\lambda(t_n)}}\right)\right.\right.$$
$$\left.\left.\tilde{P}^{\delta t}_{\lambda(t_{l+1})}\, \mathrm{A}_{t_l} e^{-\beta H_{\lambda(t_l)}} + O\left((\delta t)^2\right)\right\}\right\} \tag{8.157}$$

where A_{tn} is the error of order $O(\delta t)$ and the exponentiated work is taken as $e^{-\beta H_{\lambda(t_{n+1})}} e^{+\beta H_{\lambda(t_n)}} + A_{t_n} + O\left((\delta t)^2\right)$, and $\prod_{n=l+1}^{N-1}{}^{*}$ represents the omission of the term $\prod_{n=N}^{N-1}$. In the limit of large $N \to \infty$ the net error remains, because the average, Eq. 8.157, accumulates N terms of the order $O(\delta t) = O(1/N)$. During that procedure one might neglect the higher-order errors because they are small with an order $O(N)$ at most. The mentioned example emphasizes that caution must be paid when choosing the exponentiated infinitesimal work since even a small mistake results in a serious error at the end. Practically, one might get an error A_{t_n} of the order $O(\delta t \times (t_s)^{-1})$. If the switching time t_s is longer, the final error is not too big.

8.19 Quantum Jarzynski Equality for One-Half Spin

The obtained quantum Jarzynski equality, Eq. 8.156, can be applied to a simple quantum system like the one-half spin which is coupled to an isothermal heat bath and time-dependent magnetic field. We assume that the magnetic field which acts on the one-half spin is linearly reversed during a finite time interval which is described by the Hamiltonian $H_{\mathrm{S}}(t)$

$$H_{\mathrm{S}}(t) = -\frac{\lambda(t)}{2}\sigma^x - \frac{1}{2}\Delta\sigma^x$$
$$\lambda(t) = \lambda_0\left(1 - 2\frac{t}{t_s}\right) \tag{8.158}$$

where $\lambda(t)$ is the magnetic field at time t and $\sigma_{x,z}$ are the Pauli matrices. The level splitting Δ and the magnetic field amplitude λ_0 are constants. The one-half spin system couples to the heat bath which is described by term $\gamma\sigma^x \sum_\alpha \left(a_\alpha^\dagger + a_\alpha\right)$, where γ is a coupling constant and the bath is represented as a set of harmonic oscillators $\sum_{\alpha} \hbar\omega_{\alpha}\left(a_{\alpha}^{\dagger}a_{\alpha}+1/2\right)$. Here, $a_{\alpha}^{\dagger}$ (a_{α}) is a creation (annihilation) operator with a mode α. The time-dependent evolution of the system dynamics begins at time $t = 0$ from a thermal equilibrium state and completes at time $t = t_s$. The dynamics of the quantum system is found by implementing the projection technique. We assume that the heat bath always remains in a thermal equilibrium state characterized by an inverse temperature β. The equation of motion for the density matrix of the one-half spin system $\rho(t)$ is

$$\frac{\partial \rho}{\partial t} = -\frac{i}{\hbar}\left[H_s(t), \rho(t)\right] + \frac{\gamma^2}{\hbar^2}\Gamma(\rho) \tag{8.159}$$

The relaxation term containing Γ is introduced to ensure that the system will relax into an instantaneous thermal equilibrium state. For a stationary case described by the time-independent Hamiltonian, the equation of motion characterizes the relaxation to a thermal equilibrium state $\propto \exp(-\beta H_s(t))$. One can also write an explicit expression for Γ, which is quite cumbersome. The result for Γ contains the interaction between the spin and the bath σ_x, and also the autocorrelation function of the thermal heat bath variables. A numerical solution has been obtained in the second order of perturbation over the coupling constant γ (Yukawa 2000). The computational results are shown in Fig. 8.5, where we have used $\Delta = 0.1$ and $\lambda_0 = 1$, respectively. The expectation value of work is represented by several distinct time series using Eq. 8.148 with different values of switching time t_s. The data are compared with the difference of exact Helmholtz free energy at the corresponding time. Besides, we show the result of the quantum Jarzynski equality. One can see that the value of work is more converged into the exact Helmholtz free-energy difference when the switching time is longer. It happens because the dynamics approaches the quasistatic limit. However, if the switching time t_s is fast, the dynamics is dominated by a nonadiabatic transition. It requires an additional amount of work. We can also see that the results obtained using the quantum Jarzynski equality are in good agreement with the difference of exact Helmholtz free energy.

The quantum analogue of the Jarzynski equality (Yukawa 2000) relates the average of the exponentiated work operator with the exact Helmholtz free-energy difference. The relationship is

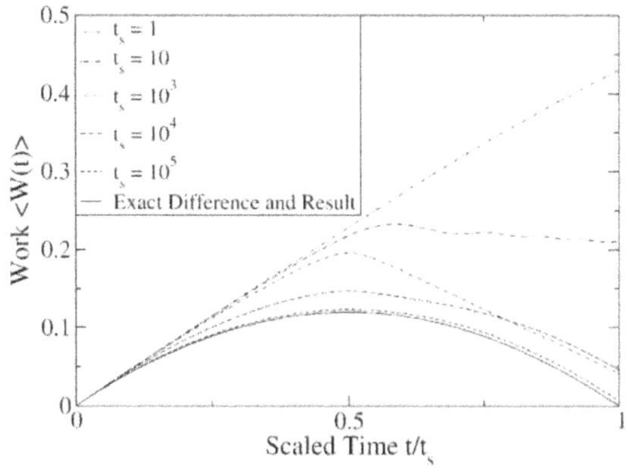

Figure 8.5 Expectation value of the work represented by the time series by a difference of the exact Helmholtz free energy and by the quantum Jarzynski equality. The results are shown versus time t in units of t_S. Adapted from Yukawa (2000).

established for the switching process between an initial thermal equilibrium state and the final thermal equilibrium state, even though the actual final state of the process is not thermal equilibrium. The practical implementation of the quantum analogue of the Jarzynski equality was illustrated by an example of the one-half spin system exposed to an alternating magnetic field. The quantum system interacts with the thermal heat bath. We see that the result coincides with the exact difference of the Helmholtz free energy. The quantum Jarzynski equality considered here includes the classical quality. The density operator becomes a diagonal matrix in the classical limit $\hbar \to 0$, that is, it coincides with the distribution function of the relevant classical system. In the classical limit $\hbar \to 0$, the transition between diagonal elements are involved only because the infinitesimal time evolution operator acts like a Fokker–Planck operator. Then, the average part of the quantum one is similar to Eq. 8.153. During the switching process one can select an arbitrarily dynamics. The key point is to enforce the property that the time-dependent thermal equilibrium state $\propto \exp(-\beta H_{\lambda(t)})$ corresponds to a *singular solution of the dynamics*. The mentioned property does not imply that the Liouvillian dynamics with the projection technique is mandatory. The dynamics with a singular solution produces the same results despite an instability of the singular solution $\exp(-\beta H_{\lambda(t)})$. Any other results derived from the dynamics are not physical. Other relevant issues involve the relationship between the quantum Jarzynski equality and the fluctuation theorem which is a quantum analogue of the classical fluctuation theorem. It establishes the correspondence between the distribution function and the entropy production rate, and is also valid in dynamics between nonequilibrium steady states. Another issue is to extend the quantum Jarzynski equality into nonequilibrium steady dynamics (Gallavotti 1995, Gallavotti and Cohen 1995a, 1995b, Crooks 1999, Hatano 1999a, 1999b).

Problems

8-1. Explain why nonequilibrium effects are important in graphene nanodevices.
8-2. Which parameters characterize the nonequilibrium state of a nanosystem?
8-3. What is the difference between the mechanisms of coherent and incoherent transmission through a nanojunction?
8-4. What is the limit of applicability of the S-matrix technique to a nonequilibrium system?
8-5. What is the basic idea of implementing piece-wise coherence in a complex system?
8-6. Explain why homogeneous approximation allows us to simplify the expressions for Green functions.
8-7. Determine why the electron distribution functions in the electrodes attached to the central island of a quantum dot can be considered equilibrium Fermi functions.
8-8. Determine how the effect of electron chirality in monolayer graphene enters the Landauer–Butticker formula for electric current and the kinetic equation for the electron distribution function.
8-9. How can one keep a small system in the state of quasistatic equilibrium during a switching process?

8-10. Explain why the presence of operators in the quantum Jarzynski equality, Eq. 8.158, instead of variables complicates the decomposition of the time variable.

References

Chiu, H. Y., V. Perebeinos, Y. M. Lin, and P. Avouris (2010). Controllable p-n junction formation in mono layer graphene using electrostatic substrate engineering. *Nano Lett*, **10**(11): 4634–4639.

Crooks, G. E. (1999). Entropy production fluctuation theorem and the nonequilibrium work relation for free energy differences. *Phys Rev E*, **60**(3): 2721–2726.

Eliashbe.Gm (1970). Film superconductivity stimulated by a high-frequency field. *JETP Lett USSR*, **11**(3): 114.

Gallavotti, G. (1995). Ergodicity, ensembles, irreversibility in Boltzmann and beyond. *J Stat Phys*, **78**(5–6): 1571–1589.

Gallavotti, G., and E. G. D. Cohen (1995a). Dynamical ensembles in nonequilibrium statistical-mechanics. *Phys Rev Lett*, **74**(14): 2694–2697.

Gallavotti, G., and E. G. D. Cohen (1995b). Dynamical ensembles in stationary states. *J Stat Phys*, **80**(5–6): 931–970.

Gorkov, L. P., and Eliashbe.Gm (1968). Generalization of Ginzburg-Landau equations for non-stationary problems in case of alloys with paramagnetic impurities. *Sov Phys JETP: USSR*, **27**(2): 328.

Hatano, T. (1999a). Heat conduction in the diatomic Toda lattice revisited. *Phys Rev E*, **59**(1): R1–R4.

Hatano, T. (1999b). Jarzynski equality for the transitions between nonequilibrium steady states. *Phys Rev E*, **60**(5): R5017–R5020.

Ivlev, B. I., and Eliashbe.Gm (1971). Influence of nonequilibrium excitations on properties of superconducting films in a high-frequency field. *JETP Lett USSR*, **13**(8): 333.

Jarzynski, C. (1997a). Berry's conjecture and information theory. *Phys Rev E*, **56**(2): 2254–2256.

Jarzynski, C. (1997b). Equilibrium free-energy differences from nonequilibrium measurements: a master-equation approach. *Phys Rev E*, **56**(5): 5018–5035.

Jarzynski, C. (1997c). Nonequilibrium equality for free energy differences. *Phys Rev Lett*, **78**(14): 2690–2693.

Keldysh, L. V. (1965). Diagram technique for nonequilibrium processes. *Sov Phys JETP: USSR*, **20**(4): 1018.

Lin, Y. M., V. Perebeinos, Z. H. Chen, and P. Avouris (2008). Electrical observation of subband formation in graphene nanoribbons. *Phys Rev B*, **78**(16).

Rinzan, M., G. Jenkins, H. D. Drew, S. Shafranjuk, and P. Barbara (2012). Carbon nanotube quantum dots as highly sensitive terahertz-cooled spectrometers. *Nano Lett*, **12**(6): 3097–3100.

Shafranjuk, S. E. (2008). Probing the intrinsic state of a one-dimensional quantum well with photon-assisted tunneling. *Phys Rev B*, **78**(23).

Shafranjuk, S. E. (2009). Directional photoelectric current across the bilayer graphene junction. *J Phys: Condens Mater*, **21**(1).

Shafranjuk, S. E. (2011a). Electromagnetic properties of the graphene junctions. *Eur Phys J B*, **80**(3): 379–393.

Shafranjuk, S. E. (2011b). Resonant transport through a carbon nanotube junction exposed to an ac field. *J Phys: Condens Mater*, **23**(49).

Tselev, A., Y. F. Yang, J. Zhang, P. Barbara, and S. E. Shafranjuk (2009). Carbon nanotubes as nanoscale probes of the superconducting proximity effect in Pd-Nb junctions. *Phys Rev B*, **80**(5).

Yang, Y. F., G. Fedorov, J. Zhang, A. Tselev, S. Shafranjuk, and P. Barbara (2012). The search for superconductivity at van Hove singularities in carbon nanotubes. *Supercond Sci Tech*, **25**(12).

Yukawa, S. (2000). A quantum analogue of the Jarzynski equality. *J Phys Soc Jpn*, **69**(8): 2367–2370.

Yukawa, S., M. Kikuchi, G. Tatara, and H. Matsukawa (1997a). Quantum ratchets. *J Phys Soc Jpn*, **66**(10): 2953–2956.

Yukawa, S., M. Kikuchi, G. Tatara, and H. Matsukawa (1997b). Quantum ratchets [**66**, 2953, (1997)]. *J Phys Soc Jpn*, **66**(12): 4055–4055.

Zhang, J., A. Tselev, Y. F. Yang, K. Hatton, P. Barbara, and S. Shafraniuk (2006). Zero-bias anomaly and possible superconductivity in single-walled carbon nanotubes. *Phys Rev B*, **74**(15).

Chapter 9

Graphene Thermoelectric Nanocoolers and Electricity Cogenerators

A direct conversion between electrical and thermal energy can be accomplished using the thermoelectric (TE) cogenerators. Among advantages of the TE convertors is the absence of mechanical components and undesirable chemical residue. In particular, a TE device contains no moving parts nor does it emit undesirable chemicals, like chlorofluorocarbons, etc. (DiSalvo 1999, Bell 2008). Implementing a TE device, the heat can be converted to electricity and vice versa. The TE coolers are consuming the electrical power which is supplied to a TE device and which causes the Peltier effect. Depending on the sign of electric current, the Peltier effect results either in cooling or in the heating of attached objects. A reciprocal phenomenon is the Seebeck effect, which can be implemented for converting the heat into usable electrical power. A serious setback on the path of a wide implementation of the TE devices is their low efficiency. Typical efficiencies of the most available TE cogenerators are far below than can be achieved using many other approaches, for example, a traditional internal combustion engine. It strongly motivates the searching of systems, mechanisms, and materials which might considerably improve the TE. A basic ingredient of such

Graphene: Fundamentals, Devices, and Applications
Serhii Shafraniuk

ISBN 978-981-4613-47-7 (Hardcover), 978-981-4613-48-4 (eBook)
www.panstanford.com

activity is a better understanding of fundamental issues related to the heat transport and the TE phenomena on the nanoscale (Dubi and Di Ventra 2011).

9.1 Thermoelectric Effects on the Nanoscale

Let us focus on the Seebeck effect, which is responsible for generation of electricity from the heat. When one side of a conducting material is hotter than the opposite side, the density of elementary excitations is larger at the hot side and smaller at the cold side. Because the inhomogeneity of density creates a difference of chemical potential $\Delta\mu = \mu_{\text{hot}} - \mu_{\text{cold}}$, it thereby forces the excitations to propagating from the hot side to the cold side. The elementary excitations in solids which, for example, are phonons, electrons, and holes, have a finite mass. Therefore their propagation induces the flow of heat and of charge between the hot and cold sides. Besides, if additionally they also carry an electric charge, like, for example, electrons and holes, their propagation induces a finite electric current across the sample. For electrically charged excitations, $\Delta\mu = V_{\text{hot}} - V_{\text{cold}} = \pm\Delta V$ where the sign of electric potential difference ΔV depends on whether the charge carriers are electrons or holes. In the isolated sample, thermally excited charge carriers propagate until the ΔV built up by the accumulation of charge is sufficient to cancel the thermoelectric (TE) current. Thus, the gradient of temperature $\Delta T = T_{\text{hot}} - T_{\text{cold}}$ across the sample induces a finite voltage $\Delta V = S\Delta T$ where the thermopower S is a linear-response, two-terminal property. The Seebeck coefficient S is determined as

$$S = \lim_{\Delta T \to 0} \left. \frac{\Delta V}{\Delta T} \right|_{I=0}. \tag{9.1}$$

One can see that the Seebeck coefficient S when defined with Eq. 9.1, corresponds to the limit of zero electric current I across the sample. The thermopower either of graphene or carbon nanotube (CNT) junction can be approximately written (Mott 1958, Cutler and Mott 1969, Lunde and Flensberg 2005) as

$$S_M(\mu, T) = -\frac{\pi^2}{3}\frac{k_B}{|e|}k_B T \frac{\partial \ln G(\mu, T)}{\partial E}, \tag{9.2}$$

where G is the electrical conductance of the junction, T is the temperature of the electrodes, μ is the chemical potential, k_B is Boltzmann's constant, and e is the electron charge, respectively. The above Eq. 9.2 is regarded as the Mott formula which stays valid under an assumption that the electron spectrum is smooth. If the last assumption is valid, Eq. 9.2 provides a fairly accurate estimation of thermopower. However, in the vicinity of resonances, one should use Eq. 9.2 with caution (Bergfield and Stafford 2009a, 2009b), because it is broken down. Keeping in mind that thermopower is a measure of coupling between the thermal and electric components, that is, between entropic and charge degrees of freedom, one arrives at a natural unit of thermopower which is entropy per unit charge (k_B/e). Other units which are frequently used during the measurements of S are $\mu V/K(\sim 0.012k_B/e)$. Because of the opposite charge of electrons and holes, the sign of thermopower can be either positive or negative that reflects the nature of the charge carriers. Therefore, S is positive when the transport is dominated by hole, whereas it becomes negative for the electron-dominated transport. It differs from the conductance, which is always positive. Dependence of the thermopower sign versus the charge carrier type serves to probe, for example, the alignment of the energy level between the graphene (or CNT) and metal electrodes (Paulsson and Datta 2003). Thus, one can accurately determine, for example, the energy mismatch if the work function (WF) of the electrodes approaches the ionization potential (or the so-called HOMO level) of the adsorbed molecule assuming a positive thermopower. This helps to adequately describe the molecular energy levels. Besides, as it follows from Eq. 9.2, if one can maximize variations in G, it would allow a considerable improving of the Seebeck coefficient S. The TE response of graphene and CNT can be studied by implementing the electron tunneling via the Schottky barrier through the interfaces metal/graphene or metal/CNT. A similar approach had been recently used to measure the TE response of a single-molecule junction. The measurements of the single-molecule conductance were reported by Reddy et al. (2007) and Baheti et al. (2008) with a modified scanning tunneling microscope (STM) similar to the device first developed by Xu and Tao (2003). The authors implemented a modified STM which was repeatedly moved up and down all over the substrate prepared with the adsorbed molecules. During each period of motion, the STM tip quickly creates, measures, and breaks thousands of molecular junctions. In addition, during each period, they measure the TE voltage ΔV induced in response to the applied temperature gradient ΔT. Obtained histograms of induced voltage ΔV versus the temperature difference ΔT allowed to extract the Seebeck coefficient S. Similar data were also obtained in other experiments (Sadat et al. 2010, Tan et al. 2010) where the authors implemented a modified atomic force microscope (AFM). The experimental approach is explained schematically in Fig. 9.1a. The plots of thermoelectic voltage measured versus spacing between the hip and surface through Au–1,4-benzenedithiol–Au junctions are represented in Fig. 9.1b. In Fig. 9.1c we show the histograms of the room-temperature (RT) data obtained in similar experimental measurements (Baheti et al. 2008). One can notice several interesting features:

(a) The TE voltages remain almost unchanged amid the Au tip deformations and moving over $\sim 50^{\circ}$A. It means that ΔV does not depend on the lead–molecule coupling details.
(b) The thermopower *is* quite sensitive to chemical structure. It is confirmed by the fact that if one replaces the thiol (SH) end groups by cyanide (CN) in Au–1,4-benzenedithiol–Au, the junction transport is adjusted from hole dominated to electron dominated.

As it follows from the experiments the chemical symmetry can be extracted immediately from thermopower and furthermore, the knowledge of chemical symmetry can be utilized to improve the efficiency of TE materials and devices. Also we would to emphasize the difference in tendencies taking place for the conductance and thermopower. As the molecular length increases, the signal for conductance decays, whereas the Seebeck coefficient *increases* (Kamenetska et al. 2010, Reddy et al. 2007, Baheti et al. 2008). The mentioned experiments (Kamenetska et al. 2010, Reddy et al. 2007, Baheti et al. 2008) suggest that in the ballistic regime the thermopower is almost insensitive to the length of molecular bridge which links the two electrodes. Therefore, on the molecular level, almost coherent transport through the TE materials proceeds in

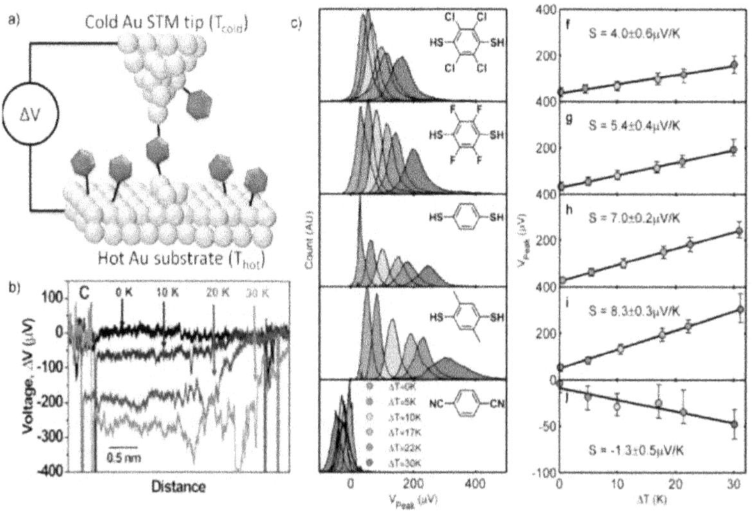

Figure 9.1 (a) A sketch of a single-molecule thermopower experiment where authors have implemented a modified Au STM tip. The Au tip is moved up and down over the Au substrate with adsorbed molecules, forming many molecular junctions. The thermoelectric voltage ΔV is measured for each junction in response to the applied temperature gradient $\Delta T = T_{\text{cold}} - T_{\text{hot}}$, and the thermopower S_{junction} is deduced. Here, T_{cold} corresponds to an ambient temperature (i.e., room temperature). (b) A trace of ΔV as a function of the tip–surface distance for several values of ΔT. Reprinted with permission from Reddy et al. (2007). Copyright 2007 American Chemical Society. (c) The molecular thermopower is obtained from the histograms which are built from the repeated junction formation and thermoelectric voltage measurement. One can see the effect of substituent and end groups in the thermoelectric response. Reprinted with permission from Baheti et al. (2008). Copyright 2008 American Chemical Society.

the form of single-molecule responses. It is quite different from the conductance which strongly depends on the molecular bridge length and is roughly proportional to the number of molecules inside the bridge connecting two electrodes. A large quantum-enhanced TE response might occur due to constructive and destructive transport interferences taking place in certain molecular junctions. Other mechanisms of enhanced conductance are due to many-body spectral features (Bergfield et al. 2011), as well as in the presence of sharp molecular spectral features (Finch et al. 2009).

9.2 Performance of a Thermoelectric Device

The dimensionless figure of merit *ZT* serves as a handy characteristic of the efficiency of a TE device. It is defined as

$$ZT = \frac{S^2 GT}{\kappa_{\mathrm{el}} + \kappa_{\mathrm{ph}}} \tag{9.3}$$

As it follows from Eq. 9.3, the efficiency of a TE is immediately related to the Seebeck coefficient *S* and to the electrical conductance *G*. On the one hand, a large magnitude of the Seebeck coefficient ensures that the coupling between electronic and thermal currents is as strong as possible. On the other hand, the electric current flows with the least Joule heating as soon as *G* achieves its maximum value. Besides, one should simultaneously maintain a finite temperature gradient across the device. It is accomplished when the thermal conductance $\kappa = \kappa_{\mathrm{el}} + \kappa_{\mathrm{ph}}$ is as small as possible (DiSalvo 1999, Bell 2008). A commercially viable green-energy solution corresponds to $ZT \geq 4$. The TE devices with large $ZT \geq 4$, if they were constructed, would resolve a huge variety of heating and cooling problems at both the macro- and nanoscales, with no mechanical parts involved, no operational carbon footprint and without toxic chemical residue output (DiSalvo 1999, Bell 2008). Using an engineering language, Eq. 9.3 briefly represents basic requirements to an efficient TE material which simultaneously must be a phonon glass and an electron crystal. In respect to the bulk materials a general strategy is to increase *G*, and *S* (by maximizing variation in *G*, see Eq. 9.3, and to reduce κ_{ph}. However, in many bulk materials the electronic part of the heat conductance always dominate over the contribution phonons, and the Wiedemann–Franz (WF) law $\kappa_{\mathrm{el}}/G = LT$ is observed where *L* is the Lorenz number. In that case the minimization of κ_{el} does not help to increase *ZT*. After many years of research, the figure of merit *ZT* has been improved only marginally (Tritt 2011). After overcoming the challenges of major material engineering, packaging and fabrication (DiSalvo 1999, Bell 2008), the best TE materials available now in the laboratory have only $ZT \approx 3$. The commercially available TE devices exhibit even more modest $ZT \approx 1$ which is still below the desired TE efficiency.

In the nanoscale graphene/CNT junctions, the excitations which involve electrons, holes and phonons, all contribute to the heat transfer. The graphene and CNT junctions suggest great opportunities for the quantum well engineering. The electron spectral singularities in those quantum wells are exploited to achieve large values of the Seebeck coefficient S and electric conductance G. However, there is a serious complication when using CNTs and graphene for the immediate TE applications. The problem is that their heat conductance $\kappa_{\rm el} + \kappa_{\rm ph}$ is large and is formed presumably by the phonon transport since $\kappa_{\rm ph} >> \kappa_{\rm el}$. Therefore, the WF law is not generally observed in the mentioned materials. The mentioned hurdle is overcome using graphene or CNT ambipolar field-effect transistors (FETs) as elements of the TE circuits. For instance, the ZT value can be greatly improved implementing metallic multilayered electrodes in sequence with graphene stripes or CNT. Another efficient solution is to decouple the phonon and electron heat flows by redirecting them along different independent paths. Using the graphene and CNT ambipolar transistors opens the path to address longstanding technological challenges in the field of thermoelectrics. The fabrication and development of graphene and CNT TE devices benefits from implementing various research methods. One example is the scanning probe calorimetric device with picowatt sensitivity and nanometer spatial resolution (Sadat et al. 2011, Canetta and Narayanaswamy 2013a, 2013b). Such calorimeter is capable of observing the energy transport through a single CNT or graphene stripe. Response of the TE system is measured in thermodynamic characteristics, for example, like the efficiency η. That characteristic η indicates which fraction of the heat is converted into usable power P and the generated power amount. One can also establish the connection between the figure of merit ZT and the efficiency η. In the linear response one finds the following expression

$$\eta_{\rm max} = \eta_{\rm C} \frac{\sqrt{1+ZT}-1}{\sqrt{1+ZT}+\frac{T_{\rm cold}}{T_{\rm hot}}}, \tag{9.4}$$

where T_α is the temperature of reservoir α and $\eta_{\rm C} = 1 - T_{\rm cold}/T_{\rm hot}$ is the Carnot efficiency. From Eq. 9.4 one can see that when

$ZT \to \infty$ one gets $\eta_{\max} \to \eta_C$. In general, one cannot accurately extend the thermodynamic response Eq. 9.4 beyond the linear approximation because ZT represents just an ad hoc rule-of-thumb metric. The nonlinear heat transport in a CNT junction has recently been studied experimentally by Small et al. (2003) and theoretically by Shafranjuk (2009).

Immediately near the metal/graphene and metal/CNT contacts, the Seebeck coefficient S might be considerably increased due to the chiral fermion interference features in conditions of Klein tunneling (Shafranjuk 2009). Such interference might provide large values for both ZT and η. Furthermore, quantum interference effects can be engineered with atomic precision via synthetic chemistry (Guedon et al. 2012). Unique aspects of the chiral transport through the graphene and CNT quantum wells, in addition to the reduced heat conductivity, high transparency and robustness of the metal/graphene and metal/CNT interfaces, and high flexibility of material properties, are very promising for the study of graphene and CNT TE materials.

9.3 Quantum Theory of Electronic Thermal Transport

Let us discus basic equations for the description of the heat transfer in graphene and CNT junction. The junction under consideration is composed of macroscopic metallic electrodes M attached to an arbitrary interacting nanoscale system. We derive an exact formula for the heat and charge current in an interacting nanostructure. Then, we discuss linear and nonlinear TE responses of the junction by implementing the corresponding expression for the electric current (Meir and Wingreen 1992, Jauho et al. 1994). An open quantum system is described by the Hamiltonian

$$H = H_{\text{mol}} + \sum_{\alpha=1}^{M} \left[H^{\alpha}_{\text{lead}} + H^{\alpha}_{T} \right], \tag{9.5}$$

where H^{α}_{lead} and H^{α}_{T} are the lead and tunneling Hamiltonians for lead α and H_{mol} is the graphene (or CNT) Hamiltonian. Each metallic electrode is represented as a bath of noninteracting fermions. We start from the fundamental thermodynamic identity at constant volume

$$TdS = dE - \mu dN, \tag{9.6}$$

where T, S, E, and N are temperature, entropy, internal energy, and particle number, respectively. We employ Eq. 9.6 to derive the heat current. When the above Eq. 9.6 is applied to electrode α, it gives

$$I_\alpha^Q \equiv T_\alpha \frac{dS_\alpha}{dt} = \frac{d}{dt}\left\langle H_{\text{lead}}^{(\alpha)}\right\rangle - \mu_\alpha \frac{d}{dt}\langle N_\alpha\rangle, \tag{9.7}$$

where T_α and μ_α are the temperature and chemical potential of electrode α respectively, and I_α^Q is the heat current flowing from the molecule into electrode α. The standard quantum mechanics allows to compute the time derivatives on the right hand side of Eq. 9.7 as

$$\begin{aligned} I_\alpha^Q &= -\frac{i}{\hbar}\left\{\left\langle\left[H_{\text{lead}}^\alpha, H\right]\right\rangle - \mu_\alpha\langle[N_\alpha, H]\rangle\right\} \\ &= \frac{i}{\hbar}\sum_{\substack{k\in\alpha \\ n,\sigma}}(\varepsilon_{k\sigma}-\mu_\alpha)\left[V_{nk}\left\langle d_{n\sigma}^\dagger c_{k\sigma}\right\rangle - V_{nk}^*\left\langle c_{k\sigma}^\dagger d_{n\sigma}\right\rangle\right]. \end{aligned} \tag{9.8}$$

The second line of the last Eq. 9.8 can be rewritten in terms of nonequilibrium Green's functions (Keldysh 1965, Meir and Wingreen 1992, Jauho et al. 1994) complemented by the Dyson's equation. The derivation yields the following general formula (Bergfield and Stafford 2009a, 2009b):

$$\begin{aligned} I_\alpha^{(\nu)} = -\frac{i}{\hbar}\int dE(E-\mu_\alpha)^\nu \\ \times \operatorname{Tr}\left\{\Gamma^\alpha(E)G^<(E) + f_\alpha(E)\left[G(E)-G^\dagger(E)\right]\right\}, \end{aligned} \tag{9.9}$$

where $f_\alpha(E)$ is the Fermi–Dirac distribution for electrode $\alpha\Gamma^\alpha(E)$ is the matrix of tunneling width which describes the bonding between the molecule and electrode $\alpha, I_\alpha^{(\nu)} = I_\alpha^Q$ represents the thermal current, and $-eI_\alpha^{(0)}$ is the Meir–Wingreen (Meir and Wingreen 1992, Jauho et al. 1994) expression for the electric current. The above formula now contains the Fourier transforms of the retarded $G(E)$ and the Keldysh "lesser" $G^<(E)$ Green's functions

$$\begin{aligned} G_{n\sigma,m\sigma}(t) &= -i\theta(t)\left\langle\{d_{n\sigma}(t), d_{m\sigma'}^\dagger(0)\}\right\rangle, \\ G_{n\sigma,m\sigma}^<(t) &= i\left\langle d_{m\sigma'}^\dagger(0)\,d_{n\sigma}(t)\right\rangle. \end{aligned} \tag{9.10}$$

Both $G(E)$ and $G^<(E)$ originate from time-ordered Green's functions on the Keldysh time contour (Keldysh 1965). Therefore, if one computes, for example, $G(E)$, then $G^{<}(E)$ is obtained immediately without further approximations. The above expression for electric current, Eq. 9.9, is exact and is valid for an arbitrary number of electrodes. The summation over the electrode's indices is performed in expressions for the tunneling self-energies when constructing the Green's functions of the junction. Because Eq. 9.9 contains exact Green functions which cannot be computed in general, one should use appropriate expressions. Instead of the exact Green functions, one uses approximate forms obtained under certain assumptions. It permits to include the quantum many-body effects while accounting for the macroscopic number of electrons and vibrations. Roughly speaking, there are two different versions of the Green function methods. One category includes the many-body perturbation theories, which are based, for example, on the effective single-particle methods related to the Kohn–Sham scheme of density functional theory (KS-DFT), the molecular Dyson equation, the Kadanoff–Baym equations, and GW (Strange et al. 2011) approximation to Hedin's equations (Hedin 1965). The method of nonequilibrium Green functions typically assumes that the Green's function $G(E)$ of the junction is represented by a perturbative series in terms of noninteracting Green's functions $G_0(E)$ and the self-energy $\Sigma(E)$. The series are obtained using an effective single-particle theory. The self-energy $\Sigma(E)$ accounts for a few physical processes to infinite order implementing the Dyson's equation, $G(E) = G_0(E) + G_0(E)\Sigma(E)G(E)$. Another method aims to solve the relevant molecular system representing it as a mathematical model. One can solve the problem either exactly or approximately. An approximate solution is obtained considering electron hopping across the graphene/metal (CNT/metal) interfaces as a perturbation. Since all the processes within the density matrix approaches are included up to infinite order, they are quoted as complementary. Considering transport through the graphene and CNT-based junctions, there are many nontrivial issues, including the resonant tunneling and Coulomb blockade which have been observed experimentally (Zhang et al. 2006, Tselev et al. 2009). Properties of the single graphene and CNT junctions involve a variety of physical phenomena. The junction's properties are evolving versus the gate and source–drain voltages, which suggests that many aspects of charge transport are important.

9.4 Electron Transport and Elastic Collisions

Electron scattering on lattice defects, atomic impurities, and tunneling through the barriers proceed mostly without energy losses. Such processes are regarded as elastic. The electron energy dissipation occurs mostly when electrons collide with the lattice oscillations, phonons, whereas the contribution of other inelastic scattering processes in the graphene and CNT junctions is typically low. For instance, a small inelastic current originating from the electron–phonon interaction is visible only at a large bias (Galperin et al. 2008). Therefore, in most cases of interest, the elastic processes dominate the electron transport through the graphene/metal and CNT/metal junctions. The elastic component of the electronic current can be computed when the Coulomb self-energy entering the electron Green function $G(E)$ is simplified neglecting the inelastic processes. Then, one arrives at the multiterminal Landauer–Buttiker formula (Buttiker 1986, Sivan and Imry 1986)

$$I_{\alpha}^{(\nu)} = \frac{1}{h} \sum_{\beta=1}^{M} dE(E - \mu_{\alpha})^{\nu} T_{\alpha\beta}(E) \left[f_{\beta}(E) - f_{\alpha}(E) \right], \tag{9.11}$$

where β is the index of the electrodes, whereas the transmission function is given by

$$T_{\alpha\beta}(E) = \mathrm{Tr}\left\{ \Gamma^{\alpha}(E) G(E) \Gamma^{\beta}(E) G^{\dagger}(E) \right\}. \tag{9.12}$$

The above Eq. 9.12 remains valid even for strongly interacting systems although it ignores an inelastic contribution to transport. One also obtains a handy expression for the usable produced power P in a two-terminal device which takes the form

$$\begin{aligned} P &= I^{(0)}(\mu_1 - \mu_2), \\ \eta &= \frac{P}{\left| I_1^{(1)} \right|} \end{aligned} \tag{9.13}$$

where η is the energy conversion efficiency which characterizes the thermodynamic response of the junction Deriving Eqs. 9.14

and 9.15, we assume that $T_1 > T_2$ In the linear response regime one assumes that $\Delta T << T$ and $|eV| << \mu$. The electron distribution function in the α electrode can be represented by

$$f_\alpha (E) \cong f_0 (E) + \left(-\frac{\partial f_0}{\partial E}\right) \left[\Delta\mu_\alpha + \frac{E-\mu}{T}\Delta T_\alpha\right], \qquad (9.14)$$

where T is the temperature, μ is the chemical potential, and $f(E)$ is the equilibrium Fermi distribution at zero bias. Further simplification of expression for the electronic current, Eq. 9.11 with $\nu = 0, 1$ gives

$$\begin{pmatrix} I_\alpha^{(0)} \\ I_\alpha^{(1)} \end{pmatrix} = \sum_\beta \begin{pmatrix} L_{\alpha\beta}^{(0)} & L(1)\alpha\beta \\ L_{\alpha\beta}^{(1)} & \frac{1}{T} L_{\alpha\beta}^{(2)} \end{pmatrix} \begin{pmatrix} \mu_\beta - \mu_\alpha \\ T_\beta - T_\alpha \end{pmatrix}, \qquad (9.15)$$

where we have used the matrix form and have introduced the Onsager linear-response function

$$L_{\alpha\beta}^{(\nu)} (\mu, T) = \frac{1}{h} \int dE (E-\mu)^\nu \left(-\frac{\partial f_0}{\partial E}\right) M_e (\varepsilon)\, T_{\alpha\beta} (E). \qquad (9.16)$$

$\alpha = 1 \ldots 3$, $M_e (E) = N_G (E) (\hbar v/L)$ is the number of modes, $N_G (E)$ is the electron density of states. Various transport coefficients are represented via the L functions in a compact fashion

$$\begin{aligned} G &= e^2 L(0), \\ S &= -\frac{1}{eT}\frac{L^{(1)}}{L^{(0)}}, \end{aligned} \qquad (9.17)$$

and

$$\begin{aligned} \kappa^{el} &= \frac{1}{T}\left(L^{(2)} - \frac{\left[L^{(1)}\right]^2}{L^{(0)}}\right), \\ ZT^{el} &= \left(\frac{L^{(0)} L^{(2)}}{\left[L^{(1)}\right]^2} - 1\right)^{-1}. \end{aligned} \qquad (9.18)$$

In the above Eqs. 9.17 and 9.18 G is the electrical conductance, S is the thermopower, κ_{el} is the electronic contribution to the thermal conductance, and ZT^{el} is the electronic contribution to ZT Performing the Sommerfeld expansion for the junction's transparency $T(E)$ around $E = \mu$ to the first order straight from Eqs. 9.16 and 9.18 one can also derive the Mott's formula Eq. 9.2.

One benefit of the approach described above is that the linear-response transport coefficients of interacting system look similar to the corresponding coefficients for noninteracting system. The difference only is that they contain the transparency matrix $T_{\alpha\beta}(E)$ which includes many-body effects The discussed above approach for computing the electronic characteristics provides a clean and convenient path to studying the heat and charge transport in the nanodevices. The derived equations are offering the convenient guidelines to accounting for the molecular vibrations, electron–electron correlations, etc. contributing into the electronic characteristics of graphene and CNT junctions with multiple electrodes.

9.5 Reversible Peltier Effect in Carbon Nanojunctions

The maximum temperature difference for a traditional cooler $ZT = (T_{\mathbf{hot}} - T_{\mathbf{cold}})_{\max}/T$

$$(T_{\text{hot}} - T_{\text{cold}})_{\max} = \frac{1}{2} Z T^2 \tag{9.19}$$

is expressed in terms of the figure of merit Z

$$Z = \frac{\left(\alpha_p - \alpha_n\right)^2}{KR} \tag{9.20}$$

where $\alpha_{p,\,n}$ are Seebeck coefficients of the p- and n-type regions,

$$K = k_p \frac{A_p}{L_p} + k_n \frac{A_n}{L_p} \tag{9.21}$$

is the thermal conductance of the p- and n-type regions connected in parallel, and

$$R = \rho_p \frac{L_p}{A_p} + \rho_n \frac{L_n}{A_n} \tag{9.22}$$

is the electrical resistance of the p- and n-type regions connected in series, and $L_{p,\,n}$ and $A_{p,\,n}$ are the corresponding length and area of the p- and n-type regions. The Seebeck coefficient α is related to the Peltier coefficient Π by

$$\Pi_{p,\,n} = T\alpha_{p,\,n} \tag{9.23}$$

where

$$\Pi = \frac{1}{q} \frac{\int \sigma(E)\left(E_f - E\right)\left(-\partial f_0/\partial E\right)}{\int \sigma(E)\left(-\partial f_0/\partial E\right)} \tag{9.24}$$

where q is the electric charge. The differential conductivity $\sigma(E)$ gives the contribution of a carrier at energy E to the overall conductivity

$$\sigma(E) = q^2\tau(E) \int\int v_x^2(E, k_y, k_z)\, dk_y dk_z \approx q^2\tau(E)\, \bar{v}_x^2(E)\, \bar{n}(E) \tag{9.25}$$

in terms of the relaxation time $\tau(E)$, the average carrier velocity $\bar{v}_x(E)$ and of the average carrier density $\bar{n}(E)$. From the above expressions one can derive the following criteria for the figure of merit ZT:

(i) Seebeck (Peltier) coefficients must differ, that is, $\alpha_{p,\,n} >> \alpha_{n,\,p}$.
(ii) High energies contribute to the transport coefficients stronger, which means that one prefers the junction transparency peaked at higher energies (within the energy region $\Delta E \approx k_B T$ indeed).
(iii) Nonequilibrium effects may enhance ZT as well. The semi-conducting CNTs have a strong potential for creating efficient nanocoolers whose electric and heat transport properties are controlled by the gate and split-gate voltages.

For a symmetric setup, the figure of merit becomes

$$Z = \frac{(\alpha_p - \alpha_n)^2}{\kappa_p + \kappa_n + \kappa_{\mathbf{ph}}} \sigma_{\text{seq}} \tag{9.26}$$

where $\alpha_{p,\,n}$ are Seebeck coefficients of the p- and n-type regions,

$$\alpha_{p,\,n} = -\frac{1}{q_{p,\,n} T} \frac{L^{(1)}}{L^{(0)}} \tag{9.27}$$

$q_p = e$, $q_n = -e$, σ_{seq} is the conductivity of two equal size conductors connected in sequence,

$$\sigma_{\text{seq}} = \frac{\sigma_p \sigma_n}{\sigma_p + \sigma_n} \tag{9.28}$$

$$\sigma_{p,\,n} = L_{p,\,n}^{(0)} \tag{9.29}$$

and $\kappa_{n,\,p}$ is the electron (hole) thermal conductivity

$$\kappa_{n,\,p} = \frac{1}{e^2 T}\left(L_{p,\,n}^{(2)} - \frac{\left[L_{p,\,n}^{(1)}\right]^2}{L_{p,\,n}^{(0)}} \right) = \frac{1}{e^2 T}\left(L_{p,\,n}^{(2)} + qT\alpha_{p,\,n} L_{p,\,n}^{(1)} \right). \tag{9.30}$$

The Onsager functions are represented as

$$L^{(\alpha)} = e^2 \int \frac{dk}{\pi a^2} \left(-\frac{\partial f}{\partial \varepsilon}\right) \tau (k)\, v (k)\, v (k)\, [\varepsilon (k) - \zeta]^\alpha \tag{9.31}$$

where f is the Fermi–Dirac distribution, $\varepsilon (k)$ is the energy dispersion relationship, $\tau (k)$ is the relaxation time, and ζ is the chemical potential. For a CNT/graphene junction, one uses the following expression for the electric current density:

$$\begin{aligned} j_\mathrm{e} =& eN (0) \int d\varepsilon \left(G_R^K (0, D, \varepsilon) - G_L^K (0, D, \varepsilon)\right) \\ =& eN (0) \int d\varepsilon \sum_k |t_{kk'}|^2 \left(\left(2n_k^R - 1\right) \delta (\varepsilon - \varepsilon_k + eV)\right. \\ & \left. - \left(2n_k^L - 1\right) \delta (\varepsilon - \varepsilon_k)\right) \\ =& 2eN (0) \int d\varepsilon |t_\varepsilon|^2 (n_\varepsilon - n_{\varepsilon - eV}) \end{aligned} \tag{9.32}$$

The corresponding heat current density j_Q is obtained as

$$\begin{aligned} j_\mathrm{Q} \propto& N (0) \int d\varepsilon (\varepsilon - eV) \left(G_R^K (0, D, \varepsilon) - G_L^K (0, D, \varepsilon)\right) \\ =& 2N (0) \int d\varepsilon |t_\varepsilon|^2 (\varepsilon - eV) (n_\varepsilon - n_{\varepsilon - eV}) \\ =& - jV + 2N (0) \int d\varepsilon |t_\varepsilon|^2 \varepsilon (n_\varepsilon - n_{\varepsilon - eV}) \end{aligned} \tag{9.33}$$

which gives

$$L_{p,n}^{(\alpha)} = 2e^2 N (0) \int (\varepsilon - \zeta)^\alpha \left|t_\varepsilon^{p,n}\right|^2 \left(-\frac{\partial n_\varepsilon}{\partial \varepsilon}\right) d\varepsilon \tag{9.34}$$

9.6 Thermoelectric Figure of Merit and Fourier's Law

For a symmetric TE setup, the figure of merit becomes

$$\begin{aligned} Z =& \frac{(\alpha_p - \alpha_n)^2 \sigma_\mathrm{seq}}{\kappa_p + \kappa_n + \kappa_L} \\ =& \frac{4}{T} \frac{L_p^{(0)} L_n^{(0)}}{L_p^{(0)} + L_n^{(0)}} \left(\frac{L_p^{(1)}}{L_p^{(0)}} + \frac{L_n^{(1)}}{L_n^{(0)}}\right)^2 \\ & \times \left(L_p^{(2)} + L_n^{(2)} - \frac{\left[L_p^{(1)}\right]^2}{L_p^{(0)}} - \frac{\left[L_n^{(1)}\right]^2}{L_n^{(0)}} + \frac{1}{3} C_L v_L l_L\right)^{-1} \end{aligned} \tag{9.35}$$

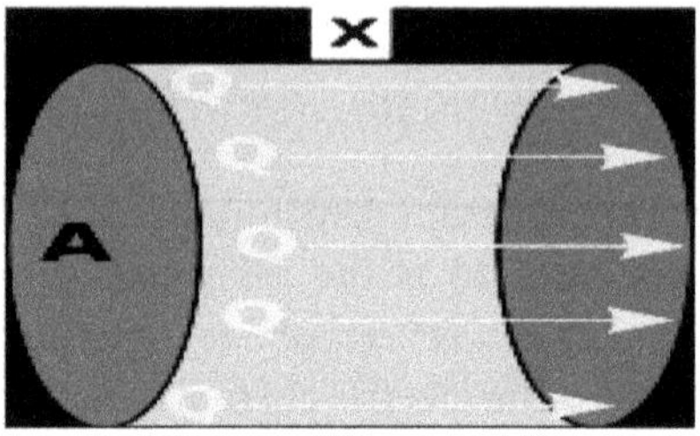

Figure 9.2 Linear case characterized by uniform temperature across equally sized end surfaces and by perfectly insulated sides.

By integrating the differential form over the material's total surface S, we arrive at the integral form of Fourier's law:

$$\frac{\partial Q}{\partial t} = -k \oint_S \nabla T \cdot dS \tag{9.36}$$

where $\partial Q/\partial t$ is the amount of heat transferred per unit time, [W] or [J s^{-1}], S is the surface through which the heat is flowing, [m^2], and k is the thermal conductivity in SI units—W/(m K)—or in English units—BTU/(hr ft. F°). For the two to be converted, one uses the relation 1 BTU/(hr ft. F°) = 1.730735 W/(m K).

9.6.1 *Linear Heat Flow*

The above differential Eq. 9.36, when integrated for a simple linear case (Fig. 9.2), where uniform temperature across equally sized end surfaces and perfectly insulated sides exist, gives the heat flow rate between the end surfaces as:

$$\frac{\Delta Q}{\Delta t} = -kA\frac{\Delta T}{\Delta x} \tag{9.37}$$

where A is the cross-sectional surface area, ΔT is the temperature difference between the ends, and Δx is the distance between the ends.

The law Eq. 9.37 forms the basis for the derivation of the heat equation. Fourier's law is the thermal heat flow analogue of the electrical Ohm's law. The above differential equation Eq. 9.36, when integrated for a simple linear situation, where uniform temperature across equally sized end surfaces and perfectly insulated sides exist, gives the rate of heat flow between the end surfaces by Eq. 9.37. The heat equation is an important partial differential equation which describes the distribution of heat (or variation in temperature) in a given region over time. For a function of three spatial variables (x, y, z) and one time variable t, the heat equation is

$$\frac{\partial T}{\partial t} - k\left(\frac{\partial^2 T}{\partial x^2} + \frac{\partial^2 T}{\partial y^2} + \frac{\partial^2 T}{\partial z^2}\right) = 0 \tag{9.38}$$

where k is a constant and T is the temperature at a given location (x, y, z).

9.6.2 *Cooling Power*

The net cooling power is determined by Fourier's law

$$\begin{aligned}\frac{\Delta Q}{\Delta t} &= -\kappa A \frac{\Delta T}{\Delta x} \\ &= -N_{\mathbf{T}} \frac{\pi d_{\mathbf{T}}^2}{8} \frac{1}{L} \left(\alpha_p - \alpha_n\right)^2 \cdot \sigma_{\text{seq}} T^2.\end{aligned} \tag{9.39}$$

The cooling power density for one CNT ($N_{\mathbf{T}} = 1$) is

$$\begin{aligned}\frac{\Delta Q}{\Delta t} &= \frac{\left(\alpha_p - \alpha_n\right)^2}{L} \cdot \sigma_{\text{seq}} T^2 \\ &= \left(\alpha_p - \alpha_n\right)^2 \frac{1}{L} \cdot \left(\frac{L}{A} \frac{2e^2}{h} \bar{T} N_{\mathbf{ch}}\right) \cdot T^2 \\ &= \left(\alpha_p - \alpha_n\right)^2 \cdot T^2 \cdot \left(\frac{2e^2}{h} \bar{T} \frac{N_{\mathbf{ch}}}{A}\right)\end{aligned} \tag{9.40}$$

where we have used that

$$\Delta T = (T_{\mathbf{hot}} - T_{\mathbf{cold}})_{\max} = \frac{1}{2} Z T^2 \tag{9.41}$$

$$Z = \frac{\left(\alpha_p - \alpha_n\right)^2 \sigma_{\text{seq}}}{\kappa_p + \kappa_n + \kappa_L}. \tag{9.42}$$

Then, one gets

$$\begin{aligned}\frac{\Delta Q}{\Delta t} &= -\kappa A \frac{\Delta T}{\Delta x} = -\kappa A \frac{1}{\Delta x}\left(\frac{1}{2} Z T^2\right) \\ &= -N_T \frac{\pi d^2}{8} \frac{1}{L} \left(\alpha_p - \alpha_n\right)^2 \sigma_{\text{seq}} T^2 \\ &= -N_T \frac{\pi d^2}{8L} \left(\frac{1}{e} \frac{L^{(1)}}{L^{(0)}}\right)^2 L_{p,n}^{(0)} \\ &= -2N(0) \cdot N_T \cdot \mathbf{K}\end{aligned} \tag{9.43}$$

where we have defined

$$\mathbf{K} = \frac{\pi d^2}{8L}\left(\frac{M^{(1)}}{M^{(0)}}\right)^2 \cdot M^{(0)} \tag{9.44}$$

$$M^{(\alpha)} = \int (\varepsilon - \zeta)^{\alpha} |t_{\varepsilon}|^2 \left(-\frac{\partial n_{\varepsilon}}{\partial \varepsilon}\right) d\varepsilon. \tag{9.45}$$

In the above formulas N_{T} is the number of nanotubes in the rope, d_{T} and L are the average tube diameter and length, respectively, and α_p and σ_{seq} are determined for a single tube. From Eq. 9.43 one can see that the heat-sucking rate depends on the ratio N_{T}/L.

9.6.3 *Seebeck Coefficient*

Seebeck coefficients of the p- and n-type regions (in units of Volt/Kelvin),

$$\alpha_{p,\,n} = \pm\frac{1}{eT}\frac{L^{(1)}_{p,\,n}}{L^{(0)}_{p.n}} = \pm\frac{1}{eT}\frac{\int (\varepsilon - \zeta)\,\left|t_{\varepsilon}^{p,\,n}\right|^2 \left(-\frac{\partial n_{\varepsilon}}{\partial \varepsilon}\right) d\varepsilon}{\int \left|t_{\varepsilon}^{p,\,n}\right|^2 \left(-\frac{\partial n_{\varepsilon}}{\partial \varepsilon}\right) d\varepsilon} \tag{9.46}$$

where $q_p = e$, $q_n = -e$. Roughly speaking, one gets

$$\alpha_{p,\,n} = \pm\tilde{\sigma}\frac{V_{\mathrm{SD}}}{T} \tag{9.47}$$

where typically $\tilde{\sigma} \simeq 10^{-3}$,

$$\alpha_{\pm} = \frac{4\rho}{T} \tag{9.48}$$

and where

$$\rho\left(\nu_{\mathrm{SD}}\right) = T\log 2 - \frac{V_{\mathrm{SD}}}{2} + T\vartheta_0\left(\frac{E - V_{\mathrm{SD}}}{T}\right) \tag{9.49}$$

9.6.4 *Electron Thermal Conductivity*

The electron (n) and hole (p) thermal conductivity is defined as

$$\kappa_{n,\,p} = \frac{1}{e^2T}\left(L^{(2)}_{p,\,n} - \frac{\left[L^{(1)}_{p,\,n}\right]^2}{L^{(0)}_{p,\,n}}\right) = \frac{1}{e^2T}\left(L^{(2)}_{p,\,n} + qT\alpha_{p,\,n}L^{(1)}_{p,\,n}\right) \tag{9.50}$$

where the auxiliary functions are

$$L^{(\alpha)}_{p,\,n} = 2e^2N(0)\int (\varepsilon - \zeta)^{\alpha}\left|t_{\varepsilon}^{p,\,n}\right|^2\left(-\frac{\partial n_{\varepsilon}}{\partial \varepsilon}\right) d\varepsilon. \tag{9.51}$$

The electron distribution factors are

$$n_0(\varepsilon) = \frac{1}{e^{\varepsilon/T}+1}; -\frac{\partial n_0(\varepsilon)}{\partial \varepsilon} = \frac{1}{T}\frac{e^{\frac{\varepsilon}{T}}}{\left(e^{\frac{\varepsilon}{T}}+1\right)^2}. \tag{9.52}$$

Also using that

$$L_{p,n}^{(0)} = 2e^2 N(0)\,\bar{T}\cdot\left(\frac{1}{2}+\vartheta_0\left(\frac{E_i-\zeta}{T}\right)\right) = \frac{L}{A}\frac{2e^2}{h}N_{\mathbf{ch}}\bar{T} \tag{9.53}$$

where ϑ_α accounts for a resonant contribution from electron spectral singularities and is defined as

$$\vartheta_\alpha(x) = \iota x^\alpha e^x \left(e^x+1\right)^{-2}, \tag{9.54}$$

where the conductance is

$$G = \frac{2e^2}{h}N_{\mathbf{ch}}\bar{T}, \tag{9.55}$$

whereas the conductivity is

$$\sigma = \frac{L}{A}\frac{2e^2}{h}N_{\mathbf{ch}}\bar{T}. \tag{9.56}$$

For $L = 1\ \mu\text{m}$ and $A = \pi d^2/4 = 3.141\cdot 4\ \text{nm}^2/4$

$$\sigma = \frac{L}{A}\frac{2e^2}{h}N_{\mathbf{ch}}\bar{T} = 2.5\times 10^7\,\frac{\text{S}}{\text{m}}. \tag{9.57}$$

We represent the energy dependence of $t_\varepsilon^\pm$ as $\left|t_\varepsilon^\pm\right|^2 \rightarrow \bar{T} + \bar{T}\iota\delta(\varepsilon - E_k)$, where the second term accounts for the resonant tunneling at the energies E_k of bound states or Van Hove singularities and ι is a dimensionless parameter, $\iota < 1$. We also find that

$$L_{p,n}^{(1)} = 2e^2 N(0)\,\bar{T}\cdot\left(\iota\frac{E_i-\zeta}{T}\frac{e^{\frac{E_i}{T}}}{\left(e^{\frac{E_i}{T}}+1\right)^2} + \left(T\log(2)-\frac{\zeta}{2}\right)\right)$$
$$= 2L_{p,n}^{(0)}\cdot\left(T\log(2)-\frac{\zeta}{2}+\vartheta_1\left(\frac{E_i-\zeta}{T}\right)\right) \tag{9.58}$$

where

$$L_{p,n}^{(2)} = 2e^2 N(0)\,\bar{T}\cdot\left(\frac{1}{6}\left(\pi^2T^2+3\zeta(\zeta-4T\log(2))\right)+\vartheta_2\left(\frac{E_i-\zeta}{T}\right)\right) \tag{9.59}$$

and

$$\kappa_{n,\,p} = \frac{1}{e^2 T}\left(L^{(2)}_{p,\,n} - \frac{1}{L^{(0)}_{p,\,n}}\left[L^{(1)}_{p,\,n}\right]^2\right)$$

$$= \frac{1}{e^2 T}\left[2e^2 N(0)\,\bar{T}\cdot\left(\frac{1}{6}\left(\pi^2 T^2 + 3\zeta(\zeta - 4T\log(2))\right)\right)\right.$$

$$\left. + \vartheta_2\left(\frac{E_i - \zeta}{T}\right) - \frac{1}{e^2 N(0)\,\bar{T}}\left[2e^2 N(0)\,\bar{T}\cdot\left(T\log(2) - \frac{\zeta}{2}\right)\right]^2\right]$$

$$= \frac{N(0)\,\bar{T}}{T}\left[\left(\frac{1}{3}\left(\pi^2 T^2 + 3\zeta(\zeta - 4T\log(2))\right)\right)\right.$$

$$\left. + \vartheta_2\left(\frac{E_i - \zeta}{T}\right) - 4\left(T\log(2) - \frac{\zeta}{2} + \vartheta_1\left(\frac{E_i - \zeta}{T}\right)\right)^2\right] \quad (9.60)$$

or

$$\kappa_{n,\,p} = \frac{L}{A}\frac{2}{h}N_{\mathbf{ch}}\bar{T}\cdot T\left[\left(\pi^2 + 3\nu(\nu - 4\log 2) + 3\vartheta_2(\nu_i - \nu)\right)/3\right.$$

$$\left. - 4\left(\log(2) - \nu/2 + \vartheta_0(\nu_i - \nu)\right)^2\right] \quad (9.61)$$

where $\nu = \zeta/T$ and $\nu_i = E_i/T$ we use a useful identity

$$N(0)\,\bar{T} = \frac{L}{A}\frac{2}{h}N_{\mathbf{ch}}\bar{T}. \quad (9.62)$$

For the electron (hole) thermal conductivity $\kappa_{n,\,p}$ one obtains

$$\kappa_{n,\,p} = \frac{2}{h}N_{\mathbf{ch}}\cdot\frac{L}{A}\cdot\frac{2}{T}\bar{T}\cdot\left[\frac{1}{6}\left(\pi^2 T^2 + 3\zeta(\zeta - 4T\log(2)\right)\right.$$

$$\left. + 3\vartheta_2\left(\frac{E_i - \zeta}{T}\right)\right)$$

$$\left. - 2\left(T\log(2) - \frac{\zeta}{2} + \vartheta_0\left(\frac{E_i - \zeta}{T}\right)\right)^2\right] \quad (9.63)$$

which is expressed in the units of W/ (mK), and for a tube with $L =$ 1 μm and $A = \pi d^2/4 = 3.141\cdot 4\,\text{nm}^2/4$ (i.e., 2 nm in diameter)

$$\kappa_{n,\,p} = \frac{1}{h}\frac{L}{A}\cdot\frac{\bar{T}}{T}\cdot(k_{\mathbf{B}}T)^2 = 3\times 10^{-4}\frac{\text{W}}{\text{m}\cdot\text{K}}. \quad (9.64)$$

The above estimation means that the thermal conductivity is determined by phonons mostly. One also finds the Seebeck coefficient and thermal conductivity in the form

$$\alpha_\pm = 4\rho/(eT)$$
$$\kappa_\pm = \left(\sigma_0/e^2 T\right)\cdot\left[\psi/3 - 4\rho^2\right] \quad (9.65)$$

where the "±" signs correspond to electrons (holes) and

$$\rho_i(\nu_{\mathbf{SD}}) = T\log 2 - \frac{V_{\mathbf{SD}}}{2} + T\vartheta_0\left(\frac{E_i - V_{\mathbf{SD}}}{T}\right)$$
$$= T\left(\log 2 - \frac{\nu_{\mathbf{SD}}}{2} + \vartheta_0(\nu_i - \nu_{\mathbf{SD}})\right), \quad (9.66)$$

$\nu_{\mathbf{SD}} = V_{\mathbf{SD}}/T$, and

$$\psi_i(\nu_{\mathbf{SD}}) = \left(\pi^2 T^2 + 3V_{\mathbf{SD}}(V_{\mathbf{SD}} - 4T\log(2)) + 3T^2\vartheta_2\left(\frac{E_i - V_{\mathbf{SD}}}{T}\right)\right)$$
$$= T^2\left(\pi^2 + 3\nu_{\mathbf{SD}}(\nu_{\mathbf{SD}} - 4\log(2)) + 3\vartheta_2(\nu_i - \nu_{\mathbf{SD}})\right). \quad (9.67)$$

Using Eqs. 9.65–9.67, one estimates the figure of merit for a symmetric TE setup as

$$Z \cdot T = \frac{32\rho_i^2(\nu_{\mathbf{SD}})}{T^2\psi_i(\nu_{\mathbf{SD}})/6 - 2\rho_i^2(\nu_{\mathbf{SD}}) + \rho_{\mathbf{ph}}}. \quad (9.68)$$

The measured RT $\kappa_{\mathbf{ph}}$ is in the range ~30–6000 W/ (mK). For 1–100 phonon modes and for the tube with $d = 10^{-9}$ m and $L = 10^{-6}$ m it gives

$$\rho_{\mathbf{ph}} = \frac{1}{2k_{\mathbf{B}}T}\frac{e^2}{\sigma_0}\kappa_{\mathbf{ph}} = 0.1 - 10. \quad (9.69)$$

One can notice that the product ZT does not depend on the properties of electrodes but instead, strongly depends on temperature T along with the source drain bias voltage $V_{\mathbf{SD}}$. For $\kappa_{\mathbf{ph}} = 10$, the product $T \cdot Z$ can be increased by 2 orders of magnitude, neglecting the phonon part in a microscopic model where the electron and hole envelope wavefunctions are obtained as solutions of the Dirac equation.

9.6.5 *Figure of Merit*

For a symmetric setup, the figure of merit becomes

$$Z \times T = \frac{(\alpha_p - \alpha_n)^2 \sigma_{\text{seq}}}{\kappa_p + \kappa_n + \kappa_{\mathbf{ph}}} \times T = 0.3 - 40. \quad (9.70)$$

For the above estimation we used the following parameters. We evaluate that $\kappa_{\mathbf{ph}} = n_q \cdot \kappa_0 = (50 - 5000)$ W/(m K) for 1–100 phonon modes and $\alpha_p = -\alpha_n \simeq 2.4 \times 10^{-4}$ V/K, $\sigma_{\text{seq}} = 2.5 \times 10^7$ S/m for one conducting channel, and use the temperature $T =$

300 K, nanotube diameter $d_T = 2$ nm, and nanotube length $L = 1$ µm. Furthermore, we also use

$$N(0) \to \frac{L}{A}\frac{2}{h}N_{\mathbf{ch}} \tag{9.71}$$

$$\alpha_{p,n} = \pm\frac{4}{e}\left(\log(2) - \frac{\nu}{2}\right) \tag{9.72}$$

which means that $Z \times T$ can be significantly increased varying the source–drain bias voltage ζ. For 1–100 phonon modes and tube with $d = 2 \cdot 10^{-9}$ m and $L = 10^{-6}$ m it gives

$$\rho_{\mathbf{ph}} = \frac{1}{2k_{\mathbf{B}}T}\frac{h}{2\bar{T}}\frac{1}{N_{\mathbf{ch}}}\frac{A}{L}\kappa_{\mathbf{ph}} \simeq 0.1 - 10, \tag{9.73}$$

$ZT = 1 - 70$, where the dimensionless figure of merit is defined by

$$\frac{1}{2}ZT = \frac{(T_{\mathbf{hot}} - T_{\mathrm{cold}})_{\max}}{T} \tag{9.74}$$

In Eq. 9.74

$$L^{(0)}_{p,n} = 2e^2 N(0)\,\bar{T}\cdot\frac{1}{2} = \frac{L}{A}\frac{2e^2}{h}N_{\mathbf{ch}}\bar{T} \tag{9.75}$$

$$L^{(1)}_{p,n} = 2e^2 N(0)\,\bar{T}\cdot\left(T\log(2) - \frac{\zeta}{2}\right) \tag{9.76}$$

$$L^{(2)}_{p,n} = 2e^2 N(0)\,\bar{T}\cdot\left(\frac{1}{6}\left(\pi^2 T^2 + 3\zeta(\zeta - 4T\log(2))\right)\right). \tag{9.77}$$

We assume that the conductance is

$$G = \frac{2e^2}{h}\bar{T}N_{\mathbf{ch}} \tag{9.78}$$

The above expression can be rewritten via the conductivity as

$$\sigma = \frac{L}{A}\frac{2e^2}{h}\bar{T}N_{\mathbf{ch}} \tag{9.79}$$

where $N_{\mathbf{ch}}$ is the voltage-dependent number of conducting channels. Then,

$$L^{(0)}_{p,n} = \frac{L}{A}\frac{2e^2}{h}\bar{T}N_{\mathbf{ch}} \tag{9.80}$$

$$\alpha_{p,n} = \frac{1}{e}\frac{L^{(1)}}{L^{(0)}} = \frac{4}{eT}\left(k_{\mathrm{B}}T\log(2) - \frac{\zeta}{2}\right) \tag{9.81}$$

$$\frac{\Delta Q}{\Delta t} = 16\left(T\log(2) - \frac{\zeta}{2}\right)^2\frac{2}{h}\frac{N_{\mathbf{ch}}}{A}. \tag{9.82}$$

9.7 Phonon Transport and Thermal Conductivity

Acoustic phonon transport is measured via the lattice thermal conductivity. Such an approach is used in the experimental technique for investigation of graphene. Here, we discuss the main contributions into the heat conductivity. Fourier's law for thermal conductivity is rewritten as

$$\vec{\phi} = -K\nabla T, \tag{9.83}$$

where ∇T is the temperature gradient, $\vec{\phi}$ is the heat flux, and $K = (K_{ij})$ is the tensor of thermal conductivity. If the environment is isotropic and the temperature variations are small, the thermal conductivity is independent on direction of the heat flow, and thus, one sets K as a constant. Generally speaking, the thermal conductivity $K \equiv K(T)$ is a function of temperature in the wide temperature range. The flow of heat in solid materials is caused by propagation of the phonons and electrons which gives $K = K_\mathrm{p} + K_\mathrm{e}$, where K_p and K_e are the phonon and electron contributions, respectively. In many conducting materials like, for example, metals or degenerately doped semiconductors, K_e dominates because of a large concentration of electric current carriers. The electronic part of thermal conductivity K_e can be obtained from the WF law using the experimental data on the electrical conductivity σ via

$$\frac{K_\mathrm{e}}{\sigma T} = \frac{\pi^2 k_\mathrm{B}^2}{3e^2} \tag{9.84}$$

where e is the electron charge and k_B is the Boltzmann's constant. The situation is different in carbon materials, where the phonons usually serve as main heat carriers. In particular, the acoustic phonons dominate the heat conduction even in the metal-like graphite (Bergman et al. 2000, Klemens 2000). It happens due to the strong covalent sp^2 bonding which leads to high values of in-plane phonon group velocities and low crystal lattice unharmonicity for in-plane vibrations. The phonon thermal conductivity is written as

$$K_\mathrm{p} = \sum_j C_J(\omega)\, v_j^2(\omega)\tau_j(\omega)\, d\omega \tag{9.85}$$

where j is the index of the phonon polarization branches, $v_j = d\omega_j/dq$ is the phonon group velocity of the jth branch, which, in

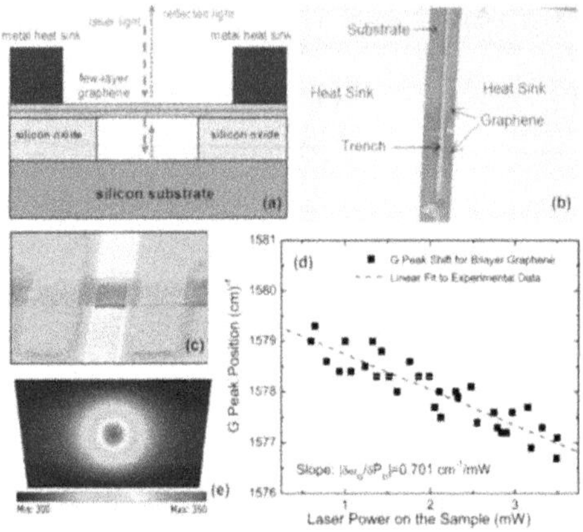

Figure 9.3 (a) Measurement of the thermal conductivity performed on suspended FLG flakes using the excitation laser light. (b) The FLG flakes attached to metal heat sinks. The images were obtained by means of optical microscopy. (c) The suspended graphene flake shown as a colored image obtained by scanning electron microscopy. (d) The position of the Raman G-peak measured versus laser power. Thus, one extracts the increase of local temperature in response to the dissipated power from the experimental data. (e) Simulation of the temperature distribution inside the flake by a finite-element method using the given geometry to extract the thermal conductivity. After Ghosh, et al. (2010), reproduced with permission from the Nature Publishing Group.

many solids, can be approximate by the sound velocity, τ_j is the phonon relaxation time, $C_j = \hbar\omega_j \partial N_0 (\omega_j) / \partial T$ is the contribution to heat capacity from the jth branch, and

$$N_0 (\omega_j) = \frac{1}{e^{\frac{\hbar\omega_j}{k_B T}} - 1} \tag{9.86}$$

is the Bose–Einstein phonon equilibrium distribution function. The phonon branches include two transverse acoustic and one longitudinal acoustic branch. If the thermal conductivity is affected by extrinsic processes like scattering of phonons on a rough boundary or by the phonon–defect scattering, it is regarded as extrinsic. Because the acoustic phonons are spatially confined in the nanostructures it causes the phonon energy spectra to be quantized. Such discreteness of the phonon energy spectra generally leads to a decreasing of the phonon group velocity. Altering of the phonon energies, group velocities and the density of states which are complemented by the phonon scattering from boundaries altogether modifies the thermal conductivity of nanostructures. On the one hand, the spatial confinement of acoustic phonons frequently leads to reduction of phonon part of the thermal conductivity (Balandin and Wang 1998, Zou and Balandin 2001). On the other hand, one can increase the thermal conductivity by embedding a nanostructure between acoustically hard barrier layers which cause spatial confinement of acoustic phonons (Balandin 2005, Pokatilov et al. 2005). One can include the mentioned factors and evaluate the boundary phonon scattering as

$$\frac{1}{\tau_{B,j}} = \frac{v_j}{D}\frac{1-p}{1+p}, \tag{9.87}$$

where p is the specularity parameter defined as a probability of specular scattering at the boundary and D is the nanostructure or grain size. The specular scattering which corresponds to $p = 1$ conserves the momentum, and therefore, it does not increase the thermal resistance. On the other hand, the diffusive scattering of phonons from rough interfaces which corresponds to $p{-}{>}0$, alters the phonon momentum and hence limits the phonon MFP. The appropriate magnitude of p is either computed from the surface roughness or is obtained from fitting the experimental data. Specularity of phonons can be mimicked by a phenomenological expression, as suggested by Ziman

$$p(\lambda) = \exp\left(-\frac{16\pi^3\eta^2}{\lambda^2}\right), \tag{9.88}$$

where λ is the wavelength of the incident phonon and η is the root-mean-square deviation of the surface height from the reference plane. If the phonon-boundary scattering is dominant, the thermal conductivity scales with the nanostructure or grain size D as

$$K_p \approx C_p v \Lambda \approx C_p v^2 \tau_B \approx C_p v D. \tag{9.89}$$

When the dimensions D of the structure are very small, $D << \Lambda$, the strong quantization of phonon energy spectra inside the spatially confined region complicates the dependence of thermal conductivity versus the physical size of the structure (Balandin and Wang 1998, Zou and Balandin 2001). Because of the dependence of specific heat C_{p} on the phonon density of states, it causes a distinction in the behavior of $C_{\mathrm{p}}(T)$ for 3D, 2D, and 1D systems. Besides, the details of the phonon spectrum influence the temperature dependence $K(T)$ at low temperatures T (Bergman et al. 2000, Klemens 2000). At low temperatures T, $K(T) \approx T^3$ in the bulk 3D samples whereas it changes to $K(T) \approx T^2$ in 2D systems. An ability of a material to conduct heat is characterized by the thermal conductivity K, whereas the speed of heating is determined by another characteristic which is the thermal diffusivity α. Thus, the latter parameter α tells how fast the material transfers the heat. The thermal diffusivity is computed as

$$\alpha = \frac{K}{C_{\mathrm{p}}\rho_{\mathrm{m}}} \tag{9.90}$$

where ρ_{m} is the mass density. Frequently, the measurement of thermal diffusivity is more straightforward than determining the thermal conductivity.

9.7.1 *Estimation of Phonon Thermal Conductivity*

The quantum of thermal conductance at $T = 300$ K is evaluated as

$$\rho_0 = \frac{\pi^2 k_{\mathrm{B}}^2 T}{3h} = 1.541\,0 \times 10^{-10}\frac{\mathrm{W}}{\mathrm{K}} \tag{9.91}$$

which corresponds to the conductivity of a tube with $L = 1\,\mu\mathrm{m}$, cross-sectional area $A = 3.141 \cdot 4\,\mathrm{nm}^2/4$ (i.e., 2 nm in diameter) and

$$\kappa_0 = \frac{\rho_0 L}{A} = 50\frac{\mathrm{W}}{\mathrm{mK}}. \tag{9.92}$$

At $T = 10$ K it is $\rho_0 = 5 \times 10^{-12}$ W/K and $\kappa_0 = 1.6$ W/(m K), while the measured RT $\kappa_{\mathbf{ph}}$ ranges between 35 and ~3000W/(m K). The authors Berber et al. (2000a, 2000b) even got an estimated value from their measurements as high as ~6000W/(m K). The corresponding electron/hole thermal conductivity is as low as

$$\kappa_{\mathrm{n,p}} = 3 \times 10^{-4}\frac{\mathrm{W}}{\mathrm{m\,K}}. \tag{9.93}$$

9.8 Recent Experiments on Measuring the Thermal Conductivity of Graphene

The heat conduction of graphene shown in Fig. 9.1d was measured by the group of researchers from UC Riverside in 2007 (Balandin et al. 2008, Calizo et al. 2008, Ghosh et al. 2008). The study of the phonon transport became possible after development of the optothermal Raman measurement technique. The experimental samples had been fabricated of large-area suspended graphene layers exfoliated from the high-quality Kish and highly ordered pyrolytic graphite. The obtained results suggest that the thermal conductivity is changed in a wide range and can exceed that of the bulk graphite, which is ~2000 W/(m K) at RT. The authors also concluded that the phonon contribution to heat conduction in the ungated graphene near RT is much larger than that of electrons, that is, $K_e << K_p$. At RT, the phonon mean free path in graphene was obtained as ~800 nm (Balandin et al. 2008, Calizo et al. 2008, Ghosh et al. 2008). Other groups implemented the Raman optothermal technique modified it via addition of a power meter under the suspended portion of graphene. They obtained rather high valves of the thermal conductivity of suspended high-quality chemical-vapor-deposited graphene. The measured K_p exceeded ~2500 W/(m K) at 350 K, and it achieved as much as $K \approx 1400$ W/(m K) at 500 K (Cai et al. 2010). The obtained value of K_p was also larger than the thermal conductivity of bulk graphite at RT. The Raman optothermal measurements performed on the suspended graphene found the thermal conductivity in the range from ~1500 to ~5000 W/(m K) (Jauregui et al. 2010). Besides, the group that aimed to reproduce the Raman-based measurements found $K \approx 630$ W/(m K) for the suspended graphene membrane (Faugeras et al. 2010). The data variations can be understood by considering the differences in the actual temperature of graphene under laser heating, strain distribution inside the suspended graphene of various sizes and geometries. Also, it is worth to mention here the experimental study which reported the thermal conductivity of graphene to be ~1800 W/(m K) at 325 K and ~710 W/(m K) at 500 K (Lee et al. 2011). The latter values are smaller than those of the bulk graphite. The discrepancy with other authors emerges probably from the assumption that the optical absorption coefficient should be 2.3% under conditions of the relevant experiment. The authors disregarded, however, that due to many-body effects, the absorption in graphene, when $\lambda > 1$ eV depends on the wavelength λ (Kim et al. 2009, Mak et al. 2011). The absorption of 2.3% is observed only in the near-infrared at ~ 1 eV. When the phonon wavelength λ decreases and energy $\hbar\omega = 2\pi\hbar/\lambda$ increases then, the absorption steadily increases. In particular, the Raman laser lines 514.5 nm and 488 nm correspond to 2.41 eV and 2.54 eV, respectively. At $\hbar\omega = 2.41$ eV the absorption is about $1.5 \times 2.3\% \approx 3.45\%$ (Kim et al. 2009). The last value of absorption 3.45% agrees with the other independently obtained data reported by Chen et al. (2011). If we replace the aforementioned value 2.3% which were assumed by Lee et al. (2011) by 3.45%, then we get $\sim$2700 W/(m K) at 325 K and 1065 W/(m K) near 500 K. Such magnitudes of thermal conductivity exceed the corresponding values for the bulk graphite and they are also consistent with the data published by other groups (Cai et al. 2010), where the measurements were conducted by the same Raman optothermal technique but with the measured light absorption. Very interesting data were obtained for suspended or partially suspended graphene where they are expected to be closer to the intrinsic thermal conductivity. Suspension of graphene sheets diminishes the heat exchange with the substrate and reduces phonon scattering on the substrate defects and impurities. The latter scattering mechanisms are suppressing the thermal conductivity of fully supported graphene. Such a tendency is confirmed with experimental measurements for exfoliated graphene on SiO_2/Si. The obtained data indicate that at RT in-plane $K \approx 600$ W/(m K) (Seol, Jo et al. 2010, Seol, Moore et al. 2010). It is much lower value than it follows from solution of the Boltzmann transport equation (BTE) and comparing with their experiments. The theoretical estimations predict that the thermal conductivity of free graphene should be $\sim$3000 W/(m K) at RT. The presently known data indicate considerable discrepancies among the experimental values of the thermal conductivity of graphene. Nevertheless, we conclude that K is very large compared to that for bulk silicon which is $K = 145$ W/(m K) at RT or bulk copper which is $K = 400$ W/(m K) at RT.

The latter materials serve as important ingredients for electronic applications. The variations in K of graphene can be caused by different lateral sizes of graphene samples, inhomogeneity of thickness, quality of materials, concentration of lattice and interface defects, contamination of surfaces, orientation and size of grains, and also by distribution of strains. Sometimes the measurements of thermal conductivity are performed at different temperatures T of the sample. In most cases, additional heating of samples is necessary for improving the spectral resolution of the Raman spectrometers involved in the temperature measurements. For these reasons, the values of thermal conductivity cannot be consistently compared because the distribution of additional heating over the sample's area is not homogeneous. The experimental data reported for thermal conductivity of CNTs vary even more than for graphene (Balandin 2011).

9.9 Microscopic Model of the Thermoelectric Effect

9.9.1 *The Electron Green Function of Infinite Space*

The electron field operator of an infinite space is

$$\psi_\sigma(x,\ t) = \sum_k a_{\sigma k} e^{ikx} e^{i\varepsilon_k t}$$

The free-electron retarded Green function is

$$\begin{aligned} G^{\mathrm{R}}(x,\ t;\ x',\ t') &= -i\theta(t-t')\left\langle\left[\psi_{L\sigma}(x,\ t)\,\psi^\dagger_{L\sigma'}(x',\ t')\right]_+\right\rangle \\ &= -i\theta(t-t')\sum_{kk'} e^{ikx-ik'x'} e^{i\varepsilon_k t - i\varepsilon_{k'} t'}\left\langle\left[a_{\sigma k},\ a^\dagger_{\sigma' k'}\right]_+\right\rangle \\ &= -i\theta(t-t')\sum_{kk'} e^{ikx-ik'x'} e^{i\varepsilon_k t - i\varepsilon_{k'} t'}\delta_{\sigma\sigma'}\delta_{kk'} \\ &= -i\theta(t-t')\sum_{kk'} e^{ikx-ik'x'} e^{i\varepsilon_k t - i\varepsilon_{k'} t'}\delta_{\sigma\sigma'}\delta_{kk'} \\ &= -i\theta(t-t')\,\delta_{\sigma\sigma'}\sum_{k} e^{ik(x-x')} e^{i\varepsilon_k (t-t')} \end{aligned} \tag{9.94}$$

The Keldysh–Green function (Keldysh 1965) is

$$\begin{aligned}G^{\mathrm{K}}(x, t; x', t') &= -i\left\langle\left[\psi_{L\sigma}(x, t)\, \psi^{\dagger}_{L\sigma'}(x', t')\right]_{-}\right\rangle \\ &= -i\sum_{kk'} e^{ikx-ik'x'} e^{i\varepsilon_k t - i\varepsilon_{k'} t'}\left\langle\left[a_{\sigma k}, a^{\dagger}_{\sigma' k'}\right]_{-}\right\rangle \\ &= -i\sum_{kk'} e^{ikx-ik'x'} e^{i\varepsilon_k t - i\varepsilon_{k'} t'}\delta_{\sigma\sigma'}\delta_{kk'}(2n_k - 1) \\ &= -i\delta_{\sigma\sigma'}\sum_{k} e^{ik(x-x')} e^{i\varepsilon_k (t-t')}(2n_k - 1) \qquad (9.95)\end{aligned}$$

Other Green functions introduced by Keldysh are

$$\begin{aligned}G^{+} &= G^{\mathrm{c}} - G^{\mathrm{A}} = \left(G^{\mathrm{R}} - G^{\mathrm{A}} + G^{\mathrm{K}}\right)/2 \\ G^{-} &= G^{\mathrm{c}} - G^{\mathrm{R}} = \left(-G^{\mathrm{R}} + G^{\mathrm{A}} + G^{\mathrm{K}}\right)/2 \qquad (9.96)\end{aligned}$$

9.9.2 *The DC Electric Current*

Direct calculation of the electric current gives

$$\begin{aligned}j_{\mathrm{e}} &\propto eN(0)\int d\varepsilon\left(G^{\mathrm{K}}_{\mathrm{R}}(0, D, \varepsilon) - G^{\mathrm{K}}_{\mathrm{L}}(0, D, \varepsilon)\right) \\ &= eN(0)\int d\varepsilon\sum_{k}|t_{kk'}|^2 \\ &\quad\times\left(\left(2n^{\mathrm{R}}_{k} - 1\right)\delta(\varepsilon - \varepsilon_k + eV) - \left(2n^{\mathrm{L}}_{k} - 1\right)\delta(\varepsilon - \varepsilon_k)\right) \\ &= 2eN(0)\int d\varepsilon|t_{\varepsilon}|^2\left(n_{\varepsilon} - n_{\varepsilon - eV}\right) \qquad (9.97)\end{aligned}$$

which is rather a standard expression.

9.9.3 *The DC Heat Current*

A straightforward calculation of the heat current density gives

$$\begin{aligned}j_{\mathrm{Q}} &\propto N(0)\int d\varepsilon(\varepsilon - eV)\left(G^{\mathrm{K}}_{\mathrm{R}}(0, D, \varepsilon) - G^{\mathrm{K}}_{\mathrm{L}}(0, D, \varepsilon)\right) \\ &= N(0)\int d\varepsilon\sum_{k}|t_{kk'}|^2(\varepsilon - eV) \\ &\quad\times\left(\left(2n^{\mathrm{R}}_{k} - 1\right)\delta(\varepsilon - \varepsilon_k + eV) - \left(2n^{\mathrm{L}}_{k} - 1\right)\delta(\varepsilon - \varepsilon_k)\right) \\ &= 2N(0)\int d\varepsilon|t_{\varepsilon}|^2(\varepsilon - eV)\left(n_{\varepsilon} - n_{\varepsilon - eV}\right) \\ &= -jV + 2N(0)\int d\varepsilon|t_{\varepsilon}|^2\varepsilon\left(n_{\varepsilon} - n_{\varepsilon - eV}\right) \qquad (9.98)\end{aligned}$$

which is a conventional expression. One evaluates the zero-temperature steady-state heat current as

$$J_{Q0} = 2N(0)\ W \int \varepsilon |t_\varepsilon|^2 (n_{\varepsilon - eV} - n_\varepsilon) d\varepsilon$$
$$= 2N(0)\ W \int_0^{eV} \varepsilon |t_\varepsilon|^2 d\varepsilon = 2N(0)\ \bar{T}\, W \frac{(eV)^2}{2} \quad (9.99)$$

where $\bar{T} = |t_\varepsilon|^2$, W is the graphene stripe width. So, the conductance G is

$$G_0 = \frac{e^2}{\hbar^2} \bar{T}\, W \sqrt{2meV} = \frac{2e^2}{h} \bar{T} \pi W k_c (V) = \frac{2e^2}{h} \bar{T}\, N_{\mathbf{ch}} (V) \quad (9.100)$$

where $N_{\mathbf{ch}}(V)$ is the voltage-dependent number of conducting channels. The above expression, Eq. 9.100, can be rewritten via the conductivity as

$$\sigma_0 = \frac{G_0}{W} = \frac{2e^2}{h} \bar{T} \frac{N_{\mathbf{ch}} (V)}{W}, \quad (9.101)$$

while the voltage-dependent electron wavevector is

$$\mathbf{k}_c (V) = \sqrt{2meV}, \quad (9.102)$$

which coincides with the classic Landauer formula. For a 2D metal

$$G_0 = \frac{e^2}{\hbar^2} \bar{T}\, W p_F = \frac{2e^2}{h} \bar{T}\, W \frac{\pi p_F}{\hbar}$$
$$= \frac{2e^2}{h} \bar{T}\, W \pi k_F = \frac{2e^2}{h} \bar{T} \frac{W}{\lambda_{\mathbf{DB}}} = \frac{2e^2}{h} \bar{T}\, N_{ch} \quad (9.103)$$

where $\lambda_{\mathbf{DB}}$ is the DeBroile wavelength.

9.9.4 *Fourier's Law*

For semiconducting nanotubes, the local-gate voltage induces a coordinate-dependent shift $eU(x)$ of the electron energy ε, which is determined by Eq. 2.90–2.95, which now is written as

$$\varepsilon - eU(x) = \hbar v \sqrt{\mathbf{q}_\nu (n)^2 + k^2} \quad (9.104)$$

where we use $\gamma = \hbar v$. The above formula is rewritten in terms of the longitudinal electron wavevector

$$\mathbf{k} = \sqrt{\left(\frac{\varepsilon - eU(x)}{\hbar v}\right)^2 - \mathbf{q}_\nu (n)^2} \quad (9.105)$$

The transversal electron wavevector q_υ is quantized, which, for example, for a CNT reads

$$q_\upsilon(n) = \frac{2}{d_T}\left(n - \frac{\nu}{3}\right) \tag{9.106}$$

where n is the integer number which also denotes the index of electron subband. It is frequently instructive to introduce dimensionless units which are convenient for numeric calculations. The 10^{-2} VHS spacing serves as a natural energy scale unit Δ_T for CNTs:

$$\Delta = \hbar v \frac{2}{d_T} = 1\,\text{eV} \tag{9.107}$$

More convenient energy units are

$$\Delta_T = \hbar v \frac{2}{d_T} \times 10^{-2} = \frac{1\,\text{eV}}{100 \times d_{\mathbf{T}}\,(\text{in nm})} \tag{9.108}$$

which, for example, for $d_{\mathbf{T}} = 1$ nm gives

$$\Delta_T = 10\,\text{meV for}\, d_{\mathbf{T}} = 1\text{ nm}. \tag{9.109}$$

Then, a new spatial scale unit is $100 \times d_{\mathbf{T}}$ (in nm). For a nanotube, for example, $L = 0.5$ μm it gives $L = 5$. Assuming a piece-wise form of $eU(x)$, for the longitudinal electron momentum inside the CNT one finds $\tilde{k} = \sqrt{(\varepsilon - eU_0/\hbar v)^2 - q_\upsilon(n)^2}$, where the electron wavevector in the transversal direction is $q_\upsilon(n) = (2\pi/L)(n - \nu/3)$, n is the electron subband index, and the tube is at $\nu = 0$ metallic and semiconducting or otherwise dielectric, when $\nu \neq 0$, and $\tilde{s} = \text{sign}(\varepsilon - eU_0)$.

According to Fourier's law, one evaluates the quantity

$$\begin{aligned}\frac{\Delta Q}{\Delta t} &= -\kappa A \frac{\Delta T}{\Delta x} = -N_T \frac{\pi d^2}{8} \frac{(\alpha_p - \alpha_n)^2}{L} \sigma_{\text{seq}} T^2 \\ &= -N_T \frac{\pi d^2}{8L}\left(\frac{1}{e}\frac{L^{(1)}}{L^{(0)}}\right)^2 L^{(0)}_{p,n} = -2N(0) \cdot N_T \cdot K\end{aligned} \tag{9.110}$$

where

$$K = \frac{\pi d^2}{8L}\left(\frac{M^{(1)}}{M^{(0)}}\right)^2 \cdot M^{(0)} \tag{9.111}$$

is measured in units of $\text{nm}/(\text{eV})^2$ and is computed numerically

$$M^{(\alpha)} = \int (\varepsilon - \zeta)^\alpha |t_\varepsilon|^2 \left(-\frac{\partial n_\varepsilon}{\partial \varepsilon}\right) d\varepsilon. \tag{9.112}$$

In the above formulas N_{T} is the number of nanotubes in the rope, d_{T} and L are the average tube diameter and length, and α_p and σ_{seq} are determined for a single tube. From Eq. 9.110 one can see that the heat-sucking rate depends on the ratio N_{T}/L. For instance, for a metallic nanotube with $\nu = 0$ and $n = 0$ one gets

$$q_1(0) = 0 \tag{9.113}$$

$$k'(\varepsilon = 0) = k_0 = \left|\frac{eU_0}{\hbar v_F}\right|. \tag{9.114}$$

For the sake of convenience, one also can introduce units where $\hbar v_{\mathrm{F}} = 1$ and $eU_0 = 1$. This means that for $eU_0 = 1\,\mathrm{eV}$ one gets the following units

$$k_0 = \frac{eU_0}{\hbar v_{\mathrm{F}}} = 1.8756 \times 10^9 \frac{1}{\mathrm{m}} \tag{9.115}$$

or the unit of length

$$\frac{1}{k_0} \simeq 0.5\,\mathrm{nm} \tag{9.116}$$

The electron density of states, for example, for Nb electrodes is $N(0) = 5.1 \times 10^{10}\,\mathrm{eV}^{-1}\,\mu\mathrm{m}^{-3}$. The rough estimation for one conducting channel, $L = 1\,\mu\mathrm{m}$ and $A = \pi d^2/4 = 3.141 \cdot 4\,\mathrm{nm}^2/4$, $\sigma_{\mathrm{seq}} = 2.5 \times 10^7$ S/m, $\kappa_{n,\,p} = 3 \times 10^{-4}$ W/(m K), $\kappa_{\mathbf{ph}} = n_q \cdot \kappa_0 = (50 - 5000)$ W/(m K) where $n_q = 1 - 100$ phonon modes, $\kappa_0 = 50$ W/(m K), and $\alpha_{p,\,n} = 2.4 \times 10^{-4}$ V/K. Then, one obtains

$$\frac{\Delta Q}{\Delta t} = \left(4\left(T\log(2) - \frac{\zeta}{2}\right)\right)^2 \frac{2}{h}\frac{N_{\mathbf{ch}}}{A} = 7 \times 10^8 \frac{\mathrm{W}}{\mathrm{cm}^2} \tag{9.117}$$

At $T = 0.003$ (15 K), $eU_0 = 0.5$ eV, $L = 50 - 150$ nm, $eV_G = 0.2$ eV, and $d_T = 1.5$ nm for just one single-walled CNT (one conducting channels) one also gets

$$\frac{\Delta Q}{\Delta t} = 2 \cdot N(0) \cdot \frac{\pi\,(1.5)^2}{8}\frac{\left(\alpha_p - \alpha_n\right)^2}{L} \cdot \sigma_{\mathrm{seq}} T^2 \simeq 1.2 \times 10^{-5}\,\mathrm{W} \tag{9.118}$$

which means that a metallic CNT with a single conducting channel gives the net cooling power

$$\frac{\Delta Q}{\Delta t} \simeq 12\,\mu\mathrm{W} \tag{9.119}$$

This provides an upper estimation of the cooling power per unit area as

$$\frac{1}{A_T}\frac{\Delta Q}{\Delta t} = 7 \times 10^8 \cdot \frac{\mathrm{W}}{\mathrm{cm}^2} \tag{9.120}$$

where the single CNT area is $A = \pi \cdot d_T^2/4 = 1.7671 \times 10^{-14}\mathrm{cm}^2$.

9.9.5 *Cooling Efficiency of a Gated Stack of Nanotubes*

A stack of 100 nanotubes has the height $H = 100 \cdot d_T \cdot 5 = 1.0 \times 10^{-6}$ m. If one nanotube can carry the maximum power $W_T = 10\ \mu\mathrm{W}$, then the 100 tube stack with width 1 cm has the area

$$A_{\mathrm{net}} = 100 \cdot d_T \cdot 2 = 2.0 \times 10^{-5}\mathrm{cm}^2 \tag{9.121}$$

and carries the power $W_{\mathrm{net}} = 5$ kW. According to Fourier's law

$$\begin{aligned} W_{\mathrm{net}} &= \frac{\Delta Q}{\Delta t} = -\kappa A_{\mathrm{net}} \frac{\Delta T}{L} = -\frac{A_{\mathrm{net}}}{L}\left(\alpha_p - \alpha_n\right)^2 \sigma_{\mathrm{seq}} \\ T^2 &= -\frac{A_{\mathrm{net}}}{L} ZT \cdot \kappa T. \end{aligned} \tag{9.122}$$

The difference of temperatures $T_{\mathbf{hot}} - T_{\mathbf{cold}}$ between the hot and cold ends of a conventional cooler

$$\frac{(T_{\mathbf{hot}} - T_{\mathbf{cold}})_{\max}}{T} = \frac{1}{2} ZT \tag{9.123}$$

is expressed in terms of the figure of merit ZT. One obtains

$$ZT \cdot \kappa = \left(\alpha_p - \alpha_n\right)^2 \sigma_{\mathrm{seq}} \tag{9.124}$$

where $\alpha_{p,n}$ are Seebeck coefficients of the p- and n-type regions. Taking, for example, $ZT = 50$, $T = 1000$ K, the thermal conductivity $\kappa = 30$–6000 W/(m K), and the length $L = 0.1$ cm one gets

$$W_{\mathrm{net}} = -\frac{A_{\mathrm{net}}}{L} ZT \cdot \kappa T = 0.5\,\mathrm{kW}. \tag{9.125}$$

9.9.6 *Cooling Power*

A metallic CNT with a single conducting channel gives the net cooling power

$$\frac{\Delta Q}{\Delta t} \simeq 12\ \mu\mathrm{W}. \tag{9.126}$$

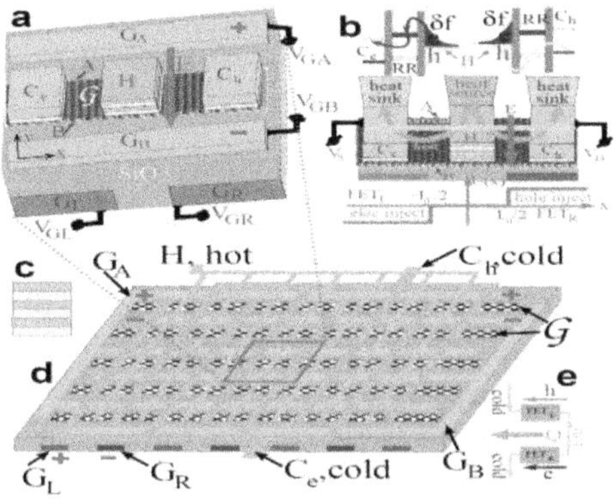

Figure 9.4 (a) The G-TEG cell which consists of a graphene G– stripe with zigzag edges deposited on a dielectric substrate. One edge of G is terminated by atoms of the A sublattice, while the opposite edge is terminated by the B sublattice atoms. The G-stripe is polarized in the y direction by the transverse electric field $|\mathrm{E}| = V_{\mathrm{SG}}/W = (V_{\mathrm{GA}} - V_{\mathrm{GB}})/W$, where W is the spacing between G_{A} and G_{B}. Thus the field is created by the split*gate electrodes $\mathrm{G_A}$ and $\mathrm{G_B}$. Two back-gate electrodes $\mathrm{G_L}$ and $\mathrm{G_R}$ having opposite polarity induce a step-wise change of the electrochemical potential $U(x)$ versus coordinate x. (b) The energy diagram (top panel), the heat and electric current diagram (central panel), and the $U(x)$ profile along the x direction inside G (bottom panel). (c) The multilayered H electrode where the phonon density of states $F_{\mathrm{q}}(\omega)$ in adjacent layers is mismatched. (d) The TEG device consists of the interdigitated split-gate combs $\mathrm{G_A(B)}$ on the one hand and the graphene comb G on the other hand. The H comb serves as the heat source, while the $\mathrm{G_e}$ and $\mathrm{G_h}$ combs represent the heat sinks. The back-gate electrode combs $\mathrm{G_L}$ and $\mathrm{G_R}$ provide the dynamic, coordinate-dependent n- and p-doping of G. (e) Heat (Q) and electric (J) currents in the G-TEG involving the field-effect transistors ($\mathrm{FETs_R}$).

The cooling power density per unit area for a CNT cooler may reach the value at least

$$P_{\mathbf{u}}^{\mathbf{CNT}} = \frac{1}{A_T}\frac{\Delta Q}{\Delta t} = 7 \times 10^8 \frac{\mathrm{W}}{\mathrm{cm}^2} \tag{9.127}$$

which is significantly better (by the factor $\sim 10^3$) as compared, for example, to literature data for SiGe/Si coolers. However, for various reasons, the performance of realistic devices similar to those shown in Fig. 9.3, is lower than the expected values $P_{\mathbf{u}}^{\mathbf{CNT}}$. We expect that optimized Peltier CNT coolers may achieve the cooling power

$$P_{\mathbf{u}}^{\mathbf{CNT}} = \left(10^4 - 10^5\right) \frac{\mathrm{W}}{\mathrm{cm}^2}. \tag{9.128}$$

9.10 Converting Heat into Electricity by a Graphene Stripe with Heavy Chiral Fermions

Thermoelectric generators (TEGs) can convert the heat energy into electricity immediately (Rowe and Min 1995a, 1995b, 1995c, 1995d, Lin et al. 2000, Small et al. 2003, Shafranjuk 2009). However, available solid state TEGs typically have a low figure of merit $Z\delta T = Q^{\mathbf{eh}}/Q^{\mathbf{ph}} << 1$ (here $Q^{\mathbf{eh}}$ is the generated electric power, $Q^{\mathbf{ph}}$ is the heat flow energy carried by the lattice oscillations, phonons; δT is the temperature difference across TEG) (Small et al. 2003, Pop et al. 2007, Shafranjuk 2009). Here we consider an approach which allows a drastic improving of TEG characteristics. The key element of the suggested TEG is the graphene stripe G with atomic zigzag edges (Brey and Fertig 2007) polarized by the transverse electric field **E** (see Fig. 9.1). The graphene TEG exploits the gate voltage-controlled heavy chiral fermion (HCF) states formed inside G and it also benefits from the electric/heat current filtering. The design strategy is illustrated by an expression for the figure of merit $Z\delta T = G_e S^2 \delta T^2/\Lambda$ for a symmetric TEG (Rowe and Min 1995a, 1995b, 1995c, 1995d). Here S is the Seebeck coefficient, G_e is the electric conductance, $\Lambda = \Lambda_\mathbf{e} + \Lambda_\mathbf{ph}$ is the thermal conductance due to the electrons (e) and phonons (ph). In typical TEG devices (Hoffmann et al. 2009, Persson et al. 2009) $\Lambda_\mathbf{ph}/\Lambda_\mathbf{e} \approx 10^3 - 10^4$, providing that $\Lambda \simeq \Lambda_\mathbf{ph} >> \Lambda_\mathbf{e}$. Since normally $\Lambda >> G_\mathrm{e} S^2 \delta T^2$, one gets $Z\delta T << 1$. In the TEG described here, instead of regular electrons and holes, we exploit HCF particles to boost the electric conductance $G_\mathrm{e} \propto \sqrt{m^*/m_\mathrm{e}}$ and the Seebeck coefficient S. In contrast to a pristine graphene, the HCF mass m^* in G-TEG exceeds the free electron mass m_e by a huge factor $m^*/m_\mathrm{e} \sim 10^2 - 10^6 >> 1$. To decimate $\Lambda_\mathbf{ph}$, we also implement special multilayered "hot" metallic electrodes. It allows reducing of the phonon part $Q^{\mathbf{ph}}$ the heat energy flow by a factor of $\sim 10^{-2} - 10^{-3}$. The combined approach yields $Z\delta T \geq 1$, or even $Z\delta T >> 1$, which manifests a significant improvement over other TE devices.

The whole TEG circuit represents a 2D array formed by three interdigitated combs, G, $\mathrm{G_A}$, and $\mathrm{G_B}$, as shown in Fig. 9.3d. The comb G consists of parallel graphene stripes alternating with metallic split-gate stripes $G_{\mathbf{A(B)}}$ which form the other two combs. The $G_{\mathbf{A}}$ and $G_{\mathbf{B}}$ combs serve as split gates to creating of the transverse electric field $\mathbf{E} \neq 0$ across G. Additionally, there are three other metallic combs, C_e, C_h, and H, which are oriented in perpendicular to the former G and $G_{A,B}$ combs. The two combs, C_e and C_h, act as heat sinks, while the H comb represents a heat source. Besides, there are two more back-gate combs, G_L and G_R, which are placed underneath of the dielectric substrate and which are also oriented in-perpendicular to the $G_{A,B}$ and G combs.

Let us consider an elementary cell of the TEG device sketched in Fig. 9.3a (we call it G-TEG) and its diagrams shown in Fig. 9.3b. The G-TEG cell consists of two FETs (Lemme et al. 2007, Chandra et al. 2011, Shafranjuk 2011, Xia et al. 2011), left (FET_L) and right (FET_R), both being fabricated on the same section of the graphene stripe G. The two transistors, FET_L and FET_R, are connected electrically in a sequence, but thermally in parallel (see Fig. 9.3e) (Shafranjuk 2009). The edges of each G-stripe have an atomic zigzag shape (Brey and Fertig 2007). The opposite edges of the same stripe are terminated by carbon atoms belonging to two distinct sublattices A and B (Soriano and Fernandez-Rossier 2010, Soriano et al. 2010, Soriano and Fernandez-Rossier 2012). The split-gate voltage $V_{\mathbf{SG}} = V_{\mathbf{GA}} - V_{\mathbf{GB}}$ polarizes the G-stripe in the transverse y direction. The left (G_L) and right (G_R) back-gate electrodes are controlling the concentration and type of the charge carriers inside the left ($x < -L_0/2$) and right ($x > L_0/2$) G-TEG sections, which are forming FET_L and FET_R, correspondingly. The local-gate voltage $V_{\mathbf{G}}(x)$ induces an x-dependent change $U(x)$ of the electrochemical potential along the G-stripe, as shown at the bottom of Fig. 9.3b.

Sharp distinctive HCF levels at energies $E_p = \pm\Delta$ result from reflections at the atomic zigzag edges (Brey and Fertig 2007) of the graphene stripes G when the G-stripes are polarized in the transverse $\hat{y}$ direction by a finite electric field $\mathbf{E} \neq 0$. The field controls the value of Δ thereby adjusting the quantized level's energy with setting to a required position. We will see that the quantized energy levels originate from splitting of a zero-energy level at $E_\mathrm{p} = 0$ when the finite transverse electric field $\mathbf{E} \neq 0$ is applied. The zero-energy level splitting manifests the quantum-confined Stark effect (see Chapter 3). The mentioned effect becomes

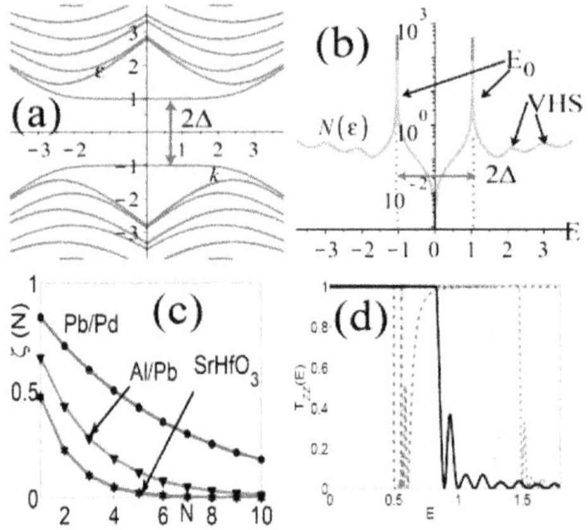

Figure 9.5 (a) The electron energy subbands in the graphene stripe G with zigzag edges. The transverse split-gate voltage V_{SG} induces an energy gap $2\Delta = eV_{SG} = e(V_{GA} - V_{GB})$, which splits the narrow zero-energy level. (b) Density of electron states N'' for three inelastic scattering rates $\gamma_{1;2;3} = 0.01$, 0.001, and 0.0001 in units of Δ. The two sharp peaks constitute the heavy chiral fermion (HCF) excitations arising at energies $\pm\Delta$. The corresponding peak height $N(\varepsilon = \Delta)$ depends on γ. It exceeds the Van Hove singularity (VHS) peaks N_{VHS} by the big factors $N(\varepsilon = \Delta)/N_{VHS} = m^*/m_e = 10, 100$, and 1000. (c) The phonon transmission coefficient ζ for multilayered metallic electrode H (shown in Fig. 9.5c) versus the number of layers N. (d) The transmission coefficient $T_{RR} = G_{zz}(\varepsilon)$ between the RR and uncovered G sections for different electron subbands.

possible due to preserving of the parity-time (PT) symmetry: Despite the fact that the time-reversal T symmetry is broken, a more general combined PT inversion symmetry is observed. According to Sections 3.3–3.5, the PT symmetry is responsible for the origin of so-called HCF resonances. Mathematical model of HCF is formulated in terms of a mere mean field approach (Ando 2005, Shafranjuk 2009). The method (Ando 2005, Shafranjuk 2009) describes the electronic excitation spectrum E_p of a graphene stripe in terms the Dirac equation for bipartite sublattices. One adds a symmetry breaking term $\propto \Delta$ into the Hamiltonian

$$\mathcal{H} = -i\hbar v\left(\left(\hat{\sigma}_x \otimes \hat{1}\right)\partial_x + \left(\hat{\sigma}_y \otimes \hat{\tau}_z\right)\partial_y\right) + U(x)\left(\hat{1} \otimes \hat{1}\right) + \Delta\left(\hat{\sigma}_z \otimes \hat{\tau}_z\right) \tag{9.129}$$

where $v = 8.1 \times 10^5$ m/s$\simeq c/300$ is the massless fermion speed, $\hat{\sigma}_i$ and $\hat{\tau}_k$ are the Pauli matrices, $\otimes$ is the Kronecker product, $\{i, k\} = 1 \ldots 3$, and $\Delta = e|\mathbf{E}|W/2 = eV_{\mathbf{SG}}/2 = e(V_{\mathbf{GA}} - V_{\mathbf{GB}})/2$, $V_{\mathbf{SG}}$ is the split-gate voltage, and W is the stripe width. The pseudospin polarization means that the electric charge is depleted in one zigzag edge while is accumulated in the other. Then an electric dipole is formed as soon as $V_{\mathbf{SG}} = 2\Delta/e \neq 0$. At $V_{\mathbf{SG}} = 0$ there are two flat bands giving rise to a large density of states at the Fermi energy and being associated to zigzag edge states (Brey and Fertig 2007). When $V_{\mathbf{SG}} \neq 0$, the resulting pseudospin polarization driven by the Δ term in Eq. 9.129 yields an excitation spectrum $\varepsilon_{\pm} = \pm v\sqrt{k^2 + q^2 + (\Delta/v)^2} \pm U$ characterized by the energy gap 2Δ. The electron wavefunction Ψ satisfies the boundary conditions $\Psi_{\mathbf{A}}(0) = \Psi_{\mathbf{A'}}(0) = 0$ at $y = 0$ and $\Psi_{\mathbf{B}}(W) = \Psi_{\mathbf{B'}}(0) = 0$ at $y = W$ for zigzag atomic edges. It gives (Brey and Fertig 2007) the transversal quantization condition (QC) in the form $k = q_{\mathrm{p}}/\tan(q_{\mathrm{p}}W)$ where $q_{\mathrm{p}} = \pm\sqrt{(E_{\mathrm{p}} \mp U)^2/v^2 - k^2 - (\Delta/v)^2}$. The QC actually determines the localized energy levels when is being solved in respect to the excitation energy E_{p}. Thus, for a finite $\Delta \neq 0$, the graphene zigzag stripes G are band insulators with the pseudospin polarization. The steady-state energy spectrum of the G-stripe features two flat-bottom bands $E_{\mathrm{p}} = \pm\Delta$, which represent the highest occupied and lowest unoccupied bands. The flat bands correspond to a very high effective electron mass $m^* = (10^2 - 10^6)m_{\mathrm{e}} >> m_{\mathrm{e}}$.

The sign of $U(x)$ determines whether the HCF charge carriers inside of each $\mathrm{FET}_{\mathbf{L,R}}$ are electrons or holes. In Fig. 9.3b, the $V_{\mathbf{G}}(x)$ profile is step-wise, so $V_{\mathbf{G}}(x) = -U_0/\alpha_{\mathbf{G}}$ for $x < -L_0/2$ (inside $\mathrm{FET}_{\mathbf{L}}$) and $V_{\mathbf{G}}(x) = U_0/\alpha_{\mathbf{G}}$ for $x > L_0/2$ (inside $\mathrm{FET}_{\mathbf{R}}$). Here U_0 is the shift of the electron electrochemical potential and $\alpha_{\mathbf{G}}$ is the back-gate efficiency. The G-stripe RR section located immediately under the central hot H electrode $(-L_0/2 < x < L_0/2)$ remains neutral, since $V_{\mathbf{G}}(x) = 0$. This part of the G-stripe serves as an electron–hole recombination region (RR).

In the TEG shown in Fig. 9.3d, the heat is supplied via the hot H comb, while it is drained by the "cold" $\mathrm{C}_{\mathbf{e}}$ and $\mathrm{C}_{\mathbf{h}}$ sink combs. The phonon fraction of the heat flow is filtered out by the multilayered H electrode (see Sections 9.11–9.12). Since the electron temperature

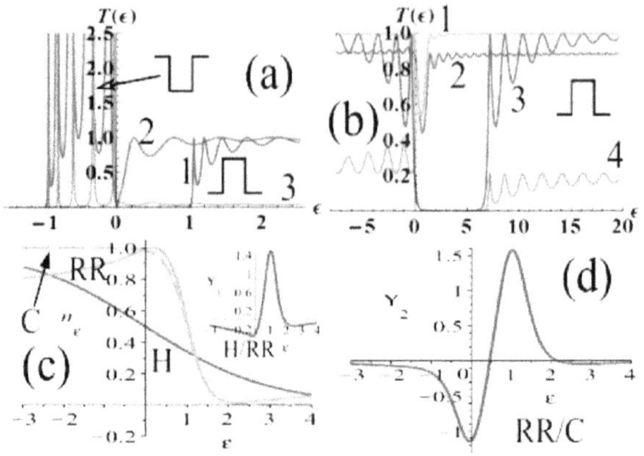

Figure 9.6 (a) The transmission coefficient $T(\varepsilon)$ for conventional electrons and conventional heavy fermions through the well and potential barrier. (b) $T(\varepsilon)$ for chiral fermions: curves 1 and 2 correspond to regular chiral fermions, while curves 3 and 4 correspond to the HCF excitations. Curves 1 and 3 are for the incident angle $\varphi = \pi/8$, while curves 2 and 4 are for $\varphi = 3\pi/8$. (c) The electron distribution functions in the cold electrode (C), recombination region (RR) and in the hot electrode (H). Inset shows the driving factor Υ_H =RR(ε). (d) The driving factor Υ_{RR} =C(ε).

$T_{\mathbf{H}}$ in H is much higher than the electron temperature in the G-stripe, T_G, one achieves the nonequilibrium thermal injection Q^{eh} of the hot electrons and holes from H into RR (see Section 9.13 for details). Although the contact the Seebeck coefficient $S_{\mathrm{H/RR}}$ and contact electric conductance $G_{\mathrm{H/RR}}$ through the H/RR interface vanish (i.e., $S_{\mathrm{H/RR}} = 0$ and $G_{\mathrm{H/RR}} = 0$, as shown in Section 9.13), the contact thermal conductance $\Lambda_{\mathbf{H/RR}}$ remains essentially finite (Hoffmann et al. 2009, Persson et al. 2009). After being thermally injected from H into RR, the electrons and holes are quickly (during time $\sim 10^{-13}$ s) converted into the HCF excitations. In the RR region, the nonequilibrium HCF electrons and holes populate the levels $\pm\Delta$ inversely: the upper $+\Delta$ level is populated by excessive HCF electrons while the lower $-\Delta$ level by the excessive HCF holes. The quasiparticle distribution function $n_{\varepsilon}^{\mathbf{RR}}$ is obtained as a solution of the quantum kinetic equation (Keldysh 1965) (see Section 9.13).

In the configuration shown in Fig. 9.3, a ballistic propagation of holes is from RR toward C_h while the electrons in the left G-section proceed from RR toward C_e. Thus, the latter stage implies a chiral transmission of the excessive nonequilibrium HCF electrons from the upper $+\Delta$ level localized in RR into the upper $\Delta - U_0$ level located in the uncovered G-section adjacent to C_e (see the diagram at the top of Fig. 9.3b) where $U_0 \leq \gamma$, γ is the localized level width. Simultaneously, the excessive holes are transmitted from the lower $-\Delta$ level in RR into the lower $-\Delta + U_0$ level localized near C_h. Thus, the full thermal flow $Q^{\mathbf{eh}}$ from H to RR is eventually split between the C_h and C_e sections of the graphene stripe G. In the uncovered G-section, broadening γ of the HCF level E_0 originates from coupling of the HCF states to the phonons. It yields $\gamma = \gamma_{\mathbf{ph}} = \hbar/\tau_{\mathbf{ph}}$ (typically $\tau_{\mathbf{ph}} \simeq 10^{-12}$ s at $T = 10$ K). In the RR section, there is an additional coupling (Xia et al. 2011) to the electron states in H which gives $\gamma = \gamma_{\mathbf{ph}} + \gamma_{\mathbf{H}}$. Both the electrons and holes are supplied into the electrically neutral RR by the thermal injection from H. Then the electric current J in the $FET_{\mathbf{R}}$ actually emerges due to the ballistic propagating of HCF holes from the RR region toward the C_h electrode. Simultaneously, the electric current J in the $FET_{\mathbf{L}}$ also consists of ballistic propagating the HCF electrons in the opposite direction from RR toward C_e. Thus, both the HCF electrons and holes transfer the heat from the hot (H) to the cold ($C_{\mathbf{e,\,h}}$) electrodes. Then the combined thermal flow $Q^{\mathbf{eh}}$ of electrons and holes generates an electric current accompanied by a finite voltage drop $V_{\mathbf{SD}} = \sum_n$. All the G-stripes are connected in parallel; thus the net electric current through the whole TEG is $I = \sum_k$, where I_k is the current through the kth G-stripe. The voltage drop on the two $FETs_{\mathbf{L,\,R}}$ is $V^n_{\mathbf{SD}} = 2S_n\delta T_n$. Because the heat flow is transferred by heavy-charged HCF particles ($m^*/m_e >> 1$) it ensures a considerable value of the Seebeck coefficient $\mathbf{S}_n$. It in turn yields a high TE voltage $V^n_{\mathbf{SD}} = 2\mathbf{S}_n\delta T_n$ generated by the heat flow when a finite temperature difference δT_n is maintained between the H and $C_{\mathbf{e(h)}}$ electrodes of the nth G-TEG. Then the G-TEG generates the electric power $Q^{\mathbf{el}}_{\mathbf{G}-TEG} = \sum_{k,\,n} V^n_{\mathbf{SD}}$, which is considerably higher than in TEG devices where conventional electrons and holes are engaged.

Another increase of $Z\delta T$ is accomplished when the phonon part of the heat conductance $\Lambda_{\mathbf{ph}}$ is strongly reduced. A visible decimation of $\Lambda_{\mathbf{ph}}$ by a few orders of magnitude is achieved, for example, if the hot H electrodes are metallic multilayers fabricated as sketched in Fig. 9.3c. The multilayered structure shown in Fig. 9.3c serves for decimating of the phonon transport through the TEG by the factor ζ which we plot in Fig. 9.4c versus the number of layers N (see Sections 9.11–9.12). Because the multilayered metal contacts H and $C_{e,h}$ are reducing the phonon component of the heat flow, the figure of merit $Z\delta T$ can additionally be increased by orders of magnitude.

One computes the net electric power generated by the whole TEG with equivalent G-stripes and equivalent G-TEGs as $Q_{\text{electr}} = N_G \times IV_{SD} = N_G \times G_e(2S\delta T)^2$ where N_G is the total number of **G**-TEG elements in the TEG array and we have omitted index n. The corresponding thermal power is $Q_{\text{heat}} = N_G \times \Lambda\delta T$. Both the quantities, Q_{electr} and Q_{heat}, are expressed in terms of the Seebeck coefficient S, temperature differences δT, electric G_e, and thermal Λ conductivities of the individual **G**-TEG sections. The TEG conversion efficiency is then estimated as $\{Z\delta T\}_{\text{TEG}} = Q_{\text{electr}}/Q_{\text{heat}} = 4N_G S^2 G_e \delta T/\Lambda$. The electron part, Λ_e, of the thermal conductance Λ is obtained as $1/\Lambda_e = 2/\Lambda_c^e + 1/\Lambda_G^{e,\text{cont}}$ and is typically $10^3 - 10^4$ times smaller than the phonon part, which is $1/\Lambda_{ph} = 2/\Lambda_c^{ph} + 1/\Lambda_G^{ph} + 1/\Lambda_{SiO_2}$ (Hoffmann et al. 2009, Persson et al. 2009). Here, Λ's indices e, ph, c, G, and SiO_2 relate it to the electrons, phonons, contacts, G-stripe, and dielectric SiO_2 substrate, respectively. There are only two conducting HCF channels per each G-stripe. For one conducting channel in a regular graphene stripe by width $W = 10$ nm, one gets the Seebeck coefficient at most $S \approx 10^{-4}$ V/K. For a graphene stripe with zigzag edges by the same width $W \simeq 10$ nm and the split-gate voltage $V_{SG} = 0.1$ V, where the HCF resonances are invented, one improves it to $S \approx 10^{-3} - 10^{-1}$ V/K (see Section 9.13). In combination with the multilayered hot electrode it yields as much as $Z\delta T \approx 10^2$.

The electric power generated by a single **G**-TEG is evaluated as $Q^{eh} \simeq \kappa_{HCF} G_q V_{SG}^2 \simeq 1$ mW, where $\kappa_{HCF} = \sqrt{m^*/m_e}(\delta T_{H/RR}/\delta T_{RR/C})$, $G_q = 2e^2/h = 7.75 \times 10^{-5}\Omega^{-1}$, and we have used $m^*/m_e = 10^2$ (which corresponds to $\gamma = 1$ meV), $\delta T_{H/RR} = 300$ K, $\delta T_{RR/C} = 30$ K, and $V_{SG} = 0.1$ V. Let us assume that the length of a single **G**-TEG element is $L_G \simeq 2.5$ μm, the G-stripe width is $W_G = 10$ nm, and the period in the y direction is $W_p = 40$ nm. Then the TEG device by size 1 cm×1 cm contains $N_G = 10^9$**G**-TEG elements which might generate the electric power about 1 MW. One can see that the **G**-TEG efficiency largely depends on the maximum value of V_{SG} and on reducing of $\Lambda_{\mathbf{ph}}$. For practical realizations, instead of DC gate voltages, one can operate the TEG at industrial AC frequency $f = 60$ or 50 Hz. Then required V_G and V_{SG} might be generated immediately during the heat to electricity conversion.

The obtained result suggests that the **G**-TEG devices which exploit filtering of the electric/heat currents and of using the voltage-controlled spectral singularities are able to improving of the TE conversion efficiency and the generated output electric power by 2–3 orders of magnitude as compared to presently known devices.

9.11 Blocking Phonon Flow with Multilayered Electrodes

The goal here is to reducing of the phonon transport considerably while stimulating the electron transport at the same time. We design the filter pads to separate the G-section both from the external electrodes and from the substrate thermally, but not electrically. We consider two different types of the heat/electric current valve pads. One design involves metallic multilayers Pb/Al with the layer thickness $\sim$ 10nm. Another method is to depositing of pads made of $SrHf0_3$ and/or $SrRu0_3$. The layered materials have an appreciable electric conductance while their thermal conductance along the c axis is remarkably low (Maekawa et al. 2005a, 2005b, Keawprak et al. 2009). Planting of the H/RR pad between the metallic electrodes and graphene stripe would reduce the effective $\Lambda_{\mathbf{ph}}$ significantly, because the phonons which provide large heat conductance $\Lambda_c^{\mathbf{ph}}$ between the hot and cold electrodes are eliminated from the TE path. Then the net heat conductance which involves the path HOT $\Rightarrow$ RR $\Rightarrow$ G $\Rightarrow$C $\Rightarrow$ COLD will be considerably diminished. Thus, placing of the H/RR pad with a sufficient number of nanolayers allows to decimate the phonon part $\Lambda_{\mathbf{ph}}$ of the whole thermal conductance Λ. Optimal **G**-TEG geometry is also determined by the electric and thermal transfer lengths which are estimated (Hoffmann et al. 2009, Persson et al. 2009) correspondingly as $L_{\mathbf{el}} \simeq$ 10–100 nm and $L_{\mathbf{th}} \simeq$ 100–250 nm.

9.12 Molecular Dynamics Simulations

The phonon part of the thermal transport through the TEG had been examined as follows. We describe the nonequilibrium thermal transport through the G-stripe in the presence of multiple scattering on lattice defects, boundaries, and electrons. A finite temperature difference δT between the opposite ends of each G-stripe induces the thermal flow given as a sum of contributions of the individual phononic subbands. The phonon density of states $F_{\beta}(\omega)$ related to the phonon subband β is mismatched in adjacent layers of the H electrode. Inside the G-stripe, the phonon distribution function $N(\omega)$ is nonequilibrium, which means that $N(\omega)$ deviates from the Bose–Einstein distribution in the hot (H) and cold (C) ends. For a "clean" graphene stripe, the phonon mean free path exceeds the G-stripe length L. Therefore, the nonequilibrium effect does not influence the final results. The equilibrium phonon distribution at the G-stripe ends is established due to a free phonon diffusion into the bulk of attached metallic contacts and dielectric substrate. The thermal conductance $\Lambda_{\mathbf{G}}^{\mathbf{ph}}$ of the G-stripe had been computed by using the phonon density of states $F_{\beta}(\omega)$ preliminary obtained for each phonon subband β.

Summarizing, the TE characteristics are found with solving the Dirac equation for chiral fermions in graphene (see above). The analytical model is verified by numeric calculations based on the density functional theory (Brey and Fertig 2007). The electron and phonon excitation spectra are obtained considering influence of the inelastic electron–phonon and elastic electron–impurity scatterings. They are taken into account along with processes of the electron tunneling through the interface barriers. The electron–impurity and electron–phonon scatterings are included within the Keldysh–Green function technique (Keldysh 1965) which allows deriving of the quantum kinetic equations (see below for details).

9.13 Nonequilibrium Thermal Injection

The heat energy flow $Q^{\mathbf{eh}}$ from the hot electrode H with temperature $T_{\mathbf{H}}$ into a much colder heat sinks $C_{\mathbf{e(h)}}$ proceeds along the G-stripe

with temperature T_G ($T_H >> T_G$). The process happens in two stages (*i*) and (*ii*) as illustrated by the energy diagrams at the top of Fig. 9.3b. In the diagram we show the distribution function of hot excessive quasiparticles $\delta n_{\mathbf{e,h}}$ in H where $\delta n_{\mathbf{e}} = [n(\varepsilon/T_{\mathbf{H}}) - n(\varepsilon/T_{\mathbf{G}})]\theta(\varepsilon - \varepsilon_{\mathbf{F}})$ corresponds to the thermally excited excessive electrons while $\delta n_{\mathbf{h}} = -[n(\varepsilon/T_{\mathbf{H}}) - n(\varepsilon/T_{\mathbf{G}})]\theta(\varepsilon_{\mathbf{F}} - \varepsilon)$ describes the excessive hole thermal excitations—here $n(x) = 1/(e^x + 1)$:
(*i*) The thermal injection H $\rightarrow$ RR: The temperature difference $\delta T = T_{\mathbf{H}} - T_{\mathbf{G}}$ initializes the nonequilibrium "thermal injection" of hot electrons and holes from H into the two levels $\pm\Delta$ localized inside the recombination region RR. As a result, the upper $+\Delta$ level becomes inversely populated by the nonequilibrium electrons, while the lower level $-\Delta$ is populated with nonequilibrium holes. The process is described by the quantum kinetic equation (Keldysh 1965)

$$\frac{\partial n_\varepsilon}{\partial t} = \mathcal{L}_{\mathbf{H/RR}}\{\varepsilon,\ n_\varepsilon\} + \mathcal{L}_{\mathbf{ep}}\{\varepsilon,\ n_\varepsilon\} + \mathcal{L}_{\mathbf{eh}}\{\varepsilon,\ n_\varepsilon\} + \mathcal{L}_{\mathbf{RR/C_{e,h}}}\{\varepsilon,\ n_\varepsilon\}. \tag{9.130}$$

In the quasistationary case, one sets $\partial n_\varepsilon/\partial t \equiv 0$, which gives

$$\Gamma_{\mathbf{H/RR}}\left(n_\varepsilon^{\mathbf{H}} - n_\varepsilon^{\mathbf{RR}}\right) + \Gamma_{\mathbf{RR/C_{e,h}}}\left(n_\varepsilon^{\mathbf{RR}} - n_\varepsilon^{\mathbf{e,h}}\right) - \frac{n_\varepsilon^{\mathbf{RR}} - n_\varepsilon^{\mathbf{F}}}{\tau_{\mathbf{eh}}} - \frac{n_\varepsilon^{\mathbf{RR}} - n_\varepsilon^{\mathbf{F}}}{\tau_{\mathbf{ep}}} = 0, \tag{9.131}$$

where we have used $\mathcal{L}_{\mathbf{H/RR}} = \Gamma_{\mathbf{H/RR}}\left(n_\varepsilon^{\mathbf{H}} - n_\varepsilon^{\mathbf{RR}}\right)$ for the nonequilibrium thermal injection, $\mathcal{L}_{\mathbf{ep}} = \left(n_\varepsilon^{\mathbf{RR}} - n_\varepsilon^{\mathbf{F}}\right)/\tau_{\mathbf{ep}}$ is the electron–phonon collision term, $\mathcal{L}_{\mathbf{eh}} = \left(n_\varepsilon^{\mathbf{RR}} - n_\varepsilon^{\mathbf{F}}\right)/\tau_{\mathbf{eh}}$ describes the electron–hole recombination, and $\mathcal{L}_{\mathbf{RR/C_{e,h}}} = \Gamma_{\mathbf{RR/C_{e,h}}}\left(n_\varepsilon^{\mathbf{RR}} - n_\varepsilon^{\mathbf{e,h}}\right)$ accounts for the HCF electron/hole escapes from RR into directions of $\mathrm{C_{e,h}}$. The corresponding electron distribution functions are approximated as $n_\varepsilon^{\mathbf{H}} = 1/\left(e^{\varepsilon/T_{\mathrm{H}}} + 1\right)$ in the H electrode and $n_\varepsilon^{\mathbf{e,h}} = 1/\left(e^{(\varepsilon \pm U_0)/T_{\mathrm{c}}} + 1\right)$ in the $\mathrm{C_{e,h}}$ shoulders where $U_0 = V_{\mathrm{G}}/\alpha$ is the back-gate-induced shift of the electron electrochemical potential, α is the back-gate efficiency, $n_\varepsilon^{\mathbf{F}}$ is the Fermi function. Besides, $T_{\mathbf{H(c)}}$ are the effective electron temperatures in the H and $\mathrm{C_{e,h}}$-adjacent regions. The nonequilibrium distribution function $n_\varepsilon^{\mathbf{RR}}$ in the RR region then is

$$n_\varepsilon^{\mathbf{RR}} = \frac{\Gamma_{\mathbf{RR/C_{e,h}}} n_\varepsilon^{\mathbf{e,h}} - \Gamma_{\mathbf{H/RR}} n_\varepsilon^{\mathbf{H}} - \Gamma_\varepsilon n_\varepsilon^{\mathbf{F}}}{\Gamma_{\mathbf{RR/C_{e,h}}} - \Gamma_{\mathbf{H/RR}} - \Gamma_\varepsilon} \tag{9.132}$$

where the electron energy broadening is $\Gamma_\varepsilon = \hbar/\tau_{\mathbf{eh}} + \hbar/\tau_{\mathbf{ep}}$. Typically one gets (Xia et al. 2011) $\Gamma_{\mathbf{H/RR}} = 5$ meV for a rough G/Pd interface, while $\Gamma_{\mathbf{H/RR}} = 60$ meV for a smooth G/Pd interface, $\Gamma_{\mathbf{RR/C_{e,h}}} = 100$ meV, and $\Gamma_\varepsilon = 1.5$ meV. The temperatures are taken as $T_{\mathrm{H}} = 630$ K, $T_{\mathrm{RR}} = 330$ K, and $T_{\mathbf{G}} = 300$ K, and the level position $\Delta = 50$ meV, which corresponds to the split-gate voltage $V_{\mathbf{SG}} = 0.1$ V. The above formulas allow computing of the electric and thermal currents. Both types of the currents are inhomogeneous in the vicinity of the H and $\mathrm{C_{e,h}}$ electrodes on the corresponding spatial lengths $L_{\mathbf{el}} \approx 10\text{–}100$ nm and $L_{\mathbf{th}} \approx 100\text{–}250$ nm (Hoffmann et al. 2009, Persson et al. 2009). The hot conventional electrons coming from H in RR are converted inside RR into the heavy HCF excitations during the short time (Xia et al. 2011) $\tau_{\mathrm{c}} \simeq \hbar/\Gamma$ (in Xia et al., 2011) $\tau_{\mathrm{c}} \simeq 10^{-13}$ s, due to the energy level broadening because the tunneling coupling between H and **G**. One finds $\Gamma = h v_{\mathbf{H}} \mathcal{T}_{\mathbf{H/RR}}/(4 d_{\mathbf{H}}) \simeq 5$ meV where $\mathcal{T}_{\mathbf{H/RR}}$ is the H/RR interface transparency, $v_{\mathbf{H}}$ is the Fermi velocity in H). (*ii*) During the next *chiral* transmission process, RR$\rightarrow$ **G**, which occurs on the longer timescale $\tau_{\mathbf{esc}} \leq L/v \simeq 10^{-12}$ s, most of the HCF electrons and holes inside RR are captured by the adjacent $\mathrm{FET_{L,R}}$. Simultaneously, minor fractions of HCF electrons and holes annihilate with each other (Rana 2007) during the time $\tau_{\mathbf{eh}} \geq 10^{-12}$ s (typically $\tau_{\mathbf{esc}} \approx (3-7) \cdot \tau_{\mathrm{eh}}$). It sets a requirement to the spatial dimension of RR as $L_0 = v\tau_{\mathbf{eh}} \leq 1$ μm. Since the thermal injection is essentially a nonequilibrium process, the transforming of electron states between the stages (*i*) and (*ii*) is incoherent.

Since the HCF distribution function $n_\varepsilon^{\mathrm{RR}}$ in the RR region is known, one can compute the electric conductance, G_{e}, the Seebeck coefficient, S, and the electronic part of the heat conductance, Λ_{e}. The electric *contact* conductance $G_{\mathbf{H/RR}}$ across the H/RR interface vanishes ($G_{\mathbf{H/RR}} \equiv 0$) because the electron part of the electric current is compensated by the hole part. The H/RR *contact* the Seebeck coefficient also vanishes since it is $S_{\mathbf{H/RR}} = -(1/eT)\left(L^{(1)}/L^{(0)}\right) = V_{\mathbf{H/RR}}/\delta T_{\mathbf{H/RR}} \equiv 0$, where $V_{\mathbf{H/RR}}$ is the bias voltage and $\delta T_{\mathbf{H/RR}}$ is the difference of temperatures across the H/RR interface. Here we have introduced auxiliary functions

$$L_{\mathrm{e}}^{(\alpha)} = \frac{2e^2}{h} \int d\varepsilon \cdot (\varepsilon - \mu)^\alpha M_{\mathrm{e}}(\varepsilon)\, \mathcal{T}_{\mathrm{e}}(\varepsilon)\, \Upsilon_{\mathbf{H/RR}}(\varepsilon) \qquad (9.133)$$

where

$$\Upsilon_{\mathbf{H/RR}}(\varepsilon) = -\frac{\partial\left[n_\varepsilon^{\mathrm{H}} - n_\varepsilon^{\mathrm{RR}}\right]}{\partial\varepsilon} \tag{9.134}$$

is the driving factor, $\alpha = 1\ldots 3$, $M_{\mathrm{e}}(E) = N_{\mathrm{G}}(E)(\hbar v/L)$ is the number of modes, $N_{\mathrm{G}}(E)$ is the electron density of states shown in Fig. 9.4b, L is the G-stripe length, and $\mathcal{T}_{\mathrm{e}}(\varepsilon)$ is the contact transparency. Then $L^{(1)}_{\mathrm{e},\,\mathbf{H/RR}} \equiv 0$, because the driving factor $\Upsilon_{\mathbf{H/RR}}(\varepsilon)$ vanishes because it is an even function of ε. In contrast to $G_{\mathbf{H/RR}}$ and $S_{\mathbf{H/RR}}$, both of which vanish at the H/RR contact, the thermal contact conductance

$$\Lambda_{\mathbf{H/RR}} = \frac{1}{e^2\delta T}\left(L^{(2)}_{\mathbf{H/RR}} - \frac{[L^{(1)}_{\mathbf{H/RR}}]^2}{L^{(0)}_{\mathbf{H/RR}}}\right) = \frac{L^{(2)}_{\mathbf{H/RR}}}{e^2\delta T} \tag{9.135}$$

remains essentially finite. For the sake of simplicity, we are using here the model form

$$M_{\mathrm{e}}(\varepsilon) = \sqrt{\frac{m^*}{m_{\mathrm{e}}}}\cdot\delta(\varepsilon-\Delta) \tag{9.136}$$

which allows to evaluate the thermal contact conductance as

$$\Lambda_{\mathbf{H/RR}} \simeq \mathcal{F}_{\mathrm{H/RR}}\sqrt{\frac{m^*}{m_{\mathrm{e}}}}\frac{2\Delta^2}{h}\frac{1}{\delta T_{\mathrm{H/RR}}} \tag{9.137}$$

where $\delta T_{\mathrm{H/RR}}$ is an effective electron temperature difference across the H/RR interface, and

$$\mathcal{F}_{\mathrm{H/RR}} = -\left.\frac{\partial[n_\varepsilon^{\mathrm{H}} - n_\varepsilon^{\mathrm{RR}}]}{\partial\varepsilon}\right|_{\varepsilon=\Delta} \approx 0.1. \tag{9.138}$$

Along the G-stripe, the TEG parameters S, G_{e}, and Λ_{e} are determined purely by the electron and hole transport. The underlying physical mechanism is the chiral transmission of the HCF electrons and holes from the neutral RR section to the voltage p- and n- doped **G**-sections. In the same approximation, one evaluates the electric conductance of the $\mathrm{FET}_{\mathbf{L,R}}$ along the G-stripe between the RR and $\mathrm{C}_{\mathbf{e,h}}$ as

$$\begin{aligned} G_{\mathbf{RR/C}} &= L^{(0)}_{\mathbf{RR/C}} = \frac{2e^2}{h}\int d\varepsilon\, M(\varepsilon)\; T_{\mathbf{RR/C_{e(h)}}}(\varepsilon)\;\Upsilon_{\mathbf{RR/C}}(\varepsilon) \\ &= \frac{2e^2}{h}\sqrt{\frac{m^*}{m_{\mathrm{e}}}} T_{\mathbf{RR/C_{e(h)}}}(\varepsilon)\;\Upsilon_{\mathbf{RR/C}}(\varepsilon)|_{\varepsilon=\Delta} \end{aligned} \tag{9.139}$$

where $\Upsilon_{\mathbf{RR/C}}(\varepsilon)$ is shown in Fig. 9.5d. Analogously, one finds the Seebeck coefficient along the *G*-stripe

$$S_{\mathbf{RR/C}} \simeq \frac{\Delta}{e}\frac{1}{\delta T_{\mathbf{RR/C}}} = \frac{V_{\mathbf{SG}}}{\delta T_{\mathbf{RR/C}}} \tag{9.140}$$

and

$$\Lambda^{\mathbf{RR/C}} = \frac{L^{(2)}_{\mathbf{RR/C}}}{e^2 T} - T S_{\mathrm{e}}^2 G_{\mathrm{e}} \simeq 2\sqrt{\frac{m^*}{m_{\mathrm{e}}}}. \tag{9.141}$$

The last result indicates that the Seebeck coefficient $S_{\mathbf{RR/C}}$ could be huge while the electron/hole part of the thermal conductance $\Lambda_{\mathbf{RR/C}}$ along the strip is typically low. The phonon part of the heat energy flow is

$$Q^{\mathbf{ph}}_{\mathbf{RR/C}} = \Lambda_{\mathbf{ph}}\delta T_{\mathbf{RR/C}} = N_{\mathbf{ph}}\kappa_0\delta T_{\mathbf{RR/C}} \tag{9.142}$$

where $\kappa_0 = 5 \times 10^{-12}$ W/K at $T = 10$ K, while the number of phonon modes $N_{\mathbf{ph}}$ also depends on the temperature and the stripe geometry. The electron/hole heat energy flow is

$$\begin{aligned} Q^{\mathrm{eh}} &= \delta T^2 \cdot S^2_{\mathrm{e(h)}} G_{\mathrm{e(h)}} = \delta T^2 \cdot S^2_{\mathrm{RR/C}} G_{\mathrm{RR/C}} \\ &\simeq \frac{2e^2}{h} V^2_{\mathrm{SG}} \cdot \left(\frac{\delta T_{\mathrm{H/RR}}}{\delta T_{\mathrm{RR/C}}}\right)^2 \sqrt{\frac{m^*}{m_{\mathrm{e}}}} = \frac{2e^2}{h} V^2_{\mathrm{SG}} \cdot \kappa_{\mathrm{HCF}} \end{aligned} \tag{9.143}$$

where we have used that $V_{\mathbf{SG}} = 2\Delta/e$ and $S_{\mathbf{RR/C}} = V_{\mathbf{SG}}/\delta T_{\mathbf{RR/C}}$, and also

$$G_{\mathbf{RR/C}} = \frac{2e^2}{h}\sqrt{\frac{m^*}{m_{\mathrm{e}}}} \tag{9.144}$$

Besides above we have defined the factor $\kappa_{\mathbf{HCF}}$. Because $\kappa_{\mathbf{HCF}}$ can be big, one might achieve huge values of $Q^{\mathbf{eh}}$. Typically $\Lambda^{\mathrm{H/RR}}_{\mathrm{ph}} << \Lambda^{\mathrm{RR/C}}_{\mathrm{ph}}$ (Hoffmann et al. 2009, Persson et al. 2009); therefore the **G**-TEG net phonon heat conductance is $\Lambda_{\mathrm{ph}} \simeq \Lambda^{\mathrm{H/RR}}_{\mathrm{ph}}$, which can be comparable to the contact electron heat conductance

$$\Lambda_{\mathbf{H/RR}} = \frac{2\Delta^2}{h}\mathcal{F}_{\mathrm{H/RR}}\sqrt{\frac{m^*}{m_{\mathrm{e}}}}\frac{1}{\delta T_{\mathbf{H/RR}}}. \tag{9.145}$$

It means that only the contact electron/hole and phonon heat conductances actually contribute into Λ. Summarizing the above estimates one arrives at $Z\delta T >> 1$.

9.13.1 *Transparency of the H/RR Interface*

The H/RR interface transparency is directly related to the thermal injection efficiency. The interface barriers, which contribute into $\Lambda_{\mathbf{H/RR}}$, originate from the difference of the work functions in the metallic H electrode on one hand and the graphene *G*-stripe right beneath of it on the other hand. The heat-conducting $C_{\mathbf{h}}$, H, and $C_{\mathbf{e}}$ electrodes are deposited at the top of the *G*-stripe, as schematically shown in Fig. 9.3. Another important factor is the change in number of the conducting channels when electrons and holes tunnel from the 3D metallic H electrode into the 2D graphene *G*-stripe. Conversion of the regular electrons and holes into the HCF excitations also contributes to $\Lambda_{\mathbf{H/RR}}$. Thus, for the G-TEG, $\Lambda_{\mathbf{H/RR}}$ depends on the 3D/2D conversion efficiency η and on the spatial distribution of charge carriers near the H/RR interface. The contact thermal conductance problem and its solution are illustrated in Fig. 9.3. In Fig. 9.3a we plot the transmission probability $\mathcal{T}(\varepsilon)$ as a function of the electron energy ε for the conventional electrons penetrating a nonchiral potential barrier (curve 1), and the quantum well (curve 2). Curve 3 shows $\mathcal{T}(\varepsilon)$ for the nonchiral heavy fermions transmitting via a potential well. One can see that $\mathcal{T}(\varepsilon)$ is strongly suppressed in the latter case. For such a reason, the contact conductance for conventional heavy fermions is low. Quite a different behavior $\mathcal{T}(\varepsilon)$ takes place if instead of the conventional heavy electrons there are the HCF particles as is evident from Fig. 9.3b. In Fig. 9.3b we compare $\mathcal{T}(\varepsilon)$ for the conventional chiral fermions penetrating the chiral barrier (curves 1 and 3) with the same characteristics for HCF particles (curves 2 and 4). One can see that $\mathcal{T}(\varepsilon)$ is fairly good for both types of the chiral particles if the incidence angle is small, that is, $\phi = \pi/8$ (curves 1 and 2). For bigger incidence angles, that is, $\phi = 3\pi/8$ (curves 3 and 4), for the HCF particles $\mathcal{T}(\varepsilon)$ becomes suppressed (curve 4). The electron thermal conductance Λ_{e} of G-TEG is determined by $2\Lambda_{\mathrm{H/RR}}$. The dominant contribution into $\Lambda_{\mathbf{H/RR}}$ comes at the low angles ϕ; therefore using HCF particles helps maintain Λ_{e} at a decent level.

9.13.2 *Perspectives of Thermoelectric Research for Graphene*

The TE effects and heat transport in graphene and CNT junctions belong to emerging fields of research. Recent experimental data obtained during the study of temperature distribution and thermopower in graphene stripes and CNTs indicate that efficient TE devices can be fabricated using the nanoscale technology. Because the elementary excitations in graphene and CNT are chiral fermions they scattering on the phonons is minimized. For such reasons, the contribution of inelastic scattering processes in the TE transport is small. On one hand, the electric charge transfer and the thermopower both are determined mostly by elastic processes without energy dissipation. An additional increase of electronic transport coefficients occurs at expense of interference effects and of electrode doping by applying of appropriate gate voltages. On the other hand, the heat conductance due to the propagation of phonons is also significant which tends to decrease the figure of merit. The last hurdle can be avoided by using multilayered metallic electrodes and multiple terminal configurations to decouple the electron and phonon components of the heat flow by redirecting them along different paths. The ability to controlling and manipulating by the electronic and phonon transport on the nanoscale suggests a bright future for a successful development of graphene and CNT electronics. A direct observation of energy transport in nonequilibrium nanoscale junctions can be accomplished by taking benefits of recent advances of thermal microscopy techniques. Thus, one can directly probe a variety of fundamental thermodynamic properties. One exciting application of the mentioned novel research method is, for example, an immediate mapping of the nonequilibrium temperature distribution in a quantum system subject to a thermal gradient or voltage bias. A closely relevant issue in this respect is how to define the local temperature on the nanoscale. Although that theoretical problem is not new, the recent experiments have sparked an impressive wave of interest to the topic and motivated intensive efforts to solve the number of fundamental issues. There are many remaining questions related to the meaning of temperature on the atomic (or subatomic) scale, the relationship between Fourier's law of heat conduction and the quantum heat transport, and the fundamental limits of the tip-environment coupling. Other topics of interest are related to quantum interference effects on the local temperature distribution. When a nanostructure is exposed to external fields and biases, the fermionic (electronic) and bosonic (phonon or photon) degrees of freedom distributions are characterized by different effective temperatures because in general they do not correspond to one another away from equilibrium. That difference is quite essential in small grapheme and CNT junctions, where the electron–phonon coupling is weak, or in scanning tunneling devices operating in the tunneling regime, where there is no phonon heat transport. Finding the correspondence among the temperatures of these distinct degrees of freedom will definitely represent new insights into the nature of the nonequilibrium transport problem.

A good progress in this important direction had been achieved with the ballistic heat transport measurements performed in mesoscopic systems and with the advent of nanometer precision, picowatt sensitive calorimeters. One can expect that similar experiments will be capable of probing the quantum regime of heat transport down to the quantum of heat conductance $\kappa_0 = \pi^2 k_B^2 T/3h \approx 0284$ nWK^{-1} at 300 K. It will allow to exploring of the relationships among different symmetry, electron correlation, molecular vibrations, noise, and quantum interference. All the mentioned phenomena can be studied when probing the interplay between the quantum heat and charge transport.

Problems

9-1. Determine how the Mott formula, Eq. 9.2, can be modified to account for sharp singularities of the excitation spectrum.

9-2. Explain why phonons, excitons, plasmons, and other massive excitations without electric charge reduce the figure of merit of a thermoelectric (TE) device?

9-3. Explain why singularities in the electric excitation spectrum can improve the performance of a TE device.

9-4. Explain why the Wiedemann–Franz (WF) law is not observed in graphene and carbon nanotubes (CNTs).

9-5. What is the connection between the figure of merit ZT and the Carnot efficiency η_C?

9-6. What is the connection between Fourier's law and the electrical Ohm law?

9-7. Explain why the main heat carriers in carbon materials are usually phonons.

9-8. Explain why the phonon group velocity is typically decreased in nanostructures.

9-9. Explain why preserving the parity-time (PT) inversion symmetry is important for the creation of heavy chiral fermion (HCF) states.

9-10. Explain why using multilayered metallic electrodes with low value of the phonon thermal conductance can improve the figure of merit of a thermoelectric generator (TEG)?

References

Ando, T. (2005). Theory of electronic states and transport in carbon nanotubes. *J Phys Soc Jpn*, **74**(3): 777–817.

Baheti, K., J. A. Malen, P. Doak, P. Reddy, S. Y. Jang, T. D. Tilley, A. Majumdar, and R. A. Segalman (2008). Probing the chemistry of molecular heterojunctions using thermoelectricity. *Nano Lett*, **8**(2): 715–719.

Balandin, A., and K. L. Wang (1998). Significant decrease of the lattice thermal conductivity due to phonon confinement in a free-standing semiconductor quantum well. *Phys Rev B*, **58**(3): 1544–1549.

Balandin, A. A. (2005). Nanophononics: phonon engineering in nanostructures and nanodevices. *J Nanosci Nanotech*, **5**(7): 1015–1022.

Balandin, A. A. (2011). Thermal properties of graphene and nanostructured carbon materials. *Nat Mater*, **10**(8): 569–581.

Balandin, A. A., S. Ghosh, W. Z. Bao, I. Calizo, D. Teweldebrhan, F. Miao, and C. N. Lau (2008). Superior thermal conductivity of single-layer graphene. *Nano Lett*, **8**(3): 902–907.

Bell, L. E. (2008). Cooling, heating, generating power, and recovering waste heat with thermoelectric systems. *Science*, **321**(5895): 1457–1461.

Berber, S., Y. K. Kwon, and D. Tomanek (2000a). Electronic and structural properties of carbon nanohorns. *Phys Rev B*, **62**(4): R2291–R2294.

Berber, S., Y. K. Kwon, and D. Tomanek (2000b). Unusually high thermal conductivity of carbon nanotubes. *Phys Rev Lett*, **84**(20): 4613–4616.

Bergfield, J. P., G. C. Solomon, C. A. Stafford, and M. A. Ratner (2011). Novel quantum interference effects in transport through molecular radicals. *Nano Lett*, **11**(7): 2759–2764.

Bergfield, J. P., and C. A. Stafford (2009a). Many-body theory of electronic transport in single-molecule heterojunctions. *Phys Rev B*, **79**(24).

Bergfield, J. P., and C. A. Stafford (2009b). Thermoelectric signatures of coherent transport in single-molecule heterojunctions. *Nano Lett*, **9**(8): 3072–3076.

Bergman, L., M. Dutta, K. W. Kim, P. G. Klemens, S. Komirenko, and M. A. Stroscio (2000). Phonons, electron-phonon interactions, and phonon-phonon interactions in III-V nitrides. *Ultrafast Phenom Semicond IV*, **3940**: 100–111.

Brey, L., and H. A. Fertig (2007). Elementary electronic excitations in graphene nanoribbons. *Phys Rev B*, **75**(12).

Buttiker, M. (1986). 4-terminal phase-coherent conductance. *Phys Rev Lett*, **57**(14): 1761–1764.

Cai, W. W., A. L. Moore, Y. W. Zhu, X. S. Li, S. S. Chen, L. Shi, and R. S. Ruoff (2010). Thermal transport in suspended and supported monolayer graphene grown by chemical vapor deposition. *Nano Lett*, **10**(5): 1645–1651.

Calizo, I., S. Ghosh, D. Teweldebrhan, W. Z. Bao, F. Miao, C. N. Lau, and A. A. Balandin (2008). Raman nanometrology of graphene on arbitrary substrates and at variable temperature. *Carbon Nanotubes Associated Devices*, **7037:** B371.

Canetta, C., and A. Narayanaswamy (2013a). Measurement of optical coupling between adjacent bi-material microcantilevers. *Rev Sci Instrum*, **84**(10).

Canetta, C., and A. Narayanaswamy (2013b). Sub-picowatt resolution calorimetry with a bi-material microcantilever sensor. *Appl Phys Lett*, **102**(10).

Chandra, B., V. Perebeinos, S. Berciaud, J. Katoch, M. Ishigami, P. Kim, T. F. Heinz, and J. Hone (2011). Low bias electron scattering in structure-identified single wall carbon nanotubes: role of substrate polar phonons. *Phys Rev Lett*, **107**(14).

Chen, S. S., A. L. Moore, W. W. Cai, J. W. Suk, J. H. An, C. Mishra, C. Amos, C. W. Magnuson, J. Y. Kang, L. Shi, and R. S. Ruoff (2011). Raman measurements of thermal transport in suspended monolayer graphene of variable sizes in vacuum and gaseous environments. *ACS Nano*, **5**(1): 321–328.

Cutler, M., and N. F. Mott (1969). Observation of Anderson localization in an electron gas. *Phys Rev*, **181**(3): 1336.

DiSalvo, F. J. (1999). Thermoelectric cooling and power generation. *Science*, **285**(5428): 703–706.

Dubi, Y., and M. Di Ventra (2011). Colloquium: heat flow and thermoelectricity in atomic and molecular junctions. *Rev Mod Phys*, **83**(1): 131–155.

Faugeras, C., B. Faugeras, M. Orlita, M. Potemski, R. R. Nair, and A. K. Geim (2010). Thermal conductivity of graphene in corbino membrane geometry. *ACS Nano*, **4**(4): 1889–1892.

Finch, C. M., V. M. Garcia-Suarez, and C. J. Lambert (2009). Giant thermopower and figure of merit in single-molecule devices. *Phys Rev B*, **79**(3).

Galperin, M., M. A. Ratner, A. Nitzan, and A. Troisi (2008). Nuclear coupling and polarization in molecular transport junctions: beyond tunneling to function. *Science*, **319**(5866): 1056–1060.

Ghosh, S., W. Z. Bao, D. L. Nika, S. Subrina, E. P. Pokatilov, C. N. Lau, and A. A. Balandin (2010). Dimensional crossover of thermal transport in few-layer graphene. *Nat Mater*, **9**(7): 555–558.

Ghosh, S., I. Calizo, D. Teweldebrhan, E. P. Pokatilov, D. L. Nika, A. A. Balandin, W. Bao, F. Miao, and C. N. Lau (2008). Extremely high thermal conductivity of graphene: prospects for thermal management applications in nanoelectronic circuits. *Appl Phys Lett*, **92**(15).

Guedon, C. M., H. Valkenier, T. Markussen, K. S. Thygesen, J. C. Hummelen, and S. J. van der Molen (2012). Observation of quantum interference in molecular charge transport. *Nat Nanotechnol*, **7**(5): 304–308.

Hedin, L. (1965). New method for calculating 1-particle greens function with application to electron-gas problem. *Phys Rev*, **139**(3A): A796.

Hoffmann, E. A., H. A. Nilsson, J. E. Matthews, N. Nakpathomkun, A. I. Persson, L. Samuelson, and H. Linke (2009). Measuring temperature gradients over nanometer length scales. *Nano Lett*, **9**(2): 779–783.

Jauho, A. P., N. S. Wingreen, and Y. Meir (1994). Time-dependent transport in mesoscopic systems: general formalism and applications. *Semicond Sci Tech*, **9**(5): 926–929.

Jauregui, L. A., Y. N. Yue, A. N. Sidorov, J. N. Hu, Q. K. Yu, G. Lopez, R. Jalilian, D. K. Benjamin, D. A. Delk, W. Wu, Z. H. Liu, X. W. Wang, Z. G. Jiang, X. L. Ruan, J. M. Bao, S. S. Pei, and Y. P. Chen (2010). Thermal transport in graphene nanostructures: experiments and simulations. *Graphene, Ge/III-V, Emerging Mater Post-CMOS Appl 2*, **28**(5): 73–83.

Kamenetska, M., S. Y. Quek, A. C. Whalley, M. L. Steigerwald, H. J. Choi, S. G. Louie, C. Nuckolls, M. S. Hybertsen, J. B. Neaton, and L. Venkataraman (2010). Conductance and geometry of pyridine-linked single-molecule junctions. *J Am Chem Soc*, **132**(19): 6817–6821.

Keawprak, N., R. Tu, and T. Goto (2009). Thermoelectric properties of alkaline earth ruthenates prepared by SPS. *Mater Sci Eng B*, **161**(1–3): 71–75.

Keldysh, L. V. (1965). Diagram technique for nonequilibrium processes. *Sov Phys JETP: USSR*, **20**(4): 1018.

Kim, K. S., Y. Zhao, H. Jang, S. Y. Lee, J. M. Kim, K. S. Kim, J. H. Ahn, P. Kim, J. Y. Choi, and B. H. Hong (2009). Large-scale pattern growth of graphene films for stretchable transparent electrodes. *Nature*, **457**(7230): 706–710.

Klemens, P. G. (2000). Role of optical modes in thermal conduction. *Therm Cond 25: Therm Expansion 13*, **25**: 291–299.

Lee, J. U., D. Yoon, H. Kim, S. W. Lee, and H. Cheong (2011). Thermal conductivity of suspended pristine graphene measured by Raman spectroscopy. *Phys Rev B*, **83**(8).

Lemme, M. C., T. J. Echtermeyer, M. Baus, and H. Kurz (2007). A graphene field-effect device. *IEEE Electron Device Lett*, **28**(4): 282–284.

Lin, Y. M., X. Z. Sun, and M. S. Dresselhaus (2000). Theoretical investigation of thermoelectric transport properties of cylindrical Bi nanowires. *Phys Rev B*, **62**(7): 4610–4623.

Lunde, A. M., and K. Flensberg (2005). On the Mott formula for the thermopower of non-interacting electrons in quantum point contacts. *J Phys: Condens Mater*, **17**(25): 3879–3884.

Maekawa, T., K. Kurosaki, H. Muta, M. Uno, and S. Yamanaka (2005a). Thermal and electrical properties of perovskite-type strontium molybdate. *J Alloy Compd*, **390**(1–2): 314–317.

Maekawa, T., K. Kurosaki, H. Muta, M. Uno, and S. Yamanaka (2005b). Thermoelectric properties of perovskite type strontium ruthenium oxide. *J Alloy Compd*, **387**(1–2): 56–59.

Mak, K. F., J. Shan, and T. F. Heinz (2011). Seeing many-body effects in single- and few-layer graphene: observation of two-dimensional saddle-point excitons. *Phys Rev Lett*, **106**(4).

Meir, Y., and N. S. Wingreen (1992). Landauer formula for the current through an interacting electron region. *Phys Rev Lett*, **68**(16): 2512–2515.

Mott, N. F. (1958). A theory of the origin of fatigue cracks. *Acta Metall*, **6**(3): 195–197.

Paulsson, M., and S. Datta (2003). Thermoelectric effect in molecular electronics. *Phys Rev B*, **67**(24).

Persson, A. I., Y. K. Koh, D. G. Cahill, L. Samuelson, and H. Linke (2009). Thermal conductance of InAs nanowire composites. *Nano Lett*, **9**(12): 4484–4488.

Pokatilov, E. P., D. L. Nika, and A. A. Balandin (2005). Acoustic-phonon propagation in rectangular semiconductor nanowires with elastically dissimilar barriers. *Phys Rev B*, **72**(11).

Pop, E., D. A. Mann, K. E. Goodson, and H. J. Dai (2007). Electrical and thermal transport in metallic single-wall carbon nanotubes on insulating substrates. *J Appl Phys*, **101**(9).

Rana, F. (2007). Electron-hole generation and recombination rates for Coulomb scattering in graphene. *Phys Rev B*, **76**(15).

Reddy, P., S. Y. Jang, R. A. Segalman, and A. Majumdar (2007). Thermoelectricity in molecular junctions. *Science*, **315**(5818): 1568–1571.

Rowe, D. M., and G. Min (1995a). Alpha-plot in sigma-plot as a thermoelectric-material performance indicator. *J Mater Sci Lett*, **14**(9): 617–619.

Rowe, D. M., and G. Min (1995b). Model for electrical power factor of silicon germanium gallium phosphide alloys. *Thirteenth Int Conf Thermoelectr*, (316): 271–276.

Rowe, D. M., and G. Min (1995c). Multiple potential barriers as a possible mechanism to increase the Seebeck coefficient and electrical power factor. *Thirteenth Int Conf Thermoelectr*, (316): 339–342.

Rowe, D. M., and G. Min (1995d). Of heat treatment on the Fermi energy and effective mass of n-type silicon germanium-gallium phosphide. *Thirteenth Int Conf Thermoelectr*, (316): 250–253.

Sadat, S., Y. J. Chua, W. Lee, Y. Ganjeh, K. Kurabayashi, E. Meyhofer, and P. Reddy (2011). Room temperature picowatt-resolution calorimetry. *Appl Phys Lett*, **99**(4).

Sadat, S., A. Tan, Y. J. Chua, and P. Reddy (2010). Nanoscale thermometry using point contact thermocouples. *Nano Lett*, **10**(7): 2613–2617.

Seol, J. H., I. Jo, A. L. Moore, L. Lindsay, Z. H. Aitken, M. T. Pettes, X. S. Li, Z. Yao, R. Huang, D. Broido, N. Mingo, R. S. Ruoff, and L. Shi (2010).

Two-dimensional phonon transport in supported graphene. *Science*, **328**(5975): 213–216.

Seol, J. H., A. L. Moore, I. S. Jo, Z. Yao, and L. Shi (2010). Thermal conductivity measurement of graphene exfoliated on silicon dioxide. *Proc ASME Int Heat Transfer Conf, 2010*, **6**: 519–523.

Shafranjuk, S. E. (2009). Reversible heat flow through the carbon tube junction. *EPL*, **87**(5).

Shafranjuk, S. E. (2011). Electromagnetic properties of the graphene junctions. *Euro Phys J B*, **80**(3): 379–393.

Sivan, U., and Y. Imry (1986). Multichannel Landauer formula for thermoelectric transport with application to thermopower near the mobility edge. *Phys Rev B*, **33**(1): 551–558.

Small, J. P., K. M. Perez, and P. Kim (2003). Modulation of thermoelectric power of individual carbon nanotubes. *Phys Rev Lett*, **91**(25).

Soriano, D., and J. Fernandez-Rossier (2010). Spontaneous persistent currents in a quantum spin Hall insulator. *Phys Rev B*, **82**(16).

Soriano, D., and J. Fernandez-Rossier (2012). Interplay between sublattice and spin symmetry breaking in graphene. *Phys Rev B*, **85**(19).

Soriano, D., F. Munoz-Rojas, J. Fernandez-Rossier, and J. J. Palacios (2010). Hydrogenated graphene nanoribbons for spintronics. *Phys Rev B*, **81**(16).

Strange, M., C. Rostgaard, H. Hakkinen, and K. S. Thygesen (2011). Self-consistent GW calculations of electronic transport in thiol- and amine-linked molecular junctions. *Phys Rev B*, **83**(11).

Tan, A., S. Sadat, and P. Reddy (2010). Measurement of thermopower and current-voltage characteristics of molecular junctions to identify orbital alignment. *Appl Phys Lett*, **96**(1).

Tritt, T. M. (2011). Thermoelectric phenomena, materials, and applications. *Ann Rev Mater Res*, **41**: 433–448.

Tselev, A., Y. F. Yang, J. Zhang, P. Barbara, and S. E. Shafranjuk (2009). Carbon nanotubes as nanoscale probes of the superconducting proximity effect in Pd-Nb junctions. *Phys Rev B*, **80**(5).

Xia, F. N., V. Perebeinos, Y. M. Lin, Y. Q. Wu, and P. Avouris (2011). The origins and limits of metal-graphene junction resistance. *Nat Nanotechnol*, **6**(3): 179–184.

Xu, B. Q., and N. J. J. Tao (2003). Measurement of single-molecule resistance by repeated formation of molecular junctions. *Science*, **301**(5637): 1221–1223.

Zhang, J., A. Tselev, Y. F. Yang, K. Hatton, P. Barbara, and S. Shafraniuk (2006). Zero-bias anomaly and possible superconductivity in single-walled carbon nanotubes. *Phys Rev B*, **74**(15).

Zou, J., and A. Balandin (2001). Phonon heat conduction in a semiconductor nanowire. *J Appl Phys*, **89**(5): 2932–2938.

Chapter 10

Sensing of Electromagnetic Waves with Graphene and Carbon Nanotube Quantum Dots

A fraction of the electromagnetic field spectrum spaced between 0.5 THz and 100 THz is regarded as the THz domain. An ability to generate and detect the waves belonging to the THz domain (so-called T-rays) attracts significant attention (Sizov and Rogalski 2010, Shafranjuk 2011a, 2011b). Visualizing the THz waves illustrated in Fig. 10.1 is enormously helpful in a variety of applications, including space research, medicine, industry, defense, and security.

10.1 Sensors of the Electromagnetic Field

The T-rays have a variety of unique properties which make them interesting (Fuse et al. 2007a, 2007b, 2007c). For instance, the THz spectrum of many chemical inorganic and organic molecules exhibits oscillation and rotation spectral features taking place within the THz domain. Hence, the sets of spectral line positions within the THz domain represent unique characteristics of the mentioned chemical analytes. It paves the path for the remote

Graphene: Fundamentals, Devices, and Applications
Serhii Shafraniuk

ISBN 978-981-4613-47-7 (Hardcover), 978-981-4613-48-4 (eBook)
www.panstanford.com

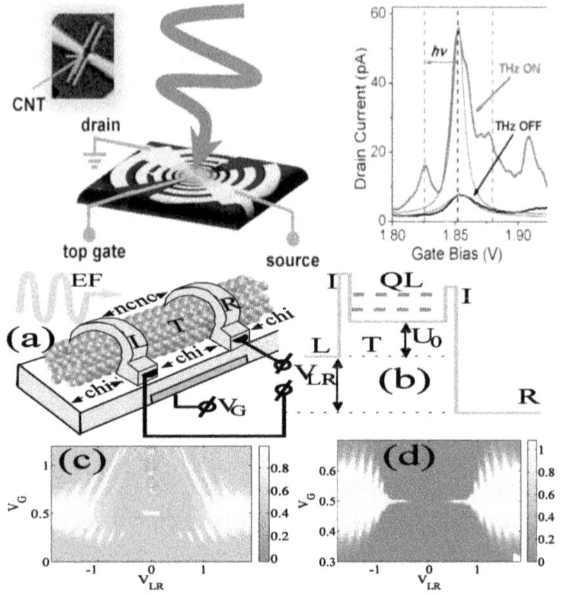

Figure 10.1 The upper panel shows the carbon nanotube quantum dot THz field sensor (Rinzan et al. 2012). (a) The CNT quantum dot design which is based on the field-effect transistor. (b) The energy diagram of the quantum dot. (c) The difference of the chiral and nonchiral tunneling contributions into the dot's conductance for the lowest-energy subband. (d) The same for the first excited energy subband. The lower panel is adapted from Shafranjuk (2011a, 2011b).

detecting, identification, and chemical analysis of the analyte of interest when measuring the reflection and transmission of the THz waves. There is another interesting application of T-rays which consists of their ability to easily penetrating through thin layers of clothes or plastic covers which reflect the conventional sunlight. That unusual property of the T-rays emerges from the fact that the submillimeter wave length is comparable to thickness of human closing and various covers. For such reasons, the T-ray spectral analyzers shown in Figs. 10.1 and 10.2 are capable of detecting the concealed weapon, hidden explosives, or toxic chemicals. That penetrative ability of T-rays is utilized in a variety of security, defense, and industrial applications (Sizov and Rogalski 2010, Shafranjuk 2011a, 2011b). However, in practice, using the quantum

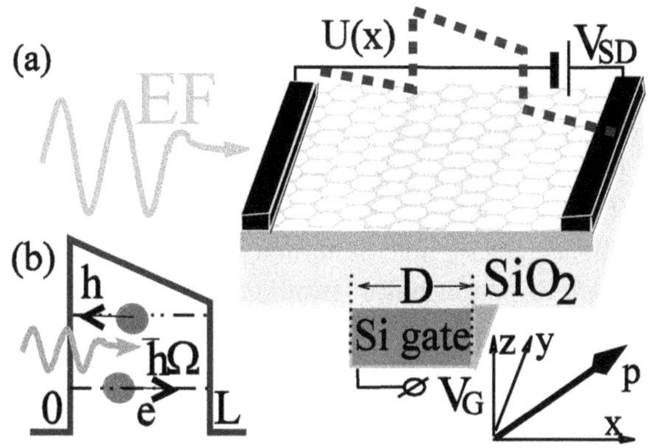

Figure 10.2 (a) The monolayer graphene junction (GJ) with metallic source and drain electrodes and a local gate electrode attached from beneath. The junction is exposed to the external electromagnetic field (EF) and is also controlled by the gate voltage V_G. The darker green hexagons mark the region of the chiral barrier, while the dashed line shows the barrier profile $U(x)$. (b) Quantum interference of electrons and holes (e–h) inside the chiral barrier. Adapted from Shafranjuk (2011a, 2011b).

dot (QD) devices for detecting T-rays is a technological and technical challenge (Maksimenko et al. 2006, Slepyan et al. 2006, Fu et al. 2008, Yngvesson et al. 2008, Seliuta et al. 2010). The energy $h\nu$ of a T-ray photon with frequency ν is low compared to typical electron energy-level spacing inside the dot. Therefore, it becomes difficult to meet the requirement

$$h\nu = \Delta E \geq kT^* \tag{10.1}$$

where T^* is the electron temperature in the dot and ΔE is the spacing between adjacent quantized electron energy levels. The above Eq. 10.1 implies that the spectral resolution of the T-ray frequency ν must exceed the level broadening kT^*, which for $\Delta E/h$ =1 THz requires that $T^* \leq$ 50 K, which is much below the room temperature 300 K. One can still satisfy the requirement (10.1) if the QD is additionally refrigerated. The T-ray sensor elements are required to be exceptionally sensitive to triggering, when absorbing the low-energy photons. Furthermore, the T-rays are not typically controlled by means of conventional optics or RF waveguides nor they are focused by conventional lenses which also introduces extra technical complications. Thus, the development of practical T-ray devices faces numerous issues yet to be solved. The aforementioned challenges nowadays are addressed by implementing of novel nanotechnological approaches, research methods, and advanced materials. The study and implementations of graphene, carbon nanotubes (CNTs), and other monoatomic materials have a strong potential in this emerging field of human activity.

Electromagnetic properties of graphene QD devices (Syzranov et al. 2008, Kienle and Leonard 2009) and CNTs (Fuse et al. 2007a, 2007b, 2007c, Rinzan et al. 2012) were examined in respect to their appropriateness for the THz sensing and emission applications. Preliminary research (Syzranov et al. 2008, Kienle and Leonard 2009, Fuse et al. 2007a, 2007b, 2007c, Rinzan et al. 2012) suggests that the novel materials are advantageous in comparison with the conventional semiconductors. A remarkable feature of graphene and CNTs is that the low-energy electrons behave like relativistic particles (Ando 2005, Katsnelson et al. 2006, Morozov et al. 2006, Novoselov et al. 2006). Furthermore, in new carbon materials, the electrons at low energies are characterized by chirality and by two one-half pseudospins per each excitation. For such reasons, the elementary excitations in the graphene and CNTs are regarded as chiral fermions, which makes them different from the electrons in conventional semiconductors where they behave as spinless excitations. The unique nature of elementary excitations in graphene and CNTs influences the whole scenario of the electron transport in carbon materials as illustrated in Fig. 10.3. During the electron scattering on impurities, phonons, and other electrons the one-half pseudospins are conserved, thus, all the electron transport coefficients are modified with special coherence factors. For these reasons, in the absence of charged impurities and ripples in the CNT/graphene devices, the electron movement is rather ballistic than diffusive. It means that the electron mean free path l_e ($l_{\mathrm{e}} = \min\{l_{\mathrm{ee}}, l_{\mathrm{ep}}, l_{\mathrm{ei}}\}$, where $l_{\mathrm{ee,ep,ei}} = v\tau_{\mathrm{ee,ep,ei}}$ and $\tau_{\mathrm{ee,ep,ei}}$ are scattering times of an electron on the other electrons, phonons, and impurities) is prolonged (Ando 2005, Shafranjuk 2009a, 2009b). Therefore, the nanotube might be regarded as being intrinsically "purified." Thus, the relativistic nature of the electrons, their one-half pseudospins and chirality are responsible for the "intrinsic purity" of graphene

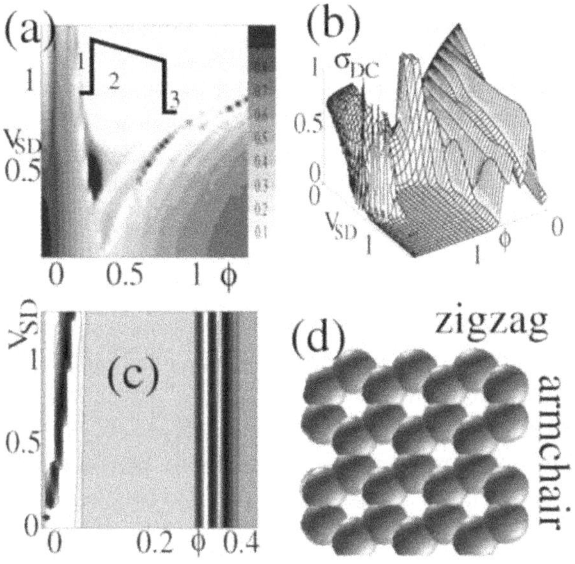

Figure 10.3 (a) The steady-state differential conductivity σ_{SD} versus the source–drain voltage V_{SD} and azimuthal angle φ (in radians) for a trapezoidal barrier formed on a wide graphene ribbon. (b) The differential conductivity σ_{SD} (in units of 8e2 $|V_{SD}|/(vh2)$) versus V_{SD} (in units of the barrier height U_0) and the azimuthal angle φ (in radians) for $D = 15$ and $U_0 = 1$. (c) Similar plot σ_{SD} (V_{SD}, φ) for a ribbon with zigzag edges for $W = 15$, $D = 20$, and $U_0 = 1$. (d) The corner of a graphene ribbon where the upper edge has a zigzag shape, while the right-side edge has the armchair shape. Adapted from Shafranjuk (2011a, 2011b).

and CNT samples (Ando 2005, Katsnelson et al. 2006, Morozov et al. 2006, Novoselov et al. 2006, Shafranjuk 2009a, 2009b). The intrinsic "purity" of graphene and CNT and ballistic transport of elementary excitations makes the difference from the electron transport in regular semiconductors. Ballistic transport in the conventional 2D semiconductors is impeded due to electron–electron collisions. Therefore, the electron system resembles a viscous fluid (Dyakonov and Shur 1996, Shur et al. 1996). In the latter case of the electronic fluid, one can create "shallow water" plasma waves (Dyakonov and Shur 1996, Shur et al. 1996). Theoretically, the "shallow water" plasma waves are described merely applying certain boundary conditions (BCs) at the edges of the conventional 2D semiconducting sample. Creation of plasma waves with acoustic dispersion law were examined for regular 2D semiconductors (Dyakonov and Shur 1996, Shur et al. 1996). The described scenario fails for graphene and CNT samples where the creation of "shallow water" acoustic plasma waves and their use for detecting the THz radiation cannot be accomplished. Because of the unique nature of the electron transport, the electron–electron scattering in CNT/graphene devices occurs on a very long length $l_{ee} = v\tau_{ee}$ (Ando 2005, Katsnelson et al. 2006, Morozov et al. 2006, Novoselov et al. 2006, Shafranjuk 2009a, 2009b). Here, we assume that the samples are not rippled and the charged impurities are removed, for example, by their annealing in vacuum. In those systems, the electrons behave rather as a diluted gas and not as a fluid. If the graphene stripe width is, for example, just a few nanometers or less, then, the contribution into the transconductance G comes just from a few conducting channels. Another contribution into G originates from the interface barriers between the metal source and drain electrodes on one side and the graphene (or CNTs) on the other side. The barriers emerge partially from imperfections and impurities located immediately at the interface, and additionally, the Schottky barriers are formed due to the difference of workfunctions of metal and graphene.

Though the task of creating the "shallow water" plasma waves in the CNT/graphene devices is problematic, the detection of T-rays can be accomplished (Fuse et al. 2007a, 2007b, 2007c, Rinzan et al. 2012) by utilizing other resonances, for example, the single-electron tunneling (SET) and the quantized states. Such resonances were observed in several independent experiments (Fuse et al. 2007a, 2007b, 2007c, Rinzan et al. 2012) with measuring DC conductance of the CNT field-effect transistor (FET) QDs under the influence of the THz field.

The QD (see Fig. 10.4) is fabricated using a section of graphene stripe or CNT. Thus, one can create quantized energy levels of chiral fermions as shown in Figs. 10.2, 10.3 inside the spatially confined region. In the absence of "shallow water" plasma waves in CNT/graphene (Akturk et al. 2007, Kienle and Leonard 2009), the quantized states have been already firmly detected (Fuse et al. 2007a, 2007b, 2007c, Rinzan et al. 2012). Sharp quantized energy levels inside the graphene or CNT sections are formed because the electron mean free path is long ($l_e > D_T$ where D_T is the section

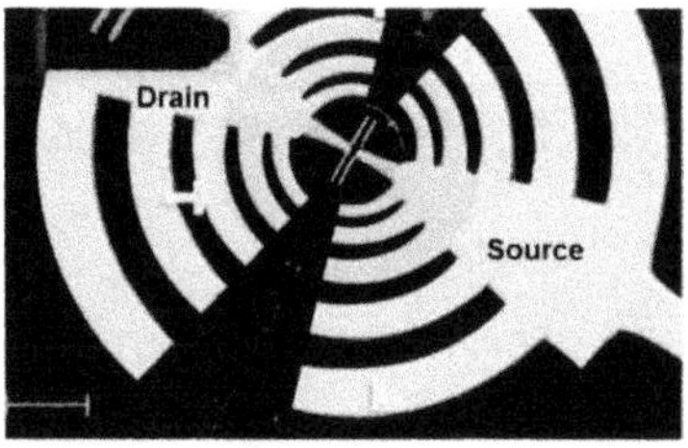

Figure 10.4 The sketch and image of the carbon nanotube quantum dot (CNT QD) sensor (Rinzan, 2012). (a) Cross-section of the CNT QD device structure. (b) The scanning tunneling microscopy (STM) image of a device. Upper-left corner: Zoom-in of the QD region. In the design, the source and drain electrodes are separated by 400 nm corresponding to energy-level spacing $\Delta E \approx 4$ meV. The PMMA gate dielectric had been deposited on top of the CNT connecting the electrodes. The CNT cannot be seen because of it. Adapted from Rinzan et al. (2012).

length) there. Positions and widths of the quantized energy levels are adjusted by applying appropriate electric potentials to the source, drain, and gate electrodes attached to CNT TEG QD (Fuse et al. 2007a, 2007b, 2007c, Rinzan et al. 2012). The electron transport through the dot is initialized applying the external source drain V_{SD} and gate V_G voltages (Zhang et al. 2006, Tselev et al. 2009). An external AC field acts on the charged particles by changing their energy and momentum. According to the pioneer work by Tien and Gordon (Tien and Gordon 1963), the electrons tunnel with absorption and emission of the AC field photons. At the THz frequencies, the amount of electron energy change $\Delta E = h\nu$ due to the absorption or emission the photons of external field corresponds to $\Delta T \sim$ tens of Kelvins. Such energy absorption by the electrons tunneling through the QD is pronounced as periodic series of steps in the I–V curves. Another implication of the AC field influence on the electric current through the QD is that the AC field not only affects the electron quantization conditions, but it also modifies the quantized states. One experimentally observes the AC field–induced changes in the system by measuring the QD conductance which is altered by several orders of magnitude against its steady-state value. That huge response of the electron transport to the external electromagnetic field is exploited in the new generation of highly sensitive detectors.

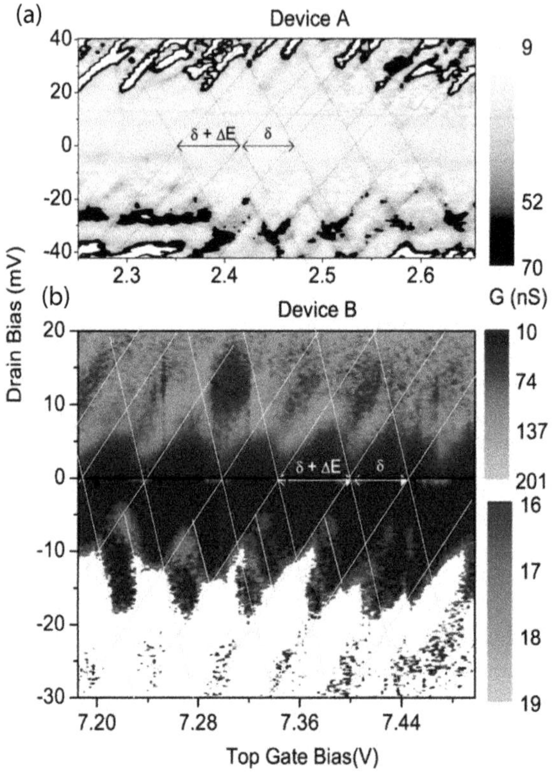

Figure 10.5 Basic experimental characteristics of the quantum dot (Rinzan, 2012). (a) Differential conductance as a function of gate voltage and source–drain bias for device A. (b) The same for device B: Because of the asymmetry of the current versus drain bias for device B, we have used two different color scales for the positive and negative source–drain bias to reveal the diamond pattern clearly. We mark the excited levels by the lines running parallel to the edges of the coulomb blockade diamonds. Characteristic parameters of the quantum dot were extracted from the slopes of diamonds. The obtained source and drain capacitances are $C_S = 3.7$ (11.8) aF and $C_D = 4.4$ (6.1) aF, the gate capacitance $C_G = 3.6$ (3.8) aF and the energy-level spacing is $\Delta E = 3.3$ (1.9) meV. The charging energy for sensor A (B) is $\delta = 13.7$ (7.4) meV. Adapted from Rinzan et al. (2012).

In this Section, we focus our attention on CNT and graphene T-ray detectors and spectral analyzers (Fuse et al. 2007a, 2007b, 2007c, Rinzan et al. 2012). We will discuss principles and basic physical mechanisms of the T-ray sensing. Apart from the T-ray sensors based on the bolometric and nonlinear effects (Slepyan et al. 2006, Fu et al. 2008, Yngvesson et al. 2008, Seliuta et al. 2010), we describe the graphene (**G**) and CNT QD THz sensing devices (Fuse et al. 2007a, 2007b, 2007c, Rinzan et al. 2012). The design technology and fabrication of such nanodevices requires special approaches. It represents a principal issue based on the simple physical reason that the T-ray photon energy is relatively low (Fuse et al. 2007a, 2007b, 2007c, Rinzan et al. 2012). There are many important requirements to the QD devices (Sizov and Rogalski 2010, Shafranjuk 2011a, 2011b) which account for not only common characteristics like the contact resistance, charge transfer length, or operational temperature, but also the nanotube diameter and chirality, as well as the shape and width of the graphene stripe edges. Such requirements are met by exploiting a higher degree of intrinsic coherence in the CNT and graphene sensors. It allows us to refine the THz sensors by making them far more capable comparing to conventional QDs which were being fabricated from regular semiconductors.

Besides, we briefly discuss available technological approaches, important experiments which along with the critical theoretical concepts are shedding the light on present results concerning the T-ray sensing by the graphene and/or CNT devices (Kral et al. 2000, Mele et al. 2000, Akturk et al. 2007, Fuse et al. 2007a, 2007b, 2007c, Shafranjuk 2008, 2009a, 2009b, 2011a, 2011b, Kawano 2009, Kienle and Leonard 2009, Rinzan et al. 2012). The physical mechanisms which are responsible for an AC field-induced change of the electron transport through the QD will be examined using the mathematical models. A consistency between the existing experimental data and theoretical predictions is discussed as well. Thus, the theory allows us to extract many useful nanodevice parameters right from the experiments. One can determine, for example, the energy-level spacing and the level width, population of the levels, the effective electron temperature, etc. Besides, a variety of other important characteristics like the noise, level broadening due to the electron–phonon interaction and the spectrum of thermal phonons generated due to electron–phonon collisions are also determined. The latter microscopic process represents a major source of the intrinsic thermal noise. An analysis of correspondence between the theory and experiment allows us to establish the basic requirements to keep the QD being "quiet." In particular, we examine Johnson–Nyquist noise and shot noise.

A straightforward detection of the THz waves is based on inducing interlevel transitions when exposing the QD to an external AC field (Kawano 2009, 2011, 2012, Rinzan et al. 2012), (Shafranjuk 2008, 2009a, 2009b, 2011a, 2011b). The external AC field inflicts changes in the electron transport through the QD which are visible in the experimental curves of differential conductance. A sharp increase of the differential conductance occurs when the electron energy in the source and drain electrodes ε matches the quantized-level (QL) position E_n, provided that $\varepsilon \rightarrow E_n \pm eV_{\mathrm{SD}}$ where V_{SD} is the source–drain bias voltage. It means that if the latter condition is satisfied, the tunneling probability through the QD is high. The mentioned sharp peaks in the DC conductance G which are related to resonances of the photon-assisted tunneling are measured experimentally. The change of electron energy ε during a single photon-assisted tunneling event corresponds to the multiples of the photon energies $\pm m\hbar\omega$ (m is the number of absorbed/emitted photons, $\hbar\omega$ is the photon energy,). The condition of energy matching is changed as $\varepsilon \pm m\hbar\omega \rightarrow E_n$ where n is the energy-level index inside the dot. The external AC field changes the resonance condition so the side peaks in the DC conductance arise at new positions $V_{\mathrm{SD}} \rightarrow V_{\mathrm{SD}} \pm m\hbar\omega/e$ versus the source–drain voltage V_{SD}. When the electron temperature is low, the side peaks are very high and narrow. As the temperature increases, the side peaks acquire a finite width Γ while their height decreases. The sharp side peaks associated with the AC field–induced photon-assisted electron tunneling in graphene and CNT QDs constitute the idea of sensing. The rapid changes in the differential conductance can be readily measured. Besides, the crisp resonances ensure the high accuracy of the QD THz detectors. By metering the width, spacing, and magnitude of the peaks, one immediately determines the frequency, intensity, and even polarization of the external AC field.

Another relevant issue is how to reduce the intrinsic and parasitic noises which reduce the performance capabilities of the device (Shafranjuk 2008, 2009a, 2009b, 2011a, 2011b). Important advantages of the graphene and CNT setups emerge from their intrinsic coherence which causes suppression of the intrinsic noise. The noise becomes much lower than in regular semiconducting devices.

In this chapter, we consider the THz sensor based on graphene and CNT FET QD. The QD is composed from the FETs and spatially confined region while the quantized states of electrons are formed. The whole QD is exposed to the THz field which induces transitions between the quantized states. We will see that the interlevel transitions change the shape of the DC differential conductance curves. In particular, the AC field causes an appearance of additional satellite peaks (so-called side peaks) which are located aside the steady-state SET peaks originating from the Coulomb blockade effect. The spacing between the side peak and the major SET peak is $V_{SD} = h\nu/e$ where V_{SD} is the source–drain bias voltage across the QD. The sharp changes of the DC differential conductance allow us to achieve a high sensitivity of the CNT/graphene FET transport versus the external THz field. Physical principles of the THz field sensing by the CNT/graphene QD utilize a variety of microscopic phenomena. Among those processes are transmissions of the particles from attached electrodes into and out the CNT/graphene QD, interelectrode transitions of the particles through the QD via the discrete energy levels, and the field-induced interlevel transitions. Below, we describe the graphene and CNT experimental setups and available experimental data. One interesting feature of the experimental QD device is its ability to work in conditions of nonequilibrium self-cooling. Such a self-cooling effect is induced by the photon-assisted single-electron tunneling (PASET) through the CNT QD. We also present a theoretical model of that phenomenon, along with a thorough discussion of basic assumptions used to formulate the theoretical model. As a steady state of the graphene and CNT QDs we consider the quantization of electron states inside the dot where the chiral transport plays a critical role. It will allow us to achieve an adequate portrayal of influence of the THz field on the quantized states. Next, we will describe and debate available numeric results related to sensing of the THz field. Furthermore, we discuss the basic criteria for improving of responsivity and quantum efficiency of the graphene and CNT QDs. The origin of intrinsic noise and evaluation of the noise-equivalent power (NEP) and sensitivity, as well as the operation temperature and the frequency range of the graphene and CNT QD THz sensor are examined as well.

10.2 THz Sensor Based on Carbon Nanotube Quantum Dots

Design, fabrication, and experimental research of the CNT QD THz detectors are based on the idea to multiply two independent resonances. Kawano (2009, 2011, 2012) and Rinzan et al. (2012) utilized two different physical phenomena occurring on the nanoscale, (*i*) creation of a spatially confined region of graphene or CNT where a geometrical quantization of the electron motion occurs and (*ii*) a SET which emerges from a discreteness of the electron charge.

Design of the QD device is established using a small section of CNT which connects two weakly coupled source (S) and drain (D) electrodes (Tans et al. 1997). The dot also is capacitively coupled to the third gate electrode. The total dot's capacitance, C_{tot}, is very low, which makes the charging energy $\delta = e^2/C_{\text{tot}}$ to be sufficient to affect the whole electron transport through the dot. Resulting charging effects are manifested as sharp peaks in the source–drain conductance versus the gate voltage at temperatures $T < \delta/k_{\text{B}}$. Thus, the resonant elastic tunneling of single electrons through the QD is represented by peak series on the experimental curves. The spacing between peaks consists of the charging energy of the dot δ and the energy-level spacing ΔE (Tans et al. 1997). By exposing the dot with an external THz field one introduces another additional quantization of the electron states. Thus, altogether there are three independent quantization mechanisms involved: (*a*) the geometrical quantization inside the CNT QD, (*b*) the quantization of the electron charge, and (*c*) the THz field of frequency ω changes the electron energy by an amount $n\hbar\omega$, where n is the integer number. The latter mechanism presumes energy change due to absorption/emission of n photons during the SET through the dot. The mentioned PASET (inelastic tunneling with absorption and emission of photons) leads to satellite peaks in the dot's DC differential conductance. The magnitude of PASET depends on the field intensity whereas the voltage spacing separating the main SET peaks from the side PASET peaks is proportional to the energy of THz field photon $h\nu$ (Kouwenhoven et al. 1994a, 1994b). It represents an advantage of the CNT QD sensor comparing to conventional bolometric sensors: The PASET sensors determine both the intensity and frequency of radiation. Another strong benefit of the PASET sensor is that it operates as a wide band spectrometer from a few hundred GHz up to frequencies which corresponds to photon energy $h\nu \approx \delta$.

Firm detecting of a weak THz field has been recently reported in the work by Kawano et al. (Fuse et al. 2007a, 2007b, 2007c, Kawano 2009, 2012). Fuse et al. (2007a, 2007b, 2007c) and Kawano (2009, 2012) implemented the aforementioned mechanism of photon assisted tunneling, while the coupling power of THz field to the CNT QD was in the femtowatt range. The experiments (Fuse et al. 2007a, 2007b, 2007c, Kawano 2009, 2012) also revealed basic issues which hindered further research: (*a*) there was too weak coupling between the detector and the 10 mW laser source, and (*b*) the results obtained on multiple samples were hard to reproduce.

The initial work (Fuse et al. 2007a, 2007b, 2007c, Kawano 2009, 2012) motivated further activity in the same field using more refined method (Rinzan et al. 2012) which involved an improved design with a local top gate for each device. Then, THz wave is firmly coupled to the CNT QD with the help of the broad band on-chip antennas. The antenna electrodes also function as source and drain electrodes. Using the improved design, (Rinzan et al. 2012) one can work with much weaker THz sources (power smaller than 10 W) within a broad band. The elevated response of the CNT QD to the THz field allows us to observe new details which had not been formerly observed because the coupling was too weak. In particular, besides the satellite peaks on the left and right sides of the main peaks, the authors (Rinzan et al. 2012) found an anomalous increase and sharpening of the main peak. Such a substantial enhancement is understood assuming that the effective electron temperature is reduced in presence of the THz irradiation, which contradicts general expectations.

CNT QDs reported in the work (Rinzan et al. 2012) had been fabricated by growing the single-wall CNT by chemical vapor deposition on undoped Si substrates. Then, the CNT was capped by 1500 nm thick thermally grown SiO_2. The broadband log-periodic antennas were patterned by e-beam lithography and simultaneously served as titanium source and drain electrodes (see Fig. 10.1b). The lower and upper threshold frequencies were equal to 680 GHz and 2.5 THz, respectively. The gate electrode was separated from the CNT section by the dielectric which was obtained by crosslinking 200 nm thick PMMA and the top gate by sputtering a Cr/Au layer (see Fig. 10.1b). The experiments were conducted in an optical cryostat at liquid He^4 temperature and above. The measurements of electric current were performed irradiating the sample with a THz laser beam which was focused via the Winston cone. It allowed to study the THz response of the CNT quantum well in a wide range of temperature, a.c. field intensity and frequency.

The measurements were performed using two devices, A and B. For each device refrigerated below 10 K, the reported data indicate Coulomb blockade oscillations. The obtained data are represented by a color map in Fig. 10.5 where we show the differential conductance versus the gate voltage and drain bias. For each device refrigerated below 10 K, the reported data indicate the Coulomb blockade oscillations. The obtained data on the differential conductance versus the gate voltage and drain bias are represented by a color map in Fig. 10.5. One can see typical Coulomb blockade diamonds, which are considerably smoothed than those usually obtained for CNT QDs using the conventional back-gate configuration when silicon dioxide serves as a gate dielectric. Nonetheless, they are quite suitable for extracting of the QD parameters. Measuring the diamond's dimensions, one determines (Rinzan et al. 2012) the values of charging energy δ and the single particle–level spacing ΔE characterizing both devices (see Fig. 10.5). The contribution of tunneling of the charge carriers through excited states of the CNT QD is marked by parallel lines drawn through the diamond's edges.

The THz beam was created by two different sources. One source was a backward wave oscillator (BWO) whose frequency was continuously tunable between 680 and 1080 GHz. The other source was a CO_2 pumped gas laser with a fixed frequency, 1.27 THz. The QD devices A and B were irradiated by the THz beams created by the two mentioned sources. Unfortunately, in the experiment with device A, the width of Coulomb blockade peak was too wide to resolve the satellite PASET peaks because it was comparable to the photon energy hv. Since the spacing between the satellite peak and the main SET peak was about the same as the width of central PASET peak, all three peaks (i.e., one central peak and the two satellite peaks) are merged into a single wider peak. Besides, at frequencies above 800 GHz, there was a significant peak broadening, accompanied by a strong enhancement of the main peak. In order to achieve a desirable peak separation $\pm n\hbar\omega$ which can be observed in our experiments we needed other devices with narrower Coulomb peaks. The narrower Coulomb peaks were obtained for device B, which allowed the PASET satellite peaks to be firmly resolved (see Figs. 10.6 and 10.7). The obtained experimental data for the satellite peak positions versus the THz field intensity and frequency are fairly consistent with the concept of photon assisted tunneling. In particular, when the THz field is stronger, the side peak height increases whereas the spacing between the side peaks and the main peaks increases proportionally to $\hbar\omega$.

Another evident tendency is that *both* the height of the side peak and the height of the main peak in device B increase, when the source intensity increases. Such behavior contradicts to general expectations and, therefore, is counterintuitive. A physical contradiction here is that the external THz field supplies energy into the CNT QD, which is then converted into the thermal energy of electrons. One should expect that the dot must heat up. The relevant increase of electron temperature would inevitably lead to diminishing of the PASET peak height and the broadening of it, as it also follows from the orthodox photon assisted tunneling theory (Tien and Gordon 1963). The last scenario with the temperature dependence of the steady-state Coulomb blockade peaks is illustrated in Figs. 10.6 and 10.7. From the plot one can see that the width of the Coulomb blockade peaks increases whereas

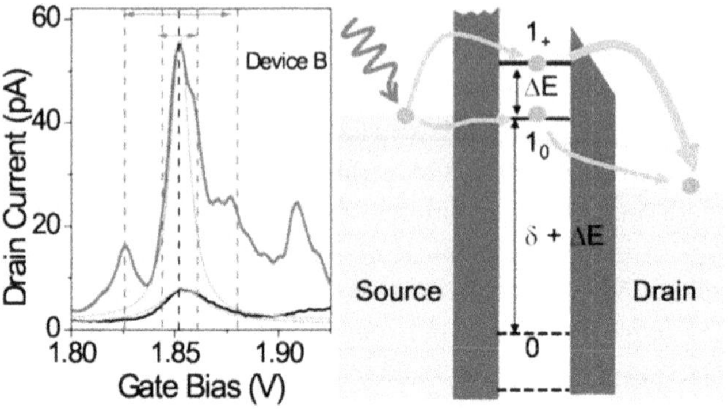

Figure 10.6 Left: Experimentally obtained DC response of carbon nanotube quantum well (device B) to a THz laser beam with $h\nu = 5.24$ meV (red curve) (Rinzan et al. 2012). The dark current of the terahertz sensor is shown as the black curve. The Lorentzian fits to the terahertz and dark Coulomb blockade peaks with widths of 11 and 28 mV are shown by the green curves. The red vertical lines indicate the satellite peak locations with spacing $2h\nu/\alpha_G$ due to PASET. They are indicated by the red arrow and the blue vertical lines. The satellite side peaks correspond to a considerable transport through excited states with spacing $2\Delta E/\alpha_G$ ($\Delta E = 1.9$ meV), sign-posted by the blue arrow. Right: A sketch of the electron tunneling processes through the carbon nanotube quantum dot. For frequencies ν lying within the terahertz domain but lower than the level spacing, $h\nu < \Delta E$, electrons tunnel from the source to the drain through level 1_0. If frequency ν exceeds the level spacing, $h\nu > \Delta E$, photon-assisted tunneling occurs through the excited level 1_+. If the tunneling rate from the dot to the drain electrode Γ_D increases when increasing the electron energy, then $\Gamma_D(E_1^+) > \Gamma_D(E_1^0)$ and the most energetic electrons leave the dot at a faster rate, causing the effective temperature of the electrons in the dot to decrease. Adapted from Rinzan et al. (2012).

their height is decreased, when the temperature is increased, whereas they become narrower and their height increases as the THz radiation is on.

The counterintuitive behavior which was observed in the experiment (Rinzan et al. 2012) can be understood considering nonequilibrium effects which complement the orthodox PASET model (Averin et al. 1991, Kouwenhoven et al. 1994a, 1994b). The nonequilibrium effects taking place in the CNT QD exposed to the external THz field are conveniently described by the quantum

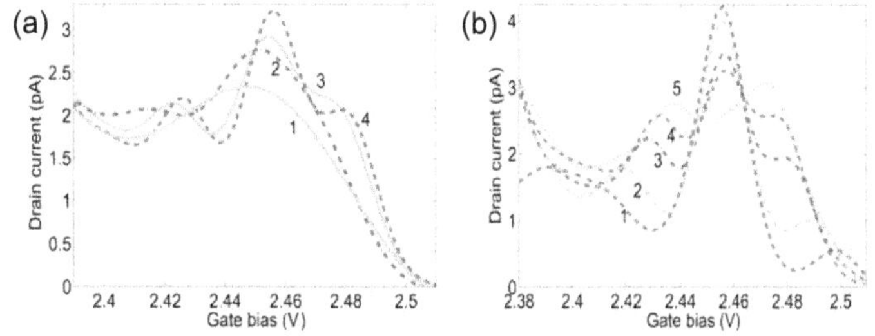

Figure 10.7 Theoretical simulations (Rinzan, 2012). Numerically computed frequency and amplitude dependence of electric current versus gate voltage. (a) Fixed frequency, $hv = 5.25$ meV, and different values of the AC field amplitude $\tilde{V}_S = 0$, 3.4, 4.4, and 5.5 mV for curves 1, 2, 3, and 4, respectively, where $\tilde{V}_S$ ($\tilde{V}_D$) is the source–dot (dot–drain) amplitude of the AC voltage drop and $\tilde{V}_D = 0.55\ \tilde{V}_S$. Curves 2–4 are related to effective electron temperature $T^*_{2\text{-}4} = 6.9$, 5.25, and 4.2 K, respectively. Curve 1 shows the "dark current" and corresponds to $T^*_1 = 9.3$ K. (b) Different frequencies, $hv = 7.31$, 6.33, 4.67, 3.31, and 3.02 meV for curves 1–5, respectively, at $\tilde{V}_S = 5.2$ mV. Curves 1–5 are related to effective electron temperatures $T^*_{1-5} = 4.5$, 4.7, 4.95, 5.15, and 6.6 K, respectively. The dot/drain tunneling rate parameter $\Gamma_D(0) = 0.6$ meV. Both graphs illustrate terahertz cooling of the dot (Rinzan, 2012).

kinetic equation. It allows us to obtain the nonequilibrium electron distribution function $g(\varepsilon_k)$, which determines the electron transport in the system under study (Keldysh 1965). The tunneling barriers are modeled by the same expressions as those suggested in the standard QD theory (Averin et al. 1991, Kouwenhoven et al. 1994a, 1994b). In particular, we use the tunneling rates $\Gamma_{\mathbf{S},\mathbf{D}}$ through source–dot and dot–drain Schottky barriers as energy dependent and model them with $\Gamma_{\mathbf{D}}(\varepsilon_k)$ as $\Gamma_D(\varepsilon_k) = \Gamma_D^{(0)}\sqrt{\varepsilon_k/\delta}$, while $\Gamma_{\mathbf{S}} \equiv$ **const**, with $\Gamma_{\mathbf{S},\mathbf{D}} \approx \delta$.

The theoretical results for the drain current versus gate voltage are shown in 10.3.1a,b. The obtained curves are presented for different field intensities and frequencies. One can see that the gate voltage dependences of drain current show distinct satellite peaks, with spacing from the main peak proportional to the photon energy $\hbar\omega$. The obtained theoretical curves confirm the experimental data. When either the field intensity or the frequency is increased, the main SET peak of the computed drain current shows an anomalous increase which is consistent with the experiment. In the calculations we assume that the lattice temperature on the dot is the same as in the source and drain electrodes, $T = 4.2$ K. The mentioned assumption is used as a BC when solving the kinetic equation for the electron distribution function. It allows us to compute the effective electron temperature (T^*) in the dot. Technically, the effective temperature T^* in the dot is obtained from the balance of the energy gain at the dot which is supplied by the THz field on the one hand and the energy taken out by escaping of the electrons from the dot. Schematic of the nontrivial cooling is presented in Fig. 10.6. Physically, the cooling originates due to the nonequilibrium extraction of excessive excitations from the dot. It becomes very efficient when the THz frequency exceeds the QL spacing ΔE, which actually determines the threshold of the nonequilibrium cooling mechanism. Below the threshold, when the photon energy *hv* is lower than the energy-level spacing, $h\nu < \Delta E$, the electrons are populating mostly the lower energy level 1_0 inside the dot, because the AC field cannot excite them to the upper 1_+ level. The upper 1_+ level is filled only when the THz photon energy exceeds the level spacing inside the dot, $h\nu \geq \Delta E$. The basic condition which determines the cooling efficiency is that the electron lifetime on the lower level 1_0 must be much longer than on the upper level 1_+. In our CNT QD the condition is fulfilled because the tunneling rate $\Gamma_D(\varepsilon_k)$ increases at $\varepsilon_k \approx E_{1_+}$, which corresponds to a large escape rate $\Gamma_D(E_{1_+}) >> \Gamma_D(E_{1_0})$ from the upper energy level into the metal drain electrode. For this reason, the electrons escape from the level 1_+ to the drain electrode much faster than they did from 1_0 (as indicated by the thicker arrow in Fig. 10.6). Because the "hottest" electrons are leaving the dot with a higher rate, the process leads to decrease of the effective electron temperature T^* inside the dot, when the THz field with $h\nu \geq \Delta E$ is applied. It results in considerable enhancement of the main peak height and to narrowing of the peak, as illustrated by Figs. 10.3a,b and 10.6. The experimental data described above suggest that CNT QDs can serve for the design of extremely sensitive quantum detectors and have a strong potential as building blocks of the THz (submillimeter) spectral analyzers. The nonequilibrium cooling method might be utilized in novel approaches to improve the overall performance and to elevate the temperature of operation, which is very desirable for practical applications.

10.3 Microscopic Model of the Carbon Nanotube Quantum Dot Sensor

10.3.1 *Carbon Nanotube Quantum Dots*

In this Subsection we discuss basic assumptions used to formulate a comprehensive theoretical model of the QD T-ray sensor. The QD is constructed from either graphene stripe or single-wall CNT resting on (or suspended over) a dielectric substrate with metallic electrodes attached (see Fig. 10.1). The section of CNT has a typical length $L_{\mathrm{T}} = 250$–5000 nm and is enclosed between two metallic electrodes S and D deposited on the top of CNT as shown in Fig. 10.1. The length D_T of CNT section is selected to obtain the QL spacing $\Delta = hv_{\mathrm{F}}/2D_{\mathrm{T}} \approx h \cdot 1$ THz, where h is the Plank constant, v_{F} is the Fermi velocity in graphene/CNT. Due to distinction between concentrations of the electric charge carriers in the metallic electrodes on the one hand and in the CNT section on the other hand, the corresponding workfuncions are different from each other, causing appearance of the interface Schottky barriers. The potential barriers are situated immediately at the interfaces between the metallic electrodes and CNTs (see inset in Fig. 10.1). Besides, we assume that our section of graphene (or CNT) contains no ripples, no mechanical defects, and no charged impurity atoms. Then, the electron propagation inside the CNT might be considered as ballistic owing to conservation of the one-half pseudospins during the processes of electron scattering on the lattice defects, phonons, and other electrons.

The incident electron waves are reflected from the ends of the graphene (or CNT) section in the form of backward electron waves. The incident and backward waves interfere with each other, which causes the electron motion to be quantized inside the graphene (or CNT) section of finite length D_{T}. The concentration of charge carriers is controlled by the gate electrode which is placed underneath the dielectric SiO_2 substrate as shown in Figs. 10.1 and 10.2. A finite electric current is induced in the section by a voltage bias applied between the source (S) and drain (D) metallic electrodes, which are attached to the ends of the section. The whole system represents a QD formed inside the graphene (CNT) section spatially confined between the two ends. The gate voltage V_G adjusts the QL positions (red dash in Fig. 10.1), while the electron injection from the metallic S and D electrodes into the graphene (or CNT) section changes the population of the levels.

10.3.2 *Role of AC Chiral Transport*

The wavefunction symmetry of the elementary excitations in graphene and CNTs is dictated by the honeycomb crystal symmetry in the materials. Therefore, in contrast to regular semiconductors or metals, the quasiparticle excitations inside the graphene (or CNT) section are not just conventional electrons or holes. The excitations here actually behave as chiral particles with two one-half spins per each particle. The directions "↑" and "↓" of one one-half pseudospin correspond to two different spectral branches. The branches are related to electron (or hole) excitations with either positive or negative electron velocity $\pm v_F$ per each. Another one-half pseudospin can also be directed either "↑" or "↓" which corresponds to zero-energy K or K′ points depending on the direction. The conservation of the one-half pseudospins during the scattering process and tunneling proclaims another quantum mechanical selection rule which is regarded as the one-half pseudospin conservation law. The one-half pseudospin conservation law is a unique property of graphene and CNTs and determines basic transport properties of those materials. The conservation law makes the transport properties to be distinct from properties of regular conductors where the charge transfer occurs due to propagation of spinless excitations. One significant consequence of the mentioned conservation laws is that the rates of electron–impurity and electron–phonon collisions in graphene and CNTs are much lower than in regular semiconductors and metals. For these reasons, the electron transport in graphene and CNTs remains ballistic up to the room temperature. The pseudospin conservation can also be viewed as an intrinsic coherence which is an immediate outcome of the phase-coherent propagation of chiral particles (Ando 2005, Shafranjuk 2009a, 2009b, 2011a, 2011b, Katsnelson 2012).

10.4 Electromagnetic Field Influence

One can model the influence of external AC field introducing time-dependent additions $V_1 \cos(\omega t)$ and $U_1 \cos(\omega t)$ to the gate V_{G} and the source drain V_{SD} voltages, respectively (Shafranjuk 2011a, 2011b). Polarization of the external field determines the relationship between V_1 and U_1. In the most trivial case (Shafranjuk 2009a, 2009b, 2011a, 2011b), the vector of external AC electric field is normal in respect to the interface separating the metal source or drain electrode on one side and the *CNT* or graphene on the other side. Then, the source–drain voltage can be represented as $V_{\mathrm{SD}} \to V_{\mathrm{SD}} + V_1 \cos(\omega t)$. Such influence of the AC voltage on electron tunneling was elaborated earlier (Tien and Gordon 1963). In this Section, we pay attention to the less trivial case which corresponds to the setup of quantum well where the external AC voltage acts only via the gate electrode, whereas $V_1 = 0$. Such a case is realized, for example, when the metallic electrodes are properly shielded (as shown in Fig. 10.1), and the AC voltage adjusts the electrochemical potential μ of electrons inside the CNT section as $\mu \to \mu + U_1 \cos(\omega t)$ (see Fig. 10.1). In other words, one can say that the AC field generates an additional time dependence of the gate voltage V_G (Shafranjuk 2009a, 2009b, 2011a, 2011b). Then, the electron excitation spectrum inside the QD is theoretically computed from solutions of the time-dependent Dirac equation complemented by a kinetic equation for the electron distribution function.

The theoretical model of the photon-assisted chiral tunneling through the graphene and CNT QDs (Shafranjuk 2009a, 2009b, 2011a, 2011b) is implemented for a thorough microscopic description of existing experimental data (Kawano 2009, 2011, 2012, Rinzan et al. 2012). The interaction of external AC field with a QD is essentially a nonlinear problem. Such a nonstationary system is generally solved by the Floquet method which is complemented by the kinetic equation and by the S-matrix technique—see, for example, Shafranjuk (2011a, 2011b) and references there in. Since the S-matrix technique was originally developed to solve the stationary steady-state problems, here we generalize it for the nonstationary case. The combined approach is capable of describing the nonlinear and nonequilibrium properties of junctions with arbitrary transparency which are subjected to influence of AC fields with arbitrary amplitude and frequency.

10.5 Key Assumptions

Considering the photon-assisted tunneling through the interface barriers whose transparency T is not very low $0.01 < T < 1$, for the sake of simplicity, one can neglect the contribution of the SET. It is the case when the Coulomb charging energy $U_Q = e^2/C$ (C is the capacitance of the Schottky contact) is lower than the thermal energy $k_B T$ and the typical electron energy ε. Then, the processes of electron tunneling between the counter electrodes S and D and the dot are correlated in phase with each other because the phase of electron is not destroyed during the tunneling. We will see that the chiral transport is strongly altered versus the source–drain and gate voltages and depends considerably on the field frequency and amplitude. By adjusting the voltages, one affects basic properties of AC transport through the graphene and CNT junctions. The source–drain and gate voltages not only modify the electronic structure and the phase-correlated transport of chiral charge carriers. They also cause injection of nonequilibrium excitations into the quantum well. The relevant changes are observed in current–voltage characteristics of the graphene and CNT junctions. In principle, one can expect that the voltage control can additionally be implemented for optimizing the general performance of THz field detectors. Considerable benefits of devices based on graphene and CNTs stem from a unique nature of those materials. The chiral tunneling indicates remarkable resonant features whereas both the inelastic and elastic scatterings are weak. It envisages achieving of the electric current density in the graphene and CNT junctions much higher than in ordinary semiconducting or even in metal devices. Similar theoretical approaches are appropriate to describe both graphene and CNT QDs. A specific topology and geometry of the graphene stripes and CNT sections are taken into account by imposing the appropriate BCs. The BCs become time dependent as soon as one applies the external AC field.

We take into account the THz field which acts on the graphene or CNT sections including an additional time-dependent term $\mu \to \mu + U_1 e^{i\omega t}$ into the Hamiltonian. The role of AC term with a finite $U_1 \neq 0$ is to adjust the electrochemical potential μ. The outcome of such an influence is learned by comparing the nonstationary effect with the steady-state characteristics. Therefore, we start from a brief discussion of the steady-state differential conductance and then compare it with our data for the case when the THz field is on. Following this path, we initially set the AC field amplitude $U_1 = 0$ and assume that the electrochemical potential μ is adjusted only by a finite DC gate voltage $V_{\mathrm{G}} \neq 0$ when the AC field is off. The obtained steady-state parameters will serve as an input on the next stage for computing of the differential conductance when the AC field is turned on. Then, we can compare the stationary and nonstationary behavior of the CNT junction to better understand the THz field influence. There are many traditional methods for studying of the influence of AC field on the transport of multisectional tunneling junctions. The applicability of any certain method is restricted. As an example one can regard a simple and spectacular model by Tien and Gordon (Tien and Gordon 1963) which describes the influence of an external AC field on the electric current through the tunneling junction. Originally, the model (Tien and Gordon 1963) was used to derive a concise expression for an electric current in conditions of photon-assisted tunneling. Unfortunately, for the multisectional graphene and CNT junctions, many assumptions of the model (Tien and Gordon 1963) cannot be fulfilled automatically. In particular, the simplicity and elegance of the model (Tien and Gordon 1963) are preserved for special cases of low-transparency junctions only. However, due to the presence of Klein tunneling, which proceeds with an ideal probability, the latter assumption of a low junction transparency fails. In principle, that complication can be avoided using another model and assuming that the junction's transparency is arbitrary, whereas the AC field intensity is low. Nonetheless, such an assumption is not well justified in many experimental situations. Indeed, a serious setback of such a straightforward generalization of the Tien and Gordon model is rather cumbersome: the cylindrical Bessel function $J_n(eU_1/h\nu)$ is not orthogonal over its index n (which physically represents itself as the number of photons which are absorbed or emitted during a tunneling process). The aforesaid motivates to use a more universal approach which is capable to describe multisectional junctions with arbitrary transparency and are exposed to an external AC field of arbitrary amplitude. Simultaneously, the approach must be simple enough to handle it mathematically.

The impact of AC field of an arbitrary amplitude on properties of the multisectional graphene and CNT junctions with arbitrary interface transparency is accounted for by the Floquet method (Aravind and Hirschfelder 1984, Son et al. 2009) combined with the time-dependent S-matrix technique and Dirac equation. The Floquet method exploits periodicity of the external field versus time which allows reducing of the nonstationary problem to a stationary task. The Floquet method represents an analogy with the Bloch theorem for the case of one-dimensional solid-state crystals. Similarly to a solid-state crystal, the nonstationary solutions of the time-dependent problem yield an energy band structure where the lattice wavevector is replaced by the field frequency ω. The Floquet technique is handy for considering the influence of an external THz field with arbitrary amplitude and frequency. We implement the Floquet method on the top of the S-matrix technique which will be appropriately generalized to a nonequilibrium case. The combined strategy conveniently describes the electron transport in presence of arbitrary interface barriers and defects, for example, atomic impurities or imperfections of crystal lattice. Here, we implement this method to study the electric transport in the graphene and/or CNT junction where two Schottky barriers of arbitrary transparency are formed at the interfaces between the source and drain electrodes and the CNT/graphene section as shown in inset in Fig. 10.1.

When a graphene or CNT junction is exposed to THz signals, one frequently observes the bolometric effect (Maksimenko et al. 2006, Slepyan et al. 2006, Fu et al. 2008, Yngvesson et al. 2008, Seliuta et al. 2010). However, at certain conditions, the heating of the electrons by THz field can be reduced. Here we implement special BCs assuming that the electrons in the source (S) and drain (D) electrodes remain in the stationary and equilibrium state. We assume that the THz field acts on the graphene (or CNT) section only. Then, the AC field influences the transport of the chiral particles inside that section. If inside the graphene (or CNT) section the dwell time $\tau_{\mathbf{D}}$ of a chiral particle is sufficiently short, $\tau_{\mathbf{D}} = D/v << \min[\tau_{\mathbf{ep}}, \tau_{\mathbf{T}}, 2\pi/\omega]$ (where $\tau_{\mathbf{T}}$ is the tunneling time between source, graphene or CNT, and drain electrodes, $\tau_{\mathbf{ep}}$ is the electron–phonon inelastic scattering time), the electron propagation is ballistic. The transmission of electrons across the source/CNT and CNT/drain interfaces are phase-correlated with each other (Datta and Mclennan 1990, Datta 1992, Datta et al. 2008, Datta et al. 2011). Besides, we assume that $\tau_{\mathbf{ep}} >> \tau_{\mathbf{T}}$, which means that electrons arrive and escape the graphene or CNT section faster than the electron–phonon scatterings occur. The last condition represents a "coherent transmission" limit when the electron distribution function n_{ε} inside CNT/graphene is almost equilibrium. It means that the Floquet state $|\kappa_m>$ of an electron inside CNT/graphene is preserved during the dwell time $\tau_{\mathbf{D}}$. Another meaning of the above assumptions is that the "time inhomogeneous" corrections become small as soon as $(\tau_{\mathbf{D}}/\tau_{\mathbf{s}})^2 << 1$ (where $\tau_{\mathbf{s}}^{-1} = \tau_{\mathbf{ep}}^{-1} + \tau_{\mathbf{T}}^{-1}$ and we take into account the typical electron–phonon relaxation time $\tau_{\mathbf{ep}}$ for "clean" semiconducting graphene and/or CNT is $\tau_{\mathbf{ep}} \geq 10^{-12}$s). The aforementioned assumptions allow us to greatly simplify the mathematical calculation of the time-averaged electric current. The above formulated conditions are accomplished, for example, for a monochromatic THz field with $\omega = 0.01\Delta$. For the sake of convenience, here and below we express the field frequency ω and other energy quantities in units of the van Hove singularity spacing $\Delta = 2\hbar v/d_{\mathbf{T}}$; $d_{\mathbf{T}}$ being the nanotube diameter. It is easy to evaluate the tunneling time through the Schottky barrier at the interfaces. For the barrier height $U_0 = 0.5\Delta$ and the nanotube diameter $d_{\mathbf{T}} = 2.5$ nm one arrives at the tunneling time $\tau_{\mathbf{T}} \geq 10^{-12}$ s. For the graphene (or CNT) section with the length $D_{\mathbf{T}} = 750$ nm one also obtains $\tau_{\mathbf{D}} = D_{\mathbf{T}}/v \leq 10^{-12}$s. While estimating the time-averaged electric current, the fulfillment of the above conditions also allows the nonequilibrium effects to be neglected. When all the above assumptions are maintained, we arrive at the conventional Landauer–Buttiker formula (Datta and Mclennan 1990, Datta 1992,

Datta et al. 2008, Datta et al. 2011)

$$I(V_{\mathrm{SD}}) = \frac{2e}{h} \int d\varepsilon M(\varepsilon) \mathbf{T}_{\varepsilon} \left(n_{\varepsilon} - n_{\varepsilon - eV_{\mathrm{SD}}} \right) \tag{10.2}$$

where $\mathbf{T}_{\varepsilon} = \kappa \sum_m |t_{\mathrm{SD}}(\varepsilon + m\hbar\omega)|^2$ is the AC field-modified transparency of the whole CNT junction, ε is the energy variable, $n_{\varepsilon} = 1/(\exp(\varepsilon/T) + 1)$ is the Fermi function, $M(\varepsilon)$ is the energy-dependent number of modes, T is the temperature, and m is an integer number of absorbed/emitted photons. The junction transparency contains the geometrical factor κ which alters in the range $\kappa = 0.3 - 1$ versus the crystallographic direction and stripe width of graphene (or nanotube rollup vector and diameter of nanotube). In the next subsections we compute the transmission coefficient $t_{\mathrm{SD}}(\varepsilon)$ in the coherent approximation.

10.6 Electron Quantization in the Steady State

As a starting point for studying of the AC transport through the graphene (or CNT) QD we use a solution for the steady state. It gives us an opportunity to compare the electric transport properties of the nanotube junction placed in the AC field against its steady-state characteristics. The transmission coefficient $t_{SD}^{(0)}(\varepsilon)$ through the CNT junction is computed in the steady state from solutions of the stationary Dirac equation for the fermion wavefunction Ψ

$$-i\hbar v \left(\left(\hat{\sigma}_x \otimes \hat{1} \right) \partial_x + \left(\hat{\sigma}_y \otimes \hat{\tau}_z \right) \partial_y \right) \Psi + eU(x, t) \Psi = \varepsilon \Psi \tag{10.3}$$

In the above Eq. 10.3 $\hat{\sigma}_i$ and $\hat{\tau}_k$ are the Pauli matrices, ε is the excitation energy of chiral fermions, $v = 8.1 \times 10^5$ m/s$\approx c/300$ is the massless fermion speed, and $\{i, k\} = 1 \ldots 3$. When an external AC field is applied across the QD, it modifies the electric potential inside the graphene (or CNT) section $U(x, t) = \mu_0 + \alpha V_{\mathrm{G}}(x, t)$ induced by the AC addition to the equilibrium gate voltage V_{G}. Thereby, we assume that the electrochemical potential inside the CNT section μ is adjusted only with the finite DC gate voltage $V_{\mathbf{G}} \neq 0$. For the gate voltage, we take a simple analytical form $V_{\mathrm{G}}^{(1)}(x, t) = U_1 \theta(x) \theta(D_{\mathrm{T}} - x) \cos \omega t$, where θ is the Heaviside step function, x is the longitudinal coordinate, and D_{T} is the CNT section length. In this subsection we consider the stationary case only and set the THz field amplitude $U_1 = 0$. The AC field influence will be accounted for in the next subsection. If $V_{\mathbf{SD}} < U_0$, the electric current in the steady state is fully suppressed. For typical values of the gate voltage $V_{\mathbf{G}} = 1$ V and the SiO_2 thickness $d = 300$ nm one finds (Katsnelson et al. 2006, Katsnelson 2012) $U_0 = 2$ meV. For an infinite single-wall CNT which is characterized by the chirality index ν the electron excitation energy ε is obtained straight as a steady-state solution of the Dirac equation as $\varepsilon = \hbar v \left(k_{\mathrm{F}m} \pm \sqrt{k^2 + q_\nu^2(n)}\right) + eU_0$. Here, we have introduced the Fermi wavevector in the S and D electrodes $k_{\mathrm{F}m}$, which for the sake of simplicity we take the same, n is the index of the electron subband, $\nu = n + m - 3N$, n, m, and N are integers related to the translation vector $\mathbf{L} = n\mathbf{a} + m\mathbf{b}$ of the graphene lattice, and $\mathbf{a}$ and $\mathbf{b}$ are primitive translation vectors. Index $\nu = 0$ when the $[m, n]$ nanotube is metallic, while $\nu \neq 0$ for semiconducting and dielectric nanotubes. The electron wavevector in the transversal direction is $q_\nu(n) = (2/d_{\mathrm{T}})(n - \nu/3)$, d_{T} is the nanotube diameter. When a finite backgate voltage V_{G} is applied, it causes a stationary shift of the electrochemical potential as $U_0 = \alpha V_{\mathrm{G}}$ where α is the gate efficiency. It is also instructive to express the electron wavevector in terms of excitation energy as $\tilde{k} = \sqrt{((\varepsilon - eU_0)/\hbar v)^2 - q_\nu(n)^2}$. The simplest case corresponds to continuous BCs with setting $V_{\mathbf{SD}} = 0$ which yields

$$t_\varepsilon = 2e^{-i\phi} k\tilde{k} s\tilde{s}/\mathbf{D} \tag{10.4}$$

where $s = 1$, $\tilde{s} = \mathbf{sign}(\varepsilon - eU_0)$, D_{T} is the tube length, $\tilde{\kappa} = \sqrt{\tilde{k}^2 + q^2}$, $\Theta = \kappa\tilde{\kappa} - 2s\tilde{s}q^2$, $\kappa = \sqrt{k^2 + q^2}$, and $\mathbf{D} = 2k\tilde{k}s\tilde{s}\cos(\tilde{k}D_{\mathrm{T}}) - i\Theta\sin(\tilde{k}D_{\mathrm{T}})$. The above expression, Eq. 10.4, describes the chiral junction with ideal interface transparency of the Schottky interface barriers positioned at the S/CNT and CNT/D interfaces. Finite Schottky barriers emerge due to the difference of work functions inside the graphene and/or CNT on the one hand and for the external S and D electrodes on the other hand. When a finite source-drain bias voltage is applied across the junction ($V_{\mathbf{SD}} \neq 0$), the electric voltage drop builds up at the interface barriers. Then, the electric field vanishes inside the CNT section and the DC differential conductance is readily determined using the trial wavefunction Ψ as a combination of the plane waves.

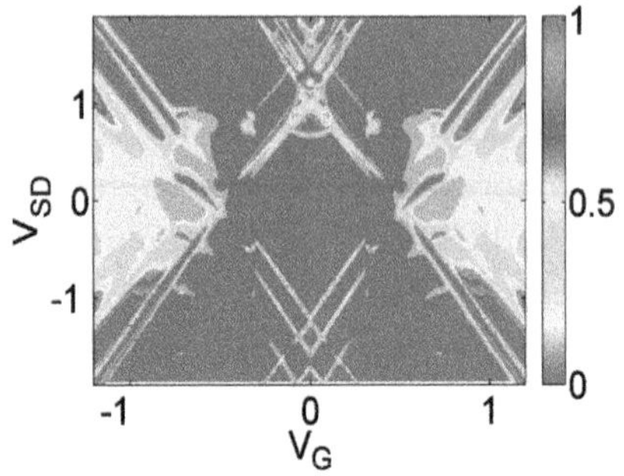

Figure 10.8 The steady-state DC differential conductance σ of the carbon nanotube junction shown as a contour plot. The scale bar is in units $G_0 = 4e^2/h$. The voltage dependence of the quantized state levels is pronounced as a pattern of parallel lines. The T section length is $D_T = 850$ nm, the barrier length is $D_B = 5$ nm, and height $U_B = 0.5\Delta$.

Let us consider the geometry with the CNT section enclosed between two Schottky barriers. The transmission amplitude via such a chiral barrier is computed using the S-matrix composition rules. One needs to know the barrier parameters and the parameters of source and drain electrodes as well. The S and D electrodes in our QD are made of ordinary metals. The pseudospins of the chiral fermions are not conserved while they tunnel through the Schottky barriers. In case of the nonchiral tunneling one implements a conventional expression for the transmission coefficient

$$t_{S(D)} = \frac{4k\tilde{k}e^{iD_B\tilde{k}}}{(k+\tilde{k})^2 - e^{2iD_B\tilde{k}}(\tilde{k}-k)^2} \tag{10.5}$$

where $\tilde{k}$ is the electron wavevector inside the Schottky barrier of length $D_\mathbf{B}$. One can see that the transmission coefficient $t_T(\varepsilon)$ acquires a different form owing to chiral features of the electron transport through the CNT section. As a first step in the block matrix equation, one computes the partial matrices. After that, one builds larger blocks by utilizing the S-matrix composition rules to obtain the steady-state transmission coefficient $t_{SD}(\varepsilon)$ of the whole sensor setup. In Fig. 10.8 we show the steady-state differential conductance of the CNT QD with the CNT section length $D_\mathbf{T} = 850$ nm, the interface I_L and I_R barrier lengths are $L_\mathbf{B} = 5$ nm for both of them, and their height is $U_\mathbf{B} = 0.5\Delta$. The van Hove singularity spacing here is $\Delta = 2\hbar v/d_\mathbf{T}$, where $d_\mathbf{T}$ is the nanotube diameter. In particular, for a nanotube with $d_T = 1$ nm, for example, one gets $\Delta = 1$ eV. The contour plot $\sigma_{SD}^{(0)}$ versus V_{SD} and versus V_G in Fig. 10.8 clearly shows sets of parallel stripes. The stripes correspond to quantized energy levels.

The QL resonances are pronounced as sharp singularities in the electric conductance. The energy positions of QLs depend on the bias source–drain and gate voltages. The bias voltages are expressed in units of Δ/e. For a nanotube with $d_T = 5$ nm, for example, one gets the energy-level spacing $\Delta = 50$ THz.

10.7 THz Field Influence on Quantum Dots

Thus, an additional time-dependent gate voltage $U_1 e^{i\omega t}$ appears when the external AC field acts on the dot. The field shifts the electrochemical potential inside the CNT section as $\mu \to \mu + U_1 e^{i\omega t}$. Then, the electron envelope wavefunction in CNT is obtained as a solution of the quasiclassical Dirac equation (Ando 2005, Katsnelson et al. 2006, Morozov et al. 2006, Novoselov et al. 2006, Shafranjuk 2009a, 2009b) completed by corresponding time-dependent BCs. We present the profile of electric potential $U(x)$ inside the quantum well by a piece-wise function composed of several coordinate-independent pieces. The electron envelope wavefunction of the entire region is constructed by matching the BCs for the electron wavefunctions from different regions as the source electrode (S), the CNT section, the drain electrode (D), and the Schottky barriers separating those regions. Complementing the Dirac equation with corresponding nonstationary BCs, one can model the influence of the external AC field which alters the electric potential profile $U(x)$ in the QD region. The nonstationary conditions allow us to compute the electron partial transmission amplitudes between each two adjacent regions (i.e., between S and CNT and between CNT and D) which are separated by Schottky barriers.

Utilizing the nonstationary interface BCs for the four-component envelope spinor wavefunction one computes the partial transmission and reflection amplitudes through the chiral barrier (Ando 2005, Katsnelson et al. 2006, Morozov et al. 2006, Novoselov et al. 2006, Shafranjuk 2009a, 2009b). The BCs at the S/CNT and CNT/D interfaces have the following form:

$$\begin{aligned}\Psi_{\mathbf{S}}(x=0,\ t) &= \Psi_{\mathbf{T}}(x=0,\ t)\\ \Psi_{\mathbf{T}}(x=D_{\mathbf{T}},\ t) &= \Psi_{\mathbf{D}}(x=D_{\mathbf{T}},\ t)\end{aligned} \tag{10.6}$$

where Ψ_{S}, Ψ_{T}, and Ψ_{D} are the time-dependent electron wavefunctions in the S, CNT, and D sections of the junction, respectively. One can factorize the piece-wise trial wavefunctions Ψ_{S}, Ψ_{T}, and Ψ_{D}, introducing the time-dependent coefficients $r_{\mathbf{T}}(t)$, $t_{\mathbf{T}}(t)$, $\alpha_{\mathbf{T}}(t)$, and $\beta_{\mathbf{T}}(t)$. This gives

$$\begin{aligned}\Psi_S &= e^{iV_Lt}\left(\chi_k e^{ikx} + r_T(t)\ \chi_{-k}e^{-ikx}\right)\\ \Psi_T &= \alpha_T(t)\sum_m e^{im\omega t}\chi_{\kappa_m}e^{i\kappa_m x} + \beta_T(t)\sum_m e^{im\omega t}\chi_{-\kappa_m}e^{-i\kappa_m x}\\ \Psi_D &= e^{iV_Rt}t_T(t)\ \chi_k e^{ik(x-D_T)}\end{aligned} \tag{10.7}$$

where κ_m is the wavevector of the so-called Floquet state and m is the index of Fourier series. The above formulas contain the time dependence which enters via the longitudinal component of the electron wavevector $\tilde{k}(t)$ inside the CNT section $\tilde{k}(t) = \sqrt{((\varepsilon - eU(t))/\hbar v)^2 - q_\nu(n)^2}$. The nonstationary BCs, Eq. 10.7, constitute a system of linear recurrence equations for the reflection $r_{\mathbf{T}}(\varepsilon)$ and transmission $t_{\mathbf{T}}(\varepsilon)$ coefficients, which read

$$\begin{aligned}\chi_k + r_T(\varepsilon_k)\ \chi_{-k} &= \sum_m \chi_{\kappa_m}\alpha_T\left(W_{\kappa_m} + \hbar m\omega - eV_{\mathrm{S}}\right)\\ &\quad + \sum_m \chi_{-\kappa_m}\beta_T\left(W_{-\kappa_m} + \hbar m\omega - eV_{\mathrm{S}}\right)\\ t_T(\varepsilon_k)\ \chi_k &= \sum_m \chi_{\kappa_m}e^{i\kappa_m D_T}\cdot\alpha_T\left(W_{\kappa_m} + \hbar m\omega - eV_{\mathrm{D}}\right)\\ &\quad + \sum_m \chi_{-\kappa_m}e^{-i\kappa_m D_T}\cdot\beta_T\left(W_{-\kappa_m} + \hbar m\omega - eV_{\mathrm{D}}\right)\end{aligned} \tag{10.8}$$

where $V_{\mathrm{S(D)}}$ are electric potentials in the S and D electrodes and W_κ is the so-called Floquet pseudoenergy. The linear system of equations, Eq. 10.8, is solved self-consistently. One can see that the recurrence in the above equations originates from the shifts $\varepsilon \to \varepsilon \pm \hbar m\omega$ of energy arguments. The above equations, Eq. 10.8, define the energy-dependent reflection and transmission coefficients. Besides, they depend on the Floquet wavevector κ_m and also on the Fourier index m. The mentioned two variables determine the nonstationary behavior of the system under the influence of an external AC field of arbitrary amplitude and frequency.

The simplest consideration corresponds to a rotating wave approximation when setting the Fourier index $m = 0, \pm 1$ (Aravind and Hirschfelder 1984, Son et al. 2009). Then, one gets the following simple expression for the chirred barrier transparency

$$t_{\mathbf{T}}^{\mathbf{RWA}} = 2k\kappa_m(\varepsilon) \cdot s\sigma_m \cdot Z_m^{-1} \tag{10.9}$$

where $\sigma_m = \pm 1$ for the conductance (valence) band and $p = \sqrt{k^2 + q^2}$ and $\pi_m = \sqrt{\kappa_m^2 + q^2}$ are absolute values of the 2D electron momentum in the source and drain electrodes and the CNT, respectively. $Z_m = k\kappa_m s\sigma_m \cos(\kappa_m D_{\mathrm{T}}) - i\left(p\pi_m - s\sigma_m q^2\right) \sin(\kappa_m D_{\mathrm{T}})$ is the denominator in Eq. 10.9, $\kappa_m = (1/2\pi) \int_{-\pi}^{\pi} \tilde{k}(t)\, e^{-imt} dt$ is the Floquet pseudostate wavevector, $\tilde{k}(t)$ is the time-dependent electron wavevector inside CNTs. One also obtains the composition rules for the time-dependent S-matrices straight from the nonstationary BCs for the four-component wavefunction. Such composition rules represent a time-dependent analog of the corresponding steady-state S-matrix composition rules (Datta and Mclennan 1990, Datta et al. 2008, Datta et al. 2011). Using the composition rules for partial transmission (and reflection) coefficients $t_{\mathrm{S(D)}}$ and $t_{\mathbf{T}}^{\mathbf{RWA}}$ one computes the transmission coefficient t_{SD} of the whole junction. The quantized energy-level positions are determined by setting the denominator Z_m^{tot} of expression for the whole transmission coefficient. It serves as a resonant transport condition for the whole QD system.

10.8 Characteristics of Electric Transport

The derived equations are used for studying of the AC electric transport through the graphene and CNT QDs exposed to the external electromagnetic field. The effect of interface barriers located between the adjacent electrodes (e.g., at the metal/graphene or metal/CNT interfaces) is accounted for using the nonstationary S-matrix technique (Datta and Mclennan 1990, Datta et al. 2008, Datta et al. 2011). An example of the influence of the external THz field on the differential conductance of a CNT junction is shown in

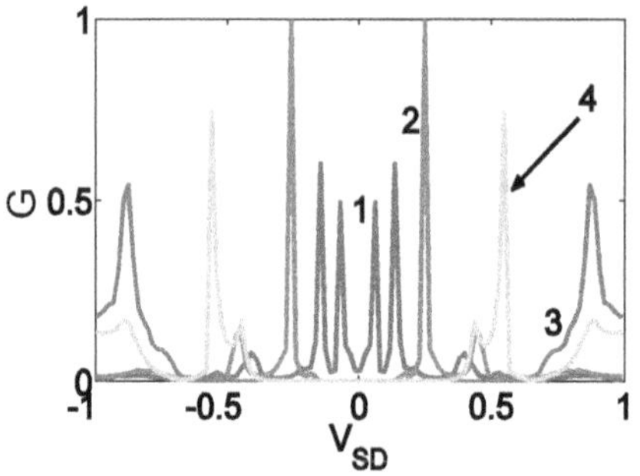

Figure 10.9 Differential conductance G (in units of $G_0 = 4e^2/h$) of a carbon nanotube quantum dot with the same parameters as in Fig. 10.2. The gate voltage for curves 1–4 is $V_G = -0.6, -0.03, 0.56$, and 1.16, respectively. The sequence of curves in the lower panel is numbered corresponding to the increase of the gate voltage.

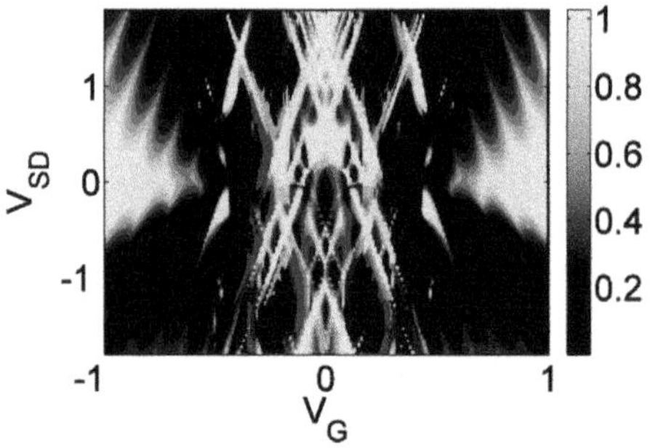

Figure 10.10 Differential DC conductance G of the carbon nanotube junction exposed to a THz field with frequency $\omega = 0.3$ and amplitude $U_1 = 0.3$ as a contour plot versus the gate and source drain voltages. The color bar is shown on the right. The diameter of the CNT $d = 2$ nm, the length of the CNT section $D = 1500$ nm, and the interface barrier width and height $d_B = 2.5$ nm and $U_0 = 5$ mV.

Fig. 10.10. The calculations were performed under the assumption that the nonequilibrium electron distribution is described in terms of an effective temperature T^*. The electron energy and the electric potential are expressed here in units of $10^{-2}\Delta$, where $\Delta = 2\hbar v/d_T$ is the Van Hove singularity spacing. Such energy scale is universal: for the nanotubes, for example, with diameter $d_T = 1$ nm one gets $10^{-2}\Delta = 10$ meV, while the tube length $L = 5$ corresponds to

0.5 μm. The contour plot in Fig. 10.10 was obtained for a setup where the CNT section is subjected to an external AC field. The plot in Fig. 10.10 shows the dependence of differential conductance $G(V_{SD}, V_G) = dI/dV_{SD}$ versus the source drain V_{SD} and gate V_G voltages. The data presented in the plot correspond to the AC field amplitude $U_1 = 0.3$ and frequency $\omega = 0.3$ (e.g., for the nanotube diameter $d_T = 2.5$ nm it corresponds to 30 THz and $U_1 = 40$ meV). We use the same junction's geometry, as shown in the previous Fig. 10.8.

As one can see from Fig. 10.10, the pattern of differential DC conductance $G(V_{SD}, V_G)$ considerably changes versus the external THz field parameters. It becomes evident when comparing the pattern in Fig. 10.10 where $U_1 \neq 0$ with the steady-state pattern shown in the former Fig. 10.8. There are several specific features of the AC field effect which are proclaimed in Fig. 10.9 where we display $G(V_{SD})$ for several different values of V_G. One can notice that spectacular crisp peaks in the plot are related to the quantized state levels. Considerable variations of $G(V_{SD})$ at different V_G, as it follows from the theoretical model, suggest that the QLs, which actually determine the resonant transport, strongly depend on the THz field amplitude U_1. The mentioned tendency is also well pronounced in the corresponding contour plot Fig. 10.10. Therefore, we study the behavior of differential electric conductance $G(V_{SD}, V_G)$ versus the external THz field frequency ω and the field amplitude U_1.

As it follows from the theoretical model, the characteristics of the electric transport through the graphene and/or CNT QDs subjected to the external THz field are sensitive to details of their electron spectrum. The more thorough study shows that the DC electric conductance $G(V_{SD}, V_G)$ of the semiconducting (with $\nu = 1$) nanotube junction depends also on the barrier height U_0. Altering the THz amplitude and frequency one affects the DC differential conductance $G(V_{SD}, V_G)$ considerably. As is evident from Fig. 10.10, a fine structure in the transport characteristics which is formed at the expense of the upper electron subbands with $n = 1, 2, 3$ is remarkably pronounced in all the $G(V_{SD}, V_G)$ curves.

The electron AC transport through the QD in the THz frequency region is essentially determined by the chiral nature of the elementary excitations. The intrinsic coherence emerges from the electron one-half pseudospin conservation in graphene and CNTs. Such conservation law restricts probabilities of certain scattering processes in graphene and CNT and is observed in the vicinity of Dirac points K and K′. The intrinsic coherence allows us to improve the overall performance of the graphene and CNT THz sensors.

10.9 Responsivity and Quantum Efficiency of the THz Detector

One defines the responsivity of a detector as a ratio of generated photovoltage to the incident (or absorbed) power of the THz beam. Such a definition assumes that the detector's response is linear and that the influence of noises is neglected. The highest responsivity is typically achieved in a region with photon energies slightly above the level energy. The responsively declines sharply in the intermediate interlevel region, where the absorption decreases. Mathematically, the THz detector *responsivity* is defined as

$$R = \frac{\Delta U}{\mathcal{S} P_s(\nu)} \tag{10.10}$$

where $\nu = \omega/2\pi$ is the AC field frequency, $\mathcal{S} = (c/\nu)^2 \mathcal{G}/4\pi$ is the antenna aperture, ΔU is the change of source–drain voltage due to the AC field influence, $\mathcal{G}$ is the antenna gain, and $P_s = c\varepsilon_0 E^2/2$ is the intensity of AC field. The calculation results for responsivity are shown in Fig. 10.11. In our CNT field-effect setup, the electric field E is associated with a drop of the gate voltage V_G on the dielectric SiO_2 substrate layer. A typical thickness of the substrate SiO_2 layer is $d_{SiO_2} = 300$ nm which, for example, for $V_G = 5$ V corresponds to the electric field $E_{SiO_2} = 1.7 \times 10^7$V/m created inside the substrate. The AC field causes a small addition to the DC gate voltage which is built up inside the gated CNT section, $E \approx 2 \times 10^3$V/m. It corresponds to the time-averaged AC field intensity

$$P_s = \frac{c\varepsilon_0 E^2}{2} = 5.4 \times 10^3 \frac{W}{m^2} \tag{10.11}$$

In the last formula, Eq. 10.11, we have used the value of the electric current $I = \left(4e^2/h\right) V_{SD} = 1.55 \times 10^{-4}\mathrm{S} \cdot V_{SD}$, and $\lambda = 3 \times 10^{-4}$ m. Besides, we use the AC field amplitude $U_1 = 4$ mV (1 THz) and

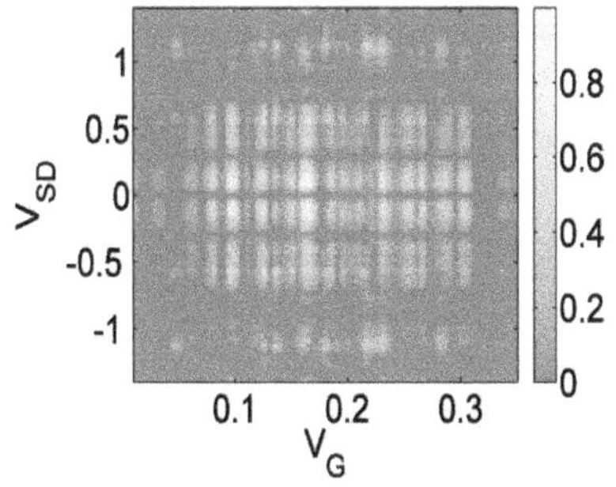

Figure 10.11 Contour plot of the CNT FET responsivity $R(V_G, V_{SD})$ (in units of 2×10^8 V/W) versus the gate and source drain voltages when the external THz field with frequency $\omega = 0.3$ and amplitude $U_1 = 0.05$ (both in units of $10^{-2}\Delta$) is applied.

$\nu = 2\pi/\omega = 1$ THz. The external AC field induces a change of the source drain voltage as

$$\Delta U = I\Delta R = I\left(\frac{1}{G_{\min}} - \frac{1}{G_0}\right) \tag{10.12}$$

where G_0 and $G_{\min}$ are the conductance of CNT FET without and in presence of the AC field, respectively, $I = G_0 V_{SD} = 6.2 \times 10^{-7}$A is a typical electric current. We also assume that the device operates in the regime of source of the current (I = const.). For certainty we set the antenna aperture as $\mathbf{S} = (c/\nu)^2\, \mathbf{G}/4\pi = 1.1 \times 10^{-8}\text{m}^2$. Using the above listed parameters we evaluate the device responsivity. At temperature $T = 50$ mK, and using the change of the AC field-induced conductance as $\delta G/G_0 = 10^4$ we get

$$R = I\left(\frac{1}{G_{\min}} - \frac{1}{G_0}\right)\frac{1}{S \cdot P_s} = 2 \times 10^8\,\frac{\text{V}}{\text{W}} \tag{10.13}$$

where we have used the valve of dipole antenna gain $\mathbf{G} = 1.5$. Furthermore, departing immediately from the quantum efficiency, for the parameters listed above, we also evaluate the number of photoelectrons excited due to absorption of $F/h\nu$ THz photons

$$\eta = \frac{i(\nu)/q}{F(\nu)/h\nu} = \frac{R(\nu)\,h\nu}{q} = 10^6\,\frac{\text{V}}{\text{A}} \tag{10.14}$$

where the photocurrent $i(\nu)$ depends on frequency ν.

10.10 Intrinsic Noise and Noise-Equivalent Power

The electron transport through the graphene and CNT QDs is suppressed because the pseudospin conservation restricts amplitudes of the electron-impurity and electron–phonon back scattering. During the backward scattering, the pseudospin rotates by the angle $+\pi\,(2m+1)$ where m is an integer number. The time reversal process causes a pseudospin rotation into the opposite direction with $-\pi\,(2m+1)$. In the latter case, the spatial part of the wavefunction remains unchanged.

After summation over all the probabilities of possible scattering processes, the factors with positive and negative phases cause a cancellation of the corresponding matrix elements. It eliminates the effect of backward scattering in the pristine graphene and in metallic CNTs. A different scenario takes place in graphene on substrate and in semiconducting nanotubes where the electron states with pseudospins "↑" and "↓" are entangled with each other. The wavefunction symmetry at the K and K' points is destroyed, because the states with the pseudospins "↑" and "↓" are hybridized. A handy description of hybridization of the electron wavefunction is available in terms of the fictitious flux $\varphi = -(\nu/3)\,\varphi_0$. After introducing the flux φ one arrives at a finite contribution of electron back scattering and obtains no perfectly conducting channel in the semiconducting nanotubes. The electron wavevector along the axis direction in the vicinity of K point for $n = 0$ and $\nu = \pm 1$ is $\mathbf{k}_{\pm} = (-2\pi\nu/3L, \pm|k|)$ where the two signs $\pm$ correspond to the right (left) moving excitations. Since the sign of the full electron momentum is not reversed now, one cannot draw a correspondence between the back scattering $\mathbf{k}_{+} \rightarrow \mathbf{k}_{-}$ and the spin rotation by $+\pi\,(2m+1)$ or $-\pi\,(2m+1)$. It explains why the semiconducting nanotubes have a finite resistance due to the finite back scattering. The electron–impurity and electron–phonon collisions also influence the electromagnetic response of the graphene and CNT QDs. The electron–impurity scattering can be described in terms of the Boltzmann transport equation, taking into account that the impurity atoms create additional potentials randomly positioned along the graphene stripe or nanotube. Because of conservation of the one-half pseudospins, the electron–phonon inelastic collision rate as well as the electron–impurity collision rate in graphene and CNTs are typically much lower than in the conventional semiconductors and metals. This expectation is confirmed by theoretical calculations made in (Shafranjuk 2009a, 2009b, 2011a, 2011b). The rate of the electron–phonon collisions remains remarkably low owing to the conservation of pseudospin during the electron transmissions. The outcome of this effect is that the QLs remain sharp even at relatively high temperatures.

Another practical consequence of the above discussion is that the intrinsic noise in graphene and CNT is remarkably low. It follows from an extremely low rate of the phonon–electric collisions in the graphene and CNTs. Therefore, the graphene and CNT QDs can be implemented as key elements of very precise THz field detectors. Such THz detectors exploit sensitivity of the electronic transport through the graphene and CNT junction (NJ) to the THz field. Details of the electron transport depend on the THz field amplitude, frequency, and polarization. Fabricating the QD arrays one constructs the spectral analyzers. In such devices, each element of the array is tuned to a certain frequency ν_{NJ} and covers a narrow frequency interval $\Delta\nu_{\mathrm{NJ}}$. The whole frequency range $\Delta\nu_{\mathrm{array}}$ corresponds to the THz domain $\Delta\nu_{\mathrm{array}} = N\Delta\nu_{\mathrm{NJ}} = 0.5$–$100$ THz, where N is a number of graphene or CNT junctions in the array, $\Delta\nu_{\mathrm{NJ}}$ is the spectral range of a single sensor element.

The lowest level of THz radiation to which the detector can respond is characterized by the noise equivalent power (NEP) (Dyakonov and Shur 1995, Dyakonov and Shur 1996, Sizov and Rogalski 2010). One defines NEP as a reference to the incident radiant power which will produce a signal just equal to the rms noise. In other terms, NEP is a radiant power which will produce the signal to noise ratio equal to 1. When considering an ideal THz photon detector, it is illustrated that shot noise is limited. It is irradiated by an incident monochromatic flux whose power density is P_s W/m^2. In that idealized case, the photocurrent is

$$i_s = RG_0 \cdot AP_s = 2.17 \times 10^{-8}\,\mathrm{A} \tag{10.15}$$

In the above Eq. 10.15, as an example, we have used the responsivity $R = 2 \times 10^8$ V/W, the CNT cross-section area $A = 10^{-16}$ m^2, the steady-state QD DC conductance $G_0 = 1.55 \times 10^{-4}$ S, and the AC field power per unit area $P_s = 7 \times 10^3$ W/m^2. For a detector limited by the shot noise, the averaged current is

$$\langle i_s(\nu)\rangle = [2qi_s(\nu)\,\Delta f]^{1/2} = [2qR(\nu)\,P_s(\nu)\,\Delta f]^{1/2} = \left[2q\frac{\eta(\nu)\,q}{h\nu}P_s(\nu)\,\Delta f\right]^{1/2} \quad (10.16)$$

where f is the chopping frequency. One also defines the signal to noise ratio as

$$\frac{S}{N} = \frac{i_s(\nu)}{\langle i_s(\nu)\rangle} = \frac{RG_0 \cdot AP_s}{[2qRG_0 \cdot AP_s\Delta f]^{1/2}} = \sqrt{\frac{AG_0RP_s}{2q\Delta f}} \quad (10.17)$$

where $R = 2 \times 10^8$ V/W is the responsivity (estimated below for $T = 50$ mK), $A = L \cdot d_{\mathrm{T}} = 1\,\mathrm{nm} \cdot 0.4\,\mu\mathrm{m} = 10^{-16}\,\mathrm{m}^2$ is the CNT cross section area, $G_0 = 1.55 \times 10^{-4}$ S is the CNT conductance with four conducting channels, $P_s = 7 \times 10^3$ W/m^2 is the AC field power density, and $\Delta f = 0.26$ THz is the chopping frequency interval.

The lowest detectable THz irradiation level is determined by the condition $S/N = 1$ which defines *sensitivity*. Then, for example, at the temperature $T = 50$ mK one gets

$$P_s^{\mathrm{min}} = \frac{2q\Delta f}{AG_0R} = 10^{-12}\,\mathrm{W} \quad (10.18)$$

The sensitivity P_s^{min} is significantly influenced by the required detection bandwidth and is actually determined by the detection noise. Another convenient characteristic is the aforementioned NEP which is defined by

$$\mathrm{NEP}_f = A\frac{P_s^{\mathrm{min}}}{\sqrt{\Delta f}} = \frac{2}{\sqrt{\Delta f}}\frac{q\Delta f}{G_0R} = 10^{-18}\frac{\mathrm{W}}{\sqrt{\mathrm{Hz}}}. \quad (10.19)$$

One can evaluate P_s^{min} from (10.18) and NEP_f from Eq. 10.19 which indicate that our graphene and CNT THz detector is very sensitive at low temperatures, $T = 50$ mK. Its key characteristics are very competitive as compared with most available THz detectors based on conventional semiconductors.

10.11 Frequency Range and Operation Temperature

The temperature of operation and frequency range of the graphene and CNT QD detector are important characteristics of the device. Ideally, the detector should be operated within a spectral region where its responsivity is close to its highest possible value. If that is the case, one gets benefits from the lowest possible detection noise which presumes a high signal-to-noise ratio and high sensitivity. Indeed, at lower temperatures, the detector responsivity is much better: that is, at $T = 50$ mK it achieves $R = 2 \times 10^8$ V/W. Nevertheless, its operation temperature may be extended even above 10 K. If the temperature increases, for example, above $T = 10$ K, the detector responsivity indeed drops down to $R = 10$ V/W which corresponds to sensitivity $P_s^{\min} = 10^{-5}$ W and $\mathrm{NEP}_f = 10^{-11}$ W/$\sqrt{\mathrm{Hz}}$. The operating frequency range $\Delta\nu$ of the available graphene and CNT QD detectors strongly depends on the fabrication quality and temperature. Typical segment length of the CNT is about $L = 0.05$–10 μm which corresponds to the level spacing 0.3–60 meV or 0.1–20 THz. The temperature broadening effects in the graphene and CNT QD T-ray detections are much lower as compared to conventional semiconducting QD devices. That is a direct consequence of "intrinsic purity" of the new carbon materials where the resonance transport can be achieved much easier than in conventional semiconductors. But even so, the temperature broadening makes the THz sensing below 0.6 THz rather problematic.

The aforesaid suggests a strong potential of graphene and CNT QD as T-ray detectors: Finding the key intrinsic mechanisms controlling sensitivity may pave the way to extremely efficient remote detectors for utilizing in various applications. The contributions into the AC transport from different electron subbands are identified using the T-rays of appropriate frequency and magnitude. The electron band structure of QDs contains sets of QL singularities whose width remains very narrow even at relatively high temperatures. It opens new opportunities for building new devices which exploit a fairly precise method for determining the THz field parameters. Therefore, the T-ray spectroscopy demonstrates a strong potential as an additional tool for study of the electron spectral characteristics of the graphene and CNT quantum multibarrier junctions. The suggested here novel spectroscopic method can be implemented in various diagnostic and sensing applications. We conclude that an array of the graphene and CNT quantum detectors has a considerable capability for the remote sensing.

Problems

10-1. Discuss why generating and detecting of THz waves represents a serious technical problem.

10-2. Explain why graphene and carbon nanotube (CNT) quantum dots (QDs) have a strong potential for THz applications.

10-3. Which physical principle is used for the detection of THz waves?

10-4. What types of quantization of electron states are used to detect THz waves by graphene and CNT QDs?

10-5. In which conditions does the AC field–induced cooling of a quantum well become possible?

10-6. Which parameters of the external AC field can be deduced from the differential conductance characteristics of a QD?

10-7. Explain why the linear equations for reflection $r_{\mathrm{T}}(\varepsilon)$ and transmission $t_{\mathrm{T}}(\varepsilon)$ coefficients have a recurrent form?

10-8. Explain why the highest responsivity corresponds to photon energies slightly above the level energy.

10-9. Discuss how the basic parameters of a THz QD detector, like sensitivity, quantum efficiency, and noise-equivalent power (NEP), depend on the type of edges of the graphene stripe or on the CNT chirality?

References

Akturk, A., N. Goldsman, G. Pennington, and A. Wickenden (2007). Terahertz current oscillations in single-walled zigzag carbon nanotubes. *Phys Rev Lett*, **98**(16).

Ando, T. (2005). Theory of electronic states and transport in carbon nanotubes. *J Phys Soc Jpn*, **74**(3): 777–817.

Aravind, P. K., and J. O. Hirschfelder (1984). 2-state systems in semiclassical and quantized-fields. *J Phys Chem,* **88**(21): 4788–4801.

Averin, D. V., A. N. Korotkov, and K. K. Likharev (1991). Theory of single-electron charging of quantum-wells and dots. *Phys Rev B,* **44**(12): 6199–6211.

Datta, S. (1992). Exclusion-principle and the Landauer-Buttiker formalism. *Phys Rev B,* **45**(3): 1347–1362.

Datta, S., and M. J. Mclennan (1990). Quantum transport in ultrasmall electronic devices. *Rep Prog Phys,* **53**(8): 1003–1048.

Datta, S., T. Saha-Dasgupta, and A. Mookerjee (2008). Recursive approach to study transport properties of atomic wire. *Eur Phys J B,* **66**(1): 57–65.

Datta, S., S. D. Wang, C. Tilmaciu, E. Flahaut, L. Marty, M. Grifoni, and W. Wernsdorfer (2011). Electronic transport properties of double-wall carbon nanotubes. *Phys Rev B,* **84**(3).

Dyakonov, M. I., and M. S. Shur (1995). 2-dimensional electronic flute. *Appl Phys Lett,* **67**(8): 1137–1139.

Dyakonov, M. I., and M. S. Shur (1996). Plasma wave electronics: novel terahertz devices using two dimensional electron fluid. *IEEE Trans Electron Devices,* **43**(10): 1640–1645.

Fu, K., R. Zannoni, C. Chan, S. H. Adams, J. Nicholson, E. Polizzi, and K. S. Yngvesson (2008). Terahertz detection in single wall carbon nanotubes. *Appl Phys Lett,* **92**(3).

Fuse, T., Y. Kawano, M. Suzuki, Y. Aoyagi, and K. Ishibashi (2007a). Coulomb peak shifts under terahertz-wave irradiation in carbon nanotube single-electron transistors. *Appl Phys Lett,* **90**(1).

Fuse, T., Y. Kawano, T. Yamaguchi, Y. Aoyagi, and K. Ishibashi (2007b). Quantum response of carbon nanotube quantum dots to terahertz wave irradiation. *Nanotechnology,* **18**(4).

Fuse, T., Y. Kawano, T. Yamaguchi, Y. Aoyagi, and K. Ishibashi (2007c). Single electron transport of carbon nanotube quantum dots under THz laser irradiation. *AIP Conf Proc,* **893**: 1013–1014.

Katsnelson, M. I. (2012). *Graphene: Carbon in Two Dimensions.* Cambridge, Cambridge University Press.

Katsnelson, M. I., K. S. Novoselov, and A. K. Geim (2006). Chiral tunnelling and the Klein paradox in graphene. *Nat Phys,* **2**(9): 620–625.

Kawano, Y. (2009). Quantum dots enable integrated terahertz imager. *Laser Focus World,* **45**(7): 45.

Kawano, Y. (2011). Highly sensitive detector for on-chip near-field THz imaging. *IEEE J Sel Top Quant,* **17**(1): 67–78.

Kawano, Y. (2012). Terahertz sensing and imaging based on nanostructured semiconductors and carbon materials. *Laser Photon Rev,* **6**(2): 246–257.

Keldysh, L. V. (1965). Diagram technique for nonequilibrium processes. *Sov Phys JETP: USSR,* **20**(4): 1018.

Kienle, D., and F. Leonard (2009). Terahertz response of carbon nanotube transistors. *Phys Rev Lett,* **103**(2).

Kouwenhoven, L. P., S. Jauhar, K. Mccormick, D. Dixon, P. L. Mceuen, Y. V. Nazarov, N. C. Vandervaart, and C. T. Foxon (1994a). Photon-assisted tunneling through a quantum-dot. *Phys Rev B,* **50**(3): 2019–2022.

Kouwenhoven, L. P., S. Jauhar, J. Orenstein, P. L. Mceuen, Y. Nagamune, I. Motohisa, and H. Sakaki (1994b). Observation of photon-assisted tunneling through a quantum-dot. *Phys Rev Lett,* **73**(25): 3443–3446.

Kral, P., E. J. Mele, and D. Tomanek (2000). Photogalvanic effects in heteropolar nanotubes. *Phys Rev Lett,* **85**(7): 1512–1515.

Maksimenko, S. A., G. Y. Slepyan, M. V. Shuba, and A. Lakhtakia (2006). Optical scattering by achiral carbon nanotubes and application as nanoantennas and composite mediums: art. no. 632807. *Nanomodeling II,* **6328**: 32807–32807.

Mele, E. J., P. Kral, and D. Tomanek (2000). Coherent control of photocurrents in graphene and carbon nanotubes. *Phys Rev B,* **61**(11): 7669–7677.

Morozov, S. V., K. S. Novoselov, M. I. Katsnelson, F. Schedin, L. A. Ponomarenko, D. Jiang, and A. K. Geim (2006). Strong suppression of weak localization in graphene. *Phys Rev Lett,* **97**(1).

Novoselov, K. S., E. McCann, S. V. Morozov, V. I. Fal'ko, M. I. Katsnelson, U. Zeitler, D. Jiang, F. Schedin, and A. K. Geim (2006). Unconventional quantum Hall effect and Berry's phase of 2 pi in bilayer graphene. *Nat Phys,* **2**(3): 177–180.

Rinzan, M., G. Jenkins, H. D. Drew, S. Shafranjuk, and P. Barbara (2012). Carbon nanotube quantum dots as highly sensitive terahertz-cooled spectrometers. *Nano Lett,* **12**(6): 3097–3100.

Seliuta, D., I. Kasalynas, J. Macutkevic, G. Valusis, M. V. Shuba, P. P. Kuzhir, G. Y. Slepyan, S. A. Maksimenko, V. K. Ksenevich, V. Samuilov, and Q. Lu (2010). Terahertz sensing with carbon nanotube layers coated on silica fibers: Carrier transport versus nanoantenna effects. *Appl Phys Lett,* **97**(7).

Shafranjuk, S. E. (2008). Probing the intrinsic state of a one-dimensional quantum well with photon-assisted tunneling. *Phys Rev B,* **78**(23).

Shafranjuk, S. E. (2009a). Directional photoelectric current across the bilayer graphene junction. *J Phys: Condens Mater,* **21**(1).

Shafranjuk, S. E. (2009b). Reversible heat flow through the carbon tube junction. *EPL,* **87**(5).

Shafranjuk, S. E. (2011a). Electromagnetic properties of the graphene junctions. *Eur Phys J B,* **80**(3): 379–393.

Shafranjuk, S. E. (2011b). Resonant transport through a carbon nanotube junction exposed to an ac field. *J Phys: Condens Mater,* **23**(49).

Shur, M. S., J. Q. Lu, and M. I. Dyakonov (1996). Plasma wave electronics: Terahertz sources and detectors using two dimensional electronic fluid in high electron mobility transistors. *Ninety Eight: 1998 IEEE Sixth Int Conf Terahertz Electron Proc,* 127–130.

Sizov, F., and A. Rogalski (2010). THz detectors. *Prog Quant Electron,* **34**(5): 278–347.

Slepyan, G. Y., M. V. Shuba, S. A. Maksimenko, and A. Lakhtakia (2006). Theory of optical scattering by achiral carbon nanotubes and their potential as optical nanoantennas. *Phys Rev B,* **73**(19).

Son, S. K., S. Y. Han, and S. I. Chu (2009). Floquet formulation for the investigation of multiphoton quantum interference in a superconducting qubit driven by a strong ac field. *Phys Rev A,* **79**(3).

Syzranov, S. V., M. V. Fistul, and K. B. Efetov (2008). Effect of radiation on transport in graphene. *Phys Rev B,* **78**(4).

Tans, S. J., M. H. Devoret, H. J. Dai, A. Thess, R. E. Smalley, L. J. Geerligs, and C. Dekker (1997). Individual single-wall carbon nanotubes as quantum wires. *Nature,* **386**(6624): 474–477.

Tien, P. K., and J. P. Gordon (1963). Multiphoton process observed in interaction of microwave fields with tunneling between superconductor films. *Phys Rev,* **129**(2): 647.

Tselev, A., Y. F. Yang, J. Zhang, P. Barbara, and S. E. Shafranjuk (2009). Carbon nanotubes as nanoscale probes of the superconducting proximity effect in Pd-Nb junctions. *Phys Rev B,* **80**(5).

Yngvesson, K. S., K. Fu, B. Fu, R. Zannoni, S. H. Adams, A. Ouarraoui, E. Carrion, J. Donovan, M. Muthee, J. Nicholson, and E. Polizzi (2008). Microwave and terahertz detection in bundles of single-wall carbon nanotubes. *33rd Int Conf Infrared, Millimeter Terahertz Waves,* **1–2**: 665–666.

Zhang, J., A. Tselev, Y. F. Yang, K. Hatton, P. Barbara, and S. Shafraniuk (2006). Zero-bias anomaly and possible superconductivity in single-walled carbon nanotubes. *Phys Rev B,* **74**(15).

Chapter 11

Other Atomic Monolayers

An impressive progress in graphene research has raised questions for other examples of 2D materials with distinct and useful properties. Preliminary results obtained in this new field of knowledge paved the way to an entire world of 2D crystals. Atomic monolayer materials (AMMs) forming 2D atomic crystals have been widely investigated owing to their remarkable mechanical, thermal, electronic, magnetic, and optical properties. Examples of 2D layered materials include hexagonal boron nitride (h-BN), transition metal dichalcogenides, the chalcogenides of group III, group IV, and group V, transition metal oxides, tertiary compounds of carbonitrides, and other traditionally nonlayered structures such as germananes (atomic layers of germanium) and silicenes (silicon-based layered structures). The corresponding atomic structure is shown in Fig. 11.1.

11.1 Atomic Monolayers

Some of the new 2D materials already find use in commercial applications (see, for example, recent work of Wilson 1969, Wilson and Yoffe 1969, Mattheis.Lf 1973, Novoselov et al. 2005, Balandin et al. 2008, Bunch et al. 2008, Castro Neto et al. 2009, Geim 2009,

Graphene: Fundamentals, Devices, and Applications
Serhii Shafraniuk

ISBN 978-981-4613-47-7 (Hardcover), 978-981-4613-48-4 (eBook)
www.panstanford.com

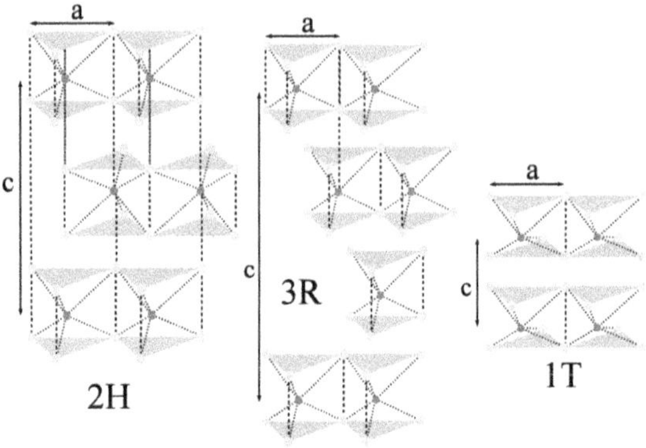

Figure 11.1 Crystal structure of TMDC corresponding to 2H (hexagonal symmetry, two layers per repeat unit, trigonal prismatic coordination), 3R (rhombohedral symmetry, three layers per unit, trigonal prismatic coordination), and 1T (tetragonal symmetry, one layer per repeat unit, octahedral coordination), lattice constant $a = 0.31$–0.37 nm, and the stacking index c corresponds to interlayer spacing 0.65 nm.

Mayorov et al. 2011, Nair et al. 2012, Zhan et al. 2012, and references therein). The unique properties of atomic monolayer materials (AMMs) stem from their peculiar electronic band structure. In particular, in graphene, a single sheet of carbon atoms arranged in a 2D honeycomb crystal lattice, the charge carriers propagating through the honeycomb lattice are described by Dirac rather than the Schrödinger equation. As a result, the charge carriers in pristine graphene obey a linear dispersion law near the Fermi level. Furthermore, they behave like massless relativistic particles with high mobility, large mean free path, long scattering times, and high coherence. This allows to observe relativistic and macroscopic quantum effects in graphene even at room temperature. In addition to the usual spin characteristic of ordinary solid state systems, electrons in graphene have two other pseudospin degrees of freedom (Osada and Sasaki 2012). Unfortunately, the absence of an energy gap in the electron energy excitation spectrum of graphene limits its potential for many nanoelectronic applications. Other AMMs, like transition metal dichalcogenides (TMDCs) (Fig. 11.1), have a significant energy gap ($\sim$1–2 eV for dichalcogenides TiS_2 or MoS_2), which makes them better candidates than graphene for many applications. The TMDCs show a variety of electronic, optical, mechanical, chemical, and thermal properties that attracted a strong interest for decades (Wilson 1969, Wilson and Yoffe 1969, Yoffe 1993). Nowadays, there is a resurgence of attention to the atomically thin 2D forms of TMDCs owing to recent advances in sample preparation, optical detection, transfer, and manipulation of 2D materials, and physical understanding of 2D materials learned from graphene. These properties of AMMs are promising for building a new class of photonic, electronic, and spintronic devices (Wilson 1969, Wilson and Yoffe 1969, Yoffe 1993, Yoon et al. 2011), and offer a convenient tool for investigating some fundamental issues of physics.

The 2D monolayers of TMDCs indicate properties that are complementary but distinct from those in graphene. On the one hand, the graphene devices encapsulated in boron nitride (BN) dielectric layers 5 display an exceptionally high carrier mobility exceeding 10^6 cm^2 V^{-1} s^{-1} at 2 K (Elias et al. 2011) which remains large even at room temperature 10^5 cm^2 V^{-1} s^{-1}. On the other hand, the pristine graphene lacks a band gap; therefore the field-effect transistors (FETs) made from graphene cannot be effectively switched off and have low on/off switching ratios. To engineer the band gaps in graphene one uses the nanostructuring (Han et al. 2007, Li et al. 2008, Lin et al. 2011, Ling et al. 2011), chemical functionalization (Balog et al. 2010) and applying a high electric field to bilayer graphene (Zhang et al. 2009). Nevertheless, the mentioned approaches inevitably increase complexity and reduce the mobility of charge carriers. In contract to graphene, several 2D TMDCs possess sizable band gaps around 1–2 eV (Wilson 1969, Wilson and Yoffe 1969, Yoffe 1993), promising new interesting FET and optoelectronic devices. A transition metal M which is an element from group IV (Ti, Zr, Hf, and so on), group V (e.g., V, Nb, or Ta), or group VI (Mo, W, and so on) and a chalcogen X (which is either S, Se, or Te) form a class of materials with the formula MX_2. These materials are quoted as TMDCs with a layered structure of the form X–M–X. The chalcogen atoms are positioned in two hexagonal planes separated by a plane of metal atoms, as shown in Fig. 11.2 on the left. A variety of bulk crystal polytypes are formed when the adjacent atomic layers are weakly held together.

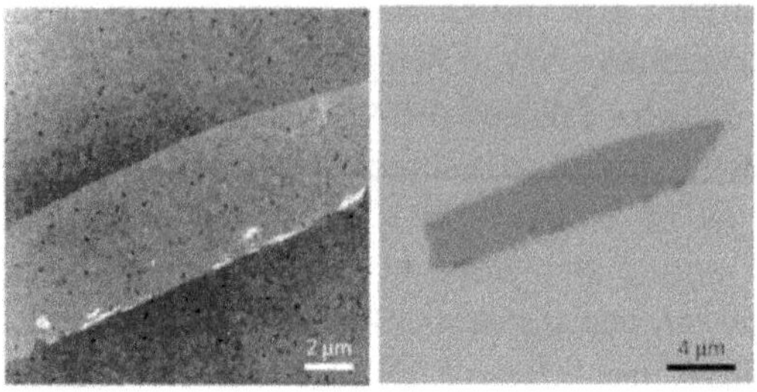

Figure 11.2 Optical microscopy image and atomic force microscopy (AFM) image of a monolayer flake of MoS_2. Adapted from Benameur et al. (2011).

The polytypes differ in the order of stacking and in the metal atom coordination, as illustrated on the right panel in Fig. 11.2. The metal atoms have octahedral or trigonal prismatic coordination, and the overall symmetry of TMDCs is either hexagonal or rhombohedral. TMDCs exhibit a variety of transport properties which are changed from metallic to semiconducting, as summarized in Table 11.1. Some TMDCs exhibit unconventional phenomena, for example, charge density waves and superconductivity (Wilson et al. 1975, Sipos et al. 2008). Much attention is paid to the layer-dependent properties of TMDC. For instance, in some semiconducting TMDCs there is transition from an indirect band gap in the bulk to a direct one in the monolayer. In particular, the indirect band gap of 1.3 eV in bulk MoS_2 transforms into a wider direct band gap of 1.8 eV in the single-layer form MoS_2(Kuc et al. 2011). The presence of direct band gap allows us to obtain the photoluminescence of the monolayer MoS_2, which opens up the possibility of many optoelectronic applications (Kuc et al. 2011). Experimental HAADF-STEM and AFM images of this material are shown in Fig. 11.3. Another interesting phenomena in the monolayer MoS_2 is the valley polarization which is caused by its peculiar electronic structure and is absent in bilayer MoS_2 (Mak et al. 2012, Zeng et al. 2012). There are a lot of exciting layer-dependent properties of the 2D materials which are not observed in the corresponding bulk materials.

Despite the fact that the bulk TMDCs were intensively explored in the last decades, the properties of corresponding monolayers and

Table 11.1 Summary of TMDC. References: [A] Beal et al. (1975); [B] Wilson et al. (1975); [C] Mak et al. (2010); [D] Liu et al. (2011); [E] Ding et al. (2011); [F] Kam and Parkinson (1982); [G] Kuc et al. (2011); (E): experimental; (T): theoretical

Summary of TMDC						
	S_2		Se_2		Te_2	
Nb	Metal Superconducting CDW	[A](E)	Metal Superconducting CDW	[A] [B](E)	Metal	[E] (T)
Ta	Metal Superconducting CDW	[A] [B](E)	Metal Superconducting CDW	[A] [B](E)	Metal	[E] (T)
Mo	Semiconducting 1L: 1.8 eV Bulk: 1.2 eV	[C](E) [F] (T)	Semiconducting 1L: 1.5 eV Bulk: 1.1 eV	[D](T) [F] (E)	Semiconducting 1L: 1.1 eV Bulk: 1.0 eV	[E] (T) 165 (E)
W	Semiconducting 1L: 2.1 eV 1L: 1.9 eV Bulk: 1.4 eV	25 (T) [D] (T) [F] (E)	Semiconducting 1L: 1.7 eV Bulk: 1.2 eV	[E] (T) [F] (E)	Semiconducting 1L: 1.1 eV	[E] (T)

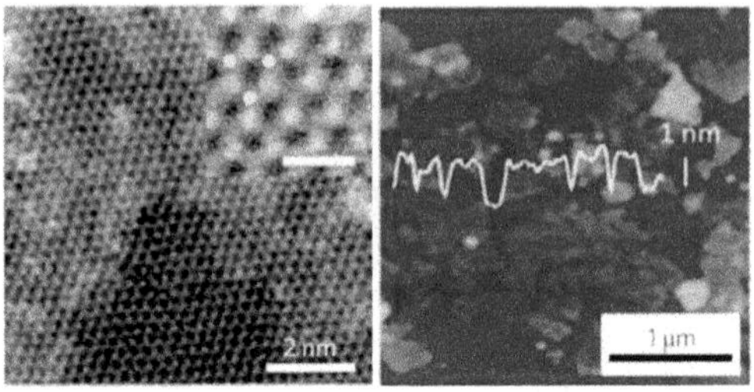

Figure 11.3 High-angle angular dark-field scanning transmission electron microscopy (HAADF-STEM) image of MoS_2 (middle) from the suspension in the left panel and atomic force microscopy (AFM) image of flakes of MoS_2 deposited on SiO_2 (right). Adapted from Eda et al. (2011), © 2011 ACS.

bilayers have not been understood yet. Therefore, it motivates a better clarifying of the mentioned issues.

Although many unusual physical effects have already been established experimentally, and prototype devices demonstrated, AMM's full potential for electronics remains yet to be revealed. We anticipate that the AMM-based devices will exhibit new unconventional transport, spectral, and magnetic properties, including long-range resonant tunneling, sub-THz collective plasma oscillations, and tunable magnetism. It is also expected that further research in this exciting field will lead to new discoveries in fundamental physics.

11.2 Monolayered and Few-Layered Materials

11.2.1 *Hexagonal Boron Nitride*

The most stable and softest among BN polymorphs is the hexagonal form corresponding to graphite. The BN nanotubes are obtained with the structure similar to that of carbon nanotubes, that is, graphene (or BN) sheets rolled on themselves. Nevertheless, properties of the BN nanotube are quite different from the carbon nanotubes which can be either metallic or semiconducting depending on the rolling direction and radius. In contrast to the latter, the properties of BN nanotube are almost independent of tube chirality and morphology because it is always an electrical insulator with a wide band gap of ~5.5 eV like diamond. The BN nanotubes are more thermally and chemically stable as compared to the carbon nanotubes, which is very beneficial for many applications.

11.2.2 *Transition Metal Dichalcogenides*

Titanium disulfide (TiS_2) belongs to TMDCs, which are ionic compounds of a transition metal atom bonded to two groups of 16 anions (usually sulfur, selenium, or tellurium). That TiS_2 material is a golden-yellow solid with high electrical conductivity. The mentioned dichalcogenide has a hexagonal close-packed (hcp) structure. The chemical structure of TiS_2 is similar to cadmium iodide (CdI_2) structure, where cations fill a half of the octahedral holes. The structure of cadmium iodide is common of d-metal chalcogenides and d-metal halides. Furthermore, there is a small overlap between the conduction band and the valence band in the electron spectrum of the titanium disulfide which means that the conductor is a semimetal. The material properties of the titanium disulfide powder are also studied implementing the high pressure synchrotron X-ray diffraction (XRD) at room temperature. Obtained at ambient pressure, the material behaves as semiconductor, whereas at high pressures of 8 GPa the material indicates semimetal properties. The transport properties change when the pressure is increased up to 15 Gpa. The density of electron states at the Fermi level is not affected by pressure up to 20 GPa, and there is no phase change below 20.7 GPa. Nevertheless, if the pressure exceeds 26.3 GPa, the structure of titanium disulfide is changed, though the new structure of titanium disulfide under high pressure is not well understood yet. Images of molybdenum disulfide are shown in Fig. 11.4. Schematics of CVD of MoS_2 from solid *S* and MoO_3 precursors are displayed in Fig. 11.5.

11.2.3 *Material Properties*

Subjecting the samples of titanium disulfide to high pressure has also allowed to study the compressibility. The unit cell size of titanium disulfide 5.695 angstroms in absence of the external pressure decreases down to 3.407 angstroms when the pressure 17.8 GPa is applied. The decrease in unit cell size of TiS_2 has exceeded the values which had been observed for both MoS_2 and

Figure 11.4 Left: Resulting MoS_2 films on SiO_2. The red dots indicate the heating elements in the furnace. In this optical microscopy image, the lighter regions are MoS_2 and the darker regions are SiO_2. Right: Resulting MoS_2 layers that are visible in optical microscopy. Adapted from Lee et al. (2012).

Figure 11.5 Schematic of CVD of MoS_2 from solid S and MoO_3 precursors.

WS_2. It suggests that titanium disulfide is softer and also is more compressible. The experiments indicate that the compression of titanium disulfide is anisotropic. The volume of the TiS_2 unit cell depends on pressure. In particular, because the van der Waals forces which bind S and Ti atoms together are weak, the axis perpendicular to S-Ti-S layers (*a* axis) is less compressible than the axis parallel to S-Ti-S layers (*c* axis). For these reasons, the *c* axis is compressed by 9.5%, whereas the *a* axis is compressed only by 4% when the external pressure 17.8 GPa is applied.

The experiments have also determined the longitudinal sound velocity as 5284 m/s in the plane parallel to S-Ti-S layers while the longitudinal sound velocity perpendicular to the layers is 4383 m/s. Molybdenum disulfide is also classified as a dichalcogenide and represents an inorganic compound with the formula MoS_2. Bulk MoS_2 is a diamagnetic with indirect-band-gap semiconductor similar to silicon and with a gap of 1.2 eV. It occurs as the mineral molybdenite and looks like a silvery black solid as the principal ore for molybdenum. The dichalcogenide MoS_2 is unaffected by dilute

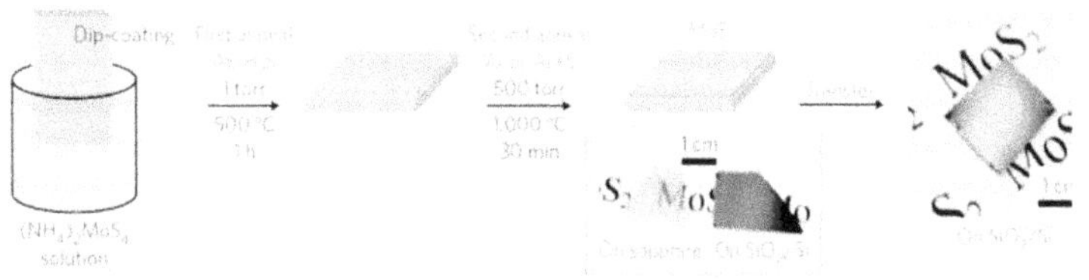

Figure 11.6 CVD growth of MoS_2 from a dip-coated precursor on the substrate and growth in the presence of Ar gas and S vapor. Adapted from Liu et al. (2012), © 2012 ACS.

acids and oxygen and is relatively unreactive. Thus, MoS_2 represents a typical example of the layered TMDC family of materials. Its crystals are composed of vertically stacked, weakly interacting layers which are bound together with the van der Waals interactions. Using the Scotch® tape or lithium-based intercalation one can extract the 0.65 nm thick single layers of MoS_2 (see Fig. 11.2). Furthermore, using MoS_2 suspensions, one can also prepare large-area thin films. Although the bulk MoS_2 is semiconducting with an indirect band gap of 1.2 eV, the single-layer MoS_2 is a direct-gap semiconductor with a band gap of 1.8 eV (Brivio et al. 2011). Other remarkable features are the absence of dangling bonds and the thermal stability up to 1100°C which make MoS_2 interesting for many nanoscale applications. The influence of quantum mechanical confinement in the electronic and optical properties of MoS_2 nanotubes and nanowires are pronounced as well. A variety of nanoelectronic applications can benefit from abundance of other features of the MoS_2 atomic monolayers (AMs). The AM MoS_2 can also be regarded as a semiconducting analogue of graphene, which does not have a band gap in its pristine form. A quantum mechanical confinement results in the appearance of band gaps up to 400 meV. The band structures for bulk and monolayer molybdenum disulfide are shown in Fig. 11.7. Similar plots for wolfram disulfide are presented in Fig. 11.8. It can be introduced by patterned or exfoliated graphene nanoribbons (Li et al. 2008), although it diminishes the mobility significantly (200 $cm^2\ V^{-1}\ S^{-1}$ for a 150 meV band gap). Other adverse effects are the loss of coherence (Sols et al. 2007) or increased off-state currents due to edge roughness (Yoon and Guo 2007). Applying a perpendicular electric field to the

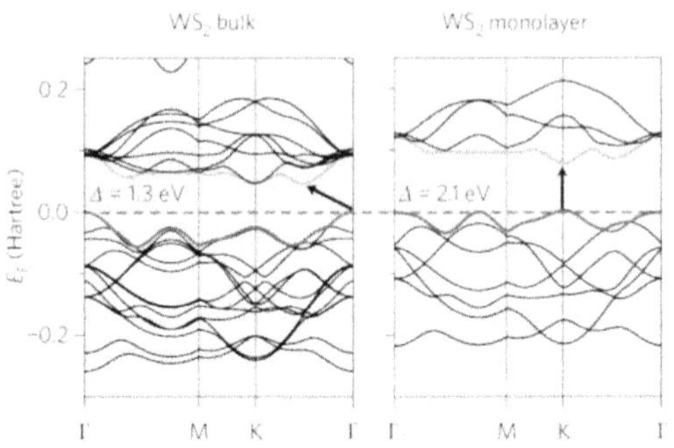

Figure 11.7 Band structures calculated from first-principle density functional theory (DFT) for bulk and monolayer MoS_2. The horizontal dashed lines indicate the Fermi level. The arrows indicate the fundamental band gap (direct or indirect). The top of the valence band (blue) and bottom of the conduction band (green) are highlighted. Adapted from Kuc et al. (2011), © 2011 APS.

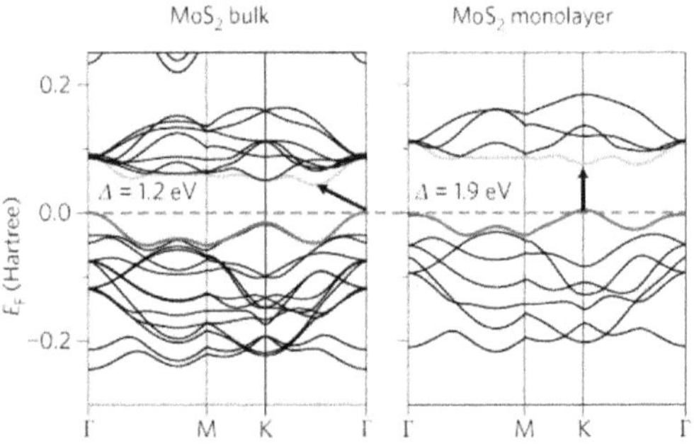

Figure 11.8 The same as in Fig. 11.7 but for WS_2. Adapted from Kuc et al. (2011), © 2011 APS.

bilayer graphene ribbons, one can induce the band gaps as well (Zhang et al. 2009). However, to open an optical gap ~250 meV, it requires the application of a sizable voltage exceeding 100 V (Zhang et al. 2009). It complicates the strategy of building the graphene logic circuits which are able to operate at room temperature with low standby power dissipation. The key requirement is that one needs a current on/off ratio I_{on}/I_{off} from 1×10^4 to 1×10^7 and a band gap bigger than 400 meV (Schwierz 2010) to accomplish any potential replacement of silicon in CMOS-like digital logic devices.

The atomic structure of molybdenum disulfide is detailed as follows. Each Mo (IV) center in MoS_2, is located in a trigonal prismatic coordination sphere, and therefore, is linked to six sulfide ligands. Besides, each sulfur center is bound to three Mo centers which are pyramidal. Such bindings form a layered structure which emerges from the interconnections of the trigonal prisms. In the structure, the molybdenum atoms are sandwiched between layers of sulfur atoms. Molybdenum disulfide has a low coefficient of friction because of the weakness of the van der Waals interaction between the sheets of sulfide atoms. That lubricating property is practically implemented in industry. A low magnitude of the friction coefficient is a common property of many other layered inorganic materials which also exhibit lubricating properties. Two good examples of the material are graphite, which requires volatile additives, and hexagonal boron nitride (h-BN). It is generally known as a solid lubricant or a dry lubricant. Though bulk materials form layered structures, the tendency becomes different for tiny nanoscale samples. In particular, instead of multilayers, nanoparticulate MoS_2 forms fullerene and nanotubular microstructures.

11.2.4 *Chalcogenides of Group III, Group IV, and Group V*

A chemical compound which consists of one chalcogen ion and at least one more electropositive element is called *chalcogenide*. The term chalcogenide is most commonly used for sulfides, selenides, and telluridesa, though all the elements of group 16 of the periodic table are defined as chalcogens rather than oxides.

One example is represented by fluorescence spectra of colloidal CdTe quantum dots with sizes between 2 and 20 nm from left to right. The fluorescence experiences a blue shift emerging from the quantum confinement. The bulk samples of CdTe are transparent in the infrared region of optical spectrum which approximately corresponds to the band-gap energy 1.44 eV at 300 K. Those energies are associated with the infrared wavelengths between 860 nm and 20 μm. The fluorescence of CdTe is observed at 790 nm. For the CdTe quantum dots fabricated of the CdTe crystals whose dimensions are smaller than a few nanometers, the fluorescence peak position moves towards through the visible range.

Transition metal oxides belong to the class of materials with both the transition elements and oxygen. They involve insulators along with the poor metals. In many instances, changing either temperature or pressure the material may experience a metal–insulator transition, thus indicating both types of transport properties. Many transition metal oxides below the critical temperature become superconductors.

Tertiary compounds of carbonitrides. A family of compounds with a general formula near C_3N_4 represents graphitic *carbon nitrides* (also denoted by g-C_3N_4). They also involve structures based on heptazine units which, depending on conditions of reaction, show a variety of condensation degrees, properties, and reactivities. Due to unique semiconductor properties of carbon nitrides, they show unusual catalytic behavior for many reactions, such as for trimerization reactions, activation of benzene, and also activation of carbon dioxide which is also regarded as artificial photosynthesis.

Beta carbon nitride (β-C_3N_4) was theoretically predicted in 1985 by Marvin Cohen and Amy Liu. According to estimations, the superhard β-C_3N_4 material should be even harder than diamond. That conclusion follows from inspecting the nature of the crystalline bonds because according to theory, carbon and nitrogen atoms could form a particularly short and strong bond in a stable crystal lattice in a ratio of 1:1.3.

Along with graphene, other layered nanomaterials which include the TMDCs also have a great potential in nanoelectronics, sensing, and energy harvesting. A new wave of interest had been motivated by recent progress in sample preparation, transfer, and manipulation of AMMs, optical detection, and physical understanding of the 2D materials. A rapid development of this field is stimulated using accumulated experience with graphene. In this emerging field, the graphene research creates a surge of scientific and engineering activity related to atomically thin 2D layers formed from TMDC materials. Because the TMDC samples are fabricated as ultrathin monolayer films with structural flexibility, many of new thin-film devices such as thin-film transistors (TFTs), sensors, and diodes can be created. The atomic layered structure is well suitable for ionic or molecular intercalation, which additionally suggests a variety of applications in ion batteries, hydrogen storage, catalysis, and electrochemical double-layer capacitors.

In this subsection, we briefly discuss electrical characteristics of AM films and flakes, which will be produced through liquid phase exfoliation processing. Obtaining of atomically monolayer samples is a crucial step forward from the investigation of individual handmade devices towards large scale processing of monolayer nanomaterials for numerous applications (Lee et al. 2010, Mak, Lee et al. 2010, Li et al. 2012). One example of AMs is the aforementioned MoS_2. In a natural form MoS_2 has a layered structure represented by a sandwich of three hexagonal atomic layers, sulfur–molybdenum–sulfur (see Fig. 11.2). The 2D layers are formed owing to a strong covalent bonding between atoms (Lee et al. 2010, Mak, Lee et al. 2010, Li et al. 2012). The interlayer coupling is rather weak since it is due to the van der Waals interactions. Anisotropic layered structure of MoS_2 causes a significant anisotropy of the electric transport. Experimental results for the carrier mobility in monolayer molybdenum disulfide are presented in Fig. 11.9. Both *p*- and *n*-types of conduction were obtained in natural MoS_2 crystals (Kuc et al. 2011). Electric transport characteristics of mono- and few-layer MoS_2 flakes fabricated by mechanical cleaving have a direct band gap (Kuc et al. 2011), whereas the bulk MoS_2 has an indirect band gap. The initial stage of the work suggests the study of electronic transport of AM obtained from liquid phase exfoliation. We are going to conduct four-probe measurements for the MoS_2 films and flakes using a source meter instrument at temperatures between the liquid nitrogen and room temperature in vacuum and ambient air. Special attention will be devoted to the electric contacts of the metal source, drains, and local gate electrodes. We expect to obtain Schottky barriers at the contacts between metal and MoS_2 films.

11.2.5 *Synthesis*

Implementing of new electronic and optical properties demonstrated by the 2D TMDCs into actual applications depends on reliable production of atomically thin uniform samples. Here, we briefly discuss the available methods for top-down exfoliation from bulk materials and for bottom-up synthesis, and evaluate their relative merits.

Top-down methods. Using adhesive tape (Novoselov et al. 2005, Alem et al. 2009, Lee et al. 2010, Mak, Lee et al. 2010, Splendiani

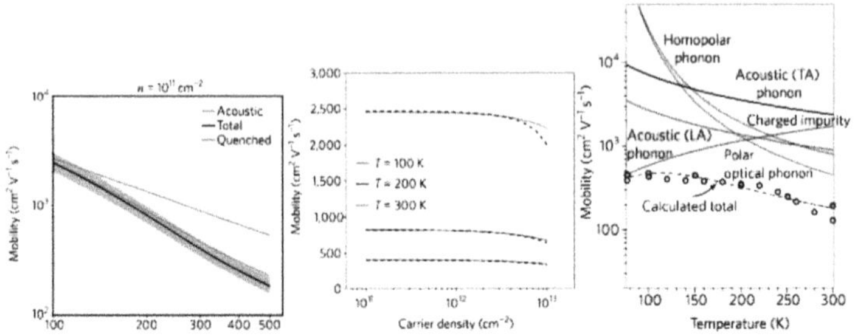

Figure 11.9 Left: Calculated carrier mobility in monolayer MoS_2 as a function of temperature, as determined by contributions from acoustic, quenches, and total phonons; the gray band shows the uncertainty in calculated mobility values due to a 10% uncertainty in computed deformation potentials associated with phonons. Center: Mobility versus the carrier density at different temperatures. Right: The calculated and measured carrier mobility in multilayer MoS_2 as a function of temperature, showing the scattering contributions from charged impurities (red line), homopolar out-of-plane phonons (green lines), and polar-optical phonons (blue line), as well as the total mobility due to the combined effects (dashed line). Figure reproduced with permission from Kaasbjerg et al. (2012, © 2012 APS.

et al. 2010, Bertolazzi et al. 2011, Radisavljevic, Radenovic et al. 2011, Radisavljevic, Whitwick et al. 2011) for a micromechanical cleavage one can peel atomically thin flakes of TMDCs from their parent bulk crystals. After the flakes are applied to substrates, they are optically identified by light interference (Benameur et al. 2011, Li et al. 2012, Li et al. 2012), implementing the same techniques that were developed for graphene. The result of micromechanical cleavage process yields an AM flake of MoS_2 derived from a bulk crystal of MoS_2 shown in Fig. 11.10. Using the same approach of the micromechanical cleavage one also exfoliates other layered materials such as BN (Novoselov et al. 2005, Alem et al. 2009, Lee et al. 2010, Mak, Lee et al. 2010, Splendiani et al. 2010, Bertolazzi et al. 2011, Radisavljevic, Radenovic et al. 2011, Radisavljevic, Whitwick et al. 2011) and oxide nanosheets (Kalantar-zadeh et al. 2010, Osada and Sasaki 2012) into single sheets. Thus, one obtains single-crystal flakes of high cleanliness and purity that are appropriate for thorough characterization and for fabrication of individual devices (Ayari et al. 2007, Novoselov et al. 2005,

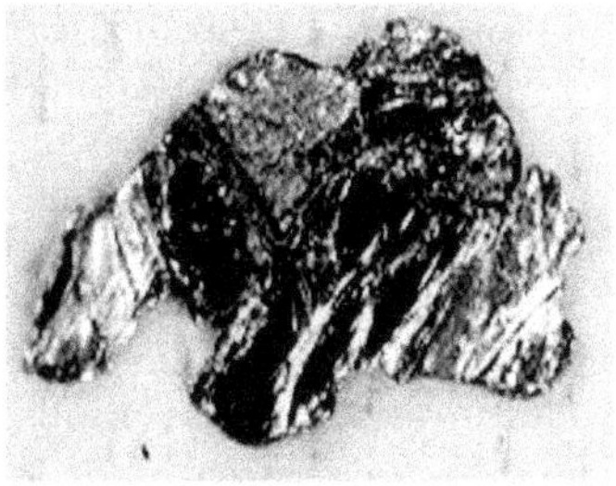

Figure 11.10 A bulk piece of molybdenum disulfide.

Alem et al. 2009, Lee et al. 2010, Mak, Lee et al. 2010, Splendiani et al. 2010, Bertolazzi et al. 2011, Radisavljevic, Radenovic et al. 2011, Radisavljevic, Whitwick et al. 2011). A disadvantage is that the described approach is not scalable and it is not possible to accomplish a systematic control of flake thickness and size. Another method is to thin MoS_2 down to monolayer thickness by thermal ablation with micrometer-scale resolution by a focused laser spot. However, the requirement for the laser raster scanning makes it impossible for scale-up (Castellanos-Gomez et al. 2012).

Large numbers of exfoliated nanosheets can be fabricated by implementing the liquid-phase preparations of TMDCs. The method can be complemented with other procedures including composites and hybrids by simple mixing of dispersions of different materials (Coleman et al. 2011, Smith et al. 2011), and thin films and coatings by filtration, inkjet printing, spray coating, and doctor blading. The use of solution-based graphene paves the way for the development of high-frequency flexible electronics with a current gain cutoff frequency of 2.2 GHz (Gordon et al. 2002). Therefore, the solution-based TMDCs are expected to have good prospects for fabricating of flexible electronics and composite materials.

The layers can be exfoliated in liquid with intercalation of TMDCs by ionic species (Bissessur et al. 1996, Hutchison et al. 1996). Although the intercalation approach was suggested about 40 years ago, nowadays such methods attract a renewed interest (Eda et al. 2011, Zeng et al. 2011). The TMDC had been first intercalated by lithium (Dines 1975), whereas the intercalation-driven exfoliation was reported by Eda et al. (2011). The process consists of submerging the bulk powder of TMDC in a solution of a lithium-containing compound such as *n*-butyllithium during a day or longer. It allows a firm intercalating of lithium ions. Then, the intercalated material is exposed to water. A vigorous reaction of water with the lithium between the layers causes the H_2 gas to evolve. Then, the layers are rapidly separated (Eda et al. 2011). The lithium-based intercalation yields the solution-phase MoS_2 flakes which are derived from the exfoliation.

The described methods of chemical exfoliation produce gram quantities of submicrometer-sized monolayers (Tsai, Heising et al. 1997, Tsai, Schindler et al. 1997), but thus obtained exfoliated material is altered electronically and structurally, as compared to the original bulk material (Eda et al. 2011). For instance, when the process is applied to MoS_2, the Mo atom coordination is changed from trigonal prismatic (2H-MoS_2) to octahedral (1T-MoS_2), whereas the electronic structure of the exfoliated nanosheets is changed from semiconducting to metallic (Gordon et al. 2002, Eda et al. 2011). A phase change from 1T-MoS_2 to 2H-MoS_2 might result from annealing at 300°C, which restores the Mo atom coordination, and also reinstates the semiconducting band gap of the pristine material. The mentioned reinstatement was confirmed by the re-emergence of band-gap photoluminescence (Gordon et al. 2002, Eda et al. 2011). The method of lithium-based chemical exfoliation has been proved to work (Gordon et al. 2002) for TMDCs such as MoS_2, WS_2, $MoSe_2$, and SnS_2.

A better controllable and faster lithiation method is based on an electrochemical cell with a lithium foil anode and TMDC-containing cathode, as was reported by Eda et al. (2011) and Zeng et al. (2011). The lithiation is controlled and metered because the intercalation occurs during a galvanic discharge in the electrochemical cell. Then, applying a sonication in water, one exfoliates the Li-intercalated material and obtains monolayer TMDC nanosheets. The method was shown to work for MoS_2, WS_2, TiS_2, TaS_2, ZrS_2, and graphene (Eda et al. 2011, Zeng et al. 2011) and later for BN, $NbSe_2$, WSe_2, Sb2Se_3, and Bi_2Te_3. The method is relatively fast, since it requires only a few hours for Li intercalation, compared with more than a day for the *n*-butyllithium method.

Another approach of the TMDC exfoliating is the ultrasonication in appropriate liquids, including organic solvents, aqueous surfactant solutions, or solutions of polymers in solvents (Coleman et al. 2011, Smith et al. 2011). Using ultrasonication, the mechanical exfoliation of layered crystals results in flakes whose size achieves a few hundred nanometres. One stabilizes the exfoliated nanosheets against reaggregation either by solvation or by steric or electrostatic repulsion which is based on adsorption of the molecules from solution. Solutions of 2D layered materials exfoliated in organic solvents are used along with thin films from filtration of solution-based material (Coleman et al. 2011, Smith et al. 2011). The most probable mechanism is that molecules from solution are adsorbed onto the TMDC nanosheets by noncovalent interactions.

One can estimate the energy which is required to exfoliate layered crystals. Such energy is related to the surface energy, which is equal to the half of energy for splitting a monolayer from the crystal per the monolayer surface area. The inverse gas chromatography and exfoliation studies show that the surface energies of MoS_2, WS_2, $MoSe_2$, and BN are falling in the range 65–75 mJ m^{-2} (Coleman et al. 2011, Smith et al. 2011, Cunningham et al. 2012). That energy range is close to the surface energy of graphene, which has been found in the range 65–120 mJ m^{-2}. It suggests that the inorganic layered compounds are exfoliated as easily as graphite, or even easier.

An evident benefit of the ion exfoliation is a high output of monolayers (Eda et al. 2011, Zeng et al. 2011). The development of Li intercalation by complementing it with an electrochemical method has made the process even more controllable and faster (Eda et al. 2011, Zeng et al. 2011). A disadvantage of that approach is the flammability of the Li compounds under the ambient conditions. Therefore, the process should proceed under inert gas, whereas Li is a quite expensive material. It motivates researchers to find alternative intercalants. Although liquid exfoliation is insensitive to ambient conditions, the concentration of obtained monolayer flakes is very low (Coleman et al. 2011, Smith et al. 2011, Cunningham et al. 2012). Therefore, for electronic or photonic applications where monolayers are required, it makes sense to implement well-annealed ion-exfoliated nanosheets. For other applications, like composite materials, where large quantities are required, one can use liquid exfoliation instead of the former. Most of nanoelectronic and optoelectronic technological processes require conducting of rational control over nanosheet thickness and size. It could be achieved with sorting of flakes by layer thickness (Green and Hersam 2009) and lateral size (O'Neill et al. 2012) as it was done for graphene where new layer-controlled chemistries (Shih et al. 2011) and postsynthesis were utilized.

11.2.6 *Bottom-Up Fabrication*

Wafer-scale fabrication of electronic devices and transparent, flexible optoelectronics critically depend on methods for synthesizing large-area and uniform layers. Fabrication of large-scale devices (Lin et al. 2011, Wu, Vendome et al. 2011, Wu, Lin et al. 2011, Grasso et al. 2012, Wu et al. 2012) became viable owing to implementation of wafer-scale synthesis methods via chemical vapour deposition (CVD) on metal substrates (Li et al. 2009) and epitaxial growth on SiC substrates (Hass et al. 2008). The growing of atomically thin films of MoS_2 on insulating substrates were accomplished using the CVD methods (Lee, Zhang et al. 2012, Peng et al. 2012, Zhan et al. 2012). The CVD technique exploits different solid precursors which are heated to high temperatures. It involves sulfur powder and MoO_3 powder which are vaporized and then subsequently codeposited onto a nearby substrate (Lee, Zhang et al. 2012, Peng et al. 2012, Zhan et al. 2012). In this scenario, a thin layer of Mo metal is deposited onto a wafer heated with solid sulfur (Lee, Zhang et al. 2012, Peng et al. 2012, Zhan et al. 2012). The substrates are then dip-coated in a solution of $(NH_4)_2MoS_4$ and are heated in the presence of sulfur gas (Liu, Zhang et al. 2012). The CVD techniques are summarized in Figs. 11.4 and 11.6. Using the CVD techniques, the final MoS_2 film thickness is frequently dependent on the concentration or thickness of the initial precursor. Nevertheless, a precise knowledge of the number of layers over a large area has not yet been obtained. Obtaining the MoS_2 with the CVD technique on Cu foil as a surface template was done similarly as for the previous CVD-grown graphene. This method yields single-crystal MoS_2 flakes with lateral size of a few micrometers (Shi et al. 2012). Although the mentioned results are preliminary, they are promising for the further work involving a wider range of materials other than MoS_2. That activity might lead to production of large-area uniform sheets of TMDCs with controllable layer number. It opens path for fabricating of high-quality top-gated monolayer molybdenum disulfide field effect transistors shown in Fig. 11.11.

Other approaches are based on chemical preparation of MoS_2 and $MoSe_2$ (Peng et al. 2001a, 2001b). They have been obtained using hydrothermal synthesis which consists of growth the single crystals from an aqueous solution in an autoclave at high temperature and pressure. Besides, there are methods (Matte et al. 2010, Matte et al. 2011) to synthesize WS_2, MoS_2, WSe_2, and $MoSe_2$, where the reaction of molybdic or tungstic acid with either thiourea or selenourea at elevated temperatures is utilized to obtain the corresponding layered TMDC materials (Matte et al. 2010, Matte et al. 2011). The mentioned procedures yield satisfactory good-quality materials with the flake dimensions typically of hundreds of nanometers to a few micrometers. Nonetheless, the flake thickness has not been determined as adequate to AMs.

11.2.7 *Electronic Band Structure*

The first-principle and tight-binding calculations (Kobayashi and Yamauchi 1995, Li and Galli 2007, Ding, Wang et al. 2011, Ding, Xiang et al. 2011, Kuc et al. 2011, Liu et al. 2011, Ataca et al. 2012) which are complemented by spectroscopic experiments have shown that the band structure of many TMDC materials is consistent with their general features (Novoselov et al. 2005, Alem et al. 2009, Lee et al. 2010, Mak, Lee et al. 2010, Splendiani et al. 2010, Bertolazzi et al. 2011, Radisavljevic, Radenovic et al. 2011, Radisavljevic, Whitwick et al. 2011). The NbX_2 and TaX_2 compounds are metallic, whereas MoX_2 and WX_2 are semiconducting (Kobayashi and Yamauchi 1995, Li and Galli 2007, Ding, Wang et al. 2011, Ding, Xiang et al. 2011, Kuc et al. 2011, Liu et al. 2011, Ataca et al. 2012). The first-principle calculations show a difference between the band structures of bulk and monolayer MoS_2 and WS_2, as illustrated by Figs. 11.7 and 11.8 (Kuc et al. 2011). One can see that at the Γ point, the band-gap transition is indirect for the bulk material. However, it gradually shifts to be direct for the monolayer. In contrast, at the K point, the direct excitonic transitions remain relatively unchanged versus the layer number (Novoselov et al. 2005, Alem et al. 2009, Lee et al.

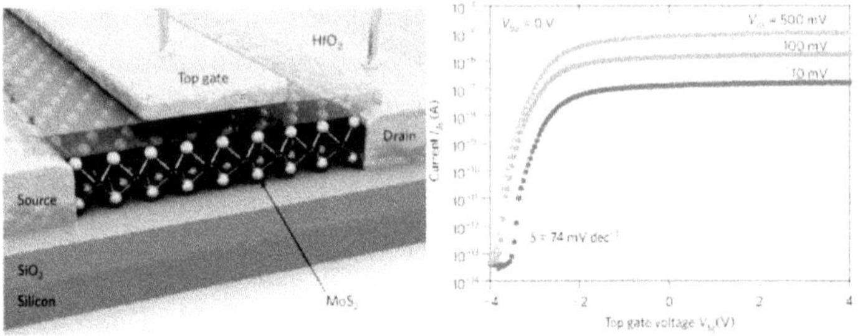

Figure 11.11 Left: Schematic illustration of HfO_2-top-gated monolayer MoS_2 FET device. Right: Source–drain current (I_{ds}) versus top-gate voltage (V_{tg}) curve recorded from the top-gated device shown on the left for a bias voltage ranging from 10 mV to 500 mV. Measurements are performed at room temperature with the back gate grounded. Top-gate width, 4 mm; top-gate length, 500 nm. The device can be completely turned off by changing the top-gate bias from −2 to −4 V. For V_{ds} = 10 mV, the I_{on}/I_{off} ratio is larger than 10^6. For V_{ds} = 500 mV, the I_{on}/I_{off} ratio $>1 \times 10^8$ in the measured range, while the subthreshold swing S = 74 mV dec^{-1}. Adapted from Benameur et al. (2011), © 2011 NPG, and Radisavljevic et al. (2011), © 2011 ACS.

2010, Mak, Lee et al. 2010, Splendiani et al. 2010, Bertolazzi et al. 2011, Radisavljevic, Radenovic et al. 2011, Radisavljevic, Whitwick et al. 2011).

The main contribution into the band structure change versus the layer number originates from the quantum confinement. Another contribution comes also from hybridization between *pz*-orbitals on S atoms and *d*-orbitals on Mo atoms (Splendiani et al. 2010). There is a spatial correlation between the electronic distributions and atomic structure (Splendiani et al. 2010). According to calculations based on the density functional theory (DFT), the conduction-band states at the K point of MoS_2 are mainly due to localized *d*-orbitals on the Mo atoms. They are located in the middle of the S–Mo–S layer sandwiches and are robust in respect to interlayer coupling. On the contrary, the states located near the Γ point are combined from the antibonding *pz*-orbitals on the S atoms and the *d*-orbitals on Mo atoms. Therefore, they are strongly influenced by the interlayer coupling (Splendiani et al. 2010). Although direct excitonic states near the K point do not depend on the layer numbers, the story differs for the transition at the Γ point which shifts considerably from an indirect one to a larger direct one (Kuc et al. 2011). All the MoX_2 and WX_2 materials undergo similar transformation from the indirect- to direct-band gap when the number of layers decreases. The transformation covers the band-gap energy range 1.1–1.9 eV. The data for bulk and monolayer band gaps which cover several TMDCs are given in Table 11.1.

The values of semiconducting energy gap for the bulk and for the monolayer TMDCs are similar to each other and are about 1.1 eV, as shown in Table 11.1. The gap value 1.1 eV is close to the band gap in silicon which makes TMDCs suitable for fabrication of digital transistors (Schwierz 2010). The transition to direct band gap in the monolayer form has important implications for photonics, optoelectronics, and sensing. The photophysical properties follow from the electronic band structure. One can see the excitonic transitions A and B, and the direct and indirect band gaps at E_g and $E_{g'}$, which are indicated on the simplified band diagram in Fig. 11.16.

11.3 Electric Transport in Nanodevices

The creation of high performance transistors for digital electronics is among the most important applications of semiconductors. The technological efforts in the digital electronics industry during last decades has been focused on scaling of the existing transistors down to the smallest dimensions. Nowadays, the best processors are composed of silicon-based metal–oxide–semiconductor field-effect transistors (MOSFETs) with dimensions of 22 nm. The further scaling down to the atomic dimensions eventually will be affected by the statistical, quantum effects, and heat dissipation issues. Altogether, it motivates the spread of novel device concepts and materials. In this respect, the 2D semiconductor materials are beneficial due to their handy processability and absence of short-channel effects that impair the overall device performance (Schwierz 2010).

The generic FET structure includes a region of the semi-conducting channel which is connected to the source and drain electrodes. That idea also has been adapted to the 2D TMDCs where the channel is separated from the gate electrode by a

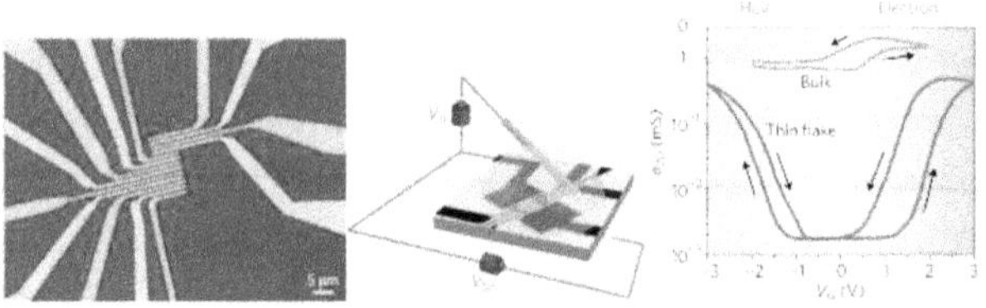

Figure 11.12 Left: Integrated circuit based on single-layer MoS_2. Center: Schematic of electric double-layer transistor (EDLT, a FET gated by ionic liquids). V_{DS} is the source–drain voltage, and V_G is the gate voltage. Right: Conductivity as a function of top-gate voltage for both bulk and thin-flake MoS_2 EDLT devices. Thin-flake devices show on/off ratios $>10^2$ for both electron and hole transport. Adapted from Lee et al. (2010), Yoon et al. (2011) and Zhang et al. (2012).

dielectric layer. It allows the electric current flowing between the source and drain electrodes to be readily controlled by the gate electrode. Thus, the gate electrode modulates conductivity of the semiconducting channel. The basic material which is used nowadays for the design of semiconducting electronic circuits is silicon. The material satisfies the present industrial requirements for the performance and manufacturability in digital logic. Although silicon is widely implemented for fabricating the digital logic, other electronic systems like light-emitting diodes (LEDs), high-power electronics, high-temperature electronics, radio-frequency electronics, and photovoltaics use other semiconductors like SiC, GaN, Ge, and GaAs which are better suitable for these purposes (Morkoc et al. 1994). Other promising nanomaterials which have been recently examined and suggested include carbon nanotubes, graphene, and semiconductor nanowires.

The key parameters which determine the fast operation of digital logic transistors are as follows. They include a high charge carrier mobility, a high on/off ratio which is actually the ratio of on-state to off-state conductance for effective switching, high conductivity which is the product of charge density and mobility, and a low value of the off-state conductance for low power consumption during operation.

In regular semiconductors, the charge carrier density can be increased by doping. However, there is an adverse effect of chemical doping which can also decrease the mobility due to a stronger scattering on atomic impurities. Therefore, while doping the semiconductors chemically, one should simultaneously maintain mobility at a sufficient level. The criterion of the tradeoff is that for digital logic, on/off ratios of 10^4–10^7 are generally required for use as switches (Schwierz 2010). Because graphene is 2D, a lot of interest has been devoted to its electronic device applications. The carrier mobility of graphene has exceptionally high value, whereas an external gate voltage can readily modulate its current flow (Schwierz 2010). The high-frequency and radio-frequency analog graphene transistors have been designed taking advantage of the high carrier mobility and high transconductances (Schwierz 2010). Their cutoff frequencies achieved hundreds of gigahertz. An absence of band gap in a pristine graphene makes it impossible to achieve low off-state current, which is a prerequisite for the design of digital logic transistor. Therefore, there is a demand for new 2D semiconductor materials with a considerable band gap to ensure high on/off ratios while simultaneously maintaining high carrier mobility and scalability to smaller dimensions.

Other properties which are very desirable for the next-generation electronics are mechanical flexibility and transparency. Attention of researchers nowadays is focused on TMDCs as promising ultrathin materials with tunable band gaps that might ensure high on/off ratios in the new generation of FETs (Benameur et al. 2011, Radisavljevic, Radenovic et al. 2011, Radisavljevic, Whitwick et al. 2011). An important advantage of the 2D semiconductors such as MoS_2 and others when compared with classical 3D electronic materials is the subnanometer thickness. An extreme thinness, when utilized jointly with a band gap of 1–2 eV, can yield high on/off ratios, allows more efficient control over switching (Colinge 2004a, 2004b, 2004c) and helps to reduce the short-channel effects and power dissipation which cause major restrictions to transistor miniaturization.

11.4 Electronic Transport versus Scattering Mechanisms

Transport and scattering of charge carriers in 2D TMDC layers occur exclusively within the plane of the material. The scattering mechanisms contributing to the mobility of carriers are as follows:

(i) electron scattering on the acoustic and optical phonons; (ii) scattering of electrons on the Coulomb potential created by randomly positioned charged impurities; (iii) interface phonon scattering on surfaces; and (iv) elastic scattering of electrons on roughness. Relative contribution of the listed scattering mechanisms and the degree to which they affect the carrier mobility depend on temperature, carrier density, layer thickness, effective mass of electrons, electronic and phonon band structures. The listed scattering mechanisms are also typical for other semiconductors and graphene.

As the temperature increases, the phonon scattering becomes more important to determine the carrier mobility. In the 2D TMDCs, where partial ionic bonds connect the atoms of metal and chalcogen, the crystal deformation creates the polarization fields which tend to alter electron scattering. The electron mobility versus temperature calculated from the first principles by Kaasbjerg et al. (2012a, 2012b) is shown in Fig. 11.9 for a single-layer MoS_2. Along with the common effect of the acoustic and optical phonon scattering, we show the mobility due to acoustic phonon scattering alone. When the temperature is low ($T < 100$ K), the major contribution comes from acoustic phonons but it switches to the optical component to prevail at higher temperatures. Quenching of the out-of-plane homopolar mode appears to be not important as one would expect for top-gated devices. It gives just a slight increase of mobility. The optical phonons restrict the value of room-temperature mobility to $\sim 4^{10}$ $cm^2\ V^{-1}\ s^{-1}$ (Kaasbjerg et al. 2012a, 2012b).

Similar to graphene, charged impurities which are randomly positioned in the 2D TMDC layer or on the top of it cause the Coulomb scattering in 2D TMDCs. It becomes a dominant scattering mechanism at low temperatures, as for graphene (Hwang et al. 2007). The mobility can be improved by engineering the dielectric environment (Jena and Konar 2007), like it was made for graphene (Konar et al. 2010a, Konar et al. 2010b) and also for MoS_2 (Radisavljevic, Radenovic et al. 2011). The electron scattering on charged impurities also generally limits the mobility of monolayer graphene to values below 10,000 $cm^2\ V^{-1}\ s^{-1}$ resting on the dielectric SiO_2 substrate (Zhu, Wang et al. 2009, Zhu, Perebeinos et al. 2009). Doping the TMDC with ionic impurities allows to tune the carrier concentration and band gap. The adverse effect of chemical doping is that the mobility worsens due to excessive scattering. It makes the choice of doping level in a particular device a tricky task which can strongly influence its performance. An illustration of influence of the carrier concentration and temperature on mobility of MoS_2 is presented in Fig. 11.9. The collective influence of phonons and charged impurities on the mobility of multilayer MoS_2 is given in Fig. 11.9 (right). A similar tendency takes place also for a monolayer MoS_2.

For the impurity scattering to prevail over the phonon scattering the impurity concentration must exceed $\sim 5 \times 10^{11}$ cm^{-2} which corresponds to a heavy doping (Kaasbjerg et al. 2012a, 2012b). It suggests that the room-temperature mobility of MoS_2 can be increased from $\sim$0.5–3 cm^2 V^{-1} s^{-1} to $\sim$200 cm^2 V^{-1} s^{-1} with tuning of the dielectric screening of Coulomb scattering on the charged impurities using gate dielectric materials with a high dielectric constant.

Since the 2D materials are extremely thin, the surface phonons and ripples scatter the charge carriers very strongly. Electron scattering on interfacial roughness can dominate in GaAs-based quantum wells (Sakaki et al. 1987). However in graphene, the electron scattering on charged impurities prevails over the short-range scattering and scattering on surface ripples (Hwang et al. 2007). When graphene is resting on a dielectric SiO_2 substrate, there is a visible contribution from the electron scattering on the surface polar phonons localized in the SiO_2 (Chen et al. 2008). The primary scattering in freely suspended graphene (Hwang et al. 2007) emerges due to out-of-plane flexural phonons (Castro et al. 2010). Besides, ripples like in graphene take place also in freestanding MoS_2 (Brivio et al. 2011), which causes an additional decrease of mobility. The theoretical value of room-temperature mobility of MoS_2 limited by scattering on phonons was obtained as $\sim$410 cm^2 V^{-1} s^{-1}, which is also the case for other single-layer TMDCs. The phonon modes and experimental Raman spectra of monatomic molybdenum disulfide are shown in Fig. 11.16.

11.5 TMDC Transistors

Unique properties of semiconducting 2D TMDCs can be exploited to fabricate channels of the FETs. The most attractive features are a high mobility comparable to Si, lack of dangling bonds, and structural stability (Fivaz and Mooser 1967). Initial experiments performed on crystals of WSe_2 which represent the premature use of TMDCs in the FETs, already indicated the level of mobility exceeding the characteristics of the best single-crystal Si FETs which for the room-temperature *p*-type conductivity was about 500 $cm^2\ V^{-1}\ s^{-1}$. Other observations were ambipolar behavior and $\sim 10^4$ on/off ratio at the temperature of 60 K (Podzorov, Gershenson et al. 2004, Podzorov, Pudalov et al. 2004). That initial work has motivated creating of devices with mobility values in the range 0.1–10 $cm^2\ V^{-1}\ s^{-1}$ (Perisanu et al. 2007) based on thin layers of MoS_2 with a back-gated configuration.

The MoS_2 top-gated transistor based on monolayers, as shown in Fig. 11.11, was first reported by Kis and coworkers (Radisavljevic, Radenovic et al. 2011). Current–voltage curves for this FET are shown in Fig. 11.13. The device demonstrated room-temperature mobility exceeding 200 $cm^2\ V^{-1}\ s^{-1}$, *n*-type conduction, high on/off current ratio ($\sim 10^8$), and subthreshold swing of 74 mV

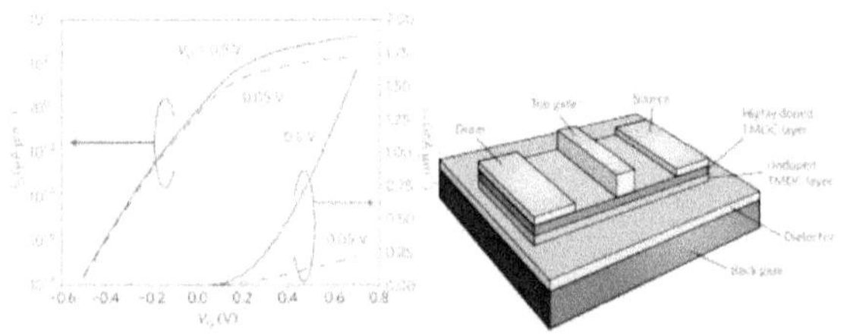

Figure 11.13 Left: Simulated device characteristics for a monolayer MoS_2 FET with 2.8 nm thick HfO_2 top-gate oxide, 15 nm gate length, and power supply voltage 0.5 V. The source–drain current (I_D) is plotted against the gate voltage (V_G) for 0.05 and 0.5 V drain voltage (V_D) on linear (right axis) and logarithmic (left axis) scales. Right: Proposed TMDC-based high-electron-mobility transistor (HEMT) device with top-gated Schottky contact and TMDC layers with different doping levels. Adapted from Yoon et al. (2011).

per decade (Radisavljevic, Radenovic et al. 2011). An advantage of the top-gated geometry is that it reduces the voltage necessary to switch the device. Furthermore, one can allocate multiple devices on the same substrate. An additional benefit emerges from the dielectric engineering, as suggested by Jena and Konar (2007). Much better mobility of the monolayer MoS_2 in this device is obtained implementing high-κdielectric HfO_2. The hole mobility $\sim$250 cm^2 V^{-1} s^{-1} at room temperatures along with the per decade subthreshold swing $\sim$60 mV and 10^6 on/off ratio (Fang et al. 2012) was achieved in the top-gated p-type FET with a high-k dielectric where the active channel was made of WSe_2 monolayer flake. Liquid exfoliation allows to fabricate the thin-film MoS_2 transistors which show similar electrical performance (Lee et al. 2011). It paves the way for creating of transparent, flexible, 2D electronic circuits. Large areas of MoS_2 and the wafer-scale fabrication of devices will become possible due to development of CVD synthesis methods.

The fabrication of single-layer MoS_2 transistors (Radisavljevic, Radenovic et al. 2011) was based on the Scotch® tape–based micromechanical exfoliation (Novoselov et al. 2004) of atomic MoS_2 monolayers. Then, the monolayers were transferred to degenerately doped silicon substrates covered with 270 nm thick SiO_2 as shown in Fig. 11.14. The mentioned oxide thickness is optimal

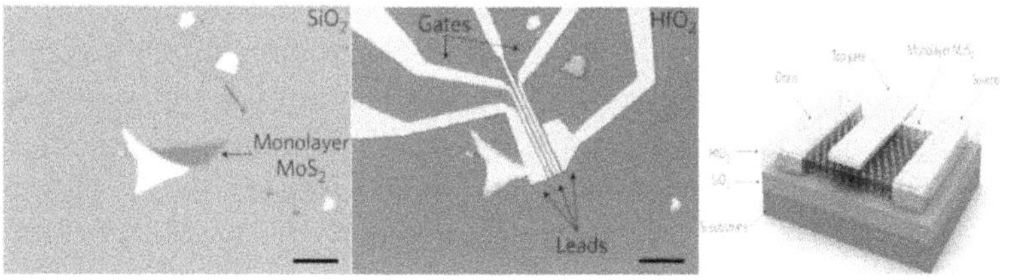

Figure 11.14 Left: A monolayer of MoS_2 with thickness 0.65 nm, resting on the 270 nm thick dielectric SiO_2 layer formed at the top of the silicon substrate. Center: The device made of the flake whose optical image is shown on the left. It represents two FETs connected in series. The three gold leads serve as source and drain electrodes for the two transistors. The 30 nm thick ALD-deposited HfO_2 layer acts both as a gate dielectric and a mobility booster in the MoS_2 monolayer. Scale bars are 10 mm. Right: 3D schematic view of one of the transistors shown in (b). Adapted from Radisavljevic et al. (2011), © 2011 NPG.

for optical detection of single-layer MoS_2 and correlates with the contrast when measured by atomic force microscopy (AFM). The electrical 50 nm thick gold electrodes were formed using electron-beam lithography and deposition. The residue of the resist was removed by annealing the device at 200°C (Ishigami et al. 2007) which also yielded a decrease of the contact resistance. Typical mobility of the above single-layer devices was in the range 0.1–10 $cm^2\ V^{-1}\ s^{-1}$, which is consistent with the formerly reported values for single layers (Novoselov et al. 2004) and thin crystals containing more than 10 layers of MoS_2 (Ayari et al. 2007). The mentioned value of mobility is smaller than the room-temperature mobility of bulk MoS_2 ~200–500 $cm^2\ V^{-1}\ s^{-1}$ (Fivaz and Mooser 1967). It motivated the further efforts to enhance the mobility by engineering the dielectric screening as it was demonstrated for graphene (Chen, Xia et al. 2009a, 2009b). A sketch of the MoS_2 FET with electrical wiring is presented in Fig. 11.14 (right). The top-gate length, source–gate spacing, and gate–drain spacing were 500 nm and the width of the top gate of that device was 4 mm. The experimental measurements were performed at room temperature. The wiring schematic of the device is shown in Fig. 11.15 (left). The measurements were conducted using the semiconductor parameter analyzer and shielded probe station with voltage sources connected, as shown in Fig. 11.15 (left). The MoS_2 transistors with 0.65 nm thick conductive channels were measured applying the drain–source bias V_{ds} via the pair of gold source and drain electrodes and gate voltage V_{bg} to the degenerately doped silicon substrate. The top gate was not fixed electrically at this time. The electric current characteristics through the MoS_2 transistors versus the back-gate voltage are shown in Figs. 11.14 and 11.15. The shown current characteristic is typical for the FETs with an *n*-type channel. Similar curves to those shown in Fig. 11.15 had been obtained for all the MoS_2 transistors which have been fabricated. They do not depend on the number of layers or contacting material and show behavior typical for the FETs with *n*-type channels. There is no visible hysteresis of the I_{ds}–V_{bg} curves even after many sweeps on the same device are performed. When the gate voltages are maintained constant the electric current I_{ds} remains constant as well. It suggests that the top gate is not likely

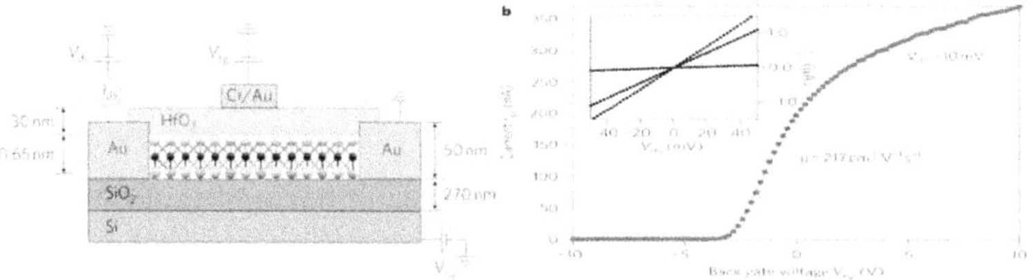

Figure 11.15 Wiring of the MoS2 monolayer transistors. Left: A sketch of the MoS2 FET complemented by electrical wiring. An atomic monolayer MoS_2 with thickness 6.5 Å was placed on a 270 nm thick SiO_2 dielectric with a degenerately doped silicon substrate under it, which serves as the back gate. The gold electrodes act as drain and source. The 30 nm ALD-grown HfO_2 separates the MoS_2 atomic sheet from the top gate with thickness 4 mm. The top-gate length, sourc–gate spacing, and gate–drain spacing are 500 nm each. Right: Transfer characteristic of the MoS_2 FET with 10 mV applied bias voltage V_{ds} at room temperature. Back-gate voltage V_{bg} is applied to the substrate and the top gate is disconnected. Inset: The I_{ds}–V_{ds} curve measured for V_{bg} values of 0, 1, and 5 V. Adapted from Radisavljevic et al. (2011), © 2011 NPG.

to accumulate the electric charge during the measurements. The constant surface charge trapped on the top gate is estimated as $n \approx 4.6 \times 1012\ \text{cm}^{-2}$. In principle, the trapped charge can shift the threshold voltage by ~1 V but it will inflict no change in the slope of the I_{ds}–V_{bg} curve in Fig. 11.15 used to estimate the channel mobility. The linear behavior of the source current versus source bias characteristics (Fig. 11.15, inset) is in the range of +50 mV of voltages which confirms that the gold contacts are ohmic.

The atomic-scale thickness allows us to improve the expected resilience of MoS_2 to short-channel effects which is confirmed by theoretical simulations of performance of the single-layer MoS_2 transistor (Liu et al. 2011, Yoon et al. 2011). Numeric simulations suggest that the top-gated MoS_2 transistors with gate lengths of 15 nm have a potential to perform in the ballistic regime with on-current as high as 1.6 mA μm^{-1}, subthreshold swing close to 60 mV per dec and current on/off ratio of 10^{10}. The computed current-voltage characteristics for a single-layer MoS_2 transistor at different operating conditions are shown in Fig. 11.13 (left) (Yoon et al. 2011). Although the mobility of MoS_2 is generally below the typical values for conventional III–V transistors, the overall electrical performance characteristics, relatively high earth abundance and a high degree of electrostatic control ensure that MoS_2 can be the best material for low-power electronics (Yoon et al. 2011).

Digital logic circuits represent another important application of the 2D TMDC transistors. A functional electronic circuit has been recently reported by Radisavljevic, Whitwick et al. (2011), where up to six independently switchable transistors were fabricated. The devices were formed by lithographical patterning of multiple sets of the electrodes on the common monolayer MoS_2, as shown in Fig. 11.13 (right) (Radisavljevic, Whitwick et al. 2011). A logical inverter was fabricated on a single flake of MoS_2 which represented an integrated circuit composed of two TMDC transistors. The device operated converting a logical 0 into a logical 1, and it also worked as a logical NOR gate (Radisavljevic, Whitwick et al. 2011). The latter logical gate can serve as an elementary building block for large digital logic circuits since it represents one of the universal gates that can be built in combinations to form all other logic operations. The complex integrated circuits have been recently built on bilayer MoS_2 also by Wang et al. (2012). The mentioned circuit included a logical NAND gate, static random access memory, an inverter, and five-stage ring oscillator.

Using ionic liquid as a gate in a thin (10 nm thick) MoS_2 electric double-layer transistor (EDTL), which is shown in Fig. 11.12 (center), one can obtain very high carrier concentrations of $\sim 10^{14}$ cm^2 (Zhang et al. 2012). The controlled switching between the *n*- and *p*-type transports regarded as ambipolarity can be utilized in applications like *p–n*-junction optoelectronics and CMOS logic. In the reported ambipolar device (Zhang et al. 2012), the on/off ratio exceeds $\sim$200 which, nonetheless, is still much below than that for the single-layer device described above (Radisavljevic, Whitwick et al. 2011). Such a relatively low on/off ratio occurs because of the bulk of flakes involved in the off-current transmission (Zhang et al. 2012). The source–drain current as a function of gate voltage characterizing a thin-flake and bulk MoS_2 ambipolar devices is plotted in Fig. 11.12 (right).

11.6 Perspectives of TMDC Electronics

High-performance flexible electronics belongs to one of the most promising near-term applications of the semiconducting 2D TMDCs. According to measurements, single-layer MoS_2 is 30 times mechanically stronger than steel and before breaking it can be deformed up to 11%. Therefore, MoS_2 is one of the strongest semiconducting materials and is very advantageous for fabrication of flexible substrates. Using the electrochemically lithiated and exfoliated MoS_2 (He et al. 2012) and the CVD-grown MoS_2 with an ionic gel gate dielectric (Pu et al. 2012), some groups have fabricated flexible transistors which work as gas sensors. The excellent mechanical properties like those observed in MoS_2 were also reported for other dichalcogenide semiconductors (Ataca et al. 2012). In principle, one can position the hybrid all-2D electronics on flexible substrates where the semiconducting 2D TMDCs are combined with other 2D materials, such as conducting graphene and insulating BN.

The high-electron-mobility transistors (HEMTs) represent another potential application of the 2D TMDC. The HEMT devices are usually fabricated as planar junctions of conventional semiconductors with distinct band gaps. One example is the planar junction GaAs/AlGaAs formed with the molecular beam evaporation techniques (Khan, Bhattarai et al. 1993, Khan, Kuznia et al. 1993a,

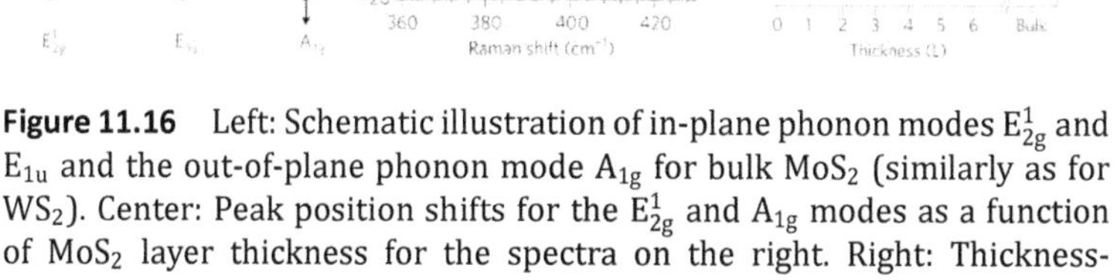

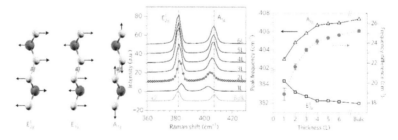

Figure 11.16 Left: Schematic illustration of in-plane phonon modes E_{2g}^1 and E_{1u} and the out-of-plane phonon mode A_{1g} for bulk MoS_2 (similarly as for WS_2). Center: Peak position shifts for the E_{2g}^1 and A_{1g} modes as a function of MoS_2 layer thickness for the spectra on the right. Right: Thickness-dependent Raman spectra for MoS_2. Adapted from Lee et al. (2010) and Molina-Sanchez and Wirtz (2011).

1993b, 1993c). In the HEMT junctions, the semiconducting electrode with the larger band gap is undoped, whereas the other electrode with the smaller band gap is highly doped. Since the scattering from dopants is minimized, the electrons move freely from the doped to the undoped layers. One can use two different TMDCs with similar lattice constants and distinct band gaps as two semiconducting electrodes in the HEMT setup, shown in Table 11.1. A strong motivation here is that the TMDCs are generally better earth-abundant and easily processable than traditional semiconductors. The described TMDC-based 2D HEMT device is sketched in Fig. 11.13 (right), where an undoped TMDC layer is interfaced with another highly doped layer. One also designs a reverse-biased Schottky contact which serves as the gate electrode modulating the signal through the device. Among the factors which restrict the use of TMDCs in nanoelectronics one should mention a relatively high effective mass of carriers and relatively low carrier mobility values (Yoon et al. 2011). Those parameters impose the limits of performance in some applications.

Another very promising activity is to construct vertical heterostructures with stacking of different 2D layered materials at the top of each other. The approach can yield a novel type of hybrid devices whose operating principles are quite different from conventional methods. A recent example of such a device is the field-effect tunneling transistor which is based on a new device architecture (Britnell et al. 2012). For that device, the authors reported an impressive value of switching ratio. The authors used two independently controlled graphene layers which are separated by thin MoS_2 or h-BN layers acting as tunneling barriers.

Optoelectronic devices are capable of generating, detecting, interacting with or controlling the light by electronic methods. Many optoelectronic applications including solar cells, optical switches, lasers, LEDs, displays, and photodetectors are based on semiconductor quantum dots, nanowires, carbon nanotubes, and other nanomaterials (Avouris et al. 2008a, 2008b, Collins and Avouris 2008, Shaver et al. 2008, Steiner et al. 2008, Xia et al. 2008). An increasingly important research direction is devoted to flexible and transparent optoelectronic devices which can be utilized in transparent displays, solar arrays, and wearable electronics. An ability of a semiconducting material or device to absorb and emit light strongly depends on its electronic band structures. If a semiconducting band gap is direct, the photons with energy exceeding the band-gap energy can be easily emitted or absorbed. However, if the band gap is indirect, efficiency of the photon absorption or emission process is strongly diminished because an additional phonon is required to supply the difference in momentum for the process to be completed. The primarily direct semiconducting band gaps in the single-layer TMDCs make them very attractive for a variety of optoelectronic applications. Because the 2D TMDC semiconductors are atomically thin and processable, they have great potential for flexible and transparent optoelectronics.

11.7 Vibrational and Optical Properties of TMDCs

The general optical properties of TMDC depend on the electronic band structures of those materials. In particular, when the band gap changes from indirect to direct and its magnitude increases, the transition is clearly observed as a change not only in photoconductivity of MoS_2, but also in absorption spectra and photoluminescence as is evident from experimental data presented in Fig. 11.17. It causes an increase by factor $\sim 10^4$ in photoluminescence quantum yield when the structure changes from bulk to monolayer MoS_2. Even higher quantum yield occurs for regions of the MoS_2 monolayer flake suspended over holes in the substrate, as visible in (Mak, Lee et al. 2010, Mak, Shan et al. 2010). However, in the latter case, the overall quantum yield is significantly lower than the near-unity values that would be expected for a direct-gap semiconductor. That magnitude for MoS_2 measured so far is about 10^{-5}–10^{-6} for few-layer samples and up to 4×10^{-3} for monolayer samples (Mak, Lee et al. 2010, Mak, Shan et al. 2010). The experimental data (Mak, Lee et al. 2010, Mak, Shan et al. 2010) suggest that the quantum yield is higher for suspended samples and samples placed on hBN substrates than on SiO_2. However, the quenching mechanisms in MoS_2 and other TMDCs is not understood well to firmly control and to increase the quantum yield for the optoelectronic applications.

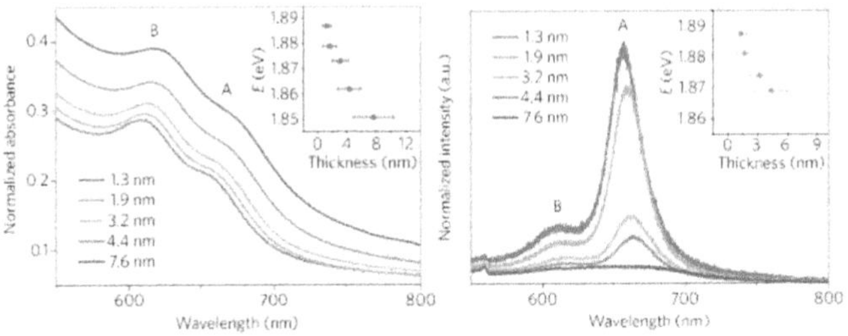

Figure 11.17 Absorption (left) and photoluminescence spectra (right) of MoS_2 thin films with average thicknesses ranging from 1.3 to 7.6 nm. Insets of left and right show the energy of the A exciton peak as a function of average film thickness. The peak energies were extracted from the absorption and photoluminescence spectra in the main panel, respectively. Adapted from Eda et al. (2011) © ACS.

A direct-gap luminescence feature at 1.9 eV (Mak, Lee et al. 2010, Mak, Shan et al. 2010) is pronounced as the main peak of the monolayer MoS_2 photoluminescence spectrum. The indirect-gap luminescence in the few-layer samples of MoS_2 exhibit additional peaks (Mak, Lee et al. 2010, Mak, Shan et al. 2010). Such band-structure peaks appear also in the photoconductivity of TMDCs, with steps of increasing the MoS_2 photocurrent as soon as the photon energy corresponds to the direct- and indirect-gap energies (Mak, Lee et al. 2010, Mak, Shan et al. 2010). There are also two main peaks corresponding to exciton bands, the so-called A and B excitons. Thus, the optical absorption spectrum of bulk MoS_2 shows the direct-gap transitions at the K point of the Brillouin zone between the maximum of split valence bands and the minimum of the conduction band (Coehoorn et al. 1987a, 1987b, Mak, Lee et al. 2010, Mak, Shan et al. 2010). A numeric study indicates that the A and B excitons are related to the expected energies of the gap energies at the K point (Ramasubramaniam 2012). The corresponding band splitting emerges due to spin–orbit coupling in monolayer MoS_2. Besides, according to the calculations, the binding energies of excitons are big because the dielectric constants are smaller than in the bulk. Another increase of the binding energy is caused by the 2D confinement which amounts ~0.897 eV for monolayer MoS_2 and 0.424 eV for bilayer (Cheiwchanchamnangij and Lambrecht 2012). Another numeric result is that the energy required to create an exciton is much lower than the band gap because the transition energy of the excitons is offset by the exciton binding energy. Therefore, the optical transition energies are not equal to the transport band-gap energies.

Ab initio calculations of phonon spectra in the 2D TMDCs have been performed by Molina-Sanchez and Wirtz (2011) and compared with the experimental Raman spectra (Lee et al. 2010). The in-plane E^1_{2g} and E_{1u} phonon modes, and the out-of-plane A_{1g} mode are responsible for the main Raman peaks as seen in Fig. 11.16 (center). When the layer becomes thinner, the frequency of A_{1g} mode near 406 cm^{-1} lowers, whereas the frequency of E^1_{2g} mode near 382 cm^{-1} increases, as shown in Fig. 11.16 (center) (Lee et al. 2010). The shifts of the corresponding peak positions is used to determine the layer thicknesses from the Raman spectroscopy data. The shifts are emerging from a variation of effective restoring forces on atoms due to the change of the layer's thickness and also due to an increase of dielectric screening of the long-range Coulomb interactions (Molina-Sanchez and Wirtz 2011).

11.7.1 *Transparent and Flexible Optoelectronics*

Many applications, like displays and wearable electronics require flexible and transparent optoelectronic nanodevices. Their properties are critical using in huge variety of transparent and flexible components including light emitters, semiconductors, conductors, dielectrics, and optical absorbers. To ensure a good performance of complex flexible and transparent nanoelectronic circuit, one should integrate diverse classes of 2D materials with distinct properties. The materials which are widely implemented nowadays like, for example, indium tin oxide, are inflexible and increasingly expensive due to the scarcity of indium. Thus, graphene whose conductivity is high while the broadband absorption is low, serves as a promising flexible and earth-abundant replacement for the currently implemented materials for the design of transparent conductors. A promising choice for the semiconducting components with tunable band gaps are the 2D TMDCs which can serve as light-absorbing components or light-emitting devices. Besides, layered perovskites and BN are promising as 2D dielectric materials.

11.7.2 *Photodetection and Photovoltaics*

Light-absorbing materials in alternative thin-film solar cells (Alharbi et al. 2011) constitute another promising application of TMDCs due to their relatively high earth abundance and their direct band gaps. Their optoelectronic properties in the visible range are suitable for fabricating flexible photovoltaic elements that can coat buildings and curved structures. The compatibility of conduction- and valence-band edges and work functions of many TMDCs with the work functions of generally used electrode materials (Beal et al. 1975) is another strong side which supports their suitability for wide applications. Furthermore, a capability for tuning the band gap of TMDCs while intercalating them with metal ions and organic molecules (Friend and Yoffe 1987) permits optical absorbances to be adjusted in the photovoltaic applications.

There are many benefits of the TMDCs functionality in photovoltaics and photodetectors. In particular, a phototransistor made from a single layer of MoS_2 has shown its potential as a photodetector and thin films of MoS_2 and WS_2 are photosensitive, as illustrated in Fig. 11.18 (Yin et al. 2012). The experimental data suggest that the photodetector's photocurrent varies versus the

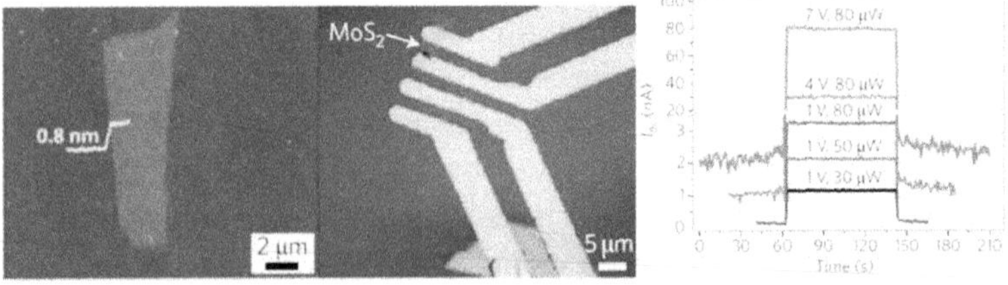

Figure 11.18 Left: Atomic force microscopy image of MoS_2 monolayer flake (left) and optical microscopy image of this flake made into a device with metal contacts (right). The white trace in the left panel shows a height profile from AFM along the MoS_2 flake edge. Right: Photoswitching characteristics of single-layer MoS_2 phototransistor at different optical power (P_{light}) and drain voltage (V_{ds}). Adapted from Yin et al. (2012).

incident light intensity, has high photoresponsivity and responds during 50 ms when the light level changes. One can also tune the photodetection of different wavelengths merely using MoS_2 layers of different thicknesses. Detecting of green light is accomplished with the single- and double-layer MoS_2 where the respective band-gap energies are 1.8 and 1.65 eV (Lee, Kim et al. 2012, Lee, Min et al. 2012). The red light is detected using the triple-layer MoS_2 with a band gap of 1.35 eV. A photoconversion efficiency 1.3% was obtained using the bulk heterojunction solar cell made from MoS_2 atomic layer nanosheets, TiO_2 nanoparticles, and poly(3-hexylthiophene) (P3HT) (Shanmugam et al. 2012a, 2012b). Besides, using the stable inorganic absorber material TiO_2, the electrochemical solar cells were sensitized with WS_2 (Thomalla and Tributsch 2006). One more application is implementation of TMDCs as conductors and electron-blocking layers in the polymer LEDs.

Combining the TMDC materials with different band gaps can yield a variety of devices. The examples include multijunction solar cells which efficiently absorb photons of different energies in the full solar spectrum, thus reducing the energy losses via thermalization, as sketched in Fig. 11.19 (Polman and Atwater 2012). Shown

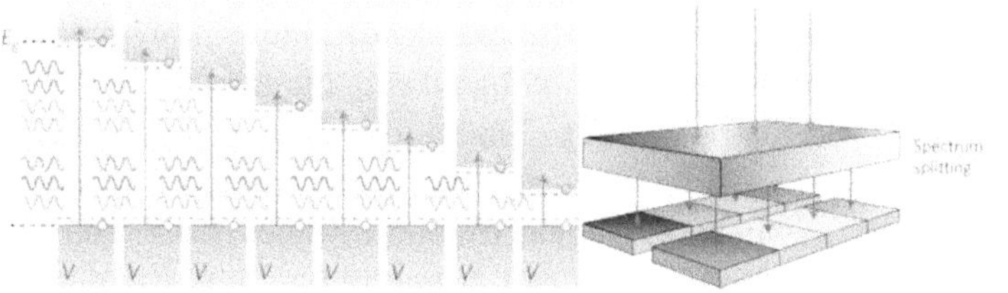

Figure 11.19 Left: Energy-level diagram of proposed multijunction solar cell made of stacked semiconductors of different band gaps to absorb different wavelengths from the solar spectrum to reduce thermalization losses. The blue dashed lines represent the quasi-Fermi levels defining the open-circuit voltage, and yellow dots represent electrons in the device. Right: Proposed solar cell device with parallel structures that may be fabricated using lift-off and printing techniques for patterning the semiconductor layers and a micro- or nanophotonic spectrum-splitting layer. Various band gaps available from different TMDCs are promising for the use in the multijunction photovoltaic devices. Adapted from Polman and Atwater (2012).

devices might be fabricated from two distinct TMDCs which are characterized by different band gaps. The bandwidth is altered between visible and near-infrared regions of spectrum, as listed in Table 11.1. The photodetector involving different layers (Lee, Kim et al. 2012, Lee, Min et al. 2012) is as a good example where tuning of band gap enables a selective absorption of light with different wavelengths.

11.7.3 *Emission of Light*

Electroluminescence and photoluminescence represent the other two important phenomena. The electroluminescence mode is utilized in LEDs and diode lasers since in those optoelectronic devices the photons are emitted in response to electrical stimulus. During the photoluminescence process, the material absorbs one photon and then re-radiates another photon which might have a different energy. The radiative recombination of electrons and holes in direct-band-gap semiconductors creates photons. The process intensity is much higher than in indirect-band-gap semiconductors. That circumstance motivates an interest to monolayer semiconducting TMDCs whose direct band gaps make them good candidates for the future flexible optoelectronics, since they can be used as active light-emitting layers. It makes TMDC more preferable than graphene, where designing of band gap requires an additional chemical procedure. The electroluminescence from MoS_2 is induced, for example, in electrically excited Au nanocontacts (Carladous et al. 2002), whereas the electroluminescence from SnS_2 was obtained on samples exfoliated from lithium intercalation and incorporated into a composite polymer matrix. The quantum yield of the photoluminescence in monolayer MoS_2, which has a direct band gap, is much higher than for bilayer and bulk MoS_2. Nevertheless, the quantum yield of photoluminescence which was measured in MoS_2 is considerably lower than was expected from the direct-gap semiconductor. It motivates further studies to understand the photoluminescence emission mechanisms and quenching processes before any feasible optoelectronic devices could be created.

11.7.4 *Orbit, Spin, and Valley Interactions*

The signals transmitted through the nanoelectronic circuits are carried either by charge or spin. An additional property of charge carriers is the valley index which refers to the confinement of electrons or holes. The index helps to distinguish between excitations positioned at the conduction-band minima or valence-band maxima at the same energies but different locations in momentum space. Using of the valley index leads to potential "valleytronic" devices (Xiao et al. 2012). Substantial splitting of spins in certain materials can emerge due to various reasons, for example, when driving the system out of equilibrium or symmetry breaking. The working principle of spintronic devices is based on the system's ability to maintain the population of spin-polarized carriers.

Special conditions can be created in the group IV semiconducting dichalcogenides (MoS_2, $MoSe_2$, WS_2, and WSe_2). They simultaneously provide a strong spin–orbit-induced electronic band splitting along with spin–valley coupling (Xiao et al. 2012). As follows from the structural diagrams in Fig. 11.1 (Xiao et al. 2012), there is no inversion symmetry in the monolayer TMDCs like MoS_2. It makes difference from graphene and from bilayer MoS_2, which are both centrosymmetric. A very strong spin–orbit splitting emerges because of absence of inversion symmetry, presence of the in-plane confinement of electron motion, and high mass of the elements in the MX_2 materials. The valence-band splitting ranges between 0.15 and 0.45 eV (Zhu et al. 2011a, 2011b). It is quite different for graphene, where a very weak spin–orbit interaction is caused by the low mass of carbon. Because the spin splitting in the MX_2 materials is very strong it makes them promising candidates for spintronic devices.

The absence of inversion symmetry and strong spin–orbit coupling introduces an interplay between the spin and valley transport (Xiao et al. 2012). The valley confinement was experimentally realized in monolayer MoS_2. Optical acting on samples using circularly polarized light allows the carrier populations in distinct valleys to be controlled, which was independently reported by Mak et al. (2012). The result represents the first step towards the building of novel valleytronic devices. A variety of quite distinctive spin, orbit, and valley properties in TMDCs contains a strong potential for their new applications.

11.7.5 *Molecular Sensing*

New opportunities in the molecular sensing applications benefit from a rapid progress of electronics, optoelectronics, and chemistry of TMDC. Because of 2D nature of the new materials, they have very high surface-to-volume ratio. Therefore, TMDCs are exceptionally sensitive to subtle details of their surroundings. Placing the 2D samples in gases and vapors, their physical properties change due to the intercalation, charge transfer, doping, lattice vibrations, and permittivity shift. The listed changes of properties are exploited designing gas sensors. The adsorbed molecules transfer the electric charge into the graphene sheet which changes its resistivity and therefore is used as the detection mechanism (Schedin et al. 2007, Fowler et al. 2009).

Presence of the adsorbates causes change of the electronic properties of TMDCs which is detectable electrically with TMDCs as elements of transistor circuits measuring their current–voltage characteristics. Furthermore, one measures the optical characteristics including the photoluminescence, absorbance, and Raman spectra. In particular, with intercalating MoS_2 with Li^+ ions one changes positions of the Raman A_{1g} and E^1_{2g} peaks, whereas the peak intensities are decreased. The mechanism is related with strain which is introduced by the Li^+ ions entering interstitial sites (Balendhran et al. 2012). The photoluminescence of monolayer MoS_2 also has a potential for applications in biosensing, where stable fluorescent markers are of importance for imaging and fluorometric assays.

Sensitive detectors of NO gas were designed using transistors fabricated from single- and few-layer MoS_2 sheets (Li, Lu et al. 2012, Li, Yin et al. 2012, Liu, Lai et al. 2012). Since the adsorbed NO induces the *p*-doping, it also changes the resistivity of the intrinsically *n*-doped MoS_2 which constitutes the probable detection mechanism. Very sensitive NO_2 gas detectors (He et al. 2012) were also fabricated implementing the flexible TFTs whose active regions were composed of MoS_2 obtained by the electrochemical lithiation and exfoliation technique. An array of thin-film humidity sensors which detect the moisture from fingers (Feng et al. 2012, Jin et al. 2012, Xiao et al. 2012) was designed using liquid-phase exfoliated nanosheets of VS_2. In the sensors, adsorbed water molecules modulate the conductivity which aims to inhibit the conduction along V atoms at flake edges (Feng et al. 2012). Since the water molecules are easily adsorbed by the hydrophilic MoS_2 surface and polarized versus the applied gate voltage, the MoS_2 transistors demonstrate the humidity-dependent hysteresis in their current–voltage characteristics (Late et al. 2012). Another sensing application of MoS_2 uses flakes incorporated into a glassy carbon electrode in an electrochemical cell. The flakes are electrochemically reduced, and the reduced material shows electrochemical sensitivity in respect of glucose and other biomolecules like dopamine (Wu et al. 2012).

11.8 Future Applications of 2D Materials

Although the layered TMDC materials have been known for decades, atomically thin 2D forms represent a relatively new and exciting subject for nanotechnology. Nowadays, there is a variety of promising applications of the materials in nanoelectronics and optoelectronics. Electronic spectra of many TMDC compounds have large band gaps which makes them suitable as channel materials in digital logic transistors. Presence of direct band gaps in a variety of single-layer TMDCs paves the way for their implementation in optoelectronics. There are many successful examples of the MoS_2 transistors with high on/off ratios and integrated circuits with digital logic operation. Furthermore, the 2D sheets of MoS_2 were used as chemical and gas sensors. Many unconventional properties related to the valley polarization and strong spin–orbit effects were also observed. The former results concerning the material processing and characterization, the bulk TMDC intercalation chemistry and methodology of device fabrication, and nanoscale characterization were utilized for the scientific research and creating of new TMDC devices. The approaches were developed previously for carbon nanotubes and graphene. However, the 2D TMDCs indicate many properties that have not been seen previously in other materials. When the researchers study them deeper, they find more distinctive, unexpected, and exciting phenomena.

Further progress in this area requires the development of scalable and controllable sample preparation which are aimed at obtaining of a considerable amount of atomically thin and uniform TMDC layers, either in solution or on substrates. There are many serious challenges when using the solution-phase preparation. They include an improvement of methods to control the area and thickness of either exfoliated or chemically grown flakes, and finding new recipes and chemicals that can efficiently and safely produce these materials in large volumes. The crystal growth of solid-state samples assumes improvement of the design of large areas, large grain sizes, uniformity, and precise control of the number of layers. Fabrication of the high-quality large samples will facilitate a more adequate understanding of physical and chemical properties of the TMDCs, as well as invent a wide variety of applications. Development and studies of the 2D TDMC represent a rapidly expanding and relatively new field of research.

Problems

11-1. Discuss how to engineer the energy gap in the electron excitation spectrum of pristine graphene.

11-2. What are the benefits of a direct band gap in the electron spectrum of monolayer MoS_2 as compared to the indirect band gap in MoS_2 multilayers?

11-3. What is the physical reason for transition metal dichalcogenides (TMDCs) to behave like dry lubricants?

11-4. How do we estimate the exfoliation energy of layered crystals?

11-5. What is the difference between the band structures of bulk and monolayer TMDCs?

11-6. What are the benefits of using 2D atomic monolayers (AMs) in field-effect transistors (FETs)?

11-7. Discuss the specific role of 2D geometry in the mechanism of electron scattering.

11-8. What are the benefits of using TMDC materials in optoelectronics?

11-9. Explain why the presence of a direct band gap in a TDMC allows us to create devices which utilize electroluminescence and photoluminescence?

References

Alem, N., R. Erni, C. Kisielowski, M. D. Rossell, W. Gannett, and A. Zettl (2009). Atomically thin hexagonal boron nitride probed by ultrahigh-resolution transmission electron microscopy. *Phys Rev B,* **80**(15).

Alharbi, F., J. D. Bass, A. Salhi, A. Alyamani, H. C. Kim, and R. D. Miller (2011). Abundant non-toxic materials for thin film solar cells: alternative to conventional materials. *Renew Energy,* **36**(10): 2753–2758.

Ataca, C., H. Sahin, and S. Ciraci (2012). Stable, single-layer MX2 transition-metal oxides and dichalcogenides in a honeycomb-like structure. *J Phys Chem C,* **116**(16): 8983–8999.

Avouris, P., M. Freitag, and V. Perebeinos (2008a). Carbon-nanotube optoelectronics. *Carbon Nanotubes,* **111**: 423–454.

Avouris, P., M. Freitag, and V. Perebeinos (2008b). Carbon-nanotube photonics and optoelectronics. *Nat Photon,* **2**(6): 341–350.

Ayari, A., E. Cobas, O. Ogundadegbe, and M. S. Fuhrer (2007). Realization and electrical characterization of ultrathin crystals of layered transition-metal dichalcogenides. *J Appl Phys,* **101**(1).

Balandin, A. A., S. Ghosh, W. Z. Bao, I. Calizo, D. Teweldebrhan, F. Miao, and C. N. Lau (2008). Superior thermal conductivity of single-layer graphene. *Nano Lett,* **8**(3): 902–907.

Balendhran, S., J. Z. Ou, M. Bhaskaran, S. Sriram, S. Ippolito, Z. Vasic, E. Kats, S. Bhargava, S. Zhuiykov, and K. Kalantar-zadeh (2012). Atomically thin layers of MoS2 via a two step thermal evaporation-exfoliation method. *Nanoscale,* **4**(2): 461–466.

Balog, R., B. Jorgensen, L. Nilsson, M. Andersen, E. Rienks, M. Bianchi, M. Fanetti, E. Laegsgaard, A. Baraldi, S. Lizzit, Z. Sljivancanin, F. Besenbacher, B. Hammer, T. G. Pedersen, P. Hofmann, and L. Hornekaer (2010). Bandgap opening in graphene induced by patterned hydrogen adsorption. *Nat Mater,* **9**(4): 315–319.

Beal, A. R., H. P. Hughes, and W. Y. Liang (1975). Reflectivity spectra of some group VA transition-metal dichalcogenides. *J Phys C Solid State,* **8**(24): 4236–4248.

Benameur, M. M., B. Radisavljevic, J. S. Heron, S. Sahoo, H. Berger, and A. Kis (2011). Visibility of dichalcogenide nanolayers. *Nanotechnology,* **22**(12).

Bertolazzi, S., J. Brivio, and A. Kis (2011). Stretching and breaking of ultrathin MoS2. *ACS Nano,* **5**(12): 9703–9709.

Bissessur, R., J. Heising, W. Hirpo, and M. Kanatzidis (1996). Toward pillared layered metal sulfides. Intercalation of the chalcogenide clusters Co(6)Q(8)(PR(3))(6) (Q=S, Se, and Te and R=alkyl) into MoS2. *Chem Mater,* **8**(2): 318.

Bohr, M. T., R. S. Chau, T. Ghani, and K. Mistry (2007). The high-k solution. *IEEE Spectrum,* **44**(10): 29–35.

Britnell, L., R. V. Gorbachev, R. Jalil, B. D. Belle, F. Schedin, A. Mishchenko, T. Georgiou, M. I. Katsnelson, L. Eaves, S. V. Morozov, N. M. R. Peres, J. Leist, A. K. Geim, K. S. Novoselov, and L. A. Ponomarenko (2012). Field-effect tunneling transistor based on vertical graphene heterostructures. *Science,* **335**(6071): 947–950.

Brivio, J., D. T. L. Alexander, and A. Kis (2011). Ripples and layers in ultrathin MoS2 membranes. *Nano Lett,* **11**(12): 5148–5153.

Bunch, J. S., S. S. Verbridge, J. S. Alden, A. M. van der Zande, J. M. Parpia, H. G. Craighead, and P. L. McEuen (2008). Impermeable atomic membranes from graphene sheets. *Nano Lett,* **8**(8): 2458–2462.

Carladous, A., R. Coratger, F. Ajustron, G. Seine, R. Pechou, and J. Beauvillain (2002). Light emission from spectral analysis of Au/MoS2 nanocontacts stimulated by scanning tunneling microscopy. *Phys Rev B,* **66**(4).

Castellanos-Gomez, A., M. Barkelid, A. M. Goossens, V. E. Calado, H. S. J. van der Zant, and G. A. Steele (2012). Laser-thinning of MoS2: on demand generation of a single-layer semiconductor. *Nano Lett,* **12**(6): 3187–3192.

Castro, E. V., H. Ochoa, M. I. Katsnelson, R. V. Gorbachev, D. C. Elias, K. S. Novoselov, A. K. Geim, and F. Guinea (2010). Limits on charge carrier mobility in suspended graphene due to flexural phonons. *Phys Rev Lett,* **105**(26).

Castro Neto, A. H., F. Guinea, N. M. R. Peres, K. S. Novoselov, and A. K. Geim (2009). The electronic properties of graphene. *Rev Mod Phys,* **81**(1): 109–162.

Cheiwchanchamnangij, T., and W. R. L. Lambrecht (2012). Quasiparticle band structure calculation of monolayer, bilayer, and bulk MoS2. *Phys Rev B,* **85**(20).

Chen, F., J. L. Xia, D. K. Ferry, and N. J. Tao (2009). Dielectric screening enhanced performance in graphene FET. *Nano Lett,* **9**(7): 2571–2574.

Chen, F., J. L. Xia, and N. J. Tao (2009). Ionic screening of charged-impurity scattering in graphene. *Nano Lett,* **9**(4): 1621–1625.

Chen, J. H., C. Jang, S. D. Xiao, M. Ishigami, and M. S. Fuhrer (2008). Intrinsic and extrinsic performance limits of graphene devices on SiO2. *Nat Nanotechnol,* **3**(4): 206–209.

Coehoorn, R., C. Haas, and R. A. Degroot (1987a). Electronic-structure of Mose2, Mos2, and Wse2 .2. The nature of the optical band-gaps. *Phys Rev B,* **35**(12): 6203–6206.

Coehoorn, R., C. Haas, J. Dijkstra, C. J. F. Flipse, R. A. Degroot, and A. Wold (1987b). Electronic-structure of Mose2, Mos2, and Wse2 .1. Band-structure calculations and photoelectron-spectroscopy. *Phys Rev B,* **35**(12): 6195–6202.

Coleman, J. N., M. Lotya, A. O'Neill, S. D. Bergin, P. J. King, U. Khan, K. Young, A. Gaucher, S. De, R. J. Smith, I. V. Shvets, S. K. Arora, G. Stanton, H. Y. Kim, K. Lee, G. T. Kim, G. S. Duesberg, T. Hallam, J. J. Boland, J. J. Wang, J. F. Donegan, J. C. Grunlan, G. Moriarty, A. Shmeliov, R. J. Nicholls, J. M. Perkins, E. M. Grieveson, K. Theuwissen, D. W. McComb, P. D. Nellist, and V. Nicolosi (2011). Two-dimensional nanosheets produced by liquid exfoliation of layered materials. *Science,* **331**(6017): 568–571.

Colinge, J. P. (2004a). Multiple-gate SOI MOSFETs. *Solid State Electron,* **48**(6): 897–905.

Colinge, J. P. (2004b). Novel gate concepts for MOS devices. *Essderc 2004: Proceedings of the 34th European Solid-State Device Research Conference,* 45–49.

Colinge, J. P. (2004c). SOI for hostile environment applications. *IEEE International SOI Conference, Proceedings,* 1–4.

Collins, P. G., and P. Avouris (2008). The electronic properties of carbon nanotubes. *Carbon Nanotubes: Quantum Cylinders of Graphene,* 49–81.

Cunningham, G., M. Lotya, C. S. Cucinotta, S. Sanvito, S. D. Bergin, R. Menzel, M. S. P. Shaffer, and J. N. Coleman (2012). Solvent exfoliation of transition metal dichalcogenides: dispersibility of exfoliated nanosheets varies only weakly between compounds. *ACS Nano,* **6**(4): 3468–3480.

Dines, M. B. (1975). Lithium Intercalation via n-butyllithium of layered transition-metal dichalcogenides. *Mater Res Bull,* **10**(4): 287–291.

Ding, Y., Y. L. Wang, J. Ni, L. Shi, S. Q. Shi, and W. H. Tang (2011). First principles study of structural, vibrational and electronic properties of graphene-like MX2 (M=Mo, Nb, W, Ta; X=S, Se, Te) monolayers. *Phys B Condens Mater,* **406**(11): 2254–2260.

Ding, Y. C., A. P. Xiang, X. J. He, and X. F. Hu (2011). Structural, elastic constants, hardness, and optical properties of pyrite-type dinitrides (CN2, SiN2, GeN2). *Phys B Condens Mater,* **406**(8): 1357–1362.

Eda, G., H. Yamaguchi, D. Voiry, T. Fujita, M. W. Chen, and M. Chhowalla (2011). Photoluminescence from chemically exfoliated MoS2. *Nano Lett,* **11**(12): 5111–5116.

Elias, D. C., R. V. Gorbachev, A. S. Mayorov, S. V. Morozov, A. A. Zhukov, P. Blake, L. A. Ponomarenko, I. V. Grigorieva, K. S. Novoselov, F. Guinea, and A. K. Geim (2011). Dirac cones reshaped by interaction effects in suspended graphene. *Nat Phys,* **7**(9): 701–704.

Fang, H., S. Chuang, T. C. Chang, K. Takei, T. Takahashi, and A. Javey (2012). High-performance single layered WSe2 p-FETs with chemically doped contacts. *Nano Lett,* **12**(7): 3788–3792.

Feng, J., L. L. Peng, C. Z. Wu, X. Sun, S. L. Hu, C. W. Lin, J. Dai, J. L. Yang, and Y. Xie (2012). Giant moisture responsiveness of VS2 ultrathin nanosheets for novel touchless positioning interface. *Adv Mater,* **24**(15): 1969–1974.

Fivaz, R., and E. Mooser (1967). Mobility of charge carriers in semiconducting layer structures. *Phys Rev,* **163**(3): 743.

Fowler, J. D., M. J. Allen, V. C. Tung, Y. Yang, R. B. Kaner, and B. H. Weiller (2009). Practical chemical sensors from chemically derived graphene. *ACS Nano,* **3**(2): 301–306.

Friend, R. H., and A. D. Yoffe (1987). Electronic-properties of intercalation complexes of the transition-metal dichalcogenides. *Adv Phys,* **36**(1): 1–94.

Geim, A. K. (2009). Graphene: status and prospects. *Science,* **324**(5934): 1530–1534.

Gordon, R. A., D. Yang, E. D. Crozier, D. T. Jiang, and R. F. Frindt (2002). Structures of exfoliated single layers of WS2, MoS2, and MoSe2 in aqueous suspension. *Phys Rev B,* **65**(12).

Grasso, C. S., Y. M. Wu, D. R. Robinson, X. H. Cao, S. M. Dhanasekaran, A. P. Khan, M. J. Quist, X. Jing, R. J. Lonigro, J. C. Brenner, I. A. Asangani, B. Ateeq, S. Y. Chun, J. Siddiqui, L. Sam, M. Anstett, R. Mehra, J. R. Prensner, N. Palanisamy, G. A. Ryslik, F. Vandin, B. J. Raphael, L. P. Kunju, D. R. Rhodes, K. J. Pienta, A. M. Chinnaiyan, and S. A. Tomlins (2012). The mutational landscape of lethal castration-resistant prostate cancer. *Nature,* **487**(7406): 239–243.

Green, A. A., and M. C. Hersam (2009). Solution phase production of graphene with controlled thickness via density differentiation. *Nano Lett,* **9**(12): 4031–4036.

Han, M. Y., B. Ozyilmaz, Y. B. Zhang, and P. Kim (2007). Energy band-gap engineering of graphene nanoribbons. *Phys Rev Lett,* **98**(20).

Hass, J., W. A. de Heer, and E. H. Conrad (2008). The growth and morphology of epitaxial multilayer graphene. *J Phys: Condens Mater,* **20**(32).

He, Q. Y., Z. Y. Zeng, Z. Y. Yin, H. Li, S. X. Wu, X. Huang, and H. Zhang (2012). Fabrication of flexible MoS2 thin-film transistor arrays for practical gas-sensing applications. *Small,* **8**(19): 2994–2999.

Hutchison, J. C., R. Bissessur, and D. F. Shriver (1996). Conductivity anisotropy of polyphosphazene-montmorillonite composite electrolytes. *Chem Mater,* **8**(8): 1597.

Hwang, E. H., S. Adam, and S. Das Sarma (2007). Carrier transport in two-dimensional graphene layers. *Phys Rev Lett,* **98**(18).

Ishigami, M., J. H. Chen, W. G. Cullen, M. S. Fuhrer, and E. D. Williams (2007). Atomic structure of graphene on SiO2. *Nano Lett,* **7**(6): 1643–1648.

Jena, D., and A. Konar (2007). Enhancement of carrier mobility in semiconductor nanostructures by dielectric engineering. *Phys Rev Lett,* **98**(13).

Jin, Y., J. Feng, X. L. Zhang, Y. G. Bi, Y. Bai, L. Chen, T. Lan, Y. F. Liu, Q. D. Chen, and H. B. Sun (2012). Solving efficiency-stability tradeoff in top-emitting organic light-emitting devices by employing periodically corrugated metallic cathode. *Adv Mater,* **24**(9): 1187–1191.

Kaasbjerg, K., K. S. Thygesen, and K. W. Jacobsen (2012a). Phonon-limited mobility in n-type single-layer MoS2 from first principles. *Phys Rev B,* **85**(11).

Kaasbjerg, K., K. S. Thygesen, and K. W. Jacobsen (2012b). Unraveling the acoustic electron-phonon interaction in graphene. *Phys Rev B,* **85**(16).

Kalantar-zadeh, K., A. Vijayaraghavan, M. H. Ham, H. D. Zheng, M. Breedon, and M. S. Strano (2010). Synthesis of atomically thin WO3 sheets from hydrated tungsten trioxide. *Chem Mater,* **22**(19): 5660–5666.

Kam, K. K., and B. A. Parkinson (1982). Detailed photocurrent spectroscopy of the semiconducting group-VI transition-metal dichalcogenides. *J Phys Chem,* **86**(4): 463–467.

Khan, M. A., A. Bhattarai, J. N. Kuznia, and D. T. Olson (1993). High-electron-mobility transistor based on a GaN—AlxGa1—xN heterojunction. *Appl Phys Lett,* **63**(9): 1214–1215.

Khan, M. A., J. N. Kuznia, A. R. Bhattarai, and D. T. Olson (1993a). Metal-Semiconductor field-effect transistor based on single-crystal gan. *Appl Phys Lett,* **62**(15): 1786–1787.

Khan, M. A., J. N. Kuznia, D. T. Olson, M. Blasingame, and A. R. Bhattarai (1993b). Schottky-Barrier photodetector based on Mg-doped P-type gan films. *Appl Phys Lett,* **63**(18): 2455–2456.

Khan, M. A., J. N. Kuznia, D. T. Olson, T. George, and W. T. Pike (1993c). Gan/Aln digital alloy short-period superlattices by switched atomic layer metalorganic chemical-vapor-deposition. *Appl Phys Lett,* **63**(25): 3470–3472.

Kobayashi, K., and J. Yamauchi (1995). Electronic-structure and scanning-tunneling-microscopy image of molybdenum dichalcogenide surfaces. *Phys Rev B,* **51**(23): 17085–17095.

Konar, A., T. A. Fang, and D. Jena (2010a). Effect of high-kappa gate dielectrics on charge transport in graphene-based field effect transistors. *Phys Rev B,* **82**(11).

Konar, A., T. A. Fang, N. Sun, and D. Jena (2010b). Anisotropic charge transport in nonpolar GaN quantum wells: polarization induced line charge, and interface roughness scattering. *Phys Rev B,* **82**(19).

Kuc, A., N. Zibouche, and T. Heine (2011). Influence of quantum confinement on the electronic structure of the transition metal sulfide TS2. *Phys Rev B,* **83**(24).

Late, D. J., B. Liu, H. S. S. R. Matte, V. P. Dravid, and C. N. R. Rao (2012). Hysteresis in single-layer MoS2 field effect transistors. *ACS Nano,* **6**(6): 5635–5641.

Lee, C., H. Yan, L. E. Brus, T. F. Heinz, J. Hone, and S. Ryu (2010). Anomalous Lattice vibrations of single- and few-layer MoS2. *ACS Nano,* **4**(5): 2695–2700.

Lee, H. S., C. J. Kim, D. Lee, R. R. Lee, K. Kang, I. Hwang, and M. H. Jo (2012). Large electroabsorption susceptibility mediated by internal photoconductive gain in Ge nanowires. *Nano Lett,* **12**(11): 5913–5918.

Lee, H. S., S. W. Min, Y. G. Chang, M. K. Park, T. Nam, H. Kim, J. H. Kim, S. Ryu, and S. Im (2012). MoS2 nanosheet phototransistors with thickness-modulated optical energy gap. *Nano Lett,* **12**(7): 3695–3700.

Lee, K., H. Y. Kim, M. Lotya, J. N. Coleman, G. T. Kim, and G. S. Duesberg (2011). Electrical characteristics of molybdenum disulfide flakes produced by liquid exfoliation. *Adv Mater,* **23**(36): 4178.

Lee, Y. H., X. Q. Zhang, W. J. Zhang, M. T. Chang, C. T. Lin, K. D. Chang, Y. C. Yu, J. T. W. Wang, C. S. Chang, L. J. Li, and T. W. Lin (2012). Synthesis of large-area MoS2 atomic layers with chemical vapor deposition. *Adv Mater,* **24**(17): 2320–2325.

Li, H., G. Lu, Z. Y. Yin, Q. Y. He, H. Li, Q. Zhang, and H. Zhang (2012). Optical identification of single- and few-layer MoS2 sheets. *Small,* **8**(5): 682–686.

Li, H., Z. Y. Yin, Q. Y. He, H. Li, X. Huang, G. Lu, D. W. H. Fam, A. I. Y. Tok, Q. Zhang, and H. Zhang (2012). Fabrication of single- and multilayer MoS2 film-based field-effect transistors for sensing NO at room temperature. *Small,* **8**(1): 63–67.

Li, T. S., and G. L. Galli (2007). Electronic properties of MOS2 nanoparticles. *J Phys Chem C,* **111**(44): 16192–16196.

Li, X. L., X. R. Wang, L. Zhang, S. W. Lee, and H. J. Dai (2008). Chemically derived, ultrasmooth graphene nanoribbon semiconductors. *Science,* **319**(5867): 1229–1232.

Li, X. S., W. W. Cai, J. H. An, S. Kim, J. Nah, D. X. Yang, R. Piner, A. Velamakanni, I. Jung, E. Tutuc, S. K. Banerjee, L. Colombo, and R. S. Ruoff (2009). Large-area synthesis of high-quality and uniform graphene films on copper foils. *Science,* **324**(5932): 1312–1314.

Lin, M. W., C. Ling, Y. Y. Zhang, H. J. Yoon, M. M. C. Cheng, L. A. Agapito, N. Kioussis, N. Widjaja, and Z. X. Zhou (2011). Room-temperature high on/off ratio in suspended graphene nanoribbon field-effect transistors. *Nanotechnology,* **22**(26).

Lin, Y. M., A. Valdes-Garcia, S. J. Han, D. B. Farmer, I. Meric, Y. N. Sun, Y. Q. Wu, C. Dimitrakopoulos, A. Grill, P. Avouris, and K. A. Jenkins (2011). Wafer-scale graphene integrated circuit. *Science,* **332**(6035): 1294–1297.

Ling, C., G. Setzler, M. W. Lin, K. S. Dhindsa, J. Jin, H. J. Yoon, S. S. Kim, M. M. C. Cheng, N. Widjaja, and Z. X. Zhou (2011). Electrical transport properties of graphene nanoribbons produced from sonicating graphite in solution. *Nanotechnology,* **22**(32).

Liu, K. K., W. J. Zhang, Y. H. Lee, Y. C. Lin, M. T. Chang, C. Su, C. S. Chang, H. Li, Y. M. Shi, H. Zhang, C. S. Lai, and L. J. Li (2012). Growth of large-area and highly crystalline MoS2 thin layers on insulating substrates. *Nano Lett,* **12**(3): 1538–1544.

Liu, L. T., S. B. Kumar, Y. Ouyang, and J. Guo (2011). Performance limits of monolayer transition metal dichalcogenide transistors. *IEEE Trans Electron Devices,* **58**(9): 3042–3047.

Liu, Y. P., T. Lai, H. L. Li, Y. Wang, Z. X. Mei, H. L. Liang, Z. L. Li, F. M. Zhang, W. J. Wang, A. Y. Kuznetsov, and X. L. Du (2012). Nanostructure formation and passivation of large-area black silicon for solar cell applications. *Small*, **8**(9): 1392–1397.

Mak, K. F., K. L. He, J. Shan, and T. F. Heinz (2012). Control of valley polarization in monolayer MoS2 by optical helicity. *Nat Nanotechnol*, **7**(8): 494–498.

Mak, K. F., C. Lee, J. Hone, J. Shan, and T. F. Heinz (2010). Atomically thin MoS2: a new direct-gap semiconductor. *Phys Rev Lett*, **105**(13).

Mak, K. F., J. Shan, and T. F. Heinz (2010). Electronic structure of few-layer graphene: experimental demonstration of strong dependence on stacking sequence. *Phys Rev Lett*, **104**(17).

Matte, H. S. S. R., A. Gomathi, A. K. Manna, D. J. Late, R. Datta, S. K. Pati, and C. N. R. Rao (2010). MoS2 and WS2 analogues of graphene. *Angew Chem, Int Ed*, **49**(24): 4059–4062.

Matte, H. S. S. R., B. Plowman, R. Datta, and C. N. R. Rao (2011). Graphene analogues of layered metal selenides. *Dalton Trans*, **40**(40): 10322–10325.

Mattheis.Lf (1973). Band structures of transition-metal-dichalcogenide layer compounds. *Phys Rev B*, **8**(8): 3719–3740.

Mayorov, A. S., R. V. Gorbachev, S. V. Morozov, L. Britnell, R. Jalil, L. A. Ponomarenko, P. Blake, K. S. Novoselov, K. Watanabe, T. Taniguchi, and A. K. Geim (2011). Micrometer-scale ballistic transport in encapsulated graphene at room temperature. *Nano Lett*, **11**(6): 2396–2399.

Molina-Sanchez, A., and L. Wirtz (2011). Phonons in single-layer and few-layer MoS2 and WS2. *Phys Rev B*, **84**(15).

Morkoc, H., S. Strite, G. B. Gao, M. E. Lin, B. Sverdlov, and M. Burns (1994). Large-band-gap sic, III-V nitride, and II-VI ZnSe-based semiconductor-device technologies. *J Appl Phys*, **76**(3): 1363–1398.

Nair, R. R., H. A. Wu, P. N. Jayaram, I. V. Grigorieva, and A. K. Geim (2012). Unimpeded permeation of water through helium-leak-tight graphene-based membranes. *Science*, **335**(6067): 442–444.

Novoselov, K. S., A. K. Geim, S. V. Morozov, D. Jiang, Y. Zhang, S. V. Dubonos, I. V. Grigorieva, and A. A. Firsov (2004). Electric field effect in atomically thin carbon films. *Science*, **306**(5696): 666–669.

Novoselov, K. S., D. Jiang, F. Schedin, T. J. Booth, V. V. Khotkevich, S. V. Morozov, and A. K. Geim (2005). Two-dimensional atomic crystals. *Proc Natl Acad Sci U S A*, **102**(30): 10451–10453.

O'Neill, A., U. Khan, and J. N. Coleman (2012). Preparation of High concentration dispersions of exfoliated MoS2 with increased flake size. *Chem Mater*, **24**(12): 2414–2421.

Osada, M., and T. Sasaki (2012). Two-dimensional dielectric nanosheets: novel nanoelectronics from nanocrystal building blocks. *Adv Mater,* **24**(2): 210–228.

Peng, C., Y. J. Zhan, and J. Lou (2012). Size-dependent fracture mode transition in copper nanowires. *Small,* **8**(12): 1889–1894.

Peng, Y. Y., Z. Y. Meng, C. Zhong, J. Lu, Z. P. Yang, and Y. T. Qian (2001a). A novel molten-salt method to produce nanocrystalline WS2. *Chem Lett,* (1): 64–65.

Peng, Y. Y., Z. Y. Meng, C. Zhong, J. Lu, W. C. Yu, Y. B. Jia, and Y. T. Qian (2001b). Hydrothermal synthesis and characterization of single-molecular-layer MoS2 and MoSe2. *Chem Lett,* (8): 772–773.

Perisanu, S., P. Vincent, A. Ayari, M. Choueib, S. T. Purcell, M. Bechelany and D. Cornu (2007). High Q factor for mechanical resonances of batch-fabricated SiC nanowires. *Appl Phys Lett,* **90**(4).

Podzorov, V., M. E. Gershenson, C. Kloc, R. Zeis, and E. Bucher (2004). High-mobility field-effect transistors based on transition metal dichalcogenides. *Appl Phys Lett,* **84**(17): 3301–3303.

Podzorov, V., V. M. Pudalov, and M. E. Gershenson (2004). Light-induced switching in back-gated organic transistors with built-in conduction channel. *Appl Phys Lett,* **85**(24): 6039–6041.

Polman, A., and H. A. Atwater (2012). Photonic design principles for ultrahigh-efficiency photovoltaics. *Nat Mater,* **11**(3): 174–177.

Pu, J., Y. Yomogida, K. K. Liu, L. J. Li, Y. Iwasa, and T. Takenobu (2012). Highly flexible MoS2 thin-film transistors with ion gel dielectrics. *Nano Lett,* **12**(8): 4013–4017.

Radisavljevic, B., A. Radenovic, J. Brivio, V. Giacometti, and A. Kis (2011). Single-layer MoS2 transistors. *Nat Nanotechnol,* **6**(3): 147–150.

Radisavljevic, B., M. B. Whitwick, and A. Kis (2011). Integrated Circuits and logic operations based on single-layer MoS2. *ACS Nano,* **5**(12): 9934–9938.

Ramasubramaniam, A. (2012). Large excitonic effects in monolayers of molybdenum, and tungsten dichalcogenides. *Phys Rev B,* **86**(11).

Sakaki, H., T. Noda, K. Hirakawa, M. Tanaka, and T. Matsusue (1987). Interface roughness scattering in GaAs/AlAs quantum-wells. *Appl Phys Lett,* **51**(23): 1934–1936.

Schedin, F., A. K. Geim, S. V. Morozov, E. W. Hill, P. Blake, M. I. Katsnelson, and K. S. Novoselov (2007). Detection of individual gas molecules adsorbed on graphene. *Nat Mater,* **6**(9): 652–655.

Schwierz, F. (2010). Graphene transistors. *Nat Nanotechnol,* **5**(7): 487–496.

Shanmugam, M., T. Bansal, C. A. Durcan, and B. Yu (2012a). Molybdenum disulphide/titanium dioxide nanocomposite-poly 3-hexylthiophene bulk heterojunction solar cell. *Appl Phys Lett,* **100**(15).

Shanmugam, M., T. Bansal, C. A. Durcan, and B. Yu (2012b). Schottky-barrier solar cell based on layered semiconductor tungsten disulfide nanofilm. *Appl Phys Lett,* **101**(26).

Shaver, J., S. A. Crooker, J. A. Fagan, E. K. Hobbie, N. Ubrig, O. Portugall, V. Perebeinos, P. Avouris, and J. Kono (2008). Magneto-optical spectroscopy of highly aligned carbon nanotubes: identifying the role of threading magnetic flux. *Phys Rev B,* **78**(8).

Shi, Y. M., W. Zhou, A. Y. Lu, W. J. Fang, Y. H. Lee, A. L. Hsu, S. M. Kim, K. K. Kim, H. Y. Yang, L. J. Li, J. C. Idrobo, and J. Kong (2012). van der Waals epitaxy of MoS2 layers using graphene As growth templates. *Nano Lett,* **12**(6): 2784–2791.

Shih, C. J., A. Vijayaraghavan, R. Krishnan, R. Sharma, J. H. Han, M. H. Ham, Z. Jin, S. C. Lin, G. L. C. Paulus, N. F. Reuel, Q. H. Wang, D. Blankschtein, and M. S. Strano (2011). Bi- and trilayer graphene solutions. *Nat Nanotechnol,* **6**(7): 439–445.

Sipos, B., A. F. Kusmartseva, A. Akrap, H. Berger, L. Forro, and E. Tutis (2008). From Mott state to superconductivity in 1T-TaS(2). *Nat Mater,* **7**(12): 960–965.

Smith, R. J., P. J. King, M. Lotya, C. Wirtz, U. Khan, S. De, A. O'Neill, G. S. Duesberg, J. C. Grunlan, G. Moriarty, J. Chen, J. Z. Wang, A. I. Minett, V. Nicolosi, and J. N. Coleman (2011). Large-Scale exfoliation of inorganic layered compounds in aqueous surfactant solutions. *Adv Mater,* **23**(34): 3944.

Sols, F., F. Guinea, and A. H. C. Neto (2007). Coulomb blockade in graphene nanoribbons. *Phys Rev Lett,* **99**(16).

Splendiani, A., L. Sun, Y. B. Zhang, T. S. Li, J. Kim, C. Y. Chim, G. Galli, and F. Wang (2010). Emerging photoluminescence in monolayer MoS2. *Nano Lett,* **10**(4): 1271–1275.

Steiner, M., F. N. Xia, H. H. Qian, Y. M. Lin, A. Hartschuh, A. J. Meixner, and P. Avouris (2008). Carbon nanotubes and optical confinement: controlling light emission in nanophotonic devices. *Carbon Nanotubes Associated Devices,* **7037**.

Thomalla, M., and H. Tributsch (2006). Photosensitization of nanostructured TiO2 with WS2 quantum sheets. *J Phys Chem B,* **110**(24): 12167–12171.

Tsai, H. L., J. Heising, J. L. Schindler, C. R. Kannewurf, and M. G. Kanatzidis (1997). Exfoliated-restacked phase of WS2. *Chem Mater,* **9**(4): 879.

Tsai, H. L., J. L. Schindler, C. R. Kannewurf, and M. G. Kanatzidis (1997). Plastic superconducting polymer-NbSe2 nanocomposites. *Chem Mater,* **9**(4): 875.

Wang, H., L. L. Yu, Y. H. Lee, Y. M. Shi, A. Hsu, M. L. Chin, L. J. Li, M. Dubey, J. Kong, and T. Palacios (2012). Integrated circuits based on bilayer MoS2 transistors. *Nano Lett,* **12**(9): 4674–4680.

Wilson, J. A. (1969). Correction. *Adv Phys,* **18**(76): 848.

Wilson, J. A., F. J. Disalvo, and S. Mahajan (1975). Charge-density waves and superlattices in metallic layered transition-metal dichalcogenides. *Adv Phys,* **24**(2): 117–201.

Wilson, J. A., and A. D. Yoffe (1969). Transition metal dichalcogenides discussion and interpretation of observed optical, electrical and structural properties. *Adv Phys,* **18**(73): 193.

Wu, S. X., Z. Y. Zeng, Q. Y. He, Z. J. Wang, S. J. Wang, Y. P. Du, Z. Y. Yin, X. P. Sun, W. Chen, and H. Zhang (2012). Electrochemically reduced single-layer MoS2 nanosheets: characterization, properties, and sensing applications. *Small,* **8**(14): 2264–2270.

Wu, Y. H., J. Vendome, L. Shapiro, A. Ben-Shaul, and B. Honig (2011). Transforming binding affinities from three dimensions to two with application to cadherin clustering. *Nature,* **475**(7357): U510–U107.

Wu, Y. Q., K. A. Jenkins, A. Valdes-Garcia, D. B. Farmer, Y. Zhu, A. A. Bol, C. Dimitrakopoulos, W. J. Zhu, F. N. Xia, P. Avouris, and Y. M. Lin (2012). State-of-the-art graphene high-frequency electronics. *Nano Lett,* **12**(6): 3062–3067.

Wu, Y. Q., Y. M. Lin, A. A. Bol, K. A. Jenkins, F. N. Xia, D. B. Farmer, Y. Zhu, and P. Avouris (2011). High-frequency, scaled graphene transistors on diamond-like carbon. *Nature,* **472**(7341): 74–78.

Xia, F. N., M. Steiner, Y. M. Lin, and P. Avouris (2008). Cavity-controlled, electrically-induced infrared emission from a single single-wall carbon nanotube (SWCT). *Conf Opt Fiber Commun/Natl Fiber_Opt Eng Conf,* **1–8**: 460–462.

Xiao, D., G. B. Liu, W. X. Feng, X. D. Xu, and W. Yao (2012). Coupled spin and valley physics in monolayers of MoS2 and other group-VI dichalcogenides. *Phys Rev Lett,* **108**(19).

Xiao, J., X. L. Chen, P. V. Sushko, M. L. Sushko, L. Kovarik, J. J. Feng, Z. Q. Deng, J. M. Zheng, G. L. Graff, Z. M. Nie, D. W. Choi, J. Liu, J. G. Zhang, and M. S. Whittingham (2012). High-performance LiNi0.5Mn1.5O4 spinel controlled by Mn3+ concentration and site disorder. *Adv Mater,* **24**(16): 2109–2116.

Yin, Z. Y., H. Li, H. Li, L. Jiang, Y. M. Shi, Y. H. Sun, G. Lu, Q. Zhang, X. D. Chen, and H. Zhang (2012). Single-layer MoS2 phototransistors. *ACS Nano,* **6**(1): 74–80.

Yoffe, A. D. (1993). Low-dimensional systems: quantum-size effects and electronic-properties of semiconductor microcrystallites (zero-dimensional systems) and some quasi-2-dimensional systems. *Adv Phys,* **42**(2): 173–266.

Yoon, Y., K. Ganapathi, and S. Salahuddin (2011). How good can monolayer MoS2 transistors be? *Nano Lett,* **11**(9): 3768–3773.

Yoon, Y., and J. Guo (2007). Effect of edge roughness in graphene nanoribbon transistors. *Appl Phys Lett,* **91**(7).

Zeng, H. L., J. F. Dai, W. Yao, D. Xiao, and X. D. Cui (2012). Valley polarization in MoS2 monolayers by optical pumping. *Nat Nanotechnol,* **7**(8): 490–493.

Zeng, Z. Y., Z. Y. Yin, X. Huang, H. Li, Q. Y. He, G. Lu, F. Boey, and H. Zhang (2011). Single-layer semiconducting nanosheets: high-yield preparation and device fabrication. *Angew Chem, Int Ed,* **50**(47): 11093–11097.

Zhan, Y. J., Z. Liu, S. Najmaei, P. M. Ajayan, and J. Lou (2012). Large-area vapor-phase growth and characterization of MoS2 atomic layers on a Si02 substrate. *Small,* **8**(7): 966–971.

Zhang, Y. B., T. T. Tang, C. Girit, Z. Hao, M. C. Martin, A. Zettl, M. F. Crommie, Y. R. Shen, and F. Wang (2009). Direct observation of a widely tunable bandgap in bilayer graphene. *Nature,* **459**(7248): 820–823.

Zhang, Y. J., J. T. Ye, Y. Matsuhashi, and Y. Iwasa (2012). Ambipolar MoS2 thin flake transistors. *Nano Lett,* **12**(3): 1136–1140.

Zhu, W., Z. F. Wang, Q. W. Shi, K. Y. Szeto, J. Chen, and J. G. Hou (2009). Electronic structure in gapped graphene with a Coulomb potential. *Phys Rev B,* **79**(15).

Zhu, W. J., V. Perebeinos, M. Freitag, and P. Avouris (2009). Carrier scattering, mobilities, and electrostatic potential in monolayer, bilayer, and trilayer graphene. *Phys Rev B,* **80**(23).

Zhu, Z. Y., Y. C. Cheng, and U. Schwingenschlogl (2011a). Giant spin-orbit-induced spin splitting in two-dimensional transition-metal dichalcogenide semiconductors. *Phys Rev B,* **84**(15).

Zhu, Z. Y., Y. C. Cheng, and U. Schwingenschlogl (2011b). Vacancy induced half-metallicity in half-Heusler semiconductors. *Phys Rev B,* **84**(11).

Problems and Hints

Problems of Chapter 1

1-1. Calculate the area of hexagonal unit cell and the area of the hexagonal first Brillouin zone of the honeycomb graphene atomic lattice.
Hint: Use the method described in the book by Kittel (2005), *Introduction to Solid State Physics.* Take into account the fact that a honeycomb lattice is characterized by primitive translation vectors $\mathbf{a} = a\,(1,\ 0)$ and $\mathbf{b} = a\left(-1/2,\ \sqrt{3}/2\right)$, where the lattice constant is $a = 0.246$ nm. An elementary lattice cell contains two carbon atoms, which are denoted as A and B. The nearest-neighbor carbon atoms are connected by the vectors $\boldsymbol{\tau}_1 = a\left(0,\ 1/\sqrt{3}\right)$, $\boldsymbol{\tau}_2 = a\left(-1/2,\ -1/2\sqrt{3}\right)$, and $\boldsymbol{\tau}_3 = a\left(1/2,\ -1/2\sqrt{3}\right)$.

1-2. Consider how the electron dispersion law (Eq. 1.12) is changed if the intersite next-nearest-neighbor hopping is not neglected, that is, if one retains terms with $t' \neq 0$ in Eqs. 1.8 and 1.9, where t' is the intersite next-nearest-neighbor hopping energy.
Hint: Retain the next-nearest-neighbor terms in Eqs. 1.8 and 1.9 which connect the sites belonging to the same sublattice, which is either A or B. Such extra terms enter the determinant given by the left side of Eq. 1.11. The roots of the equation $\det(\hat{M}) \equiv 0$ will contain extra terms with the intersite next-nearest-neighbor hopping energy t'.

1-3. Compute the density of electron states in pristine graphene without and with *trigonal warping* of the electronic spectrum. Analyze how trigonal warping affects the electron density of states.

Hint: For the case when trigonal warping is neglected $(t' \neq 0)$ use Eq. 1.34 and the linear dispersion law given by Eq. 1.33. The effect of trigonal warping is taken into account using the electron dispersion law given in Eq. 1.39 and Eq. 1.34 without the right side.

1-4. What is the difference between the mass of elementary excitation in graphene and the cyclotron mass?

Hint: Using Eqs. 1.41–1.44, discuss the experimental curve shown in Fig. 1.7 extracted from the SdH experiment. Take into account the fact that the electron cyclotron mass is constant in conventional conductors where the dispersion law is parabolic, whereas the electron wavefunction conforms to the Schrödinger equation.

1-5. Explain why the dispersion laws of electron excitations in pristine graphene and carbon nanotubes are different from each other, although the materials have the same crystal lattice symmetry, spacing, and intersite coupling. What is the origin of Van Hove singularities in carbon nanotubes?

Hint: Take into account quantization of the electron motion in the lateral direction perpendicular to the nanotube axis. It serves to create Van Hove singularities in the electron density of states. The Van Hove singularities appear due to the 1D nature of the electronic spectrum. The quantization does not occur in pristine graphene, whose dimensions are infinite. There is a similar quantization in the graphene ribbons occurring in the lateral direction to the ribbon length. Besides, quantization of the electron motion in the lateral direction causes an appearance of the subband structure and an energy gap at the Fermi energy.

1-6. Explain why the electric conductivity of pristine graphene has a finite value $G = e^2/h$ at $\varepsilon = 0$, which should not be in a zero-gap semiconductor where it is supposed to vanish.

Hint: Take into account the contribution of a topological singularity at $\mathbf{k} = 0$, causing the zero-mode anomaly in the conductivity of graphene. One performs the spin-rotation operation $R^{-1}[\theta(\mathbf{k})][s]$ with the help of Eqs. 1.52–1.57 in Eq. 1.57, including the Berry phase, continuously changing the direction of $\mathbf{k}$. The change of sign occurs only if the origin $\mathbf{k} = 0$ is enclosed inside the closed contour.

1-7. Why does the tunneling of chiral relativistic particles lead to the Klein paradox with an ideal transmission probability and angular dependence on the angle of incidence? What is the influence of the intervalley scattering and symmetry change between the two different sublattices?
Hint: Take into account the effect of the intervalley scattering and symmetry change between two different sublattices. The four-component trial wavefunction, Eqs. 1.60–1.62, should be modified appropriately. Consider the tunneling process which occurs between two graphene regions with different types of charge carriers (e.g., one region is n-doped, while the other is p-doped). The tunneling probability is computed solving the boundary conditions for the four-component trial wavefunction, Eqs. 1.60–1.62. It allows us to compute the electric conductivity, which is electron-like on one side of the barrier, whereas it is hole-like on the other side. One finds that during the chiral tunneling process, normally incident chiral electrons are fully converted into the chiral holes. Because the effective mass (or relationship between momentum and velocity) has an opposite sign for electrons and holes, the hole momentum created in the p-region is inverse to the momentum of an electron incoming from the n-region. Furthermore, if one requires the conservation of momentum parallel to the n–p interface, the velocity of the tunneling particle is inverted. Therefore, the bunch of transmitted holes in the p-region is actually focused on the source of incident electrons located in the n-region.

Problems of Chapter 2

2-1. Why are some processes of electron scattering in pristine graphene permitted, whereas others are prohibited?
Hint: Take into account the degree of deviation of the electron momentum during the scattering process: If an electron deviates by angle $\varphi = \pi/3$ emerging from the K point

and getting into the K′ point, then the intervalley one-half pseudospin is flipped. Such a process is prohibited. However, if the scattering is weaker, for example, as it happens during the electron scattering on a neutral impurity, the process is permitted since the scattering which occurs between the two different K points conserves the pseudospins.

2-2. Determine how the full charge density per unit area (or the charge imbalance) changes when a finite gate voltage is applied to a graphene sample.
Hint: Use Eqs. 2.3–2.6 for the electron and hole density per unit area to consider the case when a finite gate voltage is applied ($\mu \neq 0$). Compute the dependence of the full charge density per unit area or the charge imbalance $n_s(\mu)$ for the case when a finite gate voltage is applied.

2-3. Compute the bipolar conductivity $\sigma_0(\mu)$ using the same dependence for the total carrier density $N_s(\mu)$ of a charged sheet of pristine graphene which represents a graphene field-effect transistor (FET), that is, in conditions when a finite gate voltage is applied ($\mu \neq 0$).
Hint: Use Eq. 2.23 for the bipolar conductivity $\sigma_0(\mu)$ as the sum of electron and hole components complemented with Eq. 2.7 for the case when a finite gate voltage V_G is applied ($V_G \neq 0$) and compute the dependence $\sigma_0(V_G)$.

2-4. Compute the quantum capacitance of a single-wall carbon nanotube (CNT) FET when a finite gate voltage $V_G \neq 0$ is applied underneath the substrate. What is the effect of Van Hove singularities on the quantum capacitance?
Hint: Use Eq. 2.16 complemented with the expressions for the electron density of states, Eqs. 1.49 and 1.50, for single-wall CNTs, $\rho_{CNT}(\varepsilon)$. Deduce the effect of Van Hove singularities on the quantum capacitance of CNTs. Take into account the fact that in CNTs, the electron density of states acquires a strong dependence on the energy and is considerably enhanced in the vicinity of singularities. The quantum capacitance for such reasons is strongly dependent versus the gate voltage.

2-5. Compute the bipolar conductivity of a CNT FET when a finite gate voltage $V_G \neq 0$ is applied. What is the contribution of Van Hove singularities?

Hint: Use Eq. 2.23 complemented with the expressions for the electron density of states, Eqs. 1.49 and 1.50, for single-wall CNTs $\rho_{CNT}(\varepsilon)$. Determine the effect of Van Hove singularities on the bipolar conductivity of a CNT FET. Be advised that a strong dependence of the electron density of states on the energy in CNTs originates from the spectral singularities and can be strongly increased in their vicinity. The bipolar conductivity for this reason is strongly dependent on the gate voltage.

2-6. Explain how the basic equation of graphene planar electrostatics (Eq. 2.42) changes if the value of the gate voltage corresponding to the charge neutrality point is finite, $V_{NP} \neq 0$.
Hint: Use Eq. 2.40 under an assumption that at the neutrality point (NP), the interface trap charges are present. It causes a finite value of the gate voltage $V_{NP} \neq 0$ to be retained. The interface trap capacitance per unit area C_{it} is obtained from Eqs. 2.27 and 2.28.

2-7. Determine in which way the interface trap charges affect the dependence of the electron Fermi energy on the gate voltage.
Hint: Under an assumption $V_{NP} \neq 0$ and using Eq. 2.40, derive an explicit dependence of the electron Fermi energy ε_F versus the gate voltage at $V_G>0$. Examine the relation for the graphene charge density n_S versus the gate voltage for the case when the charge neutrality point V_{NP} voltage is nonzero (i.e., see Eq. 2.50).

2-8. What is the condition for the drift current component to dominate over the diffusion part or vice versa?
Hint: Use Eq. 2.64 for the ratio of the diffusion to the drift current $\kappa = J_{diff}/J_{dr}$ which corresponds to an idealized graphene channel when one neglects the interface trap density ($C_{it} = 0$). Consider the case when the dimensionless parameter κ remains constant along the channel for a given electric bias $V_{SD} \neq 0$. Then, κ is expressed via the ratio of characteristic capacitances. One finds that in a high-doped regime, when the capacitance C_Q is large, and simultaneously C_{ox} is low (i.e., when the gate oxide layer is thick) it presumes $C_Q >> C_{ox}$, and one obtains $\kappa << 1$. Otherwise, in the opposite limit, one gets $\kappa >> 1$.

2-9. Explain why the Dirac equation fails to describe the electron transport in conditions when the electron scattering on the phonons, atomic impurities, and lattice imperfections in the graphene and CNT samples is strong.
Hint: The electron–impurity scattering and, likewise, the electron–phonon absorption and emission occur randomly. Therefore, the phase of electron wavefunction is randomized as soon as electrons and holes scatter or emit the thermal phonons. The randomness limits the applicability of a mere Dirac equation, and it also requires a more general method for the study of electron transport. The phases of electrons become randomized either inside or outside the chiral barrier: If the ratio $\tau_{\mathbf{ep}}/\tau_{\mathbf{dw}} < 1$, where $\tau_{\mathbf{ep}}$ is the electron–phonon scattering time and τ_{dw} is the dwell time, the phonons are emitted inside the barrier and the scattering approach fails. However, if $\tau_{\mathbf{ep}}$ is long enough, ensuring that $\tau_{\mathbf{ep}}/\tau_{\mathbf{dw}} > 1$, the electron wavefunction phase is preserved inside the barrier region, while it is randomized outside the barrier. In the latter case, the electron motion remains coherent and ballistic inside the barrier, since the time dependence of the electron wavefunction is the same for all the electrons.

Problems of Chapter 3

3-1. Explain why trigonal warping in bilayer graphene occurs at low energies.
Hint: Use the effective Hamiltonian, Eq. 3.5, which is valid in the limiting case $v_F k << 1$ at zero bias $V = 0$, and in Eq. 3.5 set $\gamma_3 = 0$. Then, the resulting electron-hole spectrum consists of two parabolic bands $\varepsilon_{k,\pm} \approx \pm v^2 k^2/t_\perp$ merging with each other at $\varepsilon = 0$. In such a limiting case, the spectrum is symmetric (see Fig. 3.2, left). Besides, the two additional bands arise at finite $\varepsilon = \pm t_\perp$. However, if γ_3 is finite, it serves as a source of trigonal warping. Therefore, the energy bands rebuild qualitatively, which constitute the trigonal distortion occurring at low energies. Furthermore, instead of two bands, we obtain three sets of Dirac-like linear bands which touch each other at

$k = 0$. One Dirac point is located at $\varepsilon = 0$ and $k = 0$, and three other Dirac points are also at $\varepsilon = 0$ but with a finite momentum at three equivalent points.

3-2. Explain which properties of bilayer graphene are characterized by winding numbers.
Hint: In bilayer graphene, the winding number illustrates topological properties of the electron spectral branches. The topological arguments are used studying the stability of the Dirac points where the electron bands touch each other (see, for example, Fig. 3.2, left). The winding number is introduced for a closed curve in the plane which encircles a given point. Thus, the winding number represents the total number of times when the curve encircles the point counterclockwise with no change of the wavefunction. In particular, the winding number of the point where the two parabolic bands merge with each other for $\gamma_3 = 0$ is equal to $+2$. The term γ_3 which causes trigonal warping and splitting of the Dirac point at $k = 0$ is characterized with the winding number -1, whereas three Dirac points at $k \neq 0$ are described with winding numbers $+1$. Applying the in-plane magnetic field, or slightly rotating one graphene layer with respect to another, one can also split the $\gamma_3 = 0$ degeneracy into two Dirac points with winding numbers $+1$.

3-3. Determine what is the effect of rotation of one graphene layer on the other in bilayer graphene. How is the mutual rotating of the two atomic planes in bilayer graphene expressed in the electron excitation spectrum?
Hint: In the tight-binding scheme, we are interested in the electronic states close to the Fermi level; therefore, it is sufficient to take into account only the p_z orbitals. When the planes are rotated, the neighboring atoms are not located on the top of each other (as is the case in the Bernal AB stacking). Thus, interaction between the layers is not restricted to $pp\pi$ terms and some $pp\sigma$ terms have to be introduced. One can split the $\gamma_3 = 0$ degeneracy into two Dirac points with the winding number $+1$, applying the in-plane magnetic field, or slightly rotating the planes. Breaking the inversion symmetry and violating the equivalence between the two atomic layers in

bilayer graphene occur when the bias term in Eq. 3.3 is finite, $V \neq 0$.

3-4. Discuss the basic difference between two cases of chiral tunneling in monolayer graphene on the one hand and bilayer graphene on the other hand.
Hint: On the one hand, the chiral tunneling through the barrier in monolayer graphene has a strong angular dependence whereas the tunneling probability T for the normal incidence is ideal ($T = 1$). On the other hand, for tunneling in bilayer graphene, the chirality of the particles is revealed only at finite incidence angles $\varphi \neq 0$.

3-5. How should we measure the energy gap in the electron excitation spectrum of bilayer graphene when a finite bias voltage is applied between the atomic layers?
Hint: Use Eqs. 3.8 and 3.9 for the electron excitation spectrum in bilayer graphene. The experimental measurement of the energy gap in the bilayer graphene system under bias is performed using the dependence of the energy gap on the applied voltage V. The appearance of an energy gap in the bilayer graphene spectrum makes it very interesting for technological applications.

3-6. Explain why for chiral fermions propagating in a stripe of monolayer graphene the time reversal symmetry is violated when applying a lateral electric field. What type of symmetry is preserved instead?
Hint: It is well known that a finite electric field $\mathbf{E} = \{0, E_y, 0\} \neq 0$ does not violate the time reversal symmetry of propagating charged particles. However, a different scenario occurs in a narrow stripe of graphene subjected to a finite electric field applied perpendicular to its axis. In the latter case, the stripe of monolayer graphene behaves similarly as an electric dipole. The dipole energy $\Delta = \mathbf{D}E_y$ flips the sign under a parity transformation $\mathbf{P}$. Because the dipole moment $\mathbf{D}$ is a vector, the expectation value of it in the state $|\psi>$ is proportional to $<\psi|D|\psi>$. Therefore, when the dipole is subjected to the time reversal operation, $\mathbf{T}$, an invariant state must acquire a vanishing dipole momentum $\mathbf{D} = 0$. In different terms, a nonvanishing $D \neq 0$ suggests that the

symmetry is broken when either **P** or **T** is applied separately. On the contrary, the combined **PT** symmetry is retained. For the above reasons, a graphene stripe represents a one-dimensional quantum well polarized in the lateral y direction where quantized energy subbands arise. In the absence of an external electric field, electrons and holes can occupy only the Van Hove singularity (VHS) energy subbands inside the quantum well. A finite electric field $\mathbf{E} = \{0, E_y, 0\}$ causes the quantum-confined Stark effect pronounced in the VHS splitting. Therefore, the electron states tend to move to the higher energies, while the hole states are shifted to lower energies. Besides, the external electric field pulls electrons and holes to opposite sides perpendicular to the G stripe axis, separating them spatially. The spatial separation between the electrons and holes is limited by the presence of the confining potential barriers fringing the graphene stripe. It means that the two $\mathrm{HCF}_{\mathrm{e(h)}}$ edge states characterized by opposite $\pm$ one-half pseudospins and by opposite electric charges $\pm e$ are able to exist in the system, even under the influence of an electric field.

3-7. What makes heavy chiral fermion (HCF) singularities shown in Fig. 3.5b,c sharper than the Van Hove singularities (VHSs) in the electron density of states (DOS)?
Hint: The sharp spectral singularities caused by HCF in graphene stripes with zigzag edges originate from the very flat bottom of the conductance band and the very flat top of the valence band. The bands are separated by an energy gap 2Δ, the value of which is obtained from Eqs. 3.26 and 3.27 as a solution of the excitation energy ε. When the split gate voltage is applied, $V_{\mathrm{SG}} \neq 0$, the splitting of the HCF level 2Δ becomes finite, and the graphene zigzag stripe G is a band insulator with a pseudospin polarization. The two flat bands correspond to very sharp HCF singularities in the electron DOS which emerge at energies $\varepsilon = \pm\Delta$ (see Fig. 3.5a–c). In vicinity of the HCF singularities, $\varepsilon = \pm\Delta$, one can use an approximate analytical expression for the DOS in the form $\propto \mathrm{Re}\{1/(\varepsilon - \Delta)^{\alpha}\}$. Here, $\alpha \simeq 1.3$ is obtained using the last formula to fit the numeric solution of Eqs. 3.26 and 3.27). The

DOS divergences are smoothed with the setting $\varepsilon \rightarrow \varepsilon + i\gamma$. The effective HCF bandwidth γ at the HCF singularity is actually the rate of inelastic electron–phonon collisions which is typically low inside the graphene stripe. The flat bands correspond to a very high effective electron mass $m^* = (10^2 - 10^5) \times m_e >> m_e$, which is estimated as $m^* \simeq \hbar^2 / (\gamma a^2)$, where a is the graphene lattice constant (Kittel 2005).

Problems of Chapter 4

4-1. Explain why the Grüneisen parameter for the lowest transverse acoustic (TA) out-of-plane acoustic (ZA) modes is negative.
Hint: Take into account the "membrane effect." The partial Grüneisen parameters are negative for some phonon bands of graphene. At low temperatures, when most optical modes with positive Grüneisen parameters are not active, the contribution from the negative Grüneisen parameters will be dominant and the thermal expansion coefficient, which is directly proportional to the Grüneisen parameter, is negative. The lowest negative Grüneisen parameters are related to the lowest transversal acoustic ZA branches. The phonon frequencies of the modes are enhanced as the in-plane lattice parameter increases, because the atoms within the layer upon stretching cannot freely deviate in the z direction. This is similar to the behavior of a string, which, when it is stretched, will have vibrations with smaller amplitude and higher frequency. This phenomenon, named the "membrane effect," was predicted by Lifshitz in 1952.

4-2. Discuss why the contribution of ZA phonons to thermal transport in graphene might be neglected.
Hint: Use Eqs. 4.4–4.7 defining the Grüneisen parameter γ_s of the individual vibrational modes, thermal conductivity, and the Umklapp relaxation time $\tau_{U,s}^{K}(q)$, depending on three phonon modes. Take into account the fact that out-of-plane ZA phonons have low group velocity $v_s(q)$ and low $\tau_{U,s}^{K}(q) \propto 1/\gamma_s^2$ (because the Grüneisen parameter might be neglected).

4-3. Explain why the acoustic phonon energy spectra in nanostructures are spatially quantized, and what impact this has on the phonon group velocity.
Hint: In small nanostructures, whose dimensions L are comparable with just a few phonon wavelengths λ_s, the lattice oscillations constitute standing waves with frequencies $\Omega_s = n\omega_s$, where $n = \text{int}\ (L/\lambda_s)$. Therefore, the phonon spectrum represents a set of fixed energies $E_s = n\hbar\omega_s$, which do not depend on the phonon's wavevector $\mathbf{q}$. For these reasons, the phonon group velocity $\mathbf{v}_s(\mathbf{q}) = d\omega_s(\mathbf{q})/d\mathbf{q}$ vanishes in tiny nanostructures.

4-4. What are the basic assumptions for deriving the Klements equation, Eq. 4.2?
Hint: The assumptions are as follows: (i) only waves of a high group velocity make a substantial contribution to thermal conduction K; (ii) only acoustic modes contribute to K, while optical modes, which are the relative motions of atoms within a molecular group, are neglected; (iii) the modes of the acoustic branches, which have a high group velocity, have a frequency range from zero to ω_m and are considered in a Debye approximation; (iv) the phonon mean free path $l(\omega)$ is temperature dependent and is reduced due to the scattering of phonons by solutes and other imperfections (this sensitivity holds for bulk specimens and, to an even greater degree, for thin films; the external boundaries scatter phonons, and thin films usually contain a larger concentration of various defects); and (v) the intrinsic mean free path, limited by anharmonic energy interchange between groups of three waves, is a function of both absolute temperature T and frequency ω.

4-5. Explain why the heat transport in single-layer graphene differs from the heat transport in basal planes of bulk graphite.
Hint: Take into account the coupling between the layers, along with the contribution of ZA phonons. Experimental data suggest that there is a noticeable difference between phonons existing in 2D graphene on the one side and phonons of 3D bulk graphite on the other side. The phonon

transport in basal planes of bulk graphite is determined by a certain low-bound cut-off frequency ω_{min}, above which the heat transport is approximately 2D. On the contrary, below ω_{min}, there is a strong coupling between the cross-plane phonon modes. It leads to propagation of phonons in all directions. Furthermore, it reduces the contributions of low-energy modes to the transport of phonons within the basal planes to negligible values. In spectroscopy experiments, an onset of cross-plane coupling, which is the ZO' phonon branch near ~4 THz and which is visible in the spectrum of bulk graphite, serves as a reference point. Computing the phonon transport characteristics, one takes into account that $\omega_{\text{min}} = \omega_{\text{ZO'}}(q = 0)$ for the ZO' branch, which helps to avoid the logarithmic divergence in the Umklapp-limited thermal conductivity.

4-6. Explain the role of 2D three-phonon Umklapp processes.
Hint: Accurate computing of the thermal conductivity of graphene requires considering all the possible three-phonon Umklapp processes which obey Eqs. 4.15 and 4.16. The momentum and energy conservation laws during the three-phonon Umklapp processes are given in Eqs. 4.15 and 4.16. Accounting for all the possible three-phonon Umklapp processes in Eqs. 4.17 and 4.19 allows us to compute the total phonon relaxation rate $\tau_{\text{tot}}^{-1}(s, q) = \tau_{\text{U}}^{-1}(s, q) + \tau_{\text{B}}^{-1}(s, q)$, where the phonon relaxation times τ_{U} and τ_{B} are determined by the Umklapp and boundary scatterings, respectively. During calculations one should observe the actual phonon dispersions.

4-7. Explain why the number of vibrational modes $3N$-6 of nonlinear molecules is smaller than the number of vibrational modes $3N$-5 of linear molecules.
Hint: Take into account the fact that the positions of atoms in the rotations of a linear molecule along the bond axis do not change. Generally, a chemical compound which consists of N atoms has $3N$ degrees of freedom in total. The number $3N$ indicates an ability of each atom to be displaced in three different directions (x, y, and z). The movement of the molecule is generally regarded as a whole one. Thus, it is

instructive to distinguish different parts of the $3N$ degrees of freedom. The convention is to split the molecular degrees of freedom into translational, rotational, and vibrational motions. The translational motion of the molecule as a whole along each of the three spatial dimensions also corresponds to three degrees of freedom. Consequently, other three degrees of freedom are related to molecular rotations around the x, y, and z axes. However, there are certain exceptions to the above scheme. For instance, linear molecules can have only two rotations since the rotations along the bond axis do not change the positions of atoms in the molecule. Other remaining molecular degrees of freedom are related with vibrational modes. The latter modes correspond to molecular stretching and bending motions of chemical bonds.

4-8. Explain why the locations of Stokes and anti-Stokes peaks create a symmetric pattern around the frequency difference $\Delta\nu = 0$ between transmitted and incident photons.
Hint: The Stokes and anti-Stokes peaks are located around $\Delta\nu = 0$ symmetrically. The symmetry of frequency shifts around zero frequency $\nu = 0$ originates from the energy difference between the corresponding upper and lower resonant states. The intensities are different only for the pairs of features which are also temperature dependent. The reason is that the corresponding intensities depend on populations of initial states of the material, which are changed versus the temperature. In the steady state, the upper state is less populated than the lower state. Therefore, Stokes transitions from the lower to the upper state prevail over anti-Stokes transitions, which occur in the opposite direction. This intensifies the Stokes scattering peaks as compared to the anti-Stokes scattering peaks. The temperature dependence of the ratio of two opposite processes can be utilized for practical temperature measurement.

4-9. What is the physical difference between the Raman effect and fluorescence?
Hint: Take into account the fact that the time between absorption and the following emission is finite. Actually, both processes yield the same result. However, during fluore-

scence, the incident light is completely absorbed; therefore, the system initially goes to an excited state and, then, after a certain resonance lifetime, recombines into various states with lower energy. In either case, a photon with a frequency different from that of the incident photon is emitted, provided that a molecule is either excited to a higher-energy level or recombined to a lower-energy level.

4-10. What is the necessary condition for a Raman shift to appear? How should molecular vibrations and rotations be classified?
Hint: The basic condition for a Raman shift to be optically activated is that the molecule be polarized. The molecular distortion α in an electric field, and thus the vibrational Raman cross section, are determined with its polarizability. Molecular vibrations and rotations, which participate in Raman scattering, occur for molecules or crystals. The symmetries of the molecular vibrations and rotations are determined from the appropriate character table for the symmetry group. They are listed in many textbooks on quantum mechanics or group theory for chemistry.

4-11. How should the Fermi velocity and the electron–phonon coupling constant be measured directly from the Raman effect?
Hint: Use Eqs. 4.25, 4.26, and 4.28, which express those parameters via the features extracted from the experimental Raman curves. The strategy for the experimental determination of the electron–phonon coupling is based on density functional theory calculations. The Raman spectra anomalies are pronounced as two sharp kinks in the phonon dispersion shown in Fig. 4.13. The slope of the kinks $\eta^{\text{LO}}(\Gamma, K)$ is proportional to the ratio of the square of the electron–phonon coupling matrix element and the slope of the π bands at the K point.

4-12. What is the difference between the structure of the 2D peak in bulk graphite and graphene (Kittel 2005, Klemens 2000, Bergman 2000)?
Hint: Consider Raman spectra versus the number of monoatomic carbon layers shown in Figs. 4.15 and 4.16. On the one hand, the structure of the 2D peak in bulk graphite

consists of two components, $2D_1$ and $2D_2$. The relevant heights of $2D_1$ and $2D_2$ are measured as roughly 1/4 and 1/2 of the G peak height, respectively. On the other hand, graphene has a single, sharp 2D peak, the intensity of which is about four times stronger than that of the G peak. One can also learn the evolution of Raman spectra versus the number of atomic layers in graphene from Figs. 4.15 and 4.16. In Figs. 4.15 and 4.16 one can see the development of the 2D band versus the number of layers for 514.5 nm and 633 nm excitations. It says that the 2D band in bilayer graphene is very broad and is shifted up in energies as compared to monolayer graphene. In contrast to bulk graphite, the 2D band of graphene consists of four components: $2D_{1B}$, $2D_{1A}$, $2D_{2A}$, and $2D_{2B}$. Relative intensities of the two components $2D_{1A}$ and $2D_{2A}$ are higher than those of the other two, as seen in Fig. 4.16. If the number of layers increases, it leads to a substantial decrease of relative intensity of the lower frequency $2D_1$ peaks, as is evident from Fig. 4.16. When the number of layers in graphene exceeds five, the Raman spectrum resembles the spectrum of bulk graphite. Thus, Raman spectroscopy distinguishes graphene as consisting of a single layer, a bilayer, and a few (less than five) layers.

Problems of Chapter 5

5-1. Explain why the conservation of a one-half pseudospin eliminates the electron back scattering on impurities and phonons. Hint: For illustration use Fig. 5.1, where the forward scattering (fs), backward scattering (bs), and intervalley-scattering (is) processes in the vicinity of K and K′ points are shown. For metallic nanotubes and pristine graphene the rotation angle is equal to 180° multiplied by the odd integer number. That's why after summation over all probabilities of the possible scattering processes, the matrix elements mutually cancel each other. It results in the absence of backward scattering.

5-2. Discuss why the conservation of a one-half pseudospin is less pronounced in semiconducting carbon nanotubes (CNTs)

than in metallic nanotubes. Explain also how the difference in pseudospin conservation affects electron back scattering with impurities and phonons in the two relevant materials.
Hint: Use the illustration in the form of pseudospin rotations. In contrast to pristine graphene and metallic nanotubes, for semi-conducting nanotubes and graphene stripes, during electron–impurity and electron–phonon scattering, the rotation angle of the one-half pseudospin is different: 180° is multiplied by a noninteger number. That's why after summation over all the probabilities of the possible scattering processes, the matrix elements do not cancel each other, which causes a finite contribution of the residual backward scattering.

5-3. Explain why the phonon drag effect is minimized in pristine graphene and metallic CNTs, while it remains finite in graphene stripes and in semiconducting nanotubes.
Hint: Take into account the conservation of a one-half pseudospin during the electron scattering on atomic impurities, phonons, ripples, etc. The one-half pseudospin conservation takes place in pristine graphene and in metallic nanotubes, while it is violated in graphene stripes and in semiconducting nanotubes.

5-4. What are the necessary conditions for the plasma mode to exist?
Hint: Take into account Eqs. 5.58–5.60 defining the continuum of boson excitations in the frequency interval where the imaginary part of electron susceptibility $\kappa_2(q)$ is finite but vanishes elsewhere.

5-5. Explain why phonon frequencies in conducting materials are renormalized.
Hint: The conduction electrons screen out the long-range Coulomb forces between ions on the one hand and between ions and electrons on the other hand. The phonon frequencies are influenced by the dressed electron–phonon interaction, which largely depends on screening. The corresponding frequency renormalization effect is described in Eqs. 5.116 and 5.117.

5-6. For what reason does the electron–electron interaction become dependent on energy?

Hint: The energy dependence originates from the electron–phonon interaction. The resonance takes place at the dressed phonon frequency ω_q, which corresponds to the ionic over-screening of the bare Coulomb interaction for $q_0 < \omega$ and underscreening for $q_0 > \omega$. For high-frequency $q_0 >> \omega$, the ions do not respond and $\mathrm{Re}V\,(q)$ approaches the bare Coulomb interaction reduced by the electronic dielectric function $\kappa(\mathbf{q}, q_0)$.

5-7. Explain why phonon scattering in graphene depends on the number of atomic layers.
Hint: Experimental data indicate that heat conductance decreases for few-layer graphene samples as the number of layers increases. Saturation of heat conductance occurs in bulk graphene when the number of layers exceeds four. Such a behavior is understood if one considers the Umklapp scattering of phonons and that the phonon momentum acquires the perpendicular across the layer component, whereas the phonon dispersion is also altered.

Problems of Chapter 6

6-1. Discuss why Mott excitons require special conditions for appearing in graphene.
Hint: Take into account the fact that the energy gap in graphene is absent when the electron excitation spectrum has a conic shape. The absence of the energy gap and the linear dispersion law in pristine graphene and metallic carbon nanotubes (CNTs) make the formation of bound states impossible (see Problem 6-3 below). However, excitons can be formed when a finite energy gap is induced in graphene and the electron dispersion becomes nonlinear due to the interaction of the carbon material with the dielectric substrate or due to a spatial confinement in the lateral direction.

6-2. Explain how conductive electron screening changes the electron–electron Coulomb interaction and effective electron mass in carbon nanotubes (CNTs).

Hint: Using Fig. 6.1, compare how the band gaps and effective mass of electrons depend on the Coulomb energy. One can see that the Coulomb interaction causes a considerable increase of the band gaps. In particular, when taking into account the electron–electron interaction $(e^2/\kappa L)^2/(2\pi\gamma/L) = 1$, the first band gap becomes about 1.5 times larger than that without interaction. On the contrary, the effective electron mass in the semiconducting nanotubes diminishes only slightly; therefore the electron–electron interaction is a minor effect. The results are practically the same using either dynamic or static random phase approximation.

6-3. Explain why the binding of two particles causes no bound states if their dispersion law has a conic shape.
Hint: The moments of the particles remain constant and parallel to each other due to the equality of velocities, whereas they maintain a constant separating distance between the two particles. Indeed, such an idealized picture is related only to an infinite pristine graphene sheet and almost never occurs in realistic samples. As soon as the mass becomes finite, the particles cannot maintain the constant spatial separation any more. Thus, the propagation of particles with a finite mass is accompanied by variation of the electron–hole distance. If the mass is positive, the electron–hole pair forms a bound state.

6-4. Explain why interparticle binding occurs only in certain sectors of the momentum space, whereas bound states are prohibited in the other sectors belonging to the same space.
Hint: The dispersion law for electrons in graphene is highly anisotropic. For these reasons, exciton states can only be formed at certain angles, as shown in Fig. 6.4 (right). The sign of the exciton mass depends on the angle of the exciton momentum. Therefore, the formation of a bound state depends on the angle of the exciton momentum.

6-5. Explain why for generating Mott excitons in graphene, photons with a short wavevector are required.
Hint: Take into account the fact that in contrast to conventional semiconductors with a finite energy gap, the absence of a gap in the electron excitation spectrum of pristine graphene causes a

spread of exciton energies in the wide range between zero and several tenths of electron volts.

6-6. How does the use of phonon- or impurity-assisted absorption help to excite Mott excitons in graphene?
Hint: The photon absorption during the electron scattering on phonons or on impurities in graphene is accompanied by a large momentum change. Such a large change of electron momentum ensures that during the absorption of a photon with a large momentum (large energy) the energy and momentum conservation laws are observed. If, during the process, the energy of the absorbed photon is large, it strongly stimulates the generation of Mott excitons.

6-7. Discuss the role of screening with the charge carriers of moving excitons in graphene.
Hint: Take into account the fact that the moving excitons in graphene only partially are accompanied by charge carriers. Consider what happens to the binding energy when the screening radius of the moving electron–hole pair becomes shorter than the exciton radius.

6-8. What are the signatures of the Tomonaga–Luttinger liquid (TTL) state in narrow graphene stripes and CNTs which have been observed in experiments?
Hint: Take into account the power law dependence of the density of states of linear bands (as shown in Fig. 6.9), which is pronounced in the differential conductance and photoemission experiments.

6-9. Explain how the quantized levels of charge and spin excitations in the TTL state inside a 1D quantum well (QW) can be separated from each other and are identified in experiments.
Hint: Use Fig. 6.12 showing the difference between splitting of the quantized levels due to the photon-assisted tunneling (PAT) and the Zeeman effect. The AC field splits the charge boson energy levels due to PAT, while the DC magnetic field splits the spin boson levels due to the Zeeman effect. The splitting depends on the field parameters, which can be used in relevant experiments to identify the quantized energy levels associated with the charge and spin bosons in the TLL state. Besides,

one determines the spacing of quantized levels Δ and the TLL parameter g.

Problems of Chapter 7

7-1. Discuss a possible relationship between the intrinsic coherence of graphene and quantum coherence in conventional superconductors.
Hint: Take into account the coherence originating from conservation of the electron chirality and pseudospin in graphene, on the one hand, and the quantum coherence of the superconducting state, on the other hand. In superconductors, the superfluid condensate of Cooper pairs is manifested in a variety of unconventional phenomena. It includes, for example, the Meissner effect, Abrikosov vortices, Josephson current, etc. The basic mechanism to form the superconducting state consists of Cooper coupling between an electron state, say, with spin "up," and its time*reversed state, which is a hole with spin "down." Another type of intrinsic coherence occurs in graphene, where it emerges due to conservation of the one-half pseudospin. For these reasons, the superconducting state becomes robust and stable in respect to the electron scattering on atomic defects and lattice oscillations. Therefore, graphene has a very high charge carrier mobility. Thus, the graphene/superconducting junctions represent an attractive subject of scientific research and applications. Definitely, the two-dimensional structure of graphene suggests an unsurpassed flexibility for the control of transport properties using a variety of methods.

7-2. Which conservation laws determine the Andreev reflection (AR) from the N/S interface which separates the normal metal N and superconductor S?
Hint: Take into account the continuity of electric current across the N/S interface. The electric charge carriers in the normal material N are electrons and holes. Crossing the N/S interface, they must be converted into the superfluid condensate of Cooper pairs which carries the electric current

in a superconductor. The conversion process between two different systems of the charge carriers is described with the electric current continuity equation. The microscopic mechanism of conversion of the electric current at the N/S interface is the AR. During the elementary process of AR, the electron excitation is regarded as a filled state at positive energy $\varepsilon > 0$, which is above the Fermi energy, E_F, whereas the hole excitation is an empty state at $-\varepsilon < 0$ below E_F. Since the absolute value of the excitation energy ε is the same, AR is an elastic process. Nevertheless, since the electric charges of the electron ($-e$) and of the hole ($+e$) are opposite, an extra charge of $2e$ is lost in the conversion process. The missing charge $2e$ is transmitted to the superfluid condensate, which is interpreted as a creation of an extra Cooper pair superconductor.

7-3. Explain why, when the barrier strength Z is finite, AR occurs mostly at energies $E = \Delta$, whereas it is suppressed at lower electron energies $E < \Delta$.
Hint: Use the particle's current probabilities $A = a^*a$ and $B = b^*b$ for the reflection and transmission amplitudes complemented with Eqs. 7.18–7.23. For an ideal interface transparency, when the barrier is absent (i.e., $Z = 0$), one obtains $b = d = 0$. It means that there is no branch crossing, because all the reflections become just the ARs. However, if there is a finite barrier at the N/S interface ($Z \neq 0$), there is a conventional reflection which is stronger at energies $E < \Delta$, whereas the AR is suppressed. Nevertheless, the AR is retained at energies $E \simeq \Delta$, that is, close to the gap edge, where the singularity in the electron density of states occurs.

7-4. Explain why the differential conductance of graphene/ metal junctions depends on the bias voltage and is asymmetric.
Hint: Take into account the effect of the final width and asymmetry of Schottky-type barriers at the graphene/metal interface. Schottky barriers arise due to the difference of workfunctions of graphene and metal electrodes. Typically, the shape of a Schottky barrier is asymmetric, whereas its height is low. Therefore, the transparency of a Schottky barrier depends on the energy of tunneling particles. If the

junction is formed attaching the two or more metal electrodes to graphene, the electric current flows through two or more Schottky barriers. Because the distinct barriers of the same junction are nonequivalent and their asymmetries are also different, the differential conductance of graphene/metal junctions has an asymmetric shape.

7-5. What are the most important distinctions between the classical BTK model and a real graphene/metal junction?
Hint: Take into account the fact that particle transmission through a graphene/metal junction involves several different mechanisms. One can distinguish the presence of interface Schottky barriers (see Problem 7-4), a change of the number of conducting quantum channels during the tunneling from rt he3D metallic electrode into the 2D graphene, and the resistance of the metal electrodes and graphene sheet. Different contributions are separated experimentally along with applying a DC magnetic field, and one measures the ratio of the excess zero-voltage conductance obtained at a very low temperature to normal state conductance.

7-6. Explain why the excessive conductance due to AR appears only below the threshold voltage Δ/e.
Hint: Use Eqs. 7.12–7.14 together with Eqs. 7.18–7.23. AR is a process where the participating electron excitation is regarded as a filled state at positive energy $\varepsilon > 0$, which is above the Fermi energy, E_F, whereas the hole excitation is an empty state at $-\varepsilon < 0$ below E_F. AR is an elastic process because the absolute value of the excitation energy ε is the same. Because of the opposite electric charges of the electron ($-e$) and of the hole ($+e$) involved in the process, an extra charge of $2e$ is transmitted to the superfluid condensate, which is interpreted as the creation of an extra Cooper pair superconductor.

7-7. Explain why the reflection (r) and transmission (t) amplitudes entering Eq. 7.27 for the graphene/superconductor interface are 4×4 matrices.
Hint: Consider the S-matrix which, besides AR, also accounts for one-half pseudospin flips. The diagonal elements of such an S-matrix then correspond to conventional reflection (r)

and transmission (t) amplitudes, whereas the nondiagonal elements describe the AR processes preserving the time reversal symmetry.

7-8. Explain why conventional AR happens when electrons and holes both lie in the conduction band, whereas specular AR occurs when they are in different, conduction and valence, bands.
Hint: Take into account the fact that the wavevector of a reflected hole is directed opposite the velocity when the hole belongs to the conduction band. On the contrary, if the reflected hole belongs to the valence band, the parallel component of the velocity v_y remains the same, whereas the normal component v_x flips the sign, which is now quoted as specular reflection.

7-9. Explain why reflectionless AR can cause an increase of the N/S interface conductance more than twice.
Hint: Consider the contribution of multiple AR processes taking place while an electron dwells near the N/S interface.

7-10. Explain why the period of Rowell–McMillan oscillations is energy dependent and why it changes versus the barrier strength.
Hint: Consider the Rowell–McMillan oscillations originating from interference between the electron and hole evanescent waves in the middle section of the Pd/CNT/Pd junction. Amplitudes of the electron wavefunction inside the junctions are energy dependent and also depend on the interface barrier strength. Therefore, the resulting interference pattern depends on the electron energy as well as the interface barrier strength.

Problems of Chapter 8

8-1. Explain why nonequilibrium effects are important in graphene nanodevices.
Hint: Consider an example of a graphene quantum well in an extremely nonequilibrium state which is determined under competition between thermal relaxation and external driving.

The system represents a nonequilibrium quantum device and can be constructed using graphene field-effect transistors. The quantum well system performs in competitive situations while balancing the thermal relaxation and the external driving which emerge in nonequilibrium conditions.

8-2. Which parameters characterize the nonequilibrium state of a nanosystem?
Hint: Take into account the balance of energy supply and dissipation in the system. Such conditions imply that the sample is subjected to either an external field or heat flow. Thus, the nonequilibrium effects in graphene and carbon nanotube devices emerge if the energy supply from outside the system exceeds the dissipation and escape of energy from the system to outside. The state of the system deviates from equilibrium when energy is supplied into the system.
At first, the energy can be absorbed with a certain subsystem and only later is redistributed among other subsystems in the same system. When a graphene device is exposed to an external electromagnetic field, the field acts directly on the electrically charged particles. which are the chiral fermions.

8-3. What is the difference between the mechanisms of coherent and incoherent transmission through a nanojunction?
Hint: Compare the dwell time τ_D inside the junction with the electron–phonon inelastic scattering time, the tunneling time between the electrodes, and the escape time of an electron excitation from the nanojunction region. In the case of partial incoherence, one can implement a combined approach which takes into account the piece-wise coherence inside the system. The partial incoherence can be introduced by atomic impurities, structural defects, ripples and roughness of surfaces and interfaces, etc. As an example of the mentioned geometry, one can consider a double-barrier junction (see, for example, Fig. 8.1) where the electron propagation consists of coherent and noncoherent pieces. Using the piece-wise representation of the electron trajectories, one can compute basic transport characteristics of the complex system. In the method, the whole electron trajectory is composed from separate coherent and noncoherent pieces (see, for example,

Fig. 8.1). Different fractions of trajectories with various types of electron transport are matched with each other by sticking different coherent and noncoherent pieces. Electron propagation along coherent pieces of trajectories between the scatterers are described in terms of S-matrices, while the incoherent connecting parts of the electron propagation are determined via the probability matrices and the distribution functions previously computed.

8-4. What is the limit of applicability of the S-matrix technique to a nonequilibrium system?
Hint: Take into account the fact that a time-dependent external field breaks the energy conservation law. When the Hamiltonian depends on time, the S-matrix blocks acquire dependence on two time variables. The nonlocal time dependence introduces additional technical difficulties describing the system in terms of the S-matrix technique.

8-5. What is the basic idea of implementing piece-wise coherence in a complex system?
Hint: Consider a piece-wise representation of electron trajectories when the whole electron trajectory is composed of separate coherent and noncoherent pieces (see, for example, Fig. 8.1). Different pieces of trajectories with different coherent and noncoherent types of transport are tailored to each other, either with S-matrix amplitudes or with probabilities.

8-6. Explain why homogeneous approximation allows to simplify the expressions for Green functions.
Hint: When an external time-dependent field is applied to the electron system, the electron Green function depends on two time variables (see Eq. 8.25 as an example). To simplify the expression, one uses homogeneous approximation, which transforms the Green function into quasihomogeneous form. Thus, one obtains a Green function which is dependent just on a single time variable, $\tau = t - t'$, whereas the dependence on another time variable, $T = t+t'$, is neglected. Using Eqs. 8.27–8.29, one evaluates the shortest spatial scale which ensures conservation of the electron momentum when an external AC field acts on the nanojunction.

8-7. Determine why the electron distribution functions in the electrodes attached to the central island of a quantum dot can be considered equilibrium Fermi functions.
Hint: Consider the balance of energy supplied to the central island from the attached electrodes and from the external AC field due to photon-assisted tunneling on the one hand and energy dissipation due to inelastic electron–phonon collisions and electron escape from the island on the other hand.

8-8. Determine how the effect of electron chirality in monolayer graphene enters the Landauer–Butticker formula for electric current and the kinetic equation for the electron distribution function.
Hint: Consider the additional factors originating from the electron chirality entering Eqs. 8.58 and 8.62 and Eqs. 8.109–8.111.

8-9. How can one keep a small system in the state of quasistatic equilibrium during a switching process?
Hint: The state of quasistatic equilibrium can be maintained selecting a longer switching time t_s, thereby allowing us to alter the macroscopic parameter λ sufficiently slowly.

8-10. Explain why the presence of operators in the quantum Jarzynski equality, Eq. 8.158, instead of variables complicates the decomposition of the time variable.
Hint: To check the difference between the use of classic variables and quantum mechanical operators, one can introduce slightly different definitions of the infinitesimal exponentiated work and evaluate the errors which are accumulated at each step of the whole transition process. Consider how the error depends on the switching time duration.

Problems of Chapter 9

9-1. Determine how the Mott formula, Eq. 9.2, can be modified to account for sharp singularities of the excitation spectrum.
Hint: Use Eqs. 9.16 and 9.17, where the energy-dependent electron density of states under integral takes into account

the singularities in the excitation spectrum. A sharp spectral singularity can be approximated with a Dirac delta function, which allows us to elucidate the spectral singularity contribution into the transport coefficients.

9-2. Explain why phonons, excitons, plasmons, and other massive excitations without electric charge reduce the figure of merit of a thermoelectric (TE) device?

Hint: A temperature gradient causes a drift of the mentioned electrically neutral excitations from the "hot" to the "cold" end. Because of a finite mass, the flow of excitations is accompanied with transfer of energy through the sample. In TE devices, the charged particles drift under the influence of an electric field. Typically, in TE coolers and energy generators, the drift of charged particles occurs from the "cold" to the "hot" end. On the contrary, neutral excitations flow in the opposite direction, transferring energy from the "hot" to the "cold" end, which, according to Eq. 9.3, serves as a drawback for the overall performance of the device.

9-3. Explain why singularities in the electric excitation spectrum can improve the performance of a TE device.

Hint: Use Eq. 9.3 for the figure of merit ZT, where the numerator is $\propto S^2GT$. The last product S^2GT can be considerably increased in materials or devices with sharp peaks of the electron density of states. Both the Seebeck coefficient, S, and the differential conductance, G, are enhanced when the Fermi level is matched with a spectral singularity. An additional increase of ZT is accomplished, reducing the denominator $\kappa_{\rm el}+\kappa_{\rm ph}$ of Eq. 9.3. Both the electron ($\kappa_{\rm el}$) and the phonon ($\kappa_{\rm ph}$) heat conductance can be reduced using special materials and designs of the TE device.

9-4. Explain why the Wiedemann–Franz (WF) law is not observed in graphene and carbon nanotubes (CNTs).

Hint: Consider the contribution of the phonon part of the thermal conductance which almost always exceeds the electronic part in graphene and CNTs. Exclusions can occur at low temperatures or using a special design or in vicinity of strong singularities in the electron spectrum.

9-5. What is the connection between the figure of merit ZT and the Carnot efficiency η_C?
Hint: Use Eq. 9.4 to analyze the connection between the two characteristics.

9-6. What is the connection between Fourier's law and the electrical Ohm law?
Hint: Consider Eqs. 9.36 and 9.37, which establish the relationship between the amount of heat transferred per unit time through a surface S and the gradient of temperature. Equations 9.36 and 9.37 suggest that Fourier's law is the thermal heat flow analogue of the electrical Ohm law.

9-7. Explain why the main heat carriers in carbon materials are usually phonons.
Hint: According to Baheti et al. (2008) and Balandin and Wang (1998), acoustic phonons dominate the heat conduction, even in metal-like graphite. Furthermore, the high values of in-plane phonon group velocities and low crystal lattice unharmonicity for in-plane vibrations take place due to the strong covalent sp^2 bonding.

9-8. Explain why the phonon group velocity is typically decreased in nanostructures.
Hint: Take into account the fact that acoustic phonons are spatially confined in nanostructures. It causes quantization of the phonon energy spectra and a decrease of the phonon group velocity.

9-9. Explain why preserving the parity-time (PT) inversion symmetry is important for the creation of heavy chiral fermion (HCF) states.
Hint: Consider Section 3.5, which shows how preserving the PT symmetry leads to HCF states and to sharp distinctive quantized levels when a transverse electric field is applied.

9-10. Explain why using multilayered metallic electrodes with low value of the phonon thermal conductance can improve the figure of merit of a thermoelectric generator (TEG)?
Hint: Use Eq. 9.3 for the figure of merit ZT with the denominator $\propto \kappa_{el} + \kappa_{ph}$. Typically, in graphene and CNTs, the phonon part of the thermal conductance κ_{ph} exceeds the electron part, $\kappa_{ph} >> \kappa_{el}$. To evaluate κ_{ph} for a TE device

with multilayered electrodes, take into account Fourier's law jointly with Kirchhoff rules for the heat current. If one connects different elements with distinct thermal conductances into series, the thermal conductance of the whole device is strongly diminished, which considerably improves ZT.

Problems of Chapter 10

10-1. Discuss why generating and detecting of THz waves represents a serious technical problem.
Hint: Compare the energy $h\nu$ of a T-ray photon with frequency ν with the typical electron energy-level spacing inside the dot and express both of them in temperature units $h\nu/k$ and $\Delta E/k$. Successful detection and generation of THz waves requires that the condition $h\nu = \Delta E \geq kT^* > \Gamma$, where Γ is level broadening, be observed. It imposes a necessity of strong cooling and of heat management.

10-2. Explain why graphene and carbon nanotube (CNT) quantum dots (QDs) have a strong potential for THz applications.
Hint: Chirality of charge carriers along with one-half pseudospin conservation causes an intrinsic coherence of graphene and in CNTs. In turn, the intrinsic coherence helps to reduce the level broadening Γ. For these reasons, the generation and detection of THz waves can be accomplished at lower temperatures T of the active region.

10-3. Which physical principle is used for the detection of THz waves?
Hint: Consider a sharp increase of the differential conductance which occurs when the electron energy in the source and drain electrodes, ε, matches the quantized-level position E_n, provided that $\varepsilon \rightarrow E_n \pm eV_{SD}$, where V_{SD} is the source–drain bias voltage. The matching condition between the THz photon energy and the energy-level spacing is also enforced with the single-electron tunneling (SET) resonances occurring in conditions of Coulomb blockade.

10-4. What types of quantization of electron states are used to detect THz waves by graphene and CNT QDs?

Hint: Consider three different types of quantization: (*a*) geometrical quantization of electron propagation inside the QD, (*b*) quantization of the electron charge which results in SET under the conditions of Coulomb blockade, and (*c*) quantization of electron states with the electromagnetic THz field of frequency ω, which alters the electron energy by an amount $n\hbar\omega$, where n is an integer.

10-5. In which conditions does the AC field–induced cooling of a quantum well become possible?
Hint: Discuss the following conditions: (a) the THz frequency exceeding the quantized-level spacing ΔE, which actually determines the threshold of the nonequilibrium cooling mechanism, and (b) the energy dependence of transparency of the Schottky barrier between the QD and the metal electrode attached, which ensures that the electron lifetime on the lower level 1_0 must be much longer than on the upper level 1_+.

10-6. Which parameters of the external AC field can be deduced from the differential conductance characteristics of a QD?
Hint: Consider the origin, spacing, and amplitude of the side peaks which accompany the Coulomb blockade peaks in the curves of differential conductance when the external AC field is applied to a QD. The voltage spacing ΔV of side peaks is related to the AC field frequency ν as $\Delta V = h\nu/e$, whereas their amplitude is $\propto J_n^2(eU_1/h\nu)$, which can be used to determine both the AC field frequency ν and the amplitude U_1.

10-7. Explain why the linear equations for reflection $r_{\mathrm{T}}(\varepsilon)$ and transmission $t_{\mathrm{T}}(\varepsilon)$ coefficients have a recurrent form?
Hint: Consider two types of electron waves traveling through a QD which correspond to the left- and right-moving electrons. Furthermore, there are two types of photon-assisted tunneling processes, which are absorption and emission of m photons with energies $\pm\hbar m\omega$. It causes shifts $\varepsilon \rightarrow \varepsilon \pm \hbar m\omega$ of the energy arguments in Eq. 10.8, which take the form of recurrence equations for the reflection $r_{\mathrm{T}}(\varepsilon)$ and transmission $t_{\mathrm{T}}(\varepsilon)$ coefficients.

10-8. Explain why the highest responsivity corresponds to photon energies slightly above the level energy.
Hint: Evaluate the change of source–drain voltage due to the AC field influence ΔU using Eq. 10.2 and results of Section 10.6. An approximate magnitude of ΔU can be obtained, for example, from the contour plot in Fig. 10.10 and Eq. 10.12.

10-9. Discuss how the basic parameters of a THz QD detector, like sensitivity, quantum efficiency, and noise-equivalent power (NEP), depend on the type of edges of the graphene stripe or on the CNT chirality.
Hint: Consider the rate of electron–impurity scattering, back scattering, and intrinsic noises versus the CNT chirality and the shape of atomic edges. For pristine graphene or metallic nanotubes, electron–impurity scattering causes the pseudospin to rotate by the angle $+\pi\,(2m+1)$, where m is an integer. The time reversal scattering process causes a pseudospin rotation into the opposite direction by the angle $-\pi\,(2m+1)$. In the latter case, the spatial part of the wavefunction remains unchanged. After summation over all the scattering processes, the scattering amplitudes mutually cancel each other. It causes an eventual suppression of the intrinsic noise. The scenario is different in graphene on a substrate and in semiconducting nanotubes. In the vicinity of the K point, the electron wavevector along the axis direction for $n = 0$ and $\nu = \pm 1$ is $\mathbf{k}_{\pm} = (-2\pi\nu/3L, \pm|k|)$, where the two signs $\pm$ correspond to the right- (left-) moving excitations, respectively. When the electron momentum changes as $\mathbf{k}_{+} \rightarrow \mathbf{k}_{-}$, the sign of the full momentum is not reversed. Therefore, there is no correspondence between the back scattering and the spin rotation by $+\pi\,(2m+1)$ or $-\pi\,(2m+1)$. It explains why the intrinsic noises generated during electron scattering are increased in the latter case.

Problems of Chapter 11

11-1. Discuss how to engineer the energy gap in the electron excitation spectrum of pristine graphene.

Hint: Consider the electron excitation spectrum of graphene stripes, the spectrum of multilayered graphene, and the role of the substrate.

11-2. What are the benefits of a direct band gap in the electron spectrum of monolayer MoS_2 as compared to the indirect band gap in MoS_2 multilayers?
Hint: The presence of a direct band gap in the electron spectrum of monolayer MoS_2 allows photoluminescence, which opens up the possibility of many optoelectronic applications.

11-3. What is the physical reason for transition metal dichalcogenides (TMDCs) to behave like dry lubricants?
Hint: Take into account the weakness of the van der Waals interaction between the sheets of sulfide atoms. It causes a TMDC to have a low coefficient of friction because the bulk materials form layered structures. That lubricating property is practically implemented in industry. The low magnitude of the friction coefficient is a common property of many other layered inorganic materials which also exhibit lubricating properties.

11-4. How do we estimate the exfoliation energy of layered crystals?
Hint: Consider the fact that the exfoliation energy is related to the surface energy, which is equal to half the energy for splitting a monolayer from the crystal as per the monolayer surface area. According to experimental data obtained from inverse gas chromatography and exfoliation studies, the surface energies of MoS_2, WS_2, $MoSe_2$, and boron nitride (BN) are about 65–75 mJ m^{-2}. Such values are similar to the surface energy of graphene, which is 65–120 mJ m^{-2}. For these reasons, inorganic layered compounds are exfoliated as easily as graphite, or even more easily.

11-5. What is the difference between the band structures of bulk and monolayer TMDCs?
Hint: Take into account the first-principle calculations, which show that near the Γ point, the band-gap transition is indirect for the bulk material. However, it is eventually shifted and becomes direct for the monolayer. In contrast, near

the K point, direct excitonic transitions remain relatively independent from the layer number.

11-6. What are the benefits of using 2D atomic monolayers (AMs) in field-effect transistors (FETs)?

Hint: Discuss the following benefits: (i) an external gate voltage can readily modulate its current flow; (ii) as compared to conventional chemical doping, electrostatic doping with gate electrodes is more flexible and introduces no undesirable atomic defects and it allows us to create much "cleaner" FET structures; and (iii) an extreme thinness of TMDCs, when utilized together jointly with a band gap of 1–2 eV, can yield high on/off ratios, allowing more efficient control over switching.

11-7. Discuss the specific role of 2D geometry in the mechanism of electron scattering.

Hint: Take into account the fact that because 2D materials are extremely thin, there is strong electron scattering on surface phonons and ripples.

11-8. What are the benefits of using TMDC materials in optoelectronics?

Hint: Flexible and transparent optoelectronic devices can be utilized for generating, detecting, interacting with, or controlling light by electronic methods. Since 2D TMDC semiconductors are atomically thin and processable, they have great potential for flexible and transparent optoelectronics. The electronic band structure of a semiconducting material or device determines its ability to absorb and emit light. The primarily direct semiconducting band gaps in single-layer TMDCs makes them very attractive for a variety of optoelectronic applications. In semiconductors with a direct band gap, photons can be easily emitted or absorbed if their energy exceeds the band gap. However, in semiconductors with an indirect band gap, photon absorption (or emission) is strongly reduced because the process requires an additional phonon to supply the difference in momentum for the process to be completed.

11-9. Explain why the presence of a direct band gap in a TDMC allows us to create devices which utilize electroluminescence and photoluminescence?
Hint: Take into account the fact that electroluminescence arises in response to electrical signals applied to a semiconductor with a direct band gap. The process intensity is much higher than that in indirect-band-gap semiconductors. Photoluminescence is induced when a material absorbs one photon and then reradiates another photon with a different energy. Electroluminescence and photoluminescence represent two important phenomena. They are utilized in light-emitting diodes (LEDs) and diode lasers, since in these optoelectronic devices photons are emitted. The radiative recombination of electrons and holes in direct-band-gap semiconductors creates photons. It suggests that atomic 2D layer TDMCs with a direct band gap are good candidates for the use of electroluminescence and photoluminescence phenomena.

Index